Unitary Representations of Groups, Duals, and Characters

Mathematical
Surveys
and
Monographs
Volume 250

Unitary Representations of Groups, Duals, and Characters

Bachir Bekka
Pierre de la Harpe

AMS
American Mathematical Society
Providence, Rhode Island

2010 *Mathematics Subject Classification.* Primary 22D10, 22D25, 22D40, 46L55.

The first author has been supported by the ANR (French Agence Nationale de la Recherche) through the projects Labex Lebesgue (ANR-11-LABX-0020-01) and GAMME (ANR-14-CE25-0004).

For additional information and updates on this book, visit
www.ams.org/bookpages/surv-250

Library of Congress Cataloging-in-Publication Data

Names: Bekka, Bachir M., author. | de la Harpe, Pierre, author.
Title: Unitary representations of groups, duals, and characters / Bachir Bekka, Pierre de la Harpe.
Description: Providence, Rhode Island: American Mathematical Society, [2020] | Series: Mathematical surveys and monogrphs, 0076-5376; volume 250 | Includes bibliographical references and index.
Identifiers: LCCN 2020027691 | ISBN 9781470456276 (paperback) | ISBN 9781470462871 (ebook)
Subjects: LCSH: Representations of groups. | Unitary groups. | Topological groups. | Ergodic theory. | Operator algebras. | Characters of groups. | AMS: Topological groups, Lie groups – Locally compact groups and their algebras – Unitary representations o | Topological groups, Lie groups – Locally compact groups and their algebras – C^*-algebras and W^*- | Topological groups, Lie groups – Locally compact groups and their algebras – Ergodic theory on groups | Functional analysis – Selfadjoint operator algebras (C^*-algebras, von Neumann (W^*-) algebras, etc.)
Classification: LCC QA176 .B45 2020 | DDC 512/.55–dc23
LC record available at https://lccn.loc.gov/2020027691

10 9 8 7 6 5 4 3 2 1 25 24 23 22 21 20

Contents

Foreword ix

Acknowledgments xi

Introduction 1
- Representation theory of finite groups 1
- Representations of compact groups and abelian groups 2
- Primitive dual and locally compact groups of type I 4
- The normal quasi-dual of a locally compact group 5
- Characters of a locally compact group 6
- Thoma's dual space 7
- Historical comments on dual spaces of groups 7
- Overview 10

Chapter 1. Unitary dual and primitive dual 13
- 1.A. Definition of the unitary dual 13
- 1.B. Functions of positive type and GNS construction 19
- 1.C. Weak containment and Fell topology for representations of topological groups 27
- 1.D. Topological properties of the dual of a group 32
- 1.E. Primitive dual of a topological group 38
- 1.F. Induced representations, irreducibility, and equivalence 40
- 1.G. On decomposing representations into irreducible representations 52
- 1.H. On decomposable and diagonalisable operators 59
- 1.I. Direct integrals of von Neumann algebras 63

Chapter 2. Representations of locally compact abelian groups 67
- 2.A. Canonical representations of abelian groups 67
- 2.B. Canonical decomposition of representations of abelian groups 73
- 2.C. The SNAG Theorem 80
- 2.D. Containment and weak containment in terms of projection-valued measures 85
- 2.E. Canonical decomposition of projection-valued measures 88
- 2.F. Applying the SNAG Theorem to nonabelian groups 89

Chapter 3. Examples of irreducible representations 95
- 3.A. Infinite dihedral group 95
- 3.B. Two-step nilpotent groups 99
- 3.C. The affine group of a field 110
- 3.D. Solvable Baumslag–Solitar groups 114
- 3.E. Lamplighter group 118

3.F. General linear groups 119
3.G. Some nondiscrete examples 121

Chapter 4. Finite-dimensional irreducible representations 123
4.A. Finite-dimensional irreducible representations of some semi-direct products 123
4.B. All finite-dimensional irreducible representations for some groups 126
4.C. Classes of groups in terms of finite-dimensional representations 132

Chapter 5. Describing all irreducible representations of some semi-direct products 147
5.A. Constructing some irreducible representations 147
5.B. Cocycles over measurable group actions 149
5.C. Cocycles with values in the unitary group of a Hilbert space 151
5.D. Constructing all irreducible representations 155
5.E. Identifying induced representations 162
5.F. On the existence of infinite nonatomic ergodic invariant measures 165

Chapter 6. Types for representations, quasi-duals, groups of type I 169
6.A. Comparing representations, quasi-equivalence 170
6.B. Group representations and von Neumann algebras 176
6.C. The quasi-dual of a topological group 192
6.D. Groups of type I 196
6.E. A class of groups of type I 200

Chapter 7. Non type I groups 211
7.A. A class of non type I groups 211
7.B. Operator algebras, traces, and types 214
7.C. Types of group representations 219
7.D. Types of representations of discrete groups 221
7.E. On types of representations of locally compact groups 227
7.F. Non type I factor representations and irreducible representations 228

Chapter 8. Representations of C*-algebras and of LC groups, the Glimm Theorem 231
8.A. Spectrum and primitive ideal space of a C*-algebra 232
8.B. C*-algebras and representations of LC groups 242
8.C. Functorial properties of group C*-algebras 246
8.D. Second-countable and σ-compact LC groups 254
8.E. The central character of a representation, of a primitive ideal 256
8.F. Characterization of type I groups: The Glimm Theorem 257
8.G. The von Neumann algebra of a group representation 259
8.H. Variants 260

Chapter 9. Examples of primitive duals 263
9.A. A weak containment result for induced representations 263
9.B. Two-step nilpotent groups 265
9.C. Affine groups of infinite fields 271
9.D. Solvable Baumslag–Solitar groups 272
9.E. Lamplighter group 275

9.F. General and special linear groups 276
9.G. On the noninjectivity of the map from the dual to the primitive dual 288
9.H. Borel comparison for duals of groups 290

Chapter 10. Normal quasi-dual and characters 297
10.A. Traces and Hilbert algebras 297
10.B. The standard representation of a semi-finite von Neumann algebra 307
10.C. Group representations associated with traces 308
10.D. Normal factor representations and characters 312

Chapter 11. Finite characters and Thoma's dual 319
11.A. Factor representations of finite type and finite characters 319
11.B. GNS Construction for traces on groups 320
11.C. Thoma's dual 324
11.D. Characters and primitive duals 328

Chapter 12. Examples of Thoma's duals 331
12.A. Two-step nilpotent groups 331
12.B. Affine groups of infinite fields 336
12.C. Solvable Baumslag–Solitar groups 338
12.D. Lamplighter group 341
12.E. General linear groups 346

Chapter 13. The group measure space construction 353
13.A. Construction of factors of different types 353
13.B. Ergodic group actions without invariant measure 365

Chapter 14. Construction of factor representations for some semi-direct products 373
14.A. Crossed product von Neumann algebras for semi-direct products 373
14.B. Some factor representations of semi-direct products 376
14.C. Some normal factor representations 377
14.D. The von Neumann algebra of the regular representation as a crossed product 383
14.E. Examples of normal factor representations 387

Chapter 15. Separating families of finite type representations 397
15.A. Finite type representations and bi-invariant metrics 397
15.B. Groups for which finite type representations are separating 399
15.C. Locally compact groups with a finite von Neumann algebra 401
15.D. Connected and totally disconnected LC groups 403
15.E. Faithful traces on group C*-algebras 405
15.F. Traces and Invariant Random Subgroups 416

Appendix 421
A.1. Topology 421
A.2. Borel spaces 424
A.3. Measures on Borel spaces and σ-finiteness 425
A.4. Radon measures on locally compact spaces 429
A.5. Groups and actions 431

A.6. Locally compact groups 435
A.7. Locally compact abelian groups and duality 437
A.8. Hilbert spaces and operators 440
A.9. Projection-valued measures 441
A.10. C*-algebras 442
A.11. Von Neumann algebras 443

Bibliography 447

Notation Index 463

Index 467

Foreword

This is an expository book on unitary representations of topological groups, and of several dual spaces, which are spaces of such representations up to some equivalence. The most important notions are defined for topological groups, but special attention is paid to the case of discrete groups.

The unitary dual of a group G is the space of equivalence classes of its irreducible unitary representations; it is both a topological space and a Borel space. It is useful in important situations, but much less useful in many other situations, for example because, for an infinite discrete group, it is usually not countably separated as a Borel space.

We discuss other candidates for being a dual space of a group. The two most important ones are the primitive dual, the space of weak equivalence classes of unitary irreducible representations, and the normal quasi-dual, the space of quasi-equivalence classes of traceable factor representations. The last space is parametrized by "characters", which can be finite or infinite. The space of finite characters is an object particularly easy to describe: it is in natural bijection with the space of indecomposable central positive definite functions on the group, that we call the Thoma dual. We discuss other related spaces, such as the space of finite-dimensional irreducible unitary representations.

These spaces are much better behaved than the unitary dual. They deserve to be studied on their own, and related to each other. In contrast to the unitary dual, they can in principle be classified for every specific group. We illustrate the theory by giving an explicit description of these alternative dual spaces for a series of examples: Heisenberg groups over fields and over the integers, affine groups of infinite fields, solvable Baumslag–Solitar groups, lamplighter groups, and general (or special) linear groups over infinite fields.

Operator algebras play an important role in the discussion, in particular von Neumann algebras associated with unitary representations and C*-algebras associated with locally compact groups.

Acknowledgments

This book would not have been possible without the support of our academic institutions and the feedback of many friends and colleagues. The first author gratefully acknowledges partial support by the ANR (French Agence Nationale de la Recherche) through the projects Labex Lebesgue (ANR-11-LABX-0020-01) and GAMME (ANR-14-CE25-0004).

We are indebted to Pierre-Emmanuel Caprace for his numerous most valuable remarks and suggestions, especially in connection with traces on totally disconnected groups in Section 15.D; to Sebastien Gouëzel for his help with the construction of infinite invariant measures for Bernoulli shifts in Section 14.E; to François Maucourant for discussions concerning the nonexistence of invariant measures for boundary actions as in Section 13.B; and to Rostislav Grigorchuk for his constant interest and stimulating questions.

We would also like to thank Marcia Almeida, Brian Bartling, and Eriko Hironaka of the American Mathematical Society for their help with the preparation of the manuscript for publication. Thanks are also due to the referees of this book for their many helpful suggestions.

It is the occasion to celebrate Jacques Dixmier and the writing
of his two books [**Dixm–vN, Dixm–C***] *on operator algebras.*
More than 50 years after publication,
they remain unrivalled references on the subject.

Introduction

Unitary group representations, that is, representations of groups by linear isometries of a Hilbert space, play an important role in a large variety of subjects, including number theory, geometry, probability theory, and theoretical physics. Given a group G, a major problem is to describe the most general representation of G in terms of a dual space, that is, a set of "elementary" representations of G, up to a suitable equivalence relation. The most common dual space of G is the unitary dual $\widehat{G}$. While the choice of $\widehat{G}$ as a dual space is adequate for some classes of groups G (as compact groups, locally compact abelian groups, or semi-simple Lie groups), we will see that for other groups, especially for infinite discrete groups, it is necessary to consider a replacement space instead of $\widehat{G}$.

Representation theory of finite groups

The theory of **linear representations** of finite groups is a classical subject of fundamental importance. It was founded by Frobenius in the late 1890s and further developed by Burnside and Schur, among others (see below). Let G be a finite group and let π be a representation of G on a finite-dimensional complex vector space V. There is a decomposition $V = V_1 \oplus \cdots \oplus V_n$ in a direct sum of **irreducible** invariant subspaces V_i. Such a decomposition not usually being unique, a more canonical one can be obtained as follows. Let $\{\sigma_1, \ldots, \sigma_k\}$ be the set of equivalence classes of irreducible representations of G. The space V admits a *unique* decomposition $V = W_1 \oplus \cdots \oplus W_k$ in a sum of invariant subspaces W_j such that the restriction of π to W_j is equivalent to the direct sum of n_j copies of σ_j, for some nonnegative integer n_j; the W_j are called the **isotypical components** of π. As a result, one faces two main problems in the representation theory of G:

(A) determine the **dual space** $\widehat{G}$ of G, that is, the set of equivalence classes of irreducible representations of G;
(B) for a given representation π of G, determine the multiplicities n_σ in the isotypical decomposition $\pi \simeq \bigoplus_{\sigma\in\widehat{G}} n_\sigma\sigma$, where we use $\simeq$ to indicate equivalent representations.

One way to approach Problem (A), already initiated by Frobenius, is through the determination of the set of **characters** of irreducible representations of G. A representation π of G gives rise to its character

$$\chi_\pi : G \to \mathbf{C}, \quad g \mapsto \mathrm{Tr}(\pi(g)),$$

where Tr stands for "trace"; note that χ_π is a class function. (Recall that a class function or central function on a group G is a function f on G which is constant on conjugacy classes: $f(hgh^{-1}) = f(g)$ for all $g, h \in G$.) Every representation is determined up to equivalence by its character. The map $\sigma \mapsto \chi_\sigma$ sets up a bijection

$$\widehat{G} \overset{\approx}{\longrightarrow} E(G)$$

between $\widehat{G}$ and an appropriate set $E(G)$ of class functions on G. By $\approx$, we indicate a bijection between sets, or a homeomorphism between topological spaces.

In view of Schur's orthogonality relations, Problem (B) can also be solved through the use of characters.

Following Frobenius, Molien, Burnside, E. Noether, and others, the representation theory of G can be viewed as the study of modules over the **group algebra**

$$\mathbf{C}[G] = \Big\{ \sum_{g \in G} c_g g \;\Big|\; c_g \in \mathbf{C} \Big\}.$$

Indeed, every representation $\pi : G \to \mathrm{GL}(V)$ extends linearly to a representation

$$\widetilde{\pi} : \mathbf{C}[G] \to \mathrm{End}(V), \qquad \sum_{g \in G} c_g g \mapsto \sum_{g \in G} c_g \pi(g).$$

The map $\pi \to \widetilde{\pi}$ sets up a bijection between the representations of G and the modules over $\mathbf{C}[G]$, preserving equivalence and irreducibility.

The character χ_π of a representation $\pi : G \to \mathrm{GL}(V)$ has a natural extension $\widetilde{\chi}_\pi$ to $\mathbf{C}[G]$ given by

$$\widetilde{\chi}_\pi : \mathbf{C}[G] \to \mathbf{C}, \quad x \mapsto \mathrm{Tr}(\widetilde{\pi}(x)).$$

The map $\chi_\sigma \mapsto \widetilde{\chi}_\sigma$ sets up a bijection between $E(G)$ and an appropriate set $\mathrm{Char}(G)$ of linear functionals $\widetilde{\chi}$ on $\mathbf{C}[G]$ which satisfies the trace property:

$$\widetilde{\chi}(xy) = \widetilde{\chi}(yx) \qquad \text{for all } x, y \in \mathbf{C}[G].$$

The dual space $\widehat{G}$ can be described in terms of the ideal theory of the ring $\mathbf{C}[G]$ as follows: the map $\sigma \to \ker(\widetilde{\sigma})$ sets up a bijection

$$\widehat{G} \overset{\approx}{\longrightarrow} \mathrm{Prim}(G)$$

between $\widehat{G}$ and the space of **primitive ideals** of $\mathbf{C}[G]$. (Recall that a primitive ideal in a ring R is the annihilator of a simple R-module.)

Representations of compact groups and abelian groups

For a topological group G, representations of interest in this book are **unitary representations**, that is, homomorphisms $\pi : G \to \mathcal{U}(\mathcal{H})$ from G to the unitary group of a (not necessarily finite-dimensional) complex Hilbert space $\mathcal{H}$, which are continuous in an appropriate sense (see Section 1.A). We will deal mostly with topological groups which are *locally compact*, and also sometimes second-countable locally compact.

A locally compact group G admits a **left Haar measure**, that is, a positive measure μ_G on the Borel subsets of G which is invariant by left translations; second-countable locally compact groups are characterized among standard Borel groups by the fact that they admit a translation-invariant measure (see [**Mack–57**]). A locally compact group G has a distinguished unitary representation, the regular representation λ_G on $L^2(G) = L^2(G, \mu_G)$, which plays a central role in the theory—as it already does for finite groups. The unitary dual $\widehat{G}$ of G is the set of equivalence classes of irreducible unitary representations of G.

Compact groups. Most of Frobenius' theory extends to the case of a compact group G, as shown by Schur and Weyl between 1924 and 1927. First, every representation of G is equivalent to a unitary representation. Then:

(I) every irreducible representation of G is finite-dimensional;
(II) the map $\sigma \mapsto \chi_\sigma$ sets up a bijection between $\widehat{G}$ and the set of irreducible characters of G, which is a set of class functions on G;
(III) every representation $\pi : G \to \mathcal{U}(\mathcal{H})$ admits a canonical isotypical decomposition $\pi \simeq \bigoplus_{\sigma\in\widehat{G}} n_\sigma \sigma$;
(IV) every unitary representation π of G "extends" uniquely to a $*$-representation $\widetilde{\pi}$ of the group algebra $L^1(G) = L^1(G, \mu_G)$;
(V) the map $\rho \mapsto \ker(\widetilde{\rho})$ sets up a bijection

$$\widehat{G} \xrightarrow{\approx} \mathrm{Prim}(G)$$

between $\widehat{G}$ and the space of primitive (or maximal) ideals of $L^1(G)$.

The crucial tool for this extension is the use of averaging procedures which are made possible by the fact that the Haar measure μ_G on G is *finite* when (and only when) G is compact. One should add to the previous list one more item, a key step to finding a complete system of irreducible representations of G:

(VI) every irreducible representation of G is equivalent to a subrepresentation of the regular representation λ_G.

Locally compact abelian groups. Now let G be a locally compact abelian group. Then $\widehat{G}$ coincides with the set of unitary characters $G \to \mathbf{T}$ and is itself a locally compact abelian group, for a natural topology. By Pontryagin and van Kampen duality theory, the dual group of $\widehat{G}$ can naturally be identified, as a topological group, with G. This justifies the term "dual" used for $\widehat{G}$.

Let $\pi : G \to \mathcal{U}(\mathcal{H})$ be a unitary representation of G. In general, π does not admit a direct sum decomposition into irreducible representations. For example, when G is noncompact, its left regular representation does not contain any irreducible subrepresentation. However, π can be analyzed by means of a unique projection-valued measure on $\widehat{G}$ or a unique collection of measure classes on $\widehat{G}$ with associated multiplicities (see Chapter 2). In case of the regular representation $\pi = \lambda_G$ on $L^2(G)$, such a decomposition generalizes the Plancherel formula of classical Fourier analysis on $\mathbf{R}^n$ or $\mathbf{T}^n$.

The set $E(G)$ obviously coincides with $\widehat{G}$, and here also the map $\chi \to \ker(\widetilde{\chi})$ is a bijection between $\widehat{G}$ and the space of primitive (or maximal) ideals of $L^1(G)$, where $\widetilde{\chi} : L^1(G) \to \mathbf{C}$ is the multiplicative linear functional canonically associated with χ.

As a result, we see that the representation theory of G has the features (I)-(V) of Frobenius theory listed above, appropriately interpreted.

Even when G is noncompact, so that no unitary character of G is equivalent to a subrepresentation of the regular representation λ_G, the projection-valued measure on $\widehat{G}$ associated with λ_G has $\widehat{G}$ as support. Using a notion to be introduced in Section 1.C, it follows that every $\chi \in \widehat{G}$ is weakly contained in λ_G, and this is to be viewed as an analogue of feature (VI) in the list above.

Primitive dual and locally compact groups of type I

We now turn to the case of a *general* locally compact group G. By the Gel'fand–Raikov Theorem, G has plenty of irreducible unitary representations: the unitary dual $\widehat{G}$ separates the points of G (Theorem 1.B.19). Observe that, when G is not compact, $\widehat{G}$ will usually contain infinite-dimensional representations (for more on this, see Section 4.C).

As in the case of compact or abelian groups, the unitary representations are in bijective correspondence with the $*$-representations of the group algebra $L^1(G)$; however, a more suitable algebra is the **maximal C*-algebra** $C^*_{\text{max}}(G)$ of G which is the enveloping C*-algebra of $L^1(G)$. Every unitary representation π of G "extends" to a representation $\widetilde{\pi}$ of $C^*_{\text{max}}(G)$. The **primitive ideal space** of $C^*_{\text{max}}(G)$, or **primitive dual** of G, is a space of closed two-sided ideals in $C^*_{\text{max}}(G)$:

$$\text{Prim}(G) = \{\ker(\widetilde{\pi}) \mid \pi \in \widehat{G}\}.$$

We have a canonical surjective map

$$\kappa^{\text{d}}_{\text{prim}} : \widehat{G} \twoheadrightarrow \text{Prim}(G), \quad \pi \mapsto \ker(\widetilde{\pi}).$$

The space $\text{Prim}(G)$ is equipped with a natural topology, the Jacobson topology (see Section 8.A).

There exists a description of $\text{Prim}(G)$ at the group level: it is the space of **weak equivalence classes** of irreducible unitary representations of G. Thus $\text{Prim}(G)$ is a well-defined space for any topological group. See Section 1.C for the notion of weak equivalence, and Sections 1.E & 8.B for $\text{Prim}(G)$. Besides $\widehat{G}$, the space $\text{Prim}(G)$ is the most important dual space of a group G.

Let $\pi : G \to \mathcal{U}(\mathcal{H})$ be a unitary representation of G on a separable Hilbert space. There is a way π can be decomposed in irreducible parts. As already mentioned in the previous section, decomposition in direct sums does not suffice (unless G is compact), but there is a notion of **direct integral** of unitary representations, and

$$\pi = \int_{\widehat{G}}^{\oplus} n_\rho \, \rho \, d\mu(\rho),$$

where μ is a Borel measure on $\widehat{G}$ (equipped with a natural Borel structure), and every irreducible representation $\rho \in \widehat{G}$ occurs with a multiplicity n_ρ. However, for some groups G, an unexpected pathology appears: there exist unitary representations π of G having two direct integral decompositions

$$\pi = \int_{\Omega}^{\oplus} \rho d\mu(\rho) = \int_{\Psi}^{\oplus} \rho d\nu(\rho),$$

where Ω, Ψ are *disjoint* Borel subsets of $\widehat{G}$, and μ [respectively ν] a measure on Ω [respectively Ψ]. See Section 1.G.

It turns out that the pathology mentioned above is related to the existence of factor representations of several types. Let $\pi : G \to \mathcal{U}(\mathcal{H})$ be a unitary representation. The bicommutant $\pi(G)''$ of $\pi(G)$ is a von Neumann subalgebra of $\mathcal{L}(\mathcal{H})$ and π is said to be **factorial**, or a **factor representation**, if $\pi(G)''$ is a factor, that is, if the centre of $\pi(G)''$ consists of scalar multiples of the identity only.

Examples of factor representations include representations of the form $n\pi$, where π is an irreducible unitary representation and n a cardinal. Groups without factor representations of another type constitute an important class defined as follows: a locally compact group G is a **group of type I** if every factor representation of G is of the form $n\pi$ for some $\pi \in \widehat{G}$ and some cardinal n.

The class of second-countable locally compact groups G of type I is characterized by a number of equivalent properties:

- for every unitary representation π of G, the direct integral decomposition $\pi = \int_{\widehat{G}} n_\rho \, \rho \, d\mu(\rho)$ mentioned above is unique in an appropriate sense (Section 6.D.b);
- the dual $\widehat{G}$, equipped with its natural Borel structure, is countably separated;
- the canonical map $\kappa_{\mathrm{prim}}^{\mathrm{d}} : \widehat{G} \twoheadrightarrow \mathrm{Prim}(G)$ is injective, and therefore bijective.

The last two characterizations are part of Glimm's Theorem (see Section 8.F).

The class of locally compact groups of type I includes, among others, abelian groups, compact groups, and Lie groups with finitely many connected components which are either semisimple or nilpotent (see Section 6.D). As a rule, discrete groups are *not* of type I; more precisely, a discrete group G is of type I if and only if G is virtually abelian (Thoma's Theorem 7.D.3).

Let G be a second-countable locally compact group. Every unitary representation π of G has a canonical direct integral decomposition $\int_X \sigma_x d\mu(x)$ into *factor* representations σ_x; when G is of type I, there are multiplicities n_x and irreducible representations ρ_x such that $\sigma_x = n_x \rho_x$ for all $x \in X$, and therefore a canonical decomposition $\pi_x \simeq \int_X n_x \rho_x d\mu(x)$ into *irreducible* representations ρ_x. When G is not of type I, factor representations can be of other types, and this is a reason why direct integral decomposition in irreducible representations is not canonical.

Let G be a second-countable locally compact group which is not of type I. Since $\widehat{G}$ is not countably separated, there is no systematic procedure to encode the equivalence classes of irreducible unitary representations of G. By contrast, for the Borel structure associated with the Jacobson topology, $\mathrm{Prim}(G)$ is a standard Borel space, by a theorem of Effros (Subsection 8.A.c). The space $\mathrm{Prim}(G)$ is therefore much better behaved than $\widehat{G}$, and this justifies it as an alternative dual space for a locally compact group G.

In Chapter 6, there is a definition of being of type I which applies to topological groups which are not necessarily locally compact.

The normal quasi-dual of a locally compact group

Let G be a locally compact group. If G is not of type I, then G admits factor representations that are not of type I. It is convenient to introduce one more notion of equivalence: two unitary representations π_1, π_2 of G are **quasi-equivalent** if there are cardinals n_1, n_2 such that $n_1\pi_1$ and $n_2\pi_2$ are equivalent (more on this notion in Section 6.A). The **quasi-dual** $\mathrm{QD}(G)$ of G is the set of quasi-equivalence classes of factor representations of G. It has a natural Borel structure, and there is a natural inclusion

$$\kappa_{\mathrm{qd}}^{d} : \widehat{G} \hookrightarrow \mathrm{QD}(G)$$

of the dual as a subspace of the quasi-dual. See Section 6.C.

When G is σ-compact, given a factor representation π of G, it is a fact (see Section 8.D) that the C^*-kernel of π is a primitive ideal of $C^*_{\max}(G)$, that is, $\ker(\widetilde{\pi}) \in$

$\text{Prim}(G)$. Consequently, there is a natural extension

$$\kappa_{\text{prim}}^{\text{qd}} : \text{QD}(G) \twoheadrightarrow \text{Prim}(G)$$

of the map $\kappa_{\text{prim}}^{\text{d}} : \widehat{G} \twoheadrightarrow \text{Prim}(G)$.

When G is of type I, the inclusion $\widehat{G} \hookrightarrow \text{QD}(G)$ is a Borel isomorphism. When G is not of type I, a description of $\text{QD}(G)$ is out of reach, possibly even more than for $\widehat{G}$. But one may try to describe $\text{Prim}(G)$ as the collection of C^*-kernels of *suitable* factor representations of G. This leads to the definition of $\text{QD}(G)_{\text{norm}}$, as follows.

Let G be a locally compact group. A unitary representation π of G is **normal** if it is factorial and if the factor $\pi(G)''$ admits a (not necessarily finite) trace τ such that there exists a positive element $x \in C^*_{\text{max}}(G)$ for which $0 < \tau(x) < \infty$. The **normal quasi-dual** $\text{QD}(G)_{\text{norm}}$ is the set of normal factor representations of G, up to quasi-equivalence (see Chapter 10). It inherits from $\text{QD}(G)$ a Borel structure. When G is second-countable, the Borel space $\text{QD}(G)_{\text{norm}}$ is standard (Theorem 10.D.7).

When G is σ-compact, the restriction of the map $\kappa_{\text{prim}}^{\text{qd}}$ defined above provides a map

$$\kappa_{\text{prim}}^{\text{norm}} : \text{QD}(G)_{\text{norm}} \to \text{Prim}(G), \quad \pi \mapsto \ker(\widetilde{\pi}).$$

When G is a connected Lie group, a remarkable result of Pukánszky shows that $\kappa_{\text{prim}}^{\text{norm}}$ is a bijection (Chapter 10). This result indicates that, for such a group, normal factor representations form the natural class of unitary representations one should seek to classify. It should be mentioned that not every irreducible unitary representation of G is normal when G is not of type I (Corollary 10.D.4).

For a general locally compact group G, the map $\kappa_{\text{prim}}^{\text{norm}}$ can fail to be injective or surjective (see Section 11.D). Nevertheless, we think that $\text{QD}(G)_{\text{norm}}$ is an interesting object which deserves to be studied as a dual space of G in its own right.

Characters of a locally compact group

Let G be a second-countable locally compact group. The normal quasi-dual $\text{QD}(G)_{\text{norm}}$ admits a parametrization in terms of the set $\text{Char}(G)$ of characters of G: a normal factor representation π of G is uniquely determined, up to quasi-equivalence, by its **character** χ_π which is a (not necessarily finite) trace on the maximal C*-algebra $C^*_{\text{max}}(G)$ of G (see Section 10.D).

Assume that G is of type I. Then all the dual spaces we have considered above may be canonically identified:

$$\widehat{G} \xrightarrow{\approx} \text{Prim}(G) \xleftarrow{\approx} \text{QD}(G)_{\text{norm}} \xrightarrow{\approx} \text{Char}(G).$$

Nevertheless, something is gained: the possibility to parametrize the unitary dual $\widehat{G}$ by means of the space $\text{Char}(G)$ of characters, in analogy to the case of finite or compact groups. This programme has been carried out by Harish–Chandra for a connected semisimple Lie group, in which case the set of characters can be identified with a set of appropriate central distributions on the group [**Hari–54**].

Thoma's dual space

Let G be a second-countable locally compact group. The set $\mathrm{Char}(G)$ of characters of G consists of two parts:

- the set $E(G)$ of finite characters of G;
- the set $\mathrm{Char}(G) \smallsetminus E(G)$ of infinite characters.

The set $E(G)$ admits a simple description: it can be viewed as the set of continuous functions $\varphi : G \to \mathbf{C}$ which are central, of positive type, normalized by $\varphi(e) = 1$, and extremal with these properties. (See Section 1.B for functions of positive type, and Section 11.C for the Thoma dual.)

In general, the set $\mathrm{Char}(G) \smallsetminus E(G)$ is nonempty and seems difficult to describe (see Section 14.E, for the construction of a few infinite characters of the Baumslag–Solitar group $\mathrm{BS}(1,p)$ and of the lamplighter group). By contrast, $E(G)$ can be determined for several discrete groups (see Chapter 12). Moreover, for some classes of discrete groups G such as groups of polynomial growth, all characters are finite (that is, $\mathrm{Char}(G) = E(G)$) and the canonical map $E(G) \twoheadrightarrow \mathrm{Prim}(G)$ is a bijection (Theorem 11.D.3).

Thoma made a crucial use of the space $E(G)$ in his characterization of discrete groups which are of type I and suggested that $E(G)$ should be viewed as a dual space of discrete groups [**Thom–64a, Thom–64b**]. We will refer to $E(G)$ as the **Thoma's dual** of G. Again, Thoma's dual consists of two parts:

- the set $E(G)_{\mathrm{fd}}$ of almost periodic functions in $E(G)$, which parametrizes the space $\widehat{G}_{\mathrm{fd}}$ of equivalence classes of finite-dimensional irreducible unitary representations of G;
- the set $E(G) \smallsetminus E(G)_{\mathrm{fd}}$, of which each element gives rise to a unitary representation of G generating a factor of type II_1.

For a discrete group G which is not of type I, the set $E(G) \smallsetminus E(G)_{\mathrm{fd}}$ is nonempty (Theorem 8.F.4). By contrast, when G is connected, we have $E(G) = E(G)_{\mathrm{fd}}$, by a result of Kadison and Singer (see Corollary 15.D.3). In particular, if G is a semisimple connected Lie group without compact factor, $E(G)$ consists only of the trivial character 1_G. A similar result holds in the context of totally disconnected groups: if such a group G is compactly generated and has no proper open normal subgroups, then $E(G) = \{1_G\}$ (see Corollary 15.D.7).

Historical comments on dual spaces of groups

Unitary dual. The study of linear representations of finite groups was initiated by Frobenius in a series of articles between 1896 and 1900 (see Vol. III of [**Frob–Col**]). This work was continued, among others, by Burnside and Schur who established in particular the fact that such representations are equivalent to unitary representations.

Using invariant integration, Schur and Weyl extended between 1924 and 1927 most of Frobenius' representation theory from finite groups to compact groups (see in particular [**Schu–24**] and [**Weyl–24**], as well as [**Bore–86**]). This culminated with the Peter–Weyl Theorem [**PeWe–27**] and the classification of the unitary dual of simple compact Lie groups (see [**Weyl–Col**]).

The proof of the existence of a Haar measure [**Haa–33**] opened up the possibility to a theory of unitary representations for general locally compact groups. A strong impetus for the development of the theory was provided by applications to quantum theory. This motivated Stone, von Neumann, and Weyl to determine the unitary dual of the real Heisenberg group (see [**Ston–30**] and [**vNeu–31**]), and Wigner to study the unitary dual of the Poincaré group $\mathrm{SO}(3,1) \ltimes \mathbf{R}^4$ (see [**Wign–39**], as well as [**Rose–04**]).

A duality theory for locally compact abelian groups was developed around 1935 by Pontryagin and van Kampen, and exposed in the influential books by Pontryagin and Weil, [**Pont–39**] and [**Weil–40**]. The classical spectral theorem for a one-parameter family of unitary operators (which amounts to the classification of unitary representations of $\mathbf{R}$) was generalized to locally compact abelian groups by Segal, Naimark, Ambrose, and Godement (see the SNAG Theorem in Section 2.C).

Gel'fand and Raikov showed in [**GeRa–43**] that every locally compact group G has enough irreducible unitary representations to separate the points of G, triggering a systematic study of unitary representations for such groups. The first major results on classifications of unitary duals did concern the affine group $\mathrm{Aff}(\mathbf{R})$ of $\mathbf{R}$ [**GeNa–47a**], the special linear group $\mathrm{SL}(2,\mathbf{C})$ [**GeNa–47b**], and $\mathrm{SL}(2,\mathbf{R})$ [**Barg–47**]. Further developments include Mackey's analysis of induced representations for extensions of locally compact groups [**Mack–52, Mack–58**], Harish–Chandra's monumental work on the unitary dual of reductive Lie groups [**Hari–Col**], and Kirillov's orbit method for nilpotent Lie groups [**Kiri–62**]. For a history of the subject of unitary group representations see [**Mack–76**].

Operator algebras and primitive duals. As already mentioned above, the representation theory of a finite group G amounts to the study of modules over the group algebra $\mathbf{C}[G]$, which is a semisimple ring by Maschke's Theorem. This approach goes back to the first works on the subject by Frobenius, Molien, and Burnside, in the last years of the 19th century; see [**Hawk–71, Hawk–72, Hawk–74**].

Gel'fand developed in [**Gelf–41**] a general theory of Banach algebras and, together with Naimark, laid down the foundations for the theory of C*-algebras [**GeNa–43**]. As already mentioned above when discussing the primitive dual, the unitary representation theory of a locally compact group G can be viewed as the representation theory of a specific C*-algebra, namely the maximal C*-algebra $C^*_{\mathrm{max}}(G)$ of G.

It was soon realized by Godement, Mautner, Segal, and others that the theory of von Neumann algebras, which was developed by Murray and von Neumann in a series of papers published in the 30s (see [**vNeu–Col**]), is of fundamental importance for the study of unitary representations, in particular in questions concerning existence and uniqueness of decomposition of such representations as direct integrals of irreducible unitary representations.

Given a C*-algebra A, its dual or spectrum $\widehat{A}$ is defined as the set of irreducible representations of A in complex Hilbert spaces, modulo unitary equivalence; A is said to be of type I if every representation of A generates a von Neumann algebra of type I (see Remark 8.F.2). When A is separable, Glimm's theorem mentioned above in the group case shows that $\widehat{A}$, equipped with a natural Borel structure, is countably separated exactly when A is of type I [**Glim–61a**].

For an algebra A, Jacobson introduced in [**Jaco–45**] the set $\mathrm{Prim}(A)$ of primitive ideals of A as a dual space of A and endowed it with a topology which nowadays bears his name. For a C*-algebra A, the topological space $\mathrm{Prim}(A)$ is a quotient of $\widehat{A}$. Its topology was studied by Fell [**Fell–60a**]. In many situations, $\mathrm{Prim}(A)$

has much better topological properties than $\widehat{A}$. In particular, for a locally compact group G, Fell gives a description of the topology of $\text{Prim}(G) = \text{Prim}(C^*_{\max}(G))$ in terms of the weak containment of irreducible representations of G. The relevance of $\text{Prim}(A)$ for the description of a type I C*-algebra A as an algebra of sections of a bundle of elementary algebras was demonstrated by Kaplansky [**Kapl–51b**]. This study was further pursued in Fell [**Fell–61**]. As already mentioned in the group case, Effros [**Effr–63**] proved the important result that $\text{Prim}(A)$ is always a standard Borel space when A is separable. All these facts demonstrate the importance of the study of $\text{Prim}(G)$ for a locally compact group G.

Normal quasi-dual space and characters. Characters of finite groups were considered by Frobenius even before he defined linear representations of such groups (see Vol. III of [**Frob–Col**]). A famous formula for the characters of irreducible representations of compact Lie groups was given by Weyl (see Vol. II of [**Weyl–Col**]). Gel'fand and Naimark described the characters of irreducible unitary representations of G as conjugation-invariant distributions, for $G = \text{SL}_2(\mathbf{C})$ in [**GeNa–47b**] and other classical groups in [**GeNa–57**]. A deep and far-reaching study of characters of irreducible unitary representations of a general semisimple Lie group was undertaken by Harish–Chandra [**Hari–56**].

Godement introduced and studied a notion of normal factor representations of a unimodular locally compact group in [**Gode–54**]. This notion was extended and formulated in the context of C*-algebras (or even involutive Banach algebras) by Guichardet in [**Guic–63**]. The normal quasi-dual space $\text{QD}(A)_{\text{norm}}$ of a C*-algebra A is defined as the set of normal factor representations of A, modulo quasi-equivalence of representations. Answering of question of Dixmier [**Dixm–C***, 7.5.4]), Halpern showed in [**Halp–75**] that the normal quasi-dual space $\text{QD}(A)_{\text{norm}}$ is a standard Borel space for any separable C*-algebra A.

Let G be a connected Lie group. As already mentioned above when discussing the normal dual, Pukánszky [**Puka–74**] proved that the natural map from $\text{QD}(G)_{\text{norm}} = \text{QD}(C^*_{\max}(G))_{\text{norm}}$ to $\text{Prim}(G)$ is a bijection. A detailed analysis of $\text{QD}(G)_{\text{norm}}$ was performed by Guichardet [**Guic–63**] for the Baumslag–Solitar group $G = \text{BS}(2, 1)$.

Vershik and Kerov gave in [**VeKe–81**] a description of $\text{QD}(G)_{\text{norm}}$ for the infinite symmetric group $G = \text{Sym}_{\text{fin}}(\mathbf{N})$.

The subset $\text{QD}(G)_{\text{fin}}$ of $\text{QD}(G)_{\text{norm}}$ consisting of factor representations of finite type was studied by Godement [**Gode–51b**] for a locally compact unimodular group G. He showed in particular (see Theorem 15.B.4) that, when G is compactly generated, the family of such representations is separating if and only if G is a so-called SIN-group (that is, if G has arbitrary small conjugation-invariant neighbourhoods).

Thoma's dual (that is, the set $E(G)$ of characters of representations from $\text{QD}(G)_{\text{fin}}$) was systematically studied by Thoma in the case of a discrete group G and explicitly determined for some concrete examples [**Thom–64a**, **Thom–64b**, **Thom–64c**].

In recent years, work of Vershik ([**Vers–11**]) revealed the relevance of $E(G)$ for the study of the so-called random invariant subgroups of a discrete group G. As a consequence, information about $E(G)$ can be used to show the essential freeness of actions of G by measure preserving transformations of a probability space (see Section 15.F).

Other dual spaces. Given a locally compact group G, one may consider other dual spaces of G, besides the ones discussed above; as for instance, the space $\mathrm{Max}(C^*_{\max}(G))$ of maximal two-sided ideals of $C^*_{\max}(G)$ and the space $\mathrm{Prim}_*(L^1(G))$ of kernels of irreducible $*$-representations of $L^1(G)$. There is an injection

$$\mathrm{Max}(C^*_{\max}(G)) \hookrightarrow \mathrm{Prim}(G)$$

and a surjective canonical map

$$\mathrm{Prim}(G) \twoheadrightarrow \mathrm{Prim}_*(L^1(G)).$$

In general, these maps are not bijective (see Section 8.H), and so $\mathrm{Max}(C^*_{\max}(G))$ and $\mathrm{Prim}_*(L^1(G))$ provide only partial information on $\mathrm{Prim}(G)$.

There is no duality involved in connection with the dual spaces we introduced so far, apart from the Pontryagin–van Kampen duality in the case of an abelian group G. So, the use of the word "dual space" is questionable; it would be more accurate to speak of "structure space" when we refer to $\widehat{G}$ or $\mathrm{Prim}(G)$. Nevertheless, we will stick to the more traditional "dual space".

There are genuine duality theories for locally compact groups, such as the Tannaka–Krein duality for compact groups (see for instance [**HeRo–70**, § 30]), and various duality theories for general locally compact groups, developed by Stinespring, Kac, Ernest, Enock, and Schwartz, among others (see [**EnSc–92**]). These duality theories involve objects such as Hopf algebras, which are quite different from our dual spaces.

Overview

The first aim of this book is to provide a description of the dual spaces introduced above for a group G:

- the unitary dual $\widehat{G}$;
- the space $\widehat{G}_{\mathrm{fd}}$ of equivalence classes of finite-dimensional irreducible representations of G;
- the primitive dual $\mathrm{Prim}(G)$;
- the normal quasi-dual $\mathrm{QD}(G)_{\mathrm{norm}}$;
- the space of characters $\mathrm{Char}(G)$;
- and Thoma's dual $E(G)$.

As much as possible, basic definitions are formulated for topological groups, even if most of the interesting statements and examples are stated for locally compact groups, indeed often for second-countable ones. Systematically, the definitions will be illustrated by a family of basic examples, which are all discrete groups:

- the infinite dihedral group D_∞;
- two-step nilpotent discrete groups, in particular the Heisenberg groups $H(R)$ over a commutative ring R, and more specifically $H(\mathbf{K})$ over a field $\mathbf{K}$, and $H(\mathbf{Z})$ over the integers $\mathbf{Z}$;
- the affine group $\mathrm{Aff}(\mathbf{K})$ of a field $\mathbf{K}$;
- the solvable Baumslag–Solitar group $\mathrm{BS}(1,p)$ for a prime p;
- the lamplighter group, which is the wreath product $\mathbf{Z} \wr (\mathbf{Z}/2\mathbf{Z})$;
- and the general or special linear group $\mathrm{GL}_n(\mathbf{K})$ or $\mathrm{SL}_n(\mathbf{K})$ over an infinite field $\mathbf{K}$.

Among other things, these worked-out examples (except D_∞, the only group of type I in the list) will demonstrate that, in general:

- a complete description of $\widehat{G}$ is hopeless or useless;
- a description of the spaces $\widehat{G}_{\mathrm{fd}}$, $\mathrm{Prim}(G)$, and $E(G)$ is often possible;
- the map $\kappa^{\mathrm{d}}_{\mathrm{prim}} : \widehat{G} \twoheadrightarrow \mathrm{Prim}(G)$ is not injective, indeed it has large fibers;
- a full description of the space $\mathrm{Char}(G) \smallsetminus E(G)$ seems to be difficult in general.

Along the way, we establish results on the dual spaces of locally compact groups G which can be written as a semi-direct product of a countable discrete subgroup H and an open abelian normal subgroup N (observe that, apart from $\mathrm{GL}_n(\mathbf{K})$, every group appearing above is of this form). Let $G = H \ltimes N$ be such a group. We will

- show that the description of $\widehat{G}$ amounts to the description of G-quasi-invariant probability measures on $\widehat{N}$ and of certain cocycles $\widehat{N} \times H \to \mathcal{U}(\mathcal{K})$ with values in the unitary group of a Hilbert space $\mathcal{K}$, usually a hopeless task (Theorem 5.D.9);
- give a description of the space $\widehat{G}_{\mathrm{fd}}$, when the subgroup H is moreover abelian (Theorem 4.A.3);
- construct factor representations of G of various types (Theorem 14.B.1);
- construct normal factor representations of G and determine their characters (Theorem 14.C.2).

Other features of this book include complete proofs of the following facts:

- $\mathrm{SL}_2(\mathbf{R})$ is a group of type I (Theorem 6.E.17);
- a finitely generated group Γ which is not virtually abelian is not of type I (this is a special case of Thoma's theorem 7.D.1);
- the primitive dual of $\mathrm{PSL}_n(\mathbf{K})$) consists of only two points when $\mathbf{K}$ is an algebraic extension of a finite field and $n \geq 3$ (Corollary 9.F.19).

For a second-countable locally compact group G, the duals and sets of characters introduced in this book are organized as in the following diagram.

$$
\begin{array}{ccccccccc}
 & & & & & & & & \mathrm{Tr}_1(G) \\
 & & & & & & & & \cup \\
\widehat{G}_{\mathrm{fd}} & & \hookrightarrow & & \mathrm{QD}(G)_{\mathrm{fin}} & \approx & \mathrm{Char}(G)_{\mathrm{fin}} & \approx & E(G) \\
 & & & & \cap & & \cap & & \\
\cap & & & & \mathrm{QD}(G)_{\mathrm{norm}} & \approx & \mathrm{Char}(G) & & \\
 & & & & \cap & & & & \\
\widehat{G} & & \overset{\kappa^{\mathrm{d}}_{\mathrm{qd}}}{\hookrightarrow} & & \mathrm{QD}(G) & & & & \\
 & \searrow \kappa^{\mathrm{d}}_{\mathrm{prim}} & & \swarrow \kappa^{\mathrm{qd}}_{\mathrm{prim}} & & & & & \\
 & & \mathrm{Prim}(G) & & & & & &
\end{array}
$$

We discuss in detail some of the basic tools needed for the proofs; in particular, we will give an account on

- the Gel'fand–Naimark–Segal construction associated to arbitrary functions of positive type as well as to central functions of positive type on a group (Section 1.B, 8.A, and 11.B);

- induced representations of a locally compact group from an open subgroup, and the Mackey–Shoda irreducibility and equivalence criteria (Section 1.F);
- unitary representations of a locally compact abelian group, in terms of multiplicity-free representations, and also in terms of projection-valued measures (Chapter 2);
- von Neumann algebras associated with a representation (Chapter 6);
- C*-algebras associated with a locally compact group (Chapter 8);
- traces on C*-algebras and the theory of Hilbert algebras (Chapter 10);
- the group measure space construction of Murray and von Neumann (Chapter 13).

Various other reminders on topology, measure theory, abelian harmonic analysis, and functional analysis are collected in an appendix to this book.

Some comments are in order concerning operator algebras. On the one hand, the theory of unitary representations of groups and that of operator algebras cannot be separated, neither conceptually nor historically. On the other hand, we would find it frustrating to go through all the necessary prerequisites of operator algebras before addressing group representations. We have tried to propose a helpful selection of "reminders" on operator algebras, scattered through the chapters on group representations. For what is not explained here in detail, most citations refer to Dixmier's books [**Dixm–vN**, **Dixm–C***]. For a shorter exposition, we can also recommend [**Wall–92**, Chap. 14].

CHAPTER 1

Unitary dual and primitive dual

In Section 1.A, *we introduce unitary group representations and unitary duals. The unitary dual* $\widehat{G}$ *of a topological group* G *is the set of equivalence classes of its irreducible unitary representations; we write simply "dual" for "unitary dual".*

A unitary representation π *of a topological group* G *is cyclic if there exists a vector* ξ *in the Hilbert space of* π *such that* $\pi(G)\xi$ *generates the Hilbert space of* π. *There is a correspondence between equivalence classes of pairs* (π, ξ), *where* π *is a unitary representation with cyclic vector* ξ, *and a class of functions on* G, *called functions of positive type. Under this correspondence, irreducible representations are mapped into indecomposable functions of positive type and vice versa. This is the subject of Section* 1.B *on the Gel'fand–Naimark–Segal construction.*

In Section 1.C, *we introduce the so-called Fell topology on the dual* $\widehat{G}$ *of a topological group* G *and show how it is related to the notion of weak containment between unitary representations. Section* 1.D *is devoted to the interplay between the properties of a locally compact group* G *and the topological properties of its dual* $\widehat{G}$.

It turns out that, in general, the topological space $\widehat{G}$ *does not even have the weakest of all separation properties, the* T_0 *property. In Section* 1.E, *we introduce our second dual space: the primitive dual* $\mathrm{Prim}(G)$ *of a topological group* G *is the set of weak equivalence classes of irreducible representations of* G. *The space* $\mathrm{Prim}(G)$ *is the largest* T_0 *quotient of* $\widehat{G}$. *In case* G *is a locally compact group, we will give an alternative description of* $\mathrm{Prim}(G)$ *later in Chapter* 8.

One of the main methods for the construction of unitary representations of a group G *is induction, a way to build representations of* G *out of representations of a subgroup* H. *In Section* 1.F, *we give an account of induced representations in case the subgroup* H *is open. In particular, a treatment of the Mackey–Shoda criteria of irreducibility and equivalence of induced representations is presented.*

In Section 1.G, *we give a short introduction to the notions of direct integrals of Hilbert spaces and direct integrals of unitary representations. We discuss how every unitary representation of a second-countable locally compact group is equivalent to a direct integral of* irreducible *representations. Moreover, we show that such direct integral decompositions are not unique in general.*

As a preparation for later chapters, we give a characterization of the decomposable operators between two direct integrals of Hilbert spaces (Section 1.H*) and we discuss the notion of a direct integral of von Neumann algebras (Section* 1.I*).*

1.A. Definition of the unitary dual

Let $\mathcal{H}$ be a Hilbert space; in this book, Hilbert spaces are always assumed to be *complex*. We denote by $\mathcal{L}(\mathcal{H})$ the involutive algebra of bounded linear operators on $\mathcal{H}$ and by

$$\mathcal{U}(\mathcal{H}) = \{U \in \mathcal{L}(\mathcal{H}) \mid U^*U = UU^* = \mathrm{Id}_{\mathcal{H}}\}$$

the group of unitary operators on $\mathcal{H}$. Sometimes, the field $\mathbf{C}$ of complex numbers is viewed as a Hilbert space.

On $\mathcal{L}(\mathcal{H})$, the **strong topology** is the locally convex topology defined by the set of seminorms $x \mapsto \|x\xi\|$, for $\xi \in \mathcal{H}$, and the **weak topology** is the locally convex topology defined by the family of seminorms $x \mapsto |\langle x\xi \mid \eta\rangle|$, for $\xi, \eta \in \mathcal{H}$. The restrictions of these two topologies to $\mathcal{U}(\mathcal{H})$ coincide, and make $\mathcal{U}(\mathcal{H})$ a topological group. More on this in Appendix A.8.

Let G be a topological group. The unit element in G is denoted by e, unless tradition imposes 0 or 1.

DEFINITION 1.A.1. A **unitary representation** π of G on a Hilbert space $\mathcal{H}$ is a homomorphism $\pi : G \to \mathcal{U}(\mathcal{H})$ such that the associated map

$$G \times \mathcal{H} \to \mathcal{H}, \ (g, \xi) \mapsto \pi(g)\xi$$

is continuous; we often write $\mathcal{H}_\pi$ rather than $\mathcal{H}$ for the Hilbert space of π, and $(\pi, \mathcal{H})$ for π.

From now on in this book, all representations of groups are unitary
unless explicitly written otherwise,
and we write "**representation**" for "unitary representation".

Note also that our definition of "representation" *includes* the continuity requirement.

PROPOSITION 1.A.2. *Let G be a topological group, $\mathcal{H}$ a Hilbert space, and $\pi : G \to \mathcal{U}(\mathcal{H})$ a homomorphism. The following two conditions are equivalent:*

(i) *the map $G \times \mathcal{H} \to \mathcal{H}$, $(g, \xi) \to \pi(g)\xi$ is continuous, i.e., π is a representation of G on $\mathcal{H}$;*
(ii) *the map π is continuous with respect to the strong topology on $\mathcal{U}(\mathcal{H})$.*

PROOF. One implication is obvious and the other follows from the inequality

$$\|\pi(g)\xi - \pi(h)\eta\| \leq \|\pi(g)\xi - \pi(h)\xi\| + \|\pi(h)\xi - \pi(h)\eta\| = \|\pi(g)\xi - \pi(h)\xi\| + \|\xi - \eta\|$$

for $g, h \in G$ and $\xi, \eta \in \mathcal{H}$. □

We write 1_G for the **unit representation** of G in $\mathbf{C}$, defined by $1_G(g) = \mathrm{Id}_{\mathbf{C}}$ for all $g \in G$.

A $\pi(G)$-invariant closed subspace $\mathcal{K}$ of $\mathcal{H}$ gives rise to a **subrepresentation** $G \to \mathcal{U}(\mathcal{K})$, assigning to $g \in G$ the restriction of $\pi(g)$ to $\mathcal{K}$.

Let $(\pi_1, \mathcal{H}_1), (\pi_2, \mathcal{H}_2)$ be two representations of G. A bounded linear operator $T : \mathcal{H}_1 \to \mathcal{H}_2$ is an **intertwiner** between π_1 and π_2 if $T\pi_1(g) = \pi_2(g)T$ for all $g \in G$. Observe that the set $\mathrm{Hom}_G(\pi_1, \pi_2)$ of intertwiners between π_1 and π_2 is a linear subspace of the space $\mathcal{L}(\mathcal{H}_1, \mathcal{H}_2)$ of bounded linear operators from $\mathcal{H}_1$ to $\mathcal{H}_2$. For one representation $(\pi, \mathcal{H})$, the set $\mathrm{Hom}_G(\pi, \pi)$ is the commutant $\pi(G)'$; it is a subalgebra of $\mathcal{L}(\mathcal{H})$ which contains the set $\mathbf{C}\mathrm{Id}_{\mathcal{H}}$ of scalar multiples of the identity of $\mathcal{H}$. More generally, the **commutant** of a subset $\mathcal{S}$ of $\mathcal{L}(\mathcal{H})$ is

$$\mathcal{S}' = \{T \in \mathcal{L}(\mathcal{H}) \mid TS = ST \text{ for all } S \in \mathcal{S}\}.$$

DEFINITION 1.A.3. Two representations π_1 and π_2 of a topological group G are **equivalent**, and this is written $\pi_1 \simeq \pi_2$ if there exists an isomorphism $V : \mathcal{H}_1 \to \mathcal{H}_2$ of Hilbert spaces such that $\pi_2(g) = V\pi_1(g)V^{-1}$ for all $g \in G$.

The representations π_1 and π_2 are **disjoint** if there does not exist a subrepresentation of π_1 equivalent to a subrepresentation of π_2.

(We come back to these definitions in Chapter 6.)

The **direct sum** of a family $(\pi_\iota)_{\iota\in I}$ of representations of G is the representation $\pi = \bigoplus_{\iota\in I} \pi_\iota$ of G in the Hilbert direct sum $\bigoplus_{\iota\in I} \mathcal{H}_{\pi_\iota}$ defined by $\pi(g) = \bigoplus_{\iota\in I} \pi_\iota(g)$ for all $g \in G$.

PROPOSITION 1.A.4. *Let $(\pi, \mathcal{H})$ be a representation of a topological group G such that $\mathcal{H}$ contains a $\pi(G)$-invariant closed subspace $\mathcal{K}$.*

Then the orthogonal complement $\mathcal{K}^\perp$ is also a $\pi(G)$-invariant closed subspace of $\mathcal{H}$. It follows that π is the direct sum of the two subrepresentations of G in the subspaces $\mathcal{K}$ and $\mathcal{K}^\perp$.

PROOF. Let $\xi \in \mathcal{K}^\perp$. For all $\eta \in \mathcal{K}$, we have

$$\langle \pi(g)\xi \mid \eta\rangle \;=\; \langle \xi \mid \pi(g^{-1})\eta\rangle \;\in\; \langle \xi \mid \mathcal{K}\rangle \;=\; \{0\},$$

and hence $\pi(g)\xi \in \mathcal{K}^\perp$. This shows that $\mathcal{K}^\perp$ is $\pi(G)$-invariant. □

LEMMA 1.A.5. *Let $(\pi_1, \mathcal{H}_1)$, $(\pi_2, \mathcal{H}_2)$ be two representations of a topological group G, and $T \in \mathrm{Hom}_G(\pi_1, \pi_2)$. Set $\mathcal{K}_1 = (\ker T)^\perp$ and let $\mathcal{K}_2$ denote the closure of the image of T.*

Then $\mathcal{K}_1$ and $\mathcal{K}_2$ are closed invariant subspaces of $\mathcal{H}_1$ and $\mathcal{H}_2$ respectively, and the subrepresentation of π_1 defined by $\mathcal{K}_1$ is equivalent to the subrepresentation of π_2 defined by $\mathcal{K}_2$.

PROOF. Since T intertwines π_1 and π_2, the space $\mathcal{K}_1$ is $\pi_1(G)$-invariant, the space $T(\mathcal{K}_1)$ is $\pi_2(G)$-invariant, and therefore $\mathcal{K}_2$ is $\pi_2(G)$-invariant.

One checks immediately that $T^* \in \mathcal{L}(\mathcal{H}_2, \mathcal{H}_1)$ intertwines π_2 and π_1, and that $T^*T \in \mathcal{L}(\mathcal{H}_1)$ intertwines π_1 with itself. Since $|T| = (T^*T)^{1/2}$ is a limit in the strong operator topology of polynomials in T^*T (see Section A.9), $|T|$ also intertwines π_1 with itself.

Consider the partial isometry U of the polar decomposition $T = U|T|$; recall that $\ker U = \ker T$ and that the restriction of U to $\mathcal{K}_1$ is an isometry onto $\mathcal{K}_2$. By the definition of U (i.e., $U\xi = 0$ if $\xi \in \ker T$ and $U|T|\xi = T\xi$ if $\xi \in (\ker T)^\perp$; see Appendix A.9), it follows that U intertwines the restriction of π_1 to $\mathcal{K}_1$ and the restriction of π_2 to $\mathcal{K}_2$. Therefore these two restrictions are equivalent. □

In the following proposition, the implications (ii) $\Rightarrow$ (i) and (iii) $\Rightarrow$ (iv) are immediate consequences of Lemma 1.A.5, and the converse implications are trivial.

PROPOSITION 1.A.6. *Let $(\pi_1, \mathcal{H}_1)$ and $(\pi_2, \mathcal{H}_2)$ be two representations of a topological group G. On the one hand, the following two properties are equivalent:*

(i) *π_1 and π_2 are equivalent;*
(ii) *$\mathrm{Hom}_G(\pi_1, \pi_2)$ contains an invertible operator from $\mathcal{H}_1$ to $\mathcal{H}_2$.*

On the other hand, the following two properties are equivalent:

(iii) *π_1 and π_2 are disjoint;*
(iv) *$\mathrm{Hom}_G(\pi_1, \pi_2) = \{0\}$.*

COROLLARY 1.A.7. *Let π, and π_i for $i \in I$, be representations of G. Assume that π and π_i are disjoint for every $i \in I$.*

Then π and $\bigoplus_{i\in I} \pi_i$ are disjoint.

PROOF. In view of Proposition 1.A.6, it suffices to show that $\mathrm{Hom}_G(\pi, \bigoplus_{i\in I} \pi_i) = \{0\}$.

Let $\mathcal{H}_i$ be the Hilbert space of π_i and $\mathcal{H} = \bigoplus_{i\in I} \mathcal{H}_i$. Denote by P_i the orthogonal projection from $\mathcal{H}$ to $\mathcal{H}_i$. Let $T \in \mathrm{Hom}_G(\pi, \bigoplus_{i\in I} \pi_i)$. Then $P_iT \in \mathrm{Hom}_G(\pi, \pi_i)$, and hence $P_iT = 0$ for every $i \in I$. Therefore, $T = 0$. □

The following proposition shows that the equivalence of representations satisfies a form of the **Schröder–Bernstein property**.

PROPOSITION 1.A.8. *Let π_1, π_2 be two representations of G in Hilbert spaces $\mathcal{H}, \mathcal{K}$ respectively. Assume that π_1 is equivalent to a subrepresentation of π_2 and π_2 is equivalent to a subrepresentation of π_1.*

Then π_1 and π_2 are equivalent.

PROOF. By hypothesis, there exist

- a G-invariant closed subspace $\mathcal{K}'$ of $\mathcal{K}$ and a G-equivariant partial isometry S of initial space $\mathcal{H}$ and final space $\mathcal{K}'$,
- a G-invariant closed subspace $\mathcal{H}'$ of $\mathcal{H}$ and a G-equivariant partial isometry T of initial space $\mathcal{K}$ and final space $\mathcal{H}'$.

Set $U = TS$ and $\mathcal{H}'' = U(\mathcal{H})$. Then

$$(\sharp) \qquad \mathcal{H} \supset \mathcal{H}' \supset \mathcal{H}''$$

and U is a G-equivariant partial isometry of initial space $\mathcal{H}$ and final space $\mathcal{H}''$. For all $m \geq 0$, set

$$\mathcal{H}^{(2m)} = U^m(\mathcal{H}) \quad \text{and} \quad \mathcal{H}^{(2m+1)} = U^m(\mathcal{H}').$$

Observe that $\mathcal{H}^{(0)} = \mathcal{H}$, $\mathcal{H}^{(1)} = \mathcal{H}'$, and $\mathcal{H}^{(2)} = \mathcal{H}''$. It follows from $(\sharp)$ that

$$\mathcal{H}^{(2m)} \supset \mathcal{H}^{(2m+1)} \supset \mathcal{H}^{(2m+2)} \quad \text{for all } m \geq 0.$$

Set $\mathcal{J} = \bigcap_{n\geq 0} \mathcal{H}^{(n)}$. We have orthogonal Hilbert sums

$$\mathcal{H} = \mathcal{J} \oplus (\mathcal{H}^{(0)} \ominus \mathcal{H}^{(1)}) \oplus (\mathcal{H}^{(1)} \ominus \mathcal{H}^{(2)}) \oplus (\mathcal{H}^{(2)} \ominus \mathcal{H}^{(3)}) \oplus (\mathcal{H}^{(3)} \ominus \mathcal{H}^{(4)}) \oplus \cdots,$$

$$\mathcal{H}' = \mathcal{J} \oplus (\mathcal{H}^{(1)} \ominus \mathcal{H}^{(2)}) \oplus (\mathcal{H}^{(2)} \ominus \mathcal{H}^{(3)}) \oplus (\mathcal{H}^{(3)} \ominus \mathcal{H}^{(4)}) \oplus (\mathcal{H}^{(4)} \ominus \mathcal{H}^{(5)}) \oplus \cdots,$$

where $\mathcal{H}^{(i)} \ominus \mathcal{H}^{(j)}$ denotes the orthogonal complement of $\mathcal{H}^{(j)}$ in $\mathcal{H}^{(i)}$. Since U induces for all $m \geq 0$ a map $(\mathcal{H}^{(2m)} \ominus \mathcal{H}^{(2m+1)}) \to (\mathcal{H}^{(2m+2)} \ominus \mathcal{H}^{(2m+3)})$ which is a G-equivariant surjective isometry, we also have

$$\mathcal{H}' \simeq \mathcal{J} \oplus (\mathcal{H}^{(1)} \ominus \mathcal{H}^{(2)}) \oplus (\mathcal{H}^{(0)} \ominus \mathcal{H}^{(1)}) \oplus (\mathcal{H}^{(3)} \ominus \mathcal{H}^{(4)}) \oplus (\mathcal{H}^{(2)} \ominus \mathcal{H}^{(3)}) \oplus \cdots.$$

It follows that the representations of G in $\mathcal{H}$ and $\mathcal{H}'$ are equivalent. Since the representations of G in $\mathcal{H}'$ and $\mathcal{K}$ are equivalent by hypothesis, this concludes the proof. □

DEFINITION 1.A.9. A representation π of the topological group G is **irreducible** if its Hilbert space $\mathcal{H}_\pi$ is not $\{0\}$ and if the only $\pi(G)$-invariant closed subspaces of $\mathcal{H}_\pi$ are $\{0\}$ and $\mathcal{H}_\pi$.

REMARK 1.A.10. (1) The irreducibility of a representation is a property of its equivalence class: if π is an irreducible representation of G, then every representation which is equivalent to π is irreducible.

(2) Representations of G of dimension 1 are irreducible. They are equivalent to **unitary characters** of G, that is, to continuous homomorphisms $G \to \mathbf{T}$.

Here and below, $\mathbf{R}$ denotes the field of real numbers, and $\mathbf{T}$ the circle group, viewed sometimes as $\{z \in \mathbf{C} \mid |z| = 1\}$ and sometimes as $\mathbf{R}/\mathbf{Z}$.

In the following proposition, Claim (1) is known as Schur's lemma. For the easy proof, we refer to [**BeHV–08**, Appendix A].

PROPOSITION 1.A.11. *Let π, π_1, π_2 be representations of a topological group G:*

(1) *The representation π is irreducible if and only if* $\mathrm{Hom}_G(\pi, \pi) = \mathbf{C}\mathrm{Id}_{\mathcal{H}_\pi}$.
(2) *Suppose that π_1 and π_2 are irreducible; then:*
either π_1 and π_2 are equivalent and $\dim_{\mathbf{C}} \mathrm{Hom}_G(\pi_1, \pi_2) = 1$,
or π_1 and π_2 are not equivalent and $\mathrm{Hom}_G(\pi_1, \pi_2) = \{0\}$.

Given a representation of a group, it is often useful to consider its restriction to the centre.

DEFINITION 1.A.12. Let G be a topological group; let Z denote its centre. A representation π of G is said to **admit a central character** if, for every $g \in Z$, there exists $\lambda_\pi(g) \in \mathbf{T}$ such that $\pi(g) = \lambda_\pi(g)\mathrm{Id}_{\mathcal{H}_\pi}$. In this case, the map

$$\lambda_\pi : Z \to \mathbf{T}, \quad g \mapsto \lambda_\pi(g)$$

is a unitary character of Z called the **central character** of π. It only depends on the equivalence class of π.

By Schur's lemma, every irreducible representation admits a central character. More generally, every factor representation admits a central character (see Proposition 8.E.1). Some induced representations have central characters (see Proposition 1.F.7).

The next proposition shows that the dimensions of the Hilbert spaces in which irreducible representations of G exist are bounded by some cardinal. There is a reminder on the dimensions of Hilbert spaces in Appendix A.8.

PROPOSITION 1.A.13. *Let G be a topological group. Let $\aleph$ be a cardinal such that there exists in G a dense subset, say X, of cardinality $\aleph$.*

Then, for every irreducible representation $(\pi, \mathcal{H})$ of G, the dimension of $\mathcal{H}$ is bounded by $\aleph$.

In particular, if G is separable, then $\mathcal{H}$ is a separable Hilbert space.

PROOF. Let ξ be a nonzero vector in $\mathcal{H}$ and let $\mathcal{K}$ be the closed subspace of $\mathcal{H}$ generated by $\pi(G)\xi$. Since X is dense in G and since X has cardinality $\aleph$, it is clear that $\dim \mathcal{K} \le \aleph$. Since $\mathcal{K}$ is G-invariant and π is irreducible, we have $\mathcal{K} = \mathcal{H}$. □

The following consequence of Proposition 1.A.13 is crucial for the subject of this book.

COROLLARY 1.A.14. *The equivalence classes of irreducible representations of a topological group G constitute a set, denoted by $\widehat{G}$.*

PROOF. Let $\aleph$ be a cardinal which satisfies the condition stated in Proposition 1.A.13 and let $\mathcal{H}$ be a fixed Hilbert space of dimension $\aleph$. The family $\mathcal{R}$ of irreducible representations of G in closed linear subspaces of $\mathcal{H}$ is a set. Proposition 1.A.13 shows that every irreducible representation of G is equivalent to some representation from $\mathcal{R}$. So, $\widehat{G}$ is the quotient space $\mathcal{R}/\simeq$ of $\mathcal{R}$ by the relation of equivalence of representations. □

DEFINITION 1.A.15. The **unitary dual** $\widehat{G}$ of G is the space of equivalence classes of irreducible representations of G.

Most often below, we write simply "**dual**" for "unitary dual".

When G is abelian, its dual $\widehat{G}$ can be identified with the group $\mathrm{Hom}(G, \mathbf{T})$ of continuous homomorphisms $G \to \mathbf{T}$.

The dual of a topological group is a topological space for the Fell topology, of which we postpone a first definition until Section 1.C. For a locally compact group, an equivalent definition of the Fell topology on its dual is discussed in Section 8.B.

Next, we examine the action of automorphisms of G on the representation theory of G.

NOTATION 1.A.16. When we wish to emphasize that an action of a group G on a set X is a left action $G \times X \to X$, we refer to this action by the notation

$$G \curvearrowright X.$$

Similarly, for a right action $X \times G \to X$, we use

$$X \curvearrowleft G.$$

See also Definition A.5.7.

REMARK 1.A.17. Let $\mathrm{Aut}(G)$ be the group of bicontinuous automorphisms of G. We consider $\mathrm{Aut}(G)$ as acting on G on the left.

(1) Let $\theta \in \mathrm{Aut}(G)$. For a representation $(\pi, \mathcal{H})$ of G, the **conjugate representation of π by θ** is the representation π^θ of G which is defined on the same Hilbert space $\mathcal{H}$ by

$$\pi^\theta(g) = \pi(\theta(g)) \quad \text{for } g \in G.$$

Let $\aleph$ be a cardinal (for example, the cardinal $\aleph_0$ of the set of integers). For each cardinal $n \leq \aleph$, let $\mathcal{H}_n$ be a Hilbert space of dimension n. Denote by $\mathrm{Rep}_{\leq\aleph}(G)$ the set of representations of G in one of the Hilbert spaces $\mathcal{H}_n$, with $n \leq \aleph$. There is a natural **right** action $\mathrm{Rep}_{\leq\aleph}(G) \times \mathrm{Aut}(G) \to \mathrm{Rep}_{\leq\aleph}(G)$ defined by

$$(\pi, \theta) \mapsto \pi^\theta \quad \text{for } \pi \in \mathrm{Rep}_{\leq\aleph}(G) \text{ and } \theta \in \mathrm{Aut}(G).$$

The subset $\mathrm{Irr}_{\leq\aleph}(G)$ of $\mathrm{Rep}_{\leq\aleph}(G)$ consisting of irreducible representations is invariant by $\mathrm{Aut}(G)$, and hence the action just defined restricts to a right action of $\mathrm{Aut}(G)$ on $\mathrm{Irr}_{\leq\aleph}(G)$.

(2) Let $\aleph$ be a cardinal such that every irreducible representation of G is of dimension at most $\aleph$ (see Proposition 1.A.13). The dual $\widehat{G}$ can be identified with the quotient of $\mathrm{Irr}_{\leq\aleph}(G)$ by the relation of equivalence of representations, i.e., $\widehat{G} = \mathrm{Irr}_{\leq\aleph}(G)/\simeq$ (see also Subsection 6.C.a).

The action described in (1) induces a right action $\widehat{G} \curvearrowleft \mathrm{Aut}(G)$ of $\mathrm{Aut}(G)$ on $\widehat{G}$.

(3) Let π be a representation of G and $a \in G$. By a slight abuse of notation, we denote by π^a the conjugate representation of π by the inner automorphism defined by a. Observe that $\pi^a : g \mapsto \pi(aga^{-1})$ is equivalent to π; indeed, the unitary operator $\pi(a)$ intertwines π^a and π.

(4) We extend as follows the situation considered in Item (3). Let N be a closed normal subgroup of G and ρ a representation of N. Every element $a \in G$ defines an automorphism of N; the conjugate representation ρ^a of ρ by a is the conjugate representation of ρ by this automorphism, so that

$$\rho^a(n) = \rho(ana^{-1}), \qquad \text{for all } n \in N.$$

The map

$$\widehat{N} \times G \to \widehat{N}, \qquad (\rho, a) \mapsto \rho^a$$

defines a *right* action of G on the dual space $\widehat{N}$ of N. Since ρ^a is equivalent to ρ for $a \in N$, this action factorizes into an action $\widehat{N} \curvearrowleft G/N$ of the quotient group G/N.

1.B. Functions of positive type and GNS construction

The material of this section can be found in many places, for example in [**Dixm–C***, § 13] and [**BeHV–08**, Appendix C].

Let G be a topological group.

DEFINITION 1.B.1. A **function of positive type** on G is a continuous function $\varphi : G \to \mathbf{C}$ such that

$$\sum_{i,j=1}^{n} c_i \overline{c_j} \varphi(g_j^{-1} g_i) \geq 0 \quad \text{for all } n \geq 1, \; c_1, \ldots, c_n \in \mathbf{C}, \; \text{and } g_1, \ldots, g_n \in G.$$

Many authors use "**positive definite function**" instead of "function of positive type".

REMARK 1.B.2. (1) Let φ be such a function of positive type on G. For $n = 2$, $g_1 = e$, and $g_2 = g$, the definition implies that the matrix $\begin{pmatrix} \varphi(e) & \varphi(g^{-1}) \\ \varphi(g) & \varphi(e) \end{pmatrix}$ is positive Hermitian; it follows that

$$\varphi(g^{-1}) = \overline{\varphi(g)} \quad \text{and} \quad |\varphi(g)| \leq \varphi(e) \quad \text{for all } g \in G.$$

Note that the function $\check{\varphi}$, defined by $\check{\varphi}(g) = \varphi(g^{-1})$ for all $g \in G$, is of positive type, as well as the function $\mathrm{Re}(\varphi) = \frac{1}{2}(\varphi + \check{\varphi})$.

(2) Functions of positive type were first studied in classical harmonic analysis on groups like $\mathbf{T}$ or $\mathbf{R}$. For an historical survey, we refer to [**Stew–76**].

NOTATION 1.B.3. Denote by $P(G)$ the set of continuous functions $\varphi : G \to \mathbf{C}$ which are of positive type.

For $\varphi, \psi \in P(G)$ and $c \in \mathbf{R}_+$, it follows from the definition that $\varphi + \psi$, $c\varphi$, and $\overline{\varphi}$ are of positive type. [For $\varphi\psi$, see Example 1.B.9 (4).] Therefore $P(G)$ is a convex cone in the space $C^b(G)$ of bounded continuous complex-valued functions on G, which is closed under complex conjugation.

Let $P_{\leq 1}(G)$ be the subset of functions $\varphi \in P(G)$ such that $\sup_{g \in G} |\varphi(g)| = \varphi(e) \leq 1$, and $P_1(G)$ the subset of functions $\varphi \in P_{\leq 1}(G)$ normalized by $\varphi(e) = 1$. The subsets $P_{\leq 1}(G)$ and $P_1(G)$ are also convex.

DEFINITION 1.B.4. Let π be a representation of G on a Hilbert space $\mathcal{H}_\pi$. A **matrix coefficient** of π is a continuous function $\varphi_{\pi,\xi,\eta}$ on G defined by

$$\varphi_{\pi,\xi,\eta}(g) = \langle \pi(g)\xi \mid \eta \rangle \quad \text{for all } g \in G,$$

for some $\xi, \eta \in \mathcal{H}_\pi$. A **diagonal matrix coefficient** is a function of the form $\varphi_{\pi,\xi,\xi}$; for simplicity, we instead write $\varphi_{\pi,\xi}$.

Diagonal matrix coefficients are of positive type. Indeed, for any positive integer n and $c_1, \dots, c_n \in \mathbf{C}$, $g_1, \dots, g_n \in G$, $\xi \in \mathcal{H}_\pi$, we have

$$\sum_{i,j=1}^{n} c_i \overline{c_j} \varphi_{\pi,\xi}(g_j^{-1} g_i) = \Big\langle \sum_{i=1}^{n} c_i \pi(g_i)\xi \;\Big|\; \sum_{j=1}^{n} c_j \pi(g_j)\xi \Big\rangle \geq 0.$$

These coefficients $\varphi_{\pi,\xi}$ are the **functions of positive type associated with** π. Note that $\varphi_{\pi,\xi}(e) = \Vert\xi\Vert^2$.

Since two equivalent representations give rise to the same space of diagonal matrix coefficients, functions of positive type are associated with equivalence classes of representations. Below, we will often use (abusively) the same symbol for a representation and its equivalence class.

DEFINITION 1.B.5. Let G be a topological group and π a representation of G on a Hilbert space $\mathcal{H}$. A vector $\xi \in \mathcal{H}$ is **cyclic** if the smallest closed subspace of $\mathcal{H}$ containing $\pi(G)\xi$ is $\mathcal{H}$ itself; the representation π is **cyclic** if it has a cyclic vector.

CONSTRUCTION 1.B.6. The **Gel'fand–Naimark–Segal representation**, or for short the **GNS representation**, or the **GNS construction**, shows that the diagonal matrix coefficients of Definition 1.B.4 provide all examples of functions of positive type on a topological group G. More precisely, the construction below shows that, given $\varphi \in P(G)$, $\varphi \neq 0$, there exist a representation π_φ of G on a Hilbert space $\mathcal{H}_\varphi$, and a cyclic vector $\xi_\varphi \in \mathcal{H}_\varphi$, such that φ is a diagonal matrix coefficient of π_φ, i.e., such that $\varphi(g) = \langle \pi_\varphi(g)\xi_\varphi \mid \xi_\varphi \rangle$ for all $g \in G$.

Let $\mathbf{C}[G]$ denote the complex vector space of functions $f : G \to \mathbf{C}$ with finite support. Then $\mathbf{C}[G]$ is a unital involutive algebra for the convolution product defined by

$$(f_1 * f_2)(x) = \sum_{y \in G} f_1(xy^{-1}) f_2(y) \quad \text{for } f_1, f_2 \in \mathbf{C}[G],\ x \in G,$$

and the involution given by

$$f^*(x) = \overline{f(x^{-1})} \quad \text{for } f \in \mathbf{C}[G],\ x \in G.$$

For $g \in G$, the characteristic function δ_g of $\{g\}$ is in $\mathbf{C}[G]$. For $g, h \in G$, we have $\delta_g \delta_h = \delta_{gh}$ and $\delta_g^* = \delta_{g^{-1}}$.

Let φ be a nonzero function of positive type on G. A positive Hermitian form Φ_φ is defined on $\mathbf{C}[G]$ by

$$\Phi_\varphi(f_1, f_2) = \sum_{x,y \in G} f_1(x) \overline{f_2(y)} \varphi(y^{-1}x) \quad \text{for all } f_1, f_2 \in \mathbf{C}[G].$$

It follows from the Cauchy–Schwarz inequality that the set J_φ of all $f \in \mathbf{C}[G]$ such that $\Phi_\varphi(f, f) = 0$ is a left ideal of $\mathbf{C}[G]$. We write $[f]$ for the class in the vector space $\mathbf{C}[G]/J_\varphi$ of a function $f \in \mathbf{C}[G]$. We define $\mathcal{H}_\varphi$ as the Hilbert space completion of $\mathbf{C}[G]/J_\varphi$ for the scalar product given on $\mathbf{C}[G]/J_\varphi$ by

$$\langle [f_1] \mid [f_2] \rangle = \Phi_\varphi(f_1, f_2) \quad \text{for all } f_1, f_2 \in \mathbf{C}[G].$$

For $g \in G$, the operator on $\mathbf{C}[G]$ defined by $f \mapsto \big(x \mapsto f(g^{-1}x)\big)$ preserves the bilinear form Φ_φ, passes therefore to the quotient $\mathbf{C}[G]/J_\varphi$, and extends to a unitary operator on $\mathcal{H}_\varphi$, denoted by $\pi_\varphi(g)$. It is clear that $\pi_\varphi(gh) = \pi_\varphi(g)\pi_\varphi(h)$ for all $g, h \in G$.

Let $f \in \mathbf{C}[G]$; set $\xi = [f] \in \mathbf{C}[G]/J_\varphi$. Then

$$\begin{aligned} \langle \pi_\varphi(g)\xi \mid \xi\rangle &= \sum_{x,y\in G} f(x)\overline{f(y)}\varphi(y^{-1}gx), \\ \langle \xi \mid \pi_\varphi(g)\xi\rangle &= \sum_{x,y\in G} f(x)\overline{f(y)}\varphi(y^{-1}g^{-1}x), \end{aligned}$$

so that

$$\Vert\pi_\varphi(g)\xi - \xi\Vert^2 = 2\Vert\xi\Vert^2 - \sum_{x,y\in G} f(x)\overline{f(y)}\left(\varphi(y^{-1}gx) + \varphi(y^{-1}g^{-1}x)\right).$$

Since φ is continuous and f has finite support,

$$\lim_{g\to e} \sum_{x,y\in G} f(x)\overline{f(y)}\varphi(y^{-1}gx) = \lim_{g\to e} \sum_{x,y\in G} f(x)\overline{f(y)}\varphi(y^{-1}g^{-1}x) = \Vert\xi\Vert^2.$$

It follows that $g \mapsto \pi_\varphi(g)\xi$ is continuous for every $\xi \in \mathbf{C}[G]/J_\varphi$, and therefore for every $\xi \in \mathcal{H}_\varphi$. This shows that the assignment $g \mapsto \pi_\varphi(g)$ is continuous, i.e., that it is a representation of G on $\mathcal{H}_\varphi$. Moreover, the vector $\xi_\varphi := [\delta_e]$ is a cyclic vector of norm $\sqrt{\varphi(e)}$, and we have

$$\varphi(g) = \langle \pi_\varphi(g)\xi_\varphi \mid \xi_\varphi\rangle \quad \text{for all } g \in G.$$

The **GNS triple** associated with φ consists of the Hilbert space $\mathcal{H}_\varphi$, the representation π_φ, and the cyclic vector ξ_φ.

REMARK 1.B.7. The terminology "GNS construction", which applies both to groups and C*-algebras (see Construction 8.A.8), refers to two seminal articles: [**GeNa–48**] and [**Sega–47**].

The GNS construction described above associates a representation of a group G to a function of positive type on the group. It is closely related to the GNS construction which associates a representation of a C*-algebra A to a state on the C*-algebra. See Subsection 8.A.b, and [**Dixm–C***, § 2].

Construction 1.B.6 uses the algebra $\mathbf{C}[G]$ of finitely supported functions, as in [**Thom–64a**]; this will be suitable for our treatment of traces in Section 11.B. There are variants using, instead of $\mathbf{C}[G]$, a space of continuous functions on G, as in [**BeHV–08**, C.1–C.4], or $L^1(G, \mu_G)$ when G is locally compact with Haar measure μ_G, as in [**Foll–16**, 3.20].

DEFINITION 1.B.8. Let G be a group, H a subgroup, and $\varphi : H \to \mathbf{C}$ a function. The **trivial extension** of φ to G is the function $\widetilde{\varphi} : G \to \mathbf{C}$ defined by

$$\widetilde{\varphi}(g) = \begin{cases} \varphi(g) & \text{if } g \in H, \\ 0 & \text{if } g \in G \smallsetminus H. \end{cases}$$

Observe that if G is a topological group, H an open subgroup, and φ a continuous function, then $\widetilde{\varphi}$ is continuous.

EXAMPLE 1.B.9. (1) On a topological group G, the constant function of value 1 is of positive type, and the corresponding GNS representation is the unit representation 1_G.

More generally, every unitary character $\chi : G \to \mathbf{T}$ is a function of positive type. With the notation of Construction 1.B.6, the ideal J_χ is the kernel of the algebra homomorphism $\mathbf{C}[G] \to \mathbf{C}$, $f \mapsto \sum_{x \in G} \chi(x) f(x)$, of codimension one, and the GNS representation π_χ of G is equivalent to χ.

(2) Let H be an open subgroup of the topological group G, and $\psi : G \to \mathbf{C}$ the characteristic function of H. Then ψ is of positive type, and π_ψ is equivalent to the quasi-regular representation of G on $\ell^2(G/H)$, with cyclic vector the characteristic function of the point H in G/H (see Definition 1.F.5 for the definition).

More generally, let $\chi : H \to \mathbf{T}$ be a unitary character, and $\widetilde{\chi} : G \to \mathbf{C}$ the trivial extension of χ. Then $\widetilde{\chi}$ is of positive type, and $\pi_{\widetilde{\chi}}$ is equivalent to the representation $\mathrm{Ind}_H^G \chi$, as defined in 1.F below. (This is a particular case of what is proved in Proposition 1.F.10.)

(3) In particular, for a discrete group Γ, the characteristic function δ_e of the unit element is of positive type, and π_{δ_e} is equivalent to the left regular representation λ_Γ.

(4) Let φ, ψ be two functions of positive type on a topological group. The function $\varphi + \psi$ is of positive type, and the representation $\pi_{\varphi+\psi}$ is equivalent to a subrepresentation of the direct sum $\pi_\varphi \oplus \pi_\psi$. The function $\varphi\psi$ is of positive type, and the representation $\pi_{\varphi\psi}$ is equivalent to a subrepresentation of the tensor product $\pi_\varphi \otimes \pi_\psi$.

(5) Let G be a *locally compact abelian* group and $\widehat{G}$ the dual group of G. For a probability measure μ on the σ-algebra $\mathcal{B}(\widehat{G})$ of Borel subsets of $\widehat{G}$, let $\mathcal{F}(\mu)$ denote the Fourier–Stieltjes transform of μ, which is a bounded continuous function on G. Bochner's Theorem establishes that the map $\mu \mapsto \mathcal{F}(\mu)$ is a bijection from the set of probability measures on $\mathcal{B}(\widehat{G})$ to the set $P_1(G)$ of normalized functions of positive type on G (Theorem A.7.5).

In Construction 2.A.1 below, we associate to a probability measure μ on $\widehat{G}$ a canonical representation π_μ of G on $L^2(\widehat{G}, \mu)$. Let $1_{\widehat{G}}$ denote the constant function of value 1 on $\widehat{G}$. Then the matrix coefficient $\langle \pi_\mu(g) 1_{\widehat{G}} \mid 1_{\widehat{G}} \rangle = \int_{\widehat{G}} \chi(g) d\mu(\chi)$ is the inverse Fourier transform $\overline{\mathcal{F}}(\mu)$ of μ, and hence $\langle \pi_\mu(g) 1_{\widehat{G}} \mid 1_{\widehat{G}} \rangle = \mathcal{F}(\check{\mu})$, where $\check{\mu}$ is the image of μ by the map $G \to G$, $g \mapsto g^{-1}$ (see Appendix A.7); as a result, $(\pi_\mu, L^2(\widehat{G}, \mu), 1_{\widehat{G}})$ is a GNS-triple associated with $\mathcal{F}(\check{\mu})$.

Let us say that two cyclic representations $\pi_j : G \to \mathcal{U}(\mathcal{H}_j)$, with cyclic vector $\xi_j \in \mathcal{H}_j$ $(j = 1, 2)$, are **equivalent** if there exists a surjective isometry $U : \mathcal{H}_1 \to \mathcal{H}_2$ such that $U\pi_1(g) = \pi_2(g)U$ for all $g \in G$, and $U\xi_1 = \xi_2$. In this case, it is clear that the functions of positive type associated with ξ_1 and ξ_2 coincide, that is, $\langle \pi_1(g)\xi_1 \mid \xi_1 \rangle = \langle \pi_2(g)\xi_2 \mid \xi_2 \rangle$ for all $g \in G$. The next result shows that the converse statement is also true.

PROPOSITION 1.B.10. *Let G be a topological group. The assignments*

$$\varphi \rightsquigarrow (\pi_\varphi, \mathcal{H}_\varphi, \xi_\varphi)$$

given by the GNS construction and

$$(\pi, \mathcal{H}, \xi) \mapsto \big(\varphi_{\pi,\xi} : g \mapsto \langle \pi(g)\xi \mid \xi \rangle\big)$$

induce bijections inverse to each other between

(i) *the set $P_1(G)$ of normalized functions of positive type on G and*

(ii) *the set of triples $(\pi, \mathcal{H}, \xi)$ where π is a representation of G on $\mathcal{H}$ and $\xi \in \mathcal{H}$ a unit cyclic vector, modulo equivalence.*

PROOF. Let us verify that the assignment of the function $\varphi_{\pi,\xi}$ to an equivalence class $(\pi, \mathcal{H}, \xi)$ is injective. The other verifications are left to the reader.

Let $(\pi_1, \mathcal{H}_1, \xi_1)$ and $(\pi_2, \mathcal{H}_2, \xi_2)$ be triples consisting of cyclic representations π_1 and π_2, with cyclic vectors ξ_1 and ξ_2 respectively. Let φ_1, φ_2 be the functions on G defined by

$$\varphi_1(g) = \langle \pi_1(g)\xi_1 \mid \xi_1 \rangle \quad \text{and} \quad \varphi_2(g) = \langle \pi_2(g)\xi_2 \mid \xi_2 \rangle \quad \text{for all } g \in G.$$

Assume that $\varphi_1 = \varphi_2$, and denote this function by φ. We have to show that the triples $(\pi_1, \mathcal{H}_1, \xi_1)$ and $(\pi_2, \mathcal{H}_2, \xi_2)$ are equivalent.

For all $g_1, \dots, g_n \in G$ and $c_1, \dots, c_n \in \mathbf{C}$, we have

$$\Big\| \sum_{i=1}^{n} c_i \pi_2(g_i)\xi_2 \Big\|^2 = \sum_{i,j=1}^{n} c_i \overline{c_j} \varphi(g_j^{-1} g_i) = \Big\| \sum_{i=1}^{n} c_i \pi_1(g_i)\xi_1 \Big\|^2.$$

It follows that the map $\sum_{i=1}^n c_i\pi_1(g_i)\xi_1 \mapsto \sum_{i=1}^n c_i\pi_2(g_i)\xi_2$ is well defined on the linear span of $\pi_1(G)\xi_1$ and extends to a Hilbert space isomorphism $U : \mathcal{H}_1 \to \mathcal{H}_2$. One checks that $U\pi_1(g) = \pi_2(g)U$ for all $g \in G$, and $U\xi_1 = \xi_2$, as was to be shown. □

REMARK 1.B.11. There is a natural *right* action $P(G) \curvearrowleft \mathrm{Aut}(G)$ of the group $\mathrm{Aut}(G)$ of bicontinuous automorphisms of G on the set $P(G)$ of functions of positive type on G, given by

$$\varphi^\theta(g) = \varphi(\theta(g)) \quad \text{for } \theta \in \mathrm{Aut}(G),\ \varphi \in P(G),\ g \in G.$$

Let $\varphi \in P(G)$ and $\theta \in \mathrm{Aut}(G)$. If $(\pi, \mathcal{H}, \xi)$ is a GNS-triple associated with φ, then $(\pi^\theta, \mathcal{H}, \xi)$ is a GNS-triple associated with φ^θ, where π^θ is the conjugate representation of π by θ as defined in Remark 1.A.17.

In Proposition 1.B.14, we will characterize those $\varphi \in P(G)$ for which π_φ is irreducible. This requires a definition and two preliminary lemmas.

Let $\mathrm{Extr}(P_{\le 1}(G))$ be the set of functions φ in $P_{\le 1}(G)$ that are **indecomposable** in the following sense: if $\varphi = t\varphi_1 + (1-t)\varphi_2$ for $\varphi_1, \varphi_2 \in P_{\le 1}(G)$ and $t \in \,]0,1[$, then $\varphi_1 = \varphi_2 = \varphi$. Similarly define $\mathrm{Extr}(P_1(G))$.

As the following lemma shows, there is a simple relationship between the sets $\mathrm{Extr}(P_{\le 1}(G))$ and $\mathrm{Extr}(P_1(G))$.

LEMMA 1.B.12. *For every topological group G, we have*

$$\mathrm{Extr}(P_{\le 1}(G)) = \mathrm{Extr}(P_1(G)) \cup \{0\}.$$

PROOF. Let us first show that $\mathrm{Extr}(P_1(G)) \cup \{0\} \subset \mathrm{Extr}(P_{\le 1}(G))$.

Let $\varphi \in \mathrm{Extr}(P_1(G))$; let $\varphi_1, \varphi_2 \in P_{\le 1}(G)$ and $t \in \,]0,1[$ be such that

$$t\varphi_1 + (1-t)\varphi_2 = \varphi.$$

Then $t\varphi_1(e) + (1-t)\varphi_2(e) = 1$. Therefore $\varphi_1(e) = \varphi_2(e) = 1$, that is, $\varphi_1, \varphi_2 \in P_1(G)$, and hence $\varphi_1 = \varphi_2 = \varphi$. So, $\varphi \in \mathrm{Extr}(P_{\le 1}(G))$.

Let $\varphi_1, \varphi_2 \in P_{\le 1}(G)$ and $t \in \,]0,1[$ be such that $t\varphi_1 + (1-t)\varphi_2 = 0$. Then $\varphi_1(e) = 0$ and $\varphi_2(e) = 0$, that is, $\varphi_1 = 0$ and $\varphi_2 = 0$. Therefore $0 \in \mathrm{Extr}(P_{\le 1}(G))$.

For the opposite inclusion, consider $\varphi \in P_{\leq 1}(G)$ such that $\varphi \notin (P_1(G) \cup \{0\})$. For $t := \varphi(e)$, we have $0 < t < 1$, $\frac{1}{t}\varphi \in P_1(G)$, and $\varphi = t(\frac{1}{t}\varphi) + (1-t)0$. It follows that $\varphi \notin \mathrm{Extr}(P_{\leq 1}(G))$. □

Let G be a topological group, and $\varphi \in P(G)$. For $\psi \in P(G)$, write $\psi \leq \varphi$ if $\varphi - \psi \in P(G)$. Let $T \in \pi_\varphi(G)'$ be such that $0 \leq T \leq I$. Here $\pi_\varphi(G)'$ is the commutant of $\pi_\varphi(G)$ and I stands for the identity operator on $\mathcal{H}_\varphi$. Define $\varphi_T : G \mapsto \mathbf{C}$ by

$$\varphi_T(g) = \langle \pi_\varphi(g)T\xi_\varphi \mid \xi_\varphi \rangle = \langle \pi_\varphi(g)T^{1/2}\xi_\varphi \mid T^{1/2}\xi_\varphi \rangle \quad \text{for all } g \in G.$$

Then $\varphi_T \in P(G)$ and, since $\varphi - \varphi_T = \varphi_{I-T}$, we have $\varphi_T \leq \varphi$. (For $T \geq 0$ and $T^{1/2}$, there is a reminder in Appendix A.10.)

LEMMA 1.B.13. *The map $T \mapsto \varphi_T$ is a bijective correspondence between the set of $T \in \pi_\varphi(G)'$ such that $0 \leq T \leq I$ and the set of $\psi \in P(G)$ such that $\psi \leq \varphi$.*

PROOF. Let $T, S \in \pi_\varphi(G)'$ with $0 \leq T, S \leq I$ be such that $\varphi_T = \varphi_S$; then

$$\langle \pi_\varphi(g)\xi_\varphi \mid T\xi_\varphi \rangle = \langle \pi_\varphi(g)\xi_\varphi \mid S\xi_\varphi \rangle \quad \text{for all } g \in G.$$

Since ξ_φ is cyclic for π_φ, this implies that $T\xi_\varphi = S\xi_\varphi$. Moreover, since $T, S \in \pi_\varphi(G)'$,

$$T\pi_\varphi(g)\xi_\varphi = \pi_\varphi(g)T\xi_\varphi = \pi_\varphi(g)S\xi_\varphi = S\pi_\varphi(g)\xi_\varphi \quad \text{for all } g \in G,$$

so that $T = S$. This shows that the map $T \mapsto \varphi_T$ is injective.

Let $\psi \in P(G)$ be such that $\psi \leq \varphi$. Denote by Φ_ψ the associated positive Hermitian form on $\mathbf{C}[G]$, as in Construction 1.B.6. For $f_1, f_2 \in \mathbf{C}[G]$, we have, by the Cauchy–Schwarz inequality,

$$|\Phi_\psi(f_1, f_2)|^2 \leq \Phi_\psi(f_1, f_1)\Phi_\psi(f_2, f_2) \leq \Phi_\varphi(f_1, f_1)\Phi_\varphi(f_2, f_2).$$

This shows that Φ_ψ induces a continuous positive Hermitian form on $\mathcal{H}_\varphi$, defined by $([f_1], [f_2]) \mapsto \Phi_\psi(f_1, f_2)$. It follows that there exists $T \in \mathcal{L}(\mathcal{H}_\varphi)$, with $0 \leq T \leq I$, such that

$$\Phi_\psi(f_1, f_2) = \langle [f_1] \mid T[f_2] \rangle \quad \text{for all } f_1, f_2 \in \mathbf{C}[G].$$

Therefore

$$(\sharp) \qquad \psi(y^{-1}x) = \Phi_\psi(\delta_x, \delta_y) = \langle \pi_\varphi(x)\xi_\varphi \mid T\pi_\varphi(y)\xi_\varphi \rangle \quad \text{for all } x, y \in G.$$

For $x, y, z \in G$, we have

$$\begin{aligned}\langle \pi_\varphi(x)\xi_\varphi \mid T\pi_\varphi(z)\pi_\varphi(y)\xi_\varphi \rangle &= \psi((zy)^{-1}x) \\ &= \psi(y^{-1}(z^{-1}x)) = \langle \pi_\varphi(z^{-1}x)\xi_\varphi \mid T\pi_\varphi(y)\xi_\varphi \rangle \\ &= \langle \pi_\varphi(x)\xi_\varphi \mid \pi_\varphi(z)T\pi_\varphi(y)\xi_\varphi \rangle.\end{aligned}$$

Since ξ_φ is a cyclic vector, it follows that $T \in \pi_\varphi(G)'$. By ($\sharp$), we have $\psi = \varphi_T$. This shows that the map $T \mapsto \varphi_T$ is surjective. □

PROPOSITION 1.B.14. *Let G be a topological group. Let $\varphi \in P_1(G)$; denote by $(\mathcal{H}_\varphi, \pi_\varphi, \xi_\varphi)$ the corresponding GNS triple.*

The representation π_φ is irreducible if and only if $\varphi \in \mathrm{Extr}(P_1(G))$.

PROOF. Assume first that π_φ irreducible; let $\varphi_1, \varphi_2 \in P_1(G)$ and $t \in]0,1[$ be such that $\varphi = t\varphi_1 + (1-t)\varphi_2$. Let $j \in \{1,2\}$; since $\varphi_j \le \varphi$, there exists by Lemma 1.B.13 $T_j \in \pi_\varphi(G)'$ with $0 \le T_j \le I$ such that $\varphi_j = \varphi_{T_j}$. By irreducibility of π_φ, we then have $T_j = \lambda_j \mathrm{Id}$ for some $\lambda_j \in \mathbf{C}$; indeed, $T_j = \mathrm{Id}$, because $\varphi_j(e) = 1$. Therefore $\varphi_1 = \varphi_2 = \varphi$, and it follows that $\varphi \in \mathrm{Extr}(P_1(G))$.

Assume now that π_φ is not irreducible; there exists an orthogonal projection $P \in \pi_\varphi(G)'$ such that $0 \neq P \neq I$. Set $t := \varphi_P(e) = \|P\xi_\varphi\|^2$. We claim that $t \neq 0$. Indeed, assume that $t = 0$. Then $P\xi_\varphi = 0$ and hence $P\pi(g)\xi_\varphi = \pi(g)P\xi_\varphi = 0$ for all $g \in G$; since ξ_φ is a cyclic vector, this implies $P = 0$, in contradiction with our assumption. Similarly, we have $t \neq 1$, because $I - P \neq 0$.

Set $\varphi_1 = \frac{1}{t}\varphi_P$ and $\varphi_2 = \frac{1}{1-t}\varphi_{I-P}$; then $\varphi_1, \varphi_2 \in P_1(G)$ and $\varphi = t\varphi_1 + (1-t)\varphi_2$; so, $\varphi \notin \mathrm{Extr}(P_1(G))$. □

The case of locally compact groups. Assume from now on that G is a locally compact group. Choose a left Haar measure μ_G on G, i.e., on the σ-algebra $\mathcal{B}(G)$ of the Borel subsets of G.

The convex cone $P(G)$ of continuous functions of positive type on G and its subsets $P_{\le 1}(G)$, $P_1(G)$, $\mathrm{Extr}(P_{\le 1}(G))$, $\mathrm{Extr}(P_1(G))$ can be viewed as subspaces of $L^\infty(G, \mu_G)$. We consider these sets as topological spaces, for the topology induced by the weak*-topology of $L^\infty(G, \mu_G)$.

THEOREM 1.B.15. *Let G be a locally compact group. Let the notation be as above.*

(1) *The convex subset $P_{\le 1}(G)$ of $L^\infty(G, \mu_G)$ is compact for the weak*-topology.*
(2) *The convex hull of* $\mathrm{Extr}(P_{\le 1}(G))$ *is weak*-dense in $P_{\le 1}(G)$.*
(3) *The convex hull of* $\mathrm{Extr}(P_1(G))$ *is weak*-dense in $P_1(G)$.*
(4) *On $P_1(G)$, the weak*-topology and the topology of uniform convergence on compact subsets of G coincide.*
(5) *For Γ a discrete group, the subspace $P_1(\Gamma)$ of $\ell^\infty(\Gamma)$ is compact for the weak*-topology.*

SKETCH OF THE PROOF AND REFERENCE. Claim (1) is a consequence of the Banach–Alaoglu Theorem. Claim (2) is a consequence of the Krein–Milman Theorem, and Claim (3) follows. Claim (4) is a theorem of Raikov. It follows that the convex hull of $\mathrm{Extr}(P_1(G))$ is dense in $P_1(G)$ for the topology of uniform convergence on compact subsets of G. Claim (5) follows from Claim (1) because the evaluation map $\ell^\infty(\Gamma) \to \mathbf{C}$, $\varphi \mapsto \varphi(e)$ is continuous for the weak*-topology.

For details, we refer to [**BeHV–08**, C.5]. □

REMARK 1.B.16. (1) Note that the space $P_1(G)$ does not need to be closed in $L^\infty(G, \mu_G)$ for the weak*-topology when G is not discrete. For example, the weak*-closure of $P_1(\mathbf{T})$ in $L^\infty(\mathbf{T}, \mu_T)$ contains 0. Indeed, for $k \in \mathbf{Z}$, let $e_k \in P_1(\mathbf{T})$ be the unitary character defined on $\mathbf{T} = \{z \in \mathbf{C} \mid |z| = 1\}$ by $e_k(z) = z^k$. By the Riemann–Lebesgue lemma, we have

$$\lim_{k \to \infty} \int_{\mathbf{T}} f(z) e_k(z) d\mu_T(z) = 0 \qquad \text{for all } f \in L^1(\mathbf{T}, \mu_T),$$

i.e., the sequence $(e_k)_{k \ge 1}$ converges to $0 \in L^\infty(\mathbf{T}, \mu_T)$ for the weak*-topology. More generally, the Riemann–Lebesgue lemma holds for an arbitrary locally compact abelian group G (Proposition 3 in [**BTS1–2**, Chap. II, § 1, No 2]), and the same argument as above shows that the closure of $P_1(G)$ contains 0 if $\widehat{G}$ is not compact, that is, if G is not discrete (see Proposition 1.D.6).

(2) For a locally compact group G, the bijection established in Proposition 1.B.10 fits into a larger picture; indeed, there are canonical bijections between the following sets:

(i) $P_1(G)$,
(ii) the set of cyclic representations $(\pi, \mathcal{H}, \xi)$ of G with $\|\xi\| = 1$ modulo equivalence,
(iii) the set of positive forms of norm 1 on the group algebra $L^1(G, \mu_G)$,
(iv) the set of positive forms of norm 1 on the maximal C*-algebra $C^*_{\max}(G)$, as defined in Section 8.B.

We come back to this in Example 8.A.9(2).

(3) A nondiscrete topological group G has "discontinuous" representations, as can be checked as follows. Denote by G_{disc} the group G made discrete (that is, viewed as a discrete group). The regular representation $\lambda_{G_{\mathrm{disc}}}$ of G_{disc} on $\ell^2(G_{\mathrm{disc}})$ is not continuous when viewed as a homomorphism $G \to \mathcal{U}(\ell^2(G_{\mathrm{disc}}))$; indeed, the Dirac delta function δ supported at the origin, which is a function of positive type associated to $\lambda_{G_{\mathrm{disc}}}$, is not continuous on G.

As we now show, locally compact groups have a distinguished *faithful* representation.

EXAMPLE 1.B.17. Let G be a locally compact group and μ_G a left Haar measure on G. The **left regular representation** λ_G of G is defined on $L^2(G, \mu_G)$ by

$$\lambda_G(g)f(x) = f(g^{-1}x) \quad \text{for all } g, x \in G \text{ and } f \in L^2(G, \mu_G).$$

This representation is faithful. Indeed, let $g \in G$, $g \neq e$. Since G is a locally compact group, there exists a compact neighbourhood V of e in G such that $gV \cap V = \emptyset$. Since $\lambda_G(g)(\mathbf{1}_V) = \mathbf{1}_{gV}$ and since $\mathbf{1}_{gV} \neq \mathbf{1}_V$ (both viewed as elements in $L^2(G, \mu_G)$), we see that $\lambda_G(g) \neq \mathrm{Id}_{L^2(G,\mu_G)}$; this shows that λ_G is faithful.

The **right regular representation** ρ_G of G is defined on $L^2(G, \mu_G)$ by

$$\rho_G(g)f(x) = \Delta_G(g)^{1/2} f(xg) \quad \text{for all } g, x \in G,\ f \in L^2(G, \mu_G),$$

where Δ_G is the modular function of G (see Appendix A.6); note that this uses the *right* action of G on itself, and yet μ_G is still a *left* Haar measure on G.

The representations λ_G and ρ_G are equivalent. Indeed, if $U : L^2(G, \mu_G) \to L^2(G, \mu_G)$ is defined by

$$(Uf)(g) = \Delta_G(g)^{-1/2} f(g^{-1}) \quad \text{for all } f \in L^2(G, \mu_G) \text{ and } g \in G,$$

then U is unitary and $U\lambda_G(g) = \rho_G(g)U$ for all $g \in G$. More on this in [**BeHV–08**, A.4].

If $G \neq \{e\}$, the commutant $\lambda_G(G)'$ contains $\rho_G(G)$; in particular, $\lambda_G(G)'$ contains operators which are not multiples of the identity. It follows that the left-regular representation is not irreducible.

DEFINITION 1.B.18. A family $\mathcal{F}$ of representations of a topological group G is **separating**, or **separates the points of** G, if, for every $g \in G$, $g \neq e$, there exists in $\mathcal{F}$ a representation π on a Hilbert space $\mathcal{H}$ such that $\pi(g) \neq \mathrm{Id}_{\mathcal{H}}$.

For example, a one-element family $\{\pi\}$ is separating when the representation π is faithful.

The Gel'fand–Raikov Theorem of [**GeRa–43**] establishes that every locally compact group has a large supply of irreducible representations:

THEOREM 1.B.19 (**Gel'fand–Raikov**). *Let G be a locally compact group. The family of irreducible representations of G separates the points of G.*

PROOF. Let $g \in G, g \neq 1$. By Proposition 1.B.14, it suffices to show that there exists $\varphi \in \mathrm{Extr}(P_1(G))$ with $\varphi(g) \neq 1$.

Since the regular representation λ_G of G is faithful (Example 1.B.17), $\lambda_G(g)$ is not the identity on $L^2(G, \mu_G)$; thus, there exists $\psi \in P_1(G)$ associated with λ_G such that $\psi(g) \neq 1$. Since the convex hull of $\mathrm{Extr}(P_1(G))$ is dense in $P_1(G)$ for the topology of uniform convergence on compact subsets (Theorem 1.B.15), it follows that there exists $\varphi \in \mathrm{Extr}(P_1(G))$ with $\varphi(g) \neq 1$. □

REMARK 1.B.20. (1) For second-countable locally compact groups, there is another proof of the Gel'fand–Raikov Theorem 1.B.19, using a direct integral decomposition of a faithful representation (e.g., the left regular representation) into irreducible representations [**Mack–76**, Pages 109–110].

(2) The dual $\widehat{G}$ of a general topological group G need not contain more than the unit representation 1_G. Indeed, there are topological groups G such that, for every representation π of G on a Hilbert space $\mathcal{H}$ (irreducible or not), $\pi(g) = \mathrm{Id}_{\mathcal{H}}$ for all $g \in G$. Examples include abelian topological groups described in [**Bana–83**, **Bana–91**], the group of orientation preserving homeomorphisms of the closed unit interval $[0,1]$ with the compact-open topology (this is a special case of a result of [**Megr–01**], where isometric representations in reflexive Banach spaces are considered), the group of isometries of the universal Urysohn metric space $\mathbb{U}_1$ of diameter one with the compact-open topology [**Pest–07**, Corollary 1.4], and examples occurring in [**BYTs–16**, Corollary 4.10].

(3) There are also topological groups G that have faithful representations but no irreducible representation distinct from 1_G. This is the case of the group of measurable functions from $[0,1]$ to the unit circle, with the topology induced by the weak topology on the von Neumann algebra $L^\infty([0,1])$, that makes it an abelian Polish group [**Glas–98**] (see also [**BeHV–08**, Example C.5.10]).

1.C. Weak containment and Fell topology for representations of topological groups

In this section, we introduce a topology on the dual space $\widehat{G}$ of a topological group G as well as the related notion of weak containment for representations of G.

Weak containment of representations.

DEFINITION 1.C.1. Let G be a topological group. Let π_1, π_2 be two representations of G. Define π_1 to be **weakly contained** in π_2, and write $\pi_1 \preceq \pi_2$ if every function of positive type associated with π_1 can be approximated, uniformly on every compact subset of G, by sums of functions of positive type associated with π_2.

The relation $\pi_1 \preceq \pi_2$ only depends on the equivalence classes of π_1 and π_2. The relation of weak containment is transitive: if $\pi_1 \preceq \pi_2$ and $\pi_2 \preceq \pi_3$, then $\pi_1 \preceq \pi_3$.

(Note that there is a related *but different* notion of weak containment due to Zimmer; see [**Zimm–84**, 7.3.5].)

The representations π_1 and π_2 are **weakly equivalent**, written $\pi_1 \sim \pi_2$, if $\pi_1 \preceq \pi_2$ and $\pi_2 \preceq \pi_1$.

REMARK 1.C.2. Let G be a topological group and π_1, π_2 two representations of G.

(1) If π_1 is contained in π_2, then π_1 is weakly contained in π_2. It follows that, if π_1 and π_2 are equivalent, then they are weakly equivalent.

(2) Let n_1, n_2 be two cardinal numbers, not 0; then $\pi_1 \preceq \pi_2$ if and only if $n_1\pi_1 \preceq n_2\pi_2$. In particular, π_1 and $n_1\pi_1$ are weakly equivalent.

(3) In some particular situations, weak equivalence implies equivalence. For example, if π_1, π_2 are irreducible and π_1 is finite-dimensional, then π_1 and π_2 are weakly equivalent if and only if they are equivalent. This is a very particular case of Proposition 9.A.3.

Amenability of locally compact groups can be characterized in terms of weak containment; for the notion of amenability and for a proof of the next theorem, we refer to [**BeHV–08**, Appendix G].

THEOREM 1.C.3 (**Hulanicki–Reiter**). *For a locally compact group G, the following properties are equivalent:*

(i) *the unit representation 1_G is weakly contained in the regular representation λ_G;*
(ii) *every representation of G is weakly contained in λ_G;*
(iii) *G is amenable.*

We will need the following criterion several times for weak containment. For a proof, see [**BeHV–08**, Lemma F.1.3].

PROPOSITION 1.C.4. *Let G be a topological group and π_1, π_2 two representations of G. Assume that π_1 has a cyclic vector ξ; let φ_ξ be the corresponding function of positive type: $\varphi_\xi(g) = \langle \pi_1(g)\xi \mid \xi \rangle$ for all $g \in G$. The following conditions are equivalent:*

(i) *π_1 is weakly contained in π_2, i.e., for every function of positive type φ associated with π_1, for every $\varepsilon > 0$, and every compact subset Q of G, there exist functions of positive type $\psi_1, \dots, \psi_k$ associated with π_2 such that $\sup_{g\in Q} \left|\varphi(g) - \sum_{i=1}^k \psi_i(g)\right| < \varepsilon$.*
(ii) *For every $\varepsilon > 0$ and every compact subset Q of G, there exist functions of positive type $\psi_1, \dots \psi_k$ associated with π_2 such that $\sup_{g\in Q} |\varphi_\xi(g) - \sum_{i=1}^k \psi_i(g)| < \varepsilon$.*

Fell topology.

DEFINITION 1.C.5. Let G be a topological group. Let $\mathcal{R}$ be a set of equivalence classes of representations of G and let $\pi \in \mathcal{R}$. For a finite sequence $\varphi_1, \dots, \varphi_n$ of functions of positive type associated with π, a compact subset Q of G, and a real number $\varepsilon > 0$, define $W(\pi; \varphi_1, \dots, \varphi_n, Q, \varepsilon)$ to be the subset of those $\rho \in \mathcal{R}$ which have the following property: there exist functions $\psi_1, \dots, \psi_n$ from G to $\mathbf{C}$, each one a sum of functions of positive type associated with ρ, such that

$$|\varphi_i(g) - \psi_i(g)| < \varepsilon \quad \text{for all } i \in \{1, \dots, n\} \text{ and } g \in Q.$$

The sets $W(\pi; \varphi_1, \ldots, \varphi_n, Q, \varepsilon)$ constitute a basis of a topology on $\mathcal{R}$ called the **Fell topology**.

Among the most important cases for $\mathcal{R}$, there are the dual $\widehat{G}$, the set of equivalence classes of cyclic representations of G, and the set of equivalence classes of representations of G on Hilbert spaces of dimension at most some cardinal $\aleph$ (for example on separable Hilbert spaces when G is a second-countable group).

In case $\mathcal{R} = \widehat{G}$, the following proposition provides another description of the Fell topology.

PROPOSITION 1.C.6. *Let G be a topological group and $(\rho_\iota)_{\iota \in I}$ a net in $\widehat{G}$. Let $\pi \in \widehat{G}$ and let φ be a normalized function of positive type associated with π. The following properties are equivalent:*

(i) *the net $(\rho_\iota)_{\iota \in I}$ converges to π in the Fell topology;*
(ii) *there exists a net $(\psi_\iota)_{\iota \in I}$, where each ψ_ι is a sum of functions of positive type associated with ρ_ι, such that $\lim_\iota \psi_\iota = \varphi$ uniformly on compact subsets of G.*

Suppose moreover that G is locally compact. Properties (i) *and* (ii) *are equivalent to:*

(iii) *there exists a net $(\psi_\iota)_{\iota \in I}$ of functions of positive type associated with ρ_ι such that $\lim_\iota \psi_\iota = \varphi$ uniformly on compact subsets of G.*

PROOF. The equivalence of (i) and (ii) is an immediate consequence of Proposition 1.C.4. For (iii), we refer to [**BeHV–08**, Proposition F.1.4]. □

The following fact, which is a direct consequence of the definitions, will be needed in Section 1.E.

PROPOSITION 1.C.7. *Let G be a topological group. Let π_1, π_2 be two irreducible representations of G, viewed as points in the dual $\widehat{G}$.*

Then π_1 and π_2 are weakly equivalent if and only if the closures $\overline{\{\pi_1\}}$ and $\overline{\{\pi_2\}}$ of $\{\pi_1\}$ and $\{\pi_2\}$ in $\widehat{G}$ are equal.

The next proposition gives a description of Fell's topology on $\widehat{G}$ in terms of weak containment.

PROPOSITION 1.C.8. *Let G be a locally compact group, $(\pi_\iota)_{\iota \in I}$ a net in $\widehat{G}$, and $\pi \in \widehat{G}$. The following properties are equivalent:*

(i) *$(\pi_\iota)_{\iota \in I}$ converges to $\pi \in \widehat{G}$;*
(ii) *π is weakly contained in $\bigoplus_{\jmath \in J} \pi_\jmath$ for every subnet $(\pi_\jmath)_{\jmath \in J}$ of $(\pi_\iota)_{\iota \in I}$.*

PROOF. The fact that (i) implies (ii) is a straightforward consequence of Proposition 1.C.6 and holds for an arbitrary topological group G. For the fact that (ii) implies (i), see [**BeHV–08**, Proposition F.2.2]. □

The following proposition gives, for a locally compact group G, a description of the Fell topology on $\widehat{G}$ in terms of the weak*-topology on the space of indecomposable functions of positive type on G.

PROPOSITION 1.C.9. *Let G be a locally compact group,* $\mathrm{Extr}(P_1(G))$ *the space of indecomposable normalized functions of positive type on G, viewed as a subspace of $L^\infty(G)$ with the weak*-topology, and $\widehat{G}$ the dual of G, with the Fell topology.*

The surjective map $p : \mathrm{Extr}(P_1(G)) \twoheadrightarrow \widehat{G}$ *given by the Gelfand–Naimark construction (see Construction* 1.B.6 *and Proposition* 1.B.14*) is continuous and open. In other words, the Fell topology on* $\widehat{G}$ *is the quotient topology of the weak*-topology on* $\mathrm{Extr}(P_1(G))$ *by the map* p.

PROOF. Recall from Theorem 1.B.15 that the topology on $\mathrm{Extr}(P_1(G))$ coincides with the topology of uniform convergence on compact subsets of G. The continuity of p is therefore an immediate consequence of Proposition 1.C.6.

To show that p is open, let U be an open subset of $\mathrm{Extr}(P_1(G))$. Assume, by contradiction, that $p(U)$ is not open. Then there exists $\pi \in p(U)$ and a net $(\pi_\iota)_{\iota \in I}$ in $\widehat{G} \smallsetminus p(U)$ such that $\lim_\iota \pi_i = \pi$. Let $\varphi \in U$ be a normalized function of positive type associated with π. By Proposition 1.C.6, there exist functions of positive type φ_ι associated with π_ι such that $\lim_\iota \varphi_\iota = \varphi$. We have $\varphi_\iota \notin p^{-1}(p(U))$ and therefore $\varphi_\iota \notin U$. Since $\varphi \in U$ and U is open, this is a contradiction. □

REMARK 1.C.10. (1) For a locally compact group G, weak containment can be equivalently defined in terms of C*-algebras (Proposition 8.B.4, see also Proposition 6.E.4), and there are other equivalent definitions of the Fell topology on $\widehat{G}$. See [**Dixm–C***, §§ 3 & 18], as well as our Section 8.B.

(2) The terminology "Fell topology" for topological groups and C*-algebras refers to articles published by Fell in the early 60s, in particular to [**Fell–60a**]. For groups, this topology appears earlier in Godement's thesis [**Gode–48**, Section 16].

Functorial properties of Fell's topology. We examine how Fell's topology behaves with respect to continuous homomorphisms.

PROPOSITION 1.C.11. *Let* G *and* H *be topological groups.*

(1) *Let* $\alpha : G \to H$ *be a continuous homomorphism with* dense *image. The induced map*
$$\widehat{\alpha} : \widehat{H} \to \widehat{G}, \quad \pi \mapsto \pi \circ \alpha$$
is injective and continuous for the Fell topologies on $\widehat{H}$ *an* $\widehat{G}$.

(2) *Let* θ *be a bicontinuous automorphism of the topological group* G. *The induced map* $\widehat{\theta} : \widehat{G} \to \widehat{G}$ *is a homeomorphism.*

(3) *Let* G *be a topological group,* N *be a closed normal subgroup of* G, *and* $p : G \twoheadrightarrow G/N$ *the canonical projection. Assume that, for every compact subset* $\overline{Q}$ *of* G/N, *there exists a compact subset* Q *of* G *with* $p(Q) = \overline{Q}$. *Then* $\widehat{p}$ *is a homeomorphism between* $\widehat{G/N}$ *and a closed subspace of* $\widehat{G}$.

PROOF. Observe that (2) is an immediate consequence of (1).

To show (1), let $\pi \in \widehat{H}$. Then $\pi \circ \alpha$ is a representation of G, since α is a continuous homomorphism; moreover, $\pi \circ \alpha$ is irreducible, since α has dense image. So, $\widehat{\alpha} : \widehat{H} \to \widehat{G}$ is well defined. The injectivity of $\widehat{\alpha}$ is a consequence of the density of the image of α. The fact that $\widehat{\alpha}$ is continuous with respect to the Fell topologies is straightforward to check.

To show (3), observe that the image of $\widehat{p}$ coincides with the subset
$$X = \{\pi \in \widehat{G} \mid \pi|_N = \mathrm{Id}\}$$
of $\widehat{G}$. It is clear that X is a closed subset of $\widehat{G}$. The continuity of the inverse map $\widehat{p}^{-1} : X \to \widehat{G/N}$ follows easily from the fact that, for every compact subset $\overline{Q}$ of G/N, there exists a compact subset Q of G with $p(Q) = \overline{Q}$. □

REMARK 1.C.12. (1) In Proposition 1.C.11(1), the image of $\widehat{\alpha}$ is not necessarily closed, as shown by the following two examples.

First consider the injection $\alpha : \mathbf{R} \to \mathbf{T}^2$, $t \mapsto (e^{2\pi it}, e^{2\pi i\theta t})$, where θ is an irrational number. The dual $\widehat{\mathbf{R}}$ of $\mathbf{R}$ consists of the unitary characters $\chi_k : \mathbf{R} \to \mathbf{T}$, $t \mapsto e^{2\pi ikt}$, for all $k \in \mathbf{R}$, and the image $\widehat{\alpha}(\widehat{\mathbf{T}^2})$ is $\{\chi_{m+\theta n} \in \widehat{\mathbf{R}} \mid m, n \in \mathbf{Z}\}$, which is a dense countable subgroup of $\widehat{\mathbf{R}}$.

Let $\mathbf{Q}$ be the field of rational numbers, considered here as an additive group with the discrete topology, and consider the canonical injection $\alpha : \mathbf{Q} \to \mathbf{R}$. The image $\widehat{\alpha}(\widehat{\mathbf{R}})$ is not closed in $\widehat{\mathbf{Q}}$. Indeed, otherwise, $\widehat{\alpha}(\widehat{\mathbf{R}})$ would be a locally compact group and, hence (see [**HeRo–63**, Theorem 5.29]), the continuous homomorphism $\widehat{\alpha}$ would be a homeomorphism between $\widehat{\mathbf{R}}$ and $\widehat{\alpha}(\widehat{\mathbf{R}})$; since $\widehat{\mathbf{Q}}$ and therefore $\widehat{\alpha}(\widehat{\mathbf{R}})$ are compact, this would imply that $\mathbf{R}$, which is homeomorphic to $\widehat{\mathbf{R}}$, is compact.

(2) The condition on the pair (G, N) stated in Proposition 1.C.11(3) is always satisfied if G is a locally compact group (see Lemma B.1.1 in [**BeHV–08**]), or if G is a complete metrizable topological group (see Chap. 9, §2, Proposition 2 in [**BTG5–10**]). However, there are examples of topological groups for which this condition is not satisfied (see Remark 1.7.9 in [**BeHV–08**]).

EXAMPLE 1.C.13. Let G be a topological group. Let $[G, G]$ denote the derived group of G, also called the commutator subgroup of G, generated by its commutators $[g, h] := g^{-1}h^{-1}gh$, with $g, h \in G$. Let $\overline{[G, G]}$ be its closure, and $G_{\mathrm{ab}} := G/\overline{[G, G]}$ the Hausdorff abelianized group of G. Assume that G is either locally compact or metrizable complete.

The dual $\widehat{G}_{\mathrm{ab}}$ of G_{ab} can be identified with the closed subspace $\{\pi \in \widehat{G} \mid \dim \pi = 1\}$ of $\widehat{G}$, in other words with the closed subspace of $\widehat{G}$ of the unitary characters of G. Since $\widehat{G}_{\mathrm{ab}}$ is a Hausdorff topological space, it follows that every unitary character of G defines a closed point in $\widehat{G}$.

REMARK 1.C.14. Let G be a topological group. The dual $\widehat{G}$ has a few remarkable subspaces, besides the subspace $\widehat{G}_{\mathrm{ab}}$ of the previous example.

We denote by $\widehat{G}_{\mathrm{fd}}$ the subspace of the dual $\widehat{G}$ containing equivalence classes of finite-dimensional irreducible representations of G. This subspace will be described in Chapter 4 for our standard examples. In general, $\widehat{G}_{\mathrm{fd}}$ need not be closed in $\widehat{G}$. For example, there are discrete groups Γ which have plenty of infinite-dimensional irreducible representations and which are such that $\widehat{\Gamma}_{\mathrm{fd}}$ is dense in $\widehat{\Gamma}$; one example is a nonabelian free group of rank two; see Subsection 15.E.b.

For every integer $n \geq 1$, denote by $\widehat{G}_{\leq n}$ the subspace of $\widehat{G}$ of equivalence classes of irreducible representations of dimension at most n. Observe that $\widehat{G}_{\mathrm{fd}} = \bigcup_{n \geq 1} \widehat{G}_{\leq n}$. When G is locally compact, it is known that $\widehat{G}_{\leq n}$ is a closed subspace of $\widehat{G}$ for each integer $n \geq 1$ [**Dixm–C***, 3.6.3].

Given a representation π of G, its **support** $\mathrm{supp}(\pi)$ is the subspace of $\widehat{G}$ of irreducible representations ρ which are weakly contained in π. When G is locally compact, $\mathrm{supp}(\pi)$ is a nonempty closed subset of $\widehat{G}$. For example, the support of the left regular representation of G is the so-called reduced dual of G. More on supports in Section F.2 of [**BeHV–08**].

1.D. Topological properties of the dual of a group

We review how topological properties of a locally compact group G and of its dual space $\widehat{G}$ are related.

Properties for general locally compact groups. Recall that a topological space X is **quasi-compact** if every open covering of X contains a finite covering; equivalently, if every net $(x_\iota)_{\iota\in I}$ in X has a convergent subnet $(x_\jmath)_{\jmath\in J}$.

A topological space X is **locally quasi-compact** if every point in X has a fundamental system consisting of quasi-compact neighbourhoods (see Exercice 30 of Chap. I, §9 in [**BTG1–4**]).

PROPOSITION 1.D.1. *Let G be a locally compact group. Then $\widehat{G}$, equipped with Fell's topology, is locally quasi-compact.*

PROOF. Consider the map

$$p: \operatorname{Extr}(P_1(G)) \to \widehat{G}$$

given by the GNS construction. Recall from Proposition 1.C.9 that p is continuous and open, where $\operatorname{Extr}(P_1(G)) \subset L^\infty(G, \mu_G)$ is equipped with the weak*-topology and μ_G is a Haar measure on G.

Let $\pi \in \widehat{G}$ and let $U \subset \widehat{G}$ be an open neighborhood of π. Fix a normalized function of positive type φ_0 associated with π.

Let U^c be the complement of U in $\widehat{G}$. Denote by $\mathcal{C}_{U^c}$ the closure in $L^\infty(G, \mu_G)$, for the weak*-topology, of the convex hull of functions of positive type ψ associated to representations from U^c with $\|\psi\|_\infty = \psi(e) \le 1$. Observe that $\mathcal{C}_{U^c}$ is a compact convex subset of $P_{\le 1}(G)$ and that $0 \in \mathcal{C}_{U^c}$.

We claim that $\varphi_0 \notin \mathcal{C}_{U^c}$. Indeed, assume, by contradiction, that there exists a net $(\psi_\iota)_{\iota\in I}$ in $\mathcal{C}_{U^c}$ with $\lim_i \psi_\iota = \varphi_0$ in the weak*-topology. Then

$$1 = \|\varphi_0\|_\infty \le \liminf_\iota \|\psi_\iota\|_\infty \le \limsup_\iota \|\psi_\iota\|_\infty \le 1,$$

so that $\lim_\iota \|\psi_\iota\|_\infty = \lim_\iota \psi_\iota(e) = 1$. For $\iota \in I$, set $\widetilde{\psi}_\iota = \psi_\iota / \psi_\iota(e)$; then $\widetilde{\psi}_\iota \in P_1(G)$ and $\lim_i \widetilde{\psi}_\iota = \varphi_0$ in the weak*-topology and hence in the topology of uniform convergence on compact subsets of G, by Theorem 1.B.15(4). This implies that π is weakly contained in the direct sum $\bigoplus_{\sigma\in U^c} \sigma$. This is a contradiction, since U^c is a closed subset in $\widehat{G}$ and $\pi \notin U^c$.

It follows from the Hahn–Banach theorem (see [**Rudi–73**, Theorem 3.4]) that there exist a function $f_0 \in L^1(G, \mu_G)$ and real numbers α, β such that

$$(*) \qquad \operatorname{Re}\left(\int_G \psi(g) f_0(g) d\mu_G(g)\right) \le \alpha < \beta \le \operatorname{Re}\left(\int_G \varphi_0(g) f_0(g) d\mu_G(g)\right)$$

for all $\psi \in \mathcal{C}_{U^c}$. Observe that, since $0 \in \mathcal{C}_{U^c}$, we have $\alpha \ge 0$.

Let δ be such that $\alpha < \delta < \beta$. Consider the neighbourhood

$$V := \left\{\varphi \in \operatorname{Extr}(P_1(G)) \mid \delta \le \operatorname{Re}\left(\int_G \varphi(g) f_0(g) d\mu_G(g)\right)\right\}$$

of φ_0 in $\operatorname{Extr}(P_1(G))$ and set

$$K := p(V).$$

Since p is an open map, K is a neighbourhood of π in $\widehat{G}$.

Let $\varphi \in V$. By $(*)$, we have $\varphi \notin \mathcal{C}_{U^c}$, and, hence $\varphi' \notin \mathcal{C}_{U^c}$ for every $\varphi' \in p^{-1}(\{p(\varphi)\})$. This shows that $K \subset U$. We claim that K is a quasi-compact subset of $\widehat{G}$.

Let $(\pi_\iota)_{\iota \in I}$ be a net in K. We have to show that this net has a subnet which converges to some element of K.

For every $\iota \in I$, let $\varphi_\iota \in V$ be such that $p(\varphi_\iota) = \pi_\iota$. Since $P_{\leq 1}(G)$ is compact in the weak*-topology, there exists a subnet $(\varphi_\jmath)_{\jmath \in J}$ which is weak*-convergent to some function $\varphi_1 \in P_{\leq 1}(G)$. It is clear that

$$(**) \qquad \delta \leq \operatorname{Re}\left(\int_G \varphi_1(g) f_0(g) d\mu_G(g)\right).$$

Since $0 \leq \alpha < \delta$, we have $\varphi_1 \neq 0$.

Let π_1 be the representation of G associated to φ_1 by the GNS construction. Denote by $\mathcal{C}_{\pi_1}$ the weak*-closure of the convex hull of normalized functions of positive type associated to π_1. Let P_{π_1} be the set of extreme points of the compact convex set $\mathcal{C}_{\pi_1}$.

It is a fact (see the proof of Proposition F.2.7 in [**BeHV–08**]) that every element in P_{π_1} is an extreme point of $P_{\leq 1}(G)$, that is, in view of Lemma 1.B.12,

$$P_{\pi_1} \subset \operatorname{Extr}(P_{\leq 1}(G)) = \operatorname{Extr}(P_1(G)) \cup \{0\}.$$

Consider the continuous affine map $\Psi : \mathcal{C}_{\pi_1} \to \mathbf{R}$ defined by

$$\Psi(\varphi) = \operatorname{Re}\left(\int_G \varphi(g) f_0(g) d\mu_G(g)\right) \qquad \text{for } \varphi \in \mathcal{C}_{\pi_1}.$$

The image of Ψ is a compact interval $[a, b]$ of $\mathbf{R}$.

Since φ_1 belongs to $\mathcal{C}_{\pi_1}$, Inequality $(**)$ shows that $\delta \leq \Psi(\varphi_1)$ and it follows that $\delta \leq b$.

The set $\Psi^{-1}(b)$ is a nonempty compact convex subset of $\mathcal{C}_{\pi_1}$. Let φ be an extreme point of $\Psi^{-1}(b)$. It is clear that φ is an extreme point of $\mathcal{C}_{\pi_1}$, that is, $\varphi \in P_{\pi_1}$. We have $\varphi \neq 0$, since $b \neq 0$. It follows that $\varphi \in \operatorname{Extr}(P_1(G))$. Moreover, we have $\varphi \in V$, since

$$\delta \leq b = \Psi(\varphi) = \operatorname{Re}\left(\int_G \varphi(g) f_0(g) f(g) d\mu_G(g)\right).$$

Let $\sigma = p(\varphi) \in \widehat{G}$ be the representation associated to φ. Then $\sigma \in K$, since $\varphi \in V$. Moreover, as $\varphi \in \mathcal{C}_{\pi_1}$, we have

$$\sigma \prec \pi_1.$$

Since $(\varphi_\jmath)_{\jmath \in J}$ converges to φ_1 and $\pi_\jmath = p(\varphi_\jmath)$, the net $(\pi_\jmath)_{\jmath \in J}$ converges to π_1. It follows that, for every subnet $(\pi_{\jmath'})_{\jmath' \in J'}$ of $(\pi_\jmath)_{\jmath \in J}$, we have

$$\sigma \preceq \bigoplus_{\jmath' \in J'} \pi_{\jmath'}.$$

Therefore, $(\pi_\jmath)_{\jmath \in J}$ converges to σ, by Proposition 1.C.8. □

REMARK 1.D.2. The proof of Proposition 1.D.1 can be adapted to prove a far more general result concerning the spectrum of an arbitrary C*-algebra (see Proposition 8.A.11).

We will often deal with locally compact groups which are second-countable. The following proposition shows that this property is inherited by their duals. (For second-countable locally compact abelian groups, see also Proposition A.7.3.)

PROPOSITION 1.D.3. *Let G be a second-countable locally compact group. Then $\widehat{G}$ is second-countable.*

PROOF. Let μ_G be a Haar measure on G. Since G is second-countable, $L^1(G, \mu_G)$ is separable, and hence the unit ball of $L^\infty(G, \mu_G)$ is second-countable for the weak*-topology. It follows that the subspace $\mathrm{Extr}(P_1(G))$ of extreme points in $P_1(G)$ is second-countable for the weak*-topology. Proposition 1.C.9 shows, therefore, that $\widehat{G}$ is second-countable. □

REMARK 1.D.4. Proposition 1.D.3 is a special case of Corollary 8.A.14 which states that the spectrum of any separable C*-algebra is second-countable.

Properties for abelian groups. Let G be a locally compact abelian group. The Fell topology on $\widehat{G} = \mathrm{Hom}(G, \mathbf{T})$ is simply the topology of uniform convergence of unitary characters on compact subsets of G.

PROPOSITION 1.D.5. *Let G be a locally compact abelian group. Then $\widehat{G}$ is locally compact.*

PROOF. By Proposition 1.D.1, $\widehat{G}$ is locally quasi-compact. Since a net of complex-valued functions can have at most one limit point, $\widehat{G}$ is a Hausdorff topological space. □

The next proposition shows that there is a perfect duality between discrete and compact abelian groups.

PROPOSITION 1.D.6. *Let G be a locally compact abelian group. The following conditions are equivalent:*

(i) *G is discrete (respectively, compact);*
(ii) *$\widehat{G}$ is compact (respectively, discrete).*

PROOF. It is clear that $\widehat{G}$ is compact if G is discrete.

Assume now that G is compact. To show that $\widehat{G}$ is discrete, it suffices to show that the unit character 1_G is open in $\widehat{G}$. For any $z \in \mathbf{T} \smallsetminus \{1\}$, there exists $n \in \mathbf{Z}$ such that $|z^n - 1| \geq \sqrt{2}$. This implies that, for every $\chi \in \widehat{G} \smallsetminus \{1_G\}$, we have

$$\sup_{g \in G} |\chi(g) - 1_G(g)| \geq \sqrt{2},$$

i.e., $\{1_G\}$ is open in $\widehat{G}$. □

Properties for compact groups. We show that the dual of a compact group is a discrete space.

PROPOSITION 1.D.7. *Let G be a compact group. Then $\widehat{G}$, equipped with Fell's topology, is discrete.*

PROOF. Let $(\pi_\iota)_\iota$ be a net in $\widehat{G}$ converging to $\pi \in \widehat{G}$. Let φ be a normalized function of positive type associated with π. There exists a net $(\psi_\iota)_\iota$ of functions of positive type ψ_ι associated with π_ι such that $\lim_\iota \psi_\iota = \varphi$ uniformly on G (see Proposition 1.C.6).

Let μ_G be a Haar measure on G. Then

$$\lim_\iota \int_G \varphi(x)\overline{\psi_\iota(x)}d\mu_G(x) = \int_G |\varphi(x)|^2 d\mu_G(x) \neq 0$$

and, hence, there exists ι_0 such that

$$\int_G \varphi(x)\overline{\psi_\iota(x)}d\mu_G(x) \neq 0 \qquad \text{for all } \iota \geq \iota_0.$$

It follows from Schur's orthogonality relations (see, for example, [**Foll–16**, Theorem 5.8]) that π is equivalent to π_ι for all $\iota \geq \iota_0$. □

REMARK 1.D.8. (1) The converse of Proposition 1.D.7 holds: let G be a locally compact group such that $\widehat{G}$ is discrete; then G is compact. See [**Shte–71**, Theorem 1], and [**Bagg–72**, Theorem 3.4] for second-countable G; see also [**Wang–75**, Theorem 7.6].

(2) By Propositions 1.D.3 and 1.D.7, the dual of a second-countable compact group is discrete and countable. However, there are noncompact locally compact groups with countable dual. The following example appears in [**Bagg–72**, 4.5, 4.6], where it is attributed to Fell, and in [**Shte–71**, § 4]. For more on locally compact groups with a countable dual, see [**Wang–75**, Section 6].

EXAMPLE 1.D.9 (**A noncompact locally compact group with countable dual**). Let p be a prime, $\mathbf{Q}_p$ the field of p-adic numbers, $\mathbf{Z}_p$ its ring of integers, and $\mathbf{Z}_p^\times$ the multiplicative group of invertible elements in $\mathbf{Z}_p$. Consider the natural action of $\mathbf{Z}_p^\times$ by automorphisms of $\mathbf{Q}_p$ given by multiplication. Let $M_p := \mathbf{Z}_p^\times \ltimes \mathbf{Q}_p$ be the p-adic motion group. We claim that the dual space of $\widehat{M_p}$ is countable.

Indeed, identify $\widehat{\mathbf{Q}_p}$ with $\mathbf{Q}_p$ (see Section 3.D). Under this identification, the dual action of $\mathbf{Z}_p^\times$ on $\widehat{\mathbf{Q}_p}$ corresponds to its natural action on $\mathbf{Q}_p$. The $\mathbf{Z}_p^\times$-orbits in $\mathbf{Q}_p$ are $\{0\}$ and the sets O_j of p-adic numbers of valuation $j \in \mathbf{Z}$. Moreover, the stabilizer in $\mathbf{Z}_p^\times$ of every point in $\mathbf{Q}_p \smallsetminus \{0\}$ is trivial.

Choose, for every $j \in \mathbf{Z}$, a representative χ_j for the orbit O_j and let π_j be the induced representation $\mathrm{Ind}_{\mathbf{Q}_p}^{M_p}\chi_j$ (compare Section 1.F). Using the "Mackey machine" (see Remark 2.F.5), one can show that

$$\widehat{M_p} = \widehat{\mathbf{Z}_p^\times} \cup \{\pi_j \mid j \in \mathbf{Z}\}.$$

Since $\widehat{\mathbf{Z}_p^\times}$ is countable, $\widehat{M_p}$ is countable.

Observe that $\widehat{M_p}$ is not discrete; in fact, every $\chi \in \widehat{\mathbf{Z}_p^\times}$ belongs to the closure of $\{\pi_j \mid j \in \mathbf{Z}\}$.

Properties for discrete groups. We show that the dual of a discrete group is quasi-compact.

PROPOSITION 1.D.10. *Let Γ be a discrete group. Then $\widehat{\Gamma}$ is a quasi-compact topological space.*

PROOF. Let $(\pi_\iota)_{\iota\in I}$ be a net in $\widehat{\Gamma}$. Choose, for every $\iota \in I$, a normalized function of positive type φ_ι associated with π_ι. Since $P_1(\Gamma)$ is compact (Theorem 1.B.15), there exists a subnet $(\varphi_\jmath)_{\jmath\in J}$ which converges pointwise on Γ to some normalized function of positive type φ. Let ρ be the representation associated with φ by the GNS construction. Choose $\pi \in \widehat{G}$ with $\pi \preceq \rho$. Then $\pi \preceq \bigoplus_{\jmath'\in J'} \pi_{\jmath'}$ for every subnet $(\pi_{j'})_{j' \in J'}$ of $(\pi_j)_{j \in J}$. Therefore $(\pi_j)_{j \in J}$ converges to π, by Proposition 1.C.8. □

REMARK 1.D.11. (1) It will be shown in Proposition 9.H.1 that the dual of a countable infinite discrete group is uncountable.

(2) As the following example shows, a nondiscrete locally compact group may have a quasi-compact dual space. Another example appears in [**Bagg–72**, 4.7] and [**Shte–71**, § 4].

EXAMPLE 1.D.12 (**A nondiscrete locally compact group with quasi–compact dual**). Let $\mathbf{K}$ be an infinite field. We view its additive group $(\mathbf{K}, +)$ as a discrete abelian group. The multiplicative group $H := \mathbf{K}^\times$ acts naturally on $\mathbf{K}$. Set $N := \widehat{\mathbf{K}}$ and consider the dual action of H on the infinite compact group N. Let G be the locally compact group $H \ltimes N$, where H is equipped with the discrete topology. Observe that G is nondiscrete and that N is a compact and open normal subgroup of G. We claim that $\widehat{G}$ is quasi-compact.

Indeed, by Pontryagin duality (see Theorem A.7.2), $\widehat{N}$ can be canonically identified with $\mathbf{K}$, the action of H on $\widehat{N}$ corresponding to its natural action on $\mathbf{K}$. The H-orbits on $\widehat{N} \cong \mathbf{K}$ are $\{1_N\}$ and $\widehat{N} \smallsetminus \{1_N\}$; moreover, H acts freely on $\widehat{N} \smallsetminus \{1_N\}$.

Choose $\chi_0 \in \widehat{N} \smallsetminus \{1_N\}$ and let π_0 be the induced representation $\mathrm{Ind}_N^G \chi_0$. Then π_0 is irreducible (see Theorem 1.F.12) and, using the Mackey machine as in Example 1.D.9, one can show that

$$\widehat{G} = \widehat{H} \cup \{\pi_0\}.$$

Since $\widehat{H}$ is compact, it is obvious that the topological space $\widehat{G}$ is quasi-compact.

Separation properties of the dual space of a group. Let G be a locally compact group which is either abelian or compact. Propositions 1.D.6 and 1.D.7 show in particular that the topological space $\widehat{G}$ has the T_2 separation property (that is, $\widehat{G}$ is a Hausdorff topological space), hence it has the weaker properties T_1 and T_0. For a reminder on the properties T_0, T_1, and T_2, see Appendix A.1. As we now see, the dual space $\widehat{G}$ of a locally compact group G fails in general to have these separation properties.

We sometimes use **LC group** as shorthand for "locally compact group". The statement of the following theorem involves notions (group of type I, CCR group) to be defined in Chapter 6 and its proof uses results stated in Chapter 8.

THEOREM 1.D.13. *Let G be a second-countable locally compact group. Let $\widehat{G}$ be its dual, equipped with the Fell topology.*

(1) *$\widehat{G}$ has property T_0 if and only if G is of type I.*
(2) *$\widehat{G}$ has property T_1 if and only if G is a CCR group.*

PROOF. (1) Assume that G is of type I and let π and σ be two distinct points in $\widehat{G}$. It follows from Glimm's Theorem (Theorem 8.F.3 below) that π and σ are not weakly equivalent. We may assume that, say, π is not weakly contained in ρ. It follows from the definition of the Fell topology that there exists a neighbourhood U of π with $\rho \notin U$. This shows that $\widehat{G}$ is a T_0-space.

The converse statement is also a part of Glimm's Theorem; we prefer to give an independent proof. Assume that G is not of type I. By Corollary 7.F.4 below, there exist two (in fact, uncountably many) distinct points π and ρ in $\widehat{G}$ which are weakly equivalent. Then, every neighbourhood of π contains ρ and every neighbourhood of ρ contains π. This shows that $\widehat{G}$ is not a T_0-space.

(2) Assume that $\widehat{G}$ has property T_1. Then, by (1), G is of type I. Let π be a point in $\widehat{G}$. By Glimm's Theorem, $\pi(C^*_{\max}(G))$ contains the ideal $\mathcal{K}(\mathcal{H})$ of compact operators on the Hilbert space $\mathcal{H}$ of π, where $C^*_{\max}(G)$ denotes the maximal C*-algebra of G, to be introduced in Chapter 8. We have to show that $\pi(C^*_{\max}(G))$ coincides with $\mathcal{K}(\mathcal{H})$.

Assume, by contradiction, that $\mathcal{K}(\mathcal{H})$ is a proper ideal of $\pi(C^*_{\max}(G))$. Then $I := \pi^{-1}(\mathcal{K}(\mathcal{H}))$ is a proper ideal of $C^*_{\max}(G)$. So, the quotient $A := C^*_{\max}(G)/I$ is a nonzero C*-algebra. Let ρ be a (nonzero) irreducible representation of A. Then $\rho \circ p$ is an irreducible representation of $C^*_{\max}(G)$, where $p : C^*_{\max}(G) \twoheadrightarrow A$ is the canonical homomorphism. Denote again by ρ the representation of G corresponding to $\rho \circ p$ (see Section 8.B). Since C*ker(ρ) obviously contains C*ker(π), the representation ρ is weakly contained in π (see Proposition 8.B.4). However, $\{\pi\}$ is a closed point in $\widehat{G}$, since $\widehat{G}$ is a T_1-space. Therefore, ρ is weakly equivalent to π; this means that C*ker(ρ) = C*ker(π). Since $I \subset$ C*ker(ρ), it follows that $I \subset$ C*ker(π). This is a contradiction, since $\mathcal{K}(\mathcal{H})$ is contained in $\pi(C^*_{\max}(G))$.

Conversely, assume that G is a CCR group. Let $\pi \in \widehat{G}$. Then π is a CCR representation and hence $\{\pi\}$ is a closed point in $\widehat{G}$, by Corollary 9.A.4 below. □

REMARK 1.D.14. (1) Let G be a locally compact group. Every one-dimensional representation of G defines a closed point of $\widehat{G}$, as observed in Example 1.C.13. In particular $\widehat{G}$ always contains a closed point, 1_G. Compare with Remark 1.C.2(3).

More generally, every finite-dimensional irreducible representation π of G defines a point in the dual $\widehat{G}$ which is closed. Indeed, such a representation is CCR (see Remark 6.E.8) and therefore $\{\pi\}$ is closed in $\widehat{G}$ (Corollary 9.A.4).

(2) The converse statement to (1) holds for countable groups: let Γ be a countable group and π an irreducible representation of Γ such that $\{\pi\}$ is a closed point in $\widehat{\Gamma}$; then π is finite-dimensional.

Indeed, consider π as an irreducible representation of the C*-algebra $A := C^*_{\max}(\Gamma)$, and set $B = \pi(A)$, which is the C*-algebra generated by $\pi(\Gamma)$. The closure of $\{\pi\}$ in $\widehat{A}$ is the set of irreducible representations ρ of A such that C*ker(ρ) $\supset$ C*ker(π), which can also be seen as the set of irreducible representations of B. If $\{\pi\}$ is closed, then this set has exactly one element, i.e., the spectrum $\widehat{B}$ is a one-point space [**Dixm–C***, 2.11.2 & 3.2.1]. Since Γ is countable, B is separable, and this implies that B is isomorphic to the algebra of compact operators on a Hilbert space [**Dixm–C***, 4.7.3]. Since Γ is a discrete group, A is unital, hence B is unital, and it follows that B is finite-dimensional. As π is irreducible, the claim is proved.

(3) Let Γ be a discrete group. The following five conditions are equivalent (type I groups and CCR groups are defined in Chapter 6):

(i) Γ has an abelian subgroup of finite index,
(ii) the group Γ is type I,
(iii) the dual space $\widehat{\Gamma}$ is T_0,
(iv) the group Γ is CCR,
(v) the dual space $\widehat{\Gamma}$ is T_1.

The equivalence (i) $\Longleftrightarrow$ (ii) is Thoma's Theorem 7.D.1. Equivalences (ii) $\Longleftrightarrow$ (iii) and (iv) $\Longleftrightarrow$ (v) are those of Theorem 1.D.13. The implication (i) $\Rightarrow$ (iv) follows from (the easy part of) Thoma's Theorem 7.D.1. The implication (v) $\Rightarrow$ (iii) is obvious.

REMARK 1.D.15. We do not know of a characterization of the class $\mathcal{HD}$ of LC groups with a Hausdorff dual space. Here is a sample of results concerning this class of groups:

—Every second-countable LC group in $\mathcal{HD}$ is of type I (see Theorem 1.D.13).

— Every central-by-compact group belongs to $\mathcal{HD}$ (see [**GrMo–71b**, Corollary 5.2]). An LC group G is called **central-by-compact** if G/Z is compact, where Z is the centre of G. The class of central-by-compact groups includes direct products of a compact group and an abelian LC group; the next example shows that this inclusion is strict.

— The set $\mathbf{Z}^2 \times \{\pm 1\}$ with the multiplication defined by $(m_1, m_2, \varepsilon)(n_1, n_2, \delta) = (m_1+n_1, m_2+n_2, \varepsilon\delta(-1)^{m_1 n_2})$ is a discrete group in $\mathcal{HD}$ with centre $\{(m_1, m_2, 1) \in \mathbf{Z}^2 \times \{\pm 1\} \mid m_1 \equiv m_2 \equiv 0 \pmod 2\}$ of index 4. The group is not a direct product of this centre by a finite group. This example appears in [**Palm–78**, Page 701].

— A countable discrete group belongs to $\mathcal{HD}$ if and only if it is central-by-finite (see [**Raeb–82**, Proposition 6]).

— A connected LC group G belongs to $\mathcal{HD}$ if and only if G is compact-by-abelian (see [**BaSu–81**]).

— Let G be a semi-direct product $K \ltimes A$, with K a compact group and A a locally compact abelian group. A necessary and sufficient condition for G to belong to $\mathcal{HD}$ is given in [**Bagg–68**, Section 9]; see also [**KaTa–13**, Proposition 5.62].

—The following conjecture is stated in [**BaSu–81**, p.67]: a LC group G belongs to $\mathcal{HD}$ if and only if G is of type I and G is compact-by-abelian.

REMARK 1.D.16. Dual spaces of groups may have discrete subsets that reflect important properties of the group. In particular, a locally compact group G has **Kazhdan Property (T)** if the class of the unit representation 1_G is isolated in $\widehat{G}$; this property has shown to be a very significant one [**Kazh–67**], [**BeHV–08**].

1.E. Primitive dual of a topological group

As was seen in Theorem 1.D.13, the dual space $\widehat{G}$ of an LC group G does not have the T_0 separation property in general. Given an arbitrary topological X, one can associate to X a canonical topological space with the T_0 separation property. When applied to $\widehat{G}$, this construction leads to the following definition (see Reformulation 1.E.2 below).

DEFINITION 1.E.1. Let G be a topological group. The **primitive dual** $\mathrm{Prim}(G)$ is the space of weak equivalence classes of irreducible representations of G, with the topology defined as follows. The natural surjective map

$$(\kappa 1) \qquad \kappa^{\mathrm{d}}_{\mathrm{prim}} : \widehat{G} \twoheadrightarrow \mathrm{Prim}(G)$$

associates to an equivalence class of irreducible representations its weak equivalence class (κ stands for $\kappa\alpha\nu o\nu\iota\kappa o\sigma$, i.e., for canonical). The topology on $\mathrm{Prim}(G)$ is defined as the quotient by (κ1) of the Fell topology on the dual $\widehat{G}$.

In view of Proposition 1.C.7, the definition of $\mathrm{Prim}(G)$ and its topology can be reformulated in terms of the notion of a universal Kolmogorov quotient of a topological space, i.e., of the largest T_0 quotient of a topological space (see Appendix A.1).

REFORMULATION 1.E.2. *Let G be a topological group.*
The universal Kolmogorov quotient of the dual $\widehat{G}$ is the pair made of the primitive dual $\mathrm{Prim}(G)$ *and the natural surjective map* $\widehat{G} \twoheadrightarrow \mathrm{Prim}(G)$.

REMARK 1.E.3. Here we collect three observations about the map $\kappa^{\mathrm{d}}_{\mathrm{prim}}$.

(1) The map $\kappa^{\mathrm{d}}_{\mathrm{prim}} : \widehat{G} \twoheadrightarrow \mathrm{Prim}(G)$ is surjective, by definition. In general, it is far from being injective. Indeed, when G is a second-countable LC group, this map is injective (and therefore bijective) if and only if G is of type I. More on this in Section 9.G.

(2) The subspace $\widehat{G}_{\mathrm{fd}}$ of $\widehat{G}$, of equivalence classes of finite-dimensional irreducible representations of G, has been introduced in Remark 1.C.14. It follows from Remark 1.C.2(3) that the restriction of $\kappa^{\mathrm{d}}_{\mathrm{prim}}$ to $\widehat{G}_{\mathrm{fd}}$ is a bijection onto its image.

(3) For a locally compact group G, we will give in Section 8.B an equivalent definition of the topological space $\mathrm{Prim}(G)$ as the space of primitive ideals of the maximal C*-algebra of G, equipped with the Jacobson topology. When G is moreover σ-compact, we will extend the map $\kappa^{\mathrm{d}}_{\mathrm{prim}}$ to the quasi-dual $\mathrm{QD}(G)$ of G defined in Section 6.C.

REMARK 1.E.4. Here are some known topological properties of primitive duals:

(1) Let G be a topological group. By definition (see Reformulation 1.E.2), the primitive dual $\mathrm{Prim}(G)$ is a T_0-space.

(2) We will determine in Chapter 9 the primitive duals of some discrete groups. In most cases, they are not T_1-spaces.

(3) Here are a few results concerning the T_1 separation property for primitive duals for some classes of groups.

— Let Γ be a finitely generated solvable group. Then $\mathrm{Prim}(\Gamma)$ is a T_1-space if and only if Γ is virtually nilpotent (see [**MoRo–76**, Theorem 5]).

— Discrete nilpotent groups have T_1 primitive duals (see [**Pogu–82**]).

— Points in primitive duals of LC-groups need not be locally closed. Recall that, in a topological space X, a point x is locally closed if $\{x\}$ is open in its closure $\overline{\{x\}}$, equivalently if there exists a neighbourhood V of x such that x is closed in V [**BTG1–4**, Chap. III, § 4, no 6].

For example, let Γ be either the Baumslag-Solitar group $BS(1,p)$ or the lamplighter group (see Sections 3.D and 3.E). Then $\mathrm{Prim}(\Gamma)$ contains non locally closed points. Indeed, on the one hand, it follows from the description of $\mathrm{Prim}(\Gamma)$ in Section 9.D (or Section 9.E) that there exists $\pi \in \widehat{\Gamma}$ such that $\kappa(\pi)$ is a dense point in $\mathrm{Prim}(\Gamma)$, where $\kappa = \kappa^{\mathrm{d}}_{\mathrm{prim}} : \widehat{G} \twoheadrightarrow \mathrm{Prim}(G)$ is the canonical map. On the other hand, since Γ is an amenable residually finite group, $\kappa(\widehat{\Gamma}_{\mathrm{fd}})$ is dense in $\mathrm{Prim}(\Gamma)$, by Corollary 15.E.14. Since elements in $\kappa(\widehat{\Gamma}_{\mathrm{fd}})$ are closed (Proposition 9.A.3), $\kappa(\widehat{\Gamma}_{\mathrm{fd}})$ is contained in the complement of $\{\kappa(\pi)\}$ in $\mathrm{Prim}(\Gamma)$. It follows that the point $\{\kappa(\pi)\}$ is not locally closed in $\mathrm{Prim}(\Gamma)$.

— If G is a connected LC group, all points of $\mathrm{Prim}(G)$ are locally closed. Moreover, $\mathrm{Prim}(G)$ is a T_1-space if and only if G is a projective limit of connected

Lie groups, each of which is type R on its radical [**MoRo–76**, Theorem 1]. (A connected Lie group G is of type R on its radical if the adjoint action of G on the radical of its Lie algebra is distal.)

— We refer to Page 726 of the review article by Palmer [**Palm–78**] for more on locally compact groups G for which $\mathrm{Prim}(G)$ is a T_1-space.

(4) Concerning the Hausdorff property T_2 for primitive duals, we quote only the following result. Let G be an LC group G in which every conjugacy class has a compact closure; then the space $\mathrm{Prim}(G)$ is Hausdorff. See [**LiMo–74**] for σ-compact groups, and [**Kani–79**] in general.

(5) Let G be a locally compact group. We will show in Proposition 8.A.15 that $\mathrm{Prim}(G)$ is a Baire space.

We will continue our discussion of primitive duals, both for locally compact groups and for C*-algebras, in Chapter 8.

1.F. Induced representations, irreducibility, and equivalence

The construction of irreducible representations of a group is often based on induction of representations, which is a procedure to build representations of a group out of representations of its subgroups.

To avoid technical problems, we will restrict ourselves in this section to the case of *open* subgroups. In fact, given an LC group G, a *closed* subgroup H of G, and a representation σ of H, an induced representation $\mathrm{Ind}_H^G\sigma$ of G can always be defined. For this more general construction, see [**Mack–52**], [**Mack–58**], [**Foll–16**], [**BeHV–08**], and [**KaTa–13**].

In the particular case of inducing a *one-dimensional* representation σ of an open subgroup H of G, we will state and prove sufficient conditions for which $\mathrm{Ind}_H^G\sigma$ is irreducible, and for which two such representations are not equivalent.

Let G be a topological group, and H an open subgroup of G. A **left transversal** for H in G is a subset T of G such that G is the disjoint union $\bigsqcup_{t\in T} tH$. It is sometimes convenient, but it is not necessary, to choose $e \in T$ as a representative of the class H in G/H.

Similarly, a **right transversal** for H in G is a subset T of G such that $G = \bigsqcup_{t\in T} Ht$. Observe that if T is a left transversal for H in G, then the set T^{-1} of inverses of elements in T is a right transversal for H in G.

DEFINITION 1.F.1. Let G be a topological group, H an open subgroup, and $(\sigma, \mathcal{K}_\sigma)$ a representation of H. An **induced representation** of σ from H to G is a representation π of G on some Hilbert space $\mathcal{H}$ that satisfies the following conditions:

(1) there exists a closed subspace $\mathcal{K}$ of $\mathcal{H}$ that is invariant by $\pi(H)$;
(2) the representation $\mathrm{Res}_H^G(\pi)|_{\mathcal{K}} : H \to \mathcal{U}(\mathcal{K})$ obtained by restriction of π to H and $\mathcal{K}$ is equivalent to σ;
(3) the space of π is the orthogonal Hilbert sum $\mathcal{H} = \bigoplus_{t\in T} \pi(t)(\mathcal{K})$, for some left transversal T of H in G.

An induced representation π as above is often denoted by $\mathrm{Ind}_H^G\sigma$.

REMARK 1.F.2. Note that, in view of (1), $\pi(t)(\mathcal{K})$ only depends on the left coset in G of $t \in T$ and, hence, if (3) holds for one left transversal T for H, it holds for any such transversal. Moreover, G acts in the obvious way transitively on the set

$$\{\mathcal{L} \mid \mathcal{L} = \pi(t)(\mathcal{K}) \text{ for some } t \in T\} = \{\mathcal{L} \mid \mathcal{L} = \pi(g)(\mathcal{K}) \text{ for some } g \in G\},$$

and the stabilizer of $\mathcal{K}$ is H.

Let us check that

(4) an induced representation of σ is *unique* up to equivalence,

that is, if π_1, π_2 are representations of G which satisfy Conditions (1) to (3), then π_1 and π_2 are equivalent.

Indeed, for $j = 1, 2$, there exist by (1) and (2) a closed $\pi_j(H)$-invariant subspace $\mathcal{K}_j$ such that $\mathrm{Res}_H^G(\pi_j)|_{\mathcal{K}_j}$ is equivalent to σ. Therefore there exists a surjective isometry $V : \mathcal{K}_1 \to \mathcal{K}_2$ which intertwines $\mathrm{Res}_H^G(\pi_1)|_{\mathcal{K}_1}$ and $\mathrm{Res}_H^G(\pi_2)|_{\mathcal{K}_2}$. We claim that the map $U : \mathcal{H}_1 \to \mathcal{H}_2$ given by

$$U\pi_1(g)\eta = \pi_2(g)V\eta \quad \text{for } g \in G, \eta \in \mathcal{K}_1$$

is well-defined; indeed, let $g, g' \in G$ and $\eta, \eta' \in \mathcal{K}_1$ be such that $\pi_1(g)\eta = \pi_1(g')\eta'$. If $\eta = 0$, then $\eta' = 0$ and $\pi_2(g)V\eta = \pi_2(g')V\eta'$. So, we may assume that $\eta \neq 0$. We then have $\pi_1(g'^{-1}g)\eta = \eta' \in \mathcal{K}_1 \smallsetminus \{0\}$. It follows from (3) that $g'^{-1}g \in H$. So, $g' = gh$ and $\eta' = \pi_1(h^{-1})\eta$ for some $h \in H$, and hence

$$\pi_2(g')V\eta' = \pi_2(gh)V\pi_1(h^{-1})\eta = \pi_2(gh)\pi_2(h^{-1})V\eta = \pi_2(g)V\eta.$$

Then U is an isometry, and it is surjective because its image contains $\pi_2(t)(\mathcal{K}_2)$ for all $t \in T$. Finally, U intertwines π_1 and π_2; indeed, let $t \in T$ and $\xi \in \pi_1(t)(\mathcal{K}_1)$, say $\xi = \pi_1(t)\eta$ for $\eta \in \mathcal{K}_1$; for $g \in G$, we have

$$U\pi_1(g)\xi = U\pi_1(gt)\eta = \pi_2(g)\pi_2(t)V\eta = \pi_2(g)U\pi_1(t)\eta = \pi_2(g)U\xi.$$

Some properties of $\mathrm{Ind}_H^G\sigma$ are straightforward consequences of the definition. For example:

(5) The dimension of $\mathrm{Ind}_H^G\sigma$ is $|G/H|\dim(\sigma)$. In particular $\mathrm{Ind}_H^G\sigma$ is finite-dimensional if and only if H of finite index in G and σ is finite-dimensional.

(6) If N is a normal subgroup of G contained in H and such that $\sigma(n) = \mathrm{Id}_{\mathcal{K}}$ for all $n \in N$, then $\left(\mathrm{Ind}_H^G\sigma\right)(n) = \mathrm{Id}_{\mathcal{H}}$ for all $n \in N$.

REMARK 1.F.3. The characterization of induced representations by Conditions (1) to (3) of Definition 1.F.1 is valid for open subgroups only. In the more general case of induction from closed subgroups, these conditions need not be satisfied. Indeed, consider for instance the affine group of the real line $\mathrm{Aff}(\mathbf{R}) = \mathbf{R}^\times \ltimes \mathbf{R}$, with its natural locally compact topology (see Section 3.C). Let π be the induced representation $\mathrm{Ind}_N^{\mathrm{Aff}(\mathbf{R})}\chi$, for the normal subgroup $N = \{(1, b) \mid b \in \mathbf{R}\}$ and the unitary character of N given by $\chi(1, b) = e^{ib}$. Then (see Remark 3.C.7) π can be realized on $L^2(\mathbf{R}^\times, dt/|t|)$ by the formula

$$\pi(a, b)\xi(t) = \exp(-2\pi ibt)\,\xi(at)$$

for all $\xi \in L^2(\mathbf{R}^\times, dt/|t|)$. Using the Fourier transform on $N \approx \mathbf{R}$, we see that the restriction of π to N is equivalent to the regular representation λ_N; in particular, $\pi|_N$ contains no finite-dimensional subrepresentation.

In the general case of inducing from closed subgroups of LC groups, there is however a characterization of induced representations in terms of "transitive systems of imprimitivity". We refer to [**Mack–58**, § 6], as well as to [**Foll–16**, Chap. 6].

CONSTRUCTION 1.F.4. Let G be a topological group, H an *open* subgroup of G, and $(\sigma, \mathcal{K})$ a representation of H. In the literature, there are several models for the induced representation $\mathrm{Ind}_H^G \sigma$ as defined in Definition 1.F.1. We are going to describe three of them.

(1) Consider the Hilbert space $\mathcal{H}$ of maps $f : G \to \mathcal{K}$ with the following properties:

(a) $f(hx) = \sigma(h)f(x)$ for all $h \in H$, $x \in G$;
(b) $\sum_{Hx \in H\backslash G} \|f(x)\|^2 < \infty$.

Observe that, in view of (a) and the unitarity of σ, the number $\|f(x)\|^2$ only depends on the coset Hx; therefore, the sum in (b), which is a sum over one x for each coset in $H\backslash G$, is well defined. Observe also that, for $g \in G$ and $f \in \mathcal{H}$, the function $G \to \mathcal{K}$, $x \mapsto f(xg)$ is again in $\mathcal{H}$. The induced representation $\pi := \mathrm{Ind}_H^G \sigma$ is the representation of G defined in $\mathcal{H}$ by

$$(\pi(g)f)(x) = f(xg) \qquad \text{for all } g \in G, \ f \in \mathcal{H} \text{ and } x \in G.$$

Using another model for π, we will verify below that the homomorphism

$$G \to \mathcal{U}(\mathcal{H}), \ \ g \mapsto \pi(g)$$

is indeed continuous.

Let us check that π as defined above is indeed an induced representation of σ in the sense of Definition 1.F.1. Let T be a left transversal for H in G; denote by t_0 the element in $T \cap H$. For every $t \in T$, let $\mathcal{H}_t$ be the space of functions $f : G \to \mathcal{K}$ with support contained in Ht^{-1} (in particular of functions with support in H when $t = t_0$). Then $\mathcal{H}_{t_0}$ is a closed $\pi(H)$-invariant subspace of $\mathcal{H}$. Moreover $\mathcal{H}_t$ is closed in $\mathcal{H}$ for all $t \in T$, and $\mathcal{H} = \bigoplus_{t \in T} \mathcal{H}_t$. This shows that Conditions (1) and (3) of Definition 1.F.1 are satisfied. For $\xi \in \mathcal{K}$, define a function $f_\xi : G \to \mathcal{K}$ by $f_\xi(g) = \sigma(g)\xi$ if $g \in H$ and $f_\xi(g) = 0$ if $g \in G \smallsetminus H$. Then $f_\xi \in \mathcal{H}_{t_0}$, and $\xi \mapsto f_\xi$ defines an isometry U of $\mathcal{K}$ onto $\mathcal{H}_{t_0}$; moreover, $U\sigma(h)\xi = \pi(h)U\xi$ for all $h \in H$ and $\xi \in \mathcal{K}$, so that Condition (2) of Definition 1.F.1 is also satisfied, as was to be shown.

(2) It will be convenient for us to work with another model for induced representations, as follows. Choose a right transversal T, so that we have a disjoint union $G = \bigsqcup_{t \in T} Ht$. For every $g \in G$ and $t \in T$, there are uniquely defined elements

$$(*) \qquad \alpha(t, g) \in H \text{ and } t \cdot g \in T \text{ such that } tg = \alpha(t, g)(t \cdot g).$$

It is routine to check that the map $\alpha : T \times G \to H$ satisfies

$$\alpha(t, g_1 g_2) = \alpha(t, g_1)\alpha(t \cdot g_1, g_2) \qquad \text{for all } t \in T \text{ and } g_1, g_2 \in G,$$

i.e., it is a cocycle, see Definition 5.B.1 below, and that

$$T \times G \to T, \ \ (t, g) \mapsto t \cdot g$$

is an action of G on T *on the right*, which is equivalent to the natural action of G on $H\backslash G$ given by right multiplication.

Let $\ell^2(T, \mathcal{K})$ denote the Hilbert space of maps $f: T \to \mathcal{K}$ such that $\sum_{t\in T} \|f(t)\|^2 < \infty$, with scalar product $\langle f_1 \mid f_2\rangle := \sum_{t\in T}\langle f_1(t) \mid f_2(t)\rangle_{\mathcal{K}}$. The induced representation $\pi = \mathrm{Ind}_H^G \sigma$ of G is defined on $\ell^2(T, \mathcal{K})$ by

$$(\pi(g)f)(t) = \sigma(\alpha(t, g))(f(t \cdot g)) \quad \text{for all } g \in G, f \in \ell^2(T, \mathcal{K}) \text{ and } t \in T.$$

We leave it to the reader to check that π satisfies Conditions (1) to (3) of Definition 1.F.1.

We now show that π is indeed a representation of G, i.e., that the map

$$G \to \ell^2(T, \mathcal{K}), \quad g \mapsto \pi(g)f$$

is continuous for every $f \in \ell^2(T, \mathcal{K})$.

For $t \in T$ and $\xi \in \mathcal{K}$, let $f_{t,\xi}$ be the element of $\ell^2(T, \mathcal{K})$ defined by $f_{t,\xi}(t) = \xi$ and $f_{t,\xi}(s) = 0$ for $s \in T, s \neq t$. By definition of π, we have

$$\pi(g)f_{t,\xi} = f_{t\cdot g^{-1}, \sigma(\alpha(t\cdot g^{-1}, g))\xi}$$

for $g \in G$. Since the linear span of $\{f_{t,\xi} \mid t \in T, \ \xi \in \mathcal{K}\}$ is dense in $\ell^2(T, \mathcal{K})$, it suffices to prove that the map $g \mapsto \pi(g)f_{t,\xi}$ is continuous for $t \in T$ and $\xi \in \mathcal{K}$. Moreover, as π is a homomorphism, it suffices to prove the continuity of this map at the unit element e.

Now consider a fixed pair of $t \in T$ and $\xi \in \mathcal{K}$, and $\varepsilon > 0$. Since σ is a representation of H, there exists a neighbourhood U of e in H such that

$$\|\sigma(h)\xi - \xi\| < \varepsilon \quad \text{for all } h \in U.$$

Since H is open, U is a neighbourhood of e in G and we can find a neighbourhood V of e such that $V \subset U$ and $t^{-1}Vt \subset U$. For $g \in t^{-1}Vt$, we have $tg \in Vt \subset Ut \subset Ht$, so that $t \cdot g = t$ and $\alpha(t, g) \in U$, hence $\pi(g)f_{t,\xi} = f_{t,\sigma(\alpha(t,g))\xi}$. It follows that

$$\|\pi(g)f_{t,\xi} - f_{t,\xi}\| = \|\sigma(\alpha(t, g))\xi - \xi\| < \varepsilon$$

for every $g \in t^{-1}Vt$.

(2) Occasionally, it may be useful to realize $\mathrm{Ind}_H^G \sigma$ on $\ell^2(T, \mathcal{K})$ where T is now a *left* transversal for H in G. For $g \in G$ and $t \in T$, define $g \cdot t \in T$ and $\beta(g, t) \in H$ by the equality $gt = (g \cdot t)\beta(g, t)$. The map $\beta : G \times T \to H$ satisfies the cocycle relation

$$\beta(g_1g_2, t) = \beta(g_1, g_2 \cdot t)\beta(g_2, t) \quad \text{for all } g_1, g_2 \in G \text{ and } t \in T;$$

compare with Definition 5.A.1.

The induced representation $\pi = \mathrm{Ind}_H^G \sigma$ is defined by

$$(\pi(g)f)(t) = \sigma\left(\beta(g^{-1}, t)^{-1}\right)\left(f(g^{-1} \cdot t)\right) \quad \text{for all} mg \in G, \ f \in \ell^2(T, \mathcal{K}) \text{ and } t \in T.$$

DEFINITION 1.F.5. The **quasi-regular representation** of a topological group G associated with an open subgroup H is the induced representation $\mathrm{Ind}_H^G 1_H$.

It can be realized as the natural representation of G on $\ell^2(H\backslash G)$ given by right translations, or equivalently as the natural representation of G on $\ell^2(G/H)$ given by left translations, and then usually denoted by $\lambda_{G/H}$. There is an intertwining isometry S from $\ell^2(H\backslash G)$ onto $\ell^2(G/H)$, defined by $(Sf)(xH) = f(Hx^{-1})$ for all $f \in \ell^2(H\backslash G)$ and $x \in G$.

When G is locally compact, as already mentioned above, there is also a quasi-regular representation $\mathrm{Ind}_H^G 1_H$ for every *closed* subgroup H of G. We refer to [**BeHV–08**, Appendix B.1], or to Definition 6.B.30 for the particular case of a compact subgroup H in G.

We state here the following properties of the induction process for open subgroups, but they hold just as well for closed subgroups of locally compact groups. Property (5) is known as the continuity of induction. See, e.g., [**BeHV–08**, Appendix E].

PROPOSITION 1.F.6. *Let G be a topological group, H an open subgroup of G, and K an open subgroup of H. Let π be a representation of G; let $\sigma_1, \sigma_2, \dots$ be representations of H, and τ a representation of K.*

(1) *If σ_1 and σ_2 are equivalent representations of H, then $\mathrm{Ind}_H^G \sigma_1$ and $\mathrm{Ind}_H^G \sigma_2$ are equivalent representations of G.*
(2) *$\mathrm{Ind}_H^G(\bigoplus_{i \geq 1} \sigma_i)$ and $\bigoplus_{i \geq 1} (\mathrm{Ind}_H^G \sigma_i)$ are equivalent.*
(3) *$\mathrm{Ind}_H^G(\mathrm{Ind}_K^H \tau)$ and $\mathrm{Ind}_K^G \tau$ are equivalent.*
(4) *$\mathrm{Ind}_H^G(\mathrm{Res}_H^G \pi)$ is equivalent to the tensor product $\pi \otimes \lambda_{G/H}$ of π with the quasi-regular representation $\lambda_{G/H}$.*
(5) *If σ_1 is weakly contained in σ_2, then $\mathrm{Ind}_H^G \sigma_1$ is weakly contained in $\mathrm{Ind}_H^G \sigma_2$.*

In favourable situations, induced representations have central characters (Definition 1.A.12) which are given as follows.

PROPOSITION 1.F.7. *Let G be a topological group and H an open subgroup of G containing the centre $Z(G)$ of G. Let σ be a representation of H. Assume that σ admits a central character $\lambda_\sigma \in \widehat{Z(H)}$.*

Then $\mathrm{Ind}_H^G \sigma$ admits a central character, which is the restriction of λ_σ from $Z(H)$ to $Z(G)$.

PROOF. Write $\mathcal{K}$ for the Hilbert space of σ and π for $\mathrm{Ind}_N^G \sigma$. We realize the Hilbert space $\mathcal{H}$ of π as the space of maps $f : G \to \mathcal{K}$ such that $f(hx) = \sigma(h) f(x)$ for all $h \in H$, $x \in G$, and $\sum_{Hx \in H \backslash G} \|f(x)\|^2 < \infty$ (see Construction 1.F.4(1)). Recall that π is the representation on $\mathcal{H}$ defined by right translations.

For every $z \in Z(G)$ and $f \in \mathcal{H}$, we have

$$(\pi(z)f)(x) = f(xz) = f(zx) = \sigma(z)(f(x)) = \lambda_\sigma(z) f(x) \quad \text{for all } x \in G,$$

and the claim follows. □

Next, we identify the conjugate representation of an induced representation by an element of the group $\mathrm{Aut}(G)$ of bicontinuous automorphisms of G.

If $(\sigma, \mathcal{K})$ is a representation of a subgroup H of G and $\theta \in \mathrm{Aut}(G)$, we denote by σ^θ the representation of the subgroup $\theta^{-1}(H)$ on $\mathcal{K}$ defined by

$$\sigma^\theta(x) = \sigma(\theta(x)) \quad \text{for all } x \in \theta^{-1}(H).$$

Observe that, in case $H = G$, this representation is the conjugate representation σ^θ of G discussed in Remark 1.A.17.

PROPOSITION 1.F.8. *Let G be a topological group, H an open subgroup, and σ a representation of H on a Hilbert space $\mathcal{K}$. Let $\theta \in \mathrm{Aut}(G)$.*

The representation $(\mathrm{Ind}_H^G \sigma)^\theta$ is equivalent to the representation $\mathrm{Ind}_{\theta^{-1}(H)}^G(\sigma^\theta)$.

PROOF. Let $\mathcal{H}$ be the Hilbert space of $\pi := \mathrm{Ind}_H^G \sigma$. Choose a right transversal T for H in G and let $\mathcal{K}$ be a $\pi(H)$-invariant subspace of $\mathcal{H}$ with Properties (1), (2), and (3) of Definition 1.F.1.

Recall that the Hilbert space of π^θ is $\mathcal{H}$. The representation $\mathrm{Res}^G_{\theta^{-1}(H)}(\pi^\theta)|_\mathcal{K}$ obtained by the restriction of π^θ to $\theta^{-1}(H)$ and $\mathcal{K}$ is equivalent to σ^θ; indeed, let $U : \mathcal{K} \to \mathcal{H}_\sigma$ be a surjective isometry intertwining $\mathrm{Res}^G_H(\pi)|_\mathcal{K}$ and σ. Then

$$U(\pi^\theta(\theta^{-1}(h)))U^{-1} = U\pi(h)U^{-1} = \sigma(h) = \sigma^\theta(\theta^{-1}(h))$$

for every $h \in H$; this shows that $U : \mathcal{K} \to \mathcal{H}_{\sigma^\theta} = \mathcal{H}_\sigma$ is a surjective isometry intertwining $\mathrm{Res}^G_{\theta^{-1}(H)}(\pi^\theta)|_\mathcal{K}$ and σ^θ.

Moreover, the set $S := \theta^{-1}(T)$ is a right transversal for $\theta^{-1}(H)$ in G and we have

$$\mathcal{H} = \bigoplus_{t\in T} \pi(t)(\mathcal{K}) = \bigoplus_{s\in S} \pi(\theta(s))(\mathcal{K}) = \bigoplus_{s\in S} \pi^\theta(s)(\mathcal{K}).$$

As a result, the representation $(\pi^\theta, \mathcal{H})$ satisfies conditions (1), (2), and (3) of Definition 1.F.1, for the subspace $\mathcal{K}$ of $\mathcal{H}$ and the representation σ^θ of $\theta^{-1}(H)$. By uniqueness of induced representations (see Remark 1.F.2), it follows that π^θ is equivalent to $\mathrm{Ind}^G_{\theta^{-1}(H)}(\sigma^\theta)$. □

We will need the following elementary proposition several times. Let N be a normal subgroup of G and $(\sigma, \mathcal{K})$ a representation of N. Let $a \in G$. Recall (see Remark 1.A.17) that σ^a is the conjugate representation of σ by a, defined on $\mathcal{K}$ by

$$(\sharp) \qquad \sigma^a(n) = \sigma(ana^{-1}) \quad \text{for all } n \in N.$$

PROPOSITION 1.F.9. *Let G be a topological group, N an open normal subgroup, and σ a representation of N on a Hilbert space $\mathcal{K}$. Let $a \in G$.*

(1) *The representations $\mathrm{Ind}^G_N(\sigma^a)$ and $\mathrm{Ind}^G_N\sigma$ are equivalent.*
(2) *The restriction of $\mathrm{Ind}^G_N\sigma$ to N is equivalent to the direct sum $\bigoplus_{t\in T}\sigma^{t^{-1}}$, where $T \subset G$ is a right transversal for N in G.*
In particular, if N is contained in the centre of G, the restriction of $\mathrm{Ind}^G_N\sigma$ to N is equivalent to a multiple of σ.

PROOF. (1) By Proposition 1.F.8, the representations $(\mathrm{Ind}^G_N\sigma)^a$ and $\mathrm{Ind}^G_N(\sigma^a)$ are equivalent. Since $(\mathrm{Ind}^G_N\sigma)^a$ and $\mathrm{Ind}^G_N(\sigma)$ are equivalent, the claim is proved.

(2) Set $\pi := \mathrm{Ind}^G_N\sigma$ and $\mathcal{H} := \ell^2(T, \mathcal{K})$. By the defining property of induced representations (Definition 1.F.1), there exists a $\pi(N)$-invariant closed subspace $\mathcal{H}_e$ of $\mathcal{H}$ such that $\pi|_N$ restricted to $\mathcal{H}_e$ is equivalent to σ and such that

$$\mathcal{H} = \bigoplus_{t\in T} \pi(t)(\mathcal{H}_e).$$

For every $n \in N$, $t \in T$, and $\xi \in \mathcal{H}_e$, we have

$$\pi(n)(\pi(t)\xi) = \pi(t)(\pi(t^{-1}nt)\xi).$$

This shows that $\pi(t)(\mathcal{H}_e)$ is $\pi(N)$-invariant and that $\pi|_N$ restricted to $\pi(t)\mathcal{H}_e$ is equivalent to $\sigma^{t^{-1}}$, for every $t \in T$.

If N is contained in the centre of G, then $\sigma^g = \sigma$ for every $g \in G$ and hence $\pi|_N$ is equivalent to a multiple of σ. □

The following proposition has already been referred to in Example 1.B.9. It will be needed again, in Chapters 9 and 12.

PROPOSITION 1.F.10. *Let G be a topological group, H an open subgroup, $\varphi : H \to \mathbf{C}$ a function of positive type, $\varphi \neq 0$, and $\widetilde{\varphi} : G \to \mathbf{C}$ its trivial extension.*

Then $\widetilde{\varphi}$ is a function of positive type on G.

Denote by σ the GNS representation of H associated with φ. The GNS representation π of G associated with $\widetilde{\varphi}$ is equivalent to the induced representation $\mathrm{Ind}_H^G \sigma$. In particular, $\mathrm{Ind}_H^G \sigma$ is cyclic and has a cyclic vector ξ such that

$$\widetilde{\varphi}(g) = \langle (\mathrm{Ind}_H^G \sigma)(g)\xi \mid \xi \rangle \quad \textit{for all } g \in G.$$

NOTE. When G is locally compact, there is a version of this proposition for H closed (rather than H open) in G, for which we refer to [**Blat–63**] or [**Foll–16**, Theorem 6.13]. This involves measures of positive type (rather than functions) and modular functions on G and H.

PROOF. Denote by $(\mathcal{K}, \sigma, \eta)$ the GNS triple associated with φ, so that $\varphi(h) = \langle \sigma(h)\eta \mid \eta \rangle$ for all $h \in H$. Let $\pi := \mathrm{Ind}_H^G \sigma$ be the induced representation of σ from H to G, as in the second model of Construction 1.F.4. Recall that this involves a right transversal T, so that $G = \bigsqcup_{t \in T} Ht$, that $\pi(G)$ acts on $\mathcal{H} = \ell^2(T, \mathcal{K})$, and that we have

$$(\pi(g)f)(t) = \sigma(\alpha(t,g))f(t \cdot g) \quad \text{for all } g \in G, \ f \in \mathcal{H}, \text{ and } t \in T,$$

with $\alpha(t,g) \in H$ and $t \cdot g \in T$. We assume moreover, for simplicity, that $e \in T$.

Let $\xi \in \mathcal{H}$ be the map defined by $\xi(e) = \eta$ and $\xi(t) = 0$ for all $t \in T \smallsetminus \{e\}$. We have

$$\begin{aligned}
\langle \pi(h)\xi \mid \xi \rangle &= \sum_{t \in T} \langle \sigma(\alpha(t,h))(\xi(t \cdot h)) \mid \xi(t) \rangle = \langle \sigma(\alpha(e,h))(\xi(e \cdot h)) \mid \xi(e) \rangle \\
&= \langle \sigma(h)\eta \mid \eta \rangle = \varphi(h) \qquad \text{for all } h \in H, \\
\langle \pi(g)\xi \mid \xi \rangle &= 0 \qquad \text{for all } g \in G \smallsetminus H.
\end{aligned}$$

Therefore $\widetilde{\varphi}(g) = \langle \pi(g)\xi \mid \xi \rangle$ for all $g \in G$, i.e., $\widetilde{\varphi}$ is the function of positive type associated with π and ξ; in particular, $\widetilde{\varphi}$ is of positive type.

To show that the GNS representation associated with $\widetilde{\varphi}$ is equivalent to π, it suffices by Proposition 1.B.10 to check that ξ is a cyclic vector for π.

For every $t \in T$ and $v \in \mathcal{K}$, let $\xi_t^v \in \mathcal{H}$ be defined by $\xi_t^v(t) = v$ and $\xi_t^v(s) = 0$ when $s \in T \smallsetminus \{t\}$; observe that $\xi_e^\eta = \xi$. The linear span X of $\{\xi_t^v \in \mathcal{H} \mid t \in T, v \in \mathcal{K}\}$ is dense in $\mathcal{H}$. On the one hand, since η is cyclic for σ, the linear span of the set

$$\{\xi_t^{\sigma(h)\eta} \in \mathcal{H} \mid t \in T, \ h \in H\}$$

is dense in X, and hence in $\mathcal{H}$. On the other hand, for $t \in T$ and $h \in H$, we have

$$\pi(t^{-1}h)\xi_e^\eta = \xi_t^{\sigma(h)\eta};$$

indeed, $\alpha(t, t^{-1}h) = h$ and, for $s \in T$, we have $s \cdot t^{-1}h = e$ if and only if $s = t$. This shows that $\xi_e^\eta = \xi$ is cyclic for π. □

From now on, we restrict ourselves to **monomial representations**, that is, to representations of the form $\mathrm{Ind}_H^G \chi$, where $\chi : H \to \mathbf{T}$ is a unitary character of H. We use Model (2) in Construction 1.F.4, so that the corresponding induced representation $\pi = \mathrm{Ind}_H^G \chi$ is realized on $\ell^2(T)$ and we have

$$(\pi(g)f)(t) = \chi(\alpha(t,g))f(t \cdot g) \quad \text{for all } g \in G, \ f \in \ell^2(T), \text{ and } t \in T.$$

We assume moreover that $e \in T$, and we denote by $(\delta_t)_{t\in T}$ the orthonormal basis of $\ell^2(T)$ defined by $\delta_t(t) = 1$ and $\delta_t(s) = 0$ when $s \in T$, $s \neq t$. Observe that we have

$$\pi(t^{-1})\delta_e = \delta_t \qquad \text{for all } t \in T.$$

In particular δ_e is a cyclic vector for π.

For $i = 1, 2$, let H_i be an open subgroup of G and χ_i a unitary character of H_i. Choose a right transversal T_i for H_i in G, with $e \in T_i$. For $g \in G$ and $t \in T_i$, we have as in $(*)$ of 1.F.4 uniquely defined elements $\alpha_i(t,g) \in H_i$ and $t \cdot_i g \in T_i$ such that $tg = \alpha_i(t,g)(t \cdot_i g)$.

Our study of the monomial representations $\pi_i = \mathrm{Ind}_{H_i}^G \chi_i$, their irreducibility and their equivalence, is based on the following lemma.

Lemma 1.F.11. *We keep the notation above. Let $S \in \mathrm{Hom}_G(\pi_1, \pi_2)$. Set $f := S\delta_e \in \ell^2(T_2)$.*

(1) *We have $f = 0$ if and only if $S = 0$.*
(2) *Assume that $\pi_1 = \pi_2$; then f is a scalar multiple of δ_e if and only if S is a scalar operator.*
(3) *For every $t \in T_2$ and $h \in H_1$, we have*
$$\chi_2(\alpha_2(t,h))f(t \cdot_2 h) = \chi_1(h)f(t).$$
(4) *Let $t \in T_2$ be such that the H_1-orbit $\{t \cdot_2 h \mid h \in H_1\}$ in T_2 is infinite; then $f(t) = 0$.*
(5) *Let $t \in T_2$ and $h \in H_1$ be such that $f(t) \neq 0$ and $t \cdot_2 h = t$; then $tht^{-1} \in H_2$ and $\chi_1(h) = \chi_2(tht^{-1})$.*

Proof. Since $S \in \mathrm{Hom}_G(\pi_1, \pi_2)$, Items (1) and (2) follow from the fact, mentioned above, that δ_e is a cyclic vector for π_1.

(3) Let $h \in H_1$. Since $\alpha_1(e,h) = h$ and $e \cdot_1 h = e$, and $s \cdot_1 h \neq e$ for $s \in T_1 \smallsetminus \{e\}$, we have

$$\begin{aligned}(\pi_1(h)\delta_e)(e) &= \chi_1(\alpha_1(e,h))\delta_e(e \cdot_1 h) = \chi_1(h),\\ (\pi_1(h)\delta_e)(s) &= \chi_1(\alpha_1(s,h))\delta_e(s \cdot_1 h) = 0 \quad \text{if } s \neq e,\end{aligned}$$

so that $\pi_1(h)\delta_e = \chi_1(h)\delta_e$.

Let $t \in T_2$. We have

$$(\pi_2(h)f)(t) = (\pi_2(h)S\delta_e)(t) = (S\pi_1(h)\delta_e)(t) = \chi_1(h)(S\delta_e)(t) = \chi_1(h)f(t)$$

and

$$(\pi_2(h)f)(t) = \chi_2(\alpha_2(t,h))f(t \cdot_2 h).$$

The claim follows.

(4) For $t \in T_2$ and $h \in H_1$, it follows from (3) that $|f(t \cdot_2 h)| = |f(t)|$, i.e., that $|f|$ is constant on the H_1-orbit of t in T_2. Since $f \in \ell^2(T_2)$, this implies that $f(t) = 0$ if this orbit is infinite.

(5) Let $t \in T_2$ and $h \in H_1$ be such that $t \cdot_2 h = t$ and $f(t) \neq 0$. By (3), we have

$$\chi_2(\alpha_2(t,h)) = \chi_1(h).$$

Moreover, since

$$th = \alpha_2(t,h)(t \cdot_2 h) = \alpha_2(t,h)t,$$

we have $tht^{-1} = \alpha_2(t,h) \in H_2$ and $\chi_1(h) = \chi_2(\alpha_2(t,h)) = \chi_2(tht^{-1})$. □

If χ is a unitary character of the subgroup H and $g \in G$, recall (see the notation introduced before Proposition 1.F.8) that χ^g is the unitary character of $g^{-1}Hg$ defined by

$$\chi^g(h) = \chi(ghg^{-1}) \quad \text{for all } h \in g^{-1}Hg.$$

The **commensurator** of a subgroup H of G is the set $\mathrm{Comm}_G(H)$ of those elements $g \in G$ such that $H \cap g^{-1}Hg$ is of finite index in both H and $g^{-1}Hg$. Equivalent definitions are recorded in Proposition A.5.6.

We are now ready to formulate the irreducibility and equivalence criteria for monomial representations.

THEOREM 1.F.12 (**Irreducibility of monomial representations**). *Let G be a topological group, H an open subgroup of G, and χ a unitary character of H. Assume that*

- *for every $g \in \mathrm{Comm}_G(H) \smallsetminus H$, the restrictions of χ and χ^g to the subgroup $H \cap g^{-1}Hg$ do not coincide.*

Then the induced representation $\mathrm{Ind}_H^G \chi$ of G is irreducible.

PROOF. Let T be a right transversal for H in G, with $e \in T$.

We will show the contraposition, and therefore assume that $\pi = \mathrm{Ind}_H^G \chi$ is not irreducible. We claim that there exists $t \in T \smallsetminus \{e\}$ such that $t \in \mathrm{Comm}_G(H)$ and $\chi = \chi^t$ on $H \cap t^{-1}Ht$. This will prove the theorem because $t \notin H$.

Since π is not irreducible, there exists $S \in \mathrm{Hom}_G(\pi, \pi) = \pi(G)'$ which is not a scalar operator. On the one hand, the support of $f := S\delta_e \in \ell^2(T)$ is not reduced to $\{e\}$, by Lemma 1.F.11(2). On the other hand, by Item (4) of the same lemma, we have $f(s) = 0$ for every $s \in T$ with an infinite H-orbit. Therefore, there exists $t \in T \smallsetminus \{e\}$ with a finite H-orbit such that $f(t) \neq 0$.

Let $t^* \in T$ be such that $Ht^{-1} = Ht^*$, that is, $t^* = e \cdot t^{-1}$. We have

$$\begin{aligned}
(\pi(t)\delta_e)(t^*) &= \chi(\alpha(t^*, t))\delta_e(t^* \cdot t) = \chi(\alpha(t^*, t)), \\
(\pi(t)\delta_e)(s) &= \chi(\alpha(s, t))\delta_t(s \cdot t) = 0 \quad \text{if } s \neq t^*,
\end{aligned}$$

and therefore $\pi(t)\delta_e = \chi(\alpha(t^*, t))\delta_{t^*}$.

Consider the adjoint operator S^* of S, which is also in $\pi(G)'$, and set $f' := S^*\delta_e$. We claim that $f'(t^*) \neq 0$.

Indeed, since $\pi(t^{-1})\delta_e = \delta_t$ and $\pi(t)\delta_e = \chi(\alpha(t^*, t))\delta_{t^*}$, we have

$$\begin{aligned}
f'(t^*) &= \langle f' \mid \delta_{t^*} \rangle = \langle S^*\delta_e \mid \delta_{t^*} \rangle = \langle \delta_e \mid S\delta_{t^*} \rangle \\
&= \langle \delta_e \mid S\chi(\alpha(t^*, t))^{-1}\pi(t)\delta_e \rangle = \chi(\alpha(t^*, t))\langle \delta_e \mid S\pi(t)\delta_e \rangle \\
&= \chi(\alpha(t^*, t))\langle \delta_e \mid \pi(t)S\delta_e \rangle = \chi(\alpha(t^*, t))\langle \pi(t^{-1})\delta_e \mid f \rangle \\
&= \chi(\alpha(t^*, t))\overline{\langle f \mid \delta_t \rangle} = \chi(\alpha(t^*, t))\overline{f(t)},
\end{aligned}$$

and in particular $f'(t^*) \neq 0$, as claimed.

Lemma 1.F.11(4) applied to S^* shows that t^* has a finite H-orbit. Therefore $t \in \mathrm{Comm}_G(H)$, by Proposition A.5.6(6).

Let $h \in H \cap t^{-1}Ht$. As $f(t) \neq 0$ and $t \cdot h = t$, Lemma 1.F.11(5), applied with $H_1 = H_2 = H$ and $\chi_1 = \chi_2 = \chi$, shows that $\chi(h) = \chi(tht^{-1}) = \chi^t(h)$. So, χ and χ^t coincide on $H \cap t^{-1}Ht$. □

REMARK 1.F.13. The sufficient condition for the irreducibility of $\mathrm{Ind}_H^G \chi$ stated in Theorem 1.F.12 is also necessary; see Theorem 6' in [**Mack–51**]. Corollary

1.F.16 below provides a proof of this fact in the special case where H is a normal subgroup.

The following two corollaries particularize Theorem 1.F.12 to two opposite cases, self-commensurating subgroups and normal subgroups.

COROLLARY 1.F.14. *Let H be an open subgroup of G such that* $\mathrm{Comm}_G(H)=H$. *Then* $\mathrm{Ind}_H^G\chi$ *is irreducible for every unitary character* χ *of* H. *In particular, the quasi-regular representation* $\lambda_{G/H}$ *is irreducible.*

EXAMPLE 1.F.15. (1) Let F be the free group on two generators a and b. Let A be the subgroup generated by a; it is an infinite cyclic group. One easily checks that $\mathrm{Comm}_F(A) = A$.

Therefore $\mathrm{Ind}_A^F\chi$ is irreducible for every unitary character χ of A.

(2) Let Γ be a group of permutations of an infinite set X and let π denote the corresponding permutation representation of Γ on $\ell^2(X)$, defined by $(\pi(\gamma)f)(x) = f(\gamma^{-1}x)$ for all $\gamma \in \Gamma$, $f \in \ell^2(X)$, and $x \in X$. If the action is 2-transitive, then the representation π is irreducible.

Indeed, if x_0 denotes some point of X and H its isotropy subgroup, the hypothesis of 2-transitivity implies that H is its own commensurator in Γ (see Example A.5.4), hence $\pi = \lambda_{\Gamma/H} = \mathrm{Ind}_H^\Gamma 1_H$ is irreducible by Corollary 1.F.14.

More generally, $\mathrm{Ind}_H^\Gamma\rho$ is irreducible for any irreducible finite-dimensional representation ρ of H. This is a particular case of a result in [**Mack–51**], and also appears in several other references, for example in [**Obat–89**].

See [**BuHa–97**] for many specific examples of irreducible representations constructed in this way.

(3) In case Γ acts 2-transitively on a finite set X, the corresponding permutation representation is the direct sum of two representations, the unit representation 1_Γ and another irreducible representation of dimension $|X| - 1$. This goes back to Burnside; see [**Burn–11**, § 250].

COROLLARY 1.F.16. *Let N be an open normal subgroup of G and χ a unitary character of N. The following conditions are equivalent:*

(i) $\chi^g \neq \chi$ *for every* $g \in G \smallsetminus N$;
(ii) *the induced representation* $\mathrm{Ind}_N^G\chi$ *is irreducible.*

PROOF. The fact that (i) implies (ii) being a particular case of Theorem 1.F.12, we only have to show that (ii) implies (i).

Assume, by contraposition, that we have $\chi^{g_0} = \chi$ for some $g_0 \in G \smallsetminus N$. It will be convenient here to use Model (1) in Construction 1.F.4 for $\pi = \mathrm{Ind}_N^G\chi$; so, π is realized in the space $\mathcal{H}$ of the functions $f : G \to \mathbf{C}$ such that $f(ng) = \chi(n)f(g)$ for all $g \in G$, $n \in N$ and such that $|f| \in \ell^2(N\backslash G)$. For $g \in G$, we have $\pi(g)f(x) = f(xg)$ for all $x \in G$.

For $f \in \mathcal{H}$, define $Tf : G \to \mathbf{C}$ by $Tf(g) = f(g_0g)$. Then $Tf \in \mathcal{H}$ and $T : \mathcal{H} \to \mathcal{H}$ is a unitary operator. Indeed, for all $g \in G$, $n \in N$, we have

$$\begin{aligned} Tf(ng) &= f(g_0ng_0^{-1}g_0g) = \chi^{g_0}(n)f(g_0g) \\ &= \chi(n)Tf(g) \end{aligned}$$

and

$$\begin{aligned}\|Tf\|^2 &= \sum_{Nx\in N\backslash G} |Tf(x)|^2 = \sum_{Nx\in N\backslash G} |f(g_0x)|^2 \\ &= \sum_{Nx\in N\backslash G} |f(x)|^2 = \|f\|^2.\end{aligned}$$

Moreover, T intertwines π with itself. Indeed, for all $g \in G$, we have

$$\begin{aligned}\pi(g)(Tf)(x) &= Tf(xg) = f(g_0xg) \\ &= (\pi(g)f)(g_0x) = T(\pi(g)f)(x)\end{aligned}$$

for all $x \in G$. Since T is obviously not a scalar operator, it follows that π is not irreducible. □

Corollary 1.F.16 will be applied in Chapter 3 to produce irreducible representations for various groups.

We now turn to the question of equivalence of monomial representations.

THEOREM 1.F.17 (**Nonequivalence of monomial representations**). *Let G be a topological group, H_1, H_2 open subgroups of G, and χ_1, χ_2 unitary characters of H_1, H_2 respectively. Assume that*

- *for every $g \in G$ such that $g^{-1}H_2g \cap H_1$ has finite index in $g^{-1}H_2g$ and in H_1, the restrictions of χ_2^g and χ_1 to the subgroup $g^{-1}H_2g \cap H_1$ do not coincide.*

Then $\mathrm{Ind}_{H_1}^G\chi_1$ and $\mathrm{Ind}_{H_2}^G\chi_2$ are not equivalent.

PROOF. We use Model (2) of Construction 1.F.4. For $i = 1, 2$, let T_i be a right transversal for H_i in G, with $e \in T_i$.

We will again show the contraposition, and therefore assume that $\pi_1 := \mathrm{Ind}_{H_1}^G\chi_1$ and $\pi_2 := \mathrm{Ind}_{H_2}^G\chi_2$ are equivalent. We claim that there exists $t \in T_2$ such that $t^{-1}H_2t \cap H_1$ has finite index in $t^{-1}H_2t$ and in H_1, and such that χ_2^t and χ_1 coincide on $t^{-1}H_2t \cap H_1$. This will prove the theorem.

Since π_1 and π_2 are equivalent, there exists $S \neq 0$ in $\mathrm{Hom}_G(\pi_1, \pi_2)$. Set $f := S\delta_e \in \ell^2(T_2)$.

On the one hand, we have $f \neq 0$, by Lemma 1.F.11(1). On the other hand, by Item (4) of the same lemma, we have $f(t) = 0$ for every $t \in T_2$ with an infinite H_1-orbit. Therefore there exists $t \in T_2$ such that $f(t) \neq 0$, with a finite H_1-orbit. In particular, the stabilizer $t^{-1}H_2t \cap H_1$ of t in H_1 has finite index in H_1.

Set $t^* = e \cdot_1 t^{-1} \in T_1$. Consider the adjoint operator S^*, which is in $\mathrm{Hom}_G(\pi_2, \pi_1)$, and set $f' := S^*\delta_e \in \ell^2(T_1)$. As in the proof of Theorem 1.F.12, we have

$$f'(t^*) = \chi_1(\alpha_1(t^*, t))\overline{f(t)},$$

and hence $f'(t^*) \neq 0$.

Lemma 1.F.11(4) applied now to S^* shows that t^* has a finite H_2-orbit in T_1. Since $t^* = e \cdot_1 t^{-1}$, the stabilizer of t^* in H_2 is $tH_1t^{-1} \cap H_2$. Therefore $tH_1t^{-1} \cap H_2$ has finite index in H_2 and so $t^{-1}H_2t \cap H_1$ has finite index in $t^{-1}H_2t$.

Let $h \in t^{-1}H_2t \cap H_1$. As $f(t) \neq 0$ and $t \cdot_2 h = t$, Lemma 1.F.11(5), shows that $\chi_1(h) = \chi_2(tht^{-1})$; hence, χ_2^t and χ_1 coincide on $t^{-1}H_2t \cap H_1$. □

REMARK 1.F.18. Theorems 1.F.12 and 1.F.17 are known as the Mackey–Shoda criteria for irreducibility and equivalence for monomial representations. The terminology refers to [**Shod–33**], for the case finite groups, and [**Mack–51**], for the general case.

These criteria are also necessary conditions for irreducibility and equivalence, respectively (compare Corollaries 1.F.16, 1.F.19, and 1.F.20). Moreover, these criteria remain valid when the inducing representations are finite-dimensional [**Klep–61**, **Corw–75**]. However, the irreducibility criterion may fail when the inducing representation is infinite-dimensional [**BeCu–03**].

We will now make explicit two consequences of Theorem 1.F.17, first for self-commensurating subgroups and then for normal subgroups.

Recall that two subgroups H and L of a group G are **commensurate** if $H \cap L$ has finite index in both H and L. Two commensurate subgroups H, L of G have the same commensurator: $\mathrm{Comm}_G(H) = \mathrm{Comm}_G(L)$; see Proposition A.5.6.

COROLLARY 1.F.19 (**Equivalence of monomial irreducible representations**). *Let G be a topological group, H_1, H_2 two open subgroups, χ_1 a unitary character of H_1, and χ_2 a unitary character of H_2. Suppose that* $\mathrm{Comm}_G(H_1) = H_1$ *and* $\mathrm{Comm}_G(H_2) = H_2$. *The following properties are equivalent:*

(i) *the irreducible representations* $\mathrm{Ind}_{H_1}^G \chi_1$ *and* $\mathrm{Ind}_{H_2}^G \chi_2$ *are equivalent;*
(ii) *there exists $g \in G$ such that $g^{-1}H_2 g = H_1$ and $\chi_2^g = \chi_1$.*

PROOF. Assume that $g^{-1}H_2g = H_1$ and $\chi_2^g = \chi_1$ for some $g \in G$. Then $\mathrm{Ind}_{H_1}^G \chi_1$ is equivalent to $(\mathrm{Ind}_{H_2}^G \chi_2)^g$ (see Proposition 1.F.8) and hence to $\mathrm{Ind}_{H_2}^G \chi_2$.

Conversely, assume that $\mathrm{Ind}_{H_1}^G \chi_1$ and $\mathrm{Ind}_{H_2}^G \chi_2$ are equivalent. Then, by Theorem 1.F.17, there exists $g \in G$ such that $g^{-1}H_2g \cap H_1$ has finite index in $g^{-1}H_2g$ and in H_1, and such that χ_2^g and χ_1 coincide on $g^{-1}H_2g \cap H_1$.

In particular, H_1 and $g^{-1}H_2g$ are commensurate. Since $\mathrm{Comm}_G(H_1) = H_1$ and

$$\mathrm{Comm}_G(g^{-1}H_2g) = g^{-1}\mathrm{Comm}_G(H_2)g = g^{-1}H_2g,$$

it follows that $g^{-1}H_2g = H_1$ and $\chi_2^g = \chi_1$. □

Note that there exist triples (G, H_1, H_2), where G is a group and H_1, H_2 subgroups which are not conjugate in G, such that the representations $\mathrm{Ind}_{H_1}^G 1_{H_1}$ and $\mathrm{Ind}_{H_2}^G 1_{H_2}$ are equivalent (and not irreducible). For examples with G finite, we quote [**Suna–86**] and [**Broo–88**].

The following corollary is a direct consequence of Theorem 1.F.17 and Proposition 1.F.9.

COROLLARY 1.F.20. *Let N be an open normal subgroup of G and χ_1, χ_2 unitary characters of N.*

Then $\mathrm{Ind}_N^G \chi_1$ *and* $\mathrm{Ind}_N^G \chi_2$ *are nonequivalent representations of G if and only if $\chi_1^g \neq \chi_2$ for every $g \in G$.*

Later (Proof of Theorem 9.B.1), we will need the following weak containment result involving induced representations.

Given an open subgroup H of a topological group G, we say that G/H is an **amenable homogeneous space** in the sense of Eymard, if the unit representation 1_G is weakly contained in the quasi-regular representation $(\lambda_{G/H}, \ell^2(G/H))$; for more on this notion, see [**Eyma–72**]. When H is normal in G, the amenability of the homogeneous space G/H is equivalent to the amenability of the (discrete) group G/H (see Theorem 1.C.3).

PROPOSITION 1.F.21. *Let G be a topological group, H an open subgroup of G, and $(\pi, \mathcal{H})$ a representation of G.*

(1) *Assume that G/H is amenable. Then π is weakly contained in the induced representation $\mathrm{Ind}_H^G(\pi|_H)$.*
(2) *Assume that H is normal in G and that G/H is abelian. Then $\mathrm{Ind}_H^G(\pi|_H)$ is weakly equivalent to the direct sum*

$$\bigoplus_{\chi \in \widehat{G/H}} \pi\chi$$

where $\pi\chi \simeq \pi \otimes \chi$ is the representation of G on $\mathcal{H}$ given by $(\pi\chi)(g) = \pi(g)\chi(gH)$ for all $g \in G$.

PROOF. Recall that $\mathrm{Ind}_H^G(\pi|_H)$ is equivalent to the tensor product $\pi \otimes \lambda_{G/H}$ (Proposition 1.F.6(4)).

(1) Since $1_G \prec \lambda_{G/H}$ by hypothesis,

$$\pi = \pi \otimes 1_G \prec \pi \otimes \lambda_{G/H} \simeq \mathrm{Ind}_H^G(\pi|_H)$$

by continuity of the tensor product operation [**BeHV–08**, Proposition F.3.2].

(2) Since G/H is an abelian group, $\lambda_{G/H}$ is weakly equivalent to $\bigoplus_{\chi \in \widehat{G/H}} \chi$; see [**BeHV–08**, Proposition F.2.7]. It follows that $\mathrm{Ind}_H^G(\pi|_H) \simeq \pi \otimes \lambda_{G/H}$ is weakly equivalent to $\pi \otimes \left(\bigoplus_{\chi \in \widehat{G/H}} \chi\right) \simeq \bigoplus_{\chi \in \widehat{G/H}} \pi\chi$, again by continuity of the tensor product. □

1.G. On decomposing representations into irreducible representations

For a second-countable locally compact group G, arbitrary representations can be decomposed into irreducible representations. This appears in several situations:

- Decomposing representations into irreducible components is already of fundamental importance in classical Fourier analysis, when G is a torus $\mathbf{T}^n$ or the translation group $\mathbf{R}^n$ of a Euclidean space.
- The theory of unitary representations of groups, and in particular decompositions into irreducible components, provide an organizing principle for the study of special functions, which appear as solutions of many partial differential equations of mathematical physics [**Vile–68**].
- In case G is of type I, its regular representation can be decomposed into irreducible representations using the so-called Plancherel measure on $\widehat{G}$ [**Dixm–C***, 18.8].
- Decomposition into irreducible representations relates to spectral analysis. For example, if Γ is a discrete subgroup of $\mathrm{SL}_2(\mathbf{R})$ such that $\Gamma\backslash\mathrm{SL}_2(\mathbf{R})$ is compact, understanding the irreducible subrepresentations of the natural representation of $\mathrm{SL}_2(\mathbf{R})$ in $L^2(\Gamma\backslash\mathrm{SL}_2(\mathbf{R}))$ is essentially equivalent to understanding the spectrum of the Laplace–Beltrami operator on the upper half-plane; see [**GGPS–69**, Chap. I, § 4] and [**Bump–97**, in particular Theorem 2.7.1].

For compact groups, decompositions in direct sums are appropriate: every representation of such a group is a direct sum of irreducible representations. But, for noncompact groups, including **Z** and **R** (see Example 1.G.10), it is in general necessary to use the more general notion of direct integral decomposition.

The theory of direct integral decompositions was founded by von Neumann in the late 1930s [**vNeu–49**], and made explicit for group representations in the early 1950s [**Maut–50a**, Theorem 1.2], [**Gode–51a**, Chapitre IV]. We give a brief account of this theory and refer to [**Dixm–vN**, Chap. II] and [**Dixm–C***, § 8] for a comprehensive exposition.

Direct integrals of representations. Let $(X, \mathcal{B}, \mu)$ be a measure space, where μ is a σ-finite measure on $(X, \mathcal{B})$. A **field of Hilbert spaces over** X is a family $(\mathcal{H}_x)_{x\in X}$, where $\mathcal{H}_x$ is a Hilbert space for each $x \in X$. Elements in $\prod_{x\in X} \mathcal{H}_x$ are called **vector fields over** X.

A **fundamental family of measurable vector fields** for a field $(\mathcal{H}_x)_{x\in X}$ over X is a sequence $(e^n)_{n\geq 1}$ of vector fields over X, where $e^n = (e^n_x)_{x\in X}$ and $e^n_x \in \mathcal{H}_x$ for all $x \in X$, with the following properties:

- the map $x \mapsto \langle e^m_x \mid e^n_x\rangle$ is measurable, for every $m, n \geq 1$,
- the linear span of $\{e^n_x : n \geq 1\}$ is dense in $\mathcal{H}_x$, for every $x \in X$.

Note that, for such a fundamental family to exist, the second property implies that the Hilbert space $\mathcal{H}_x$ is separable for every $x \in X$.

Fix a fundamental family of measurable vector fields $(e^n)_{n\geq 1}$. A vector field $\xi \in \prod_{x\in X} \mathcal{H}_x$ is said to be a **measurable vector field** if the functions $x \mapsto \langle \xi_x \mid e^n_x\rangle$ are measurable for all $n \geq 1$. The set M of measurable vector fields is a linear subspace of $\prod_{x\in X} \mathcal{H}_x$. Moreover, for $\xi, \eta \in M$, the function $X \to \mathbf{C}$, $x \mapsto \langle \xi_x \mid \eta_x\rangle$ is measurable. The pair $((\mathcal{H}_x)_{x\in X}, M)$ is called a **measurable field of Hilbert spaces** over X.

Two measurable vector fields are equivalent if they are equal μ-almost everywhere. From now on, we write (abusively) "measurable vector fields" instead of "equivalence classes of measurable vector fields" for this equivalence relation. A measurable vector field ξ is a **square-integrable vector field** if

$$\int_X \|\xi_x\|^2 d\mu(x) < \infty.$$

Equipped with the obvious inner product, the set $\mathcal{H}$ of all square-integrable vector fields is a Hilbert space called the **direct integral** of the field $(\mathcal{H}_x)_{x\in X}$ of Hilbert spaces over X and denoted by

$$\mathcal{H} = \int_X^{\oplus} \mathcal{H}_x d\mu(x).$$

EXAMPLE 1.G.1. (1) Let X be a countable set and μ the counting measure. Let $(\mathcal{H}_x)_{x\in X}$ be a field of separable Hilbert spaces over X. An appropriate fundamental family of measurable vector fields in this case is a family $(e^n)_{n\geq 1}$ such that $(e^n_x)_{n\geq 1}$ contains an orthonormal basis of $\mathcal{H}_x$ for all $x \in X$. Then every vector field is measurable and

$$\int_X^{\oplus} \mathcal{H}_x d\mu(x) = \bigoplus_{x\in X} \mathcal{H}_x.$$

(2) Let $\mathcal{K}$ be a fixed separable Hilbert space. Set $\mathcal{H}_x = \mathcal{K}$ for every $x \in X$. Choose an orthonormal basis $(\xi_n)_{n\geq 1}$ of $\mathcal{K}$ and define a fundamental family of measurable vector fields $(e^n)_{n\geq 1}$ over X by $e^n_x = \xi_n$ for every $x \in X$ and $n \geq 1$. The measurable vector fields over X are the mappings $X \to \mathcal{K}$ which are measurable with respect to the Borel structure on $\mathcal{K}$ given by the weak topology. Then $\int_X^{\oplus} \mathcal{H}_x d\mu(x)$ coincides with the Hilbert space $L^2(X, \mu, \mathcal{K})$ of square-integrable measurable maps $X \to \mathcal{K}$, where two such functions are identified if they agree μ-almost everywhere, equipped with the scalar product

$$\langle F_1 \mid F_2 \rangle := \int_X \langle F_1(x) \mid F_2(x) \rangle d\mu(x) \qquad \text{for all } F_1, F_2 \in L^2(X, \mu, \mathcal{K}).$$

The field $x \mapsto \mathcal{K}$ is called a **constant field** of Hilbert spaces.

In the special case of a countable set X with the counting measure, this space coincides with the space denoted by $\ell^2(X, \mathcal{K})$ in Construction 1.F.4(2).

The following proposition is often used to reduce questions involving direct integrals of Hilbert spaces to the situation of constant fields of Hilbert spaces. For the proof, we refer to Chap. II, §1, Proposition 2 in [**Dixm–vN**] or Proposition 7.21 in [**Foll–16**].

PROPOSITION 1.G.2. *Let $\mathcal{H} = \int_X^{\oplus} \mathcal{H}_x d\mu(x)$ be a direct integral of Hilbert spaces over the measure space (X, μ). For every $n = 1, 2, \ldots, \infty$, let*

$$X_n = \{x \in X \mid \dim \mathcal{H}_x = n\},$$

and let $\mathcal{K}_n$ be a fixed Hilbert space of dimension n. Then, for every n, the following holds:

(i) *X_n is measurable;*
(ii) *the direct integral $\int_{X_n}^{\oplus} \mathcal{H}_x d\mu(x)$ of Hilbert spaces is isomorphic to $L^2(X, \mathcal{K}_n)$ over X_n, that is, there exists a family $(U_x)_{x\in X_n}$ of Hilbert space isomorphisms $U_x : \mathcal{H}_x \to \mathcal{K}_n$ such that $\xi = (\xi_x)_{x\in X_n}$ is a square-integrable vector field over X_n if and only if $x \mapsto U_x\xi_x$ is a square-integrable map from X_n to $\mathcal{K}_n$.*

As a consequence, we have an isomorphism of Hilbert spaces over X:

$$\mathcal{H} = \oplus_n \int_{X_n}^{\oplus} \mathcal{H}_x d\mu(x) \approx \oplus_n L^2(X_n, \mu, \mathcal{K}_n).$$

DEFINITION 1.G.3. Let $\mathcal{H} = \int_X^{\oplus} \mathcal{H}_x d\mu(x)$ be a direct integral of Hilbert spaces over a measure space (X, μ). For every $x \in X$, let $T_x \in \mathcal{L}(\mathcal{H}_x)$. Then $x \mapsto T_x$ is a **measurable field of operators** over X if the map $x \mapsto \langle T_x\xi_x \mid \eta_x \rangle$ is measurable for all $\xi, \eta \in \mathcal{H}$. Assume that $x \mapsto T_x$ is μ-essentially bounded, that is, $x \mapsto \|T_x\|$ is a μ-essentially bounded function; then a bounded operator $T : \mathcal{H} \to \mathcal{H}$ is defined by

$$(T\xi)_x := T_x(\xi_x) \qquad \text{for } \xi \in \mathcal{H},\ x \in X$$

and we write

$$T = \int_X^{\oplus} T_x d\mu(x).$$

In this case, the norm $\|T\|$ is the μ-essential supremum of the function $x \mapsto \|T_x\|$ (see [**Dixm–vN**, Chap. II, § 2, no 3]). Bounded operators on $\mathcal{H}$ of this form are called **decomposable** over X. This notion will be slightly generalized in Definition 1.H.1.

DEFINITION 1.G.4. Let $\mathcal{H} = \int_X^{\oplus} \mathcal{H}_x d\mu(x)$ be as in the previous definition. To every $\varphi \in L^\infty(X,\mu)$ is associated an operator $m(\varphi)$ on $\mathcal{H}$ defined by

$$m(\varphi) := \int_X^{\oplus} \varphi(x)\mathrm{Id}_{\mathcal{H}_x} d\mu(x).$$

Note that

$$(m(\varphi)\xi)_x = \varphi(x)\xi_x \quad \text{for all } \xi \in \mathcal{H},\ x \in X.$$

Bounded operators on $\mathcal{H}$ of the form $m(\varphi)$ are called **diagonalisable** over X.

The set $\{m(\varphi) \mid \varphi \in L^\infty(X,\mu)\}$ of diagonalisable operators on $\mathcal{H}$ is obviously a commutative selfadjoint subalgebra of $\mathcal{L}(\mathcal{H})$.

Note that diagonalisable operators are decomposable, and that the converse holds when $\dim \mathcal{H}_x = 1$ for all $x \in X$.

EXAMPLE 1.G.5. Let $(X,\mathcal{B},\mu)$ be a measure space and $\mathcal{K}$ a separable Hilbert space. Consider the constant field of Hilbert spaces $x \mapsto \mathcal{K}$ and the corresponding Hilbert space of square integrable vector fields $L^2(X,\mu,\mathcal{K})$.

The decomposable operators are the operators $\widetilde{T}$ on $L^2(X,\mu,\mathcal{K})$, defined for $T \in L^\infty(X,\mu,\mathcal{L}(\mathcal{K}))$ by $(\widetilde{T}\xi)(x) = T(x)\big(\xi(x)\big)$ for all $x \in X$ and $\xi \in L^2(X,\mu,\mathcal{K})$. Here, $L^\infty(X,\mu,\mathcal{L}(\mathcal{K}))$ denote the space of weakly measurable maps $\widetilde{T} : X \to \mathcal{L}(\mathcal{K})$ such that the map $x \mapsto \|\widetilde{T}(x)\|$ is essentially bounded, up to equality almost everywhere.

The diagonalisable operators are the multiplication operators $m(\varphi)$, defined for $\varphi \in L^\infty(X,\mu)$ by $(m(\varphi)\xi)(x) = \varphi(x)\xi(x)$ for all $x \in X$ and $\xi \in L^2(X,\mu,\mathcal{K})$.

Let G be a second-countable locally compact group. For every $x \in X$, let π_x be a representation of G on $\mathcal{H}_x$. Assume that $x \mapsto \pi_x$ is a **measurable field of representations** of G, that is, there is a measurable field $x \mapsto \mathcal{H}_x$ of Hilbert spaces over X and, for every $g \in G$, a measurable field $x \mapsto \pi_x(g)$ of operators on the $\mathcal{H}_x$'s over X such that $g \mapsto \pi_x(g)$ is a representation of G for all $x \in X$. Then

$$\pi(g) := \int_X^{\oplus} \pi_x(g) d\mu(x)$$

is a unitary operator on $\mathcal{H}$ for every $g \in G$. It is clear that $g \mapsto \pi(g)$ is a homomorphism from G to the unitary group of $\mathcal{H} = \int_X^{\oplus} \mathcal{H}_x d\mu(x)$. It can be shown [**Dixm–C***, Proposition 18.7.4] that the map $g \mapsto \pi(g)$ is strongly continuous, so that π is a representation of G on $\mathcal{H}$. We write

$$\pi = \int_X^{\oplus} \pi_x d\mu(x)$$

and π is called the **direct integral** of the family $(\pi_x)_{x\in X}$.

Let H be a closed subgroup of G and σ_1, σ_2 representations of H. The following proposition generalizes the fact (see Proposition 1.F.6) that the direct sum $\mathrm{Ind}_H^G \sigma_1 \oplus \mathrm{Ind}_H^G \sigma_2$ is equivalent to the induced representation $\mathrm{Ind}_H^G(\sigma_1 \oplus \sigma_2)$.

PROPOSITION 1.G.6. *Let H be a closed subgroup of G and $\sigma = \int_X^{\oplus} \sigma_x d\mu(x)$ a direct integral of a measurable field $(\sigma_x)_{x\in X}$ of representations of H over X.*

Then $\mathrm{Ind}_H^G \sigma$ is equivalent to the direct integral $\int_X^{\oplus} (\mathrm{Ind}_H^G \sigma_x) d\mu(x)$ of the family $(\mathrm{Ind}_H^G \sigma_x)_{x\in X}$.

For a proof, we refer to [**Mack–52**, Theorem 10.1].

Direct integral decompositions of representations. We are going to describe how a representation of a second-countable locally compact group G admits decompositions as direct integrals of representations, in several ways. The basic tool for this is the following result; for a proof, see Theorem 8.3.2 and Section 18.7 in [**Dixm–C***], or Theorem 7.29 in [**Foll–16**].

Let $(\pi, \mathcal{H})$ be a representation of a group G. The **commutant** of $\pi(G)$, that is,

$$\pi(G)' = \{T \in \mathcal{L}(\mathcal{H}) \mid T\pi(g) = \pi(g)T \text{ for all } g \in G\},$$

is a von Neumann subalgebra of $\mathcal{L}(\mathcal{H})$.

THEOREM 1.G.7. *Let G be a second-countable locally compact group and let π be a representation of G on a separable Hilbert space $\mathcal{H}$. Let $\mathcal{A}$ be an abelian von Neumann subalgebra of the commutant $\pi(G)'$.*

Then there exist a standard Borel space X, a σ-finite measure μ on X, a measurable field of Hilbert spaces $x \mapsto \mathcal{H}_x$ over X, a measurable field of representations $x \mapsto \pi_x$ of G in the $\mathcal{H}_x$'s over X, and an isomorphism of Hilbert spaces $U : \mathcal{H} \to \int_X^\oplus \mathcal{H}_x d\mu(x)$, with the following properties:

(1) *$U\pi(g)U^{-1} = \int_X^\oplus \pi_x(g)d\mu(x)$for all $g \in G$;*
(2) *$U\mathcal{A}U^{-1}$ coincides with the algebra of diagonalisable operators on $\int_X^\oplus \mathcal{H}_x d\mu(x)$.*

There are two important choices of the algebra $\mathcal{A}$ in Theorem 1.G.7 which lead to two integral decompositions with particular properties:

- $\mathcal{A}$ is a maximal abelian subalgebra of $\pi(G)'$;
- $\mathcal{A}$ is the centre of $\pi(G)'$.

In the first case, the representations π_x in Theorem 1.G.7 are irreducible for μ-almost every $x \in X$. More precisely, we have the following Proposition 1.G.8; for a proof, see [**Dixm–vN**, Chap. II, § 3, no 1, corollaire 1] and also [**Dixm–C***, 8.5.1]. (In the second case, when $\mathcal{A}$ is the centre of $\pi(G)'$, the π_x's in Theorem 1.G.7 are factor representations for μ-almost every $x \in X$ and are mutually disjoint. This leads to the central decomposition of π, to be treated later; see Definition 6.A.8 for factor representations and Theorem 6.C.6 for the decomposition result.)

PROPOSITION 1.G.8. *Let G be a second-countable locally compact group. Consider a standard Borel space X, a σ-finite measure μ on X, a measurable field $x \mapsto \mathcal{H}_x$ of Hilbert spaces over X, and a measurable field $x \mapsto \pi_x$ of representations of G in the $\mathcal{H}_x$'s over X. Set*

$$\mathcal{H} = \int_X^\oplus \mathcal{H}_x d\mu(x) \quad \text{and} \quad \pi = \int_X^\oplus \pi_x d\mu(x).$$

Let $\mathcal{A}$ be the algebra of diagonalisable operators in $\mathcal{L}(\mathcal{H})$. The following properties are equivalent:

(i) *$\mathcal{A}$ is a maximal abelian subalgebra of $\pi(G)'$;*
(ii) *π_x is irreducible for μ-almost every $x \in X$.*

Since, by Zorn's lemma, maximal abelian von Neumann subalgebras of $\pi(G)'$ always exist, the following result is an immediate consequence of Theorem 1.G.7 and Proposition 1.G.8.

THEOREM 1.G.9. *Let G be a second-countable locally compact group and let π be a representation of G on a separable Hilbert space $\mathcal{H}$.*

Then there exist a standard Borel space X, a σ-finite measure μ on X, a measurable field $x \mapsto \mathcal{H}_x$ of Hilbert spaces over X, and a measurable field $x \mapsto \pi_x$ of irreducible *representations of G in the $\mathcal{H}_x$'s, such that π is equivalent to the direct integral $\int_X^\oplus \pi_x d\mu(x)$.*

EXAMPLE 1.G.10. Let G be a second-countable locally compact abelian group. Choose the normalizations of the Haar measures μ_G on G and $\mu_{\widehat{G}}$ on the dual group $\widehat{G}$ so that Plancherel's Theorem holds (Theorem A.7.6). Denote by π the representation of G on $L^2(\widehat{G}, \mu_{\widehat{G}})$ defined by

$$(\pi(g)\xi)(\chi) = \chi(g)\xi(\chi) \quad \text{for all } \xi \in L^2(\widehat{G}, \mu_{\widehat{G}}),\ \chi \in \widehat{G},\ g \in G.$$

Then the Fourier transform

$$\mathcal{F} : L^2(G, \mu_G) \to L^2(\widehat{G}, \mu_{\widehat{G}})$$

is an isomorphism of Hilbert spaces which intertwines the regular representation λ_G on $L^2(G, \mu_G)$ of Example 1.B.17 and the representation π:

$$\mathcal{F}\lambda_G(g) = \pi(g)\mathcal{F} \quad \text{for all } g \in G.$$

For every $\chi \in \widehat{G}$, set $\mathcal{H}_\chi = \mathbf{C}$; then

$$L^2(\widehat{G}, \mu_{\widehat{G}}) = \int_{\widehat{G}}^\oplus \mathcal{H}_\chi d\mu(\chi) \quad \text{and} \quad \pi = \int_{\widehat{G}}^\oplus \chi d\mu(\chi),$$

so that λ_G is equivalent to the direct integral of representations $\int_{\widehat{G}}^\oplus \chi d\mu(\chi)$.

Nonuniqueness of irreducible decompositions. Let π be a representation of a second-countable locally compact group G on a separable Hilbert space. By Theorem 1.G.9, π is equivalent to a direct integral $\int_X^\oplus \pi_x d\mu(x)$ of irreducible representations π_x over a standard Borel space X. It is clear that no uniqueness can be expected for the choice of the measure space (X, μ). For instance, if a measure ν on X is equivalent to μ, then $\int_X^\oplus \pi_x d\nu(x)$ is equivalent to $\int_X^\oplus \pi_x d\mu(x)$, as is easily checked.

One may ask the following more reasonable question. Assume that $\int_Y^\oplus \sigma_y d\nu(y)$ is another direct integral decomposition π into irreducible representations over a standard Borel space Y. Do there exist measurable subsets N, N' of X and Y such that $\mu(N) = \nu(N') = 0$ and

$$\{\pi_x \mid x \in X \smallsetminus N\} = \{\sigma_y \mid y \in Y \smallsetminus N'\},$$

where these two last spaces are viewed as subspaces of $\widehat{G}$?

This question has a positive answer for some groups, those of type I; see Section 6.D for this notion. In fact, for groups of type I, one has a refinement of Theorem 1.G.9 with a stronger uniqueness statement (Theorem 6.D.7 below). On the contrary, for groups which are not of type I, the question always has a negative answer, as the following result of Dixmier shows. The dual $\widehat{G}$ is considered here with its Mackey Borel structure; see Definition 6.C.3 below.

THEOREM 1.G.11. *Let G be a second-countable locally compact group which is not of type I.*

There exist a representation π of G on a separable Hilbert space and two direct integral decompositions into irreducible representations

$$\pi = \int_X^{\oplus} \pi_x d\mu(x) = \int_Y^{\oplus} \sigma_y d\mu(y)$$

such that π_x and σ_y are nonequivalent for all $(x, y) \in X \times Y$.

The proof of Theorem 1.G.11 can be found [**Dixm–64b**, Corollaire 2], where X and Y are moreover disjoint Borel subsets of the dual $\widehat{G}$; Dixmier proves a more general result, about representations of non type I separable C*-algebras. We verify below this nonuniqueness property for two concrete examples of representations; the first example appeared in [**Yosh–51**] and the second one in [**Mack–51**].

EXAMPLE 1.G.12. (1) Let F be the free group on two generators a and b. Let A be the subgroup generated by a and B the subgroup generated by b.

For the regular representation λ_F of F, we have (by induction in stages)

$$\lambda_F \simeq \mathrm{Ind}_{\{e\}}^F 1_{\{e\}} \simeq \mathrm{Ind}_A^F(\mathrm{Ind}_{\{e\}}^A 1_{\{e\}}) \simeq \mathrm{Ind}_A^F(\lambda_A),$$

and similarly

$$\lambda_F \simeq \mathrm{Ind}_B^F(\lambda_B).$$

We have

$$\lambda_A \simeq \int_{\widehat{A}}^{\oplus} \chi d\mu(\chi),$$

where μ is a Haar measure on $\widehat{A}$; see Example 1.G.10. Therefore, by Proposition 1.G.6, we obtain a first integral decomposition

$$\lambda_F \simeq \int_{\widehat{A}} (\mathrm{Ind}_A^F \chi) d\mu(\chi).$$

Similarly, we have a second integral decomposition

$$\lambda_F \simeq \int_{\widehat{B}} (\mathrm{Ind}_B^F \psi) d\nu(\psi),$$

where ν is a Haar measure on $\widehat{B}$.

Now, the induced representations $\mathrm{Ind}_A^F \chi$ and $\mathrm{Ind}_B^F \psi$ are irreducible for every $\chi \in \widehat{A}$ and $\psi \in \widehat{B}$ (Example 1.F.15). Moreover, since $g^{-1}Bg \cap A$ has infinite index in A (in fact, $g^{-1}Bg \cap A = \{e\}$) for every $g \in F$, the representations $\mathrm{Ind}_A^F \chi$ and $\mathrm{Ind}_B^F \psi$ are not equivalent for $\chi \in \widehat{A}$ and $\psi \in \widehat{B}$, by Theorem 1.F.17.

(2) Let Γ be the affine group of the rational field $\mathbf{Q}$, that is,

$$\Gamma = \begin{pmatrix} \mathbf{Q}^\times & \mathbf{Q} \\ 0 & 1 \end{pmatrix}$$

(see Section 3.C). Consider

the normal subgroup $N = \begin{pmatrix} 1 & \mathbf{Q} \\ 0 & 1 \end{pmatrix}$ and the subgroup $H = \begin{pmatrix} \mathbf{Q}^\times & 0 \\ 0 & 1 \end{pmatrix}$

of Γ. As in (1), we have two direct integral decompositions of the regular representation λ_Γ of Γ:

$$\lambda_\Gamma \simeq \int_{\widehat{N}} (\mathrm{Ind}_N^\Gamma \chi) d\mu(\chi) \quad \text{and} \quad \lambda_\Gamma \simeq \int_{\widehat{H}} (\mathrm{Ind}_H^\Gamma \psi) d\nu(\psi).$$

It follows from the Mackey–Shoda criteria that the induced representations $\mathrm{Ind}_N^\Gamma \chi$ and $\mathrm{Ind}_H^\Gamma \psi$ are irreducible and nonequivalent for $\chi \in \widehat{N}$ and $\psi \in \widehat{H}$ (compare with Theorem 3.C.3).

1.H. On decomposable and diagonalisable operators

In this section, we expose some results on decomposable and diagonalisable operators on fields of Hilbert spaces, as defined in Section 1.G. The results go back to [**vNeu–49**] and [**Gode–51a**]; we follow closely §2 in Chapter II of [**Dixm–vN**]. The results will be needed in Section 2.B, in particular for Theorem 2.B.9 on the canonical decomposition of representations of abelian groups, in Section 5.D for our construction of representations of some semi-direct products $H \ltimes N$, and in later chapters.

We first generalize the notion of a decomposable operator introduced in Definition 1.G.3.

DEFINITION 1.H.1. Let $(X, \mathcal{B}, \mu)$ be a measure space, where μ is a σ-finite measure on $(X, \mathcal{B})$. Let $x \mapsto \mathcal{H}_x$ and $x \mapsto \mathcal{K}_x$ be measurable fields of Hilbert spaces over $(X, \mathcal{B}, \mu)$. Set

$$\mathcal{H} = \int_X^\oplus \mathcal{H}_x d\mu(x) \quad \text{and} \quad \mathcal{K} = \int_X^\oplus \mathcal{K}_x d\mu(x).$$

Let $(T_x)_x$ be a field of operators, $T_x \in \mathcal{L}(\mathcal{H}_x, \mathcal{K}_x)$ for each $x \in X$, such that the map

$$x \mapsto \langle T_x \xi_x \mid \eta_x \rangle$$

is measurable for all $\xi \in \mathcal{H}, \eta \in \mathcal{K}$, and such that $x \mapsto \|T_x\|$ is a μ-essentially bounded function. Let $T : \mathcal{H} \to \mathcal{K}$ be the bounded operator defined by

$$(T\xi)_x = T_x(\xi_x) \quad \text{for } \xi \in \mathcal{H},\ x \in X;$$

we write

$$T = \int_X^\oplus T_x d\mu(x).$$

The norm of T coincides with the μ-essential supremum of the function $x \mapsto \|T_x\|$ (see [**Dixm–vN**, Chap. II, §2, no 3]). Bounded operators $\mathcal{H} \to \mathcal{K}$ obtained in this way are said to be **decomposable** over X.

Let $\varphi \in L^\infty(X, \mu)$. Let $m_1(\varphi) \in \mathcal{L}(\mathcal{H})$ and $m_2(\varphi) \in \mathcal{L}(\mathcal{K})$ be the associated diagonalisable operators on respectively $\mathcal{H}$ and $\mathcal{K}$ (see Definition 1.G.4). It is obvious that

$$m_2(\varphi)T = Tm_1(\varphi)$$

for every decomposable operator $T \in \mathcal{L}(\mathcal{H}, \mathcal{K})$. The main result of this section is that this property characterizes the decomposable operators.

THEOREM 1.H.2. *Let*

$$\mathcal{H} = \int_X^\oplus \mathcal{H}_x d\mu(x) \quad \textit{and} \quad \mathcal{K} = \int_X^\oplus \mathcal{K}_x d\mu(x)$$

be two direct integrals of Hilbert spaces over a Borel space $(X, \mathcal{B})$, equipped with a σ-finite measure μ. Let $T \in \mathcal{L}(\mathcal{H}, \mathcal{K})$ be such that

$$m_2(\varphi)T = Tm_1(\varphi) \quad \textit{for all } \varphi \in L^\infty(X, \mu).$$

Then T is a decomposable operator.

PROOF. In case $x \mapsto \mathcal{H}_x$ and $x \mapsto \mathcal{K}_x$ are constant fields of Hilbert spaces over (X, μ), that is (see Example 1.G.1), in case $\int_X^{\oplus} \mathcal{H}_x d\mu(x) = L^2(X, \mu, \mathcal{K}_1)$ and $\int_X^{\oplus} \mathcal{K}_x d\mu(x) = L^2(X, \mu, \mathcal{K}_2)$, for fixed Hilbert spaces $\mathcal{K}_1$ and $\mathcal{K}_2$, Theorem 1.H.2 reduces to Theorem 1.H.5, proved below. The general case follows by Proposition 1.G.2; see Théorème 1, Chap. II, §2 in [**Dixm–vN**]. □

Let $\mathcal{K}$ be a Hilbert space. For $\varphi \in L^\infty(X, \mu)$, we now denote by $m(\varphi)$ the corresponding diagonalisable operator on $L^2(X, \mu, \mathcal{K})$. The set

$$\mathcal{A} := \{m(\varphi) \mid \varphi \in L^\infty(X, \mu)\}$$

of diagonalizable operators on $L^2(X, \mu, \mathcal{K})$ is an abelian $*$-subalgebra of $\mathcal{L}(L^2(X, \mu, \mathcal{K}))$. The following result shows in particular that $\mathcal{A}$ is a von Neumann algebra.

For general facts about von Neumann algebras, see Appendix A.11.

PROPOSITION 1.H.3. *Let $(X, \mathcal{B})$ be a Borel space equipped with a σ-finite measure μ, and let $\mathcal{K}$ be a Hilbert space.*

(1) *Let $L^\infty(X, \mu)$ be equipped with the weak*-topology and $\mathcal{L}(L^2(X, \mu, \mathcal{K}))$ with the weak operator topology. The map*

$$L^\infty(X, \mu) \to \mathcal{L}(L^2(X, \mu, \mathcal{K})), \quad \varphi \mapsto m(\varphi)$$

is continuous, of image $\mathcal{A}$.

(2) *The subalgebra $\mathcal{A}$ of diagonalisable operators is closed in $\mathcal{L}(L^2(X, \mu, \mathcal{K}))$ for the weak operator topology and is therefore an abelian von Neumann algebra.*

PROOF. (1) For $\varphi \in L^\infty(X, \mu)$ and $F_1, F_2 \in L^2(X, \mu, \mathcal{K})$, we have

$$\langle m(\varphi)F_1 \mid F_2 \rangle = \int_X \varphi(x) \langle F_1(x) \mid F_2(x) \rangle \, d\mu(x).$$

Since the function $x \mapsto \langle F_1(x) \mid F_2(x) \rangle$ belongs to $L^1(X, \mu)$, this shows that, for fixed $F_1, F_2 \in L^2(X, \mu, \mathcal{K})$, the map

$$\varphi \mapsto \langle m(\varphi)F_1 \mid F_2 \rangle$$

is continuous, when $L^\infty(X, \mu)$ is equipped with the weak*-topology.

(2) The unit ball of $L^\infty(X, \mu)$ is weak*-compact. It follows from (1) that the unit ball of $\mathcal{A}$ is compact in the weak operator topology. This implies that $\mathcal{A}$ is closed in the weak operator topology (see [**Dixm–vN**, Chap. I, § 3, Th. 2]). □

For $\varphi \in L^2(X, \mu)$ and $\xi \in \mathcal{K}$, let $\varphi \otimes \xi$ denote the vector-valued function in $L^2(X, \mu, \mathcal{K})$ defined by

$$\varphi \otimes \xi(x) = \varphi(x)\xi \quad \text{for all } x \in X.$$

LEMMA 1.H.4. *Let $(X, \mathcal{B})$ and μ be as above. Assume that $\mathcal{K}$ is a separable Hilbert space and let $\{\xi_i \mid i \geq 1\}$ be a countable total subset of $\mathcal{K}$.*

For every total subset D of $L^2(X, \mu)$, the set

$$\{\varphi \otimes \xi_i \mid \varphi \in D, \;\; i \geq 1\}$$

is a total subset of $L^2(X, \mu, \mathcal{K})$.

PROOF. Let $F \in L^2(X, \mu, \mathcal{K})$ be such that

$$\langle \varphi \otimes \xi_i \mid F \rangle = 0 \qquad \text{for all } \varphi \in D, \ i \geq 1.$$

It suffices to show that $F = 0$ in $L^2(X, \mu, \mathcal{K})$.

Let $i \geq 1$. Then

$$\int_X \varphi(x) \, \langle \xi_i \mid F(x) \rangle \, d\mu(x) = 0 \qquad \text{for all } \varphi \in D.$$

Observe that, by Cauchy–Schwarz inequality, the function $x \mapsto \langle \xi_i \mid F(x) \rangle$ belongs to $L^2(X, \mu)$. Since D is a total subset of $L^2(X, \mu)$, there exists a subset $N_i \in \mathcal{B}$ such that $\mu(N_i) = 0$ and $\langle \xi_i \mid F(x) \rangle = 0$ for every $x \in X \smallsetminus N_i$. Let $N = \bigcup_{i \geq 1} N_i$. Then $\mu(N) = 0$ and we have

$$\langle \xi_i \mid F(x) \rangle = 0 \qquad \text{for all } x \in X \smallsetminus N, \ i \geq 1.$$

Since $\{\xi_i \mid i \geq 1\}$ is a total subset of $\mathcal{K}$, we therefore have $F(x) = 0$ for every $x \in X \smallsetminus N$. □

We now state and prove Theorem 1.H.2 for constant fields of Hilbert spaces.

Let $\mathcal{K}_1, \mathcal{K}_2$ be Hilbert spaces. Let $T : X \to \mathcal{L}(\mathcal{K}_1, \mathcal{K}_2), \ x \mapsto T(x)$ be a map with the following properties:

- T is measurable, that is, $x \mapsto \langle T(x)\xi \mid \eta \rangle$ is measurable for all $\xi \in \mathcal{K}_1, \eta \in \mathcal{K}_2$;
- $x \mapsto \|T(x)\|$ belongs to $L^\infty(X, \mu)$.

The operator $\widetilde{T}$, defined by

$$(\widetilde{T}F)(x) = T(x)F(x) \qquad \text{for all } F \in L^2(X, \mu, \mathcal{K}_1) \text{ and } x \in X,$$

is a decomposable operator from $L^2(X, \mu, \mathcal{K}_1)$ to $L^2(X, \mu, \mathcal{K}_2)$, with norm equal to the μ-essential supremum of $x \mapsto \|T(x)\|$.

Every decomposable operator from $L^2(X, \mu, \mathcal{K}_1)$ to $L^2(X, \mu, \mathcal{K}_2)$ is of this form.

THEOREM 1.H.5. *Let $(X, \mathcal{B})$ be a Borel space, equipped with a σ-finite measure μ, and let $\mathcal{K}_1, \mathcal{K}_2$ be* separable *Hilbert spaces. Let*

$$\widetilde{T} : L^2(X, \mu, \mathcal{K}_1) \to L^2(X, \mu, \mathcal{K}_2)$$

be a bounded linear operator such that

$$\widetilde{T} m_1(\varphi) = m_2(\varphi) \widetilde{T} \qquad \text{for all } \varphi \in L^\infty(X, \mu).$$

Then $\widetilde{T}$ is a decomposable operator.

PROOF. Let ν be a probability measure on $(X, \mathcal{B})$ which is equivalent to μ (see Proposition A.3.5). For $i = 1, 2$, the map $U_i : L^2(X, \mu, \mathcal{K}_i) \to L^2(X, \nu, \mathcal{K}_i)$, given by

$$U_i(F) = \left(\frac{d\mu}{d\nu} \right)^{1/2} F \qquad \text{for all } F \in L^2(X, \mu, \mathcal{K}_i),$$

is an isomorphism of Hilbert spaces. Set

$$\widetilde{S} = U_2 \widetilde{T} U_1^{-1} : L^2(X, \nu, \mathcal{K}_1) \to L^2(X, \nu, \mathcal{K}_2).$$

It is clear that $\widetilde{S} m_1(\varphi) = m_2(\varphi) \widetilde{S}$ for every $\varphi \in L^\infty(X, \nu)$ and that $\widetilde{S}$ is a decomposable operator if and only if $\widetilde{T}$ is a decomposable operator.

Therefore, we can assume without loss of generality that μ is a probability measure on X.

Since $\mathcal{K}_1$ is separable, there exists a countable total subset $\{\xi_j \mid j \geq 1\}$ of $\mathcal{K}_1$. For every $\varphi \in L^\infty(X, \mu)$, we have

$$\widetilde{T}(\varphi \otimes \xi_j) = \widetilde{T}(m_1(\varphi)(\mathbf{1}_X \otimes \xi_j)) = m_2(\varphi)\widetilde{T}(\mathbf{1}_X \otimes \xi_j).$$

Since $L^\infty(X, \mu)$ is dense in $L^2(X, \mu)$ and in view of Lemma 1.H.4, this shows that $\widetilde{T}$ is determined by its values on the set

$$\{\mathbf{1}_X \otimes \xi_j \mid j \geq 1\}.$$

For every $j \geq 1$, set

$$F_j := \mathbf{1}_X \otimes \xi_j \in L^2(X, \mu, \mathcal{K}_1) \quad \text{and} \quad G_j := \widetilde{T}(F_j) \in L^2(X, \mu, \mathcal{K}_2).$$

Let Q be a countable dense subset of $\mathbf{C}$. For $n \geq 1$ and for every $\lambda = (\lambda_1, \dots, \lambda_n) \in Q^n$, set

$$F_\lambda := \sum_{j=1}^n \lambda_j F_j \quad \text{and} \quad G_\lambda := \sum_{j=1}^n \lambda_j G_j.$$

Let $n \geq 1$ and $\lambda \in Q^n$. For every $\varphi \in L^\infty(X, \mu)$, we have

$$\left(\widetilde{T} m_1(\varphi)\right) F_\lambda = \left(m_2(\varphi)\widetilde{T}\right) F_\lambda = m_2(\varphi) G_\lambda,$$

and therefore

$$\begin{aligned}\int_X |\varphi(x)|^2 \|G_\lambda(x)\|^2 d\mu(x) &= \|m_2(\varphi)G_\lambda\|^2 = \|\left(\widetilde{T} m_1(\varphi)\right) F_\lambda\|^2 \\ &\leq \|\widetilde{T}\|^2 \|m_1(\varphi)F_\lambda\|^2 = \|\widetilde{T}\|^2 \int_X |\varphi(x)|^2 \|F_\lambda(x)\|^2 d\mu(x).\end{aligned}$$

Since this inequality holds for every $\varphi \in L^\infty(X, \mu)$, there exists a subset $N_\lambda \in \mathcal{B}$ with $\mu(N_\lambda) = 0$ such that

$$\|G_\lambda(x)\| \leq \|\widetilde{T}\| \|F_\lambda(x)\|$$

for all $x \in X \smallsetminus N_\lambda$.

Set $N := \bigcup_\lambda N_\lambda$, where λ runs over the countable set $\bigcup_{n\geq 1} Q^n$. Then $\mu(N) = 0$ and

$$(*) \qquad \left\| \sum_{j=1}^n \lambda_j G_j(x) \right\| \leq \|\widetilde{T}\| \left\| \sum_{j=1}^n \lambda_j F_j(x) \right\|$$

for all $x \in X \smallsetminus N$, all $n \geq 1$, and all $\lambda_1, \dots, \lambda_n \in Q$.

Let $x \in X \smallsetminus N$. Since $F_j(x) = \xi_j$ and since $\{\xi_j \mid j \geq 1\}$ is a total subset of $\mathcal{K}_1$, Inequality $(*)$ implies that we can define a bounded linear map $T(x) : \mathcal{K}_1 \to \mathcal{K}_2$ such that $T(x)\xi_j = G_j(x)$ for every $j \geq 1$, and then we have $\|T(x)\| \leq \|\widetilde{T}\|$.

Setting $T(x) = 0$ for $x \in N$, we obtain a measurable map

$$T : X \to \mathcal{L}(\mathcal{K}_1, \mathcal{K}_2), \quad x \mapsto T(x)$$

such that

$$\|T(x)\| \leq \|\widetilde{T}\| \quad \text{for } \mu\text{-almost all } x \in X.$$

Let $\widetilde{T}' : L^2(X, \mu, \mathcal{K}_1) \to L^2(X, \mu, \mathcal{K}_2)$ be the decomposable operator defined by the map T. For every $j \geq 1$, we have $\widetilde{T}'(F_j) = G_j$ by definition of F_j and $\widetilde{T}'$; hence $\widetilde{T}'(F_j) = \widetilde{T}(F_j)$, that is,

$$\widetilde{T}'(\mathbf{1}_X \otimes \xi_j) = \widetilde{T}(\mathbf{1}_X \otimes \xi_j).$$

Since, as observed above, $\widetilde{T}'$ and $\widetilde{T}$ are determined by their values on the set $\{\mathbf{1}_X \otimes \xi_j \mid j \geq 1\}$, we have $\widetilde{T} = \widetilde{T}'$. $\square$

We draw an immediate consequence of Theorem 1.H.5. Recall from Proposition 1.H.3 that, for a Hilbert space $\mathcal{K}$ and a measure space (X, μ), the algebra $\mathcal{A}$ of diagonalisable operators in $\mathcal{L}(L^2(X, \mu, \mathcal{K}))$ is a von Neumann algebra.

COROLLARY 1.H.6. *Let $(X, \mathcal{B})$ be a Borel space equipped with a σ-finite measure μ. Let $\mathcal{K}$ be a separable Hilbert space and $\mathcal{A}$ the von Neumann algebra of diagonalisable operators in $\mathcal{L}(L^2(X, \mu, \mathcal{K}))$.*

The commutant $\mathcal{A}'$ of $\mathcal{A}$ in $\mathcal{L}(L^2(X, \mu, \mathcal{K}))$ is an abelian algebra if and only if $\mathcal{K}$ is one-dimensional. In particular, the von Neumann algebra $\mathcal{A}$ is maximal abelian, *that is, $\mathcal{A} = \mathcal{A}'$, if and only if $\mathcal{K}$ is one-dimensional.*

PROOF. Assume that $\mathcal{K}$ is not one-dimensional. Viewing every operator in $\mathcal{L}(\mathcal{K})$ as a constant-valued decomposable operator in $\mathcal{L}(L^2(X, \mu, \mathcal{K}))$, we see that $\mathcal{L}(\mathcal{K})$ is isomorphic to a subalgebra of $\mathcal{A}'$. Therefore $\mathcal{A}'$ is not abelian.

Assume that $\mathcal{K}$ is one-dimensional. Then the algebra of decomposable operators in $\mathcal{L}(L^2(X, \mu, \mathcal{K})) = \mathcal{L}(L^2(X, \mu))$ coincides with the algebra $\mathcal{A}$ of diagonalisable operators. By Theorem 1.H.5, we therefore have $\mathcal{A}' = \mathcal{A}$. $\square$

1.I. Direct integrals of von Neumann algebras

Given a direct integral $\pi = \int_X^{\oplus} \pi_x d\mu(x)$ of representations of a group G, we are interested in the relationship between the von Neumann algebra $\pi(G)''$ and the family of von Neumann algebras $(\pi_x(G)'')_{x \in X}$. This will be needed in Section 6.C.

Let $\mathcal{H} = \int_X^{\oplus} \mathcal{H}_x d\mu(x)$ be a direct integral of Hilbert spaces over a measure space (X, μ), as in Section 1.G.

DEFINITION 1.I.1. A **field of von Neumann algebras** over X is an assignment $x \mapsto \mathcal{M}_x$ of a von Neumann subalgebra of $\mathcal{L}(\mathcal{H}_x)$ for every $x \in X$.

The field $x \mapsto \mathcal{M}_x$ of von Neumann algebras over X is **measurable** if there exists a sequence

$$x \mapsto T_x^{(1)}, \quad x \mapsto T_x^{(2)}, \quad \ldots$$

of measurable fields of operators over X such that, for μ-almost every $x \in X$, the von Neumann algebra $\mathcal{M}_x$ is generated by the sequence $T_x^{(1)}, T_x^{(2)}, \ldots$.

EXAMPLE 1.I.2. (1) The field $x \mapsto \mathbf{C}\mathrm{Id}_x$ of all diagonalisable operators is obviously a measurable field of von Neumann algebras.

The field $x \mapsto \mathcal{L}(\mathcal{H}_x)$ of all decomposable operators is also a measurable field of von Neumann algebras. Indeed, in view of Proposition 1.G.2, we may assume that $x \mapsto \mathcal{H}_x$ is a constant field corresponding to a fixed separable Hilbert space $\mathcal{K}$. Let $(S_n)_{n \geq 1}$ be a sequence generating $\mathcal{L}(\mathcal{K})$. It suffices to consider the sequence of constant fields $(x \mapsto S_n)_{n \geq 1}$ over X.

(2) Let G be a second countable locally compact group and let

$$\pi = \int_X^{\oplus} \pi_x d\mu(x)$$

be a direct integral of representations of G. Then $x \mapsto \pi_x(G)''$ is a measurable field of von Neumann algebras. Indeed, let $(g_n)_{n \geq 1}$ be a dense sequence in G. For every $x \in X$, the sequence $(\pi_x(g_n))_{n \geq 1}$ generates $\pi_x(G)''$.

In the following proposition, we state some crucial properties of measurable fields of von Neumann algebras; for the proof, see Chap. II, § 3, Proposition 1 and Lemme 1 in [**Dixm–vN**].

PROPOSITION 1.I.3. *Let $x \mapsto \mathcal{M}_x$ be a measurable field of von Neumann algebras over X.*

(1) *The set $\mathcal{M}$ of all decomposable operators of the form $T = \int_X^{\oplus} T_x d\mu(x)$, with $T_x \in \mathcal{M}_x$ for μ-almost every $x \in X$, is a von Neumann subalgebra of $\mathcal{L}(\mathcal{H})$.*
(2) *Assume that X is a standard Borel space. Then the field $x \mapsto \mathcal{M}'_x$ is measurable, where the von Neumann algebra $\mathcal{M}'_x$ is the commutant of $\mathcal{M}_x$ in $\mathcal{L}(\mathcal{H}_x)$.*

Next, we introduce the notion of a decomposable von Neumann algebra.

DEFINITION 1.I.4. (1) Let $x \mapsto \mathcal{M}_x$ be a measurable field of von Neumann algebras over X. The von Neumann algebra of all decomposable $T = \int_X^{\oplus} T_x d\mu(x)$ with $T_x \in \mathcal{M}_x$ for μ-almost every $x \in X$ is called the **direct integral of the von Neumann algebras** $(\mathcal{M}_x)_{x\in X}$ and denoted by $\int_X^{\oplus} \mathcal{M}_x d\mu(x)$.

(2) A von Neumann algebra $\mathcal{M}$ of $\mathcal{L}(\mathcal{H})$ is **decomposable** if there exists a measurable field $x \mapsto \mathcal{M}_x$ of von Neumann algebras of X such that

$$\mathcal{M} = \int_X^{\oplus} \mathcal{M}_x d\mu(x).$$

EXAMPLE 1.I.5. (1) The decomposable von Neumann algebra $\int_X^{\oplus} \mathbf{C}\mathrm{Id}_x d\mu(x)$ is the algebra $\mathcal{A}$ of diagonalizable operators.

The decomposable von Neumann algebra $\int_X^{\oplus} \mathcal{L}(\mathcal{H}_x)_x d\mu(x)$ is the algebra of all decomposable operators in $\mathcal{L}(\mathcal{H})$.

(2) Let $\pi = \int_X^{\oplus} \pi_x d\mu(x)$ be a direct integral of representations of a second countable locally compact group G. The decomposable algebra $\int_X^{\oplus} \pi_x(G)'' d\mu(x)$ need not coincide with $\pi(G)''$. For example, consider the case where $x \mapsto \mathcal{H}_x$ is a constant field $x \mapsto \mathcal{K}$, for some nontrivial measure space (X, μ) and some Hilbert space $\mathcal{K}$, and where π_x is the identity representation of G on $\mathcal{K}$ for every $x \in X$. Then $\pi(G)'' = \mathbf{C}\mathrm{Id}$ is strictly smaller than $\int_X^{\oplus} \pi_x(G)'' d\mu(x)$.

The commutant of a decomposable von Neumann algebra admits a neat description.

THEOREM 1.I.6. *Assume that X is a standard Borel space and let $\mathcal{M} = \int_X^{\oplus} \mathcal{M}_x d\mu(x)$ be a decomposable von Neumann algebra.*

Then $\mathcal{M}' = \int_X^{\oplus} \mathcal{M}'_x d\mu(x)$

PROOF. Let $(T^{(n)})_{n\geq 1}$ be a sequence in $\mathcal{M}$ which generates $\mathcal{M}$; upon enlarging this sequence, we may assume that $\mathcal{M}_x$ is generated by $(T_x^{(n)})_{n\geq 1}$, for μ-almost every $x \in X$.

Observe that, since $\mathcal{M}$ contains the algebra of diagonalisable operators, every operator in $\mathcal{M}'$ is decomposable (see Theorem 1.H.2).

Let $T = \int_X^{\oplus} T_x d\mu(x)$ be a decomposable operator. Then $T \in \mathcal{M}'$ if and only if

$$TT^{(n)} = T^{(n)}T \quad \text{and} \quad TT^{(n)^*} = T^{(n)^*}T$$

for every $n \geq 1$. Hence, $T \in \mathcal{M}'$ if and only if

$$T_x T_x^{(n)} = T_x^{(n)} T_x \quad \text{and} \quad T_x T_x^{(n)^*} = T_x^{(n)^*} T_x$$

for every $n \geq 1$ and μ-almost every $x \in X$. This shows that $T \in \mathcal{M}'$ if and only if $T_x \in \mathcal{M}'_x$ for μ-almost every $x \in X$. Since $x \mapsto \mathcal{M}'_x$ is measurable (Proposition 1.I.3), it follows that

$$\mathcal{M}' = \int_X^{\oplus} \mathcal{M}'_x d\mu(x).$$

□

We give a necessary and sufficient condition for a von Neumann subalgebra of $\mathcal{L}(\mathcal{H})$ to be decomposable.

THEOREM 1.I.7. *Assume that X is a standard Borel space and let $\mathcal{M}$ be a von Neumann subalgebra of $\mathcal{L}(\mathcal{H})$. The following conditions are equivalent:*

(i) *$\mathcal{M}$ is decomposable over X.*
(ii) *We have $\mathcal{A} \subset \mathcal{M} \subset \mathcal{B}$, where $\mathcal{A}, \mathcal{B}$ are respectively the algebras of diagonalizable and decomposable operators on $\mathcal{H} = \int_X^{\oplus} \mathcal{H}_x d\mu(x)$.*

PROOF. It is obvious that (i) implies (ii). Assume that (ii) holds. Recall from Theorem 1.H.2 that $\mathcal{B} = \mathcal{A}'$ and hence $\mathcal{B}' = \mathcal{A}$. Since $\mathcal{A} \subset \mathcal{M}$, we have therefore $\mathcal{M}' \subset \mathcal{B}$. So, both $\mathcal{M}$ and $\mathcal{M}'$ consist of decomposable operators.

Let $(T^{(n)})_{n \geq 1}$ and $(S^{(n)})_{n \geq 1}$ be sequences generating $\mathcal{M}$ and $\mathcal{M}'$ respectively. Write

$$T^{(n)} = \int_X^{\oplus} T_x^{(n)} d\mu(x) \quad \text{and} \quad S^{(n)} = \int_X^{\oplus} S_x^{(n)} d\mu(x).$$

For every $x \in X$, let $\mathcal{M}_x$ denote the von Neumann algebra generated by $(T_x^{(n)})_{n \geq 1}$. We claim that $\mathcal{M} = \int_X^{\oplus} \mathcal{M}_x d\mu(x)$.

Observe first that $S_x^{(n)}$ and $S_x^{(n)^*}$ commute with $T_x^{(m)}$ and $T_x^{(m)^*}$ for every $n, m \geq 1$ and for μ-almost every $x \in X$.

Let $T = \int_X^{\oplus} T_x d\mu(x)$ with $T_x \in \mathcal{M}_x$ for μ-almost every $x \in X$. Then, for every $n \geq 1$, the operator T_x commutes with $S_x^{(n)}$ and $S_x^{(n)^*}$ for μ-almost every $x \in X$; hence, T commutes with $S^{(n)}$ and $S^{(n)^*}$ for every $n \geq 1$. Therefore, $T \in (\mathcal{M}')' = \mathcal{M}$. So, we have shown that

$$\int_X^{\oplus} \mathcal{M}_x d\mu(x) \subset \mathcal{M}.$$

By Proposition 1.I.3, $x \mapsto \mathcal{M}'_x$ is a measurable field of von Neumann algebras; moreover, by Theorem 1.I.6, the commutant of $\int_X^{\oplus} \mathcal{M}_x d\mu(x)$ is $\int_X^{\oplus} \mathcal{M}'_x d\mu(x)$. So, to prove that $\mathcal{M}$ is contained in $\int_X^{\oplus} \mathcal{M}_x d\mu(x)$, it suffices to show that

$$\int_X^{\oplus} \mathcal{M}'_x d\mu(x) \subset \mathcal{M}'.$$

Let $S = \int_X^{\oplus} S_x d\mu(x)$ with $S_x \in \mathcal{M}'_x$ for μ-almost every $x \in X$. Then, for every $n \geq 1$, the operator S_x commutes with $T_x^{(n)}$ and $T_x^{(n)^*}$ for μ-almost every $x \in X$. It follows that S commutes with $T^{(n)}$ and $T^{(n)^*}$ for every $n \geq 1$. Therefore, $S \in \mathcal{M}'$. □

Recall from Example 1.I.5 that, if $\int_X^{\oplus} \pi_x d\mu(x)$ is a direct integral of representations of a second countable locally compact group G, then $x \mapsto \pi_x(G)''$ is a measurable field of von Neumann algebras. The following corollary is therefore an immediate consequence of Theorem 1.I.7 and Theorem 1.I.6.

COROLLARY 1.I.8. *Assume that X is a standard Borel space and let $\pi = \int_X^{\oplus} \pi_x d\mu(x)$ be a direct integral of representations of a second countable locally compact group G on $\mathcal{H} = \int_X^{\oplus} \mathcal{H}_x d\mu(x)$. The following conditions are equivalent:*

(i) $\pi(G)'' = \int_X^{\oplus} \pi_x(G)'' d\mu(x)$.

(ii) $\pi(G)''$ *contains the algebra $\mathcal{A}$ of diagonalisable operators on $\mathcal{H}$.*

Moreover, when these conditions hold, we have

$$\pi(G)' = \int_X^{\oplus} \pi_x(G)' d\mu(x).$$

REMARK 1.I.9. We will apply Corollary 1.I.8 in the case where $\int_X^{\oplus} \pi_x d\mu(x)$ is the central decomposition of a representation π of G (see Subsection 6.C.b).

CHAPTER 2

Representations of locally compact abelian groups

In this chapter, we describe models for the most general representation of a second-countable locally compact abelian group G.

In Section 2.A, we introduce the canonical representation π_μ of G attached to a probability measure μ on $\widehat{G}$, and show that every representation of G is equivalent to a direct sum of such representations (Corollary 2.A.7).

This result is refined in Section 2.B, where we establish that a representation of G on a separable Hilbert space is equivalent to a representation determined by a set of multiplicities $I \subset \overline{\mathbf{N}^*}$ and a sequence $(\mu_m)_{m \in I}$ of mutually singular probability measures on $\widehat{G}$ (Theorem 2.B.9). This refinement is a much deeper result; on the one hand, it can be seen as a generalization of the Multiplicity Theorem of Hahn–Hellinger, which classifies selfadjoint operators up to unitary equivalence (see Remark 2.B.10); on the other hand, it admits an extension to representations of type I groups (Theorem 6.D.7).

The SNAG Theorem, which is the theme of Section 2.C, shows that representations of G are in bijective correspondence with projection-valued measures on $\widehat{G}$. In Sections 2.D and 2.E, we reformulate containment and weak containment properties, as well as the canonical decomposition result of representations of G, in terms of projection-valued measures on $\widehat{G}$.

These facts are useful to understand irreducible representations of some non-abelian groups having abelian normal subgroups (Section 2.F)

2.A. Canonical representations of abelian groups

Let G be a locally compact abelian group, for short a **LCA group**. Let $\widehat{G}$ be its dual group, which is itself an LCA group. From now on in this chapter and until Section 2.E,

(∗) we assume that G is a *second-countable* locally compact abelian group.

For some reminder on LCA groups, see Appendix A.7.

In this chapter, a "measure" on G or on $\widehat{G}$ is always meant to be a measure on the Borel σ-algebra, $\mathcal{B}(G)$ or $\mathcal{B}(\widehat{G})$ respectively, which is finite on compact subsets. In particular, because of (∗), such a measure is a Radon measure (see Theorem A.4.1).

Construction 2.A.1. Let μ be a Radon measure on $\widehat{G}$. Note that the Hilbert space $L^2(\widehat{G}, \mu)$ is separable since $\widehat{G}$ is second-countable (see Proposition A.7.3 and Theorem A.4.1).

The **canonical representation** of G on $L^2(\widehat{G}, \mu)$ is the representation π_μ defined by

$$(\sharp) \qquad (\pi_\mu(g)f)(\chi) = \chi(g)f(\chi) \quad \text{for all } g \in G,\ f \in L^2(\widehat{G}, \mu), \text{ and } \chi \in \widehat{G}.$$

For a bounded continuous function φ on $\widehat{G}$, we will most often make no notational distinction between φ and its equivalence class in $L^\infty(\widehat{G}, \mu)$. As in Definition 1.G.4 we denote by $m(\varphi)$ the multiplication operator

$$m(\varphi) :\ L^2(\widehat{G}, \mu) \to L^2(\widehat{G}, \mu), \quad f \mapsto \varphi f$$

associated with φ. In particular, for $g \in G$, the map $\widehat{g}$ defined on $\widehat{G}$ by $\widehat{g}(\chi) = \chi(g)$ for $\chi \in \widehat{G}$ is a unitary character on $\widehat{G}$ and we have

$$\pi_\mu(g) = m(\widehat{g}).$$

Let us check that π_μ is indeed a representation of G. It is clear that $\pi_\mu(g)$ is a unitary operator on $L^2(\widehat{G}, \mu)$ for every $g \in G$ and that the map $g \mapsto \pi_\mu(g)$ is a group homomorphism. The issue is to check that the map $\pi_\mu : G \to \mathcal{U}(L^2(\widehat{G}, \mu))$ is continuous.

Let $f \in L^2(\widehat{G}, \mu)$. By polarization it suffices to show that $g \mapsto \langle \pi_\mu(g)f \mid f \rangle$ is a continuous function on G. For every $g \in G$, we have

$$\langle \pi_\mu(g)f \mid f \rangle = \int_{\widehat{G}} \chi(g)|f(\chi)|^2 d\mu(\chi) = \int_{\widehat{G}} \widehat{g}(\chi)|f(\chi)|^2 d\mu(\chi) = \overline{\mathcal{F}}(\mu_f)(g),$$

where μ_f is the finite positive measure on $\widehat{G}$ defined by $d\mu_f(\chi) = |f(\chi)|^2 d\mu(\chi)$ and $\overline{\mathcal{F}}(\mu_f)$ its inverse Fourier–Stieltjes transform (see Appendix A.7). Since $\overline{\mathcal{F}}(\mu_f)$ is a continuous function on G, this concludes the proof.

PROPOSITION 2.A.2. *Let μ, ν be two Radon measures on $\widehat{G}$ which are equivalent to each other.*

Then the canonical representations π_μ, π_ν are equivalent to each other.

About the converse, see Proposition 2.A.9(2).

PROOF. Consider the Radon–Nikodym derivative $\frac{d\nu}{d\mu}$ of ν with respect to μ. Define a linear operator

$$T :\ L^2(\widehat{G}, \mu) \to L^2(\widehat{G}, \nu)$$

by

$$T(f)(\chi) = f(\chi)\sqrt{\frac{d\mu}{d\nu}(\chi)} \quad \text{for all } f \in L^2(\widehat{G}, \mu),\ \chi \in \widehat{G}.$$

The map T is invertible; indeed, T^{-1} is given by the analogous formula, with $\frac{d\nu}{d\mu}(\chi)$ instead of $\frac{d\mu}{d\nu}(\chi)$. The map T is also isometric: for every $f \in L^2(\widehat{G}, \mu)$, we have

$$\|T(f)\|^2 = \int_{\widehat{G}} |f(\chi)|^2 \frac{d\mu}{d\nu}(\chi) d\nu(\chi) = \int_{\widehat{G}} |f(\chi)|^2 d\mu(\chi) = \|f\|^2.$$

Moreover, T intertwines the representations π_μ and π_ν:

$$\begin{aligned}(T\pi_\mu(g)(f))(\chi) &= (\pi_\mu(g)(f))(\chi)\sqrt{\frac{d\mu}{d\nu}(\chi)} = \chi(g)f(\chi)\sqrt{\frac{d\mu}{d\nu}(\chi)} \\ &= \chi(g)(T(f))(\chi) = (\pi_\nu(g)(T(f)))(\chi)\end{aligned}$$

for $g \in G$ and $\chi \in \widehat{G}$. □

EXAMPLE 2.A.3. (1) If $\mu_{\widehat{G}}$ is a Haar measure on $\widehat{G}$, then $\pi_{\mu_{\widehat{G}}}$ is equivalent to the regular representation λ_G of G. More precisely, let μ_G be the Haar measure on G for which the Fourier transform $\mathcal{F} : L^2(G, \mu_G) \to L^2(\widehat{G}, \mu_{\widehat{G}})$ is unitary; as already noted in Example 1.G.10, we have $\pi_{\mu_{\widehat{G}}}(g)\mathcal{F} = \mathcal{F}\lambda_G(g)$ for all $g \in G$.

(2) Let $\chi_0 \in \widehat{G}$, and μ be the Dirac measure at χ_0. Then $L^2(\widehat{G}, \mu) = \mathbf{C}$ and $\pi_\mu = \chi_0$.

REMARK 2.A.4. Let G be as above a second-countable LCA group. In the rest of this chapter,

$$(**)\qquad \begin{array}{c}\text{we will consider canonical representations } \pi_\mu \text{ of } G \\ \text{for } \textit{probability} \text{ measures on } \widehat{G} \text{ only.}\end{array}$$

This is because we are interested in representations of G up to equivalence only, and because every Radon measure on $\widehat{G}$ is equivalent to a probability Radon measure on $\widehat{G}$ (Proposition A.3.5).

As we will see, representations of the form π_μ are the building blocks of the most general representations of G, in the sense that every representation of G on a separable Hilbert space is equivalent to a direct sum $\bigoplus_k \pi_{\mu_k}$.

Our first goal is to show that, given a sequence $(\mu_k)_k$ of probability measures on $\widehat{G}$ singular with each other, the direct sum $\bigoplus_k \pi_{\mu_k}$ is itself a representation of the form π_μ, for some probability measures μ on $\widehat{G}$.

PROPOSITION 2.A.5. *Let $(\mu_k)_{k\geq 1}$ be a (possibly finite) sequence of probability measures on $\widehat{G}$, singular with each other. Let $(c_k)_{k\geq 1}$ be a sequence of positive real numbers such that $\sum_{k\geq 1} c_k = 1$; set $\mu = \sum_{k\geq 1} c_k\mu_k$.*

Then π_μ is equivalent to $\bigoplus_{k\geq 1} \pi_{\mu_k}$.

PROOF. Since the μ_k's are mutually singular probability measures, we can find a sequence $(A_k)_{k\geq 1}$ of pairwise disjoint subsets in $\mathcal{B}(\widehat{G})$ such that $\mu_k(A_k) = 1$ for every $k \geq 1$. Let

$$T : L^2(\widehat{G}, \mu) \to \bigoplus_{k\geq 1} L^2(\widehat{G}, \mu_k), \quad f \mapsto \bigoplus_{k\geq 1} \sqrt{c_k}\mathbf{1}_{A_k} f.$$

Observe that, since every μ_k is absolutely continuous with respect to μ, the map T is well-defined. Denoting by $\|\cdot\|_\mu$ and $\|\cdot\|_{\mu_k}$ the norms on $L^2(\widehat{G}, \mu)$ and $L^2(\widehat{G}, \mu_k)$, we have

$$\begin{aligned}\|f\|_\mu^2 &= \int_{\widehat{G}} |f(\chi)|^2 d\mu(\chi) = \sum_{k\geq 1} \int_{A_k} |\sqrt{c_k} f(\chi)|^2 d\mu_k(\chi) \\ &= \sum_{k\geq 1} \|\sqrt{c_k}\mathbf{1}_{A_k} f\|_{\mu_k}^2 = \|Tf\|^2,\end{aligned}$$

hence T is an isometry.

Moreover, it is straightforward that T intertwines π_μ and $\bigoplus_{k\geq 1} \pi_{\mu_k}$.

Finally, T is surjective; more precisely, one checks that the operator given by

$$\bigoplus_{k\geq 1} L^2(\widehat{G}, \mu_k) \to L^2(\widehat{G}, \mu), \quad \bigoplus_{k\geq 1} f_k \mapsto \sum_{k\geq 1} \frac{1}{\sqrt{c_k}}\mathbf{1}_{A_k} f_k$$

is bounded, and is the inverse T^{-1} of T. $\square$

We now show that the representations of the form π_μ are exactly the cyclic representations of G (up to equivalence).

PROPOSITION 2.A.6. *Let G be as above a second-countable LCA group.*

(1) *Let μ be a probability measure on $\widehat{G}$. Then π_μ is a cyclic representation, with cyclic vector $\mathbf{1}_{\widehat{G}}$.*
(2) *Every cyclic representation of G is equivalent to a representation π_μ for a probability measure μ on $\widehat{G}$.*

PROOF. (1) Since

$$\{\pi_\mu(g)\mathbf{1}_{\widehat{G}} \mid g \in G\} = \{\widehat{g} \mid g \in G\},$$

the claim follows from Proposition A.7.7(1), case $p = 2$.

(2) Let $(\pi, \mathcal{H})$ be a cyclic representation of G and let $\xi \in \mathcal{H}$ be a cyclic unit vector for π. The function

$$\varphi : G \to \mathbf{C}, \ \ g \mapsto \langle \pi(g)\xi \mid \xi \rangle$$

is a normalized function of positive type on G (see Section 1.B). Therefore, by Bochner's Theorem A.7.5, there exists a probability measure μ on $\widehat{G}$ such that $\varphi = \overline{\mathcal{F}}(\mu)$.

We claim that π is equivalent to π_μ. Indeed, we have

$$\langle \pi_\mu(g)\mathbf{1}_{\widehat{G}} \mid \mathbf{1}_{\widehat{G}} \rangle = \int_{\widehat{G}} \widehat{g}(\chi)d\mu(\chi) = \overline{\mathcal{F}}(\mu) = \varphi(g), \qquad \text{for every } \ g \in G,$$

i.e., φ coincides with the function of positive type $\langle \pi_\mu(\cdot)\mathbf{1}_{\widehat{G}} \mid \mathbf{1}_{\widehat{G}} \rangle$. Since ξ and $\mathbf{1}_{\widehat{G}}$ are cyclic vectors for respectively π and π_μ, the claim follows from Proposition 1.B.10. □

COROLLARY 2.A.7. *Let π be a representation of G on a separable Hilbert space $\mathcal{H}$.*

There exists a (possibly finite) sequence $(\mu_k)_{k \geq 1}$ of probability measures on $\widehat{G}$ such that π is equivalent to the direct sum $\bigoplus_{k \geq 1} \pi_{\mu_k}$.

PROOF. Since the corollary is trivial when $\mathcal{H} = \{0\}$, we assume from now on that $\mathcal{H}$ is not 0-dimensional. By Proposition 2.A.6(2), it suffices to show that $\mathcal{H}$ can be written as a direct sum $\bigoplus_{k \geq 1} \mathcal{K}_k$ of $\pi(G)$-invariant closed subspaces $\mathcal{K}_k$ such that the restriction of π to $\mathcal{K}_k$ is cyclic.

Let $\{\xi_n \mid n \geq 1\}$ be an infinite countable total subset of unit vectors in $\mathcal{H}$. We construct by induction on $n \geq 1$ a strictly increasing sequence $(k_n)_{n \geq 1}$ of positive integers, and a sequence $(\mathcal{K}_n)_{n \geq 1}$ of $\pi(G)$-invariant cyclic and mutually orthogonal subspaces of $\mathcal{H}$ such that, for all $n \geq 1$, we have $\xi_k \in \bigoplus_{i=1}^n \mathcal{K}_i$ for all $k \leq k_n$.

Let $\mathcal{K}_1$ be the closed subspace of $\mathcal{H}$ generated by $\pi(G)\xi_1$; set $k_1 = 1$. Then $\mathcal{K}_1$ is $\pi(G)$-invariant, it contains ξ_1, and the restriction of π to $\mathcal{K}_1$ is cyclic.

Assume now that, for $n \geq 1$, sequences $(k_1 < \ldots < k_n)$ and $(\mathcal{K}_1, \ldots, \mathcal{K}_n)$ have been constructed, with the desired properties. Two cases may occur.

- The space $\bigoplus_{i=1}^n \mathcal{K}_i$ coincides with $\mathcal{H}$. In this case, define $\mathcal{K}_m = \mathcal{K}_n$ and $k_m = k_n + m$ for every $m > n$.
- There exists an integer $k > k_n$ such that the projection of ξ_k on the orthogonal complement of $\bigoplus_{i=1}^n \mathcal{K}_i$ is nonzero. Let k_{n+1} be the smallest such integer and let $\eta_{k_{n+1}}$ be the projection of $\xi_{k_{n+1}}$ on the orthogonal complement of $\bigoplus_{i=1}^n \mathcal{K}_i$.

Then let $\mathcal{K}_{n+1}$ be the closed subspace of $\mathcal{H}$ generated by $\pi(G)\eta_{k_{n+1}}$. Then $\mathcal{K}_{n+1}$ is $\pi(G)$-invariant, cyclic, and orthogonal to $\bigoplus_{i=1}^{n} \mathcal{K}_i$. The projection of $\xi_{k_{n+1}}$ on the orthogonal complement of $\bigoplus_{i=1}^{n} \mathcal{K}_i$, which is $\eta_{k_{n+1}}$, as well as the projection of $\xi_{k_{n+1}}$ on $\bigoplus_{i=1}^{n} \mathcal{K}_i$, belong both to $\bigoplus_{i=1}^{n+1} \mathcal{K}_i$. Therefore, $\xi_{k_{n+1}} \in \bigoplus_{i=1}^{n+1} \mathcal{K}_i$. We also have $\xi_k \in \bigoplus_{i=1}^{n+1} \mathcal{K}_i$ for every $k < k_{n+1}$, by definition of k_{n+1}.

This provides the construction of the desired sequences.

Since $\xi_k \in \bigoplus_{i\geq 1} \mathcal{K}_i$ for every $k \geq 1$, and since $\{\xi_k \mid k \geq 1\}$ is total in $\mathcal{H}$, we have

$$\mathcal{H} = \bigoplus_{i\geq 1} \mathcal{K}_i$$

and the corollary follows.

NOTE. Using Zorn's lemma, it is easy to show that any representation of a topological group in any Hilbert space is a direct sum of cyclic representations [**Dixm–C***, 2.2.7]. □

REMARK 2.A.8. (1) Concerning the equivalence classes of the probability measures μ_k of Corollary 2.A.7, no uniqueness can be expected. For instance, let μ_1 and μ_2 be probability measures on $\widehat{G}$ with disjoint supports; let $\mu = (\mu_1 + \mu_2)/2$. Then π_μ is equivalent to $\pi_{\mu_1} \oplus \pi_{\mu_2}$, by Proposition 2.A.5.

We will give in Theorem 2.B.9 a refinement of Corollary 2.A.7, with a uniqueness property. For this, we will need to consider versions "with multiplicities" of the canonical representations π_μ of Construction 2.A.1.

(2) In the special case $G = \mathbf{R}$, Corollary 2.A.7 is equivalent to the "multiplication operator form" of the spectral theorem for a (possibly unbounded) selfadjoint operator on a separable Hilbert space (see [**Halm–51**]). In fact, Corollary 2.A.7 as well as its refined version (Theorem 2.B.9 below) can both be seen as particular cases of theorems for representations of abelian C*-algebras, for which we refer to Chapter 2 of [**Arve–76**], or Section 3.5 of [**Sund–97**].

We investigate when two representations of the form π_μ have a common subrepresentation.

PROPOSITION 2.A.9. *Let μ_1, μ_2 be probability measures on $\widehat{G}$.*

(1) *There exists a nonzero bounded operator $L^2(\widehat{G}, \mu_1) \to L^2(\widehat{G}, \mu_2)$ intertwining π_{μ_1} and π_{μ_2} if and only if μ_1 and μ_2 are not singular with each other.*

    In particular, if μ_1 and μ_2 are singular with each other, then the representations π_{μ_1} and π_{μ_2} are disjoint.

(2) *The representations π_{μ_1} and π_{μ_2} are equivalent if and only if the measures μ_1 and μ_2 are equivalent.*

PROOF. (1) Assume that there exists a nonzero bounded operator

$$T : L^2(\widehat{G}, \mu_1) \to L^2(\widehat{G}, \mu_2)$$

which intertwines π_{μ_1} and π_{μ_2}. Then $\mathcal{H}_1 := (\ker T)^\perp$ is $\pi_{\mu_1}(G)$-invariant, the closure $\mathcal{H}_2$ of the image of T is $\pi_{\mu_2}(G)$-invariant, and there exists an isometric bijective linear map $U : \mathcal{H}_1 \to \mathcal{H}_2$ intertwining π_{μ_1} and π_{μ_2} (see Lemma 1.A.5).

Let $P : L^2(\widehat{G}, \mu_1) \twoheadrightarrow \mathcal{H}_1$ be the orthogonal projection; set $f_1 := P(\mathbf{1}_{\widehat{G}})$. By Proposition 2.A.6, $\mathbf{1}_{\widehat{G}}$ is a cyclic vector for π_{μ_1}. Since P intertwines π_{μ_1} with itself, f_1 is a cyclic vector for the restriction of π_{μ_1} to $\mathcal{H}_1$.

For $i = 1, 2$ and $\varphi \in C^b(\widehat{G})$, denote by $m_i(\varphi)$ the multiplication operator by φ on $L^2(\widehat{G}, \mu_i)$. We have

$$Um_1(\widehat{g}) = m_2(\widehat{g})U \quad \text{for all } g \in G.$$

Set

$$f_2 := U(f_1) \in L^2(\widehat{G}, \mu_2).$$

For every $g \in G$, we have

$$U(\widehat{g}f_1) = U(m_1(\widehat{g})(f_1)) = m_2(\widehat{g})(U(f_1)) = \widehat{g}f_2.$$

Consider the nonzero finite measures ν_1 and ν_2 on $\widehat{G}$ given by

$$d\nu_1(\chi) := |f_1(\chi)|^2 d\mu_1(\chi) \quad \text{and} \quad d\nu_2(\chi) := |f_2(\chi)|^2 d\mu_2(\chi).$$

For $i = 1, 2$, denote by $\langle \cdot \mid \cdot \rangle_{L^2(\widehat{G}, \mu_i)}$ the scalar product on $L^2(\widehat{G}, \mu_i)$. For every $g \in G$, we have

$$\overline{\mathcal{F}}(\nu_1)(g) = \int_{\widehat{G}} \widehat{g}(\chi)|f_1(\chi)|^2 d\mu_1(\chi) = \langle \widehat{g}f_1 \mid f_1 \rangle_{L^2(\widehat{G}, \mu_1)}.$$

Since U is an isometry, it follows that

$$\begin{aligned}\overline{\mathcal{F}}(\nu_1)(g) &= \langle \widehat{g}f_1 \mid f_1 \rangle_{L^2(\widehat{G}, \mu_1)} = \langle U(\widehat{g}f_1) \mid U(f_1) \rangle_{L^2(\widehat{G}, \mu_2)} \\ &= \langle \widehat{g}f_2 \mid f_2 \rangle_{L^2(\widehat{G}, \mu_2)} = \int_{\widehat{G}} \widehat{g}(\chi)|f_2(\chi)|^2 d\mu_2(\chi) = \overline{\mathcal{F}}(\nu_2)(g).\end{aligned}$$

Therefore we have

$$\overline{\mathcal{F}}(\nu_1) = \overline{\mathcal{F}}(\nu_2),$$

hence $\nu_1 = \nu_2$, by injectivity of the Fourier–Stieltjes transform (Proposition A.7.4).

Since ν_1 is absolutely continuous with respect to μ_1, and ν_2 is absolutely continuous with respect to μ_2, it follows that μ_1 and μ_2 are not singular with each other.

Conversely, assume that μ_1 and μ_2 are not singular with each other. There exists a probability measure μ on $\widehat{G}$ and nonnegative functions ψ_1 in $L^1(\widehat{G}, \mu_1)$ and ψ_2 in $L^1(\widehat{G}, \mu_2)$ such that

$$d\mu(\chi) = \psi_1(\chi)d\mu_1(\chi) \quad \text{and} \quad d\mu(\chi) = \psi_2(\chi)d\mu_2(\chi).$$

Upon replacing ψ_1 by $\psi_1 / \max_{\chi \in \widehat{G}} \psi_1(\chi)$, we can assume that $\psi_1 \leq \mathbf{1}_{\widehat{G}}$.

Define a linear operator

$$T : L^2(\widehat{G}, \mu_1) \to L^2(\widehat{G}, \mu_2)$$

by

$$T(f)(\chi) = f(\chi)\psi_2(\chi)^{1/2} \quad \text{for all } f \in L^2(\widehat{G}, \mu_1),\ \chi \in \widehat{G}.$$

Then T is well defined and bounded; indeed, for every $f \in C^b(\widehat{G})$, we have

$$\begin{aligned}\|T(f)\|^2 &= \int_{\widehat{G}} |f(\chi)|^2 \psi_2(\chi) d\mu_2(\chi) = \int_{\widehat{G}} |f(\chi)|^2 d\mu(\chi) \\ &= \int_{\widehat{G}} |f(\chi)|^2 \psi_1(\chi) d\mu_1(\chi) \leq \int_{\widehat{G}} |f(\chi)|^2 d\mu_1(\chi) = \|f\|^2.\end{aligned}$$

Moreover, T intertwines the representations π_{μ_1} and π_{μ_2}:

$$\begin{aligned}(T\pi_{\mu_1}(g)(f))(\chi) &= (\pi_{\mu_1}(g)(f))(\chi)\psi_2(\chi)^{1/2} = \widehat{g}(\chi)f(\chi)\psi_2(\chi)^{1/2} \\ &= \widehat{g}(\chi)(T(f))(\chi) = (\pi_{\mu_2}(g)T(f))(\chi)\end{aligned}$$

for $g \in G$, $f \in L^2(\widehat{G}, \mu_1)$, and $\chi \in \widehat{G}$. Finally, $T \neq 0$, since $\mu \neq 0$, and hence $\psi_2 \neq 0$.

(2) If μ_1 and μ_2 are equivalent, then π_{μ_1} and π_{μ_2} are equivalent, as was shown in Proposition 2.A.2.

Conversely, assume that there exists a bijective isometry

$$T : L^2(\widehat{G}, \mu_1) \to L^2(\widehat{G}, \mu_2)$$

which intertwines π_{μ_1} and π_{μ_2}. Then $\ker T = \{0\}$ and the first part of the proof of (1) shows that μ_1 is absolutely continuous with respect to μ_2. The same proof for T^{-1} in place of T shows that μ_2 is absolutely continuous with respect to μ_1. It follows that μ_1 and μ_2 are equivalent. □

2.B. Canonical decomposition of representations of abelian groups

Let G be a second-countable locally compact abelian group. Refining Corollary 2.A.7, we will show in this section a canonical decomposition for every representation of G on a separable Hilbert space, in terms of the representations $\pi_\mu^{(n)}$ of the following construction.

We denote by $\overline{\mathbf{N}^*}$ the set $\{1, 2, \ldots, \infty\}$ of extended positive integers.

Construction 2.B.1. Let μ be a probability measure on $\widehat{G}$ and $n \in \overline{\mathbf{N}^*}$ an extended positive integer. Let $\mathcal{K}$ be a Hilbert space of dimension n. Let $L^2(\widehat{G}, \mu, \mathcal{K})$ denote the Hilbert space of L^2-functions from $\widehat{G}$ to $\mathcal{K}$ (already introduced in Example 1.G.1). A representation $\pi_\mu^{(n)}$ of G on $L^2(\widehat{G}, \mu, \mathcal{K})$ is defined by

$$(\pi_\mu^{(n)}(g)F)(\chi) = \chi(g)F(\chi) \quad \text{for all } g \in G,\ F \in L^2(\widehat{G}, \mu, \mathcal{K}),\ \text{and } \chi \in \widehat{G}.$$

The representation $\pi_\mu^{(n)}$ is equivalent to the multiple

$$n\pi_\mu = \underbrace{\pi_\mu \oplus \cdots \oplus \pi_\mu}_{n\text{-times}}$$

of the canonical representation π_μ of Construction 2.A.1.

Note that, for $g \in G$, we have a unitary character $\widehat{g} : \widehat{G} \to \mathbf{T}$ defined by $\widehat{g}(\chi) = \chi(g)$ for all $\chi \in \widehat{G}$ and a diagonalisable operator $m(\widehat{g})$ on $L^2(\widehat{G}, \mu, \mathcal{K})$, as in Definition 1.G.4. The definition of $\pi_\mu^{(n)}$ can also be written as

$$\pi_\mu^{(n)}(g) = m(\widehat{g}) \quad \text{for all } g \in G.$$

We draw from Section 1.H some consequences on common subrepresentations between representations of the form $\pi_\mu^{(n)}$. The following two propositions generalize Proposition 2.A.9.

Proposition 2.B.2. *Consider two probability measures μ_1, μ_2 on $\widehat{G}$, two extended positive integers n_1, n_2 in $\overline{\mathbf{N}^*}$, and two Hilbert spaces $\mathcal{K}_1, \mathcal{K}_2$, respectively of dimension n_1, n_2.*

(1) *Assume that μ_1, μ_2 are singular with each other. Then the representations $\pi_{\mu_1}^{(n_1)}$ and $\pi_{\mu_2}^{(n_2)}$ are disjoint: there exists no nonzero bounded operator between $L^2(\widehat{G}, \mu_1, \mathcal{K}_1)$ and $L^2(\widehat{G}, \mu_2, \mathcal{K}_2)$ which intertwines $\pi_{\mu}^{(n_1)}$ and $\pi_{\mu}^{(n_2)}$.*
(2) *Assume that the representations $\pi_{\mu_1}^{(n_1)}$ and $\pi_{\mu_2}^{(n_2)}$ are equivalent. Then the measures μ_1 and μ_2 are equivalent.*

PROOF. (1) Assume, by contradiction, that there exists a nonzero bounded operator

$$\widetilde{T}:\ L^2(\widehat{G}, \mu_1, \mathcal{K}_1) \to L^2(\widehat{G}, \mu_2, \mathcal{K}_2)$$

intertwining $\pi_{\mu_1}^{(n_1)}$ and $\pi_{\mu_2}^{(n_2)}$. Recall that the representation $\pi_{\mu_i}^{(n_i)}$ is equivalent to $n_i \pi_{\mu_i}$, for $i = 1, 2$.

There exists a $\pi_{\mu_1}^{(n_1)}(G)$-invariant closed subspace $\mathcal{H}_1$ of $L^2(\widehat{G}, \mu_1, \mathcal{K}_1)$ such that the restriction of $\pi_{\mu_1}^{(n_1)}$ to $\mathcal{H}_1$ is equivalent to $\pi_{\mu_1} = \pi_{\mu_1}^{(1)}$, and the restriction of $\widetilde{T}$ to $\mathcal{H}_1$ is not zero. There also exists a $\pi_{\mu_2}^{(n_2)}(G)$-invariant closed subspace $\mathcal{H}_2$ of $L^2(\widehat{G}, \mu_2, \mathcal{K}_2)$ such that the restriction of $\pi_{\mu_2}^{(n_2)}$ to $\mathcal{H}_2$ is equivalent to $\pi_{\mu_2} = \pi_{\mu_2}^{(1)}$, and such that the map

$$P \circ (\widetilde{T}|_{\mathcal{H}_1}):\ \mathcal{H}_1 \to \mathcal{H}_2$$

is not zero, where $P : L^2(\widehat{G}, \mu_2, \mathcal{K}_2) \twoheadrightarrow \mathcal{H}_2$ is the orthogonal projection. Since $P \circ \widetilde{T}$ intertwines $\pi_{\mu_1}^{(n_1)}$ and $\pi_{\mu_2}^{(n_2)}$ and since the restrictions of $\pi_{\mu_1}^{(n_1)}$ and $\pi_{\mu_2}^{(n_2)}$ to $\mathcal{H}_1$ and $\mathcal{H}_2$ are equivalent to π_{μ_1} and π_{μ_2}, this contradicts Proposition 2.A.9(1).

(2) The proof is an extension of the proof of Proposition 2.A.9(2). Let

$$\widetilde{T}:\ L^2(\widehat{G}, \mu_1, \mathcal{K}_1) \to L^2(\widehat{G}, \mu_2, \mathcal{K}_2)$$

be a bijective linear isometry intertwining $\pi_{\mu_1}^{(n_1)}$ and $\pi_{\mu_2}^{(n_2)}$.

For $\varphi \in L^\infty(\widehat{G}, \mu_i)$ and $\xi \in \mathcal{K}_i$, let $\varphi \otimes \xi$ denote the vector-valued function in $L^2(\widehat{G}, \mu_i, \mathcal{K}_i)$ defined as in Section 1.H by $(\varphi \otimes \xi)(\chi) = \varphi(\chi)\xi$. Fix a unit vector $\xi \in \mathcal{K}_1$ and set

$$F := \widetilde{T}(\mathbf{1}_{\widehat{G}} \otimes \xi) \in L^2(\widehat{G}, \mu_2, \mathcal{K}_2).$$

On the one hand, we have

$$\begin{aligned}\overline{\mathcal{F}}(\mu_1)(g) &= \int_{\widehat{G}} \widehat{g}(\chi) d\mu_1(\chi) = \int_{\widehat{G}} \widehat{g}(\chi) \|\xi\|^2 d\mu_1(\chi) \\ &= \langle m_1(\widehat{g})(\mathbf{1}_{\widehat{G}} \otimes \xi) \mid \mathbf{1}_{\widehat{G}} \otimes \xi \rangle \quad \text{for all } g \in G.\end{aligned}$$

On the other hand, consider the bounded Borel measure μ on $\widehat{G}$, given by

$$d\mu(\chi) = \|F(\chi)\|^2 d\mu_2(\chi).$$

Then, since $\widetilde{T}$ is an isometry, we have

$$\begin{aligned}\overline{\mathcal{F}}(\mu)(g) &= \int_{\widehat{G}} \widehat{g}(\chi) \|F(\chi)\|^2 d\mu_2(\chi) = \langle m_2(\widehat{g}) F \mid F \rangle \\ &= \langle m_2(\widehat{g}) \widetilde{T}(\mathbf{1}_{\widehat{G}} \otimes \xi) \mid \widetilde{T}(\mathbf{1}_{\widehat{G}} \otimes \xi) \rangle \\ &= \langle \widetilde{T}(m_1(\widehat{g})(\mathbf{1}_{\widehat{G}} \otimes \xi)) \mid \widetilde{T}(\mathbf{1}_{\widehat{G}} \otimes \xi) \rangle \\ &= \langle m_1(\widehat{g})(\mathbf{1}_{\widehat{G}} \otimes \xi)) \mid \mathbf{1}_{\widehat{G}} \otimes \xi \rangle \quad \text{for all } g \in G.\end{aligned}$$

Therefore,

$$\overline{\mathcal{F}}(\mu_1) = \overline{\mathcal{F}}(\mu).$$

By injectivity of the Fourier–Stieltjes transform, it follows that

$$d\mu_1(\chi) = d\mu(\chi) = \|F(\chi)\|^2 d\mu_2(\chi).$$

So, μ_1 is absolutely continuous with respect to μ_2.

The same argument, applied to T^{-1} in place of T, shows that μ_2 is absolutely continuous with respect to μ_1. □

Next, we identify the intertwining operators between two representations of the form $\pi_\mu^{(n_1)}$ and $\pi_\mu^{(n_2)}$ for a given probability measure μ on $\widehat{G}$.

PROPOSITION 2.B.3. *Consider a probability measure μ on $\widehat{G}$, two extended positive integers n_1, n_2 in $\overline{\mathbf{N}^*}$, and two Hilbert spaces $\mathcal{K}_1, \mathcal{K}_2$ respectively of dimension n_1, n_2.*

The space of bounded linear operators $L^2(\widehat{G}, \mu, \mathcal{K}_1) \to L^2(\widehat{G}, \mu, \mathcal{K}_2)$ intertwining $\pi_\mu^{(n_1)}$ and $\pi_\mu^{(n_2)}$ consists of the decomposable operators. In particular, $\pi_\mu^{(n_1)}$ and $\pi_\mu^{(n_2)}$ are equivalent if and only if $n_1 = n_2$.

PROOF. For $i = 1, 2$ and $\varphi \in L^\infty(\widehat{G}, \mu)$, denote by $m_i(\varphi)$ the corresponding multiplication operator on $L^2(\widehat{G}, \mu, \mathcal{K}_i)$.

Suppose first that $\widetilde{T}$ is decomposable: $(\widetilde{T}F)(\chi) = T(\chi)F(\chi)$ for some μ-essentially bounded measurable map $T: \widehat{G} \to \mathcal{L}(\mathcal{K}_1, \mathcal{K}_2)$ and for all $F \in L^2(\widehat{G}, \mu, \mathcal{K}_1)$. Then

$$(\widetilde{T}\pi_\mu^{(n_1)}(g)F)(\chi) = T(\chi)\chi(g)F(\chi) = \chi(g)T(\chi)F(\chi) = (\pi_\mu^{(n_2)}(g)\widetilde{T}F)(\chi)$$

for all $g \in G$ and $F \in L^2(\widehat{G}, \mu, \mathcal{K}_1)$, so that $\widetilde{T}$ intertwines $\pi_\mu^{(n_1)}$ and $\pi_\mu^{(n_2)}$.

Conversely, suppose now that $\widetilde{T}$ intertwines $\pi_\mu^{(n_1)}$ and $\pi_\mu^{(n_2)}$, i.e., that

$$\widetilde{T}m_1(\widehat{g}) = m_2(\widehat{g})\widetilde{T} \quad \text{for all } g \in G.$$

We also have

$$\widetilde{T}m_1(\varphi) = m_2(\varphi)\widetilde{T}$$

for all trigonometric polynomials $\varphi \in \mathrm{Trig}(\widehat{G})$, by linearity, and indeed for all $\varphi \in L^\infty(\widehat{G}, \mu)$ because $\mathrm{Trig}(\widehat{G})$ is weak*-dense in $L^\infty(\widehat{G}, \mu)$; see Lemma A.7.7(2). It follows from Theorem 1.H.5 that $\widetilde{T}$ is decomposable. □

REMARK 2.B.4. Consider a representation $\pi_\mu^{(n)}$ of G in $L^2(\widehat{G}, \mu, \mathcal{K})$, as in Construction 2.B.1. On the one hand, the particular case of Proposition 2.B.3 for $\mathcal{K}_1 = \mathcal{K}_2 = \mathcal{K}$ shows that the commuting algebra

$$\pi_\mu^{(n)}(G)' = \{T \in \mathcal{L}(L^2(\widehat{G}, \mu, \mathcal{K})) \mid T\pi_\mu^{(n)}(g) = \pi_\mu^{(n)}(g)T \text{ for all } g \in G\}$$

is the von Neumann algebra of decomposable operators on $L^2(\widehat{G}, \mu, \mathcal{K})$. On the other hand, it follows from the definition (i.e., from $\pi_\mu^{(n)}(g) = m(\widehat{g})$ for all $g \in G$) and from Proposition A.7.7 that the von Neumann algebra $\pi_\mu^{(n)}(G)''$ generated by the image of the representation is the von Neumann algebra of diagonalisable operators on $L^2(\widehat{G}, \mu, \mathcal{K})$. These two von Neumann algebras appear in Example 1.G.5.

Suppose moreover that $n = 1$ and $\mathcal{K} = \mathbf{C}$. Then $\pi_\mu(G)' = \pi_\mu(G)''$, and $\pi_\mu(G)''$ is an abelian von Neumann subalgebra of $\mathcal{L}(L^2(X, \mu))$. With a terminology to be introduced in Section 6.A, the canonical representation π_μ is **multiplicity-free** (see Theorem 6.B.21).

We will need the following two corollaries of Propositions 2.B.3 and2.B.2. Given a probability measure μ on $\widehat{G}$, a Hilbert space $\mathcal{K}$, and a Borel subset B in $\widehat{G}$, we denote by $E_\mu(B)$ the projection operator on $L^2(\widehat{G}, \mu, \mathcal{K}))$ defined by

$$(E_\mu(B)F)(\chi) = \mathbf{1}_B(\chi)F(\chi) \quad \text{for all } F \in L^2(\widehat{G}, \mu, \mathcal{K}),\ \chi \in \widehat{G}.$$

COROLLARY 2.B.5. *Let μ be a probability measure on $\widehat{G}$, and $\mathcal{K}$ a separable Hilbert space of some dimension $n \in \overline{\mathbf{N}^*}$. Let $\pi_\mu^{(n)}$ be as in Construction 2.B.1. For a closed subspace $\mathcal{H}$ of $L^2(\widehat{G}, \mu, \mathcal{K})$, the following properties are equivalent:*

(i) *$\mathcal{H}$ is invariant under the commutant $\pi_\mu^{(n)}(G)'$ of $\pi_\mu^{(n)}(G)$;*
(ii) *$\mathcal{H}$ is the range of a projection $E_\mu(B)$ for $B \in \mathcal{B}(\widehat{G})$.*

PROOF. Assume that $\mathcal{H}$ is the range of a projection $E_\mu(B)$ for $B \in \mathcal{B}(\widehat{G})$. It is clear that $E_\mu(B)$ commutes with every decomposable operator on $L^2(\widehat{G}, \mu, \mathcal{K})$, that is, with every element in $\pi_\mu^{(n)}(G)'$, by Proposition 2.B.3. Therefore, $\mathcal{H}$ is invariant under $\pi_\mu^{(n)}(G)'$.

Conversely, assume that $\mathcal{H}$ is invariant under $\pi_\mu^{(n)}(G)'$. The orthogonal projection P from $L^2(\widehat{G}, \mu, \mathcal{K})$ onto $\mathcal{H}$ therefore belongs to the von Neumann algebra $\pi_\mu^{(n)}(G)''$. As mentioned in Remark 2.B.4, $\pi_\mu^{(n)}(G)''$ coincides with the algebra $\{m(\varphi) \mid \varphi \in L^\infty(\widehat{G}, \mu)\}$ of diagonalisable operators on $L^2(\widehat{G}, \mu, \mathcal{K})$. Therefore, $P = m(\mathbf{1}_B) = E_\mu(B)$ for some $B \in \mathcal{B}(\widehat{G})$. □

NOTATION 2.B.6. For the rest of this section, we denote for each $n \in \overline{\mathbf{N}^*}$ by $\mathcal{K}_n$ a Hilbert space of dimension n, fixed once and for all.

COROLLARY 2.B.7. *Let I be a subset of $\overline{\mathbf{N}^*}$, and $(\mu_n)_{n\in I}$ a sequence of mutually singular probability measures on $\widehat{G}$. Consider the representation*

$$\pi = \bigoplus_{n\in I} \pi_{\mu_n}^{(n)} \quad \textit{of } G \textit{ in the Hilbert space} \quad \mathcal{H} = \bigoplus_{n\in I} L^2(\widehat{G}, \mu_n, \mathcal{K}_n).$$

For every $n \in I$, the orthogonal projection P_n of $\mathcal{H}$ onto $L^2(\widehat{G}, \mu_n, \mathcal{K}_n)$ belongs to the centre of the von Neumann algebra $\pi(G)'$, that is, to $\pi(G)''$.

PROOF. Since $L^2(\widehat{G}, \mu_n, \mathcal{K}_n)$ is invariant under $\pi(G)$, we have $P_n \in \pi(G)'$ for every $n \in I$.

Let $T \in \pi(G)'$. Then, $P_k \circ T \circ P_\ell \in \pi(G)'$ for every $k, \ell \in I$. For $k \neq \ell$, the measures μ_k, μ_ℓ are singular with each other and it follows from Proposition 2.B.2 that $P_k \circ T \circ P_\ell = 0$. Therefore

$$TP_n = \Big(\sum_{m\in I} P_m\Big)TP_n = P_nTP_n = P_nT\Big(\sum_{m\in I} P_m\Big) = P_nT,$$

for every $n \in I$. This shows that $P_n \in \pi(G)''$ for every $n \in I$. □

In the sequel we will use the following notation. Let μ be a probability measure on a Borel space $(X, \mathcal{B})$ and let $B \in \mathcal{B}$ be such that $\mu(B) > 0$. We denote by $\mu|B$ the probability measure on X defined by

$$(\mu|B)(C) = \frac{\mu(C \cap B)}{\mu(B)} \quad \text{for every} \quad C \in \mathcal{B}.$$

Observe that $\mu|B$ is absolutely continuous with respect to μ and that the Radon–Nikodym derivative $\frac{d(\mu|B)}{d\mu}$ vanishes outside B. When X is a disjoint union $\bigsqcup_{n\geq 1} B_n$ of measurable subsets $B_n \in \mathcal{B}$, we have

$$\mu = \sum_{n\geq 1, \mu(B_n)>0} \mu(B_n)\, \mu|B_n.$$

LEMMA 2.B.8. *Let μ be a probability measure on $(X, \mathcal{B})$ and $f : X \to \mathbf{R}_+$ a measurable function such that $\int_X f d\mu = 1$. Set $B_f = \{x \in X \mid f(x) > 0\}$.*

Then $\mu(B_f) > 0$ and $\mu|B_f$ is equivalent to the probability measure $\nu: = f\mu$ on X.

PROOF. We have $\nu(B_f) = \int_{B_f} f(x)d\mu(x) = 1$, and therefore $\mu(B_f) > 0$. For $C \in \mathcal{B}$, we have

$$\nu(C) = \int_X \mathbf{1}_C(x) f(x) d\mu(x) = \int_{C\cap B_f} f(x) d\mu(x).$$

Since $f > 0$ on B_f, it follows that $\nu(C) = 0$ if and only if $\mu(C \cap B_f) = 0$. □

The main result in this chapter is the following theorem, which shows that every representation of G on a separable Hilbert space admits a decomposition in terms of representations of the form $\pi_\mu^{(n)}$, *with uniqueness up to the appropriate equivalence.*

THEOREM 2.B.9 (**Canonical decomposition of representations of abelian groups**). *Let G be a second-countable locally compact abelian group and π a representation of G on a separable Hilbert space $\mathcal{H}$.*

There exist a set I of extended positive integers and a sequence $(\mu_n)_{n\in I}$ of probability measures on $\widehat{G}$ with the following properties:

(1) *the measures μ_n are singular with each other;*
(2) *the representation π is equivalent to the direct sum $\bigoplus_{n\in I} \pi_{\mu_n}^{(n)}$.*

Moreover, this decomposition is unique in the following sense: assume that there exist a set I' of extended positive integers and a sequence $(\mu'_n)_{n\in I'}$ of probability measures on $\widehat{G}$ which are mutually singular with each other, such that π is equivalent to the direct sum $\bigoplus_{n\in I'} \pi_{\mu'_n}^{(n)}$. Then $I = I'$, and μ_n and μ'_n are equivalent for every $n \in I$.

PROOF. Since the theorem is trivial when $\mathcal{H} = \{0\}$, we assume from now on that $\mathcal{H}$ is not 0-dimensional. By Corollary 2.A.7, there exists a (possibly finite) sequence $(\lambda_i)_{i\geq 1}$ of probability measures on $\widehat{G}$ such that π is equivalent to $\bigoplus_{i\geq 1} \pi_{\lambda_i}$. These probability measures are not necessarily mutually singular; the strategy is to replace them by probability measures μ_n which are mutually singular, at the cost of considering representations $\pi_{\mu_n}^{(n)}$ with multiplicities n. Our proof of the existence part of Theorem 2.B.9 is patterned after that of Part (1) of Theorem 3.5.3 in [**Sund–97**].

• *First step.* Choose a sequence of positive real numbers $(c_i)_{i\geq 1}$ such that $\sum_{i\geq 1} c_i = 1$ and define a probability measure μ on $\widehat{G}$ by

$$\mu := \sum_{i\geq 1} c_i \lambda_i.$$

For every $i \geq 1$, note that λ_i is absolutely continuous with respect to μ and set

$$B_i = \left\{ \chi \in \widehat{G} \;\middle|\; \frac{d\lambda_i}{d\mu}(\chi) > 0 \right\} \in \mathcal{B}(\widehat{G}).$$

Note that $\mu(B_i) > 0$ and that the probability measure $\mu|B_i$ is equivalent to $\lambda_i = \frac{d\lambda_i}{d\mu}\mu$ (Lemma 2.B.8). Since the representation π is equivalent to $\bigoplus_{i \geq 1} \pi_{\lambda_i}$, it is also equivalent to $\bigoplus_{i \geq 1} \pi_{\mu|B_i}$ (see Proposition 2.A.2).

Set

$$B := \bigcup_{i \geq 1} B_i \in \mathcal{B}(\widehat{G}).$$

Observe that $\mu(B) = 1$. For $n \in \overline{\mathbf{N}^*}$ and $i \geq 1$, set

$$A_n := \left\{ \chi \in B \;\middle|\; \sum_{j \geq 1} \mathbf{1}_{B_j}(\chi) = n \right\}$$

and

$$B_{i,n} := B_i \cap A_n.$$

In other words, A_n [respectively $B_{i,n}$] is the set of $\chi \in B$ [respectively $\chi \in B_i$] which belong to exactly n of the sets B_j. The A_n's are clearly disjoint, and so are the $B_{i,n}$'s for fixed $i \geq 1$, so that we have disjoint unions

$$B = \bigsqcup_{n \in \overline{\mathbf{N}^*}} A_n \tag{1}$$

and

$$B_i = \bigsqcup_{n \in \overline{\mathbf{N}^*}} B_{i,n} \qquad \text{for all } i \geq 1. \tag{2}$$

(Note that one may have $A_n = \emptyset$ for several values of n: for example, if all B_i coincide, then all but one of the A_n's are empty; similarly, one may have $B_{i,n} = \emptyset$.) If $n \neq n'$, then $B_{i,n}$ and $B_{i,n'}$ are disjoint; however, if $i \neq i'$, the sets $B_{i,n}$ and $B_{i',n}$ need not be disjoint.

Let $i \geq 1$ and $n \in \overline{\mathbf{N}^*}$. For $\ell \in \mathbf{N}$ such that $0 \leq \ell \leq n$ if $n < \infty$ (and no condition on ℓ if $n = \infty$), let

$$B_{i,n,\ell} := \left\{ \chi \in B_{i,n} \;\middle|\; \sum_{j=1}^{n} \mathbf{1}_{B_j}(\chi) = \ell \right\}$$

be the set of these $\chi \in B_{i,n}$ which belong to exactly ℓ of $B_1, B_2, \dots, B_n$. We have disjoint unions

$$B_{i,n} = \bigsqcup_{\ell=0}^{n} B_{i,n,\ell} \tag{3}$$

and, for every $\ell \in \mathbf{N}$ such that $0 \leq \ell \leq n$ if $n < \infty$ (and no condition if $n = \infty$),

$$A_n = \bigsqcup_{i \geq 1} B_{i,n,\ell}. \tag{4}$$

Let us check (4). By definition, the union $\bigcup_{i \geq 1} B_{i,n,\ell}$ is contained in A_n for every $\ell \leq n$. To show the reverse inclusion, let $\chi \in A_n$. Then the set $\{j \geq 1 \mid \chi \in B_j\}$ consists of n indices, say $i_1 < i_2 < \cdots$, a finite sequence with last index i_n if $n < \infty$, and an infinite sequence if $n = \infty$. Fix an integer ℓ with $1 \leq \ell \leq n$ in case $n < \infty$ and $1 \leq \ell$ in case $n = \infty$. Then $\chi \in B_{i_\ell}$ and hence $\chi \in B_{i_\ell,n,\ell}$; moreover $\chi \notin B_{i,n,\ell}$

for every $i \neq i_\ell$. Therefore $A_n = \bigcup_{i \geq 1} B_{i,n,\ell}$; moreover this is a disjoint union, hence (4) holds.

• *Second step.* For a Borel set $C \subset \widehat{G}$ such that $\mu(C) = 0$, we set $\mu|C = 0$. Whenever a representation $\pi_{\mu|C}$ for such a μ-null set C appears below, it stands for the representation on the 0-dimensional Hilbert space.

Let I be the (possibly finite) sequence of those extended positive integers n such that $\mu(A_n) > 0$. For all $n \in I$, define

$$\mu_n := \mu|A_n.$$

Fix $i \geq 1$ and $n \in I$. Define

$$\mu_{i,n} := \mu|B_{i,n}.$$

It follows from (2) and from Proposition 2.A.5 that the representation $\pi_{\mu|B_i}$ is equivalent to $\bigoplus_{n \in I} \pi_{\mu_{i,n}}$. Now consider $\ell \in \mathbf{N}$ such that $0 \leq \ell \leq n$ if $n < \infty$ (and no condition if $n = \infty$); define

$$\mu_{i,n,\ell} := \mu|B_{i,n,\ell}.$$

It follows from (3) and again from Proposition 2.A.5 that $\pi_{\mu_{i,n}}$ is equivalent to $\bigoplus_{\ell \leq n} \pi_{\mu_{i,n,\ell}}$. Therefore,

$$\pi_{\mu|B_i} \quad \text{is equivalent to} \quad \bigoplus_{n \in I} \bigoplus_{\ell \leq n} \pi_{\mu_{i,n,\ell}},$$

for every $i \geq 1$. (When $n = \infty$, the notation $\ell \leq n$ should be understood as $\ell \in \mathbf{N}$.) Since π is equivalent to $\bigoplus_{i \geq 1} \pi_{\mu|B_i}$, it follows that

$$\pi \quad \text{is equivalent to} \quad \bigoplus_{i \geq 1} \bigoplus_{n \in I} \bigoplus_{\ell \leq n} \pi_{\mu_{i,n,\ell}} = \bigoplus_{n \in I} \bigoplus_{\ell \leq n} \bigoplus_{i \geq 1} \pi_{\mu_{i,n,\ell}}.$$

By (4) and Proposition 2.A.5, $\bigoplus_{i \geq 1} \pi_{\mu_{i,n,\ell}}$ is equivalent to π_{μ_n} for *every* $\ell \leq n$. Therefore, $\bigoplus_{\ell \leq n} \bigoplus_{i \geq 1} \pi_{\mu_{i,n,\ell}}$ is equivalent to $\pi_{\mu_n}^{(n)}$. It follows that

$$\pi \quad \text{is equivalent to} \quad \bigoplus_{n \in I} \pi_{\mu_n}^{(n)}.$$

We have thus proved the existence part of the theorem.

• *Third step.* We prove the uniqueness property. By the existence part, we may assume that π is the representation $\bigoplus_{n \in I} \pi_{\mu_n}^{(n)}$ on the Hilbert space

$$\mathcal{H} = \bigoplus_{n \in I} L^2(\widehat{G}, \mu_n, \mathcal{K}_n),$$

where I is a subset of $\overline{\mathbf{N}^*}$, $(\mu_n)_{n \in I}$ is a sequence of mutually singular probability measures on $\widehat{G}$, and $\mathcal{K}_n$ is the fixed Hilbert space of dimension n (see Notation 2.B.6). Set $\mathcal{H}_n := L^2(\widehat{G}, \mu_n, \mathcal{K}_n)$ for $n \in I$.

For a subset J of $\overline{\mathbf{N}^*}$, let

$$\mathcal{H} = \bigoplus_{m \in J} \mathcal{H}'_m$$

be an orthogonal decomposition of $\mathcal{H}$ into $\pi(G)$-invariant closed subspaces $\mathcal{H}'_m$ such that the restriction of π to $\mathcal{H}'_m$ is equivalent to the representation $\pi_{\mu'_m}^{(m)}$ on $L^2(\widehat{G}, \mu'_m, \mathcal{K}_m)$, for a sequence of mutually singular probability measures $(\mu'_m)_{m \in J}$ on $\widehat{G}$. We claim that $J = I$ and that $\mathcal{H}'_n = \mathcal{H}_n$ for every $n \in I$. Once established, this will finish the proof.

For $n \in I$ and $m \in J$, denote by P_n and P'_m the orthogonal projections from $\mathcal{H}$ to $\mathcal{H}_n$ and $\mathcal{H}'_m$.

Fix $n \in I$ and $m \in J$ with $P'_m \circ P_n \neq 0$. By Corollary 2.B.7, $P'_m \circ P_n$ belongs to the centre of the von Neumann algebra $\pi(G)''$. Therefore the orthogonal complement $\mathcal{K}$ of $\ker(P'_m \circ P_n)$ and the closure $\mathcal{K}'$ of the image of $P'_m \circ P_n$ are nonzero closed subspaces of $\mathcal{H}_n$ and $\mathcal{H}'_m$ respectively, which are invariant under $\pi(G)$ as well as under $\pi(G)'$. It follows, on the one hand, that the representations of G on $\mathcal{K}$ and $\mathcal{K}'$ defined by π are equivalent (see Lemma 1.A.5) and, on the other hand, that $\mathcal{K}$ and $\mathcal{K}'$ are the ranges of projections $E_{\mu_n}(B)$ and $E_{\mu'_m}(B')$ for some Borel subsets B and B' of $\widehat{G}$ (see Corollary 2.B.5). We have $\mu_n(B) > 0$ and $\mu'_m(B') > 0$ since $\mathcal{K}$ and $\mathcal{K}'$ are nonzero. This shows that the representations of G on $\mathcal{K}$ and $\mathcal{K}'$ defined by π are equivalent to the representations $\pi^{(n)}_{\mu_n|B}$ and $\pi^{(m)}_{\mu'_m|B'}$ respectively. It follows from Proposition 2.B.2 that $\mu_n|B$ and $\mu'_m|B'$ are equivalent and from Proposition 2.B.3 that $n = m$.

So, we have proved that

$$P'_m \circ P_n = P_n \circ P'_m = 0 \qquad \text{whenever } n \neq m.$$

Since $\mathrm{Id}_{\mathcal{H}} = \sum_{n \in I} P_n = \sum_{m \in J} P'_m$, it follows that $I = J$ and that $\mathcal{H}_n = \mathcal{H}'_n$ for every $n \in I$. □

REMARK 2.B.10. (1) Theorem 2.B.9 gives a complete set of invariants for equivalence classes of representations of G on separable Hilbert spaces: a subset I of $\overline{\mathbf{N}^*}$ of multiplicities and a sequence $([\mu_n])_{n \in I}$ of equivalence classes of mutually singular probability measures on $\widehat{G}$.

(2) In the special case $G = \mathbf{R}$, Theorem 2.B.9 is equivalent to the classical Hahn–Hellinger Theorem which gives a complete set of invariants for the unitary equivalence classes of selfadjoint operators on separable Hilbert spaces. See [**Halm–51**]; see also [**Sund–97**] for a proof using a C*-algebra approach.

(3) Anticipating on a notion to be introduced in Chapter 6, we observe that an LCA group is a group of type I (see Theorem 6.E.19). Theorem 2.B.9 is a special case of a more general decomposition result for representations of such groups (Theorem 6.D.7).

The case for which the set I of extended positive integers in Theorem 2.B.9 consists of a single number will be of special interest for us (see Proposition 2.F.3).

DEFINITION 2.B.11. A representation π of a second-countable LCA group G in a separable Hilbert space $\mathcal{H}$ is said to be **homogeneous**, or more precisely n-**homogeneous**, if the set I of extended positive integers attached to π by Theorem 2.B.9(1) consists of a single number $n \in \overline{\mathbf{N}^*}$; equivalently, if π is equivalent to the representation $\pi_\mu^{(n)}$, for some probability measure μ on $\widehat{G}$ and some extended positive integer $n \in \overline{\mathbf{N}^*}$.

REMARK 2.B.12. A representation π of G is homogeneous if and only if its commutant $\pi(G)'$ is a homogeneous von Neumann algebra, in the sense of Définition 1 in [**Dixm–vN**, Chap. III, § 3].

2.C. The SNAG Theorem

Let G be a second-countable locally compact abelian group. In Sections 2.A and 2.B, we described the representations of G in terms of probability measures on $\widehat{G}$ and multiplicities. As with the spectral theorem for a selfadjoint operator on a Hilbert space (see Remark 2.A.8), it is convenient to give another description of representations of G, in terms of projection-valued measures; for these, see the reminder in Appendix A.9.

EXAMPLE 2.C.1 (**Projection-valued measures on** $\widehat{G}$). Let $\mathcal{B}(\widehat{G})$ be the σ-algebra of Borel subsets of the dual group $\widehat{G}$. Let μ be a probability measure on $\widehat{G}$.

(1) For every $B \in \mathcal{B}(\widehat{G})$, define a projection $E_\mu(B)$ on $L^2(\widehat{G},\mu)$ by

$$(E_\mu(B)f)(\chi) = \mathbf{1}_B(\chi)f(\chi) \quad \text{for all } f \in L^2(\widehat{G},\mu),\ \chi \in \widehat{G}.$$

The map

$$E_\mu : B \mapsto E_\mu(B)$$

is easily checked to be a projection-valued measure on $\widehat{G}$ and $L^2(\widehat{G},\mu)$; see Appendix A.9.

Let π_μ be the canonical representation of G on $L^2(\widehat{G},\mu)$, as in Construction 2.A.1; recall that, for all $g \in G$ and $f_1, f_2 \in L^2(\widehat{G},\mu)$, we have

$$\langle \pi_\mu(g)f_1 \mid f_2\rangle = \int_{\widehat{G}} (\pi_\mu(g)f_1)(\chi)\overline{f_2(\chi)}d\mu(\chi) = \int_{\widehat{G}} \chi(g)f_1(\chi)\overline{f_2(\chi)}d\mu(\chi).$$

Therefore

$$\pi_\mu(g) = \int_{\widehat{G}} \chi(g)dE_\mu(\chi) \quad \text{for all } g \in G.$$

The measure μ_{f_1,f_2} on $\widehat{G}$ defined by E_μ and $f_1, f_2 \in L^2(\widehat{G},\mu)$, see Appendix A.9, is given by

$$\mu_{f_1,f_2}(B) = \int_B f_1(\chi)\overline{f_2}(\chi)d\mu(\chi) \quad \text{for all } B \in \mathcal{B}(\widehat{G}),$$

i.e., by $\mu_{f_1,f_2} = f_1\overline{f_2}\mu$. Note that μ_{f_1,f_2} is a complex measure.

(2) More generally, let $n \in \overline{\mathbf{N}^*}$ and $\mathcal{K}$ a Hilbert space of dimension n. Let $\pi_\mu^{(n)}$ be the associated representation of G on $L^2(\widehat{G},\mu,\mathcal{K})$, as in Construction 2.B.1. A projection-valued measure

$$E_\mu^{(n)} : \mathcal{B} \to \mathrm{Proj}(L^2(\widehat{G},\mu,\mathcal{K}))$$

is defined by the same formula (which already appeared before Corollary 2.B.5):

$$(E_\mu^{(n)}(B)F)(\chi) = \mathbf{1}_B(\chi)F(\chi) \quad \text{for all } B \in \mathcal{B},\ F \in L^2(\widehat{G},\mu,\mathcal{K}),\ \chi \in \widehat{G}.$$

One shows as above that

$$\pi_\mu^{(n)}(g) = \int_{\widehat{G}} \chi(g)dE_\mu^{(n)}(\chi) \quad \text{for all } g \in G.$$

Observe that, for $B \in \mathcal{B}(\widehat{G})$, we have $E_\mu^{(n)}(B) = 0$ if and only if $\mu(B) = 0$.

Let $U : \mathcal{H} \to \mathcal{H}'$ be an isomorphism between two Hilbert spaces $\mathcal{H}$ and $\mathcal{H}'$ and let $E : \mathcal{B}(\widehat{G}) \to \mathrm{Proj}(\mathcal{H})$ be a projection-valued measure defined on $\widehat{G}$ and $\mathcal{H}$. As one immediately checks, the map

$$\mathcal{B}(\widehat{G}) \to \mathrm{Proj}(\mathcal{H}'), \qquad B \mapsto UE(B)U^{-1}$$

is a projection-valued measure defined on $\widehat{G}$ and $\mathcal{H}'$. This gives rise to a notion of equivalence of projection-valued measures.

DEFINITION 2.C.2. Let $\mathcal{H}, \mathcal{H}'$ be Hilbert spaces. Two projection-valued measures

$$E : \mathcal{B}(\widehat{G}) \to \text{Proj}(\mathcal{H}) \quad \text{and} \quad E' : \mathcal{B}(\widehat{G}) \to \text{Proj}(\mathcal{H}')$$

are **equivalent** if there exists a Hilbert space isomorphism $U : \mathcal{H} \to \mathcal{H}'$ such that

$$E'(B) = UE(B)U^{-1} \quad \text{for all } B \in \mathcal{B}(\widehat{G}).$$

The following result shows that there is a bijective correspondence between equivalence classes of representations of the LCA group G and equivalence classes of projection-valued measures on $\widehat{G}$. The name "SNAG Theorem" refers to Stone, Naimark, Ambrose, and Godement [**Ston–30, Naim–43, Ambr–44, Gode–44**].

THEOREM 2.C.3 (**SNAG Theorem**). *Let G be a second-countable locally compact abelian group.*

(1) *Let $\mathcal{H}$ be a Hilbert space and E a projection-valued measure on $\widehat{G}$ and $\mathcal{H}$. The formula*

$$\pi_E(g) := \int_{\widehat{G}} \chi(g) dE(\chi) \quad \text{for all } g \in G$$

defines a representation π_E of G on $\mathcal{H}$.
For every $\xi, \eta \in \mathcal{H}$, we have

$$\langle \pi_E(g)\xi \mid \eta \rangle = \overline{\mathcal{F}}(\mu_{\xi,\eta})(g) \quad \text{for all } g \in G,$$

where $\mu_{\xi,\eta}$ is the complex Borel measure on $\widehat{G}$ defined by E and ξ, η.

(2) *Let $\mathcal{H}$ be a Hilbert space and E a projection-valued measure on $\widehat{G}$ and $\mathcal{H}$. A representation $(\pi', \mathcal{H}')$ of G is equivalent to the representation π_E as in* (1) *if and only if $\pi' = \pi_{E'}$ for an equivalent projection-valued measure E' on $\widehat{G}$ and $\mathcal{H}'$.*

(3) *Let π be a representation of G on a Hilbert space $\mathcal{H}$. There exists a unique projection-valued measure E on $\widehat{G}$ and $\mathcal{H}$ such that $\pi = \pi_E$.*

(4) *Let $(\pi, \mathcal{H})$ and E be as in* (3)*. The commutant $\pi(G)'$ of $\pi(G)$ coincides with the commutant of the set $\{E(B) \mid B \in \mathcal{B}(\widehat{G})\}$. In particular, $E(B)$ belongs to the von Neumann algebra $\pi(G)''$ and the range of $E(B)$ is $\pi(G)$-invariant, for every $B \in \mathcal{B}(\widehat{G})$.*

(5) *The map $E \mapsto \pi_E$ from* (1) *induces a bijection between equivalence classes of projection-valued measures on $\widehat{G}$ and equivalence classes of representations of G.*

PROOF. (1) Let $E : \mathcal{B}(\widehat{G}) \to \text{Proj}(\mathcal{H})$ be a projection-valued measure. The map

$$f \mapsto \int_{\widehat{G}} f(\chi) dE(\chi)$$

is a $*$-homomorphism from $\text{Bor}^b(\widehat{G})$ to $\mathcal{L}(\mathcal{H})$; for the involutive algebra $\text{Bor}^b(\widehat{G})$ of bounded Borel functions from $\widehat{G}$ to $\mathbf{C}$ and for this $*$-homomorphism, see Construction A.9.1. In particular,

$$\pi_E(g) := \int_{\widehat{G}} \widehat{g}(\chi) dE(\chi) = \int_{\widehat{G}} \chi(g) dE(\chi)$$

defines a unitary operator on $\mathcal{H}$ for every $g \in G$ and

$$\pi_E(g_1)\pi_E(g_2) = \int_{\widehat{G}} \chi(g_1)\chi(g_2)dE(\chi) = \int_{\widehat{G}} \chi(g_1g_1)dE(\chi) = \pi_E(g_1g_2)$$

for all $g_1, g_2 \in G$. So, $g \mapsto \pi_E(g)$ is a homomorphism from G to the unitary group of $\mathcal{H}$. For every $\xi, \eta \in \mathcal{H}$, we have

$$\langle \pi_E(g)\xi \mid \eta\rangle = \int_{\widehat{G}} \widehat{g}(\chi)d\mu_{\xi,\eta}(\chi) = \overline{\mathcal{F}}(\mu_{\xi,\eta})(g)$$

for all $g \in G$. Since $\overline{\mathcal{F}}(\mu_{\xi,\eta})$ is a continuous function on $\widehat{G}$, it follows that $g \mapsto \pi_E(g)$ is a representation of G.

(2) Let $\mathcal{H}$ be a Hilbert space and E a projection-valued measure on $\widehat{G}$ and $\mathcal{H}$. Let $U : \mathcal{H} \to \mathcal{H}'$ be a Hilbert space isomorphism and let E' be the equivalent projection-valued measure on $\widehat{G}$ and $\mathcal{H}'$ defined by U.

Let $\xi, \eta \in \mathcal{H}'$. Let $\mu_{\xi,\eta}$ and $\mu_{U^{-1}\xi, U^{-1}\eta}$ be the bounded measures on $\widehat{G}$ defined by E' and E and associated to (ξ, η) and $(U^{-1}\xi, U^{-1}\eta)$, respectively. Since $E'(B) = UE(B)U^{-1}$ for every $B \in \mathcal{B}$, we have $\mu_{\xi,\eta} = \mu_{U^{-1}\xi,U^{-1}\eta}$. It follows with (1) that

$$\langle \pi_{E'}(g)\xi \mid \eta\rangle = \overline{\mathcal{F}}(\mu_{\xi,\eta}) = \overline{\mathcal{F}}(\mu_{U^{-1}\xi,U^{-1}\eta}) = \langle \pi_E(g)U^{-1}\xi \mid U^{-1}\eta\rangle$$

and hence that

$$\pi_{E'}(g) = U\pi_E(g)U^{-1} \quad \text{for every} \quad g \in G.$$

This shows, at once, that $\pi_{E'}$ is equivalent to π_E for every projection-valued measure E' which is equivalent to E and that every representation which is equivalent to π_E is of the form $\pi_{E'}$ for a projection-valued measure E' which is equivalent to E.

(3) We first prove the uniqueness of E. Let $E, E' : \mathcal{B}(\widehat{G}) \to \mathrm{Proj}(\mathcal{H})$ be two projection-valued measures on $\widehat{G}$ and $\mathcal{H}$ such that $\pi = \pi_E$ and $\pi = \pi_{E'}$. Let $\xi, \eta \in \mathcal{H}$. Denote by $\mu_{\xi,\eta}$ and $\mu'_{\xi,\eta}$ the associated bounded measures on $\widehat{G}$ defined by E and E', respectively.

By (1), we have

$$\overline{\mathcal{F}}(\mu_{\xi,\eta})(g) = \langle \pi_E(g)\xi \mid \eta\rangle = \langle \pi_{E'}(g)\xi \mid \eta\rangle = \overline{\mathcal{F}}(\mu'_{\xi,\eta})(g)$$

for all $g \in G$. By injectivity of the Fourier–Stieltjes transform, this implies $\mu_{\xi,\eta} = \mu'_{\xi,\eta}$. Since this holds for all $\xi, \eta \in \mathcal{H}$, it follows that $E = E'$.

Next, we show the existence of a projection-valued measure with the desired property.

By Corollary 2.A.7 there exists a family $(\mu_i)_{i\in I}$ of probability measures on $\mathcal{B}(\widehat{G})$ such that π is equivalent to the direct sum $\bigoplus_{i\in I} \pi_{\mu_i}$ (for the case of $\mathcal{H}$ not separable, see the note at the end of the proof of Corollary 2.A.7). It follows from (2) that it suffices to prove the existence of a projection-valued measure $E : \mathcal{B}(\widehat{G}) \to \mathrm{Proj}(\mathcal{H})$ with the desired property in the case of

$$\mathcal{H} = \bigoplus_{i\in I} L^2(\widehat{G}, \mu_i) \quad \text{and} \quad \pi = \bigoplus_{i\in I} \pi_{\mu_i}.$$

For every $i \in I$, let

$$E_{\mu_i} : \mathcal{B}(\widehat{G}) \to \mathrm{Proj}(L^2(\widehat{G}, \mu_i))$$

be the projection-valued measure as in Example 2.C.1. As noted there, we have

$$\pi_{\mu_i}(g) = \int_{\widehat{G}} \chi(g) dE_{\mu_i}(\chi) \quad \text{for all } g \in G.$$

So, with the notation as in (1), we have $\pi_{\mu_i} = \pi_{E_{\mu_i}}$.

Let $E : \mathcal{B}(\widehat{G}) \to \mathrm{Proj}(\mathcal{H})$ be the direct sum of the E_{μ_i}'s:

$$E(B) = \bigoplus_{i \in I} E_{\mu_i}(B) \quad \text{for all } B \in \mathcal{B}(\widehat{G}).$$

Then E is a projection-valued measure and

$$\pi(g) = \bigoplus_{i \in I} \int_{\widehat{G}} \chi(g) dE_{\mu_i}(\chi) = \int_{\widehat{G}} \chi(g) d(\bigoplus_{i \in I} E_{\mu_i})(\chi) = \int_{\widehat{G}} \chi(g) dE(\chi),$$

for all $g \in G$. So, $\pi = \pi_E$.

(4) Let $T \in \mathcal{L}(\mathcal{H})$. For $g \in G$ and $\xi, \eta \in \mathcal{H}$, we have, by (1),

$$\langle \pi(g) T\xi \mid \eta \rangle = \langle \pi_E(g) T\xi \mid \eta \rangle = \overline{\mathcal{F}}(\mu_{T\xi,\eta})(g)$$

and

$$\langle T\pi(g)\xi \mid \eta \rangle = \langle \pi_E(g)\xi \mid T^*\eta \rangle = \overline{\mathcal{F}}(\mu_{\xi,T^*\eta})(g).$$

Therefore $T \in \pi(G)'$ if and only if $\overline{\mathcal{F}}(\mu_{T\xi,\eta})(g) = \overline{\mathcal{F}}(\mu_{\xi,T^*\eta})(g)$ for every $g \in G$ and $\xi, \eta \in \mathcal{H}$. By injectivity of the Fourier–Stieltjes transform, it follows that $T \in \pi(G)'$ if and only if $\mu_{T\xi,\eta} = \mu_{\xi,T^*\eta}$ for every $\xi, \eta \in \mathcal{H}$, hence if and only if

$$\langle E(B) T\xi \mid \eta \rangle = \langle E(B)\xi \mid T^*\eta \rangle = \langle TE(B)\xi \mid \eta \rangle$$

for all $B \in \mathcal{B}(\widehat{G})$ and $\xi, \eta \in \mathcal{H}$. Therefore, $T \in \pi(G)'$ if and only $TE(B) = E(B)T$ for every $B \in \mathcal{B}(\widehat{G})$.

(5) The claim follows immediately from (2) and (3). □

Next, we show how the projection-valued measures associated with a representation of G and with its conjugate by an automorphism are related.

Recall from Remark 1.A.17 that $\mathrm{Aut}(G)$ acts on the right by automorphisms of $\widehat{G}$:

$$(\chi, \theta) \mapsto \chi^\theta \quad \text{for } \chi \in \widehat{G},\ \theta \in \mathrm{Aut}(G),$$

with $\chi^\theta = \chi \circ \theta$.

Corollary 2.C.4. *Let $(\pi, \mathcal{H})$ be a representation of G, and $\theta \in \mathrm{Aut}(G)$. Let $E : \mathcal{B}(\widehat{G}) \to \mathrm{Proj}(\mathcal{H})$ be the projection-valued measure associated with π.*

Then the projection-valued measure associated with the conjugate representation π^θ is the map $E^\theta : \mathcal{B}(\widehat{G}) \to \mathrm{Proj}(\mathcal{H})$ defined by

$$E^\theta(B) = E(B^{\theta^{-1}}) \quad \textit{for } B \in \mathcal{B}(\widehat{G}),$$

where $B^{\theta^{-1}} = \{\chi^{\theta^{-1}} \mid \chi \in B\}$.

Proof. One immediately checks that E^θ is indeed a projection-valued measure on $\widehat{G}$ and $\mathcal{H}$.

Let $\xi, \eta \in \mathcal{H}$, and let $\mu_{\xi,\eta}$ and $\nu_{\xi,\eta}$ be the associated bounded measures on $\widehat{G}$ defined by E and E^θ. Denote by $\theta_*(\mu_{\xi,\eta})$ the measure on $\widehat{G}$ which is the image of $\mu_{\xi,\eta}$ under the map

$$\widehat{G} \to \widehat{G}, \ \ \chi \mapsto \chi^\theta.$$

For every $B \in \mathcal{B}(\widehat{G})$, we have

$$\nu_{\xi,\eta}(B) = \langle E^\theta(B)\xi \mid \eta\rangle = \langle E(B^{\theta^{-1}})\xi \mid \eta\rangle = \mu_{\xi,\eta}(B^{\theta^{-1}}) = \theta_*(\mu_{\xi,\eta})(B).$$

Therefore, we have

$$\nu_{\xi,\eta} = \theta_*(\mu_{\xi,\eta}).$$

It follows that, for every $g \in G$, we have

$$\begin{aligned}\langle \pi^\theta(g)\xi \mid \eta\rangle &= \langle \pi(\theta(g))\xi \mid \eta\rangle = \int_{\widehat{G}} \widehat{\theta(g)}(\chi) d\mu_{\xi,\eta}(\chi)\\ &= \int_{\widehat{G}} \widehat{g}(\chi^\theta) d\mu_{\xi,\eta}(\chi) = \int_{\widehat{G}} \widehat{g}(\chi) d\theta_*(\mu_{\xi,\eta})(\chi) = \int_{\widehat{G}} \widehat{g}(\chi) d\nu_{\xi,\eta}(\chi).\end{aligned}$$

Therefore, we have

$$\pi^\theta(g) = \int_{\widehat{G}} \chi(g) dE^\theta(\chi) \qquad \text{for all } g \in G$$

and the uniqueness part of the SNAG Theorem shows that E^θ is the projection-valued measure associated with π^θ. □

2.D. Containment and weak containment in terms of projection-valued measures

Let G be a second-countable LCA group, $\mathcal{H}$ a Hilbert space, π a representation of G on $\mathcal{H}$, and E the corresponding projection-valued measure on $\widehat{G}$ and $\mathcal{H}$. Containment and weak containment of unitary characters of G in π can be expressed neatly in terms of the atoms and of the support of E (atoms and supports are defined in Appendix A.9).

PROPOSITION 2.D.1. *Let $\mathcal{H}$ be a Hilbert space, π a representation of G on $\mathcal{H}$, and E the projection-valued measure associated with π as in the SNAG Theorem 2.C.3. Let $\chi_0 \in \widehat{G}$. Set*

$$\mathcal{H}^{\chi_0} = \{\xi \in \mathcal{H} \mid \pi(g)\xi = \chi_0(g)\xi \text{ for all } g \in G\}.$$

(a) *The following properties are equivalent:*
  (i) *χ_0 is contained in π, i.e., $\mathcal{H}^{\chi_0} \neq \{0\}$;*
  (ii) *χ_0 is an atom of E;*
moreover, $E(\{\chi_0\})$ is the orthogonal projection on $\mathcal{H}^{\chi_0}$.

(b) *The following properties are equivalent:*
  (i) *χ_0 is weakly contained in π;*
  (ii) *χ_0 belongs to the support of E.*

PROOF. (a) Assume that χ_0 is contained in π. Let ξ be a unit vector in $\mathcal{H}^{\chi_0}$ and μ_ξ the probability measure on $\widehat{G}$ associated with E and ξ. For every $g \in G$, we have

$$\chi_0(g) = \langle \pi(g)\xi \mid \xi\rangle = \int_{\widehat{G}} \chi(g) d\mu_\xi(\chi);$$

this means that

$$\overline{\mathcal{F}}(\delta_{\chi_0}) = \overline{\mathcal{F}}(\mu_\xi)$$

and hence that $\mu_\xi = \delta_{\chi_0}$ by injectivity of the Fourier–Stieltjes transform. So, we have

$$\langle E(\{\chi_0\})\xi \mid \xi\rangle = \mu_\xi(\{\chi_0\}) = 1.$$

Therefore $E(\{\chi_0\})\xi = \xi$, that is, ξ is in the range of $E(\{\chi_0\})$, and in particular $E(\{\chi_0\}) \neq \{0\}$.

Assume now that $E(\{\chi_0\}) \neq 0$. Let ξ be a unit vector in the range of the projection $E(\{\chi_0\})$. Since $\langle E(\{\chi_0\})\xi \mid \xi\rangle = 1$, the probability measure on $\widehat{G}$ associated with E and ξ is the Dirac measure δ_{χ_0}. Therefore, for every $g \in G$, we have

$$\langle \pi(g)\xi \mid \xi\rangle = \int_{\widehat{G}} \chi(g) d\delta_{\chi_0}(\chi) = \chi_0(g).$$

By the equality case of the Cauchy–Schwarz inequality, we have

$$\pi(g)\xi = \chi_0(g)\xi \quad \text{for all } g \in G,$$

that is, $\xi \in \mathcal{H}^{\chi_0}$, and so $\mathcal{H}^{\chi_0} \neq \{0\}$.

Observe that we have shown in the course of the proof that $\mathcal{H}^{\chi_0}$ coincides with the range of $E(\{\chi_0\})$.

(b) Assume that χ_0 is weakly contained in π. Then there exists a sequence $(\xi_i)_i$ of unit vectors in $\mathcal{H}$ such that

$$\lim_i \langle \pi(g)\xi_i \mid \xi_i\rangle = \chi_0(g)$$

uniformly on compact subsets of G (see [**BeHV–08**, F.1.4]). Denote by μ_i the probability measure on $\widehat{G}$ associated with E and ξ_i (with the notation of Appendix A.9, μ_i should be μ_{ξ_i}). We therefore have

$$(*) \qquad \lim_i \overline{\mathcal{F}}(\mu_i)(g) = \lim_i \int_{\widehat{G}} \chi(g) d\mu_i(\chi) = \chi_0(g)$$

uniformly on compact subsets of G.

Let U be an arbitrary compact neighbourhood of χ_0. We claim that $\lim_i \mu_i(U) = 1$; once proved, it will follow that

$$\mu_i(U) = \langle E(U)\xi_i \mid \xi_i\rangle \neq 0$$

for i large enough and therefore $E(U) \neq 0$, showing that χ_0 is in the support of E.

Denote by μ_G a Haar measure on G. As is well known (see Theorem (31.34) in [**HeRo–70**]), there exists a function $\varphi : G \to [0,1]$ in $L^1(G, \mu_G)$ with the following properties: $\overline{\mathcal{F}}(\varphi)(\chi_0) = 1$, $\overline{\mathcal{F}}(\varphi)(\chi) \leq 1$ for all $\chi \in U$, and $\overline{\mathcal{F}}(\varphi)(\chi) = 0$ for all $\chi \in G \smallsetminus U$.

For every probability measure μ on $\widehat{G}$, we have, by Fubini's Theorem (which can be applied, because $\widehat{G}$ is second-countable, and therefore μ and μ_G are σ-finite measures),

$$\begin{aligned}
\int_{\widehat{G}} \overline{\mathcal{F}}(\varphi)(\chi) d\mu(\chi) &= \int_{\widehat{G}} \int_G \chi(g)\varphi(g) d\mu_G(g) d\mu(\chi) \\
&= \int_G \varphi(g) \left(\int_{\widehat{G}} \chi(g) d\mu(\chi) \right) d\mu_G(g) \\
&= \int_G \varphi(g) \overline{\mathcal{F}}(\mu)(g) d\mu_G(g).
\end{aligned}$$

As

$$|\varphi(g)\overline{\mathcal{F}}(\mu_i)(g)| \leq |\varphi(g)| \quad \text{for all } g \in G,$$

it follows from Lebesgue's dominated convergence theorem that

$$\begin{aligned}\lim_i \int_{\widehat{G}} \overline{\mathcal{F}}(\varphi)(\chi) d\mu_i(\chi) &= \int_{\widehat{G}} \lim_i \varphi(g) \overline{\mathcal{F}}(\mu_i)(g) d\mu_G(g) \\ &= \int_G \varphi(g)\chi_0(g) d\mu_G(g) = \overline{\mathcal{F}}(\varphi)(\chi_0) = 1.\end{aligned}$$

Since $\overline{\mathcal{F}}(\varphi) \leq \mathbf{1}_U$ and hence

$$\int_{\widehat{G}} \overline{\mathcal{F}}(\varphi)(\chi) d\mu_i(\chi) \leq \int_{\widehat{G}} \mathbf{1}_U(\chi) d\mu_i(\chi) = \mu_i(U) \leq 1,$$

it follows that $\lim_i \mu_i(U) = 1$ and the claim is proved.

Conversely, assume that $E(U) \neq 0$ for every neighbourhood U of χ_0. Let $(U_i)_i$ be a sequence of decreasing open neighbourhoods of χ_0 such that

$$\bigcap_i U_i = \{\chi_0\}.$$

For every i, let ξ_i be a unit vector in the range of the projection $E(U_i)$. Since

$$E(B)\xi_i = E(B)E(U_i)\xi_i = E(B \cap U_i)\xi_i = 0$$

for every $B \in \mathcal{B}(\widehat{G})$ such that $B \cap U_i = \emptyset$, the support of the corresponding probability measure $\mu_i := \mu_{\xi_i}$ on $\widehat{G}$ is contained in U_i.

Let K be a compact subset of G and $\varepsilon > 0$. Then the family $\{\widehat{g} \mid g \in K\} \subset C(\widehat{G})$ is equicontinuous and we therefore have, for i large enough,

$$|\widehat{g}(\chi) - \widehat{g}(\chi_0)| \leq \varepsilon \quad \text{for all } \chi \in U_i \text{ and } g \in K.$$

Since the support of μ_i is contained in U_i, it follows that, for i large enough,

$$\left| \int_{\widehat{G}} \chi(g) d\mu_i(\chi) - \chi_0(g) \right| = \left| \int_{\widehat{G}} \widehat{g}(\chi) d\mu_i(\chi) - \widehat{g}(\chi_0) \right| \leq \int_{U_i} |\widehat{g}(\chi) - \widehat{g}(\chi_0)| d\mu_i(\chi) \leq \varepsilon$$

for all $g \in K$. This shows that

$$\lim_i \langle \pi(g)\xi_i \mid \xi_i \rangle = \lim_i \int_{\widehat{G}} \chi(g) d\mu_i(\chi) = \chi_0(g)$$

uniformly on K. Therefore, χ_0 is weakly contained in π. □

For later reference, we note the following consequence of Proposition 2.D.1.

Corollary 2.D.2. *Let G, $\mathcal{H}$, π, and E be as in Proposition* 2.D.1. *Assume that the support of E is a finite subset $\{\chi_1, \dots, \chi_n\}$ of $\widehat{G}$. For $i \in \{1, \dots, n\}$, define $\mathcal{H}^{\chi_i}$ as in Proposition* 2.D.1.

Then $\mathcal{H}^{\chi_i} \neq \{0\}$ for every $i \in \{1, \dots, n\}$, and $\mathcal{H}$ decomposes as the direct sum

$$\mathcal{H} = \mathcal{H}^{\chi_1} \oplus \cdots \oplus \mathcal{H}^{\chi_n}.$$

In particular, π is equivalent to $k_1\chi_1 \oplus \cdots \oplus k_n\chi_n$, for multiplicities $k_1, \dots, k_n \in \{1, 2, 3, \dots, \infty\}$.

Proof. The complement U of $\{\chi_1, \dots, \chi_n\}$ in $\widehat{G}$ is an open set with $E(U) = 0$. For $i \in \{1, \dots, n\}$, we have $E(\{\chi_i\}) \neq 0$, since χ_i is in the support of E. Moreover, by the proof of Proposition 2.D.1, $E(\{\chi_i\})$ is the orthogonal projection on $\mathcal{H}^{\chi_i}$. Since

$$\mathrm{Id}_{\mathcal{H}} = E(U) + E(\{\chi_1, \dots, \chi_n\}) = E(\{\chi_1\}) \oplus \cdots \oplus E(\{\chi_n\}),$$

the claim follows. □

REMARK 2.D.3. Let π be a representation of G. By the existence and the uniqueness part of Theorem 2.B.9, there exists a *unique* sequence $([\mu_k])_k$ of equivalence classes of probability measures μ_k on $\widehat{G}$ attached to π.

For every k, denote by A_k the set of atoms and by S_k the support of μ_k. Obviously, A_k and S_k only depend on the class of μ_k. Let E be the projection-valued measure associated with π. As is easily checked, the set of atoms of E is $\bigcup_k A_k$ and the support of E is the closure of $\bigcup_k S_k$.

2.E. Canonical decomposition of projection-valued measures

Let G be a second-countable locally compact abelian group. Theorem 2.B.9 shows that every representation of G on a separable Hilbert space admits a canonical decomposition as a direct sum of representations of the form $\pi_{\mu_k}^{(n_k)}$. Using the SNAG Theorem, we will reformulate this result in terms of projection-valued measures: every projection-valued measure on $\widehat{G}$ is equivalent, in a canonical way, to a direct sum of projection-valued measures associated with representations of the form $\pi_{\mu_k}^{(n_k)}$.

THEOREM 2.E.1 (**Canonical decomposition of projection-valued measures of abelian groups**). *Let G be a second-countable locally compact abelian group. Let*

$$E : \mathcal{B}(\widehat{G}) \to \mathrm{Proj}(\mathcal{H})$$

be a projection-valued measure on $\widehat{G}$ and a separable Hilbert space $\mathcal{H}$. Denote by $\mathcal{M} \subset \mathcal{L}(\mathcal{H})$ the von Neumann algebra $\{E(B) \mid B \in \widehat{G}\}''$ generated by the image of E.

There exist a set $I \subset \overline{\mathbf{N}^*}$ of extended positive integers, a sequence $(P_n)_{n\in I}$ of mutually orthogonal projections in $\mathcal{M}$, and a sequence $(\mu_n)_{n\in I}$ of mutually singular probability measures on $\widehat{G}$ with the following properties:

(1) *$E = \bigoplus_{n\in I} E_n$, where E_n is the projection-valued measure $E_n = P_n \circ E$.*
(2) *For every $n \in I$, the projection-valued measure E_n is equivalent to the projection-valued measure $E_{\mu_n}^{(n)}$ as in Example 2.C.1.*

Moreover, this decomposition is unique in the following sense: assume that there exists a subset $J \subset \overline{\mathbf{N}^*}$, a sequence $(Q_m)_{m\in J}$ of mutually orthogonal projections in $\mathcal{M}$, and a sequence $(\mu'_m)_{m\in J}$ of mutually singular probability measures on $\widehat{G}$, with properties analogous to (1) and (2). Then $J = I$, and $Q_n = P_n$, and μ_n and μ'_n are equivalent for every $n \in I$.

PROOF. Let π be the representation of G on $\mathcal{H}$ defined by E. In view of Theorem 2.B.9, we can assume that

$$\pi = \bigoplus_{n\in I} \pi_{\mu_n}^{(n)}$$

for a subset I of $\overline{\mathbf{N}^*}$ and a sequence $(\mu_n)_{n\in I}$ of mutually singular probability measures on $\widehat{G}$. So, we have

$$\mathcal{H} = \bigoplus_{n\in I} L^2(\widehat{G}, \mu_n, \mathcal{K}_n),$$

where $\mathcal{K}_n$ is a Hilbert space of dimension n.

For every $n \in I$, denote by P_n the orthogonal projection of $\mathcal{H}$ onto $L^2(\widehat{G}, \mu_n, \mathcal{K}_n)$. The mutually orthogonal projections P_n belong to $\pi(G)''$, by Corollary 2.B.7. By Theorem 2.C.3(4), we have $\mathcal{M} = \pi(G)''$, and hence $P_n \in \mathcal{M}$ for every $n \in I$.

We claim that the sequence $(P_n)_{n\in I}$ satisfies Properties (1) and (2) of the theorem.

(1) Since $\sum_{n\in I} P_n = I$, we have $E = \bigoplus_{n\in I} E_n$ for $E_n = P_n \circ E$.

(2) It is clear that E_n, restricted to $L^2(\widehat{G}, \mu_n, \mathcal{K}_n)$, coincides with the projection-valued measure $E^{(n)}_{\mu_n}$ associated with $\pi^{(n)}_{\mu_n}$, for every $n \in I$.

It remains to check the uniqueness assertion. Let $J \subset \overline{\mathbf{N}^*}$, $(Q_m)_{m\in J}$ be a sequence of mutually orthogonal projections in $\mathcal{M}$, and $(\mu'_m)_{m\in I}$ a sequence of mutually singular probability measures on $\widehat{G}$, with properties (1) and (2). As π is determined by E, it follows from the uniqueness part of Theorem 2.B.9 that $J = I$ and that μ'_n is equivalent to μ_n for every $n \in I$.

For every $n \in I$, the range of Q_n is $\pi(G)$-invariant; indeed, since G is abelian, we have $\mathcal{M} = \pi(G)'' \subset \pi(G)'$, and hence $Q_n \in \pi(G)'$. Moreover, the subrepresentation of π defined by the range of Q_n is equivalent to $\pi^{(n)}_{\mu_n}$, since μ'_n is equivalent to μ_n. Since $\pi^{(n)}_{\mu_n}$ and $\pi^{(m)}_{\mu_m}$ are disjoint (see Proposition 2.B.2), we therefore have $Q_m P_n = 0$ for all $m, n \in I$ with $m \neq n$. As $\sum_{m\in I} Q_m = \sum_{n\in I} P_n = I$, it follows that $Q_n = P_n$ for every $n \in I$. □

2.F. Applying the SNAG Theorem to nonabelian groups

We will be interested in representations of groups G which are nonabelian but which have a "large" abelian normal subgroup N (this holds for the examples of Sections 3.A to 3.E). We will analyze a given representation of G by applying the SNAG Theorem and the canonical decomposition Theorem 2.E.1 to its restriction to N. This procedure is the first step of the so-called Mackey machine; see Remark 2.F.5.

In order to state our next proposition, we need one more notion. A representation $(\pi, \mathcal{H})$ of a group G is said to be **factorial**, or a **factor representation**, if the von Neumann algebra $\pi(G)'' \subset \mathcal{L}(\mathcal{H})$ generated by $\pi(G)$ is a factor, that is, if the centre of $\pi(G)''$ consists of the scalar operators on $\mathcal{H}$ only. We come back to this notion in Chapter 6; see in particular Definition 6.A.8 and Proposition 6.B.6.

Let N be a closed abelian normal subgroup of G. Recall from Remark 1.A.17 the natural right action $\widehat{N} \curvearrowleft G$, defined by $\chi^g(n) = \chi(gng^{-1})$ for $\chi \in \widehat{N}$, $g \in G$, and $n \in N$. For a subset B of $\widehat{N}$, we write B^g for the image of B under the action of $g \in G$, that is, $B^g = \{\chi^g \mid \chi \in B\}$.

PROPOSITION 2.F.1. *Let G be a locally compact group, N a closed second-countable* abelian *normal subgroup of G, and $(\pi, \mathcal{H})$ a representation of G. Let E be the projection-valued measure on $\widehat{N}$ associated with the restriction $\pi|_N$ of π to N.*

(1) *For every $g \in G$, we have*

$$\pi(g)E(B)\pi(g)^{-1} = E(B^{g^{-1}}) \quad \text{for all } B \in \mathcal{B}(\widehat{N}).$$

(2) *Assume that π is factorial. Then the support of E coincides with the closure of a G-orbit in $\widehat{N}$.*

PROOF. (1) Set $\sigma := \pi|_N$. For $g \in G$, let

$$E^g : \mathcal{B}(\widehat{N}) \to \mathrm{Proj}(\mathcal{H})$$

be the projection-valued measure associated with the conjugate representation σ^g of σ.

On the one hand, since

$$\sigma^g(n) = \pi(g)\sigma(n)\pi(g^{-1}) \quad \text{for all } n \in N,$$

we have

$$E^g(B) = \pi(g)E(B)\pi(g^{-1}) \quad \text{for all } B \in \mathcal{B}(\widehat{N}).$$

On the other hand, by Corollary 2.C.4, we have

$$E^g(B) = E(B^{g^{-1}}) \quad \text{for all } B \in \mathcal{B}(\widehat{N}).$$

Claim (1) follows from the uniqueness part of the SNAG Theorem.

(2) The support S of E is a closed subset of $\widehat{N}$, which is G-invariant, by (1). Since N is a second-countable LCA group, $\widehat{N}$ is also second-countable (Proposition A.7.3). Let $(U_i)_{i \in \mathbf{N}}$ be a countable basis for the topology of $\widehat{N}$. Denote by $I \subset \mathbf{N}$ the subset of integers i such that $U_i \cap S \neq \emptyset$.

Let $i \in I$. Then

$$U_i^G := \{\chi^g \mid \chi \in U_i, \ g \in G\}$$

is a G-invariant open subset of $\widehat{N}$. For the projection $P_i = E(U_i^G)$, we have $P_i \neq 0$, since U_i^G is an open set with a nonempty intersection with S.

On the one hand, by the SNAG Theorem, P_i belongs to the von Neumann algebra $\pi(N)''$ generated by N; in particular, $P_i \in \pi(G)''$. On the other hand, since U_i^G is G-invariant, it follows from (1) that P_i commutes with $\pi(g)$ for every $g \in G$, that is, $P_i \in \pi(G)'$. Therefore,

$$P_i \in \pi(G)'' \cap \pi(G)',$$

that is, P_i belongs to the centre of the von Neumann algebra $\pi(G)''$. Since $\pi(G)''$ is a factor and $P_i \neq 0$, it follows that $P_i = I$. We have, therefore, for every $i \in I$,

$$E(\widehat{N} \smallsetminus U_i^G) = I - P_i = 0.$$

Since I is a countable set and by the properties of a projection-valued measure, we have

$$E\Big(\bigcup_{i \in I}(\widehat{N} \smallsetminus U_i^G)\Big) = 0.$$

It follows that $E\big(\bigcap_{i \in I} U_i^G\big) = I$ and therefore

$$E\Big(S \cap \big(\bigcap_{i \in I} U_i^G\big)\Big) = E(S)E\Big(\bigcap_{i \in I} U_i^G\Big) = I,$$

because $E(S) = I$. In particular, we have

$$S \cap \big(\bigcap_{i \in I} U_i^G\big) \neq \emptyset.$$

Let $\chi_0 \in S \cap \big(\bigcap_{i \in I} U_i^G\big)$. We claim that the G-orbit of χ_0 is dense in S. Indeed, let $\chi \in S$ and let U be an open neighbourhood of χ. Then, $U_i \subset U$ for some $i \in I$. Since $\chi_0 \in U_i^G$, the G-orbit of χ_0 has a nonempty intersection with U_i and hence with U. $\square$

REMARK 2.F.2. The formula of (1) in Proposition 2.F.1 shows that $B \mapsto E(B)$ is a system of imprimitivity for $(\pi, \mathcal{H})$ based on the G-space $\widehat{N}$, in the sense of [**Mack–58**].

We will later need (proof of Theorem 5.D.9) the following refinement of Proposition 2.F.1(2).

PROPOSITION 2.F.3. *Let G be a locally compact group, N a closed second-countable abelian normal subgroup of G, and π a representation of G on a separable Hilbert space $\mathcal{H}$. Let I be the subset of $\overline{\mathbf{N}^*}$ and let $(\mu_n)_{n\in I}$ be a sequence of probability measures on $\widehat{N}$ attached to the restriction $\pi|_N$ of π to N, as in Theorem 2.B.9.*

(1) *Every probability measure μ_n is quasi-invariant under G.*
(2) *Assume that π is factorial. Then $\pi|_N$ is a homogeneous representation of N, that is, $\pi|_N$ is equivalent to the representation $(\pi_\mu^{(n)}, L^2(\widehat{N}, \mu, \mathcal{K}))$ for a probability measure μ on $\widehat{N}$ which is quasi-invariant by G and a Hilbert space $\mathcal{K}$ of some dimension $n \in \overline{\mathbf{N}^*}$. Moreover, μ is ergodic under the G-action.*

PROOF. (1) Set $\sigma = \pi|_N$ and let $E : \mathcal{B}(\widehat{N}) \to \mathrm{Proj}(\mathcal{H})$ be the projection-valued measure associated with σ. Let $(P_n)_{n\in I}$ be the sequence of orthogonal projections in the centre of $\sigma(N)''$ attached to E as in Theorem 2.E.1, so that $E = \bigoplus_{n\in I} P_n \circ E$.

Fix $g \in G$ and denote by E^g the projection-valued measure associated with the conjugate representation σ^g.

On the one hand, we have

$$E^g(B) = \pi(g)E(B)\pi(g^{-1}) \qquad \text{for all } B \in \mathcal{B}(\widehat{N}).$$

It follows that I, $(\mu_n)_{n\in I}$, and $(\pi(g)P_n\pi(g^{-1}))_{n\in I}$ are respectively a subset of $\overline{\mathbf{N}^*}$, a sequence of mutually singular probability measures on $\widehat{N}$, and a sequence of mutually orthogonal projections in the centre of $\sigma^g(N)''$, together satisfying Properties (1) and (2) of Theorem 2.E.1 for the projection-valued measure E^g.

On the other hand, by Proposition 2.F.1, we have

$$E^g(B) = E(B^{g^{-1}}) \qquad \text{for all } B \in \mathcal{B}(\widehat{N}).$$

In particular, this implies (see Theorem 2.C.3(4) that $\sigma^g(N)'' = \sigma(N)''$.

For every $n \in I$, consider the conjugate representation $(\pi_{\mu_n}^{(n)})^g$ of N on the space $L^2(\widehat{N}, \mu_n, \mathcal{K}_n)$ that is the conjugate of $\pi_{\mu_n}^{(n)}$ by g; we have

$$\left((\pi_{\mu_n}^{(n)})^g(x)F\right)(\chi) = \left(\pi_{\mu_n}^{(n)}(gxg^{-1})F\right)(\chi) = \chi(gxg^{-1})F(\chi) = \chi^g(x)F(\chi),$$

for all $x \in N$, $\chi \in \widehat{N}$, and $F \in L^2(\widehat{N}, \mu_n, \mathcal{K}_n)$. It follows that $(\pi_{\mu_n}^{(n)})^g$ is (equivalent to) the representation of N corresponding to the projection-valued measure $P_n \circ E^g$. Since $(\pi_{\mu_n}^{(n)})^g$ is equivalent to $\pi_{g_*(\mu_n)}^{(n)}$, it follows that I, $(g_*(\mu_n))_{n\in I}$, and $(P_n)_{n\in I}$ are another subset of $\overline{\mathbf{N}^*}$, a sequence of mutually singular probability measures on $\widehat{N}$, and a sequence of mutually orthogonal projections in the centre of $\sigma(N)'' = \sigma^g(N)''$, together satisfying Properties (1) and (2) of Theorem 2.E.1 for the projection-valued measure E^g.

It follows from the uniqueness part of Theorem 2.E.1 that:

- $g_*(\mu_n)$ and μ_n are equivalent for every $n \in I$;
- $P_n = \pi(g)P_n\pi(g^{-1})$ for every $n \in I$.

Since this holds for every $g \in G$, and since P_n belongs to $\pi(N)''$ and hence to $\pi(G)''$, this implies that

- every measure μ_n is quasi-invariant by G;
- every projection P_n belongs to the centre of $\pi(G)''$.

(2) Assume now that π is factorial. Since every P_n belongs to the centre of $\pi(G)''$, and $\sum_{n \in I} P_n = I$, it follows that the subset I of $\overline{\mathbf{N}^*}$ consists of a single element $n \in \overline{\mathbf{N}^*}$, that is, π is n-homogeneous for this n.

Now denote by μ the probability measure on $\widehat{N}$ associated with n. Let $B \in \mathcal{B}(\widehat{N})$ be G-invariant. The projection $E(B)$ belongs to $\pi(G)''$, and also to $\pi(G)'$, because

$$E(B) = E(B^g) = \pi(g^{-1})E(B)\pi(g) \quad \text{for all } g \in G.$$

It follows that $E(B)$ belongs to the centre of $\pi(G)''$, and we have $E(B) = 0$ or $E(B) = I$. This implies that $\mu(B) = 0$ or $\mu(B) = 1$, and shows that μ is ergodic. □

COROLLARY 2.F.4. *Let G be a locally compact group, N a closed second-countable abelian normal subgroup of G, and π a* factor *representation of G on a separable Hilbert space $\mathcal{H}$. Let E be the projection-valued measure associated with $\pi|_N$.*

(1) *The equivalence class of the ergodic G-quasi-invariant probability measure μ on $\widehat{N}$ attached to $\pi|_N$ as in Proposition* 2.F.3(2) *is unique. More specifically, the sets $B \in \mathcal{B}(\widehat{N})$ such that $\mu(B) = 0$ for any such measure μ are exactly the sets $B \in \mathcal{B}(\widehat{N})$ such that $E(B) = 0$.*
(2) *The support (respectively, the set of atoms) of every probability measure μ on $\widehat{N}$ attached to $\pi|_N$ as in Proposition* 2.F.3 *coincides with the support (respectively, the set of atoms) of E.*

PROOF. Let μ be an ergodic G-quasi-invariant probability measure $\widehat{N}$ attached to $\pi|_N$.

(1) By Proposition 2.F.3 (2), we can assume that $\mathcal{H} = L^2(\widehat{N}, \mu, \mathcal{K})$ for a Hilbert space $\mathcal{K}$ of some dimension $n \in \overline{\mathbf{N}^*}$ and that $\pi|_N$ coincides with the canonical representation $\pi_\mu^{(n)}$ of N in $L^2(\widehat{N}, \mu, \mathcal{K})$. As we have already observed in the beginning of Section 2.E, concerning the projection-valued measure $E_\mu^{(n)}$ associated with $\pi_\mu^{(n)}$, we have, for every $B \in \mathcal{B}(\widehat{N})$:

$$E_\mu^{(n)}(B) = 0 \iff \mu(B) = 0.$$

This shows Item (1)

Item (2) is an immediate consequence of Item (1). □

REMARK 2.F.5 (**Mackey machine**). We give some indications on the **Mackey machine**, which is a procedure for producing and analyzing representations of a locally compact group G in terms of the representations of a closed normal subgroup N and of subgroups of G/N. The terminology refers to [**Mack–58**]. For details and proofs, see also [**Foll–16**, Section 6.4] and/or [**KaTa–13**, Section 4.3]. The procedure can be adapted to a variety of situations, with various degrees of sophistication.

Here, we only address the easiest situation, and assume that the three following properties hold.

(a) The group G is a *semi-direct product* $G = H \ltimes N$ of a closed subgroup H of G with an *abelian* closed normal subgroup N of G.
(b) The groups H and N are second-countable, hence so is G.
(c) The semi-direct product is *regular*, i.e., there exists a Borel subset $B \subset \widehat{N}$ such that B intersects each H-orbit in $\widehat{N}$ in exactly one point, for the natural right action $\widehat{N} \curvearrowleft H$ (this is not the orthodox notion of a regular semi-direct product, but our Condition (c) implies the usually defined property).

The machine produces irreducible representations, indeed all of them, as follows. For $\chi \in \widehat{N}$, denote by $\mathcal{O}_\chi$ its H-orbit and by $H_\chi = \{h \in H \mid \chi^h = \chi\}$ its isotropy in H. For any representation σ of H_χ, we have a representation $\sigma\chi$ defined by

$$(\sigma\chi)(h, n) \;=\; \sigma(h)\chi(n) \qquad \text{for all } (h, n) \in H_\chi \ltimes N.$$

Set $\pi_{\sigma,\chi} = \operatorname{Ind}_{H_\chi \ltimes N}^{G}(\sigma\chi)$. Then

(1) Let $\chi \in \widehat{N}$ and $\sigma \in \widehat{H_\chi}$ be an irreducible representation of H_χ; the representation $\pi_{\sigma,\chi}$ of G is irreducible.
(2) Every irreducible representation of G is equivalent to a representation $\pi_{\sigma,\chi}$, for some $\chi \in \widehat{N}$ and $\sigma \in \widehat{H_\chi}$.
(3) Two irreducible representations of G of the previous form, say π_{σ_1,χ_1} and π_{σ_2,χ_2}, are equivalent if and only if χ_1, χ_2 are in the same H-orbit, i.e., there exists $h \in H$ such that $\chi_2 = (\chi_1)^h$, and the representations $x \mapsto \sigma_1(x)$ and $x \mapsto \sigma_2(h^{-1}xh)$ of H_{χ_1} are equivalent.

There are refined versions of the machine, to deal with locally compact groups G having a closed normal subgroup N which is of type I, and such that the extension $N \to G \to G/N$ satisfies appropriate conditions; but sophistication increases, among other reasons, because projective representations have to be considered.

CHAPTER 3

Examples of irreducible representations

Using the irreducibility and equivalence criteria from Section 1.F*, we construct families of nonequivalent irreducible representations for various discrete groups:*

- *the infinite dihedral group* D_∞ *(Section* 3.A*),*
- *two-step nilpotent discrete groups, in particular Heisenberg groups over fields and over the integers (Section* 3.B*),*
- *the affine group* $\mathrm{Aff}(\mathbf{K})$ *over a field* $\mathbf{K}$ *(Section* 3.C*),*
- *the Baumslag–Solitar group* $\mathrm{BS}(1,p)$ *for a prime* p *(Section* 3.D*),*
- *the lamplighter group* $\mathbf{Z} \ltimes \bigoplus_{k\in\mathbf{Z}} \mathbf{Z}/2\mathbf{Z}$ *(Section* 3.E*),*
- *and the general linear group* $\mathrm{GL}_n(\mathbf{K})$ *over* $\mathbf{K}$ *(Section* 3.F*).*

For D_∞*, in this way we obtain a complete description of the dual* $\widehat{D_\infty}$*. Moreover, this space coincides with the primitive dual* $\mathrm{Prim}(D_\infty)$*, and also with the quasi-dual* $\mathrm{QD}(D_\infty)$ *introduced in Section* 6.C.

By contrast, for any group Γ *of Sections* 3.B *to* 3.F*, we will see that it is impossible to select in these families, in a measurable way, a set of pairwise inequivalent irreducible representations, and a fortiori impossible to parametrize in a measurable way the dual* $\widehat{\Gamma}$ *of* Γ*. This is a manifestation of the fact that these groups are not of type I (a notion defined in Section* 6.D*). Moreover, as checked below (Section* 9.G*), each group of Sections* 3.B *to* 3.F *has an uncountable family of representations which are pairwise nonequivalent and nevertheless weakly equivalent to each other.*

In later chapters, we will indicate further constructions providing more irreducible representations of the groups listed above.

The final Section 3.G *refers briefly to the duals of some nondiscrete groups.*

3.A. Infinite dihedral group

DEFINITION 3.A.1. The **infinite dihedral group** D_∞ is the semi-direct product $\mathbf{Z}/2\mathbf{Z} \ltimes \mathbf{Z}$, for the nontrivial action of the two-element group $\mathbf{Z}/2\mathbf{Z} = \{1, \varepsilon\}$ on $\mathbf{Z}$, that is, for the action given by $\varepsilon \cdot n = -n$ for all $n \in \mathbf{Z}$. The group has a matrix form: $D_\infty = \begin{pmatrix} \{\pm 1\} & \mathbf{Z} \\ 0 & 1 \end{pmatrix}$. It is well known that D_∞ is also isomorphic to the free product of two groups of order two.

Observe that the commutator subgroup $[D_\infty, D_\infty]$ is the subgroup $2\mathbf{Z}$ of $\mathbf{Z}$, and that the abelianized group $(D_\infty)_{\mathrm{ab}} = D_\infty/[D_\infty, D_\infty]$ is isomorphic to the Vierergruppe $\mathbf{Z}/2\mathbf{Z} \times \mathbf{Z}/2\mathbf{Z}$.

In this section, we write Γ for D_∞ and N for its normal subgroup $\mathbf{Z}$ of index 2.

As usual, we identify $\widehat{N}$ with the circle group $\mathbf{T} = \{z \in \mathbf{C} \mid |z| = 1\}$ through the map

$$\mathbf{T} \xrightarrow{\approx} \widehat{N}, \quad z \mapsto \chi_z,$$

where

$$\chi_z(n) = z^n \qquad \text{for all } n \in \mathbf{Z},$$

and we view $\widehat{N}$ as a right Γ-space. The action of ε on $\widehat{N}$ is described by

$$\chi_z^\varepsilon = \chi_{\overline{z}} \qquad \text{for all } z \in \mathbf{T}.$$

In particular, χ_1 and χ_{-1} are Γ-fixed. For $z \neq \pm 1$, the Γ-orbit of χ_z consists of the two-element set $\{\chi_z, \chi_{\overline{z}}\}$. The space of Γ-orbits in $\widehat{N}$ can be identified with the quotient space $\mathbf{T}/\sim$ for the equivalence relation on $\mathbf{T}$ defined by $z \sim \overline{z}$.

For every $z \in \mathbf{T}$, define a representation ρ_z of Γ by

$$\rho_z = \operatorname{Ind}_N^\Gamma \chi_z.$$

It is of dimension 2. Observe that $\rho_{\overline{z}}$ is equivalent to ρ_z, since $\chi_{\overline{z}}$ is in the Γ-orbit of χ_z (see Proposition 1.F.9).

PROPOSITION 3.A.2. *Let $\Gamma = D_\infty$. The map*

$$\Phi : \widehat{\Gamma_{\mathrm{ab}}} \sqcup ((\mathbf{T} \smallsetminus \{1, -1\})/\sim) \longrightarrow \widehat{\Gamma},$$

defined by $\Phi(\chi) = \chi$ for $\chi \in \widehat{\Gamma_{\mathrm{ab}}}$ and $\Phi(\{z, \overline{z}\}) = \rho_z$ for $z \in \mathbf{T} \smallsetminus \{1, -1\}$, is a bijection.

Note that $\widehat{\Gamma_{\mathrm{ab}}}$ has four elements, and $(\mathbf{T} \smallsetminus \{1, -1\})/\sim$ is in natural bijection with the upper open half-circle $\{e^{i\theta} \in \mathbf{T} \mid 0 < \theta < \pi\}$.

PROOF. The method of proof of Proposition 3.A.2 is a special case of the Mackey machine, mentioned in Remark 2.F.5.

• *First step.* We claim that the range of Φ is contained in $\widehat{\Gamma}$ and that Φ is injective.

Indeed, for $z \in \mathbf{T} \smallsetminus \{1, -1\}$, the stabilizer of χ_z in Γ is N, so that the representation ρ_z is irreducible, by the Mackey–Shoda irreducibility criterion (Corollary 1.F.16). Moreover, if $z, w \in \mathbf{T}$ are not in the same equivalence class, then ρ_z and ρ_w are not equivalent by the Mackey–Shoda equivalence criterion (Corollary 1.F.20). This proves the nontrivial part of the first step.

Let $(\rho, \mathcal{H})$ be an irreducible representation of Γ. Assume that ρ is not one-dimensional. To prove that Φ is surjective, we have to show that ρ is equivalent to ρ_z for some $z \in \mathbf{T} \smallsetminus \{1, -1\}$. We denote by E the projection-valued measure on $\widehat{N}$ associated with $\rho|_N$, and by S the support of E.

• *Second step.* We claim that S consists of exactly one Γ-orbit in $\widehat{N}$.

Indeed, by Proposition 2.F.1(2), S is the closure of a Γ-orbit in $\widehat{N}$. Since the Γ-orbits are finite and hence closed, the claim follows.

• *Third step.* We claim that $S \neq \{\chi_1\}$ and $S \neq \{\chi_{-1}\}$.

Indeed, assume that $S = \{\chi_1\}$ or $S = \{\chi_{-1}\}$. Then $\mathcal{H} = \mathcal{H}^{\chi_1}$ or $\mathcal{H} = \mathcal{H}^{\chi_{-1}}$, by Corollary 2.D.2. Observe that χ_1 and χ_{-1} are trivial on the commutator subgroup $[\Gamma, \Gamma] = 2\mathbf{Z} \subset N$. Therefore ρ factorizes by Γ_{ab} and hence is one-dimensional. This is a contradiction.

• *Fourth step.* We have $\rho = \operatorname{Ind}_N^\Gamma \chi_z$ for some $z \in \mathbf{T} \smallsetminus \{-1, 1\}$.

Indeed, by the second and the third steps, we have $S = \{z, \overline{z}\}$ for some $z \in \mathbf{T} \smallsetminus \{-1, 1\}$. By Corollary 2.D.2 again, we have

$$\mathcal{H} = \mathcal{H}^{\chi_z} \bigoplus \mathcal{H}^{\chi_{\overline{z}}}.$$

The formula of Proposition 2.F.1(1) shows that

$$\mathcal{H}^{\chi_{\overline{z}}} = \rho(\varepsilon)\mathcal{H}^{\chi_z}.$$

Since $\{e, \varepsilon\}$ is a transversal for N in Γ, we see that $\rho = \mathrm{Ind}_N^\Gamma \sigma$, where the representation σ of N is the multiple of χ_z defined on $\mathcal{H}^{\chi_z}$ (see Definition 1.F.1). Since ρ is irreducible, σ is irreducible and hence $\sigma = \chi_z$ (that is, $\dim \mathcal{H}^{\chi_z} = 1$). □

Proposition 3.A.2 describes the dual $\widehat{D_\infty}$ of the infinite dihedral group as a set. We now describe this dual as a topological space.

Set $C = \{e^{i\theta} \in \mathbf{T} \mid 0 \leq \theta \leq \pi\} \times \{1, -1\}$; it is a compact topological space, homeomorphic to the disjoint union of two closed intervals. Define an equivalence relation on C by $(e^{i\theta}, j) \sim (e^{i\theta'}, j')$ if either $0 < \theta = \theta' < \pi$, or $\theta = \theta' \in \{0, \pi\}$ and $j = j'$; we write $[e^{i\theta}, j] \in C/\sim$ for the equivalence class of $(e^{i\theta}, j)$ in C. Note that $C/\sim$ is a non-Hausdorff quasi-compact topological space, that can be described as an interval having pairs of nonseparated points at each of its ends.

For the infinite dihedral group, we now choose the matrix model

$$D_\infty = \begin{pmatrix} \{\pm 1\} & \mathbf{Z} \\ 0 & 1 \end{pmatrix} \quad \text{and} \quad (\varepsilon, n) \text{ stands for } \begin{pmatrix} \varepsilon & n \\ 0 & 1 \end{pmatrix} \in D_\infty.$$

The four unitary characters of D_∞ are denoted by $\rho_{0,1}, \rho_{0,-1}, \rho_{\pi,1}, \rho_{\pi,-1}$ and can be defined by

$$\begin{aligned} \rho_{0,1}(\varepsilon, n) &= 1, & \rho_{0,-1}(\varepsilon, n) &= \varepsilon, \\ \rho_{\pi,1}(\varepsilon, n) &= (-1)^n, & \rho_{\pi,-1}(\varepsilon, n) &= (-1)^n \varepsilon \end{aligned}$$

for all $(\varepsilon, n) \in D_\infty$. Let $z = e^{i\theta} \in \mathbf{T}$ be such that $0 \leq \theta \leq \pi$. For $\mathrm{Ind}_N^\Gamma \chi_z$, we now write ρ_θ, rather than ρ_z as above. According to Definition 1.F.1, the induced representation ρ_θ can be viewed as acting on the Hilbert space $\mathbf{C}^2$, and defined by

$$\rho_\theta(1, n) = \begin{pmatrix} e^{in\theta} & 0 \\ 0 & e^{-in\theta} \end{pmatrix} \quad \text{and} \quad \rho_\theta(-1, n) = \begin{pmatrix} 0 & e^{in\theta} \\ e^{-in\theta} & 0 \end{pmatrix}$$

for all $n \in \mathbf{Z}$. The normalized function $\varphi_{\rho_\theta, s, t}$ of positive type associated with ρ_θ and defined by a unit vector $\xi = \begin{pmatrix} s \\ t \end{pmatrix} \in \mathbf{C}^2$ (see Definition 1.B.4) is given by

$$\text{(PT)} \qquad \varphi_{\rho_\theta, s, t}(\varepsilon, n) = \begin{cases} e^{in\theta} \, |s|^2 + e^{-in\theta} \, |t|^2 & \text{if } \varepsilon = 1, \\ e^{in\theta} \, \overline{s} \, t + e^{-in\theta} s\overline{t} = 2\mathrm{Re}(e^{in\theta}\overline{s}t) & \text{if } \varepsilon = -1, \end{cases}$$

for all $n \in \mathbf{Z}$.

Note that the representation ρ_θ is cyclic. When $0 < \theta < \pi$, it is irreducible. When $\theta = 0$, we have a direct sum $\rho_0 = \rho_{0,1} \oplus \rho_{0,-1}$ in a direct sum of Hilbert spaces $\mathbf{C}^2 = \mathbf{C}\begin{pmatrix} 1/\sqrt{2} \\ 1/\sqrt{2} \end{pmatrix} \oplus \mathbf{C}\begin{pmatrix} 1/\sqrt{2} \\ -1/\sqrt{2} \end{pmatrix}$. When $\theta = \pi$, we have $\rho_\pi = \rho_{\pi,1} \oplus \rho_{\pi,-1}$, in the same decomposition of $\mathbf{C}^2$ in a direct sum of two one-dimensional Hilbert spaces.

PROPOSITION 3.A.3. *Let the notation be as above. Define a map* $\Psi \colon C/\sim \to \widehat{D_\infty}$ *by*

$$\begin{aligned} \Psi([e^{i\theta}, j]) &= \rho_\theta \quad \textit{for } [e^{i\theta}, j] \in C/\sim, & \Psi([1, 1]) &= \rho_{0,1}, \\ \Psi([1, -1]) &= \rho_{0,-1}, \quad \Psi([-1, 1]) = \rho_{\pi,1}, & \Psi([-1, -1]) &= \rho_{\pi,-1}. \end{aligned}$$

Then Ψ *is a homeomorphism.*

PROOF. By Proposition 3.A.2, the map Ψ is a bijection.

• *First step.* The map Ψ is continuous.

Since the space $C/\sim$ is first countable, it suffices to consider sequences, rather than nets. Consider a point $[e^{i\theta}, j]$ in $C/\sim$. We have to check that, for any sequence $([e^{i\theta_k}, j_k])_{k\geq 1}$ converging to $[e^{i\theta}, j]$ in $C/\sim$, the sequence of representations $(\Psi([e^{i\theta}, j]))_{k\geq 1}$ converges to $\Psi([e^{i\theta}, j])$ in $\widehat{D_\infty}$, i.e., that, for each $u, v \in \mathbf{C}$, there exists a sequence $(\varphi_{\rho_{\theta_k}, u_k, v_k})_{k\geq 1}$ of functions of positive type associated with the ρ_{θ_k}'s such that

$$\text{(Lim)}\quad \lim_{k\to\infty} \varphi_{\rho_{\theta_k}, u_k, v_k} = \varphi_{\rho_\theta, u, v} \quad \text{(limit in the topology of simple convergence).}$$

Assume first that $0 < \theta < \pi$. Then $\Psi([e^{i\theta}, j]) = \rho_\theta$. It is obvious from the explicit formulas (PT) that $\lim_{k\to\infty} \varphi_{\rho_{\theta_k}, u, v} = \varphi_{\rho_\theta, u, v}$ for all $u, v \in \mathbf{C}$.

Assume next that $\theta = 0$ and $j = -1$. Then $\Psi([1, -1]) = \rho_{0,-1}$. For any vector $\begin{pmatrix} u \\ -u \end{pmatrix}$ in the space $\mathbf{C}\begin{pmatrix} 1/\sqrt{2} \\ -1/\sqrt{2} \end{pmatrix}$ of $\rho_{0,-1}$, the function of positive type $\varphi_{\rho_0, u, -u}$ is a constant multiple of the unitary character $\rho_{0,-1}$. It is again obvious from the explicit formulas (PT) that $\lim_{k\to\infty} \varphi_{\rho_{\theta_k}, u, -u} = \varphi_{\rho_0, u, -u}$ for all $u \in \mathbf{C}$. The arguments for the three other cases $(\theta, j) = (0, 1), (\pi, 1), (\pi, -1)$ are similar.

• *Second step.* The map $\Psi^{-1} : \widehat{D_\infty} \to C/\sim$ is continuous.

Let A be a subset of $\widehat{D_\infty}$ and $\rho \in \overline{A}$. We have to show that $\Psi^{-1}(\rho) \in \overline{\Psi^{-1}(A)}$. Two cases may occur.

∘ *First case.* ρ is in the closure of $A \cap \{\rho_{0,1}, \rho_{0,-1}, \rho_{\pi,1}, \rho_{\pi,-1}\}$. Then ρ belongs to $A \cap \{\rho_{0,1}, \rho_{0,-1}, \rho_{\pi,1}, \rho_{\pi,-1}\}$, since every unitary character of D_∞ is a closed point in $\widehat{D_\infty}$ (see Example 1.C.13). In particular, $\Psi^{-1}(\rho) \in \overline{\Psi^{-1}(A)}$.

∘ *Second case.* ρ is in the closure of $A \cap \{\rho_\theta \mid 0 < \theta < \pi\}$.

Let ξ be a unit vector in the Hilbert space of ρ and $\varphi_{\rho,\xi}$ the corresponding normalized function of positive type associated with ρ. Then there exists a sequence $(\theta_k)_{k\geq 1}$ in $(0, \pi)$ with $\rho_{\theta_k} \in A$ and sequences $(s_k)_{k\geq 1}, (t_k)_{k\geq 1}$ in $\mathbf{C}$ with $|s_k|^2 + |t_k|^2 = 1$ such that

$$\lim_k \varphi_{\rho_{\theta_k}, s_k, t_k}(1, 1) = \varphi_{\rho,\xi}(1, 1).$$

Therefore, we have (see the formulas PT)

$$(*)\qquad \lim_k \left(e^{i\theta_k}|s_k|^2 + e^{-i\theta_k}|t_k|^2\right) = \varphi_{\rho,\xi}(1, 1).$$

Assume first that $\rho = \rho_\theta$ for $0 < \theta < \pi$. Choose $\xi = \begin{pmatrix} 1 \\ 0 \end{pmatrix}$. Then $\varphi_{\rho,\xi}(1, 1) = e^{i\theta}$ and it follows from $(*)$ and from the uniform convexity of the unit disc in $\mathbf{R}^2$ that $\lim_k \theta_k = \theta$. So, $\Psi^{-1}(\rho) = [e^{i\theta}, 1] = \lim_k [e^{i\theta_k}, 1]$ is in the closure of $\Psi^{-1}(A)$.

Assume next that $\rho = \rho_{0,1}$ or $\rho = \rho_{0,-1}$. With $\xi = 1$, we have $\varphi_{\rho,\xi}(1, 1) = 1$, and therefore

$$\lim_k \left(e^{i\theta_k}|s_k|^2 + e^{-i\theta_k}|t_k|^2\right) = 1.$$

As before, this implies that $\lim_k \theta_k = 0$. So, $\overline{\Psi^{-1}(A)}$ contains $[1, 1]$ and $[1, -1]$ and this shows that $\Psi^{-1}(\rho) \in \overline{\Psi^{-1}(A)}$.

Finally, assume that $\rho = \rho_{\pi,1}$ or $\rho = \rho_{\pi,-1}$. With $\xi = 1$, we have $\varphi_{\rho,\xi}(1, 1) = -1$, and therefore

$$\lim_k \left(e^{i\theta_k}|s_k|^2 + e^{-i\theta_k}|t_k|^2\right) = -1.$$

As before, this implies that $\lim_k \theta_k = \pi$. So, $\overline{\Psi^{-1}(A)}$ contains $[-1,1]$ and $[-1,-1]$ and this shows that $\Psi^{-1}(\rho) \in \overline{\Psi^{-1}(A)}$. $\square$

3.B. Two-step nilpotent groups

Let Γ be a **two-step nilpotent group**, that is, Γ is a group of which the derived group $[\Gamma, \Gamma]$ is contained in the centre $Z = Z(\Gamma)$. Observe that Γ could be abelian; we do not exclude this case, as we sometimes have to consider abelian quotients of Γ.

Theorem 3.B.4 provides a procedure constructing irreducible representations of Γ, which is reminiscent of Kirillov's orbit method for nilpotent connected real Lie groups [**Kiri–62**]. Our exposition is inspired in part by [**Howe–77b**], which provides a description of $\mathrm{Prim}(\Gamma)$ for finitely generated discrete nilpotent groups. We will come back to representations of two-step nilpotent groups in Sections 4.B, 9.B, and 12.A.

DEFINITION 3.B.1. Let Δ be a group. A **bihomomorphism** from Δ to $\mathbf{T}$ is a map $\omega : \Delta \times \Delta \to \mathbf{T}$ such that the partial maps

$$\omega_a : \Delta \to \mathbf{T}, \ \ \delta \mapsto \omega(a, \delta) \qquad \text{and} \qquad \omega^b : \Delta \to \mathbf{T}, \ \ \delta \mapsto \omega(\delta, b)$$

are homomorphisms for every $a, b \in \Delta$. For such a bihomomorphism ω, the map

$$\Delta \to \mathrm{Hom}(\Delta, \mathbf{T}), \ \ a \mapsto \omega_a$$

is a homomorphism; when Δ is abelian, this is a homomorphism $\Delta \to \widehat{\Delta}$.

LEMMA 3.B.2. *Let Γ be a two-step nilpotent group. Let Z be its centre.*

(1) *Every maximal abelian subgroup of Γ is normal in Γ and contains Z.*
(2) *For every $\gamma \in \Gamma$, the maps*

$$\Gamma \to Z, \ \ a \mapsto [a, \gamma] \qquad \text{and} \qquad \Gamma \to Z, \ \ a \mapsto [\gamma, a] = [a, \gamma]^{-1}$$

are group homomorphisms.

(3) *Let $\psi \in \widehat{Z}$ be a unitary character of Z. The map*

$$\omega : \Gamma/Z \times \Gamma/Z \to \mathbf{T}, \ \ (aZ, bZ) \mapsto \psi([b, a])$$

is a bihomomorphism from Γ/Z to $\mathbf{T}$.

(4) *Let $\psi \in \widehat{Z}$ and ω be as in (3); assume moreover that ψ is a faithful character. Then*

$$\Gamma/Z \longrightarrow \widehat{\Gamma/Z}, \ \ aZ \mapsto \omega_a$$

is an injective homomorphism.

PROOF. (1) Maximal abelian subgroups exist in Γ, by Zorn's lemma; let N be such a subgroup. Since N is its own centralizer, Z is contained in N. Therefore $\gamma n \gamma^{-1} n^{-1} \in [N, N] \subset Z \subset N$, hence $\gamma n \gamma^{-1} \in N$, for all $\gamma \in \Gamma$ and $n \in N$. It follows that N is normal in Γ.

(2) Since $[\Gamma, \Gamma] \subset Z$, we have, for all $a, b \in \Gamma$,

$$\begin{aligned} [ab, \gamma] &= b^{-1}a^{-1}\gamma^{-1}ab\gamma = b^{-1}(a^{-1}\gamma^{-1}a\gamma)\gamma^{-1}b\gamma \\ &= (a^{-1}\gamma^{-1}a\gamma)b^{-1}\gamma^{-1}b\gamma = [a, \gamma][a, \gamma]. \end{aligned}$$

This shows that the map $\Gamma \to Z, \ a \mapsto [a, \gamma]$ is a group homomorphism. The argument for $\Gamma \to Z, \ a \mapsto [\gamma, a]$ is similar.

Observe that these homomorphisms factorize through Γ/Z.

Claim (3) is a straightforward consequence of (2), and Claim (4) is equally straightforward. □

We continue with a nonabelian two-step nilpotent group Γ, its centre Z, and a maximal abelian subgroup N of Γ. We have a continuous homomorphism

$$r : \widehat{N} \to \widehat{Z}, \quad \chi \mapsto \chi|_Z,$$

given by restriction to Z of unitary characters of N. By Pontryagin theory, unitary characters extend from closed subgroups of LCA groups (see, e.g., [**BTS1–2**, Chap. II, § 1, no 7, th. 4]). In particular, the map r is surjective. Since Z is central, the action of Γ on $\widehat{Z}$ is trivial, and hence

$$r(\chi^\gamma) = r(\chi) \qquad \text{for all } \chi \in \widehat{N}, \gamma \in \Gamma.$$

For $\psi \in \widehat{Z}$, set

$$\widehat{N}(\psi) := r^{-1}(\psi) = \left\{ \chi \in \widehat{N} \mid \chi|_Z = \psi \right\}.$$

If we agree to identify $\widehat{N/Z}$ with a subgroup of $\widehat{N}$, we also have for any $\chi_0 \in \widehat{N}(\psi)$

$$\widehat{N}(\psi) = \left\{ \chi \in \widehat{N} \mid \chi = \chi_0 \rho \text{ for some } \rho \in \widehat{N/Z} \right\}.$$

Observe that $\widehat{N}(\psi)$ is a Γ-invariant subset of $\widehat{N}$.

LEMMA 3.B.3. *Let Γ be a nonabelian two-step nilpotent group, Z its centre, N a maximal abelian subgroup of Γ, and $\psi \in \widehat{Z}$ a* faithful *unitary character of Z. Let $\widehat{N}(\psi)$ be as above, and, for $a \in \Gamma$, let ω_a be as in Definition* 3.B.1.

(1) *Let $a \in \Gamma \smallsetminus N$. The restriction of ω_a to N/Z, that is the map*

$$\omega_a|_{N/Z} : N/Z \to \mathbf{T}, \quad nZ \mapsto \psi([n,a]),$$

is a unitary character of N/Z that is distinct from $1_{N/Z}$.

(2) *Let $\chi \in \widehat{N}(\psi)$ and $a \in \Gamma$. Then $\chi^{a^{-1}} = (\omega_a|_{N/Z})\chi \in \widehat{N}(\psi)$, where $\omega_a|_{N/Z}$ is viewed as a unitary character of N.*

(3) *The group Γ/N acts freely on the set $\widehat{N}(\psi)$.*

(4) *The Γ-orbit of every element $\chi \in \widehat{N}(\psi)$ is dense in $\widehat{N}(\psi)$.*

PROOF. (1) Since Γ is not abelian, $N \subsetneqq \Gamma$. Since $a \notin N$, there exists $n \in N$ such that $[n,a] \neq 1$. Since ψ is faithful, we have $(\psi_a|_{N/Z})(nZ) = \psi([n,a]) \neq 1$. Therefore $\psi_a|_{N/Z} \neq 1_{N/Z}$.

(2) Since $\chi|_Z = \psi$, we have, for every $n \in N$,

$$\chi^{a^{-1}}(n) = \chi(a^{-1}na) = \chi(n[n,a]) = \psi([n,a])\chi(n) = \omega_a(nZ)\chi(n).$$

(3) Let $a \in \Gamma \smallsetminus N$ and $\chi \in \widehat{N}(\psi)$. By (1), we have $\omega_{a^{-1}}|_{N/Z} \neq 1_{N/Z}$. Therefore $\chi^a \neq \chi$, by (2).

(4) Any two elements in $\widehat{N}(\psi)$ differ by a unitary character of N that is trivial on Z. In view of (2), it is enough to show that the subgroup

$$H := \{\rho \in \widehat{N/Z} \mid \rho = \omega_a|_{N/Z} \text{ for some } a \in \Gamma\}$$

is dense in $\widehat{N/Z}$. By Pontryagin duality, it is equivalent to show that $H^\perp = \{e\}$, where

$$H^\perp = \{nZ \in N/Z \mid \rho(nZ) = 1 \text{ for all } \rho \in H\}$$

(see Corollaire 6 of Théorème 4 in [**BTS1–2**, Chap. II, § 1, no 7]).

Let $n \in N$ be such that $\omega_a(nZ) = 1$ for all $a \in \Gamma$, i.e., such that $\psi([n,a]) = 1$ for all $a \in \Gamma$. Then $n \in Z$, because ψ is faithful, i.e., nZ is the neutral element of N/Z, as was to be shown. □

Recall (see Definition 1.A.12) that every irreducible representation of Γ admits a central character.

Given a *faithful* unitary character ψ of Z, the following result provides a procedure for the construction of irreducible representations of Γ with ψ as central character. For the case where ψ is not faithful, see Remark 3.B.5 below.

THEOREM 3.B.4. *Let Γ be a nonabelian two-step nilpotent discrete group with centre Z and N a maximal abelian subgroup of Γ. Let ψ be a* faithful *unitary character of Z; set $\widehat{N}(\psi) = \{\chi \in \widehat{N} \mid \chi\vert_Z = \psi\}$.*

(1) *Let $\chi \in \widehat{N}(\psi)$. Then $\mathrm{Ind}_N^\Gamma \chi$ is an irreducible representation of Γ, with central character ψ.*
(2) *Let $\chi_1, \chi_2 \in \widehat{N}(\psi)$ be such that $\chi_1^\gamma \neq \chi_2$ for every $\gamma \in \Gamma$. Then $\mathrm{Ind}_N^\Gamma \chi_1$ and $\mathrm{Ind}_N^\Gamma \chi_2$ are nonequivalent irreducible representations of Γ.*

PROOF. (1) By Lemma 3.B.3(3) we have $\chi^\gamma \neq \chi$ for every $\gamma \in \Gamma \smallsetminus N$. Therefore $\mathrm{Ind}_N^\Gamma \chi$ is irreducible by the Mackey–Shoda irreducibility criterion (Corollary 1.F.16), and Ind_N^Γ has central character ψ by Proposition 1.F.7.

(2) It follows from (1) that the representations $\mathrm{Ind}_N^\Gamma \chi_1$ and $\mathrm{Ind}_N^\Gamma \chi_2$ are irreducible and from the Mackey–Shoda equivalence criterion (Corollary 1.F.20) that they are not equivalent. □

REMARK 3.B.5. The central characters of the irreducible representations of Γ in Theorem 3.B.4 are faithful. (Note that the centre of a group need not have any faithful unitary character; this is, for example, the case of groups containing noncyclic finite central subgroups.)

Let ψ be a nonfaithful unitary character of the centre Z of Γ. Here is a procedure to construct irreducible representations of Γ having ψ as central character. Observe that $\ker \psi$ is a central and hence normal subgroup of Γ.

- Consider the quotient group $\Gamma_1 = \Gamma / \ker \psi$, and the canonical projection $p : \Gamma \twoheadrightarrow \Gamma_1$. View ψ as a unitary character of the subgroup $H_1 = p(Z)$ of Γ_1. Observe that H_1 is a (possibly proper) subgroup of the centre Z_1 of Γ_1; observe also that $[\Gamma_1, \Gamma_1] \subset H_1$, since $[\Gamma, \Gamma] \subset Z$.
- Extend ψ to a unitary character $\widetilde{\psi}$ of Z_1. Consider the quotient group $\overline{\Gamma} = \Gamma_1 / \ker \widetilde{\psi}$.
- The centre of $\overline{\Gamma}$ is the image $Z_1 / \ker \widetilde{\psi}$ of Z_1 in $\overline{\Gamma}$.

  Indeed, it is clear that $Z_1 / \ker \widetilde{\psi}$ is contained in the centre of $\overline{\Gamma}$. Conversely, let $\gamma \in \Gamma_1$ be such that $[\gamma, x] \in \ker \widetilde{\psi}$, for all $x \in \Gamma_1$. Since $[\Gamma_1, \Gamma_1] \subset H_1$ and $\widetilde{\psi}\vert_{H_1} = \psi$, it follows that $[\gamma, x] \in \ker \psi$ for all $x \in \Gamma_1$. As ψ is faithful on H_1, this means that $\gamma \in Z_1$.
- It is clear that the unitary character $\widetilde{\psi}$, viewed as a character of the centre $Z_1 / \ker \widetilde{\psi}$ of $\overline{\Gamma}$, is faithful.
- Theorem 3.B.4 applies to $\widetilde{\psi}$ and provides an irreducible representation π of $\overline{\Gamma}$ having $\widetilde{\psi}$ as central character.
- Lift π to Γ to obtain an irreducible representation of Γ having ψ as central character.

Heisenberg group over a commutative ring. Next, we treat Heisenberg groups, which are prominent examples of two-step nilpotent groups. We will deal with them in a way independent of Theorem 3.B.4 (which concerns faithful central characters) and Remark 3.B.5 (which concerns nonfaithful central characters).

DEFINITION 3.B.6. Let R be a commutative ring with unit. The **Heisenberg group** over R is the group

$$H(R) = \begin{pmatrix} 1 & R & R \\ 0 & 1 & R \\ 0 & 0 & 1 \end{pmatrix}.$$

We identify $H(R)$ with R^3, by writing (a, b, c) for the matrix

$$\begin{pmatrix} 1 & a & c \\ 0 & 1 & b \\ 0 & 0 & 1 \end{pmatrix}.$$

The group law on $H(R)$ is given by

$$(a, b, c)(a', b', c') = (a + a', b + b', c + c' + ab') \quad \text{for } (a, b, c),\ (a', b', c') \in R^3.$$

The centre of $H(R)$ is $Z = \{(0, 0, c) \mid c \in R\} \approx R$. Since

$$[(a, b, c), (a', b', c')] = (0, 0, ab' - a'b) \in Z \quad \text{for } (a, b, c), (a', b', c') \in R^3,$$

we see that $H(R)$ is two-step nilpotent; moreover, $[H(R), H(R)] = Z$. The map associated with a unitary character ψ of Z as in Lemma 3.B.2(3) is the bihomomorphism $\omega : R^2 \times R^2 \to \mathbf{T}$ given by

$$\omega((a, b), (a', b')) = \psi(ab' - a'b) \quad \text{for } (a, b), (a', b') \in R^2.$$

The subgroup of $H(R)$,

$$N = \{(0, b, c) \in \Gamma \mid b, c \in R\} \approx R^2,$$

is maximal abelian.

In this section, we write Γ for $H(R)$.

Fix a unitary character ψ of the centre $Z \approx R$ of Γ. Let I_ψ be the **ideal** of R defined by

$$I_\psi := \{a \in R \mid aR \subset \ker \psi\};$$

in other words, I_ψ is the largest ideal of R contained in $\ker \psi$. Define

$$\begin{aligned} Z_\psi &:= \{(a, b, c) \in \Gamma \mid a, b \in I_\psi,\ c \in R\}, \\ N_\psi &:= \{(a, b, c) \in \Gamma \mid a \in I_\psi,\ b, c \in R\}. \end{aligned}$$

As is easily checked, Z_ψ and N_ψ are normal subgroups of Γ, and $Z_\psi \subset N_\psi$. The subgroup $\ker \psi$ of Z is central, and therefore normal, in Γ; denote by $\overline{\Gamma}$ the quotient group $\Gamma / \ker \psi$ and by $p : \Gamma \twoheadrightarrow \overline{\Gamma}$ the canonical projection.

Beware that the groups Z_ψ and N_ψ need not be abelian.

LEMMA 3.B.7. *For $\psi \in \widehat{Z}$, let Z_ψ, N_ψ, and $p : \Gamma \twoheadrightarrow \overline{\Gamma}$ be as above.*

(1) *The subgroup Z_ψ is the inverse image in Γ of the centre $\overline{Z}$ of $\overline{\Gamma}$.*
(2) *The subgroup N_ψ is the inverse image in Γ of the subgroup $p(N_\psi)$, which is a maximal abelian subgroup in $\overline{\Gamma}$.*

In particular, if ψ is faithful, then $Z_\psi = Z$ and $N_\psi = N$.

PROOF. Let $\gamma = (a, b, c) \in \Gamma$.

(1) We have $[\gamma, \gamma'] = (0, 0, ab' - a'b) \in \ker \psi$ for all $\gamma' = (a', b', c') \in \Gamma$ if and only if $ab' \in \ker \psi$ and $ba' \in \ker \psi$ for all $a', b' \in R$, that is, if and only $a \in I_\psi$ and $b \in I_\psi$. Therefore, $[\gamma, \gamma'] \in \ker \psi$ for all $\gamma' \in \Gamma$ if and only if $\gamma \in Z_\psi$. This shows that $Z_\psi = p^{-1}(\overline{Z})$.

(2) We have $[\gamma, \gamma'] = (0, 0, ab' - a'b) \in \ker \psi$ for all $\gamma' = (a', b', c') \in N_\psi$ if and only if $ab' \in \ker \psi$ for all $b' \in R$, that is, if and only $a \in I_\psi$. Therefore, $[\gamma, \gamma'] \in \ker \psi$ for all $\gamma' \in N_\psi$ if and only if $\gamma \in N_\psi$. This shows that $\overline{N_\psi} := p(N_\psi)$ is a maximal abelian subgroup in $\overline{\Gamma}$ and that $N_\psi = p^{-1}(\overline{N_\psi})$. □

Let $\psi \in \widehat{Z}$; view the ring R and the ideal I_ψ of R defined above as abelian groups. For $(\alpha, \beta) \in \widehat{I_\psi} \times \widehat{R}$, define $\chi_{\alpha,\beta,\psi} : N_\psi \to \mathbf{T}$ by the formula

$$\chi_{\alpha,\beta,\psi}(a, b, c) = \alpha(a)\beta(b)\psi(c) \quad \text{for } a \in I_\psi, b, c \in R.$$

The following lemma gives a concrete description of the sets

$$\widehat{Z_\psi}(\psi) := \{\chi \in \mathrm{Hom}(Z_\psi, \mathbf{T}) \mid \chi|_Z = \psi\}$$

and

$$\widehat{N}_\psi(\psi) := \{\chi \in \mathrm{Hom}(N_\psi, \mathbf{T}) \mid \chi|_Z = \psi\}.$$

LEMMA 3.B.8. *Let $\psi \in \widehat{Z}$. With the notation introduced above, the maps*

$$\widehat{I_\psi} \times \widehat{I_\psi} \to \widehat{Z_\psi}(\psi), \qquad (\alpha, \beta) \mapsto \chi_{\alpha,\beta,\psi},$$
$$\widehat{I_\psi} \times \widehat{R} \to \widehat{N_\psi}(\psi), \qquad (\alpha, \beta) \mapsto \chi_{\alpha,\beta,\psi}$$

are bijections.

PROOF. We give the proof for $\widehat{N_\psi}(\psi)$. That for $\widehat{Z_\psi}(\psi)$ is similar.

Let $(\alpha, \beta) \in \widehat{I_\psi} \times \widehat{R}$. Then $\chi_{\alpha,\beta,\psi}$ is a unitary character of N_ψ. Indeed, let $\gamma = (a, b, c)$ and $\gamma' = (a', b', c')$ in N_ψ. Since $ab' \in \ker \psi$, we have

$$\begin{aligned}
\chi_{\alpha,\beta,\psi}(\gamma\gamma') &= \chi_{\alpha,\beta,\psi}(a + a', b + b', c + c' + ab') \\
&= \alpha(a + a')\beta(b + b')\psi(c + c')\psi(ab') \\
&= \alpha(a + a')\beta(b + b')\psi(c + c') \\
&= \alpha(a)\alpha(a')\beta(b)\beta(b')\psi(c)\psi(c') \\
&= \chi_{\alpha,\beta,\psi}(\gamma)\chi_{\alpha,\beta,\psi}(\gamma').
\end{aligned}$$

It is obvious that $\chi_{\alpha,\beta,\psi}|_Z = \psi$.

Conversely, let χ be a unitary character of N_ψ such that $\chi|_Z = \psi$. Let $a, a' \in I_\psi$ and $b, b' \in R$. Since $\chi(0, 0, ab') = \psi(ab') = 1$, we have

$$\begin{aligned}
\chi(a + a', b + b', 0) &= \chi((a + a', b + b', 0)(0, 0, ab')) \\
&= \chi(a + a', b + b', ab') \\
&= \chi((a, b, 0)(a', b', 0)) \\
&= \chi(a, b, 0)\chi(a', b', 0).
\end{aligned}$$

This shows that $(a, b) \mapsto \chi(a, b, 0)$ is a unitary character of the group $I_\psi \times R$, which can therefore be written as $(a, b) \mapsto \alpha(a)\beta(b)$ for a uniquely defined pair $(\alpha, \beta) \in \widehat{I_\psi} \times \widehat{R}$. Moreover, since $\chi|_Z = \psi$ and since Z is contained in the centre of N_ψ, it follows that $\chi = \chi_{\alpha,\beta,\psi}$. □

We are ready to construct a family of irreducible representations of $H(R)$.

THEOREM 3.B.9. *Let R be a commutative ring with unit, Γ the Heisenberg group $H(R)$, and Z the centre of Γ. Let $\psi \in \widehat{Z}$ be a unitary character; let N_ψ be the normal subgroup of Γ defined just before Lemma* 3.B.7.

(1) *Let χ be a unitary character of N_ψ such that $\chi|_Z = \psi$. Then $\mathrm{Ind}_{N_\psi}^\Gamma \chi$ is an irreducible representation of Γ, and its central character is ψ.*
(2) *Let χ_1, χ_2 be unitary characters of N_ψ such that $\chi_1|_Z = \chi_2|_Z = \psi$, and $\chi_1^\gamma \neq \chi_2$ for every $\gamma \in \Gamma$. Then $\mathrm{Ind}_{N_\psi}^\Gamma \chi_1$ and $\mathrm{Ind}_{N_\psi}^\Gamma \chi_2$ are nonequivalent irreducible representations of Γ.*
(3) *Let $\psi' \in \widehat{Z}$, with $\psi' \neq \psi$. Let $\chi \in \widehat{N_\psi}$ be such that $\chi|_Z = \psi$ and $\chi' \in \widehat{N_{\psi'}}$ be such that $\chi'|_Z = \psi'$. Then the irreducible representations $\mathrm{Ind}_{N_\psi}^\Gamma \chi$ and $\mathrm{Ind}_{N_{\psi'}}^\Gamma \chi'$ of Γ are not equivalent.*

PROOF. (1) To show that $\mathrm{Ind}_{N_\psi}^\Gamma \chi$ is irreducible, by the Mackey–Shoda criterion (Corollary 1.F.16), it suffices to check that $\chi^{\gamma^{-1}} \neq \chi$ for all $\gamma \in \Gamma \smallsetminus N_\psi$. An argument similar to that of Lemma 3.B.3(3) shows that this is the case.

Indeed, let $\gamma \in \Gamma$ be such that $\chi^{\gamma^{-1}} = \chi$. Since $\chi|_Z = \psi$, we have

$$\chi(\gamma') = \chi^{\gamma^{-1}}(\gamma') = \chi(\gamma'[\gamma', \gamma]) = \psi([\gamma', \gamma])\chi(\gamma') \quad \text{for all } \gamma' \in N_\psi.$$

Therefore $[N_\psi, \gamma] \subset \ker \psi$, and hence $\gamma \in N_\psi$ since the image of N_ψ in $\Gamma / \ker \psi$ is a maximal abelian subgroup (Lemma 3.B.7).

For the claim concerning the central character, see the proof of Theorem 3.B.4(1).

(2) The claim is a direct consequence of the Mackey–Shoda equivalence criterion (Corollary 1.F.20).

(3) The representations $\mathrm{Ind}_{N_\psi}^\Gamma \chi$ and $\mathrm{Ind}_{N_{\psi'}}^\Gamma \chi'$ are not equivalent because their central characters are not equal. □

In the situation of Theorem 3.B.9, the representations $\mathrm{Ind}_{N_\psi}^\Gamma \chi$ and $\mathrm{Ind}_{N_\psi}^\Gamma \chi^\gamma$ are equivalent for all $\gamma \in \Gamma$, by Proposition 1.F.9. The theorem can therefore be restated as follows.

COROLLARY 3.B.10. *Let the notation be as in Theorem* 3.B.9. *For $\psi \in \widehat{Z}$, let the normal subgroup N_ψ of Γ and the subset $\widehat{N_\psi}(\psi) = \{\chi \in \widehat{N_\psi} \mid \chi|_Z = \psi\}$ of the dual of N_ψ be as above. Define*

$$X := \bigsqcup_{\psi \in \widehat{Z}} \{\psi\} \times \widehat{N_\psi}(\psi),$$

a set equipped with the right Γ-action

$$X \times \Gamma \to X, \quad ((\psi, \chi), \gamma) \mapsto (\psi, \chi^\gamma).$$

Then, for every $(\psi, \chi) \in X$, the representation $\mathrm{Ind}_{N_\psi}^\Gamma \chi$ is irreducible and has central character ψ. The map

$$X \to \widehat{\Gamma}, \quad (\psi, \chi) \mapsto \mathrm{Ind}_{N_\psi}^\Gamma \chi$$

factorizes into an injective map from the space of Γ-orbits in X to the dual $\widehat{\Gamma}$ of Γ.

We will consider two particular cases: $R = \mathbf{K}$, a field, and $R = \mathbf{Z}$, the ring of integers.

Heisenberg group over a field. Let $\mathbf{K}$ be a field and $\Gamma = H(\mathbf{K})$. The centre of Γ is $Z = \{(0,0,c) \mid c \in \mathbf{K}\} \approx \mathbf{K}$ and the subgroup $N = \{\{(0,b,c) \mid b,c \in \mathbf{K}\} \approx \mathbf{K}^2$ is maximal abelian. Fix a unitary character $\psi \in \widehat{Z} \approx \widehat{\mathbf{K}}$; let I_ψ be defined as before Lemma 3.B.7. Two cases may occur:

- If $\psi = 1_Z$, then $I_\psi = \mathbf{K}$ and hence $N_\psi = \Gamma$.
- If $\psi \neq 1_Z$, then $I_\psi = \{0\}$ and hence $N_\psi = N$.

Corollary 3.B.10 therefore takes the following form.

COROLLARY 3.B.11. *Let $\mathbf{K}$ be a field, $\Gamma = H(\mathbf{K})$, with centre Z, and $p : \Gamma \twoheadrightarrow \Gamma/Z$ the canonical projection. Let N be the maximal abelian subgroup $\{(0,b,c) \in \Gamma \mid b,c \in \mathbf{K}\}$. For $\psi \in \widehat{Z}$, set $\widehat{N}(\psi) = \{\chi \in \widehat{N} \mid \chi|_Z = \psi\}$. Define*

$$X := \widehat{\Gamma/Z} \ \sqcup \ \left(\sqcup_{\psi \in \widehat{Z} \smallsetminus \{1_Z\}} \{\psi\} \times \widehat{N}(\psi)\right).$$

Then for all $(\psi,\chi) \in \bigsqcup_{\psi \in \widehat{Z} \smallsetminus \{1_Z\}} \{\psi\} \times \widehat{N}(\psi)$, the representation $\mathrm{Ind}_N^\Gamma \chi$ is irreducible and has central character ψ. The map $X \to \widehat{\Gamma}$, defined by

$$\chi \mapsto \chi \circ p \quad \textit{for } \chi \in \widehat{\Gamma/Z},$$

and

$$(\psi,\chi) \mapsto \mathrm{Ind}_N^\Gamma \chi \quad \textit{for } \psi \in \widehat{Z} \smallsetminus \{1_Z\},\ \chi \in \widehat{N}(\psi),$$

factorizes into an injective map from the space of Γ-orbits in X to the dual $\widehat{\Gamma}$ of Γ.

REMARK 3.B.12. Let $(\psi,\chi) \in \bigsqcup_{\psi \in \widehat{Z} \smallsetminus \{1_Z\}} \{\psi\} \times \widehat{N}(\psi)$. The irreducible representation $\mathrm{Ind}_N^\Gamma \chi$ is finite-dimensional if $\mathbf{K}$ (and therefore Γ) is finite, and infinite-dimensional if $\mathbf{K}$ is infinite.

When $\mathbf{K}$ is finite, the map $X \to \widehat{\Gamma}$ in Corollary 3.B.11 is surjective, onto $\widehat{H(\mathbf{K})}$; see Corollary 4.B.3. When $\mathbf{K}$ is infinite, this is far from being true; see Remark 5.E.2(1).

REMARK 3.B.13. Fields, considered as discrete abelian groups for the addition, have duals which are compact abelian groups; these duals appear above in Corollary 3.B.11, since $Z \approx \mathbf{K}$, and again below, in Remarks 3.C.6 and 5.E.2(2) concerning $\mathrm{Aff}(\mathbf{K})$, and in Lemma 9.F.1 concerning $\mathrm{GL}_n(\mathbf{K})$. They can be described as follows.

The dual $\widehat{\mathbf{Q}}$ of $\mathbf{Q}$ is isomorphic to the quotient $\mathbf{A_Q}/\mathbf{Q}$ of the rational adele ring by its cocompact discrete subgroup $\mathbf{Q}$. There is a nice exposition of this fact in [**Conr**].

Let $\mathbf{K}$ be a field of characteristic 0, viewed as a discrete abelian group. Since $\mathbf{K}$ is a vector space over $\mathbf{Q}$, it is a direct sum of copies of $\mathbf{Q}$, hence its dual $\widehat{\mathbf{K}}$ is isomorphic to a direct product of copies of $\mathbf{A_Q}/\mathbf{Q}$.

Let p be a prime. The dual of the finite field $\mathbf{F}_p$ of order p is isomorphic (noncanonically) to $\mathbf{F}_p$ itself.

Let $\mathbf{K}$ be a field of characteristic p, viewed as a discrete abelian group. Since $\mathbf{K}$ is a direct sum of copies of $\mathbf{F}_p$, the dual $\widehat{\mathbf{K}}$ is isomorphic to a direct product of copies of $\mathbf{F}_p$.

Multiplicative groups of fields also appear below, for example in the description of the primitive dual of the affine group of an infinite field (Section 9.C).

For the field of rational, the fundamental theorem of arithmetics shows that any $x \in \mathbf{Q}^\times$ is of the form $\pm p_1^{n_1} \cdots p_k^{n_k}$, for some $k \geq 0$, distinct primes $p_1, \dots, p_k$, and integers $n_1, \dots, n_k \in \mathbf{Z} \smallsetminus \{0\}$. It follows that the group $\mathbf{Q}^\times$ is isomorphic to the direct sum of $\mathbf{Z}/2\mathbf{Z}$ and a free abelian group of countable infinite rank, hence that $\widehat{\mathbf{Q}^\times}$ is isomorphic to the direct product of $\mathbf{Z}/2\mathbf{Z}$ and a countable infinite number of copies of $\mathbf{T}$. For $\mathbf{K}$ a finite extension of $\mathbf{Q}$ (a number field), we similarly have $\mathbf{K}^\times$ isomorphic to the direct sum of a finite cyclic group of even order $\mathbf{Z}/2m\mathbf{Z}$ and a free abelian group of countable infinite rank [**Fuc73**, Chapter XVIII], hence $\widehat{\mathbf{K}^\times}$ is isomorphic to the direct product of $\mathbf{Z}/2m\mathbf{Z}$ and a countable infinite number of copies of $\mathbf{T}$.

Let $\mathbf{K}$ be an infinite algebraic extension of $\mathbf{F}_p$, for some prime p. There exists an infinite sequence of positive integers $(d_j)_{j\geq 1}$ such that $d_1 = 1$ and $d_j \geq 2$ for all $j \geq 2$, such that the following holds; we set $n_k = \prod_{j=1}^k d_j$ and $m_k = p^{n_k} - 1$ for all $k \geq 1$. The field $\mathbf{K}$ is the union of a nested sequence of finite field $F_p \leq F_{p^{n_2}} \leq \cdots \leq F_{p^{n_k}} \leq \cdots \leq \mathbf{K} = \bigcup_{k=1}^\infty F_{p^{n_k}}$. It follows that the multiplicative group $\mathbf{K}^\times$ is a direct union of an increasing sequence of cyclic groups, of the form

$$\mathbf{Z}/m_1\mathbf{Z} \leq \mathbf{Z}/m_2\mathbf{Z} \leq \cdots \leq \mathbf{Z}/m_k\mathbf{Z} \leq \cdots \leq \mathbf{K}^\times = \bigcup_{k=1}^\infty (\mathbf{Z}/m_k\mathbf{Z}),$$

hence $\widehat{\mathbf{K}^\times}$ is a projective limit of cyclic groups $\widehat{\mathbf{K}^\times} = \operatorname{proj\,lim}_k (\mathbf{Z}/m_k\mathbf{Z})$.

Heisenberg group over the integers. Let $\Gamma = H(\mathbf{Z})$ be the Heisenberg group over $\mathbf{Z}$. The irreducible representations of this group which appear below can also be found in [**Foll–16**, Pages 210–218]. Further irreducible representations of $H(\mathbf{Z})$ will be constructed in Chapter 5 (see Remark 5.E.2).

The centre of Γ is $Z = \{(0,0,c) \in \Gamma \mid c \in \mathbf{Z}\} \approx \mathbf{Z}$ and we have

$$\widehat{Z} = \{\psi_\theta \mid \theta \in [0,1[\} \approx \mathbf{T},$$

where, for $\theta \in [0,1[$, the character ψ_θ is defined by

$$\psi_\theta(0,0,c) = e^{2\pi i\theta c} \quad \text{for all } (0,0,c) \in Z.$$

We denote by $\widehat{Z}_\infty$ the subset of $\widehat{Z}$ of elements of infinite order, i.e., of faithful characters of $\mathbf{Z}$, i.e.,

$$\widehat{Z}_\infty = \{\psi_\theta \mid \theta \in]0,1] \text{ and } \theta \notin \mathbf{Q}\}.$$

Fix $\psi = \psi_\theta \in \widehat{Z}_\infty$. Since ψ is faithful, the corresponding ideal is $I_{\psi_\theta} = \{0\}$ (see the notation as before Theorem 3.B.9), so that N_ψ coincides with the maximal abelian subgroup $N = \{(0,b,c) \mid b,c \in \mathbf{Z}\}$. Define $\widehat{N}(\psi)$ as in Theorem 3.B.4, i.e., $\widehat{N}(\psi) = \{\chi \in \widehat{N} \mid \chi|_Z = \psi\}$. Then

$$\widehat{N}(\psi) = \{\chi_{\beta,\theta} \mid \beta \in [0,1[\} \approx \mathbf{T},$$

where, for $\beta, \theta \in [0,1[$, the character $\chi_{\beta,\theta} \in \widehat{N}$ is defined by

$$\chi_{\beta,\theta}(0,b,c) = e^{2\pi i(\beta b + \theta c)} \quad \text{for all } (0,b,c) \in N.$$

Consider $\chi = \chi_{\theta,\beta} \in \widehat{N}(\psi)$. For every $\gamma = (a,b,c) \in \Gamma$ and $\gamma' = (0,b',c') \in N$, we have

$$\begin{aligned}
\chi^\gamma(\gamma') &= \chi(\gamma\gamma'\gamma^{-1}) = \chi(\gamma')\chi([\gamma',\gamma^{-1}]) = \chi(\gamma')\psi([\gamma',\gamma]) \\
&= e^{2\pi i(\beta b' + \theta c')} e^{2\pi i\theta(ab')} = e^{2\pi i[(\beta+\theta a)b' + \theta c')]} \\
&= \chi_{\beta + a\theta, \theta}(0,b',c').
\end{aligned}$$

Therefore the Γ-orbit of $\chi_{\beta,\theta}$ is

$$\{\chi_{\beta+a\theta,\theta} \in \widehat{N} \mid a \in \mathbf{Z}\}.$$

Observe that $\{\beta + \theta a \pmod 1) \mid a \in \mathbf{Z}\}$ is the orbit of β under the action of $\mathbf{Z}$ on $[0,1[$ by the powers of the irrational rotation $R_\theta : \mathbf{T} \to \mathbf{T}$.

The following result is a direct consequence of Corollary 3.B.10.

COROLLARY 3.B.14. *Consider the Heisenberg group $\Gamma = H(\mathbf{Z})$ over $\mathbf{Z}$, its centre $Z = \{(0,0,c) \in \Gamma \mid c \in \mathbf{Z}\}$, and its maximal abelian subgroup $N = \{(0,b,c) \in \Gamma \mid b,c \in \mathbf{Z}\}$. For each $\theta \in [0,1[$ irrational, consider as above*
the unitary character $\psi_\theta \in \widehat{Z}_\infty$,
the subset $\widehat{N}(\psi_\theta) = \{\chi_{\beta,\theta} \in \widehat{N} \mid \beta \in [0,1[\} \approx \mathbf{T}$ of the character group $\widehat{N}$,
the irrational rotation $R_\theta : [0,1[\to [0,1[$, $\beta \mapsto \beta + \theta \pmod 1$,
and the orbit space $\mathcal{O}_\theta = [0,1[/R_\theta$.
Then:

1. *For each $\beta \in [0,1[$ and irrational $\theta \in [0,1[$, the representation $\mathrm{Ind}_N^\Gamma \chi_{\beta,\theta}$ of Γ is irreducible, infinite-dimensional, with central character ψ_θ.*
2. *For each irrational $\theta \in [0,1[$, the map $[0,1[\mapsto \widehat{\Gamma}$, $\beta \mapsto \mathrm{Ind}_N^\Gamma \chi_{\beta,\theta}$ factorizes into an injective map from $\mathcal{O}_\theta$ to $\widehat{\Gamma}$.*
3. *For $\theta, \theta' \in [0,1[$ irrational and distinct, the images of the corresponding maps in (2) are disjoint, so that these maps provide an injective map*

$$\bigsqcup_{\theta \in [0,1[, \theta \notin \mathbf{Q}} \mathcal{O}_\theta \hookrightarrow \widehat{\Gamma}.$$

REMARK 3.B.15. (1) For an irrational $\theta \in [0,1[$ and for $\beta \in [0,1[$, let $\pi_{\beta,\theta} := \mathrm{Ind}_N^\Gamma \chi_{\beta,\theta}$ be the irreducible representation of Γ of Corollary 3.B.14.

We choose $H = \{(a,0,0) \in \Gamma \mid a \in \mathbf{Z}\} \approx \mathbf{Z}$ as a right transversal for N in Γ. Construction 1.F.4(2), of which we keep the notation here, shows that the induced representation $\pi_{\beta,\theta}$ can be realized on the Hilbert space $\ell^2(\mathbf{Z})$ as follows. For $t = (k,0,0)$ in the transversal and $g = (a,b,c) \in \Gamma$, we have

$$(k,0,0)(a,b,c) \;=\; (k+a, b, c+kb) \;=\; (0,b,c+kb)(k+a,0,0),$$

hence

$$\alpha(t,g) \;=\; (0,b,c+kb) \quad \text{and} \quad t \cdot g \;=\; (k+a,0,0).$$

Thus

$$(\pi_{\beta,\theta}(a,b,c)f)(k) \;=\; \chi_{\beta,\theta}(0,b,c+kb)f(a+k) \;=\; e^{2\pi i\theta c}e^{2\pi i(\beta+\theta k)b}f(a+k)$$

for $f \in \ell^2(\mathbf{Z})$, $(a,b,c) \in \Gamma$, and $k \in \mathbf{Z}$.

(2) The map of Corollary 3.B.14(3) above is far from being surjective (see Remarks 5.E.2).

(3) Given a unitary character ψ_θ of the centre Z of $\Gamma = H(\mathbf{Z})$ with irrational θ, Corollary 3.B.14 shows that every Γ-orbit in $\widehat{N}(\psi_\theta)$ provides one equivalence class of irreducible representations of Γ with central character ψ_θ. Moreover, these orbits are in a one-to-one correspondence with the orbits of the irrational rotation R_θ on $\mathbf{T}$. Since there are uncountably many R_θ-orbits, there are uncountably many nonequivalent irreducible representations of Γ with central character ψ_θ.

We will show in Section 6.C that the dual space $\widehat{\Gamma}$ of Γ is equipped with a natural Borel structure, called the Mackey–Borel structure. We now deduce from Corollary 3.B.14 that the Borel space $\widehat{\Gamma}$ is not countably separated. This is a manifestation of the fact that Γ is not type I (see Section 6.D for this notion) and is a particular instance of Glimm's Theorem of Chapter 8.

COROLLARY 3.B.16. *Let $\Gamma = H(\mathbf{Z})$ be the Heisenberg group over $\mathbf{Z}$. The Borel space $\widehat{\Gamma}$ is not countably separated.*

PROOF. Fix an irrational number $\theta \in [0, 1[$. It is well known that the rotation R_θ is ergodic with respect to the Lebesgue measure μ on $[0, 1[$ (see for instance [**BeMa–00**, Chapter I, Example 1.4.ii]). Since μ is not supported by an R_θ-orbit, this implies that the orbit space $\mathcal{O}_\theta = [0, 1[/R_\theta$ is not countably separated for its canonical quotient Borel structure (see [**Zimm–84**, Proposition 2.1.10]). In order to show that $\widehat{\Gamma}$ is not countably separated, it suffices therefore to show that the map

$$[0, 1[\to \widehat{\Gamma}, \ \ \beta \mapsto \mathrm{Ind}_N^\Gamma \chi_{\beta,\theta}$$

of Corollary 3.B.14(2) is measurable.

As shown in Remark 3.B.15, the representations $\pi_{\beta,\theta} = \mathrm{Ind}_N^\Gamma \chi_{\beta,\theta}$ are realized in $\ell^2(\mathbf{Z})$, for all $\beta \in [0, 1[$. By definition of the Mackey–Borel structure on $\widehat{\Gamma}$, see Section 6.C, it suffices to prove that the functions $\beta \mapsto \langle \pi_{\beta,\theta}(a, b, c) f_1 \mid f_2 \rangle$ are Borel measurable, for all $(a, b, c) \in \Gamma$ and $f_1, f_2 \in \ell^2(\mathbf{Z})$.

Let $(a, b, c) \in \Gamma$ and $f_1, f_2 \in \ell^2(\mathbf{Z})$. As measurability is inherited by linear combinations and pointwise limits, we can assume that $f_1 = \delta_{k_1}$ and $f_2 = \delta_{k_2}$ for $k_1, k_2 \in \mathbf{Z}$. In this case, by the formula of Remark 3.B.15 for $\pi_{\theta,\beta}$, we have

$$\langle \pi_{\beta,\theta}(a, b, c)\delta_{k_1} \mid \delta_{k_2} \rangle = \begin{cases} e^{2\pi i \theta c} e^{2\pi i(\beta + \theta(k_1 - a))b} & \text{if } k_2 = k_1 - a, \\ 0 & \text{if } k_2 \neq k_1 - a. \end{cases}$$

Clearly, this shows that $\beta \mapsto \langle \pi_{\beta,\theta}(a, b, c)\delta_{k_1} \mid \delta_{k_2} \rangle$ is measurable. □

We now turn to the construction of irreducible representations of $\Gamma = H(\mathbf{Z})$ with a nonfaithful central character.

For this, fix a character $\psi = \psi_\theta \in \widehat{Z}$ for $\theta \in [0, 1[$ **rational**. Write

$$\theta = \frac{p}{n} \quad \text{where} \quad n \in \mathbf{N}^*, \ p \in \{0, 1, \dots, n-1\} \ \text{and} \ (p, n) = 1,$$

so that ψ is of order n. Then (with the notation as before Theorem 3.B.9), we have

$$I_{\psi_\theta} = \ker \psi_\theta = n\mathbf{Z},$$

and N_ψ coincides with the normal subgroup

$$\Gamma(n) = \{(a, b, c) \in \Gamma \mid a \in n\mathbf{Z}, \ b \in \mathbf{Z}, \ c \in \mathbf{Z}\}$$

of Γ.

We parametrize the dual of $n\mathbf{Z}$ by $[0, 1/n[$: every element of $\widehat{n\mathbf{Z}}$ is of the form $n\mathbf{Z} \ni a \mapsto e^{2\pi i \alpha a} \in \mathbf{T}$ for a unique $\alpha \in [0, 1/n[$.

By Lemma 3.B.8, every unitary character χ of $\Gamma(n)$ such that $\chi|_Z = \psi_\theta$ is of the form $\chi_{\alpha,\beta,\theta}$ for a unique pair $(\alpha, \beta) \in [0, 1/n[\times [0, 1[$, where

$$\chi_{\alpha,\beta,\theta}(a, b, c) = e^{2\pi i(\alpha a + \beta b + \theta c)} \quad \text{for } a \in n\mathbf{Z}, \ b \in \mathbf{Z}, \ c \in \mathbf{Z}.$$

For every $\gamma' = (a', b', c') \in \Gamma$ and $\gamma = (a, b, c) \in \Gamma(n)$, we have $\theta ab' \in \mathbf{Z}$ and

$$\begin{aligned}\chi^{\gamma'}_{\alpha,\beta,\theta}(\gamma) &= \chi_{\alpha,\beta,\theta}(\gamma)\psi_\theta([\gamma,(\gamma')^{-1}]) \\ &= e^{2\pi i(\alpha a+\beta b+\theta c)}e^{2\pi i\theta(a'b-ab')} \\ &= e^{2\pi i(\alpha a+\beta b+\theta c)}e^{2\pi i\theta(a'b)} \\ &= e^{2\pi i[\alpha a+(\beta+a'\theta)b+\theta c]} \\ &= \chi_{\alpha,\beta+a'\theta,\theta}(\gamma).\end{aligned}$$

Therefore the Γ-orbit of $\chi_{\alpha,\beta,\theta}$ is

$$\{\chi_{\alpha,\beta+a'\theta,\theta} \in \widehat{N} \mid a' \in \mathbf{Z}\} = \{\chi_{\alpha,\beta+\frac{k}{n},\theta} \in \widehat{N} \mid k \in \mathbf{Z}\}$$

and so contains a character $\chi_{\alpha,\beta',\theta}$ for a unique pair (α, β') in $[0, 1/n[\times [0, 1/n[$.

In view of the previous remarks, the following result is a direct consequence of Corollary 3.B.10.

COROLLARY 3.B.17. *Let n be a positive integer, $\psi = \psi_\theta$ a unitary character of Z of order n, and $(\alpha, \beta) \in [0, 1/n[\times [0, 1/n[$. Let $\Gamma(n)$ and $\chi_{\alpha,\beta,\theta}$ be defined as above.*

The induced representation

$$\pi_{\alpha,\beta,\theta} := \mathrm{Ind}_{\Gamma(n)}^{\Gamma}\chi_{\alpha,\beta,\theta}$$

is an irreducible representation of $\Gamma = H(\mathbf{Z})$ of dimension n.

Moreover, for another triple $\theta' = \frac{p'}{n'}$ and $(\alpha', \beta') \in [0, 1/n'[\times [0, 1/n'[$ of the same kind, the representations $\pi_{\alpha,\beta,\theta}$ and $\pi_{\alpha',\beta',\theta'}$ are equivalent if and only if $(\alpha, \beta, \theta) = (\alpha', \beta', \theta')$.

REMARK 3.B.18. We will see later (Corollary 4.B.5) that **every** finite-dimensional irreducible representation of $\Gamma = H(\mathbf{Z})$ is equivalent to one of the representations $\pi_{\alpha,\beta,\theta}$ of Corollary 3.B.17.

REMARK 3.B.19. Consider the real Heisenberg group $H(\mathbf{R})$, now with its natural connected and locally compact topology. Up to isomorphism, it is the only nonabelian nilpotent connected simply connected real Lie group of dimension 3. In $H(\mathbf{R})$, we consider the abelian normal closed subgroup

$$N = \begin{pmatrix} 1 & 0 & \mathbf{R} \\ 0 & 1 & \mathbf{R} \\ 0 & 0 & 1 \end{pmatrix} \subset H(\mathbf{R}) = \begin{pmatrix} 1 & \mathbf{R} & \mathbf{R} \\ 0 & 1 & \mathbf{R} \\ 0 & 0 & 1 \end{pmatrix}.$$

Again write (x, y, z) for $\begin{pmatrix} 1 & x & z \\ 0 & 1 & y \\ 0 & 0 & 1 \end{pmatrix}$. The unitary characters of $H(\mathbf{R})$ and N are

$$\begin{aligned}\chi_{u,v} &: H(\mathbf{R}) \to \mathbf{T}, \qquad (x, y, z) \mapsto e^{2\pi i(ux+vy)} \qquad \text{for } u, v \in \mathbf{R}, \\ \chi_{v,w} &: N \to \mathbf{T}, \qquad (0, y, z) \mapsto e^{2\pi i(vy+wz)} \qquad \text{for } v, w \in \mathbf{R}.\end{aligned}$$

It is a version of the Stone–von Neumann Theorem [**Ston–30, vNeu–31**] that the dual of $H(\mathbf{R})$ can be seen as a disjoint union:

$$\widehat{H(\mathbf{R})} = \mathbf{R}^2 \sqcup \mathbf{R}^\times,$$

where

- $\mathbf{R}^2$ is identified with the space of unitary characters of $H(\mathbf{R})$, i.e., the dual of the abelian group $H(\mathbf{R})_{\mathrm{ab}} = H(\mathbf{R})/[H(\mathbf{R}), H(\mathbf{R})] \approx \mathbf{R}^2$;
- $\mathbf{R}^\times$ is identified with the space of equivalence classes of infinite-dimensional representations $\mathrm{Ind}_N^{H(\mathbf{R})} \chi_{0,w}$, for $w \in \mathbf{R}^\times$.

For the Fell topology, the induced topology on either $\mathbf{R}^2$ or $\mathbf{R}^\times$ is the usual one, but a sequence of points in $\mathbf{R}^\times$ that tends to 0 in the usual sense tends to all points of $\mathbf{R}^2$ in $\mathbf{R}^2 \sqcup \mathbf{R}^\times = \widehat{H(\mathbf{R})}$.

One method to obtain these results is to use the Mackey machine. The group $H(\mathbf{R})$ acts naturally on the dual of N, that is, on

$$\widehat{N} = \{\chi_{v,w} \mid (v,w) \in \mathbf{R}^2\}.$$

The set of orbits splits into two parts:

- the set of one-point orbits, i.e., of orbits $\{\chi_{v,0}\}$, where $v \in \mathbf{R}$,
- the set of line-orbits $\mathcal{O}_w = \{\chi_{v,w} \mid v \in \mathbf{R}\}$, where $w \in \mathbf{R}^\times$.

The one-point orbits provide the unitary characters of $H(\mathbf{R})$, and the line-orbits provide the infinite-dimensional irreducible representations of $H(\mathbf{R})$. Details can be found in [**Fell–62**]; see also [**Foll–16**, Section 6.7].

More generally, let G be a nilpotent, simply connected, connected, real Lie group with Lie algebra $\mathfrak{g}$; the natural action of G on the dual space $\mathfrak{g}^*$ is called the coadjoint representation. Kirillov has constructed a map from the space of coadjoint orbits $\mathfrak{g}^*/G$ to the dual $\widehat{G}$, has shown that it is a continuous bijection, and has conjectured that it is a homeomorphism [**Kiri–62**]; the conjecture has been established by Brown [**Brow–73b**]. This has been generalized to soluble, simply connected, connected, real Lie groups for which the exponential map is surjective [**LeLu–94**]. Indeed, the orbit method has shown to be a fruitful inspiration for very large classes of groups and various problems [**Kiri–99**].

3.C. The affine group of a field

DEFINITION 3.C.1. Let $\mathbf{K}$ be field. The **affine group** over $\mathbf{K}$ is the group $\mathrm{Aff}(\mathbf{K})$ of affine transformations of $\mathbf{K}$, in matrix form

$$\mathrm{Aff}(\mathbf{K}) = \begin{pmatrix} \mathbf{K}^\times & \mathbf{K} \\ 0 & 1 \end{pmatrix} \approx \mathbf{K}^\times \ltimes \mathbf{K}.$$

The semi-direct product refers to the action $\mathbf{K}^\times \curvearrowright \mathbf{K}$ defined by $(a,t) \mapsto at$ for all $a \in \mathbf{K}^\times$ and $t \in \mathbf{K}$. The derived group and the abelianized group,

$$[\mathrm{Aff}(\mathbf{K}), \mathrm{Aff}(\mathbf{K})] = \begin{pmatrix} 1 & \mathbf{K} \\ 0 & 1 \end{pmatrix} \approx \mathbf{K} \quad \text{and} \quad \mathrm{Aff}(\mathbf{K})/[\mathrm{Aff}(\mathbf{K}), \mathrm{Aff}(\mathbf{K})] \approx \mathbf{K}^\times,$$

are both abelian, and we have a short exact sequence

$$\mathbf{K} \approx \begin{pmatrix} 1 & \mathbf{K} \\ 0 & 1 \end{pmatrix} \hookrightarrow \mathrm{Aff}(\mathbf{K}) = \begin{pmatrix} \mathbf{K}^\times & \mathbf{K} \\ 0 & 1 \end{pmatrix} \twoheadrightarrow \mathrm{Aff}(\mathbf{K})/[\mathrm{Aff}(\mathbf{K}), \mathrm{Aff}(\mathbf{K})] \approx \mathbf{K}^\times.$$

In this section, we write Γ for the group $\mathrm{Aff}(\mathbf{K})$, and N for its derived group. We identify N with $\mathbf{K}$ and Γ/N with $\mathbf{K}^\times$.

The action $\mathbf{K}^\times \curvearrowright \mathbf{K}$, $(a,t) \mapsto at$, extends to an action $\Gamma \curvearrowright \mathbf{K}$ defined by

$$\begin{pmatrix} a & s \\ 0 & 1 \end{pmatrix} \cdot t = at \quad \text{for all} \quad \begin{pmatrix} a & s \\ 0 & 1 \end{pmatrix} \in \Gamma,\ t \in \mathbf{K},$$

with kernel $[\Gamma, \Gamma]$. By duality, there is a corresponding action $\widehat{\mathbf{K}} \curvearrowleft \mathbf{K}^\times$, $(\chi, a) \mapsto \chi^a$, defined by

$$\chi^a(t) = \chi(at) \qquad \text{for all } \chi \in \widehat{\mathbf{K}},\ a \in \mathbf{K}^\times, t \in \mathbf{K}.$$

As above, it extends to an action $\widehat{\mathbf{K}} \curvearrowleft \Gamma$ with kernel $[\Gamma, \Gamma]$.

LEMMA 3.C.2. *We keep the notation above.*

(1) *The group Γ/N acts freely on $\widehat{N} \smallsetminus \{1_N\}$.*
(2) *If $\mathbf{K}$ is finite, the action of Γ on $\widehat{N} \smallsetminus \{1_N\}$ is transitive.*
(3) *If $\mathbf{K}$ is infinite, the Γ-orbit of every $\chi \in \widehat{N} \smallsetminus \{1_N\}$ is dense in $\widehat{N}$.*

PROOF. (1) Let $\chi \in \widehat{N}$, $\chi \neq 1_N$, and $a \in \mathbf{K}^\times$. Since $\chi^a(t) = \chi(at)$ for all $t \in \mathbf{K}$, we have $\chi^a = \chi$ if and only $(a-1)t \in \ker(\chi)$ for all $t \in \mathbf{K}$. Since $\ker(\chi) \neq \mathbf{K}$, this holds if and only if $a = 1$. Therefore Γ/N acts freely on $\widehat{N} \smallsetminus \{1_N\}$.

(2) Let $\chi \in \widehat{N}$, $\chi \neq 1_N$. If $\mathbf{K}$ is finite of cardinality q, it follows from (1) that the Γ-orbit of χ has cardinality $q-1$ and the action of Γ on $\widehat{N} \smallsetminus \{1_N\}$ is therefore transitive.

(3) Write χ^0 for 1_N, so that χ^a is now defined for all $a \in \mathbf{K}$. Then

$$H := \{\chi^a \mid a \in \mathbf{K}\}$$

is a subgroup of $\widehat{N}$. Indeed, we have, for all $a, b, t \in \mathbf{K}$,

$$(\chi^a \chi^b)(t) = \chi(at)\chi(bt) = \chi((a+b)t) = \chi^{a+b}(t).$$

Assume, by contradiction, that the Γ-orbit

$$\mathcal{O} := \{\chi^a \mid a \in \mathbf{K}^\times\}$$

is not dense in $\widehat{N}$. We have $H = \mathcal{O} \cup \{1_N\}$; since $N = \mathbf{K}$ is infinite, its Pontryagin dual $\widehat{N}$ is not discrete. It follows that H is not dense in $\widehat{N}$, so that the quotient group $\widehat{N}/\overline{H}$ is a nontrivial LCA group. By Pontryagin duality there exists $t_0 \in \mathbf{K} \smallsetminus \{0\}$ such that $\chi^a(t_0) = 1$ for all $\chi^a \in H$, that is, such that $\chi(at_0) = 1$ for all $a \in \mathbf{K}$. This implies that $\chi(t) = 1$ for all $t \in \mathbf{K}$, which contradicts the fact that $\chi \neq 1_N$. □

THEOREM 3.C.3. *Let $\Gamma = \mathrm{Aff}(\mathbf{K})$ be the group of affine transformations of a field $\mathbf{K}$ and $N \approx \mathbf{K}$ the derived group of Γ, as above.*

(1) *For every $\chi \in \widehat{\mathbf{K}} \smallsetminus \{1_\mathbf{K}\}$, the induced representation $\mathrm{Ind}_N^\Gamma \chi$ is irreducible.*
(2) *Let $\chi_1, \chi_2 \in \widehat{\mathbf{K}} \smallsetminus \{1_\mathbf{K}\}$ be such that $\chi_1^\gamma \neq \chi_2$ for every $\gamma \in \Gamma$. Then $\mathrm{Ind}_N^\Gamma \chi_1$ and $\mathrm{Ind}_N^\Gamma \chi_2$ are nonequivalent irreducible representations of Γ.*

Concerning (2) above, observe that, for $\chi \in \widehat{K}$ and $\gamma = (b, t) \in \Gamma$, we have $\chi^\gamma = \chi^b$.

PROOF. Claim (1) follows from the Mackey–Shoda criterion for irreducibility of monomial representations (Corollary 1.F.16) and from Lemma 3.C.2(1). Claim (2) follows from the Mackey–Shoda criterion for equivalence of monomial representations (Corollary 1.F.20). □

Theorem 3.C.3 can be restated as follows.

PROPOSITION 3.C.4. *Let* $\Gamma = \mathrm{Aff}(\mathbf{K})$ *and* $N \approx \mathbf{K}$ *be as in Theorem* 3.C.3. *The map*

$$\widehat{\mathbf{K}} \to \widehat{\Gamma}, \ \chi \mapsto \begin{cases} \mathrm{Ind}_N^\Gamma \chi & \text{if } \chi \in \widehat{\mathbf{K}} \smallsetminus \{1_\mathbf{K}\}, \\ 1_\Gamma & \text{if } \chi = 1_\mathbf{K} \end{cases}$$

factorizes into an injective map from the space of Γ*-orbits in* $\widehat{\mathbf{K}}$ *to the dual* $\widehat{\Gamma}$ *of* Γ.

REMARK 3.C.5. Let $\chi \in \widehat{\mathbf{K}} \smallsetminus \{1_\mathbf{K}\}$. Following Construction 1.F.4(2), the induced representation $\mathrm{Ind}_N^\Gamma \chi$ can be realized as follows.

Write (a, b) for a matrix $\begin{pmatrix} a & b \\ 0 & 1 \end{pmatrix}$ in $\mathrm{Aff}(\mathbf{K})$. The subgroup

$$\left\{(a, 0) \in \mathrm{Aff}(\mathbf{K}) \mid a \in \mathbf{K}^\times\right\} \approx \mathbf{K}^\times$$

is a right transversal for N in Γ, i.e.,

$$\Gamma = \bigsqcup_{a \in \mathbf{K}^\times} N(a, 0).$$

Therefore the induced representation $\pi_\chi := \mathrm{Ind}_N^\Gamma \chi$ acts on $\ell^2(\mathbf{K}^\times)$, and is given by

$$(\pi_\chi(a, b)f)(t) = \chi(tb)f(at) \quad \text{for all } (a, b) \in \mathrm{Aff}(\mathbf{K}), \ f \in \ell^2(\mathbf{K}^\times), \ \text{and } t \in \mathbf{K}^\times.$$

Note that, when $\mathbf{K}$ is finite, π_χ is the unique irreducible representation of $\mathrm{Aff}(\mathbf{K})$ of dimension greater than one up to equivalence (Corollary 4.B.6).

REMARK 3.C.6. Theorem 3.C.3 shows that every $\mathbf{K}^\times$-orbit in $\widehat{\mathbf{K}}$ provides one equivalence class of irreducible representations of $\Gamma = \mathrm{Aff}(\mathbf{K})$.

Assume from now on that $\mathbf{K}$ is infinite. Let $\widehat{\mathbf{K}}$ be equipped with its normalized Haar measure μ. By the uniqueness of μ as a translation-invariant probability measure, the action of $\mathbf{K}^\times$ on $\widehat{\mathbf{K}}$ preserves μ. Moreover this action is ergodic, since $\{0\}$ is the only finite orbit for the action of $\mathbf{K}^\times$ on $\widehat{\widehat{\mathbf{K}}} \approx \mathbf{K}$ (see [**BeMa–00**, Chapter I, Proposition 1.5]).

As in Corollary 3.B.16, we conclude that the Borel space $\widehat{\Gamma}$ is not countably separated. This is another illustration of Glimm's Theorem of Chapter 8.

REMARK 3.C.7. Let $\mathbf{K}$ be a topological field. The affine group $\mathrm{Aff}(\mathbf{K}) = \mathbf{K}^\times \ltimes \mathbf{K}$ has a topology as a subspace of the product $\mathbf{K} \times \mathbf{K}$ which makes it a topological group. When $\mathbf{K}$ is moreover a local field, $\mathrm{Aff}(\mathbf{K})$ is a second-countable locally compact group, of which the dual can be described as follows. The results can be obtained by using the Mackey machine.

Consider first the **orientation-preserving affine group** of the real line

$$\mathrm{Aff}(\mathbf{R})_0 = \begin{pmatrix} \mathbf{R}_+^\times & \mathbf{R} \\ 0 & 1 \end{pmatrix},$$

which is a connected locally compact group; here $\mathbf{R}_+^\times := \{a \in \mathbf{R} \mid a > 0\}$. The abelian normal closed subgroup of interest is the translation group $N = \begin{pmatrix} 1 & \mathbf{R} \\ 0 & 1 \end{pmatrix} \approx \mathbf{R}$. The dual of $\mathrm{Aff}(\mathbf{R})_0$ can be described as

$$\widehat{\mathrm{Aff}(\mathbf{R})_0} = \mathbf{R} \sqcup \{\pi_{-1}\} \sqcup \{\pi_1\} = \{\chi_t\}_{t \in \mathbf{R}} \sqcup \{\pi_{-1}\} \sqcup \{\pi_1\}.$$

The three parts correspond to the three orbits of the natural action of $\text{Aff}(\mathbf{R})_0$ [equivalently of $\mathbf{R}^\times_+$] on $\widehat{N} \approx \mathbf{R}$: the orbit $\{0\}$ corresponds to the family of unitary characters

$$\chi_t : \text{Aff}(\mathbf{R})_0 \to \mathbf{T}, \quad \begin{pmatrix} a & b \\ 0 & 1 \end{pmatrix} \mapsto a^{it}, \qquad \text{where } t \in \mathbf{R},$$

and the orbits $]-\infty, 0[$ and $]0, \infty[$ correspond to $\pi_\varepsilon = \text{Ind}_N^{\text{Aff}(\mathbf{R})_0} \rho_\varepsilon$, where $\varepsilon \in \{-1, 1\}$ and

$$\rho_\varepsilon : N \to \mathbf{T}, \quad \begin{pmatrix} 1 & b \\ 0 & 1 \end{pmatrix} \mapsto e^{\varepsilon i b}.$$

For the Fell topology, the induced topology on $\mathbf{R}$ is the usual one, the closure of $\{\pi_1\}$ is $\mathbf{R} \sqcup \{\pi_1\}$, and the closure of $\{\pi_{-1}\}$ is $\mathbf{R} \sqcup \{\pi_{-1}\}$. We refer to [**Mack–52**, Page 132] or [**Fell–62**, Page 263], or [**Foll–16**, § 6.7].

The affine group $\text{Aff}(\mathbf{R}) = \begin{pmatrix} \mathbf{R}^\times & \mathbf{R} \\ 0 & 1 \end{pmatrix}$ has two connected components, one being $\text{Aff}(\mathbf{R})_0$. The abelian normal closed subgroup to consider is the same translation group N as above. The action of $\text{Aff}(\mathbf{R})$ [equivalently of $\mathbf{R}^\times$] now has *two* orbits. There exists an infinite-dimensional irreducible representation π of $\text{Aff}(\mathbf{R})$, unique up to equivalence. The dual of $\text{Aff}(\mathbf{R})$ can be described as

$$\widehat{\text{Aff}(\mathbf{R})} = \widehat{\mathbf{R}^\times} \sqcup \{\pi\} = (\mathbf{R} \times \{0, 1\}) \sqcup \{\pi\}.$$

For $\mathbf{R}^\times = \{a \in \mathbf{R} \mid a \neq 0\}$, the dual $\widehat{\mathbf{R}^\times}$ can be identified to $\mathbf{R} \times \{0, -1\}$; a point $(t, \delta) \in \mathbf{R} \times \{0, 1\}$ is the equivalence class of the one-dimensional representation

$$\begin{pmatrix} a & b \\ 0 & 1 \end{pmatrix} \mapsto |a|^{it} \text{sign}(a)^\delta.$$

The representation π is that induced from ρ_1 (or equivalently from ρ_{-1}); equivalently, π is the representation of $\text{Aff}(\mathbf{R})$ on $L^2(\mathbf{R}^\times, dt/|t|)$ defined by

$$\left(\pi \begin{pmatrix} a & b \\ 0 & 1 \end{pmatrix} \right) \xi(t) = \exp(-2\pi i b t)\, \xi(at)$$

for all $\begin{pmatrix} a & b \\ 0 & 1 \end{pmatrix} \in \text{Aff}(\mathbf{R})$, $\xi \in L^2(\mathbf{R}^\times, dt/|t|)$, and $t \in \mathbf{R}^\times$. See [**GGPS–69**, Page 224], where the setting is more generally that of a nondiscrete locally compact field $\mathbf{K}$, rather than $\mathbf{R}$ as here. For the Fell topology, the subspace $\mathbf{R} \times \{0, -1\}$ of $\widehat{\text{Aff}(\mathbf{R})}$ has its usual topology and the closure of $\{\pi\}$ is the whole of $\widehat{\text{Aff}(\mathbf{R})}$.

It is known that the regular representation λ of $\text{Aff}(\mathbf{R})$ is equivalent to a multiple of the infinite-dimensional irreducible representation π [**KlLi–72**, Example 1, Page 510]. In other words, anticipating on the terminology of Chapter 6, λ is a factor representation of type I_∞, equivalent to $\infty\pi$, and therefore weakly equivalent to π.

A similar description holds for the dual of the affine group $\text{Aff}(\mathbf{K}) = \begin{pmatrix} \mathbf{K}^\times & \mathbf{K} \\ 0 & 1 \end{pmatrix}$ of a non-Archimedean locally compact field $\mathbf{K}$. Its dual is the union of the Pontryagin dual of $\mathbf{K}^\times$ and *one* infinite-dimensional representation π, dense in the dual for the Fell topology.

3.D. Solvable Baumslag–Solitar groups

DEFINITION 3.D.1. Let p be a prime. The corresponding **Baumslag–Solitar group** is the solvable group $\mathrm{BS}(1,p)$ defined by the presentation

$$\text{(BS)} \qquad \mathrm{BS}(1,p) \;=\; \langle t, x \mid txt^{-1} = x^p \rangle.$$

Let

$$\mathbf{Z}[1/p] \;=\; \{kp^n \mid k, n \in \mathbf{Z}\}$$

be the subring of $\mathbf{Q}$ generated by $1/p$. The group $\mathrm{BS}(1,p)$ is isomorphic to the group of matrices

$$\begin{pmatrix} p^{\mathbf{Z}} & \mathbf{Z}[1/p] \\ 0 & 1 \end{pmatrix} = \left\{ \begin{pmatrix} p^n & b \\ 0 & 1 \end{pmatrix} \in \mathrm{GL}_2(\mathbf{Z}[1/p]) \;\middle|\; n \in \mathbf{Z}, b \in \mathbf{Z}[1/p] \right\}.$$

There is a decomposition in semi-direct product

$$\mathrm{BS}(1,p) \;=\; A \ltimes N \;\approx\; \mathbf{Z} \ltimes \mathbf{Z}[1/p],$$

where

$$A = \left\{ \begin{pmatrix} p^n & 0 \\ 0 & 1 \end{pmatrix} \;\middle|\; n \in \mathbf{Z} \right\} \approx \mathbf{Z} \quad \text{and} \quad N = \left\{ \begin{pmatrix} 1 & b \\ 0 & 1 \end{pmatrix} \;\middle|\; b \in \mathbf{Z}[1/p] \right\} \approx \mathbf{Z}[1/p],$$

and where elements of $A \approx \mathbf{Z}$ act on $N \approx \mathbf{Z}[1/p]$ by multiplication by powers of p. In particular, the group $\mathrm{BS}(1,p)$ is solvable.

Two remarks are in order. The first is that the presentation (BS) defines a solvable Baumslag–Solitar groups for any $p \in \mathbf{Z}$, not only for p prime. Though part of what follows carries over to $\mathrm{BS}(1,p)$ for $p \in \mathbf{Z}$, we will stick to the examples with p prime, for simplicity. The second is that nonsolvable Baumslag–Solitar groups, despite their importance from many points of view, do not appear in this book.

Dual of the group $\mathbf{Z}[1/p]$ and p-adic solenoid. We need to determine the dual group $\widehat{N}$ of $N \approx \mathbf{Z}[1/p]$. For this, we introduce the field $\mathbf{Q}_p$ of p-adic numbers. On the one hand, we identify the ring $\mathbf{Z}[1/p]$ with its image in the product $\mathbf{Q}_p \times \mathbf{R}$, by the diagonal embedding:

$$\mathbf{Z}[1/p] \;=\; \{(a,a) \in \mathbf{Q}_p \times \mathbf{R} \mid a \in \mathbf{Z}[1/p]\}.$$

As is well known, $\mathbf{Z}[1/p]$ is a cocompact discrete subring of $\mathbf{Q}_p \times \mathbf{R}$; see, e.g., [**CoHa–16**, Example 5.C.10(2)]. On the other hand, we introduce the subgroup

$$\Delta \;:=\; \{(a,-a) \in \mathbf{Q}_p \times \mathbf{R} \mid a \in \mathbf{Z}[1/p]\} \;\approx\; \mathbf{Z}[1/p].$$

Recall that $\widehat{\mathbf{R}}$ is isomorphic to $\mathbf{R}$ through the map

$$\text{(dual } \mathbf{R}) \qquad \mathbf{R} \to \widehat{\mathbf{R}}, \quad y \mapsto e_y$$

where e_y is the unitary character defined by $e_y(t) = e^{2\pi i y t}$ for all $t \in \mathbf{R}$.

We will identify $\widehat{\mathbf{Q}_p}$ with $\mathbf{Q}_p$, in the following way. Every $s \neq 0$ in $\mathbf{Q}_p$ can be uniquely written as

$$s \;=\; \sum_{j=m}^{\infty} a_j p^j$$

for integers $m \in \mathbf{Z}$ and $a_j \in \{0, \dots, p-1\}$ for $j \geq m$, with $a_m \neq 0$. The "fractional part" $\{s\}$ of $s \in \mathbf{Q}_p$ is the number in $[0,1[\cap \mathbf{Z}[1/p]$ defined by

$$\{s\} = \begin{cases} \sum_{j=m}^{-1} a_j p^j & \text{if } m < 0, \\ 0 & \text{if } m \geq 0 \end{cases}$$

if $s \neq 0$, and by $\{s\} = 0$ if $s = 0$. Then $s = \{s\} + [s]$, where the "integer part" $[s] = s - \{s\}$ belongs to the ring $\mathbf{Z}_p$ of p-adic integers. Observe that, for $s \in \mathbf{Q}_p$, we have $s - \{s\} \in \mathbf{Z}$ if and only if $s \in \mathbf{Z}[1/p]$. A special unitary character χ of $\mathbf{Q}_p$ is defined by

$$\chi \colon \mathbf{Q}_p \to \mathbf{T}, \quad s \mapsto e^{2\pi i \{s\}}.$$

Note that $\ker(\chi) = \mathbf{Z}_p$, so that χ is locally constant, and in particular continuous. For $x \in \mathbf{Q}_p$, define $\chi_x \in \widehat{\mathbf{Q}_p}$ by $\chi_x(s) = \chi(xs)$. The map

(dual $\mathbf{Q}_p$) $$\mathbf{Q}_p \to \widehat{\mathbf{Q}_p}, \quad x \mapsto \chi_x$$

is a topological group isomorphism (see [**BTS1–2**, Chapitre 2, § 1, No 9]).

We therefore have an isomorphism

$$\Phi : \mathbf{Q}_p \times \mathbf{R} \to \widehat{\mathbf{Q}_p \times \mathbf{R}}, \quad (x,y) \mapsto \chi_x e_y.$$

Observe that $\mathbf{Z}$ acts on $\mathbf{Q}_p \times \mathbf{R}$ by multiplications by powers of p, that we have

$$\Phi(p^n x, p^n y) = \Phi(x,y)^{p^n} \quad \text{for all } n \in \mathbf{Z} \text{ and } (x,y) \in \mathbf{Q}_p \times \mathbf{R},$$

and that the closed subring Δ of $\mathbf{Q}_p \times \mathbf{R}$ is invariant by this action.

For the dual of the subgroup $\mathbf{Z}[1/p]$ of $\mathbf{Q}_p \times \mathbf{R}$, we have by Pontryagin duality

(∗) $$\widehat{\mathbf{Z}[1/p]} \approx \widehat{\mathbf{Q}_p \times \mathbf{R}} / \mathbf{Z}[1/p]^{\perp},$$

where

$$\mathbf{Z}[1/p]^{\perp} := \left\{ \chi \in \widehat{\mathbf{Q}_p \times \mathbf{R}} \mid \chi(a) = 1 \text{ for all } a \in \mathbf{Z}[1/p] \right\}.$$

LEMMA 3.D.2. *By restriction to Δ, the isomorphism $\Phi : \mathbf{Q}_p \times \mathbf{R} \to \widehat{\mathbf{Q}_p \times \mathbf{R}}$ provides an isomorphism $\Delta \to \mathbf{Z}[1/p]^{\perp}$.*

PROOF. Let $(x,y) \in \mathbf{Q}_p \times \mathbf{R}$. We have to prove that

$$(x,y) \in \Delta \Longleftrightarrow \Phi(x,y) \in \mathbf{Z}[1/p]^{\perp}.$$

Assume that $(x,y) \in \Delta$, that is, $x \in \mathbf{Z}[1/p]$ and $y = -x$. Then, for every $a \in \mathbf{Z}[1/p]$, we have

$$\Phi(x,y)(a,a) = \chi_x(a) e_{-x}(a) = e^{2\pi i \{xa\}} e^{-2\pi i xa} = e^{2\pi i (\{xa\} - xa)} = 1,$$

since $\{xa\} - xa \in \mathbf{Z}$. Therefore, $\Phi(x,y) \in \mathbf{Z}[1/p]^{\perp}$.

Conversely, assume that $\Phi(x,y) \in \mathbf{Z}[1/p]^{\perp}$, i.e.,

$$\Phi(x,y)(a,a) = e^{2\pi i (\{xa\} + ya)} = 1 \quad \text{for all } a \in \mathbf{Z}[1/p],$$

i.e., $\{xa\} + ya \in \mathbf{Z}$ for every $a \in \mathbf{Z}[1/p]$. Note that this implies $y \in \mathbf{Z}[1/p]$. Since $xa - \{xa\} \in \mathbf{Z}_p$ for all $a \in \mathbf{Q}_p$, we have

$$(x+y)a = (xa - \{xa\}) + \{xa\} + ya \in \mathbf{Z}_p \quad \text{for all } a \in \mathbf{Z}[1/p].$$

We have in particular

$$\frac{x+y}{p^n} \in \mathbf{Z}_p \quad \text{for all } n \in \mathbf{N},$$

and therefore $x + y = 0$. It follows that $x \in \mathbf{Z}[1/p]$, and $(x,y) \in \Delta$. □

The compact group

$$\mathbf{Sol}_p := (\mathbf{Q}_p \times \mathbf{R})/\Delta$$

is known as the p**-adic solenoid**. Recall that a solenoid is a topological group G such that there exists a continuous homomorphism $\mathbf{R} \to G$ with dense image. A locally compact solenoid is abelian, and either topologically isomorphic to $\mathbf{R}$ or compact. More on this particular solenoid $\mathbf{Sol}_p$ in Appendix to Chapter I of [**Robe–00**].

The following proposition shows that $\mathbf{Sol}_p$ can be identified with the dual of $N \approx \mathbf{Z}[1/p]$. It is a consequence of Lemma 3.D.2 and of $(*)$; see also Exercice 9 in [**BTS1–2**, Chapitre 2, § 2].

We write $[(x, y)]$ for the class in $\mathbf{Sol}_p$ of $(x, y) \in \mathbf{Q}_p \times \mathbf{R}$, and $[\chi_x e_y]$ for the class in $\widehat{\mathbf{Q}_p \times \mathbf{R}}/\mathbf{Z}[1/p]^\perp$ of $\chi_x e_y \in \widehat{\mathbf{Q}_p \times \mathbf{R}}$.

PROPOSITION 3.D.3. *For a prime p, the map*

$$(\text{dual } \mathbf{Z}[1/p]) \qquad \begin{cases} \mathbf{Sol}_p & \overset{\approx}{\to} \quad \widehat{\mathbf{Z}[1/p]} \approx \widehat{\mathbf{Q}_p \times \mathbf{R}}/\mathbf{Z}[1/p]^\perp, \\ [(x,y)] & \mapsto \qquad [\chi_x e_y] \end{cases}$$

is an isomorphism from the p-adic solenoid to the dual of the group $\mathbf{Z}[1/p]$.

The action of $\mathbf{Z}$ on $\widehat{\mathbf{Z}[1/p]}$ by multiplications by powers of p corresponds to the action of $\mathbf{Z}$ on $\mathbf{Sol}_p$ for which the generator $1 \in \mathbf{Z}$ acts by the map

$$T_p : \begin{cases} \mathbf{Sol}_p & \to \ \mathbf{Sol}_p, \\ [(x,y)] & \mapsto [(px, py)]. \end{cases}$$

We will need to identify the set $\mathrm{Per}(T_p)$ of periodic points of T_p in $\mathbf{Sol}_p$.

LEMMA 3.D.4. *The set $\mathrm{Per}(T_p)$ of T_p-periodic points in $\mathbf{Sol}_p$ is the countable subset*

$$\mathrm{Per}(T_p) = \left\{ \left[\left(\frac{a}{p^n - 1}, \frac{-a}{p^n - 1} \right) \right] \in \mathbf{Sol}_p \ \middle| \ n \in \mathbf{N}^*, a \in \mathbf{Z}[1/p] \right\}$$

of $\mathbf{Sol}_p$.

PROOF. For $(x, y) \in \mathbf{Q}_p \times \mathbf{R}$ and $n \in \mathbf{N}^*$, we have $T_p^n[(x,y)] = [(x,y)]$ if and only if $(p^n - 1)x \in \mathbf{Z}[1/p]$, $(p^n - 1)y \in \mathbf{Z}[1/p]$, and $(p^n - 1)(x + y) = 0$, if and only if $x = a/(p^n - 1)$ and $y = -a/(p^n - 1)$ for some $a \in \mathbf{Z}[1/p]$. □

Some irreducible representations of the group $\mathrm{BS}(1, p)$**.** We return to the Baumslag–Solitar group

$$\mathrm{BS}(1,p) = A \ltimes N, \ \text{ where } A = \begin{pmatrix} p^{\mathbf{Z}} & 0 \\ 0 & 1 \end{pmatrix} \approx \mathbf{Z} \text{ and } N = \begin{pmatrix} 1 & \mathbf{Z}[1/p] \\ 0 & 1 \end{pmatrix} \approx \mathbf{Z}[1/p],$$

and where the semi-direct product refers to the left action $A \curvearrowright N$, $(p^n, t) \mapsto p^n t$. The dual right action $\widehat{N} \curvearrowleft A$, $(\chi, p^n) \mapsto \chi^{p^n}$ is defined by $\chi^{p^n}(t) = \chi(p^n t)$ for all $\chi \in \widehat{N}$, $n \in \mathbf{Z}$, and $t \in N$. By the isomorphism of Proposition 3.D.3, this action is equivalent to the action of $\mathbf{Z}$ induced by the transformation $T_p : \mathbf{Sol}_p \to \mathbf{Sol}_p$. Moreover, the set

$$\widehat{N}_{\mathrm{per}} = \{\chi \in \widehat{N} \mid \text{ the orbit } \widehat{N} \cdot A \text{ is finite}\}$$

corresponds to the set $\mathrm{Per}(T_p)$ of T_p-periodic points in $\mathbf{Sol}_p$.

PROPOSITION 3.D.5. *Let p be a prime, $\Gamma = \mathrm{BS}(1,p) = A \ltimes N$ the corresponding solvable Baumslag–Solitar group, and $\widehat{N}_{\mathrm{per}} \subset \widehat{N}$ as above.*

(1) *For every $\chi \in \widehat{N} \smallsetminus \widehat{N}_{\mathrm{per}}$, the induced representation $\pi_\chi := \mathrm{Ind}_N^\Gamma \chi$ is irreducible.*
(2) *Let $\chi_1, \chi_2 \in \widehat{N} \smallsetminus \widehat{N}_{\mathrm{per}}$ be such that $\chi_1^\gamma \neq \chi_2$ for every $\gamma \in A$. Then π_{χ_1} and π_{χ_2} are nonequivalent irreducible representations of Γ.*

PROOF. Since A obviously acts freely on $\widehat{N} \smallsetminus \widehat{N}_{\mathrm{per}}$, Items (1) and (2) follow from Mackey–Shoda criteria for irreducibility and equivalence of monomial representations (Corollaries 1.F.16 and 1.F.20). □

Proposition 3.D.5 can be restated as follows.

PROPOSITION 3.D.6. *For $s \in \mathbf{Sol}_p$, denote by χ_s the unitary character of N under the identification $\widehat{N} = \widehat{\mathbf{Z}[1/p]} \approx \mathbf{Sol}_p$ of Proposition* 3.D.3.

The map

$$\mathbf{Sol}_p \smallsetminus \mathrm{Per}(T_p) \to \widehat{\Gamma}, \quad s \mapsto \mathrm{Ind}_N^\Gamma \chi_s$$

factorizes into an injective map from the space of nonperiodic T_p-orbits in $\mathbf{Sol}_p$ to the dual $\widehat{\Gamma}$ of $\Gamma = \mathrm{BS}(1,p)$.

REMARK 3.D.7. Proposition 3.D.6 shows that every nonperiodic T_p-orbit in $\mathbf{Sol}_p$ provides one equivalence class of irreducible representations of $\mathrm{BS}(1,p)$.

Equip the compact group $\mathbf{Sol}_p$ with the normalized Haar measure μ. Then, by uniqueness of μ as a translation-invariant probability measure, μ is T_p-invariant. Moreover, the action of T_p on $(\mathbf{Sol}_p, \mu)$ is ergodic. Indeed, the character group of $\mathbf{Sol}_p$ can be identified with $\mathbf{Z}[1/p]$, with the dual action of T_p on $\mathbf{Z}[1/p]$ corresponding to multiplication by p. Since $\{0\}$ is the only T_p-periodic point in $\mathbf{Z}[1/p]$, the transformation T_p of $\mathbf{Sol}_p$ is ergodic (see [**BeMa–00**, Chap I, Proposition 1.5]).

As in Corollary 3.B.16 and Remark 3.C.6, we conclude that the Borel space $\widehat{\mathrm{BS}(1,p)}$ is not countably separated.

We now proceed with the construction of irreducible representations of $\mathrm{BS}(1,p)$ associated with the periodic orbits in $\mathbf{Sol}_p$. It will turn out that these representations are finite-dimensional. We will see later (Corollary 4.B.7) that they exhaust the space $\widehat{\mathrm{BS}(1,p)}_{\mathrm{fd}}$ of finite-dimensional irreducible representations of $\mathrm{BS}(1,p)$.

For $n \in \mathbf{N}^*$, we denote by $\mathbf{Sol}_p(n)$ the elements in $\mathbf{Sol}_p$ with T_p-period n. For $k \in \mathbf{Z}$ and $b \in \mathbf{Z}[1/p]$, we write (k,b) for the element $\begin{pmatrix} p^k & b \\ 0 & 1 \end{pmatrix}$ of $\mathrm{BS}(1,p)$. Fix an integer $n \in \mathbf{N}^*$ and a point $s \in \mathbf{Sol}_p(n)$. The stabilizer of s in A is the subgroup

$$A(n) \;=\; \{(k,0) \in \mathrm{BS}(1,p) \mid k \in n\mathbf{Z}\} \;\approx\; n\mathbf{Z}.$$

For $\Gamma = \mathrm{BS}(1,p)$, the set

$$\Gamma(n) \;:=\; A(n) \ltimes N \;=\; \{(k,b) \in \Gamma \mid k \in n\mathbf{Z},\ b \in \mathbf{Z}[1/p]\}$$

is a normal subgroup of index n in Γ. As in Section 3.B, we parametrize the dual of $n\mathbf{Z}$ by $[0, 1/n[$; for $\theta \in [0, 1/n[$, we define $\chi_{s,\theta} : \Gamma(n) \to \mathbf{T}$ by

$$\chi_{s,\theta}(k,b) \;=\; \chi_\theta(k)\chi_s(b) \qquad \text{for all } (k,b) \in \Gamma(n).$$

Since χ_s is $A(n)$-invariant, $\chi_{s,\theta}$ is a unitary character of $\Gamma(n)$; obviously, we have $\chi_{s,\theta}|_N = \chi_s$. Conversely, it is clear that every unitary character χ of $\Gamma(n)$ such that $\chi|_N = \chi_s$ is of the form $\chi_{s,\theta}$ for a unique $\theta \in [0, 1/n[$.

PROPOSITION 3.D.8. *Let $\Gamma = \mathrm{BS}(1,p)$ be the Baumslag–Solitar group. Fix an integer $n \geq 1$.*

(1) *Let $s \in \mathbf{Sol}_p(n)$ and $\theta \in [0, 1/n[$. Then*
$$\pi_{s,\theta} := \mathrm{Ind}_{\Gamma(n)}^{\Gamma} \chi_{s,\theta}$$
is an irreducible representation of Γ of dimension n, where $\Gamma(n)$ is the normal subgroup of Γ defined above.

(2) *Let $(s,\theta), (s',\theta') \in \mathbf{Sol}_p(n) \times [0, 1/n[$. The representations $\pi_{s,\theta}$ and $\pi_{s',\theta'}$ are equivalent if and only if s and s' are in the same T_p-orbit and $\theta = \theta'$.*

PROOF. (1) It is clear that $\chi_{s,\theta}^{\gamma} \neq \chi_{s,\theta}$ for all $\gamma \in \Gamma \smallsetminus \Gamma(n)$. Therefore $\pi_{s,\theta}$ is irreducible, by the Mackey–Shoda criterion (Corollary 1.F.16). Moreover, since $\Gamma(n)$ has index n in Γ and since $\chi_{s,\theta}$ is one-dimensional, the induced representation $\pi_{s,\theta}$ is finite-dimensional, of dimension n.

(2) Let $(s,\theta), (s',\theta') \in \mathbf{Sol}_p(n) \times [0, 1/n[$. The unitary characters $\chi_{s,\theta}$ and $\chi_{s',\theta'}$ of $\Gamma(n)$ are conjugate under Γ if and only if s and s' are in the same T_p-orbit and $\theta = \theta'$. Therefore, the claims follow from the Mackey–Shoda criterion for equivalence of monomial representations (Corollary 1.F.20). □

3.E. Lamplighter group

DEFINITION 3.E.1. The **lamplighter group** is the solvable group Γ defined by the presentation
$$\Gamma = \langle t, x \mid x^2 = 1,\ [t^k x t^{-k}, t^l x t^{-l}] = 1 \quad \text{for all } k, l \in \mathbf{Z} \rangle.$$
The group Γ can also be described as the wreath product
$$\mathbf{Z} \wr (\mathbf{Z}/2\mathbf{Z}) = A \ltimes N, \qquad \text{where} \qquad A = \mathbf{Z} \text{ and } N = \bigoplus_{k \in \mathbf{Z}} \mathbf{Z}/2\mathbf{Z},$$
where the action of $\mathbf{Z}$ on $\bigoplus_{k \in \mathbf{Z}} \mathbf{Z}/2\mathbf{Z}$ defining the semi-direct product is given by shifting the coordinates.

The dual group of N can be identified with the compact group
$$X := \prod_{k \in \mathbf{Z}} \{0,1\}.$$
Under this identification, the action of $\mathbf{Z}$ on $\widehat{N}$ is given by the shift transformation
$$T \colon \prod_{k \in \mathbf{Z}} \{0,1\} \to \prod_{k \in \mathbf{Z}} \{0,1\}, \ \ (x_n)_{n \in \mathbf{Z}} \mapsto (x_{n+1})_{n \in \mathbf{Z}}.$$
Moreover, the normalized Haar measure on $\widehat{N} = X$ is the measure $\mu = \bigotimes^{\mathbf{Z}} \nu$, where ν is the uniform probability measure on $\{0,1\}$.

Observe that μ is T-invariant (as it should be, since T is an automorphism of X). The action of T on (X, μ) is ergodic, since $\{0\}$ is the only T-periodic point in $\widehat{X} = N$ (see [**BeMa–00**, Chap I, Proposition 1.5]).

For an integer $n \geq 1$, the set $X(n)$ of elements in X with T-period n consists of the sequences $(x_m)_{m \in \mathbf{Z}} \in \{0,1\}^{\mathbf{Z}}$ with
$$x_{j+kn} = x_j \quad \text{for all } j = 0, \ldots, n-1 \text{ and } k \in \mathbf{Z}.$$

So, $X(n)$ has exactly 2^n elements, and the set

$$\mathrm{Per}(T) = \bigcup_{n\geq 1} X(n)$$

of periodic points of T is a countable dense subset of X.

We can construct a family of irreducible representations of the lamplighter group associated with the nonperiodic T-orbits, as we did in Section 3.D for the Baumslag–Solitar group $\mathrm{BS}(1,p)$. The proof of Proposition 3.D.5 carries over *mutatis mutandis* and yields the following result.

PROPOSITION 3.E.2. *Let* $\Gamma = A \ltimes N = \mathbf{Z} \ltimes \bigoplus_{k\in\mathbf{Z}} \mathbf{Z}/2\mathbf{Z}$ *and* $X = \prod_{k\in\mathbf{Z}}\{0,1\}$ *be as above. For* $x \in X$, *denote by* χ_x *the corresponding unitary character of* N *under the identification* $\widehat{N} \approx X$.

For every $x \in X \smallsetminus \mathrm{Per}(T)$, *the representation* $\mathrm{Ind}_N^\Gamma \chi_x$ *is irreducible. The map*

$$X \smallsetminus \mathrm{Per}(T) \to \widehat{\Gamma}, \quad x \mapsto \mathrm{Ind}_N^\Gamma \chi_x$$

factorizes into an injective map from the space of nonperiodic T*-orbits in* X *to the dual* $\widehat{\Gamma}$ *of* Γ.

As for the Baumslag–Solitar group $\mathrm{BS}(1,p)$, we will also construct irreducible representations of $\Gamma = A \ltimes N$ associated with periodic orbits in X. It will turn out that these representations are finite-dimensional, and we will see later (Corollary 4.B.8) that they exhaust the space $\widehat{\Gamma}_{\mathrm{fd}}$ of finite-dimensional irreducible representations of Γ.

Fix an integer $n \geq 1$. Let $x \in X(n)$ be an element of X of T-period n, and $\chi_x \in \widehat{N}$ the corresponding unitary character of N. The stabilizer of χ_x in A is the subgroup $A(n) := n\mathbf{Z}$ of $A = \mathbf{Z}$. Then $\Gamma(n) := A(n) \ltimes N$ is a normal subgroup of index n in Γ. As in Section 3.D, we parametrize $\widehat{A(n)}$ by the interval $[0, 1/n[$: for every $\theta \in [0, 1/n[$, we have a unitary character $\chi_{x,\theta}$ of $\Gamma(n)$ defined by

$$\chi_{x,\theta}(k,b) = \chi_\theta(k)\chi_x(b) \qquad \text{for all } (k,b) \in \Gamma(n) = n\mathbf{Z} \ltimes N.$$

The following proposition is proved along the same lines as the corresponding result (Proposition 3.D.8) for the Baumslag–Solitar group.

PROPOSITION 3.E.3. *Let* $\Gamma = A \ltimes N$ *be the lamplighter group; set* $X = \widehat{N}$. *Fix an integer* $n \geq 1$; *let* $X(n)$ *be the subset of* X *and* $\Gamma(n)$ *the subgroup of index* n *of* Γ *defined above.*

(1) *Let* $x \in X(n)$ *and* $\theta \in [0, 1/n[$, *and* $\chi_{x,\theta} \in \widehat{\Gamma(n)}$ *as above. Then*

$$\pi_{x,\theta} := \mathrm{Ind}_{\Gamma(n)}^\Gamma \chi_{x,\theta}$$

is an irreducible representation of Γ *of dimension* n.

(2) *Let* $(x,\theta), (x',\theta') \in X(n) \times [0, 1/n[$. *The representations* $\pi_{x,\theta}$ *and* $\pi_{x',\theta'}$ *are equivalent if and only if* x *and* x' *are in the same* T*-orbit and* $\theta = \theta'$.

3.F. General linear groups

Let $\mathbf{K}$ be a field and n an integer, $n \geq 2$. In this section, we denote by Γ the **general linear group** $\Gamma = \mathrm{GL}_n(\mathbf{K})$ over $\mathbf{K}$, defined as usual as the group of invertible n-by-n matrices with coefficients in $\mathbf{K}$.

Basic representations of Γ are the **principal series representations**, which are defined as follows. Let

$$B = \begin{pmatrix} \mathbf{K}^* & \mathbf{K} & \cdots & \mathbf{K} \\ 0 & \mathbf{K}^* & \cdots & \mathbf{K} \\ \vdots & \vdots & \ddots & \vdots \\ 0 & 0 & \cdots & \mathbf{K}^* \end{pmatrix}$$

be the subgroup of upper-triangular matrices in Γ. Then $B = A \ltimes N$ is the semi-direct product of the subgroup

$$A = \left\{ \begin{pmatrix} a_1 & 0 & \cdots & 0 \\ 0 & a_2 & \cdots & 0 \\ \vdots & \vdots & \ddots & \vdots \\ 0 & 0 & \cdots & a_n \end{pmatrix} \;\middle|\; a_1, \ldots, a_n \in \mathbf{K}^\times \right\} \approx (\mathbf{K}^\times)^n$$

of diagonal matrices with the subgroup

$$N = \begin{pmatrix} 1 & \mathbf{K} & \cdots & \mathbf{K} \\ 0 & 1 & \cdots & \mathbf{K} \\ \vdots & \ddots & \ddots & \vdots \\ 0 & 0 & \cdots & 1 \end{pmatrix}$$

of upper-triangular matrices in Γ with 1's on the diagonal.

A fundamental fact about B is the following result.

LEMMA 3.F.1. *Assume that* $\mathbf{K}$ *is infinite. Then* B *coincides with its own commensurator* $\mathrm{Comm}_\Gamma(B)$.

PROOF. Let X denote the set of complete flags in $\mathbf{K}^n$ and x_0 the complete flag

$$V_1 = \langle e_1 \rangle \subset V_2 = \langle e_1, e_2 \rangle \subset \cdots \subset V_n = \langle e_1, \cdots, e_n \rangle,$$

where $(e_1, e_2, \ldots, e_n)$ is the canonical basis of $\mathbf{K}^n$.

The group $\Gamma = GL_n(\mathbf{K})$ acts transitively on X and B is the stabilizer of x_0. The claim is that x_0 is the only point in X with a finite B-orbit; equivalently, the claim is that, for every subgroup H of B of finite index, x_0 is the unique H-fixed point. (On commensurators, see Proposition A.5.6.)

Let H be a subgroup of finite index in B. It suffices to prove that $V_1, \ldots, V_n$ are the unique nonzero H-invariant subspaces of $\mathbf{K}^n$.

Let V be a nonzero H-invariant subspace of $\mathbf{K}^n$. Let $j \in \{1, \ldots, n\}$ be minimal such that $V \subset V_j$. There exists $v \in V$ such that $v = \sum_{k=1}^{j} \alpha_k e_k$ for scalars $\alpha_k \in \mathbf{K}$, with $\alpha_j \neq 0$. If $j = 1$, then $V = V_1$ and the claim is proved. So, we can assume that $j \geq 2$.

Let $i \in \{1, \ldots, j-1\}$. Consider the subgroup $B_{i,j} \cong \mathbf{K}$ of B consisting of the elementary matrices $E_{i,j}(\alpha)$, for $\alpha \in \mathbf{K}$. (Thus, $E_{i,j}(\alpha)$ is the matrix $(\alpha_{k,l})_{1 \leq k,l \leq n}$ with $\alpha_{k,k} = 1$ for all k, with $\alpha_{i,j} = \alpha$ and with $\alpha_{k,l} = 0$ otherwise.) Then $H_{i,j} := B_{i,j} \cap H$ is a finite index subgroup of $B_{i,j}$. Observe that $H_{i,j}$ is infinite, since $\mathbf{K}$ is infinite; in particular, $H_{i,j}$ is not the trivial group, and so there exists $\alpha \neq 0$ such that $E_{i,j}(\alpha) \in H$. On the one hand,

$$E_{i,j}(\alpha)v - v \in V,$$

since V is H-invariant. On the other hand,

$$E_{i,j}(\alpha)v - v = E_{i,j}(\alpha)\Big(\sum_{k=1}^{j} \alpha_k e_k\Big) - \sum_{k=1}^{j} \alpha_k e_k = \alpha\alpha_j e_i.$$

Since $\alpha\alpha_j \neq 0$, this shows that $e_i \in V$ for all $i \in \{1, \dots, j-1\}$. Therefore we also have $e_j \in V$, since $v \in V$. It follows that $V = V_j$ and the claim is proved. $\square$

Let $\chi = (\chi_1, \dots, \chi_n) \in (\mathbf{K}^\times)^n$. Define a representation of B, again denoted by χ, by lifting the unitary character of A defined by χ to B, that is,

$$\chi \begin{pmatrix} a_1 & * & \cdots & * \\ 0 & a_n & \cdots & * \\ \vdots & \vdots & \ddots & \vdots \\ 0 & 0 & \cdots & a_n \end{pmatrix} = \chi_1(a_1) \cdots \chi_n(a_n).$$

The following result is a consequence of Lemma 3.F.1 in combination with the Mackey–Shoda criteria (Theorems 1.F.12 and 1.F.17).

THEOREM 3.F.2. *Let $\Gamma = \mathrm{GL}_n(\mathbf{K})$ for an infinite field $\mathbf{K}$.*

(1) *For every $\chi \in (\mathbf{K}^\times)^n$, the representation $\mathrm{Ind}_B^\Gamma \chi$ is irreducible.*
(2) *For $\chi, \chi' \in (\mathbf{K}^\times)^n$ with $\chi \neq \chi'$, the representations $\mathrm{Ind}_B^\Gamma \chi$ and $\mathrm{Ind}_B^\Gamma \chi'$ are not equivalent.*

3.G. Some nondiscrete examples

There are some well-studied examples of topological groups, nondiscrete ones, of which the dual has been determined, either as a set, or more rarely as a topological space.

We have already described the dual of the real Heisenberg group $H(\mathbf{R})$ in Remark 3.B.19, and the dual of the affine group $\mathrm{Aff}(\mathbf{R})$ in Remark 3.C.7.

For the simple complex Lie group $G = \mathrm{SL}_2(\mathbf{C})$, the dual as a set has been determined as recently as 1947, independently by Bargmann, Gel'fand and Naimark, and Harish–Chandra [**Barg–47**, **GeNa–47b**, **Hari–47**]. They showed that $\widehat{G}$ consists of three parts, the first is the so-called principal series (a disjoint union of a half-line with an infinite number of copies of $\mathbf{R}$), the second the supplementary series (an open interval), and the third a point (the representation 1_G). The Fell topology on $\widehat{G}$ has been described by Fell [**Fell–61**]. See also [**Fell–60a**] for the dual of $\mathrm{SL}_n(\mathbf{C})$.

For $G = \mathrm{SL}_2(\mathbf{R})$, we refer to [**Foll–16**, Section 7.6] for a short description of the dual and its topology, without proofs; see also [**Vale–84**]. The determination of the dual as a set was obtained in [**Barg–47**], and the Fell topology in [**Mili–71**]. For $\mathrm{SO}(n) \ltimes \mathbf{R}^n$, we refer to [**KaTa–13**, Example 5.59].

For semisimple groups in general, there are descriptions of the Fell topology on parts of the dual, in particular on parts defined by the irreducible principal series, or by the discrete series. For the simple linear groups of so-called split-rank 1, there is also a description of the Fell topology on the reduced dual, i.e., on that part of the dual consisting of irreducible representations that are weakly contained in the regular representation. See [**Lips–70**].

EXAMPLE 3.G.1 (**some groups which are not locally compact**). Now consider the following non-LC topological groups. Let $\mathcal{H}$ be an infinite-dimensional separable Hilbert space.

First let $\mathcal{U}_\infty(\mathcal{H})$ be the group of unitary operators x on $\mathcal{H}$ such that $x - \mathrm{Id}_{\mathcal{H}}$ is compact, with the topology inherited from the distance defined by $d(x,y) = \|x - y\|$. The dual of $\mathcal{U}_\infty(\mathcal{H})$ is countable and discrete [**Kiri–73, Ol's–78**]. Moreover, every representation of $\mathcal{U}_\infty(\mathcal{H})$ is a direct sum of irreducible subrepresentations [**Ol's–78**, Theorem 1.11].

Representations of infinite-dimensional unitary groups have been studied earlier in [**Sega–57**].

Then let $\mathcal{U}(\mathcal{H})_{\mathrm{str}}$ denote the unitary group of $\mathcal{H}$ with the strong operator topology. The inclusion $i : \mathcal{U}_\infty(\mathcal{H}) \to \mathcal{U}(\mathcal{H})_{\mathrm{str}}$ is continuous with dense image, so that there is an embedding $\pi \mapsto \pi \circ i$ from the dual of $\mathcal{U}(\mathcal{H})_{\mathrm{str}}$ into that of $\mathcal{U}_\infty(\mathcal{H})$. Moreover, the explicit description by Kirillov and Ol'shanskii of the dual of $\mathcal{U}_\infty(\mathcal{H})$ shows that every irreducible representation of $\mathcal{U}_\infty(\mathcal{H})$ is continuous when $\mathcal{U}_\infty(\mathcal{H})$ is equipped with the strong operator topology; it follows that the restriction of irreducible representations provides a bijection from the dual of $\mathcal{U}(\mathcal{H})_{\mathrm{str}}$ onto that of $\mathcal{U}_\infty(\mathcal{H})$. This also shows that every representation of $\mathcal{U}(\mathcal{H})_{\mathrm{str}}$ is a direct sum of irreducible representations.

With the terminology of Chapter 6, this shows in particular that the groups $\mathcal{U}_\infty(\mathcal{H})$ and $\mathcal{U}(\mathcal{H})_{\mathrm{str}}$ are of type I (see Theorem 6.E.20).

CHAPTER 4

Finite-dimensional irreducible representations

This chapter is dedicated to the study of the space $\widehat{G}_{\mathrm{fd}}$ of equivalence classes of finite-dimensional irreducible representations of a group G. We describe $\widehat{G}_{\mathrm{fd}}$ for a locally compact group which is a semi-direct product $G = H \ltimes N$, where

- *H is an abelian discrete subgroup of G;*
- *N is a locally compact abelian normal closed subgroup of G.*

It turns out that every finite-dimensional irreducible representation π of G is equivalent to a representation of the form $\mathrm{Ind}_M^G \chi$, where M is a finite index subgroup of G which contains N and χ a unitary character of M (Theorem 4.A.3).

This implies that the finite-dimensional irreducible representations constructed in Chapter 3 exhaust the space $\widehat{\Gamma}_{\mathrm{fd}}$ for the groups Γ considered there, i.e., for the Heisenberg group over a commutative ring R, the affine group $\mathrm{Aff}(\mathbf{K})$ over a field $\mathbf{K}$, the Baumslag–Solitar group $\mathrm{BS}(1,p)$, and the lamplighter group — the case of $\mathrm{GL}_n(\mathbf{K})$ over an infinite field $\mathbf{K}$ is treated separately (Corollary 4.C.22). For all these groups (apart from the special case of finite groups), $\widehat{\Gamma}_{\mathrm{fd}}$ is a much smaller space than $\widehat{\Gamma}$.

The set $\widehat{\Gamma}_{\mathrm{fd}}$ separates the points of Γ for some of these groups: for the Heisenberg group $H(\mathbf{Z})$ over the integers, and for the Baumslag–Solitar group $\mathrm{BS}(1,p)$. By contrast, $\widehat{\Gamma}_{\mathrm{fd}}$ consists only of one-dimensional representations for other groups: for $H(\mathbf{K})$ or $\mathrm{GL}_n(\mathbf{K})$ over an infinite field $\mathbf{K}$.

There are several interesting classes of groups G defined in terms of the space $\widehat{G}_{\mathrm{fd}}$. We first survey results concerning the so-called Moore groups, which are those groups G such that $\widehat{G}_{\mathrm{fd}} = \widehat{G}$ (Section 4.C.a). Maximally almost periodic groups are topological groups G such that $\widehat{G}_{\mathrm{fd}}$ separates the points of G (Section 4.C.b), and minimally almost periodic groups are these G such that $\widehat{G}_{\mathrm{fd}} = \{1_G\}$ (Section 4.C.d). These classes can be characterized in terms of properties of the Bohr compactification $\mathrm{Bohr}(G)$ of G, a compact group which encodes the space $\widehat{G}_{\mathrm{fd}}$ (Section 4.C.c).

Wc discuss the class of discrete groups Γ with the property that every finite-dimensional representation of Γ has finite image; these are the groups Γ for which the Bohr compactification $\mathrm{Bohr}(\Gamma)$ coincides with another compactification of Γ, the profinite completion $\mathrm{Prof}(\Gamma)$ of Γ (Proposition 4.C.14). A prominent example of a group with this property is the group $\mathrm{SL}_n(\mathbf{Z})$ for $n \geq 3$ (Theorem 4.C.16).

4.A. Finite-dimensional irreducible representations of some semi-direct products

CONVENTION 4.A.1 (**semi-direct product**). We need to specify the definition we will adopt for a semi-direct product. Let H be a group acting on the left by automorphisms on another group N; let $\varphi : H \to \mathrm{Aut}(N)$ denote the corresponding group homomorphism. The **semi-direct product** $G = H \ltimes N$ is defined as the group with underlying set $H \times N$ and with the product

$$(h_1, n_1)(h_2, n_2) = (h_1 h_2, n_1 \varphi(h_1)(n_2)) \quad \text{for } (h_1, n_1), (h_2, n_2) \in G.$$

When φ is understood, we also write $h \cdot n$ for $\varphi(h)(n)$ for $h \in H$ and $n \in N$.

The obvious maps $H \to G$ and $N \to G$ are injective group homomorphisms; we can therefore identify H with the subgroup $H \times \{e\}$ of G and N with the normal subgroup $\{e\} \times N$ of G. Observe that we way write $nh = (h, n)$ but that, in general, $hn \neq (h, n)$; indeed, we have $hn = (h, h \cdot n)$ and $(h, n) = h(h^{-1} \cdot n)$.

The following identities will often be used below; we use the identifications noted above. For $h \in H$, $n' \in N$, we have $hn'h^{-1} = h \cdot n'$. When N is moreover abelian, for $g = (h, n) \in G$ and $n' \in N$, we have $gn'g^{-1} = hn'h^{-1} = h \cdot n'$.

Beware that, elsewhere and for example in [**BA1–3**], the semi-direct product of H and N is defined as the set $N \times H$, with the same law as above, and is then denoted by $N \rtimes H$.

Suppose moreover that H and N are locally compact groups and that φ is a continuous homomorphism from H to the group of bicontinuous automorphisms of N. Then $H \ltimes N$ is a locally compact group for the product topology on $H \times N$.

Given a closed normal subgroup N of G, recall from Section 1.F that G acts (on the right) on the dual space $\widehat{N}$ by $(\pi, g) \to \pi^g$, where π^g is given $\pi^g(n) = \pi(gng^{-1})$ for $\pi \in \widehat{N}$, $g \in G$, and $n \in N$.

Assume from now on that $G = H \ltimes N$ is a semi-direct product of an abelian discrete subgroup H and a locally compact abelian normal closed subgroup N. For $\chi \in \widehat{N}$, let

$$H_\chi = \{h \in H \mid \chi^h = \chi\}$$

be the stabilizer of χ in H. Observe that $G_\chi = H_\chi \ltimes N$ is the stabilizer of χ in G.

LEMMA 4.A.2. *Consider $G = H \ltimes N$ and $\chi \in \widehat{N}$, with $H_\chi \subset H$ and $G_\chi \subset G$, as above.*

(1) *For every $\alpha \in \widehat{H_\chi}$, the map*

$$\alpha \otimes \chi : H_\chi \ltimes N \to \mathbf{T}, \quad (h, n) \mapsto \alpha(h)\chi(n)$$

is a unitary character of G_χ which extends χ.

(2) *Let π be an irreducible representation of G_χ such that $\pi|_N$ is a multiple of χ. Then π is a unitary character of G_χ and there exists $\alpha \in \widehat{H_\chi}$ such that $\pi = \alpha \otimes \chi$.*

PROOF. (1) It is obvious that $\alpha \otimes \chi$ is continuous and that it extends χ. We have to check that $\alpha \otimes \chi$ is a homomorphism. This is indeed the case: for $h_1, h_2 \in H_\chi$ and $n_1, n_2 \in N$, we have

$$\begin{aligned}
(\alpha \otimes \chi)\big((h_1, n_1)(h_2, n_2)\big) &= (\alpha \otimes \chi)\big((h_1 h_2, n_1(h_1 \cdot n_2))\big) \\
&= \alpha(h_1)\alpha(h_2)\chi(n_1)\chi(h_1 \cdot n_2) \\
&= \alpha(h_1)\alpha(h_2)\chi(n_1)\chi(n_2) \\
&= (\alpha \otimes \chi)(h_1, n_1)\, (\alpha \otimes \chi)(h_2, n_2).
\end{aligned}$$

(2) For $g_1 = (h_1, n_1),\ g_2 = (h_2, n_2) \in G_\chi$, we have

$$\begin{aligned}\pi(g_1g_2) &= \pi(g_1)\pi(g_2)\\ &= \pi(h_1, n_1)\pi(h_2, n_2)\\ &= \pi(h_1(h_1^{-1}\cdot n_1))\pi(h_2(h_2^{-1}\cdot n_1))\\ &= \pi(h_1)\chi(h_1^{-1}\cdot n_1)\pi(h_2)\chi(h_2^{-1}\cdot n_2)\\ &= \pi(h_1)\pi(h_2)\chi(n_1)\chi(n_2)\\ &= \pi(h_1h_2)\chi(n_1n_2),\end{aligned}$$

and hence, since H is abelian, $\pi(g_1g_2) = \pi(g_2g_1)$. This shows that π is trivial on the commutator subgroup of G_χ, that is, π is a unitary character of G_χ. Moreover, $\pi = \alpha \otimes \chi$ for $\alpha := \pi|_{H_\chi} \in \widehat{H_\chi}$. □

Set

$$\widehat{N}_{\mathrm{per}} = \{\chi \in \widehat{N} \mid \text{the orbit } \ \chi\cdot H \ \text{ is finite}\}$$

(equivalently, the orbit $\chi \cdot G$ is finite). Thus, $\widehat{N}_{\mathrm{per}}$ is the set $\chi \in \widehat{N}$ such that H_χ is a subgroup of finite index in H. Observe that, since H is abelian, we have

$$H_\chi = H_{\chi^h} \qquad \text{for all } \ h \in H.$$

Set

$$X_{\mathrm{fd}} = \bigsqcup_{\chi\in\widehat{N}_{\mathrm{per}}} \{\chi\} \times \widehat{H_\chi}.$$

There is a natural action $X_{\mathrm{fd}} \curvearrowleft H$ of H on X_{fd} given by

$$(X_{\mathrm{fd}}, H) \to X_{\mathrm{fd}}, \qquad ((\chi, \alpha), h) \mapsto (\chi^h, \alpha).$$

THEOREM 4.A.3. *Let $G = H \ltimes N$ be the semi-direct product of a discrete abelian subgroup H and a locally compact abelian normal subgroup N of G. Let X_{fd} be as above.*

The map $\Phi : X_{\mathrm{fd}} \to \widehat{G}_{\mathrm{fd}}$, defined by

$$\Phi : (\chi, \alpha) \mapsto \mathrm{Ind}_{H_\chi\ltimes N}^G(\alpha\otimes\chi),$$

induces a bijection between the space of H-orbits in X_{fd} and the space $\widehat{G}_{\mathrm{fd}}$ of finite-dimensional irreducible representations of G.

PROOF. • *First step.* We claim that the range of Φ is contained in $\widehat{G}_{\mathrm{fd}}$.

Indeed, let $\chi \in \widehat{N}_{\mathrm{per}}$ and $\alpha \in \widehat{H}_\chi$. Then

$$(\alpha\otimes\chi)^h = \alpha\otimes\chi^h \neq \alpha\otimes\chi \qquad \text{for all } \ h \in H \smallsetminus H_\chi.$$

Therefore $\mathrm{Ind}_{H_\chi\ltimes N}^G(\alpha\otimes\chi)$ is irreducible, by the Mackey–Shoda criterion (Corollary 1.F.16). Moreover, since H_χ has finite index in Γ and since $\alpha\otimes\chi$ is finite-dimensional, the induced representation $\mathrm{Ind}_{H_\chi\ltimes N}^G(\alpha\otimes\chi)$ is finite-dimensional.

• *Second step.* We claim that Φ factorizes into an injective map from the space of H-orbits in X_{fd} to $\widehat{G}_{\mathrm{fd}}$.

Indeed, let (χ, α) and (χ', α') in X_{fd}.

Assume that (χ, α) and (χ', α') are in the same H-orbit. Then χ and χ' are conjugate under H and $\alpha = \alpha'$. Therefore, $H_\chi = H_{\chi'}$ and the unitary characters $\alpha\otimes\chi$ and $\alpha'\otimes\chi'$ of $H_\chi \ltimes N$ are conjugate under H. It follows that $\mathrm{Ind}_{H_\chi\ltimes N}^G(\alpha\otimes\chi)$ and $\mathrm{Ind}_{H_\chi\ltimes N}^G(\alpha'\otimes\chi')$ are equivalent (Proposition 1.F.9).

Assume now that (χ, α) and (χ', α') are not in the same H-orbit. Two cases may occur:

∘ *First case:* χ and χ' are conjugate under H. Then $H_\chi = H_{\chi'}$ and $\alpha \neq \alpha'$. Therefore the unitary characters $\alpha \otimes \chi$ and $\alpha' \otimes \chi'$ of $H_\chi \ltimes N$ are not in the same H-orbit. It follows from the Mackey–Shoda criterion for equivalence of monomial representations (Corollary 1.F.20) that $\mathrm{Ind}_{H_\chi \ltimes N}^G(\alpha \otimes \chi)$ and $\mathrm{Ind}_{H_\chi \ltimes N}^G(\alpha' \otimes \chi')$ are not equivalent.

∘ *Second case:* χ and χ' are not conjugate under H. By Proposition 1.F.9(2), the restriction of $\mathrm{Ind}_{H_\chi \ltimes N}^G(\alpha \otimes \chi)$ to $H_\chi \ltimes N$ is equivalent to a direct sum of conjugates of $\alpha \otimes \chi$; similarly, the restriction of $\mathrm{Ind}_{H_{\chi'} \ltimes N}^G(\alpha' \otimes \chi')$ to $H_{\chi'} \ltimes N$ is equivalent to a direct sum of conjugates of $\alpha' \otimes \chi'$. Therefore, the restriction of $\mathrm{Ind}_{H_\chi \ltimes N}^G(\alpha \otimes \chi)$ to N is equivalent to a direct sum of conjugates of χ and the restriction of $\mathrm{Ind}_{H_{\chi'} \ltimes N}^G(\alpha' \otimes \chi')$ to N is equivalent to a direct sum of conjugates of χ'. Since χ and χ' are not conjugate, it follows that $\mathrm{Ind}_{H_\chi \ltimes N}^G(\alpha \otimes \chi)$ and $\mathrm{Ind}_{H'_\chi \ltimes N}^G(\alpha' \otimes \chi')$ cannot be equivalent.

• *Third step.* We claim that Φ is surjective.

Indeed, let $(\pi, \mathcal{H})$ be a finite-dimensional irreducible representation of G. The restriction $\pi|_N$ is a finite-dimensional representation of the abelian normal subgroup N. So, there exists a finite subset S of $\widehat{N}$ such that $\mathcal{H} = \bigoplus_{\chi \in S} \mathcal{H}^\chi$, where

$$\mathcal{H}^\chi := \{\xi \in \mathcal{H} \mid \pi(n)\xi = \chi(n)\xi \quad \text{for all } \ n \in N\}$$

is nonzero for every $\chi \in S$.

Since N is a normal subgroup of G, we have

$$(*) \qquad \pi(g)\mathcal{H}^\chi = \mathcal{H}^{\chi^{g^{-1}}} \qquad \text{for all } \ g \in G, \chi \in S.$$

Therefore S is H-invariant. Since π is irreducible, it follows that S is a single H-orbit. Since S is finite, S is the H-orbit of some $\chi \in \widehat{N}_{\mathrm{per}}$.

By $(*)$, we have

$$\pi(h)\mathcal{H}^\chi = \mathcal{H}^\chi$$

for every h in the stabilizer H_χ of χ in H.

Let T be a right transversal for $H_\chi \ltimes N$ in G, with $e \in T$. Since S is the G-orbit of χ, we have

$$\mathcal{H} = \bigoplus_{t \in T} \pi(t)\mathcal{H}^\chi.$$

This shows that π is equivalent to the induced representation $\mathrm{Ind}_{H_\chi \ltimes N}^G \sigma$, where σ is the subrepresentation of $\pi|_{H_\chi \ltimes N}$ defined on the $(H_\chi \ltimes N)$-invariant subspace $\mathcal{H}^\chi$ (see Definition 1.F.1).

We claim that $\sigma = \alpha \otimes \chi$ for some $\alpha \in \widehat{H_\chi}$. Indeed, σ is an irreducible representation of $G_\chi = H_\chi \ltimes N$, since $\mathrm{Ind}_{H_\chi \ltimes N}^G \sigma \simeq \pi$ is irreducible (see Proposition 1.F.6(2)). Moreover, $\sigma|_N$ is a multiple of χ. So, the claim follows from Lemma 4.A.2 and we conclude that π is equivalent to $\mathrm{Ind}_{H_\chi \ltimes N}^G(\alpha \otimes \chi)$. □

4.B. All finite-dimensional irreducible representations for some groups

We are going to apply Theorem 4.A.3 to Heisenberg groups, affine groups, solvable Baumslag–Solitar groups, and the lamplighter group.

4.B.a. Heisenberg groups over rings. Let $H(R)$ be the Heisenberg group over a unital commutative ring R, as in Section 3.B. Recall that we may identify $H(R)$ with R^3, equipped with the group law

$$(a,b,c)(a',b',c') \ = \ (a+a', b+b', c+c'+ab').$$

The centre of $H(R)$ is $Z = \{(0,0,c) \mid c \in R\} \approx R$, and $N = \{(0,b,c) \mid b,c \in R\} \approx R^2$ is a maximal abelian subgroup of $H(R)$. Observe that $H(R)$ is a semi-direct product $H \ltimes N$ for $H = \{(a,0,0) \mid a \in R\} \approx R$.

Let χ be a unitary character of N. Then $\chi = \chi_{\beta,\psi}$ for a unique pair $(\beta,\psi) \in (\widehat{R})^2$, where $\chi_{\beta,\psi}$ is defined by

$$\chi_{\beta,\psi}(0,b,c) \ = \ \beta(b)\psi(c) \qquad \text{for } \ b,c \in R.$$

For $h = (a,0,0) \in H$, we have

$$\chi^h_{\beta,\psi}(0,b,c) \ = \ \beta(b)\psi(ab)\psi(c) \ = \ \chi_{\beta\psi^a,\psi}(0,b,c) \qquad \text{for } \ b,c \in R,$$

where $\psi^a \in \widehat{R}$ is defined by $\psi^a(b) = \psi(ab)$ for $b \in R$. It follows that the H-orbit of $\chi_{\beta,\psi}$ is

$$\{\chi_{\beta\psi^a,\psi} \mid a \in R\},$$

and that the stabilizer of $\chi_{\beta,\psi}$, which only depends on ψ, is

$$H_\psi \ = \ \{(a,0,0) \mid a \in I_\psi\},$$

where I_ψ is the ideal

$$I_\psi \ = \ \{a \in R \mid aR \subset \ker\psi\}$$

of R, already considered in Section 3.B.

Denote by $\widehat{R}_{\text{per}}$ the set of $\psi \in \widehat{R}$ such that I_ψ has **finite index** in R. We see that the set $\widehat{N}_{\text{per}}$ of unitary characters of N with a finite H-orbit is

$$\widehat{N}_{\text{per}} \ = \ \{\chi_{\beta,\psi} \mid \psi \in \widehat{R}_{\text{per}}, \ \beta \in \widehat{R}\}.$$

Recall from Lemma 4.A.2 that, for $\alpha \in \widehat{I_\psi}$, the formula

$$\alpha \otimes \chi_{\beta,\psi} : \ (a,b,c) \mapsto \alpha(a)\beta(b)\psi(c)$$

defines a unitary character of $H_\psi \ltimes N$. Set

$$X_{\text{fd}} \ = \ \bigsqcup_{\psi \in \widehat{R}_{\text{per}}} \{\psi\} \times \{(\alpha,\beta) \mid \alpha \in \widehat{I_\psi}, \ \beta \in \widehat{R}\},$$

with the R-action $X_{\text{fd}} \curvearrowleft R$ on X_{fd} given by

$$X_{\text{fd}} \times R \to X_{\text{fd}}, \ \ ((\alpha,\beta,\psi),a) \mapsto (\alpha, \beta\psi^a, \psi).$$

In view of these remarks, the following result is a direct consequence of Theorem 4.A.3.

COROLLARY 4.B.1. *Let Γ be the Heisenberg group $H(R)$ over a unital commutative ring R. We keep the notation above.*

The map $X_{\text{fd}} \to \widehat{\Gamma}_{\text{fd}}$, defined by

$$(\alpha,\beta,\psi) \mapsto \operatorname{Ind}^{\Gamma}_{H_\psi \ltimes N}(\alpha \otimes \chi_{\beta,\psi}),$$

induces a bijection between the space of R-orbits in X_{fd} and the space $\widehat{\Gamma}_{\text{fd}}$ of equivalence classes of finite-dimensional irreducible representations of Γ.

• First particular case: $R = \mathbf{K}$ is a field. As already noted in Section 3.B, for $\psi \in \widehat{\mathbf{K}} \smallsetminus \{1_{\mathbf{K}}\}$, we have $I_\psi = \{0\}$ and hence $H_\psi = \{0\}$; for $\psi = 1_{\mathbf{K}}$, we have $I_\psi = \mathbf{K}$ and hence $H_\psi = H$.

∘ Assume first that $\mathbf{K}$ is infinite. Then $\widehat{\mathbf{K}}_{\text{per}} = \{1_{\mathbf{K}}\}$ and the following result is a direct consequence of Corollary 4.B.1.

COROLLARY 4.B.2. *Let $\Gamma = H(\mathbf{K})$ for an infinite field $\mathbf{K}$.*
The unitary characters of Γ are the only finite-dimensional irreducible representations of Γ, that is, the map

$$\widehat{\Gamma/Z} \to \Gamma_{\text{fd}}, \qquad \chi \mapsto \chi \circ p$$

is a bijection, where Z denotes the centre of Γ and $p : \Gamma \twoheadrightarrow \Gamma/Z$ the canonical projection.

∘ Assume now that $\mathbf{K}$ is finite. Let $\psi \in \widehat{\mathbf{K}} \smallsetminus \{1_{\mathbf{K}}\}$. Recall that, for $\beta_0 \in \widehat{\mathbf{K}}$, the H-orbit of $\chi_0 = \chi_{\beta_0,\psi}$ is

$$\{\chi_{\beta_0\psi^a,\psi} \mid a \in \mathbf{K}\}.$$

Therefore the H-orbit of χ_0 coincides with $\{\chi_{\beta,\psi} \mid \beta \in \widehat{\mathbf{K}}\}$, by Lemma 3.C.2(2). In particular, $\chi_{1_{\mathbf{K}},\psi}$ belongs to the H-orbit of χ_0.

In view of these remarks, the following result is again a direct consequence of Corollary 4.B.1.

COROLLARY 4.B.3. *Let $\Gamma = H(\mathbf{K})$ for a finite field $\mathbf{K}$. Let Z denote the centre of Γ and $p : \Gamma \twoheadrightarrow \Gamma/Z$ the canonical projection.*
The map

$$\left(\widehat{\mathbf{K}} \smallsetminus \{1_{\mathbf{K}}\}\right) \sqcup \widehat{\Gamma/Z} \to \widehat{\Gamma},$$

defined by

$$\psi \mapsto \mathrm{Ind}_N^\Gamma \chi_{1_{\mathbf{K}},\psi} \qquad \textit{for } \psi \in \widehat{\mathbf{K}} \smallsetminus \{1_{\mathbf{K}}\}$$

and

$$\chi \mapsto \chi \circ p \qquad \textit{for} \qquad \chi \in \widehat{\Gamma/Z},$$

is a bijection.

REMARK 4.B.4. Let $\Gamma = H(\mathbf{K})$ for a finite field $\mathbf{K}$ and let $\psi \in \widehat{Z} \smallsetminus \{1_Z\}$. Using the model of Construction 1.F.4(2), we give an explicit formula for the irreducible representation $\pi_\psi := \mathrm{Ind}_N^\Gamma \chi_{1_{\mathbf{K}},\psi}$ of Γ of Corollary 4.B.3: for every (a, b, c) in Γ, the operator $\pi_\psi(a, b, c)$ acts on $\ell^2(\mathbf{K})$ by

$$(\pi_\psi(a, b, c)f)(x) \;=\; \psi(c)\psi(bx)f(x + a) \qquad \text{for } f \in \ell^2(\mathbf{K}),\ x \in \mathbf{K}.$$

Observe that a consequence of Corollary 4.B.3 is that any two irreducible representations of Γ with central character ψ are equivalent. This result may be viewed as an analog for Heisenberg groups over finite fields of the classical Stone–von Neumann Theorem on the uniqueness of the commutation relations (see [**Foll–16**, Theorem 6.50], and Remark 3.B.19).

• Second particular case: $R = \mathbf{Z}$ is the ring of rational integers. Recall that $\widehat{Z} \approx \widehat{\mathbf{Z}} = \{\psi_\theta \mid \theta \in [0, 1[\}$, with $\psi_\theta(n) = e^{2\pi i \theta n}$ for $n \in \mathbf{Z}$.

Let $\psi = \psi_\theta$. We have $I_\psi = \{0\}$ if ψ has infinite order, that is, if θ is irrational. We have $I_\psi = n\mathbf{Z}$ if ψ has order $n \geq 1$, that is, if $\theta = p/n$ for $p \in \{0, 1, \ldots, n-1\}$, with p and n coprime. As a consequence, denoting by $\widehat{Z}_n$ the elements of order $n \geq 1$ in $\widehat{Z}$, we have

$$\widehat{Z}_{\text{per}} = \bigsqcup_{n \geq 1} \widehat{Z}_n.$$

Fix $n \geq 1$ and $\psi = \psi_\theta \in \widehat{Z}_n$. We have

$$H_\psi = \{(a, 0, 0) \mid a \in n\mathbf{Z}\}.$$

So, the stabilizer $H_\psi \ltimes N$ of ψ in $\Gamma = H(\mathbf{Z})$ only depends on n and coincides with the normal subgroup

$$\Gamma(n) = \{(a, b, c) \mid a \in n\mathbf{Z}, b, c \in \mathbf{Z}\},$$

which was already introduced in Section 3.B. Every unitary character χ of $\Gamma(n)$ with $\chi|_Z = \psi_\theta$ is of the form $\chi_{\alpha,\beta,\theta}$ for a unique pair $(\alpha, \beta) \in [0, 1/n[\times [0, 1[$, where

$$\chi_{\alpha,\beta,\theta}(a, b, c) = e^{2\pi i(\alpha a + \beta b + \theta c)} \qquad \text{for } a \in n\mathbf{Z},\ b \in \mathbf{Z},\ c \in \mathbf{Z}.$$

The H-orbit of $\chi_{\alpha,\beta,\theta}$ is

$$\{\chi_{\alpha,\beta+a\theta,\theta} \in \widehat{N} \mid a \in \mathbf{Z}\} = \{\chi_{\alpha,\beta+\frac{k}{n},\theta} \in \widehat{N} \mid k \in \mathbf{Z}\},$$

and so contains a character $\chi_{\alpha,\beta',\theta}$ for a unique $\beta' \in [0, 1/n[$. Set

$$X = \bigsqcup_{n \geq 1} \bigsqcup_{\psi \in \widehat{Z}_n} \{\psi\} \times [0, 1/n[^2.$$

The following result is therefore a direct consequence of Corollary 4.B.1.

COROLLARY 4.B.5. *Let $\Gamma = H(\mathbf{Z})$. We keep the notation above.*

For each $n \geq 1$ and $(\alpha, \beta, \psi) \in X$, with $\psi = \psi_\theta \in \widehat{Z}_n$, the representation $\mathrm{Ind}_{\Gamma(n)}^{\Gamma} \chi_{\alpha,\beta,\theta}$ of Γ is irreducible of dimension n.

The map $X \to \widehat{\Gamma}_{\text{fd}}$ defined by $(\psi, \alpha, \beta) \mapsto \mathrm{Ind}_{\Gamma(n)}^{\Gamma} \chi_{\alpha,\beta,\theta}$ is a bijection.

This corollary shows that the set $\widehat{H(\mathbf{Z})}_{\text{fd}}$ can be seen as a disjoint union parametrized by $[0, 1[\cap \mathbf{Q}$ of two-dimensional tori.

4.B.b. Affine group of arbitrary fields. Let $\Gamma = \mathrm{Aff}(\mathbf{K})$ be the affine group of a field $\mathbf{K}$, as in Section 3.C. Recall that $\Gamma = \mathbf{K}^\times \ltimes \mathbf{K}$, for the action of $\mathbf{K}^\times$ on $\mathbf{K}$ defined by $(a, t) \mapsto at$ for all $a \in \mathbf{K}^\times$ and $t \in \mathbf{K}$. We set

$$H = \mathbf{K}^\times \quad \text{and} \quad N = \mathbf{K}.$$

COROLLARY 4.B.6. *Let $\Gamma = \mathrm{Aff}(\mathbf{K})$ be the group of affine transformations of a field $\mathbf{K}$.*

1. *Assume that $\mathbf{K}$ is infinite. The unitary characters of Γ are the only finite-dimensional irreducible representations of Γ, that is, the map*

$$\widehat{\Gamma/N} \to \widehat{\Gamma}_{\text{fd}}, \quad \chi \mapsto \chi \circ p$$

is a bijection, where $p : \Gamma \twoheadrightarrow \Gamma/N$ is the canonical projection.

2. *Assume that $\mathbf{K}$ is finite of cardinality q. Fix any $\chi \in \widehat{\mathbf{K}} \smallsetminus \{1_{\mathbf{K}}\}$. Up to equivalence, the irreducible representations of Γ are the representations of dimension one and the representation $\mathrm{Ind}_N^{\Gamma} \chi$ of dimension $q - 1$.*

PROOF. For $\chi \in \widehat{\mathbf{K}} \smallsetminus \{1_{\mathbf{K}}\}$, we have $H_\chi = \{1\}$ by Lemma 3.C.2(1).

In particular, if $\mathbf{K}$ is infinite, then $\widehat{N}_{\text{per}} = \{1_N\}$ and (1) is a direct consequence of Theorem 4.A.3.

(2) If $\mathbf{K}$ is finite of cardinality q, then Γ has $|\Gamma/[\Gamma,\Gamma]| = q-1$ representations of degree 1, and one representation of degree $q-1$ denoted by $\mathrm{Ind}_N^\Gamma \chi$ (where $\chi \neq 1_{\mathbf{K}}$) in Proposition 3.C.4. Since

$$(q-1)^2 + (q-1) \times 1^2 \; = \; (q-1)q \; = \; |\Gamma|,$$

it follows from a well-known fact (see for instance [**AlBe–95**, Corollary 2, §14]) that any irreducible representation of Γ is equivalent to one of these. □

4.B.c. Solvable Bausmslag–Solitar group. Let p be a prime. Recall from Section 3.D that the Baumslag–Solitar $\Gamma = \mathrm{BS}(1,p)$ is the semi-direct product

$$\mathrm{BS}(1,p) \; = \; A \ltimes N, \quad \text{where } A = \begin{pmatrix} p^{\mathbf{Z}} & 0 \\ 0 & 1 \end{pmatrix} \approx \mathbf{Z} \text{ and } N = \begin{pmatrix} 1 & \mathbf{Z}[1/p] \\ 0 & 1 \end{pmatrix} \approx \mathbf{Z}[1/p],$$

and the generator $\begin{pmatrix} p & 0 \\ 0 & 1 \end{pmatrix}$ of A acts on N by multiplication by p. We often denote by (k,b) the element $\begin{pmatrix} p^k & b \\ 0 & 1 \end{pmatrix}$ of Γ, where $k \in \mathbf{Z}$ and $b \in \mathbf{Z}[1/p]$.

Recall also that we can identify $\widehat{N}$ with the p-adic solenoid

$$\mathbf{Sol}_p \; = \; (\mathbf{Q}_p \times \mathbf{R})/\Delta \qquad \text{for } \Delta \; = \; \{(a,-a) \mid a \in \mathbf{Z}[1/p]\};$$

the action of A on $\widehat{N}$ corresponds to the action of $\mathbf{Z}$ on $\mathbf{Sol}_p$ for which the generator $1 \in \mathbf{Z}$ acts by the transformation $T_p : \mathbf{Sol}_p \to \mathbf{Sol}_p$ of multiplication by p. For $s \in \mathbf{Sol}_p$, denote by χ_s the corresponding unitary character of N under the identification $\widehat{N} \approx \mathbf{Sol}_p$. The set $\widehat{N}_{\text{per}}$ of characters of N with finite A-orbit coincides with

$$\{\chi_s \mid s \in \mathrm{Per}(T_p)\},$$

where $\mathrm{Per}(T_p)$ is the set of T_p-periodic points in $\mathbf{Sol}_p$.

For $n \in \mathbf{N}^*$, let $s \in \mathbf{Sol}_p(n)$, where $\mathbf{Sol}_p(n)$ is the set of elements in $\mathbf{Sol}_p$ with T_p-period n. The stabilizer of χ_s in A is

$$A(n) \; = \; \{(k,0) \in \mathrm{BS}(1,p) \mid k \in n\mathbf{Z}\} \; \approx \; n\mathbf{Z},$$

and so the stabilizer of χ_s in Γ is the normal subgroup

$$\Gamma(n) \; := \; A(n) \ltimes N \; = \; \{(k,b) \in \Gamma \mid k \in n\mathbf{Z}, \; b \in \mathbf{Z}[1/p]\}$$

of index n in Γ. We parametrize the dual of $n\mathbf{Z}$ by $[0,1/n[$ through the map

$$[0,1/n[\; \to \widehat{n\mathbf{Z}}, \;\; \theta \mapsto \chi_\theta,$$

with χ_θ defined by $\chi_\theta(k) = e^{2\pi i\theta k}$ for all $k \in n\mathbf{Z}$.

Every unitary character χ of $\Gamma(n)$ with $\chi|_N = \chi_s$ is of the form $\chi_{s,\theta}$ for a unique $\theta \in [0,1/n[$, where $\chi_{s,\theta} : \Gamma(n) \to \mathbf{T}$ is defined by

$$\chi_{s,\theta}(k,b) \; = \; \chi_\theta(k)\chi_s(b) \qquad \text{for all } (k,b) \in \Gamma(n).$$

We obtain the following direct consequence of Theorem 4.A.3.

COROLLARY 4.B.7. *Let p be a prime and let $\Gamma = \mathrm{BS}(1,p)$ be a corresponding Baumslag–Solitar group. Set*

$$X_{\mathrm{fd}} = \bigsqcup_{n \geq 1} \left(\mathbf{Sol}_p(n) \times [0, 1/n[\right)$$

and let $\widetilde{T_p} : X_{\mathrm{fd}} \to X_{\mathrm{fd}}$ be defined by $\widetilde{T_p}(s,\theta) = (T_p s, \theta)$.

The map $X_{\mathrm{fd}} \to \widehat{\Gamma}_{\mathrm{fd}}$, defined by

$$(s,\theta) \mapsto \mathrm{Ind}_{\Gamma(n)}^{\Gamma} \chi_{s,\theta} \quad \text{for } n \geq 1,\ (s,\theta) \in \mathbf{Sol}_p(n) \times [0, 1/n[,$$

induces a bijection between the space of $\widetilde{T_p}$-orbits in X_{fd} and the space $\widehat{\Gamma}_{\mathrm{fd}}$ of equivalence classes of finite-dimensional irreducible representations of Γ.

4.B.d. Lamplighter group. Recall from Section 3.E that the lamplighter group is the semi-direct product $\Gamma = A \ltimes N$ of $A = \mathbf{Z}$ with $N = \bigoplus_{k \in \mathbf{Z}} \mathbf{Z}/2\mathbf{Z}$, where the action of $\mathbf{Z}$ on $\bigoplus_{k \in \mathbf{Z}} \mathbf{Z}/2\mathbf{Z}$ is given by shifting the coordinates. Recall also that $\widehat{N}$ can be identified with $X = \prod_{k \in \mathbf{Z}} \{0,1\}$, the dual action of $\mathbf{Z}$ on $\widehat{N}$ being given by the shift transformation T on X.

For $x \in X$, we denote by χ_x the corresponding element in $\widehat{N}$ under the identification $\widehat{N} \approx X$. The set $\widehat{N}_{\mathrm{per}}$ of characters of N with finite A-orbits coincides with

$$\{\chi_x \mid x \in \mathrm{Per}(T)\},$$

where $\mathrm{Per}(T)$ is the set of T-periodic points in X.

For $n \in \mathbf{N}^*$, let $x \in X(n)$, where $X(n)$ is the set of elements in X with T-period n. The stabilizer of χ_x in A is $n\mathbf{Z}$ and so the stabilizer of χ_x is the normal subgroup

$$\Gamma(n) = n\mathbf{Z} \ltimes N$$

of index n in Γ.

Parametrizing the dual of $n\mathbf{Z}$ by $[0, 1/n[$ as before, every unitary character χ of $\Gamma(n)$ with $\chi|_N = \chi_x$ is of the form $\chi_{x,\theta}$ for a unique $\theta \in [0, 1/n[$, where $\chi_{x,\theta} : \Gamma(n) \to \mathbf{T}$ is defined by

$$\chi_{x,\theta}(k,b) = \chi_\theta(k)\chi_x(b) \qquad \text{for all } (k,b) \in \Gamma(n).$$

We obtain the following direct consequence of Theorem 4.A.3.

COROLLARY 4.B.8. *Let $\Gamma = A \ltimes N$ be the lamplighter group. Set*

$$X_{\mathrm{fd}} = \bigsqcup_{n \geq 1} \left(X(n) \times [0, 1/n[\right)$$

and let $\widetilde{T} : X_{\mathrm{fd}} \to X_{\mathrm{fd}}$ be defined by $\widetilde{T}(x,\theta) = (Tx, \theta)$.

The map $X_{\mathrm{fd}} \to \widehat{\Gamma}_{\mathrm{fd}}$, defined by

$$(s,\theta) \mapsto \mathrm{Ind}_{\Gamma(n)}^{\Gamma} \chi_{x,\theta} \quad \text{for } n \geq 1,\ (x,\theta) \in X(n) \times [0, 1/n[,$$

induces a bijection between the space of $\widetilde{T}$-orbits in X_{fd} and the space $\widehat{\Gamma}_{\mathrm{fd}}$ of finite-dimensional irreducible representations of Γ.

4.C. Classes of groups in terms of finite-dimensional representations

Given a topological group G, one may ask how large is the set $\widehat{G}_{\mathrm{fd}}$ compared to $\widehat{G}$. More specifically, consider the following properties that G may possess:

- every irreducible representation of G is finite-dimensional, that is, $\widehat{G} = \widehat{G}_{\mathrm{fd}}$;
- G has sufficiently many finite-dimensional representations, in the sense that $\widehat{G}_{\mathrm{fd}}$ separates the points of G;
- every finite-dimensional representation of G is trivial, that is, $\widehat{G}_{\mathrm{fd}} = \{1_G\}$;
- the set $\widehat{G}_{\mathrm{fd}}$ is dense in $\widehat{G}$.

We will discuss in the following subsections the classes of LC groups singled out by either one of the three first properties. The class of discrete groups with the fourth property will be treated in Section 15.E.

4.C.a. Moore groups. A locally compact group G is a **Moore group** if all irreducible representations of G are finite-dimensional. LCA groups and compact groups are clearly Moore groups. Other obvious examples are products $K \times A$, where K is a compact group and A an LCA group. More generally, using the Mackey machine (see Remark 2.F.5), it can be shown that a central-by-compact LC group is a Moore group. As indicated by the following result, the class of Moore groups is a rather restricted class of LC groups.

For the next theorem, a Lie group is an LC group in which the connected component of the identity is open and is a connected Lie group.

THEOREM 4.C.1. (1) *A Lie group is a Moore group if and only if it has an open subgroup of finite index that is central-by-compact.*

(2) *A connected locally compact group is a Moore group if and only if it is a Cartesian product $K \times \mathbf{R}^n$, with K compact and $n \geq 0$.*

(3) *A locally compact group is a Moore group if and only if it is a projective limit of Moore Lie groups. In particular:*

(4) *A discrete group is a Moore group if and only if it has an abelian subgroup of finite index.*

(5) *The class of Moore groups is stable under subgroups, quotient groups, inverse limits, and finite extensions.*

REFERENCES FOR THE PROOF. See [**Moor–72**], as well as [**Robe–69**]. Observe that (4) is an easy consequence of Thoma's Theorem 7.D.1. □

REMARK 4.C.2. (1) Note that (1) in Theorem 4.C.1 *does not* carry over to locally compact groups: the example in [**Moor–72**, Section 5] is a Moore group without any central-by-compact finite index subgroup.

(2) Note also that an abelian-by-compact locally compact group need not be a Moore group. For example, the Euclidean group $\mathrm{SO}(2) \ltimes \mathbf{R}^2$ has a well-known family of infinite-dimensional irreducible representations indexed by the orbits of $\mathrm{SO}(2)$ in $\mathbf{R}^2 \smallsetminus \{0\}$, i.e., indexed by $\mathbf{R}_+^\times$; see, e.g., [**Mack–76**, Pages 195–196].

(3) Concerning (1) and (2) in Theorem 4.C.1, note that a connected locally compact group G is central-by-compact if and only if G is isomorphic to a direct product $K \times \mathbf{R}^n$ for some compact group K and some integer $n \geq 0$ (see [**Dixm–C***, 16.4.6] and Example 4.C.7 below).

4.C.b. MAP groups and Bohr compactification. A topological group G is **maximally almost periodic**, or **MAP**, if its finite-dimensional representations separate its points, i.e., if for any $g \neq e$ in G there exists a finite-dimensional representation π of G in some Hilbert space $\mathcal{H}$ such that $\pi(g) \neq \mathrm{Id}_{\mathcal{H}}$. This class is strictly larger than the class of Moore groups. It was introduced by von Neumann [**vNeu–34**]; see also [**Weil–40**, Chap. VII] and [**Dixm–C***, § 16].

We will put this class in perspective with Proposition 4.C.6. For this, we need the notion of Bohr compactification.

THEOREM 4.C.3. *Let G be a topological group.*

There exists a compact group K and a continuous homomorphism $\beta : G \to K$ with dense image, with the following universal property: for every compact group L and every continuous homomorphism $\alpha : G \to L$, there exists a continuous homomorphism $\alpha' : K \to L$ such that $\alpha = \alpha' \circ \beta$.

The pair (K, α) is unique in the following sense: let (K', β') be a pair consisting of a compact group and a continuous homomorphism $\beta' : G \to K'$ with dense image satisfying the same universal property. Then there exists an isomorphism $\alpha : K \to K'$ of topological groups such that $\beta' = \alpha \circ \beta$.

PROOF. • *Existence part.* Here is one way to prove the existence of K. Choose a family $(\pi_i, \mathcal{H}_i)_{i \in I}$ of representatives for the set of equivalence classes of finite-dimensional representations of G. Let

$$\beta : G \to \prod_{i \in I} \mathcal{U}(\mathcal{H}_i), \qquad g \mapsto \bigoplus_{i \in I} \pi_i(g).$$

Let K be the closure of $\beta(G)$ in the compact group $\prod_{i \in I} \mathcal{U}(\mathcal{H}_i)$. Then β is a continuous homomorphism and K is a compact group.

Let L be a compact group and $\alpha : G \to L$ a continuous homomorphism. Let $(\sigma_j, \mathcal{K}_j)_{j \in J}$ be a family of representatives for the set of equivalence classes of finite-dimensional representations of L. By the Peter–Weyl theory, the continuous homomorphism

$$\prod_{j \in J} \sigma_j : L \to \prod_{j \in J} \mathcal{U}(\mathcal{K}_j)$$

is injective. So, L is topologically isomorphic to its image in the compact group $\widetilde{L} := \prod_{j \in J} \mathcal{U}(\mathcal{K}_j)$. We may therefore assume that L is a closed subgroup of $\widetilde{L}$.

Let $j \in J$ and let $p_j : \widetilde{L} \twoheadrightarrow \mathcal{U}(\mathcal{K}_j)$ be the canonical projection. The composition $p_j \circ \alpha : G \to \mathcal{U}(\mathcal{K}_j)$ is a finite-dimensional representation of G. Therefore $p_j \circ \alpha$ is equivalent to $\pi_{i(j)}$ for a unique $i(j) \in I$. Let $V_j : \mathcal{H}_{i(j)} \to \mathcal{K}_j$ be a unitary operator intertwining $\pi_{i(j)}$ and $p_j \circ \alpha$, so that

$$V_j \pi_{i(j)}(g) V_j^{-1} = p_j \circ \alpha(g) \quad \text{for all } g \in G.$$

Let $\alpha' : K \to \widetilde{L}$ be the continuous homomorphism defined by

$$\alpha'\big((U_i)_{i \in I}\big) = \big(V_j U_{i(j)} V_j^{-1}\big)_{j \in J}$$

for $(U_i)_{i \in I} \in K \subset \prod_{i \in I} \mathcal{U}(\mathcal{H}_i)$. Then

$$\alpha'(\beta(g)) = \big(V_j \pi_{i(j)}(g) V_j^{-1}\big)_{j \in J} = \big(p_j \circ \alpha(g)\big)_{j \in J} = \alpha(g)$$

for every $g \in G$. Therefore $\alpha = \alpha' \circ \beta$.

• *Uniqueness part.* Let (K', β') be a pair consisting of a compact group and a continuous homomorphism $\beta' : G \to K'$ with dense image and satisfying the same universal property as (K, β). Then, by the universal property, there exist continuous homomorphisms $\alpha : K \to K'$ and $\alpha' : K' \to K$ such that $\beta' = \alpha \circ \beta$ and $\beta = \alpha' \circ \beta'$. Then

$$\alpha'(\alpha(\beta(g))) = \alpha'(\beta'(g)) = \beta(g) \quad \text{for all } g \in G$$

and hence $\alpha' \circ \alpha = \mathrm{Id}_K$, since $\beta(G)$ is dense in K. Similarly, we have $\alpha \circ \alpha' = \mathrm{Id}_{K'}$ and this shows that α is an isomorphism. □

The compact group associated with G as in Theorem 4.C.3 is denoted by $\mathrm{Bohr}(G)$ and the corresponding homomorphism by $\beta_B : G \to \mathrm{Bohr}(G)$; the group $\mathrm{Bohr}(G)$, or more precisely the pair $(\mathrm{Bohr}(G), \beta_B)$, is called the **Bohr compactification** of G. It follows from the definition that $\mathrm{Bohr}(\cdot)$ is a covariant functor: every continuous group homomorphism $G_1 \to G_2$ gives rise to a continuous homomorphism $\mathrm{Bohr}(G_1) \to \mathrm{Bohr}(G_2)$.

We are going to describe the Bohr compactification of an LCA group. For a topological group G, we denote by G_{disc} the group G made discrete (that is, viewed as a discrete group).

Let G be a LCA group. Denote by $\Gamma = \widehat{G}_{\mathrm{disc}}$ the dual of G, viewed as a discrete group. Observe that $\widehat{\Gamma}$ is a compact group, and that we have an injective continuous homomorphism $i : G \to \widehat{\Gamma}$, $g \mapsto \widehat{g}$, defined by $\widehat{g}(\chi) = \chi(g)$ for all $g \in G$ and $\chi \in \Gamma$.

PROPOSITION 4.C.4. *Let G be a locally compact abelian group and let $\Gamma = \widehat{G}_{\mathrm{disc}}$ be the dual group of G made discrete.*

The Bohr compactification of G coincides with the compact group $\widehat{\Gamma}$, together with the embedding $i : G \to \widehat{\Gamma}$ as above.

PROOF. Observe first that $i(G)$ is dense in $\widehat{\Gamma}$. Indeed, by Pontryagin duality, every element in $\widehat{\widehat{\Gamma}}$ is of the form $\widehat{\chi}$ for some $\chi \in \Gamma = \widehat{G}_{\mathrm{disc}}$. The annihilator of $i(G)$ in $\widehat{\widehat{\Gamma}}$ therefore consists of the unit character 1_G. It follows that $i(G)$ is dense in $\widehat{\Gamma}$.

To check that the pair $(\widehat{\Gamma}, i)$ has the universal property of a Bohr compactification, let K be a compact group and $\alpha : G \to K$ a continuous homomorphism.

Let $\alpha^t : \widehat{K} \to \widehat{G}_{\mathrm{disc}} = \Gamma$ be the homomorphism obtained by duality from α and which is defined by $\alpha^t(\psi) = \psi \circ \alpha$ for every $\psi \in \widehat{K}$. Since K is compact, $\widehat{K}$ is discrete, and so α^t is obviously continuous. Let $(\alpha^t)^t : \widehat{\Gamma} \to \widehat{\widehat{K}}$ be the dual homomorphism of α^t. Observe that $(\alpha^t)^t$ is continuous and is defined by

$$(\alpha^t)^t(\chi)(\psi) = \chi(\psi \circ \alpha) \quad \text{for } \chi \in \widehat{\Gamma}, \psi \in \widehat{K}.$$

Let

$$f : \widehat{\widehat{K}} \to K, \quad \widehat{x} \mapsto x$$

be the topological group isomorphism given by Pontryagin duality and define a continuous homomorphism $\alpha' : \widehat{\Gamma} \to K$ by $\alpha' = f \circ (\alpha^t)^t$. We claim that $\alpha' \circ i = \alpha$.

Indeed, let $g \in G$. For every $\psi \in \widehat{K}$, we have

$$(\alpha^t)^t(i(g))(\psi) = i(g)(\psi \circ \alpha) = (\psi \circ \alpha)(g) = \psi(\alpha(g)) = \widehat{\alpha(g)}(\psi),$$

and therefore $f((\alpha^t)^t(i(g))) = \alpha(g)$. So, we have $\alpha' \circ i = \alpha$ as claimed. □

Let G be a topological group. The kernel of the continuous homomorphism $\beta_B : G \to \mathrm{Bohr}(G)$ is the **von Neumann kernel** N_G of G; it is a topologically characteristic closed subgroup of G. There is a natural bijection from the subset $\widehat{G}_{\mathrm{fd}}$ of $\widehat{G}$ introduced above onto the dual $\widehat{\mathrm{Bohr}(G)}$ of the Bohr compactification of G. (Observe however that the Fell topology on $\widehat{\mathrm{Bohr}(G)}$ is discrete, but that $\widehat{G}_{\mathrm{fd}}$ in general is not a discrete subspace of $\widehat{G}$.)

Next, we characterize the functions on G which "extend" to continuous functions on $\mathrm{Bohr}(G)$. For a proof, we refer to [**Dixm–C***, 16.2.1].

THEOREM 4.C.5. *Let G be a topological group. For a continuous function $f : G \to \mathbf{C}$, the following properties are equivalent:*

(i) *there exists a continuous function $f' : \mathrm{Bohr}(G) \to \mathbf{C}$ such that $f = f' \circ \beta_B$;*
(ii) *the set of left translates $\{g \mapsto f(a^{-1}g)\}_{a \in G}$ is relatively compact in the Banach space $C^b(G)$ of bounded continuous functions on G;*
(iii) *f is a uniform limit of linear combinations of coefficients of finite-dimensional irreducible representations of G.*

A function on the topological group G which satisfies the equivalent properties of Theorem 4.C.5 is called **almost periodic**. For example, a linear combination of unitary characters on G is an almost periodic function. The space $AP(G)$ of almost periodic functions on G is a closed subalgebra of the Banach algebra $C^b(G)$, for the norm $f \mapsto \sup_{g \in G} |f(g)|$ of uniform convergence on G.

Maximally almost periodic groups can be characterized in terms of almost periodic functions.

PROPOSITION 4.C.6. *For a topological group G, the following conditions are equivalent:*

(i) *G is MAP;*
(ii) *there exist a compact group K and an injective continuous homomorphism $G \to K$;*
(iii) *the homomorphism $\beta_G : G \to \mathrm{Bohr}(G)$ of G in its Bohr compactification is injective, that is, the von Neumann kernel of G is trivial;*
(iv) *almost periodic functions separate the points of G, i.e., for $g, g' \in G$ with $g \neq g'$, there exists $f \in AP(G)$ such that $f(g) \neq f(g')$.*

PROOF. Assume that G is MAP and let $(\pi_i, \mathcal{H}_i)_{i \in I}$ be a family of representatives for the equivalence classes in $\widehat{G}_{\mathrm{fd}}$. The continuous homomorphism

$$\prod_{i \in I} \pi_i : G \to K := \prod_{i \in I} \mathcal{U}(\mathcal{H}_i)$$

is injective, and this shows that (i) implies (ii).

The fact that (ii) implies (iii) follows from the universal property of the Bohr compactification (Bohr, β_G) of G.

The characterization (i) in Theorem 4.C.5 of functions in $AP(G)$ shows that (iii) implies (iv).

Assume that (iv) holds. It follows from Theorem 4.C.5(iii) that finite-dimensional irreducible representations of G separate the points of G. Therefore G is MAP. □

EXAMPLE 4.C.7. (1) A Moore group is clearly MAP. In particular, locally compact abelian groups and compact compact groups are MAP.

(2) Let G be a connected locally compact group. The following five properties are equivalent:

(i) G is a direct product $K \times \mathbf{R}^n$, with K a compact group and $n \geq 0$;
(ii) G is MAP;
(iii) G is a Moore group;
(iv) the quotient of G by its centre is compact;
(v) every neighborhood of the identity in G contains a compact neighborhood which is invariant under all inner automorphisms, i.e., G is a SIN group (more on SIN groups in Section 15.B).

For a proof, we refer to [**Dixm–C***, 16.4.6]. The equivalence of (i) and (v), due to Iwasawa [**Iwas–51**, Theorem 3], is repeated below in our Theorem15.D.1. The implication (ii) $\Rightarrow$ (v) is a particular case of Corollary 15.B.5. The equivalence of (i) and (ii) is due to Freudenthal for second-countable connected LC groups [**Freu–36**] and to Weil for connected LC groups [**Weil–40**, § 32]. For more equivalences, see Diagram 3 in [**Palm–78**].

The equivalence of (ii) and (iii) carries over to connected-by-compact locally compact groups [**GrMo–71a**, Theorem 2.18].

(3) For every topological group G, the quotient G/N_G by its von Neumann kernel is MAP.

(4) Let G be a MAP group; every subgroup of G is MAP. Let G_1, G_2 be topological groups; the direct product $G_1 \times G_2$ is MAP if and only if G_1 and G_2 are MAP. Let G be an LC group and N a closed normal subgroup which is either compact or equal to the centre of G; if G is MAP, so is G/N [**LeRo–68**, Proposition 1]. Every locally compact MAP group is unimodular [**LeRo–68**, Theorem 2].

There is an interesting class of MAP groups among discrete groups. Let Γ be a **residually finite** discrete group, i.e., a group with the following property: for every $\gamma \in \Gamma \smallsetminus \{e\}$, there exists a finite group F and a homomorphism $\varphi : \Gamma \to F$ such that $\varphi(\gamma) \neq e$.

It is clear that residually finite groups are MAP. The converse does not hold, as witnessed by the additive group $\mathbf{Q}$ of rational numbers, and indeed by any nontrivial divisible abelian group, since these are MAP and are not residually finite.

However, among finitely generated groups, the MAP groups are exactly the residually finite ones.

PROPOSITION 4.C.8. *For a finitely generated group Γ, the following properties are equivalent:*

(i) Γ *is MAP;*
(ii) Γ *is residually finite.*

PROOF. Indeed, suppose Γ is MAP. Then Γ can be viewed as a subgroup of a compact group Σ. For $\gamma \in \Gamma$, there exists a representation $\pi : \Sigma \to \mathcal{U}(n)$ such that $\pi(\gamma) \neq \mathrm{Id}_{\mathbf{C}^n}$. Since finitely generated groups that are linear [in particular finitely generated groups that are subgroups of $\mathcal{U}(n)$ for some n] are residually finite by the Mal'cev Theorem [**Mal'c–40**], there exists a finite group F and a homomorphism $\rho : \pi(\Gamma) \to F$ such that $\rho(\pi(\gamma)) \neq e$; hence Γ is residually finite. As mentioned above, the converse implication is obvious. $\square$

EXAMPLE 4.C.9. (1) Finitely generated groups which are residually finite, and therefore MAP, include linear groups (see the proof of Proposition 4.C.8) and polycyclic groups, in particular nilpotent groups (a result of K. Hirsch for which we refer to [**Robi–96**, 5.14.17]).

(2) Discrete groups which are *not* MAP include the following groups, where $\mathbf{K}$ stands for an infinite field:

- Heisenberg groups $H(\mathbf{K})$, of which the von Neumann kernel coincides with the centre, by Corollary 4.B.2;
- affine groups $\mathrm{Aff}(\mathbf{K})$, for which the von Neumann kernel coincides with the translation group $\mathbf{K}$, by Corollary 4.B.6;
- general linear groups $\mathrm{GL}_n(\mathbf{K})$ for $n \geq 2$, of which the von Neumann kernel is $\mathrm{SL}_n(\mathbf{K})$, and special linear groups $\mathrm{SL}_n(\mathbf{K})$, by Corollary 4.C.22 and Proposition 4.C.20 below;
- and of course the m.a.p. groups defined below (except the one-element group).

4.C.c. Bohr compactification and profinite completion. A compact group G is **profinite** if G is topologically isomorphic to the projective limit of finite groups. Let us recall this last notion.

Let $(F_i)_{i\in I}$ be a family of finite groups indexed by a directed set I, with respect to a collection of homomorphisms $p_{i,j} : F_i \to F_j$ for $j \leq i$ such that $p_{i,i} = \mathrm{Id}_{F_i}$ and $p_{i,j} \circ p_{j,k} = p_{i,k}$ for $k \leq j \leq i$. The associated **projective limit** is the closed subgroup $\varprojlim_i F_i$ of the compact group $\prod_{i\in I} F_i$ defined by

$$\varprojlim_i F_i = \left\{ (g_i)_{i\in I} \in \prod_{i\in I} F_i \mid g_j = p_{i,j}(g_i) \ \text{ for all } \ i,j \in I, j \leq i \right\}.$$

Among compact groups, the profinite groups admit the following nice characterization (for more on this class of groups, see [**Wils–98**]).

PROPOSITION 4.C.10. *For a compact group G, the following properties are equivalent:*

(i) *G is profinite;*

(ii) *G is topologically isomorphic to a closed subgroup of a product of finite groups;*

(iii) *G is totally disconnected.*

PROOF. It is obvious that (i) implies (ii).

Assume that G is a closed subgroup of a product $K = \prod_{i\in I} F_i$ of finite groups. Then G is compact since K is compact. For every $i \in I$, the canonical projection $p_i : K \twoheadrightarrow F_i$ is continuous. Therefore the image of the connected component G_0 of the identity of G under p_i is connected, and therefore reduced to the identity, since F_i is finite. It follows that G_0 is trivial, that is, G is totally disconnected. This shows that (ii) implies (iii).

To show that (iii) implies (i), assume that G is totally disconnected. Let $\mathcal{N}$ be the family of open normal subgroups of G.

For every $N \in \mathcal{N}$, the quotient G/N is a finite group, since G is compact and G/N is discrete. Let $K = \varprojlim_N G/N$ be the projective limit of the family $(G/N)_{N\in\mathcal{N}}$, with respect to the canonical homomorphisms

$$p_{MN} : G/M \to G/N \quad \text{ if } M \subset N.$$

We claim that the natural homomorphism

$$\Phi : G \to K = \varprojlim_N G/N$$

is a topological isomorphism. Since every quotient homomorphism $G \to G/N$ is continuous, the map Φ is continuous. Therefore, since G is compact, it suffices to show that Φ is bijective.

Let us show that Φ is injective. It is well known that the connected component of an LC group H coincides with the intersection of all open subgroups of H ([**BTG1–4**, Chap. III, § 4, no 6]). Since G is totally disconnected, it follows that

$$\ker(\Phi) = \bigcap_{N \in \mathcal{N}} N = \{e\}.$$

Therefore Φ is injective.

To show that Φ is surjective, let $(g_N N)_{N \in \mathcal{N}} \in K$. We claim that the intersection $\bigcap_{N \subset \mathcal{N}} g_N N$ of cosets in G is nonempty.

Indeed, let $\mathcal{F}$ be a finite subset of $\mathcal{N}$. Then, $M := \bigcap_{N \in \mathcal{F}} N$ is an open normal subgroup of G. Since $(g_N N)_{N \in \mathcal{N}} \in \varprojlim_N G/N$, we have $g_N N = p_{M,N}(g_M M)$ for every $N \in \mathcal{F}$. Therefore $g_M \in \bigcap_{N \in \mathcal{F}} g_N N$. We have shown that $\bigcap_{N \in \mathcal{F}} g_N N$ is nonempty for every finite subset $\mathcal{F}$ of $\mathcal{N}$. Observe that every coset $g_N N$ is closed, since open subgroups are closed. As G is compact, it follows form the finite intersection property that $\bigcap_{N \in \mathcal{N}} g_N N$ is nonempty. For $g \in \bigcap_{N \in \mathcal{N}} g_N N$, we have $\Phi(g) = (g_N N)_{N \in \mathcal{N}}$. Therefore Φ is surjective. □

For a discrete group Γ, we now discuss the relationship of $\mathrm{Bohr}(G)$ with another completion of Γ, the profinite completion of Γ.

THEOREM 4.C.11. *Let Γ be a discrete group.*

There exists a profinite group G and a homomorphism $\beta : \Gamma \to G$ with dense image, with the following universal property: for every profinite group K and every homomorphism $\alpha : \Gamma \to K$, there exists a continuous homomorphism $\alpha' : G \to K$ such that $\alpha = \alpha' \circ \beta$.

The pair (G, β) is unique in the following sense: let (G', β') be a pair consisting of a profinite group and a homomorphism $\beta' : \Gamma \to G'$ with dense image satisfying the same universal property. Then there exists an isomorphism $\alpha : G \to G'$ of topological groups such that $\beta' = \alpha \circ \beta$.

PROOF. Here is one way to construct G. Let G be the closure of the image of Γ in the compact group $\prod_N \Gamma/N$, where N runs over the collection of all normal subgroups of finite index in Γ. Let $\beta : \Gamma \to G$ be the natural map.

It follows from Proposition 4.C.10 that G is a profinite group. One checks as in the proof of Theorem 4.C.3 that (G, β) has the desired universal property and that it is the unique pair with this property. □

The profinite group associated with Γ as in Theorem 4.C.11 is denoted by $\mathrm{Prof}(\Gamma)$ and the corresponding homomorphism by $\beta_P : \Gamma \to \mathrm{Prof}(\Gamma)$. The pair $(\mathrm{Prof}(\Gamma), \beta_P)$ is called the **profinite completion** of Γ.

REMARK 4.C.12. (1) As shown by the universal property, $\mathrm{Prof}(\Gamma)$ may also be realized as the projective limit $\varprojlim \Gamma/N$ of the family $(\Gamma/N)_N$, where the N's are subgroups of finite index in Γ, with respect to the canonical maps $\Gamma/N \twoheadrightarrow \Gamma/M$

when $N \subset M$; and the homomorphism β is defined in terms of the projections $\Gamma \twoheadrightarrow \Gamma/N$.

Alternatively, one may define $\mathrm{Prof}(\Gamma)$ as the largest Hausdorff quotient of the completion of Γ in the group topology for which the collection of all finite index normal subgroups in Γ is a fundamental system of open neighborhoods of the identity.

(2) Note that $\mathrm{Prof}(\cdot)$ is a covariant functor: every group homomorphism $\Gamma_1 \to \Gamma_2$ gives rise to a continuous homomorphism $\mathrm{Prof}(\Gamma_1) \to \mathrm{Prof}(\Gamma_2)$. For properties of this functor, see for example [**RiZa–10**, Section 3.2].

(3) Observe that the group Γ is residually finite if and only if the homomorphism $\beta_P : \Gamma \to \mathrm{Prof}(\Gamma)$ is injective.

Next, we determine the profinite completion of $\mathbf{Z}$. (In the literature, this completion is often denoted by $\widehat{\mathbf{Z}}$, a notation which conflicts with our use of $\widehat{\Gamma}$ as the dual of a group Γ.)

Example 4.C.13. We may realize the profinite completion $\mathrm{Prof}(\mathbf{Z})$ of $\mathbf{Z}$ as the projective limit

$$\mathrm{Prof}(\mathbf{Z}) = \varprojlim_{n} \mathbf{Z}/n\mathbf{Z}$$

with respect to the canonical maps $\mathbf{Z}/n\mathbf{Z} \twoheadrightarrow \mathbf{Z}/m\mathbf{Z}$ when $m|n$, together with the obvious injection

$$\mathbf{Z} \hookrightarrow \varprojlim_{n} \mathbf{Z}/n\mathbf{Z}.$$

We give another realization $\mathrm{Prof}(\mathbf{Z})$ as the product $\prod_{p \in P} \mathbf{Z}_p$ over the set P of all primes of the groups $\mathbf{Z}_p$ of p-adic integers, together with the natural injection $\beta : \mathbf{Z} \hookrightarrow \prod_{p \in P} \mathbf{Z}_p$. Indeed, to check this, it suffices to prove that the pair $(\prod_{p \in P} \mathbf{Z}_p, \beta)$ satisfies the universal property of Theorem 4.C.11.

First, observe that $\prod_{p \in P} \mathbf{Z}_p$ is a totally disconnected compact group and hence a profinite group. Next, we claim that $\beta(\mathbf{Z})$ is dense in $\prod_{p \in P} \mathbf{Z}_p$. Indeed, let $a = (a_p)_p \in \prod_{p \in P} \mathbf{Z}_p$. A basis of neighbourhoods of a is given by sets of the form

$$U^{e_1, \dots, e_k}_{p_1, \dots, p_k}(a) = \left\{ (x_p)_{p \in P} \in \prod_{p \in P} \mathbf{Z}_p \;\middle|\; x_{p_i} - a_{p_i} \in p_i^{e_i} \mathbf{Z}_{p_i} \text{ for } i = 1, \dots, k \right\},$$

for pairwise distinct primes $p_1, \dots, p_k$ and integers $e_1, \dots, e_k \geq 1$. Given such a neighbourhood, for every $i = 1, \dots, k$, we can write

$$a_{p_i} = m_i + b_{p_i} \quad \text{with } m_i \in \mathbf{Z} \text{ and } b_{p_i} \in p_i^{e_i} \mathbf{Z}_{p_i}.$$

By the Chinese Remainder Theorem, there exists an integer $m \in \mathbf{Z}$ such that

$$m - m_i \in p_i^{e_i} \mathbf{Z} \quad \text{for every } i = 1, \dots, k.$$

It is clear that $\beta(m) \in U^{e_1, \dots, e_k}_{p_1, \dots, p_k}(a)$. So, $\beta(\mathbf{Z})$ is dense in $\prod_{p \in P} \mathbf{Z}_p$, as claimed.

Let K be a profinite group and $\alpha : \mathbf{Z} \to K$ a homomorphism. We claim that there exists a continuous homomorphism $\alpha' : \prod_{p \in P} \mathbf{Z}_p \to K$ such that $\alpha = \alpha' \circ \beta$. Since K is a subgroup of a product of finite groups, we may assume that $K = \prod_{i \in I} F_i$ for a family $(F_i)_{i \in I}$ of finite groups and $\alpha = (\alpha_i)_{i \in I}$ for homomorphisms $\alpha_i : \mathbf{Z} \to F_i$. Therefore it suffices to show that, for every $i \in I$, there exists a continuous homomorphism $\alpha'_i : \prod_{p \in P} \mathbf{Z}_p \to F_i$ such that $\alpha_i = \alpha'_i \circ \beta$.

Fix $i \in I$ and let $n \geq 1$ be such that $\ker \alpha_i = n\mathbf{Z}$. Write $n = p_1^{e_1} \cdots p_k^{e_k}$ for pairwise distinct primes $p_1, \ldots, p_k$ and for integers $e_1, \ldots, e_k \geq 1$. Then

$$\prod_{j=1}^{k} \mathbf{Z}_{p_j}/p_j^{e_j}\mathbf{Z}_{p_j} \approx \prod_{j=1}^{k} \mathbf{Z}/p_j^{e_j}\mathbf{Z} \approx \mathbf{Z}/n\mathbf{Z}.$$

More precisely, let

$$f_n : \prod_{p \in P} \mathbf{Z}_p \twoheadrightarrow \prod_{j=1}^{k} \mathbf{Z}_{p_j}/p_j^{e_j}\mathbf{Z}_{p_j}$$

be the canonical projection. Then

$$f_n \circ \beta : \mathbf{Z} \to \prod_{j=1}^{k} \mathbf{Z}_{p_j}/p_i^{e_j}\mathbf{Z}_{p_j}$$

is a surjective homomorphism with kernel $n\mathbf{Z}$. Therefore the map $\alpha_i' : \prod_{p \in P} \mathbf{Z}_p \to F_i$, given by

$$\alpha_i'(x) = \alpha_i(m),$$

for $x \in \prod_{p \in P} \mathbf{Z}_p$ and for any $m \in \mathbf{Z}$ such that $f_n(\beta(m)) = f_n(x)$, is a well defined homomorphism. Moreover, we have

$$\alpha_i'(\beta(m)) = \alpha_i(m) \quad \text{for all } m \in \mathbf{Z},$$

that is, $\alpha_i = \alpha_i' \circ \beta$.

Observe that, by uniqueness of the profinite completion, it follows in particular that there is an isomorphism of topological groups

$$\Phi : \varprojlim_n \mathbf{Z}/n\mathbf{Z} \to \prod_{p \in P} \mathbf{Z}_p,$$

which is uniquely determined by the conditions

$$\Phi(\beta_1(m)) = \beta_2(m) \quad \text{for } m \in \mathbf{Z},$$

where, $\beta_1 : \mathbf{Z} \hookrightarrow \varprojlim_n \mathbf{Z}/n\mathbf{Z}$ and $\beta_2 : \mathbf{Z} \hookrightarrow \prod_{p \in P} \mathbf{Z}_p$ denote the canonical injections.

□

Let Γ be a discrete group. Observe that, by the universal property, there is a continuous homomorphism $c_P^B : \mathrm{Bohr}(\Gamma) \to \mathrm{Prof}(\Gamma)$ such that $\beta_P = c_P^B \circ \beta_B$. It follows that c_P^B has dense image and is therefore surjective, since c_P^B is continuous and since $\mathrm{Bohr}(\Gamma)$ is compact.

In general, the surjective homomorphism c_P^B is not injective; for example, when $\Gamma = \mathbf{Z}$ is infinite cyclic, we have (see Proposition 4.C.4)

$$\mathrm{Bohr}(\mathbf{Z}) \approx \left((\widehat{\mathbf{Z}})_{\mathrm{disc}}\right)^{\widehat{}} \approx \left((\mathbf{R}/\mathbf{Z})_{\mathrm{disc}}\right)^{\widehat{}},$$

and so $\mathrm{Bohr}(\mathbf{Z})$ is not second-countable, whereas (see Example 4.C.13)

$$\mathrm{Prof}(\mathbf{Z}) \approx \prod_{p \in P} \mathbf{Z}_p$$

is a second-countable compact abelian group.

One may ask which groups Γ have the property that the homomorphism c_P^B is an isomorphism. This property admits equivalent reformulations.

PROPOSITION 4.C.14. *Let Γ be a discrete group. The following properties are equivalent:*

(i) *the continuous surjective homomorphism $c_P^B : \mathrm{Bohr}(\Gamma) \twoheadrightarrow \mathrm{Prof}(\Gamma)$ is an isomorphism;*
(ii) *the image of every finite-dimensional representation $\Gamma \to \mathcal{U}(n)$ is finite;*
(iii) *the compact group $\mathrm{Bohr}(\Gamma)$ is totally disconnected.*

PROOF. Assume that (i) holds. Let $\pi : \Gamma \to \mathcal{U}(n)$ be a finite-dimensional representation of Γ. By the universal property of $\mathrm{Bohr}(\Gamma)$, there is a continuous homomorphism $\pi' : \mathrm{Bohr}(\Gamma) \to \mathcal{U}(n)$ such that $\pi = \pi' \circ \beta_B$. The map

$$f \;=\; \pi' \circ (c_P^B)^{-1} \;:\; \mathrm{Prof}(\Gamma) \to \mathcal{U}(n)$$

is a continuous homomorphism. Therefore there exists a neighbourhood U of e in $\mathrm{Prof}(\Gamma)$ such that $\|f(g) - I\| < 1$ for every $g \in U$, where $\| \cdot \|$ is the operator norm on the ambient algebra $\mathcal{L}(\mathbf{C}^n)$ of $\mathcal{U}(n)$, and I stands for $\mathrm{Id}_{\mathbf{C}^n}$. There exists a normal subgroup N of finite index in $\mathrm{Prof}(\Gamma)$ such that $N \subset U$. We then have

$$\|f(g^n) - I\| \;=\; \|f(g)^n - I\| < 1$$

for every $g \in N$ and every $n \in \mathbf{Z}$. It follows that 1 is the only eigenvalue of the unitary matrix $f(g)$ and hence that $f(g) = I$ for every $g \in N$. So, f is trivial on N and hence has finite image. It follows that $\pi \;=\; f \circ c_P^B \circ \beta_B$ has finite image. Therefore (ii) holds.

Assume that (ii) holds. Recall that $\mathrm{Bohr}(\Gamma)$ can be realized as the closure of the image of the homomorphism

$$\prod_{i \in I} \pi_i \,:\, \Gamma \to \prod_{i \in I} \mathcal{U}(\mathcal{H}_i),$$

where $(\pi_i, \mathcal{H}_i)_{i \in I}$ is a set of representatives for the equivalence classes of finite-dimensional representations of Γ. It follows that $\mathrm{Bohr}(\Gamma)$ is a closed subgroup of a Cartesian product of finite groups, and therefore is totally disconnected. This proves (iii).

Assume that (iii) holds. Then $\mathrm{Bohr}(\Gamma)$ is profinite, by Proposition 4.C.10. It follows from the universal property of $\mathrm{Prof}(\Gamma)$ that there exists a continuous map $c_B^P : \mathrm{Prof}(\Gamma) \to \mathrm{Bohr}(\Gamma)$ such that $c_B^P \circ \beta_P = \beta_B$. By density of $\beta_P(\Gamma)$ in $\mathrm{Prof}(\Gamma)$ and of $\beta_B(\Gamma)$ in $\mathrm{Bohr}(\Gamma)$, we see that c_B^P is the inverse map to c_P^B. □

We characterize abelian groups for which c_P^B is an isomorphism.

COROLLARY 4.C.15. *Let Γ be a discrete* abelian *group. The following properties are equivalent:*

(i) *the map $c_P^B : \mathrm{Bohr}(\Gamma) \twoheadrightarrow \mathrm{Prof}(\Gamma)$ is an isomorphism;*
(ii) *the group Γ is a direct sum $\bigoplus_{i \in I} C_i$ of finite cyclic groups C_i of bounded orders.*

PROOF. By Prüfer's Theorem (see [**Robi–96**, 4.35]), the abelian group Γ is periodic of bounded exponent (that is, there exists an integer $N \geq 1$ such that $\gamma^N = e$ for every $\gamma \in \Gamma$) if and only if Γ is a direct sum of finite cyclic groups of bounded orders. So, it suffices to show that c_P^B is an isomorphism if and only if Γ is periodic of bounded exponent.

Assume that Γ is periodic of bounded exponent N. Then, $\chi(\gamma)^N = \chi(\gamma^N) = 1$ for all $\gamma \in \Gamma, \chi \in \widehat{\Gamma}$. So, every $\chi \in \widehat{\Gamma}$ has finite image and hence c_P^B is an isomorphism, by Proposition 4.C.14.

Conversely, assume that c_P^B is an isomorphism, that is, every $\chi \in \widehat{\Gamma}$ has finite image. So, G_{disc} is a periodic group, where G is the compact group $\widehat{\Gamma}$. Then $G = \bigcup_{n \geq 1} G_n$, where G_n is the closed subset $\{\chi \in G \mid \chi^n = 1\}$ of G. By Baire's Theorem, there exists a nonempty open subset U of G which is contained in G_n for some $n \geq 1$. Then UU^{-1} is an open neighbourhood of e contained in G_n. The subgroup H generated by UU^{-1} is contained in G_n. Moreover, H is open in G and therefore has finite index, since G is compact. It follows that $G = G_N$ for $N = n|G/H|$, that is, G_{disc} is periodic of exponent N. As $\chi(\gamma^N) = \chi^N(\gamma) = 1$ for all $\gamma \in \Gamma$ and $\chi \in \widehat{\Gamma} = G$, this implies that Γ is periodic of exponent N. □

Observe that we recover from Corollary 4.C.15 the following known fact (see [**Rudi–62**, 4.35]): if Γ is a discrete abelian group of finite exponent, then $\text{Bohr}(\Gamma)$ is totally disconnected.

Next, for $\Gamma = \text{SL}_n(\mathbf{Z})$ and $n \geq 3$, we show that the map c_P^B is an isomorphism. Combining this result with the congruence subgroup theorem, we will give a precise description of $\text{Bohr}(\Gamma)$.

Recall (see Example 4.C.13) that $\text{Prof}(\mathbf{Z})$ can be identified with the additive group of the ring $\prod_{p \in P} \mathbf{Z}_p$.

THEOREM 4.C.16. *Let n be an integer with $n \geq 3$.*

(1) *The map* $c_P^B : \text{Bohr}(\text{SL}_n(\mathbf{Z})) \twoheadrightarrow \text{Prof}(\text{SL}_n(\mathbf{Z}))$ *is an isomorphism.*
(2) *The Bohr compactification* $\text{Bohr}(\text{SL}_n(\mathbf{Z}))$ *coincides with* $\text{SL}_n(\text{Prof}(\mathbf{Z}))$, *together with the homomorphism* $\text{SL}_n(\mathbf{Z}) \to \text{SL}_n(\text{Prof}(\mathbf{Z}))$ *induced by the natural embedding*

$$\mathbf{Z} \to \text{Prof}(\mathbf{Z}) \approx \prod_{p \in P} \mathbf{Z}_p.$$

PROOF. (1) Consideration of the homomorphisms of reduction modulo N, for $N \geq 1$, shows that $\Gamma = \text{SL}_n(\mathbf{Z})$ is residually finite for every $n \geq 2$. We claim that every finite-dimensional representation of Γ has finite image, when $n \geq 3$.

To prove the claim, we assume first that $n = 3$. Let $\pi : \Gamma \to \mathcal{U}(d)$ be a finite-dimensional representation of $\Gamma = \text{SL}_3(\mathbf{Z})$. Consider the restriction of π to the subgroup

$$H := \begin{pmatrix} 1 & \mathbf{Z} & \mathbf{Z} \\ 0 & 1 & \mathbf{Z} \\ 0 & 0 & 1 \end{pmatrix} \subset \Gamma,$$

which is a copy of the Heisenberg group $H(\mathbf{Z})$ over the integers. By Corollary 4.B.5, there exists an integer $N \geq 1$ such that π is trivial on the elementary matrix

$$E_{1,3}(N) = \begin{pmatrix} 1 & 0 & N \\ 0 & 1 & 0 \\ 0 & 0 & 1 \end{pmatrix}.$$

Now, the smallest normal subgroup Δ_N of Γ containing $E_{1,3}(N)$ has finite index in Γ; see Corollary 3 in [**Hump–80**, § 17.2] for a proof of this elementary but nontrivial fact, which indeed holds for any $n \geq 3$. Since π factorizes by the quotient Γ/Δ_N, the claim is proved for $n = 3$.

We assume now that $n \geq 3$. Let σ denote the restriction of π to the upper-left corner subgroup $\mathrm{SL}_3(\mathbf{Z})$ of $\Gamma = \mathrm{SL}_n(\mathbf{Z})$. By the previous argument, there exists $N \geq 1$ such that $\sigma(\gamma^N) = I$ for all $\gamma \in \mathrm{SL}_3(\mathbf{Z})$. Therefore $\ker(\pi)$ contains $E_{1,3}(N)$, and hence the smallest normal subgroup of Γ containing $E_{1,3}(N)$, which is a group of finite index in Γ. This concludes the proof of the claim for all $n \geq 3$.

Therefore the Bohr compactification $\mathrm{Bohr}(\mathrm{SL}_n(\mathbf{Z}))$ is isomorphic to the profinite completion of $\mathrm{SL}_n(\mathbf{Z})$.

(2) By the congruence subgroup theorem [**BaLS–64, Menn–65**], the homomorphism

$$\mathrm{Prof}(\mathrm{SL}_n(\mathbf{Z})) \to \mathrm{SL}_n(\mathrm{Prof}(\mathbf{Z}))$$

defined by the universal property is an isomorphism for $n \geq 3$. The claim follows from this fact and from Item (1). □

REMARK 4.C.17. (1) In contrast to the result in Theorem 4.C.16, the group $\mathrm{SL}_2(\mathbf{Z})$ has finite-dimensional representations with infinite images. Indeed, there is an observation of Hausdorff at the heart of the Hausdorff–Banach–Tarski paradox: the rotation group SO(3) contains subgroups isomorphic to the free product $(\mathbf{Z}/2\mathbf{Z}) * (\mathbf{Z}/3\mathbf{Z})$, i.e., to the quotient $\mathrm{PSL}_2(\mathbf{Z})$ of $\mathrm{SL}_2(\mathbf{Z})$ by its centre of order 2. This is in the appendix to §X.1 in [**Haus–14**]; see also, for example, [**OsAd–76**].

In particular, $\mathrm{SL}_2(\mathbf{Z})$ has infinite three-dimensional representations, and the epimorphism $\mathrm{Bohr}(\mathrm{SL}_2(\mathbf{Z})) \twoheadrightarrow \mathrm{Prof}(\mathrm{SL}_2(\mathbf{Z}))$ has a nontrivial kernel.

(2) Let Γ be a group which has a quotient which is infinite, abelian, and not a direct sum of cyclic groups of bounded orders (see Corollary 4.C.15); for example, let Γ be a group that has a quotient isomorphic to one of $\mathbf{Z}, \mathbf{K}, \mathbf{K}^\times$, with $\mathbf{K}$ an infinite field of characteristic 0. Then Γ has an infinite one-dimensional representation, so that, again, the epimorphism $\mathrm{Bohr}(\Gamma) \twoheadrightarrow \mathrm{Prof}(\Gamma)$ has a nontrivial kernel.

This applies to many of our favourite examples: to $H(\mathbf{K})$, $H(\mathbf{Z})$, $\mathrm{BS}(1,p)$, $\mathrm{Aff}(\mathbf{K})$, and $\mathrm{GL}_n(\mathbf{K})$, with again $\mathbf{K}$ an infinite field of characteristic 0.

4.C.d. Minimally almost periodic groups. A topological group G is **minimally almost periodic**, or **m.a.p.**, if every finite-dimensional representation of G is the identity, that is, if $\widehat{G}_{\mathrm{fd}} = \{1_G\}$.

The straightforward proof of the following characterizations of m.a.p. groups is left to the reader (compare with the chacterizations of MAP groups in Proposition 4.C.6).

PROPOSITION 4.C.18. *For a topological group G, the following properties are equivalent:*

(i) *G is a m.a.p. group;*
(ii) *every continuous homomorphism from G to a compact group is the identity;*
(iii) *the Bohr compactification of G is the group with one element;*
(iv) *every almost periodic continuous function on G is constant.*

The following characterization of finitely generated groups which are m.a.p. should be compared with Proposition 4.C.8.

PROPOSITION 4.C.19. *For a finitely generated group Γ, the following properties are equivalent:*

(i) *Γ is m.a.p.;*
(ii) *Γ has no other finite quotient than the one-element group.*

PROOF. Indeed, Proposition 4.C.18(ii) shows that (i) implies (ii). To show the converse, assume Γ is not m.a.p. Therefore there exists a nontrivial representation $\pi : \Gamma \to \mathcal{U}(n)$. Since finitely generated groups which are linear are residually finite (Mal'cev's Theorem), there exists a finite group F with more than one element and a surjective homomorphism $\rho : \pi(\Gamma) \to F$. Then F is a nontrivial finite quotient of Γ. Therefore, (ii) implies (i). □

The following result in the case of $\mathbf{K} = \mathbf{R}$ appears in the original article of von Neumann [**vNeu–34**, Page 483]. See also Example $\mathcal{G}^{(\gamma)}$ in Section 5 of [**vNWi–40**].

PROPOSITION 4.C.20. *Let* $\mathbf{K}$ *be an infinite field and* n *an integer,* $n \geq 2$. *The special linear group* $\mathrm{SL}_n(\mathbf{K})$ *is minimally almost periodic.*

PROOF. We have to show that, for every integer $d \geq 1$ and every finite-dimensional representation $\pi : \mathrm{SL}_n(\mathbf{K}) \to \mathcal{U}(d)$, we have $\ker(\pi) = \mathrm{SL}_n(\mathbf{K})$. Since proper normal subgroups of $\mathrm{SL}_n(\mathbf{K})$ are central (see for example Theorem 4.9 in [**Arti–57**]), it suffices to show that $\ker(\pi)$ contains a matrix γ which is not central in $\mathrm{SL}_n(\mathbf{K})$. Moreover, if $\alpha : \mathrm{SL}_2(\mathbf{K}) \to \mathrm{SL}_n(\mathbf{K})$ is any injective homomorphism, $\pi \circ \alpha$ is a finite-dimensional representation of $\mathrm{SL}_2(\mathbf{K})$; so it suffices to prove the claim in the case $n = 2$.

For $x \in \mathbf{K}$ and $a \in \mathbf{K}^\times$, consider the elements

$$\gamma_x = \begin{pmatrix} 1 & x \\ 0 & 1 \end{pmatrix} \quad \text{and} \quad \delta_a = \begin{pmatrix} a & 0 \\ 0 & a^{-1} \end{pmatrix}$$

of $\mathrm{SL}_2(\mathbf{K})$. Observe that the γ_x's commute with each other and that

$$\delta_a \gamma_x \delta_a^{-1} = \gamma_{a^2 x}.$$

It follows that $\{\pi(\gamma_{a^2}) \mid a \in \mathbf{K}^\times\}$ is a family of conjugate and commuting matrices in $\mathcal{U}(d)$. Therefore, the $\pi(\gamma_{a^2})$'s all have the same set $\{\lambda_1, \ldots, \lambda_d\}$ of eigenvalues and, moreover, we can find a common basis in $\mathbf{C}^d$ for which they are represented by diagonal matrices. However, there are only finitely many diagonal matrices with diagonal entries from the set $\{\lambda_1, \ldots, \lambda_d\}$, and the set $\{a^2 \mid a \in \mathbf{K}^\times\}$ is infinite, since $\mathbf{K}$ is infinite. It follows that there exist $a, b \in \mathbf{K}^\times$ with $a^2 \neq b^2$ such that $\pi(\gamma_{a^2}) = \pi(\gamma_{b^2})$. The element

$$\gamma_{a^2} \gamma_{b^2}^{-1} = \gamma_{a^2 - b^2} \in \mathrm{SL}_n(\mathbf{K})$$

therefore belongs to the kernel of π and is clearly not central. □

REMARK 4.C.21. (1) The result of Proposition 4.C.20 can be extended to other groups, as follows: let Γ be a group generated by its subgroups isomorphic to either $\mathrm{SL}_2(\mathbf{K})$ or $\mathrm{PSL}_2(\mathbf{K})$, for some infinite field $\mathbf{K}$; then every finite-dimensional representation of Γ is a multiple of the identity 1_Γ.

For $\mathbf{K} = \mathbf{R}$, this extended result and structure theory imply that every connected real Lie group which is semisimple, without compact factors and with finite centre, is minimally almost periodic, as a discrete group, and therefore *a fortiori* as a Lie group (see [**Vale–86**, Remark 3]). In fact, an even stronger result is true [**SevN–50**]: such a group has no nontrivial representation in the unitary group of a factor of finite type (see Section 7.E).

(2) By contrast, a second-countable locally compact group that is solvable and noncompact is not minimally almost periodic [**Schm–84**, Theorems 3.4 & 4.1].

We can now easily determine all finite-dimensional representations of $\mathrm{GL}_n(\mathbf{K})$ for an infinite field.

COROLLARY 4.C.22. *Let* $\mathbf{K}$ *be an infinite field and* n *an integer,* $n \geq 2$.
Every irreducible finite-dimensional representation of $\Gamma = \mathrm{GL}_n(\mathbf{K})$ *is a unitary character of* Γ*; more precisely, the map*

$$\widehat{\mathbf{K}} \to \widehat{\Gamma}_{\mathrm{fd}}, \quad \chi \mapsto \varphi_\chi,$$

where $\varphi_\chi(\gamma) = \chi(\det(\gamma))$ *for* $\gamma \in \Gamma$*, is a bijection*

PROOF. Let π be an irreducible finite-dimensional representation of Γ. By Proposition 4.C.20, $\mathrm{SL}_n(\mathbf{K}) \subset \ker \pi$. Since $[\Gamma, \Gamma] = \mathrm{SL}_n(\mathbf{K}) = \ker \det$, the claim follows. □

EXAMPLE 4.C.23. There are several other known classes of m.a.p. groups: The group $\mathrm{Alt}(\mathbf{N})$ of even finitely supported permutations of the integers (a simple group), more generally every countable group which is the union of a strictly increasing sequence of simple subgroups, as well as every countable infinite group which is finitely generated and simple [**BCRZ–16**, Examples 2.12 & 2.13]. Also, there exists a countable infinite m.a.p. group that is orderable, locally solvable, and perfect; for this and other examples, we refer to [**GrKO–15**].

CHAPTER 5

Describing all irreducible representations of some semi-direct products

Let Γ be a countable discrete group. The procedure of induction of representations from subgroups of Γ, as described in Section 1.F, *does produce irreducible representations of Γ, but usually does not produce all of them. We will now describe a general construction of representations which produces* all *irreducible representations, up to equivalence, of a second-countable LC group G of the following form: $G = H \ltimes N$ is a semi-direct product, where*

- *H is a discrete subgroup of G;*
- *N is a locally compact abelian normal closed subgroup of G.*

In fact, this procedure will give a description of an arbitrary factor *representation of G, up to equivalence (Theorem* 5.D.9*). The description of a factor representation or an irreducible representation of G involves an equivalence class of probability measures μ on $\widehat{N}$ and a cocycle $\widehat{N} \times H \to \mathcal{U}(\mathcal{K})$ over $(\widehat{N}, \mu)$ with values in the unitary group of some Hilbert space $\mathcal{K}$.*

One should mention that, even when H is abelian, the procedure we are going to describe does not allow in general to establish a concrete list of all irreducible representations of G. See the comments in Remark 5.D.10.

This procedure is an extension of the Mackey machine to the context of ergodic nontransitive actions and can be cast in the more general framework of groupoid representations [**Rams–76**].

5.A. Constructing some irreducible representations

Let $G = H \ltimes N$ be a semi-direct product of a countable *discrete* group H with a normal abelian subgroup N which is locally compact and second-countable; note that N need not be discrete. In this section, we exhibit a family of representations of G attached to H-quasi-invariant measures on the second-countable LCA group $\widehat{N}$.

DEFINITION 5.A.1. Let $(X, \mathcal{B})$ be a Borel space, μ a σ-finite measure on $(X, \mathcal{B})$, and $H \curvearrowright X$, or

$$H \times X \to X, \quad (h, x) \mapsto hx,$$

a **left** action of a group H on $(X, \mathcal{B})$, preserving the class of μ. For $h \in H$, denote by $(h^{-1})_*(\mu)$ the image of μ by the action of h^{-1}, and consider the Radon–Nikodym derivative

$$c(h, x) := \frac{d(h^{-1})_*(\mu)}{d\mu}(x) \quad \text{for all } x \in X$$

of $(h^{-1})_*(\mu)$ with respect to μ. Then (see Proposition A.3.4) we have the cocycle relation

$$c(h_1 h_2, x) = c(h_1, h_2 x) c(h_2, x) \quad \text{for all } h_1, h_2 \in H, \ x \in X.$$

The map $c : H \times X \to \mathbf{R}_+^\times$ is called the **Radon–Nikodym cocycle** of the action $H \curvearrowright X$.

In case of a **right** group action $X \curvearrowleft H$, the value of the cocycle for every element $h \in H$ at a point $x \in X$ is exactly the same, that is, $c(h, x) = \frac{d(h^{-1})_*(\mu)}{d\mu}(x)$; however, we prefer to denote this value by $c(x, h)$ instead of $c(h, x)$ so that the cocycle relation has the following more natural form:

$$\text{(RN)} \qquad c(x, h_1h_2) = c(x, h_1)c(xh_1, h_2) \quad \text{for all } x \in X,\ h_1, h_2 \in H.$$

CONSTRUCTION 5.A.2. Let $G = H \ltimes N$ be a semi-direct product of a countable discrete subgroup H with a second-countable locally compact abelian normal subgroup N, with respect to a left action $H \curvearrowright N$ of H on N; see Convention 4.A.1. Recall that we identify the groups H and N to subgroups of G. For $h \in H$ and $n \in N$, we have $(h, e)(e, n)(h, e)^{-1} = (h, h \cdot n)(h^{-1}, e) = (e, h \cdot n)$, so that the left action of H on N can then be written as $H \times N \to N$, $(h, n) \mapsto hnh^{-1}$; it extends naturally to the G-action $G \times N \to N$, $(g, n) \mapsto ghg^{-1}$.

This left action $H \curvearrowright N$ corresponds to a right action $\widehat{N} \curvearrowleft H$, $(\chi, h) \mapsto \chi^h$, defined by

$$\chi^h(n) = \chi(hnh^{-1}) \quad \text{for all } \chi \in \widehat{N},\ h \in H,\ n \in N.$$

The latter extends naturally to a right action of $\widehat{N} \curvearrowleft G$, defined by $\chi^g(n) = \chi(gng^{-1})$ for all $\chi \in \widehat{N}$, $g \in G$, and $n \in N$.

Let μ be a probability measure on $\widehat{N}$ which is quasi-invariant by G (equivalently by H). For every $g = (h, n) \in G$, we define an operator $\pi_\mu(g)$ on the Hilbert space $L^2(\widehat{N}, \mu)$ by

$$(\pi_\mu(g)f)(\chi) = \chi(n)c(\chi, h)^{1/2}f(\chi^h) \quad \text{for all } f \in L^2(\widehat{N}, \mu),\ \chi \in \widehat{N}.$$

By definition of the Radon–Nikodym derivative, $\pi_\mu(g)$ is a unitary operator on $L^2(\widehat{N}, \mu)$; the cocycle identity (RN) implies that

$$G \to \mathcal{U}(L^2(\widehat{N}, \mu)), \quad g \mapsto \pi_\mu(g)$$

is a group homomorphism. Moreover, since $\widehat{N}$ is second-countable, the Hilbert space $L^2(\widehat{N}, \mu)$ is separable (Theorem A.4.1(2)) and hence π_μ is continuous, when $\mathcal{U}(L^2(\widehat{N}, \mu))$ is equipped with the strong operator topology (see Proposition A.6.1 in [**BeHV–08**]). So, $(\pi_\mu, L^2(\widehat{N}, \mu))$ is a representation of G.

REMARK 5.A.3. Let $G = H \ltimes N, \mu$, and π_μ be as in Construction 5.A.2.

(1) The representation π_μ is an extension to G of the canonical representation of the LCA group N associated with μ considered in Section 2.A (and also denoted by π_μ there).

In case $H = \{e\}$ and $G = N$, the representation π_μ is precisely that of Construction 2.A.1.

(2) The restriction of π_μ to H is the Koopman representation of H associated with the action $(\widehat{N}, \mu) \curvearrowleft H$. See Section 13.A, where Koopman representations are introduced in a more general context.

(3) Assume that μ is *ergodic* under the action of H. Then π_μ is an *irreducible* representation of G. This is a particular case of Corollary 5.D.6 below.

The construction of the representations π_μ above can be considerably generalized by adding an operator-valued cocycle over $(\widehat{N}, \mu)$ as a further ingredient; see Construction 5.C.1.

5.B. Cocycles over measurable group actions

Let μ be a σ-finite measure on a Borel space $(X, \mathcal{B})$. Let G be a topological group equipped with a **right** group action $(X, \mathcal{B}) \curvearrowleft G$ which is measurable and which preserves the class of μ.

DEFINITION 5.B.1. Let L be a topological group. A **cocycle** α **over** (X, μ) **with values in** L is a map $\alpha : X \times G \to L$ which satisfies the identity

$$\text{(Coc)} \qquad \alpha(x, g_1 g_2) = \alpha(x, g_1)\alpha(xg_1, g_2),$$

for μ-almost all $x \in X$ and all $g_1, g_2 \in G$, and which is measurable when G and L are equipped with the Borel structures underlying their respective topologies. Note that (Coc) implies that $\alpha(x, e) = e$ and $\alpha(x, g)^{-1} = \alpha(xg, g^{-1})$ for μ-almost all $x \in X$ and all $g \in G$.

Observe that, if $\alpha_1 : X \times G \to L$ is a cocycle over (X, μ) and $\varphi : X \to L$ is a measurable map, then the map $\alpha_2 : X \times G \to L$, defined by

$$\text{(Coh)} \qquad \alpha_2(x, g) = \varphi(x)\alpha_1(x, g)\varphi(xg)^{-1},$$

for all $x \in X$ and $g \in G$, is also a cocycle over (X, μ). Two cocycles α_1, α_2 over (X, μ) with values in L are **cohomologous**, or **equivalent**, if there exists a measurable map $\varphi : X \to L$ such that (Coh) holds for μ-almost all $x \in X$ and all $g \in G$.

A cocycle $\alpha : X \times G \to L$ over (X, μ) which is cohomologous to the identity cocycle $I : (x, g) \mapsto e$ is called a **coboundary**. Thus, α is a coboundary if and only if there exists a measurable map $\varphi : X \to L$ such that

$$\alpha(x, g) = \varphi(x)\varphi(xg)^{-1}$$

for μ-almost all $x \in X$ and all $g \in G$.

Observe that, if μ_1 and μ_2 are two σ-finite measures on X which are quasi-invariant by G and absolutely continuous with each other, a map $\alpha : X \times G \to L$ is a cocycle over (X, μ_1) if and only if it is a cocycle over (X, μ_2).

REMARK 5.B.2. Let $(X, \mathcal{B})$, μ, G, $(X, \mathcal{B}) \curvearrowleft G$, and L be as above. We describe a construction which provides a justification for the defining relations (Coc) and (Coh).

Let $(Y, \mathcal{C})$ be a measure space. Denote by $\mathcal{M}(X, Y)$ the space of measurable maps from X to Y, modulo the equivalence relation of equality μ-almost everywhere. There is a natural left action $G \curvearrowright \mathcal{M}(X, Y)$ defined by $(g * f)(x) = f(xg)$ for all $g \in G$, $f \in \mathcal{M}(X, Y)$, and $x \in X$.

Assume that, moreover, L acts on $(Y, \mathcal{C})$, say on the left for notational convenience. Let $\alpha : X \times H \to L$ be a cocycle. There is a **twisted action** of G on $\mathcal{M}(X, Y)$ defined by

$$(g *_\alpha f)(x) = \alpha(x, g)f(xg) \qquad \text{for all } \ g \in G, \ f \in \mathcal{M}(X, Y), \ x \in X.$$

The cocycle identity (Coc) is precisely what is needed for this to be an action, i.e., for $g_1 *_\alpha (g_2 *_\alpha f) = (g_1 g_2) *_\alpha f$ for all $g_1, g_2 \in G$ and $f \in \mathcal{M}(X, Y)$.

If α and β are two cocycles over (X, μ) with values in L which are cohomologous, then, because of (Coh), the corresponding twisted actions $*_\alpha$ and $*_\beta$ are equivalent in an appropriate sense (see [**Zimm–84**, Section 4.2]).

EXAMPLE 5.B.3. (1) Let H be a subgroup of a group G and T a right transversal, so that $G = \bigsqcup_{t \in T} Ht$. With the notation of Construction 1.F.4(2), the map

$$T \times G \to H, \quad (t, g) \mapsto \alpha(t, g)$$

is a cocycle over T (with the counting measure) with values in H.

(2) Consider a Borel space $(X, \mathcal{B})$, a σ-finite measure μ on it, and a group G acting on X on the right by measurable transformations preserving the class of μ. Denote by $g_*(\mu)$ the direct image of μ by the action of $g \in G$. The Radon–Nikodym cocycle $c_\mu : X \times G \to \mathbf{R}_+^\times$ is given by

$$c_\mu(x, g) := \frac{d(g^{-1})_*(\mu)}{d\mu}(x)$$

as in Definition 5.A.1. Let ν be another σ-finite measure on $(X, \mathcal{B})$ which is equivalent to μ. Then $\nu = \varphi\mu$ for a measurable function $\varphi : X \to \mathbf{R}_+^\times$ which is locally μ-integrable. We have

$$\begin{aligned} c_\nu(x, g) &= \frac{d(g^{-1})_*(\nu)}{d\mu}(x) = \frac{d(g^{-1})_*(\nu)}{d(g^{-1})_*(\mu)}(x) \, \frac{d(g^{-1})_*(\mu)}{d\mu}(x) \, \frac{d\mu}{d\nu}(x) \\ &= \varphi(xg) \, c_\mu(x, g) \, \frac{1}{\varphi(x)}, \end{aligned}$$

that is,

$$c_\mu(x, g) = \varphi(x) \, c_\nu(x, g) \, \varphi(xg)^{-1} \qquad \text{for all } x \in X \text{ and } g \in G.$$

This shows that the cocycles c_ν and c_μ are cohomologous.

In particular, there exists a σ-finite measure ν on $(X, \mathcal{B})$ which is equivalent to μ and G-invariant, i.e., and such that $c_\nu(x, g) = 1$ for all $x \in X$ and $g \in G$, if and only if c_μ is a coboundary, that is, if and only if c_μ is of the form

$$c_\mu(x, g) = \frac{\varphi(x)}{\varphi(xg)},$$

for some measurable and locally μ-integrable function $\varphi : X \to \mathbf{R}_+^\times$.

(3) Let $G = H \ltimes N$ be a semi-direct product where H is countable discrete and N locally compact abelian. Let μ be a probability measure on $\widehat{N}$ which is quasi-invariant by G (equivalently: by H). The map

$$\alpha : \widehat{N} \times G \to \mathbf{T}, \quad (\chi, (h, n)) \mapsto \chi(n)$$

is a cocycle, which satisfies the identity (Coc). In Construction 5.A.2, the definition of the representation π_μ on $L^2(\widehat{N}, \mu)$ involves the product of two cocycles, one is α, the other the Radon–Nikodym cocycle. They will both appear again in the definition of the representation of Construction 5.D.1.

(4) Let G, L be topological groups, $f : G \to L$ a measurable group homomorphism, and $(X, \mathcal{B}, \mu)$ a σ-finite measure space on which G acts by measurable transformations preserving the class of μ. Then

$$\alpha : X \times G \to L, \quad (x, g) \mapsto f(g)$$

is a cocycle over (X, μ) with values in L.

More generally, let G, L, M be topological groups, $\alpha : X \times G \to L$ a cocycle over (X, μ) with values in L, and $f : L \to M$ a measurable group homomorphism. Then

$$f \circ \alpha : X \times G \to M, \quad (x, g) \mapsto f(\alpha(x, g))$$

is a cocycle over (X, μ) with values in M.

(5) Let $(X, \mathcal{B}, \mu)$ be a σ-finite measure space and $T : X \to X$ a Borel isomorphism preserving the class of μ. Consider $(X, \mathcal{B}, \mu)$ as a $\mathbf{Z}$-space, the generator 1 of $\mathbf{Z}$ acting by T. Let L be a topological group and $f : X \to L$ a measurable map. Define a measurable map $\alpha_f : X \times \mathbf{Z} \to L$ by

$$\alpha_f(x, n) = \begin{cases} \prod_{i=0}^{n-1} f(T^i x) & \text{if } n > 0, \\ e & \text{if } n = 0, \\ \prod_{i=1}^{|n|} f(T^{-i}x)^{-1} & \text{if } n < 0. \end{cases}$$

One checks that α_f is a cocycle.

Every cocycle $\alpha : X \times \mathbf{Z} \to L$ is of the form α_f for some measurable map $f : X \to L$. Indeed, if $f : X \to L$ is defined by $f(x) = \alpha(x, 1)$ for all $x \in X$, then $\alpha = \alpha_f$, by the cocycle relation (Coc).

A cocycle $\alpha = \alpha_f$ is a coboundary if and only if there exists a measurable map $\varphi : X \to L$ such that

$$f(x) = \varphi(x)^{-1}\varphi(Tx)$$

for μ-almost all $x \in X$.

5.C. Cocycles with values in the unitary group of a Hilbert space

Let $(X, \mathcal{B})$ be a Borel space equipped with a σ-finite measure μ. Let G be a topological group equipped with a measurable right group action $(X, \mathcal{B}) \curvearrowleft G$ which preserves the class of μ. Let c_μ be the corresponding Radon–Nikodym cocycle, defined by $c_\mu(x, g) = \frac{d(g^{-1})_*(\mu)}{d\mu}(x)$ for $g \in G$ and $x \in X$ (as in Definition 5.A.1).

Let $\mathcal{K}$ be a Hilbert space. The unitary group $\mathcal{U}(\mathcal{K})$ of $\mathcal{K}$, equipped with the strong operator topology, is a topological group. In this section, we will be interested in cocycles over (X, μ) with values in $\mathcal{U}(\mathcal{K})$. We are going to associate to such a cocycle a representation of G.

Construction 5.C.1. Let $\mathcal{K}$ be a Hilbert space. Let $\alpha : X \times G \to \mathcal{U}(\mathcal{K})$ be a cocycle over (X, μ) with values in $\mathcal{U}(\mathcal{K})$. For every $g \in G$, we define an operator $\pi_{\mu,\alpha}(g)$ on the Hilbert space $L^2(X, \mu, \mathcal{K})$ by

$$(\pi_{\mu,\alpha}(g)F)(x) = c_\mu(x, g)^{1/2}\, \alpha(x, g)\, (F(xg))$$

for all $F \in L^2(X, \mu, \mathcal{K})$ and $x \in X$. Using the cocycle identities (RN) and (Coc), one checks as in Section 5.A that

$$G \to \mathcal{U}(L^2(X, \mu, \mathcal{K})), \quad g \mapsto \pi_{\mu,\alpha}(g)$$

is a group homomorphism.

If G is a σ-compact locally compact group and $L^2(X, \mu, \mathcal{K})$ is separable, then $\pi_{\mu,\alpha}$ is continuous and so is representation of G (see [**BeHV–08**, Lemma A.6.2]).

Example 5.C.2. Let H be an open subgroup of a topological group G and $(\sigma, \mathcal{K})$ a representation of H. Let T be a right transversal for H in G and $(t, g) \mapsto$

$\alpha(t, g)$ the cocycle over (T, μ) with values in H as in Construction 1.F.4(2), where μ is the counting measure μ. The map

$$T \times G \to \mathcal{U}(\mathcal{K}), \quad (t, g) \mapsto \sigma(\alpha(t, g))$$

is a cocycle over (T, μ) with values in $\mathcal{U}(\mathcal{K})$. As Construction 1.F.4(2) shows, the associated representation $\pi_{\mu,\alpha}$ on $\ell^2(T, \mathcal{K})$ is equivalent to the induced representation $\mathrm{Ind}_H^G \sigma$.

The previous fact is valid more generally for any second countable LC group G and any *closed* subgroup H of G: for a representation σ of H, the induced representation $\mathrm{Ind}_H^G \sigma$ of G is equivalent to a representation of the form $\pi_{\mu,\alpha}$ as in Construction 5.C.1 (see [**KaTa–13**, § 2.4] or [**Foll–16**, Remark 2, p.177]).

REMARK 5.C.3. A representation $\pi_{\mu,\alpha}$ as in Construction 5.C.1 is part of a "system of imprimitivity" $(\pi_{\mu,\alpha}, E)$ based on X in the sense of [**Mack–58**, § 5]; the projection-valued measure (see Section A.9) E on X and $L^2(X, \mu, \mathcal{K})$ is the map

$$E : \mathcal{B} \to \mathrm{Proj}(L^2(X, \mu, \mathcal{K})), \qquad B \mapsto m(\mathbf{1}_B).$$

In some sense, the most general system of imprimitivity is of the form $(\pi_{\mu,\alpha}, E)$ as above (for more details, see [**Mack–58**, Theorem 5.6]).

Next, we introduce the notions of intertwiners and irreducibility for cocycles with values in unitary groups of Hilbert spaces.

DEFINITION 5.C.4. Let

$$\alpha_1 : X \times G \to \mathcal{U}(\mathcal{K}_1) \quad \text{and} \quad \alpha_2 : X \times G \to \mathcal{U}(\mathcal{K}_2)$$

be measurable cocycles over (X, μ) with values in the unitary groups of Hilbert spaces $\mathcal{K}_1$ and $\mathcal{K}_2$. An **intertwiner** between α_1 and α_2 is a decomposable operator

$$\widetilde{T} : L^2(X, \mu, \mathcal{K}_1) \to L^2(X, \mu, \mathcal{K}_2),$$

associated with an essentially bounded measurable map

$$T : X \to \mathcal{L}(\mathcal{K}_1, \mathcal{K}_2), \quad x \mapsto T(x)$$

as in Example 1.G.5, i.e., defined by

$$(\widetilde{T}F)(x) = T(x)F(x) \quad \text{for all } F \in L^2(X, \mu, \mathcal{K}_1) \text{ and } x \in X,$$

such that the relation

$$\alpha_2(x, g)T(xg) = T(x)\alpha_1(x, g)$$

holds for all $g \in G$ and μ-almost all $x \in X$.

Though α_1 and α_2 do not have values in the same unitary group, we extend (slightly) the terminology defined earlier in this section: the cocycles α_1 and α_2 are **cohomologous** (or **equivalent**) if there exists an intertwiner $\widetilde{T}$ such that $T(x) : \mathcal{K}_1 \to \mathcal{K}_2$ is an isomorphism of Hilbert spaces for μ-almost all $x \in X$.

A measurable cocycle $\alpha : X \times G \to \mathcal{U}(\mathcal{K})$ over (X, μ) is said to be **irreducible** if every intertwiner $L^2(X, \mu, \mathcal{K}) \to L^2(X, \mu, \mathcal{K})$ between α and itself belongs to the subalgebra

$$\mathcal{A} = \{m(\varphi) \mid \varphi \in L^\infty(X, \mu)\}$$

of diagonalisable operators in $\mathcal{L}(L^2(X, \mu, \mathcal{K}))$. For diagonalisable operators in this context, again see Example 1.G.5.

REMARK 5.C.5. (1) It is obvious that every cocycle with values in the circle group $\mathbf{T} = \mathcal{U}(1)$ is irreducible.

Example 5.C.8 provides an irreducible cocycle with values in $\mathcal{U}(2)$.

(2) Assume that the action of G on (X, μ) is *ergodic* and let $\alpha : X \times G \to \mathcal{U}(\mathcal{K})$ be an irreducible cocycle over (X, μ). Then the only intertwiners between α and itself are the constant diagonalisable operators, that is, the operators of the form $m(c\mathbf{1}_X)$ for some $c \in \mathbf{C}$. Indeed, an intertwiner between α and itself is a diagonalisable operator $m(\varphi)$ for $\varphi \in L^\infty(X, \mu)$; the intertwiner relation then shows that $\varphi(xg) = \varphi(x)$ for all $g \in G$ and μ-almost all $x \in X$. It follows from the ergodicity of the action that φ is constant μ-almost everywhere (see Proposition A.5.9).

(3) We will see below (Theorem 5.D.4) instances where the irreducibility of a cocycle $\alpha : X \times G \to \mathcal{U}(\mathcal{K})$ is equivalent to the irreducibility of the associated representation $\pi_{\mu,\alpha}$ of Construction 5.C.1.

As the next lemma shows, the representation $\pi_{\mu,\alpha}$ depends, up to equivalence, only on the *classes* of μ and α.

LEMMA 5.C.6. *Let μ_1 and μ_2 be two probability measures on X which are quasi-invariant by G and absolutely continuous with each other. For $i = 1, 2$, let $\alpha_i : X \times G \to \mathcal{U}(\mathcal{K}_i)$ be a cocycle over (X, μ_1), and therefore also a cocycle over (X, μ_2), with values in the unitary group of a Hilbert space $\mathcal{K}_i$. Assume that α_1 and α_2 are cohomologous.*

Then the representations π_{μ_1,α_1} and π_{μ_2,α_2} are equivalent.

PROOF. We have $\mu_1 = \varphi\mu_2$, where $\varphi := \dfrac{d\mu_1}{d\mu_2}$ is the Radon–Nikodym derivative of μ_1 with respect to μ_2. By assumption, there exists a measurable map

$$T : X \to \mathcal{L}(\mathcal{K}_1, \mathcal{K}_2), \quad x \mapsto T(x)$$

such that $T(x)$ is a Hilbert space isomorphism and such that the relation

$$\alpha_2(x, g)T(xg) = T(x)\alpha_1(x, g)$$

holds for all $g \in G$ and μ_1-almost all $x \in X$.

The operator

$$U : L^2(X, \mu_1, \mathcal{K}_1) \to L^2(X, \mu_2, \mathcal{K}_2),$$

defined by

$$U(F)(x) = \varphi(x)^{1/2}T(x)(F(x)) \quad \text{for all } F \in L^2(X, \mu_1, \mathcal{K}_1),\ x \in X,$$

is an isomorphism of Hilbert spaces. Moreover, T intertwines the representations π_{μ_1,α_1} and π_{μ_2,α_2}; indeed, if $c_{\mu_i}(x, g) = \dfrac{d(g^{-1})_*\mu_i}{d\mu_i}(x)$ for $i = 1, 2$, we have (see Example 5.B.3(2))

$$c_{\mu_2}(x, g) = \frac{\varphi(x)}{\varphi(xg)} c_{\mu_1}(x, g),$$

and hence

$$\begin{aligned}\big(U\pi_{\mu_1,\alpha_1}(g)F\big)(x) &= \varphi(x)^{1/2}T(x)\big((\pi_{\mu_1,\alpha_1}(g)(F))(x)\big)\\ &= (\varphi(x)c_{\mu_1}(x,g))^{1/2}T(x)\alpha_1(x,g)\big(F(xg)\big)\\ &= (c_{\mu_2}(x,g)\varphi(xg))^{1/2}\alpha_2(x,g)T(xg)\big(F(xg)\big)\\ &= c_{\mu_2}(x,g)^{1/2}\alpha_2(x,g)\big(U(F)(xg)\big)\\ &= \big(\pi_{\mu_2,\alpha_2}(g)U(F)\big)(x)\end{aligned}$$

for $g \in G$, $F \in L^2(X, \mu_1, \mathcal{K}_1)$, and $x \in X$. It follows that π_{μ_1,α_1} and π_{μ_2,α_2} are equivalent. □

We give examples of irreducible cocycles over the circle for an irrational rotation. We start with examples with values in $\mathbf{T}$.

EXAMPLE 5.C.7. Let $X = \mathbf{T}$ be the circle, equipped with the normalized Lebesgue measure μ. Consider the action of $G = \mathbf{Z}$ given on $\mathbf{T}$ by an irrational rotation R_θ.

Let

$$f : \mathbf{T} \to \mathbf{T}, \ \ e^{2\pi it} \mapsto e^{2\pi it}$$

denote the identity map. We claim that the associated cocycle $\alpha_f : \mathbf{T} \times \mathbf{Z} \to \mathbf{T}$ (see Example 5.B.3(5)) with values in the circle group $\mathbf{T}$ is a not a coboundary.

Indeed, assume, by contradiction, that a measurable function $\varphi : \mathbf{T} \to \mathbf{T}$ exists such that

$$f(e^{2\pi it}) = \varphi(e^{2\pi it})^{-1}\varphi(R_\theta(e^{2\pi it})),$$

that is,

$$e^{2\pi it}\varphi(e^{2\pi it}) = \varphi(e^{2\pi i(\theta+t)})$$

for almost every $t \in \mathbf{R}$. For $n \in \mathbf{Z}$, let

$$c_n(\varphi) = \int_0^1 \varphi(e^{2\pi it})e^{-2\pi int}dt$$

be the n-th Fourier coefficient of $\varphi \in L^\infty(\mathbf{T}, \mu)$. The n-th Fourier coefficient of the function $t \mapsto \varphi(e^{2\pi i(\theta+t)})$ is $e^{-2\pi in\theta}c_n(\varphi)$ and the n-th Fourier coefficient of the function $t \mapsto e^{2\pi it}\varphi(e^{2\pi it})$ is $c_{n-1}(\varphi)$. Therefore we have

$$c_n(\varphi) = e^{2\pi in\theta}c_{n-1}(\varphi) \qquad \text{for all } n \in \mathbf{Z}.$$

This implies that $(|c_n(\varphi)|)_{n\in\mathbf{Z}}$ is a constant sequence. However, $(|c_n(\varphi)|)_{n\in\mathbf{Z}} \in \ell^2(\mathbf{Z})$, since $\varphi \in L^2(\mathbf{T}, \mu)$. It follows that $c_n(\varphi) = 0$ for all $n \in \mathbf{Z}$, that is, $\varphi = 0$. This is impossible, since $f(e^{2\pi it}) = \varphi(e^{2\pi it})^{-1}\varphi(R_\theta(e^{2\pi it}))$ and $f \neq 0$.

More generally, one can show along the same lines that, for $k \in \mathbf{Z}$, $k \neq 0$, the cocycles $\alpha_{f_k} : \mathbf{T} \times \mathbf{Z} \to \mathbf{T}$, defined by the functions

$$f_k : \mathbf{T} \to \mathbf{T}, \ \ e^{2\pi it} \mapsto e^{2\pi ikt},$$

are pairwise noncohomologous. For a more general result, see [**Furs–61**, Lemma 2.2].

Next, we give an example of an irreducible two-dimensional unitary cocycle for an irrational rotation of the circle, following [**Isma–69**]. The first such example is due to Gleason and appears in [**Kiri–67**].

EXAMPLE 5.C.8. As in Example 5.C.7, we consider the action of $\mathbf{Z}$ by an irrational rotation R_θ on the circle $\mathbf{T}$, equipped with the normalized Lebesgue measure μ. Let

$$f : \mathbf{T} \to \mathcal{U}(2), \quad z \mapsto \begin{pmatrix} 0 & 1 \\ z & 0 \end{pmatrix}.$$

We claim that the associated cocycle (see Example 5.B.3(5)) $\alpha_f : \mathbf{T} \times \mathbf{Z} \to \mathcal{U}(2)$ with values in the unitary group $\mathcal{U}(2)$ is irreducible.

Indeed, write R instead of R_θ and let $z_0 = e^{2\pi i\theta}$ so that $R(z) = z_0 z$ for every $z \in \mathbf{T}$. Let

$$T : \mathbf{T} \to \mathcal{U}(2), \quad z \mapsto T(z) = \begin{pmatrix} a(z) & b(z) \\ c(z) & d(z) \end{pmatrix}$$

be a measurable map such that

$$(*) \qquad \alpha_f(z,k)T(R^k z) = T(z)\alpha_f(z,k)$$

holds for all $k \in \mathbf{Z}$ and μ-almost all $z \in \mathbf{T}$. Since

$$\alpha_f(z,2) = f(z)f(R(z)) = \begin{pmatrix} R(z) & 0 \\ 0 & z \end{pmatrix}$$

(see Example 5.B.3(5)), it follows from $(*)$ in the case $k = 2$ that

$$\begin{pmatrix} R(z) & 0 \\ 0 & z \end{pmatrix} \begin{pmatrix} a(R^2(z)) & b(R^2(z)) \\ c(R^2(z)) & d(R^2(z)) \end{pmatrix} = \begin{pmatrix} a(z) & b(z) \\ c(z) & d(z) \end{pmatrix} \begin{pmatrix} R(z) & 0 \\ 0 & z \end{pmatrix}$$

for μ-almost all $z \in \mathbf{T}$. So, we have the relations

$$a(R^2(z)) = a(z) \quad \text{and} \quad d(R^2(z)) = d(z)$$

as well as

$$b(R^2(z)) = \overline{z_0} b(z) \quad \text{and} \quad c(R^2(z)) = z_0 c(z)$$

for μ-almost all $z \in \mathbf{T}$. Since R^2 is ergodic, it follows from the first two relations that a and d are constant μ-almost everywhere on $\mathbf{T}$.

The spectrum of the Koopman operator U_{R^2} on $L^2(\mathbf{T})$ associated to R^2 consists of the set of eigenvalues $\{z_0^{2k} \mid k \in \mathbf{Z}\}$; in particular, $\overline{z_0}$ and z_0 are not eigenvalues of U_{R^2}. It follows from the last two relations that the functions b and c, which are bounded and hence belong to $L^2(\mathbf{T})$, must be zero μ-almost everywhere on $\mathbf{T}$.

Observe that $(*)$ in the case $k = 1$ shows that the constant functions a and b must be equal.

As a result, we have shown that T is a constant multiple of the identity matrix μ-almost everywhere on $\mathbf{T}$.

5.D. Constructing all irreducible representations

Let us go back to a locally compact group $G = H \ltimes N$, as in Section 5.A. Let μ be a probability measure on $\widehat{N}$, which is quasi-invariant by the action of G (equivalently, by the action of H). We will consider cocycles over $(\widehat{N}, \mu)$ with values in the unitary group of a Hilbert space.

CONSTRUCTION 5.D.1. Let $\mathcal{K}$ be a Hilbert space. Let $\alpha : \widehat{N} \times H \to \mathcal{U}(\mathcal{K})$ be a cocycle over $(\widehat{N}, \mu)$ with values in $\mathcal{U}(\mathcal{K})$. Then $\widetilde{\alpha} : \widehat{N} \times G \to \mathcal{U}(\mathcal{K})$, defined by

$$\widetilde{\alpha}(\chi, g) = \chi(n)\alpha(\chi, h) \quad \text{for all } g = (h,n) \in G,\ \chi \in \widehat{N},$$

is easily checked to be a cocycle for the action $(\widehat{N}, \mu) \curvearrowleft G$. The associated representation of G on $L^2(\widehat{N}, \mu, \mathcal{K})$ as defined in Construction 5.C.1, which we will denote by $\pi_{\mu,\alpha}$ instead of $\pi_{\mu,\widetilde{\alpha}}$, is given as follows:

$$(\pi_{\mu,\alpha}(g)F)(\chi) = \chi(n)\, c_\mu(\chi, h)^{1/2}\, \alpha(\chi, h)\, (F(\chi^h))$$

for all $g = (h, n) \in G$, $F \in L^2(\widehat{N}, \mu, \mathcal{K})$, and $\chi \in \widehat{N}$. (Recall that c_μ is the $\mathbf{R}_+^\times$-valued Radon–Nikodym cocycle, defined by $c_\mu(\chi, h) := \frac{d(h^{-1})_*(\mu)}{d\mu}(\chi)$ for all $\chi \in \widehat{N}$ and $h \in H$.)

REMARK 5.D.2. Observe that $\pi_{\mu,\alpha}$ is an extension to G of the canonical representation $\pi_\mu^{(k)}$ of the LCA group N associated with $(\mu, \mathcal{K})$ and considered in Sections 2.A and 2.B, where k is the dimension of $\mathcal{K}$. The results obtained in these sections about the representations $\pi_\mu^{(k)}$ will be useful in our study of the representations $\pi_{\mu,\alpha}$.

We now proceed towards Theorem 5.D.4, concerning irreducibility and equivalence of representations of the form $\pi_{\mu,\alpha}$.

Given a probability measure μ on $\widehat{N}$ quasi-invariant by H, the following lemma shows that the intertwiners between two cocycles α_1, α_2 over $(\widehat{N}, \mu)$ as introduced in Definition 5.C.4 coincide with the intertwiners between the associated representations $\pi_{\mu,\alpha_1}, \pi_{\mu,\alpha_2}$ of Construction 5.D.1.

LEMMA 5.D.3. *Let $G = H \ltimes N$ be a semi-direct product of a countable discrete subgroup H with a second-countable locally compact abelian normal subgroup N. Let μ be a probability measure on $\widehat{N}$ which is quasi-invariant by the H-action. For $i = 1, 2$, let $\mathcal{K}_i$ be a separable Hilbert space, and let $\alpha_i : \widehat{N} \times H \to \mathcal{U}(\mathcal{K}_i)$ be a cocycle over $(\widehat{N}, \mu)$.*

For an operator $\widetilde{T} : L^2(\widehat{N}, \mu, \mathcal{K}_1) \to L^2(\widehat{N}, \mu, \mathcal{K}_2)$, the following properties are equivalent:

(i) *$\widetilde{T}$ is an intertwiner between the cocycles α_1 and α_2;*
(ii) *$\widetilde{T}$ is an intertwiner between the representations π_{μ,α_1} and π_{μ,α_2}.*

PROOF. For $i = 1, 2$, set $\mathcal{H}_i = L^2(\widehat{N}, \mu, \mathcal{K}_i)$. For $\varphi \in L^\infty(\widehat{N}, \mu)$, denote by $m_i(\varphi)$ the operator on $\mathcal{H}_i$ given by multiplication by φ, as in Section 1.H. Observe that, for $n \in N$, we have

$$\pi_{\mu_i,\alpha_i}(n) = m_i(\widehat{n}).$$

Recall that $\widehat{n}$ denotes the function in $C^b(\widehat{N})$ defined by $\widehat{n}(\chi) = \chi(n)$ for $\chi \in \widehat{N}$.

(i) $\Longrightarrow$ (ii) Let $\widetilde{T} : \mathcal{H}_1 \to \mathcal{H}_2$ be an intertwiner between α_1 and α_2, associated with an essentially bounded measurable map $T : \widehat{N} \to \mathcal{L}(\mathcal{K}_1, \mathcal{K}_2)$. As a decomposable operator, $\widetilde{T}$ is such that

$$\widetilde{T} m_1(\widehat{n}) = m_2(\widehat{n}) \widetilde{T} \quad \text{for all } n \in N$$

(see Section 1.H), so that $\widetilde{T}$ intertwines $\pi_{\mu,\alpha_1}|_N$ and $\pi_{\mu,\alpha_2}|_N$.

Let $h \in H$. We have

$$\begin{aligned}
\big(\pi_{\mu,\alpha_2}(h, e)\widetilde{T}(F)\big)(\chi) &= c_\mu(\chi, h)^{1/2}\alpha_2(\chi, h)\big(\widetilde{T}(F)(\chi^h)\big) \\
&= c_\mu(\chi, h)^{1/2}\alpha_2(\chi, h)\big(T(\chi^h)(F(\chi^h))\big) = c_\mu(\chi, h)^{1/2}\big(T(\chi)\alpha_1(\chi, h)\big)(F(\chi^h)) \\
&= T(\chi)\big((\pi_{\mu,\alpha_1}(h, e)F)(\chi)\big) = \big(\widetilde{T}\pi_{\mu,\alpha_1}(h, e)(F)\big)(\chi)
\end{aligned}$$

for all $F \in \mathcal{H}_1$ and μ-almost all $\chi \in \widehat{N}$. Therefore

$$\pi_{\mu,\alpha_2}(h,e)\widetilde{T} = \widetilde{T}\pi_{\mu,\alpha_1}(h,e) \quad \text{for all } h \in H.$$

It follows that $\widetilde{T}$ intertwines π_{μ,α_1} and π_{μ,α_2}.

(ii) $\Longrightarrow$ (i) Conversely, let $\widetilde{T} \in \mathcal{L}(\mathcal{H}_1, \mathcal{H}_2)$ be an intertwiner between π_{μ,α_1} and π_{μ,α_2}. Since $\widetilde{T}$ intertwines $\pi_{\mu,\alpha_1}|_N$ and $\pi_{\mu,\alpha_2}|_N$, and since $\pi_{\mu,\alpha_1}|_N$ and $\pi_{\mu,\alpha_2}|_N$ coincide with the canonical representations $\pi_\mu^{(n_1)}$ and $\pi_\mu^{(n_2)}$ of Section 2.B, where n_1 and n_2 are the dimensions of $\mathcal{K}_1$ and $\mathcal{K}_2$, it follows from Proposition 2.B.3 that $\widetilde{T}$ is a decomposable operator, associated with an essentially bounded measurable map

$$\widehat{N} \to \mathcal{L}(\mathcal{K}_1, \mathcal{K}_2), \quad \chi \mapsto T(\chi).$$

Let $h \in H$. Since $\pi_{\mu,\alpha_2}(h,e)\widetilde{T} = \widetilde{T}\pi_{\mu,\alpha_1}(h,e)$, we have as above

$$\begin{aligned} c_\mu(\chi,h)^{1/2}T(\chi)\alpha_1(\chi,h)(F(\chi^h)) &= T(\chi)\Big((\pi_{\mu,\alpha_1}(h,e)F)(\chi)\Big) \\ &= \Big(\widetilde{T}\pi_{\mu,\alpha_1}(h,e)(F)\Big)(\chi) = \big(\pi_{\mu,\alpha_2}(h,e)\widetilde{T}(F)\big)(\chi) \\ &= c_\mu(\chi,h)^{1/2}\alpha_2(\chi,h)(\widetilde{T}(F))(\chi^h) = c_\mu(\chi,h)^{1/2}\alpha_2(\chi,h)T(\chi^h)(F(\chi^h)) \end{aligned}$$

for all $F \in L^2(\widehat{N}, \mu, \mathcal{K}_1)$ and μ-almost all $\chi \in \widehat{N}$. It follows that

$$T(\chi)\alpha_1(\chi,h) = \alpha_2(\chi,h)T(\chi^h)$$

for every $h \in H$ and μ-almost all $\chi \in \widehat{N}$; hence, $\widetilde{T}$ is an intertwiner between α_1 and α_2. $\square$

The first main result in this chapter concerns the irreducibility of representations of the form $\pi_{\mu,\alpha}$ and equivalence between such representations.

THEOREM 5.D.4. *Let $G = H \ltimes N$ be a semi-direct product of a countable discrete subgroup H with a second-countable locally compact abelian normal subgroup N.*

Let μ, μ_1, μ_2 be probability measures on $\widehat{N}$ which are quasi-invariant by the H-action and $\mathcal{K}$, $\mathcal{K}_1$, $\mathcal{K}_2$ separable Hilbert spaces. Let α, α_1, and α_2 be cocycles over $(\widehat{N}, \mu)$, $(\widehat{N}, \mu_1)$, and $(\widehat{N}, \mu_2)$ with values in $\mathcal{U}(\mathcal{K})$, $\mathcal{U}(\mathcal{K}_1)$, and $\mathcal{U}(\mathcal{K}_2)$, respectively. Let $\pi_{\mu,\alpha}$, π_{μ_1,α_1}, and π_{μ_2,α_2} be the corresponding representations, as in Construction 5.D.1.

(1) *Suppose that μ is ergodic for the H-action. Then the representation $\pi_{\mu,\alpha}$ is irreducible if and only if the cocycle α is irreducible.*
(2) *If the representations π_{μ_1,α_1} and π_{μ_2,α_2} are equivalent, then the measures μ_1 and μ_2 are equivalent.*
(3) *Suppose that $\mu_1 = \mu_2 = \mu$. Then π_{μ,α_1} and π_{μ,α_2} are equivalent if and only if the cocycles α_1, α_2 are cohomologous.*

PROOF. (1) Set $\mathcal{H} = L^2(\widehat{N}, \mu, \mathcal{K})$. Assume that α is irreducible and let $\widetilde{T} \in \mathcal{L}(\mathcal{H})$ be an intertwining operator for π with itself. It follows from Lemma 5.D.3 that $\widetilde{T}$ is an intertwiner between α and itself. Since α is irreducible, $\widetilde{T}$ is a diagonalisable operator: there exists $\varphi \in L^\infty(\widehat{N}, \mu)$ such that

$$\widetilde{T} = m(\varphi).$$

The fact $\widetilde{T}$ is an intertwiner between α and itself means that

$$\varphi(\chi^h)\alpha(\chi,h) = \varphi(\chi)\alpha(\chi,h),$$

that is, since $\alpha(\chi, h)$ is a nonzero operator,

$$\varphi(\chi^h) = \varphi(\chi),$$

for every $h \in H$ and μ-almost all $\chi \in \widehat{N}$. Since the H-action on $\widehat{N}$ is ergodic, it follows that φ is constant μ-almost everywhere (see Proposition 13.A.3), and hence $\widetilde{T}$ is a scalar operator. This shows that $\pi_{\mu,\alpha}$ is irreducible.

Conversely, assume that $\pi_{\mu,\alpha}$ is irreducible and let $\widetilde{T}$ be an intertwiner between α and itself. It follows from Lemma 5.D.3 that $\widetilde{T}$ intertwines π with itself. Therefore $\widetilde{T}$ is a scalar operator, by irreducibility of π. This implies that α is irreducible.

(2) Assume that π_{μ_1,α_1} and π_{μ_2,α_2} are equivalent. Then, in particular, $\pi_{\mu_1,\alpha_1}|_N$ and $\pi_{\mu_2,\alpha_2}|_N$ are equivalent. Therefore μ_1 and μ_2 are equivalent, by Proposition 2.B.2.

(3) For $i = 1, 2$, set $\mathcal{H}_i = L^2(\widehat{N}, \mu, \mathcal{K}_i)$. If α_1 and α_2 are cohomologous, then Lemma 5.C.6 shows that π_{μ,α_1} and π_{μ,α_2} are equivalent.

Conversely, assume that π_{μ,α_1} and π_{μ,α_2} are equivalent. There exists a Hilbert space isomorphism $\widetilde{T} : \mathcal{H}_1 \to \mathcal{H}_2$ which intertwines π_{μ,α_1} and π_{μ,α_2}. It follows from Lemma 5.D.3 that $\widetilde{T}$ is an intertwiner between α_1 and α_2. In particular, $\widetilde{T}$ is decomposable. Since $\widetilde{T}$ is an isomorphism between $\mathcal{H}_1$ and $\mathcal{H}_2$, we see that $T(\chi)$ is an isomorphism between $\mathcal{K}_1$ and $\mathcal{K}_2$ for μ-almost all $\chi \in \widehat{N}$, where $\chi \mapsto T(\chi)$ is the field of operators from $\mathcal{K}_1$ to $\mathcal{K}_2$ associated with $\widetilde{T}$. This shows that α_1 and α_2 are cohomologous. □

REMARK 5.D.5. The ergodicity assumption in Theorem 5.D.4(1) is necessary. More precisely: let μ be a probability measure on $\widehat{N}$ which is quasi-invariant by the H-action, and not ergodic, and let $\alpha : \widehat{N} \times H \to \mathcal{U}(\mathcal{K})$ be a cocycle over $(\widehat{N}, \mu)$; then $\pi_{\mu,\alpha}$ is not irreducible.

Indeed, we can write $\mu = (\mu_1 + \mu_2)/2$ for H-quasi-invariant and mutually singular probability measures μ_1 and μ_2. We have a decomposition

$$L^2(\widehat{N}, \mu, \mathcal{K}) = L^2(\widehat{N}, \mu_1, \mathcal{K}) \oplus L^2(\widehat{N}, \mu_2, \mathcal{K}),$$

into $\pi_{\mu,\alpha}(G)$-invariant nontrivial subspaces.

Since every cocycle with values in the circle group $\mathbf{T}$ is irreducible, the following corollary is a straightforward consequence of Theorem 5.D.4.

COROLLARY 5.D.6. *Let $G = H \ltimes N$ be as in Theorem 5.D.4. Let μ be a probability measure on $\widehat{N}$ which is quasi-invariant and ergodic by the H-action and let $\alpha : \widehat{N} \times H \to \mathbf{T}$ be a cocycle over $(\widehat{N}, \mu)$ with values in the circle group.*

(1) *The representation $\pi_{\mu,\alpha}$ of G on $L^2(\widehat{N}, \mu)$ is irreducible.*
    Moreover, if $\beta : \widehat{N} \times H \to \mathbf{T}$ is another cocycle over $(\widehat{N}, \mu)$ which is not cohomologous to α, then $\pi_{\mu,\alpha}$ and $\pi_{\mu,\beta}$ are not equivalent.
(2) *Let ν be another probability measure on $\widehat{N}$ which is quasi-invariant and ergodic by the H-action, and not equivalent to μ. For any cocycle $\beta : \widehat{N} \times H \to \mathbf{T}$ over $(\widehat{N}, \nu)$, the irreducible representations $\pi_{\mu,\alpha}$ and $\pi_{\nu,\beta}$ are not equivalent.*

EXAMPLE 5.D.7. Let $\Gamma = H(\mathbf{Z})$ be the Heisenberg group over $\mathbf{Z}$. Recall that Γ is the semi-direct product $H \ltimes N$ and that its centre is Z, where

$$\begin{aligned} H &= \{(a,0,0) \in \Gamma \mid a \in \mathbf{Z}\}, \\ N &= \{(0,b,c) \in \Gamma \mid b,c \in \mathbf{Z}\}, \\ Z &= \{(0,0,c) \in \Gamma \mid c \in \mathbf{Z}\}. \end{aligned}$$

Fix $\theta \in [0,1[$ irrational. Set

$$\omega := e^{2\pi i\theta} \in \mathbf{T},$$

so that the irrational rotation $R_\theta : \mathbf{T} \to \mathbf{T}$ is given by $z \mapsto z\omega$. Consider the character

$$\psi_\theta : Z \to \mathbf{T}, \qquad (0,0,c) \mapsto \omega^c = e^{2\pi i\theta c}.$$

The subset $\widehat{N}(\psi_\theta)$ of $\widehat{N}$ is the set of unitary characters χ of N with restriction ψ_θ on Z. For each $z \in \mathbf{T}$, we have a character

$$\chi_z \in \widehat{N}(\psi_\theta), \qquad \chi_z(0,b,c) = z^b\omega^c \qquad \text{for all } \ b,c \in \mathbf{Z}.$$

The spaces $\mathbf{T}$ and $\widehat{N}(\psi_\theta)$ can be identified by means of the map

$$\mathbf{T} \to \widehat{N}(\psi_\theta), \quad z \mapsto \chi_z.$$

The right action $\widehat{N}(\psi_\theta) \curvearrowleft H$, $(\chi_z, h) \mapsto \chi_z^h$ is given by

$$\chi_z^h = \chi_{z\omega^a} = \chi_{R_\theta^a z} \in \widehat{N}(\psi_\theta) \qquad \text{for all } \ h = (a,0,0) \in H \ \text{ and } \ z \in \mathbf{T}.$$

Indeed, we have

$$\begin{aligned} \chi_z^h(0,b,c) &= \chi_z(h(0,b,c)h^{-1}) = \chi_z(0,b,c+ab) \\ &= z^b\omega^{c+ab} = (z\omega^a)^b\omega^c = \chi_{z\omega^a}(0,b,c). \end{aligned}$$

Let μ be the image of the Lebesgue measure $\mu_{\mathbf{T}}$ on $\mathbf{T}$ under the map $z \mapsto \chi_z$; we view μ as a measure on $\widehat{N}$ which vanishes on Borel sets disjoint from $\widehat{N}(\psi_\theta)$, and we identify the probability space $(\widehat{N}, \mu)$ with $(\mathbf{T}, \mu_{\mathbf{T}})$. Then μ is an H-invariant ergodic probability measure on $\widehat{N}$, since $\mu_{\mathbf{T}}$ is an R_θ-invariant ergodic probability measure on $\mathbf{T}$. The representation π_μ of Γ associated with μ as in Construction 5.A.2 is given on $L^2(\mathbf{T}, \mu_{\mathbf{T}})$ by the formula

$$(\pi_\mu(a,b,c)f)(z) = \chi_z(0,b,c)f(z\omega^a) = z^b\omega^c f(z\omega^a)$$

for all $f \in L^2(\mathbf{T}, \mu_{\mathbf{T}})$, $(a,b,c) \in \Gamma$, and $z \in \mathbf{T}$. By Corollary 5.D.6(1), the representation π_μ is irreducible.

We claim that π_μ is not equivalent to any one of the irreducible representations of Γ constructed in Section 3.B (compare also Proposition 5.E.1).

Indeed, since π_μ is infinite-dimensional, π_μ is not equivalent to any one of the finite-dimensional representations of Corollary 3.B.17.

Let $\pi_{\beta,\theta} = \mathrm{Ind}_N^\Gamma \chi_{\beta,\theta}$ be one of the infinite-dimensional irreducible representations of Γ of Corollary 3.B.14. On the one hand, the restriction of $\pi_{\beta,\theta}$ to the subgroup

$$B = \{(0,b,0) \in \Gamma \mid b \in \mathbf{Z}\} \approx \mathbf{Z}$$

contains the one-dimensional representation $\chi_{\beta,\theta}|_B : b \mapsto e^{2\pi i\beta b}$, as follows from the general properties of an induced representation (or from the explicit formula for $\pi_{\beta,\theta}$ from Remark 3.B.15). On the other hand, using Fourier transform, it is clear that the restriction of π_μ to B is equivalent to the regular representation λ_B of B. Since λ_B has no finite-dimensional subrepresentation, the claim is proved.

EXAMPLE 5.D.8. In the situation of the previous example, consider moreover the cocycle $\alpha : \mathbf{T}\times\mathbf{Z} \to \mathbf{T}$ over $(\widehat{N}, \mu)$ associated with the identity map $f : \mathbf{T} \to \mathbf{T}$, as in Example 5.B.3(5). Then α is given by

$$\alpha(z,a) = \begin{cases} \prod_{i=0}^{a-1} R_\theta^i(z) = z^a\omega^{a(a-1)/2} & \text{if } a > 0, \\ 1 & \text{if } a = 0, \\ \prod_{i=1}^{|a|} R_\theta^{-i}(z)^{-1} = z^a\omega^{a(a-1)/2} & \text{if } a < 0. \end{cases}$$

The representation $\pi_{\mu,\alpha}$ of Γ on $L^2(\mathbf{T}, \mu_\mathbf{T}) = L^2(\widehat{N}, \mu)$ associated with μ and α is given by

$$\begin{aligned} (\pi_{\mu,\alpha}(a,b,c)f)(z) &= \chi_z(0,b,c)\alpha(z,a)f(z\omega^a) = z^b\omega^c\alpha(z,a)f(z\omega^a) \\ &= z^{a+b}\omega^{a(a-1)/2+c}f(z\omega^a) \end{aligned}$$

for all $(a, b, c) \in \Gamma$, $f \in L^2(\mathbf{T}, \mu_\mathbf{T})$, and $z \in \mathbf{T}$. The representation $\pi_{\mu,\alpha}$ is irreducible, again by Corollary 5.D.6. It is not equivalent to π_μ, since α is not a coboundary (Example 5.C.7).

As in Example 5.D.7, one checks that $\pi_{\mu,\alpha}$ is not equivalent to any one of the irreducible representations of Γ of Section 3.B (see also Remark 5.E.2).

More generally, for an integer $i \neq 0$, let $\alpha_i : \mathbf{T} \times \mathbf{Z} \to \mathbf{T}$ be the cocycle over $(\widehat{N}, \mu)$ associated with the map $\mathbf{T} \to \mathbf{T}$, $z \mapsto z^i$. The associated representations π_{μ,α_i} of $H(\mathbf{Z})$ are irreducible and are mutually nonequivalent, since the α_i's are mutually noncohomologous. Moreover, π_{μ,α_i} is not equivalent to any one of the irreducible representations of Γ of Section 3.B.

For more representations of $H(\mathbf{Z})$, see Remark 5.E.2(1) below.

We now show the second main result of this chapter: every factor representation of a semi-direct product $G = H \ltimes N$ as above is equivalent to a representation $\pi_{\mu,\alpha}$ of Construction 5.D.1.

THEOREM 5.D.9. *Let $G = H\ltimes N$ be a semi-direct product of a countable discrete subgroup H with a second-countable locally compact abelian normal subgroup N. Let π be a* factor *representation of G on a separable Hilbert space $\mathcal{H}$.*

Then π is equivalent to a representation of the form $\pi_{\mu,\alpha}$, for a probability measure μ on $\widehat{N}$ which is quasi-invariant and ergodic by the H-action, and a cocycle $\alpha : \widehat{N} \times H \to \mathcal{U}(\mathcal{K})$ with values in the unitary group of a separable Hilbert space $\mathcal{K}$.

PROOF. By Proposition 2.F.3, there exists a probability measure μ on $\widehat{N}$ which is quasi-invariant and ergodic by the H-action and a Hilbert space $\mathcal{K}$ such that the restriction $\pi|_N$ of π to N is equivalent to the representation $\pi_{\mu,I}|_N$ on $L^2(\widehat{N}, \mu, \mathcal{K})$, where I is the trivial cocycle on $(\widehat{N}, \mu)$, defined by $I(\chi, h) = \mathrm{Id}_\mathcal{K}$ for all $\chi \in \widehat{N}$ and $h \in H$. By definition, $\pi_{\mu,I}$ is given by

$$(\pi_{\mu,I}(h,n)F)(\chi) = \chi(n)c(\chi,h)^{1/2}F(\chi^h)$$

for all $(h, n) \in H \ltimes N$ and $F \in L^2(\widehat{N}, \mu, \mathcal{K})$. We can therefore assume without loss of generality that $\mathcal{H} = L^2(\widehat{N}, \mu, \mathcal{K})$ and $\pi|_N = \pi_{\mu,I}|_N$.

Set

$$\widetilde{T_h} := \pi(h)\pi_{\mu,I}(h^{-1}) \in \mathcal{U}(\mathcal{H}) \quad \text{for } h \in H.$$

For every $n \in N$, we have $\pi(h^{-1}nh) = \pi_{\mu,I}(h^{-1}nh)$ since $h^{-1}nh \in N$, and therefore

$$\begin{aligned}\widetilde{T_h}\pi_{\mu,I}(n) &= \pi(h)\pi_{\mu,I}(h^{-1})\pi_{\mu,I}(n) = \pi(h)\pi_{\mu,I}(h^{-1}nh)\pi_{\mu,I}(h^{-1}) \\ &= \pi(h)\pi(h^{-1}nh)\pi_{\mu,I}(h^{-1}) = \pi(n)\pi(h)\pi_{\mu,I}(h^{-1}) = \pi(n)\widetilde{T_h} \\ &= \pi_{\mu,I}(n)\widetilde{T_h}.\end{aligned}$$

It follows from Proposition 2.B.3 that $\widetilde{T_h}$ is a decomposable operator. Let

$$T_h : \widehat{N} \to \mathcal{L}(\mathcal{K}), \quad \chi \mapsto T_h(\chi)$$

be an essentially bounded measurable map to which $\widetilde{T_h}$ is associated.

Since $\widetilde{T_h} \in \mathcal{U}(\mathcal{H})$ for every $h \in H$ and since H is countable, there exists a Borel subset X of $\widehat{N}$ with $\mu(X) = 1$ such that $T_h(\chi) \in \mathcal{U}(\mathcal{K})$ for every $h \in H$ and every $\chi \in X$. Upon modifying $\widetilde{T_h}$ by setting $\widetilde{T_h} = I_{\mathcal{K}}$ on $\widehat{N} \smallsetminus X$, we can assume that $T_h(\chi) \in \mathcal{U}(\mathcal{K})$ for every $h \in H$ and every $\chi \in X$.

Define

$$\alpha : \widehat{N} \times H \to \mathcal{U}(\mathcal{K}), \quad (\chi, h) \mapsto T_h(\chi).$$

Observe that α is measurable. We claim that α is a cocycle over $(\widehat{N}, \mu)$. Indeed, let $h_1, h_2 \in H$. Since π and $\pi_{\mu,I}$ are representations of G, we have

$$\begin{aligned}\widetilde{T_{h_1h_2}} &= \pi(h_1)\pi(h_2)\pi_{\mu,I}(h_2^{-1})\pi_{\mu,I}(h_1^{-1}) \\ &= \pi(h_1)\pi_{\mu,I}(h_1^{-1})\pi_{\mu,I}(h_1)\pi(h_2)\pi_{\mu,I}(h_2^{-1})\pi_{\mu,I}(h_1^{-1}) \\ &= \widetilde{T_{h_1}}\pi_{\mu,I}(h_1)\widetilde{T_{h_2}}\pi_{\mu,I}(h_1^{-1}).\end{aligned}$$

It follows that, for every $F \in L^2(\widehat{N}, \mu, \mathcal{K})$ and for μ-almost every $\chi \in \widehat{N}$, we have

$$\begin{aligned}T_{h_1h_2}(\chi)(F(\chi)) &= (\widetilde{T_{h_1h_2}}F)(\chi) \\ &= \Big(\widetilde{T_{h_1}}\pi_{\mu,I}(h_1)\widetilde{T_{h_2}}\pi_{\mu,I}(h_1^{-1})F\Big)(\chi) \\ &= T_{h_1}(\chi)\Big(\pi_{\mu,I}(h_1)\widetilde{T_{h_2}}\pi_{\mu,I}(h_1^{-1})F\Big)(\chi) \\ &= c(\chi, h_1)^{1/2}T_{h_1}(\chi)\Big((\widetilde{T_{h_2}}\pi_{\mu,I}(h_1^{-1})F)(\chi^{h_1})\Big) \\ &= c(\chi, h_1)^{1/2}T_{h_1}(\chi)\Big(T_{h_2}(\chi^{h_1})(\pi_{\mu,I}(h_1^{-1})F)(\chi^{h_1})\Big) \\ &= c(\chi, h_1)^{1/2}c(\chi^{h_1}, h_1^{-1})^{1/2}T_{h_1}(\chi)\Big(T_{h_2}(\chi^{h_1})(F(\chi))\Big) \\ &= c(\chi, h_1h_1^{-1})^{1/2}T_{h_1}(\chi)\Big(T_{h_2}(\chi^{h_1})(F(\chi))\Big) \\ &= T_{h_1}(\chi)T_{h_2}(\chi^{h_1})(F(\chi)).\end{aligned}$$

This implies that, for μ-almost all $\chi \in \widehat{N}$, we have

$$T_{h_1h_2}(\chi) = T_{h_1}(\chi)T_{h_2}(\chi^{h_1}),$$

and therefore

$$\alpha(\chi, h_1h_2) = T_{h_1h_2}(\chi) = T_{h_1}(\chi)T_{h_2}(\chi^{h_1}) = \alpha(\chi, h_1)\alpha(\chi^{h_1}, h_2).$$

Therefore α satisfies the cocycle relation.

Next, we claim that π coincides with the representation $\pi_{\mu,\alpha}$ on $L^2(\widehat{N},\mu,\mathcal{K})$. Indeed, since $\pi(h) = \widetilde{T_h}\pi_{\mu,I}(h)$, we have, for every $h \in H$ and $F \in L^2(\widehat{N},\mu,\mathcal{K})$,

$$\begin{aligned}(\pi(h)F)(\chi) &= T_h(\chi)((\pi_{\mu,I}(h)F)(\chi)) = c(\chi,h)^{1/2}T_h(\chi)(F(\chi^h)) \\ &= c(\chi,h)^{1/2}\alpha(\chi,h)(F(\chi^h)) = (\pi_{\mu,\alpha}(h)F)(\chi).\end{aligned}$$

Therefore $\pi|_H = \pi_{\mu,\alpha}|_H$. Together with $\pi|_N = \pi_{\mu,\alpha}|_N$, this implies that $\pi = \pi_{\mu,\alpha}$. □

REMARK 5.D.10. (1) Let $G = H \ltimes N$ be as in Theorem 5.D.9. In view of Theorems 5.D.4 and 5.D.9, the description of the dual $\widehat{G}$ of G is reduced to the following two problems. For every extended positive integer n in $\{1, \dots, +\infty\}$, we choose a Hilbert space $\mathcal{H}_n$ of dimension n.

(a) Find a collection $\mathcal{P}$ of probability measures on $\widehat{N}$ which are quasi-invariant and ergodic for the H-action, with the following properties:
  - every probability measure on $\widehat{N}$ which is quasi-invariant and ergodic by the H-action is equivalent to a measure in $\mathcal{P}$,
  - two measures in $\mathcal{P}$ are not equivalent to each other.

(b) For every μ in $\mathcal{P}$, and every extended positive integer n, find a collection $\mathcal{C}_{\mu,n}$ of irreducible cocycles $\alpha : H \times \widehat{N} \to \mathcal{U}(\mathcal{H}_n)$ over $(\widehat{N},\mu)$, with the following properties:
  - every such cocycle is cohomologous to a cocycle in $\mathcal{C}_{\mu,n}$,
  - two cocycles in $\mathcal{C}_{\mu,n}$ are not cohomologous.

In general, as shown by the examples discussed in Remark 5.E.2 below, both Problem (a) and Problem (b) seem intractable.

For instance, assume that G is a discrete amenable group and that N is infinite. Then the normalized Haar measure on $\widehat{N}$ is a nonatomic H-invariant probability measure and it follows from Corollary 5.F.3 below that there exist uncountably many nonequivalent probability measures on $\widehat{N}$ which are quasi-invariant and ergodic by the H-action. Their classification seems to be out of reach, even in classical situations such as the one given by an irrational rotation of the circle equipped with Lebesgue measure (Remark 5.E.2).

Similarly, the problem of classifying the cocycles over $(\widehat{N},\mu)$ for a fixed probability measure μ seems to be intractable.

(2) Theorem 5.D.9 is known to several experts. It is a special case of results on groupoid representations in [**Rams–76**] (see especially §10 there). It is also mentioned in [**Suth–83**, § 3] in case H acts freely on $\widehat{N} \smallsetminus \{1\}$.

5.E. Identifying induced representations

As in Sections 5.A and 5.D, we consider a locally compact group of the form $G = H \ltimes N$, where N is a second-countable locally compact abelian normal subgroup and H a countable discrete subgroup of G.

Among the irreducible representations $\pi_{\mu,\alpha}$ of G of Theorem 5.D.4, we will identify the ones which can be obtained by inducing a unitary character of N. As a result, the representations constructed in Sections 3.B, 3.C, and 3.D correspond to representations of the form $\pi_{\mu,\alpha}$ for very special probability measures μ and trivial cocycles α.

PROPOSITION 5.E.1. *Let $G = H \ltimes N$ be as above, μ a probability measure on $\widehat{N}$ which is quasi-invariant and ergodic by the H-action, $\mathcal{K}$ a Hilbert space, and $\alpha : H \times \widehat{N} \to \mathcal{U}(\mathcal{K})$ an irreducible cocycle over $(\widehat{N}, \mu)$. Let $\chi_0 \in \widehat{N}$ and let X be the H-orbit of χ_0 in $\widehat{N}$. The following properties are equivalent:*

(i) *the representation $\pi_{\mu,\alpha}$ is equivalent to the induced representation $\mathrm{Ind}_N^G \chi_0$;*
(ii) *the stabilizer of χ_0 in H is trivial, μ is atomic, $\mu(X) = 1$, the Hilbert space $\mathcal{K}$ is one-dimensional, and α is cohomologous to the trivial cocycle $H \times \widehat{N} \to \mathbf{T}$ of constant value 1.*

PROOF. We assume first that (ii) holds. Then $\pi_{\mu,\alpha}$ is equivalent to $\pi_\mu = \pi_{\mu,I}$, by Theorem 5.D.4. We have to show that $\pi := \mathrm{Ind}_N^G \chi_0$ is equivalent to π_μ.

We choose H as a right transversal for N in G, so that we have a disjoint union $G = \bigsqcup_{h \in H} Nh$. Let $(h_1, n_1) \in G$ and $h \in H$; if we use the notation of Construction 1.F.4(2) for

$$(h, e)(h_1, n_1) \, = \, \alpha(h, (h_1, n_1))(h \cdot (h_1, n_1)) \, \in \, NH,$$

and $h_1 \cdot_G n_1$ for the action of H on N defining the semi-direct product as in Convention 4.A.1, we have

$$\alpha(h, (h_1, n_1)) \, = \, h \cdot_G n_1 \, = \, h n_1 h^{-1} \qquad \text{and} \qquad h \cdot (h_1, n_1) \, = \, h h_1,$$

and in particular $\chi_0(\alpha(h, (h_1, n_1))) = \chi_0^h(n_1)$. The induced representation π is therefore realized on $\ell^2(H)$ by

$$(\pi(h_1, n_1) f)(h) \, = \, \chi_0^h(n_1) f(h h_1) \quad \text{ for all } f \in \ell^2(H), \ h_1, h \in H, \ n_1 \in N.$$

Since μ is atomic and $\mu(X) = 1$, we have $L^2(\widehat{N}, \mu) = \ell^2(X, \mu)$. For $h \in H$ and $\chi \in \widehat{N}$, we have

$$\mu(\{\chi^h\}) \, = \, c_\mu(\chi, h) \mu(\{\chi\}),$$

by definition of the Radon–Nikodym derivative c_μ (see Theorem A.3.2). We define a bijective linear map $T : \ell^2(X, \mu) \to \ell^2(H)$ by

$$(Tf)(h) \, = \, f(\chi_0^h) \mu(\{\chi_0^h\})^{1/2} \, = \, f(\chi_0^h) c_\mu(\chi_0, h)^{1/2} \mu(\{\chi_0\})^{1/2}$$

for all $f \in \ell^2(X, \mu)$ and $h \in H$.

Then T is unitary; indeed, since χ_0 has trivial stabilizer in H and since X is the H-orbit of χ_0, we have

$$\|Tf\|^2 \, = \, \sum_{h \in H} |(Tf)(h)|^2 \, = \, \sum_{h \in H} |f(\chi_0^h)|^2 \mu(\{\chi_0^h\}) \, = \, \sum_{\chi \in X} |f(\chi)|^2 \mu(\{\chi\}) \, = \, \|f\|^2,$$

for every $f \in \ell^2(X, \mu)$. Moreover, for $(h_1, n_1) \in G$, $f \in \ell^2(X, \mu)$, and $h \in H$, we have

$$\begin{aligned} T(\pi_\mu(h_1, n_1) f)(h) \, &= \, (\pi_\mu(h_1, n_1) f)(\chi_0^h) \mu(\{\chi_0^h\})^{1/2} \\ &= \, \chi_0^h(n_1) c_\mu(\chi_0^h, h_1)^{1/2} f(\chi_0^{h h_1}) \mu(\{\chi_0^h\})^{1/2} \\ &= \, \chi_0^h(n_1) \frac{\mu(\{\chi_0^{h h_1}\})^{1/2}}{\mu(\{\chi_0^h\})^{1/2}} f(\chi_0^{h h_1}) \mu(\{\chi_0^h\})^{1/2} \\ &= \, \chi_0^h(n_1) \mu(\{\chi_0^{h h_1}\})^{1/2} f(\chi_0^{h h_1}) \\ &= \, \chi_0^h(n_1) Tf(h h_1) \\ &= \, (\pi(h_1, n_1) Tf)(h). \end{aligned}$$

This shows that T intertwines π_μ and π; hence, π_μ and $\pi = \mathrm{Ind}_N^G \chi_0$ are equivalent, i.e., (i) holds.

Next, we show that (i) implies (ii). Assume that $\pi_{\mu,\alpha}$ and $\mathrm{Ind}_N^G \chi_0$ are equivalent. In particular, $\mathrm{Ind}_N^G \chi_0$ is irreducible, since $\pi_{\mu,\alpha}$ is irreducible by Theorem 5.D.4. It then follows from the Mackey–Shoda Criterion (Corollary 1.F.16) that the stabilizer of χ_0 in H is trivial.

We write π for $\mathrm{Ind}_N^G \chi_0$. By Proposition 1.F.9, $\pi|_N$ is equivalent to the representation

$$\bigoplus_{h\in H} \chi_0^h = \bigoplus_{\chi\in X} \chi.$$

So, $\pi_{\mu,\alpha}|_N$ is equivalent to $\bigoplus_{\chi\in X} \chi$.

The measure class attached to $\pi|_N$ as in Proposition 2.F.3 coincides with the class of μ. Since $\pi_{\mu,\alpha}|_N$ is equivalent to $\bigoplus_{\chi\in X} \chi$, it follows that X is contained in the set of atoms of μ. Therefore, we have $\mu(X) = 1$, by the ergodicity of μ. The fact (proved above) that (ii) implies (i) then shows that $\pi_\mu = \pi_{\mu,I}$ is equivalent to the induced representation π.

As a consequence, $\pi_{\mu,\alpha}$ is equivalent to $\pi_{\mu,I}$. It follows from Theorem 5.D.4 that α is cohomologous to the trivial cocycle with values in $\mathbf{T}$ and hence that $\mathcal{K}$ is of dimension one. □

REMARK 5.E.2. Using Theorem 5.D.4, we can construct uncountably many irreducible representations of the countable discrete groups $\Gamma = H \ltimes N$ appearing in Sections 3.B to 3.E, which are not equivalent to the representations constructed there by induction.

(1) Let $\Gamma = H(\mathbf{Z}) = H \ltimes N$ be the Heisenberg group over $\mathbf{Z}$, where

$$H = \{(a,0,0) \in \Gamma \mid a \in \mathbf{Z}\} \quad \text{and} \quad N = \{(0,b,c) \in \Gamma \mid b,c \in \mathbf{Z}\}$$

as before. Fix $\theta \in [0,1[$ irrational and let as before $\widehat{N}(\psi_\theta)$ be the H-invariant set in $\widehat{N}$ of unitary characters χ of N such that $\chi(0,0,c) = e^{2\pi i\theta c}$ for all $c \in \mathbf{Z}$. Then $\widehat{N}(\psi_\theta)$ is homeomorphic to $\mathbf{T}$, and the action of H on $\widehat{N}(\psi_\theta)$ corresponds to the action of the irrational rotation R_θ on the circle group $\mathbf{T}$ (see Corollary 3.B.14 and Example 5.D.7).

We constructed in Example 5.D.8 a countably infinite family of irreducible representations of Γ which are nonequivalent to those appearing in Corollaries 3.B.14 and 3.B.17. We now show that there are *uncountably many* such representations.

It is known that there are uncountably many pairwise nonequivalent R_θ-quasi-invariant and ergodic probability measures on $\mathbf{T}$ which are nonatomic. (See [**Kean–71**] and [**KaWe–72**]; this follows also from Corollary 5.F.3 below.)

In fact, a more precise result is true, as follows from [**Krie–71**, Theorem 4.11]: since the Lebesgue measure is the unique R_θ-invariant probability measure on $\mathbf{T}$, the set of equivalence classes of R_θ-quasi-invariant and ergodic probability measures on $\mathbf{T}$, equipped with a suitable Borel structure, is not countably separated. This shows that the classification of equivalence classes of R_θ-quasi-invariant and ergodic probability measures on $\mathbf{T}$ is hopeless.

In particular, Theorem 5.D.4 provides us with uncountably many irreducible representations of Γ which are nonequivalent to those appearing in Corollary 3.B.14 and Corollary 3.B.17.

We can construct further irreducible representations of Γ. Indeed, as shown in Example 5.C.7, there are several pairwise noncohomologous measurable cocycles

$$\alpha : \mathbf{T} \times \mathbf{Z} \to \mathbf{T}$$

for R_θ over $(\mathbf{T}, \mu)$, where μ is the Lebesgue measure on $\mathbf{T}$ (see also [**Furs–61**], [**Veec–69**], and [**Brow–73a**]). By Corollary 5.D.6, every such cocycle α defines an irreducible representation $\pi_{\mu,\alpha}$ as in Theorem 5.D.4. More generally, noncohomologous irreducible cocycles

$$\alpha : \mathbf{T} \times \mathbf{Z} \to \mathcal{U}(n)$$

with values in the unitary group $\mathcal{U}(n)$ exist for every $n \geq 1$ (see [**Isma–69**] and [**BaMe–86**]; see also Example 5.C.8). In this way, we obtain new representations $\pi_{\mu,\alpha}$ of Γ. This fact is also mentioned in Theorem 6.67 and the remarks on Page 218 of [**Foll–16**].

The previous remarks show that Theorem 10 in [**GoSV–16**] is erroneous, where it is claimed that every irreducible representation of $H(\mathbf{Z})$ is equivalent to a representation of the form $\pi_\mu = \pi_{\mu,I}$, for an ergodic quasi-invariant measure μ on $\widehat{N}$.

(2) Let $\Gamma = \mathrm{Aff}(\mathbf{K}) = \mathbf{K}^\times \ltimes \mathbf{K}$ be the affine group of a countable infinite field $\mathbf{K}$. The normalized Haar measure on the compact group $\widehat{\mathbf{K}}$ is a $\mathbf{K}^\times$-invariant and ergodic probability measure. Therefore, by Theorem 5.F.2, there exist uncountably many pairwise nonequivalent σ-finite $\mathbf{K}^\times$-invariant and ergodic measures on $\widehat{\mathbf{K}}$ which are nonatomic. It follows that there exist uncountably many pairwise nonequivalent $\mathbf{K}^\times$-quasi-invariant and ergodic probability measures on $\widehat{\mathbf{K}}$ which are nonatomic. As above, we obtain uncountably many irreducible representations of Γ which are nonequivalent to those of Theorem 3.C.3.

(3) Similarly, we obtain from Theorem 5.D.4 uncountably many irreducible representations of the Baumslag–Solitar group $\Gamma = \mathrm{BS}(1, p)$ and of the lamplighter group which are nonequivalent to those given in Section 3.D and Section 3.E.

5.F. On the existence of infinite nonatomic ergodic invariant measures

Let Γ be a countable *amenable* group, $(X, \mathcal{B})$ a standard Borel space, μ a nonatomic probability measure on $(X, \mathcal{B})$, and $\Gamma \curvearrowright (X, \mathcal{B}, \mu)$ an action for which the measure μ is invariant and ergodic. Following [**Schm–77b**], we will show that there are uncountably many infinite σ-finite nonatomic measures on $(X, \mathcal{B})$ which are invariant and ergodic under the action of Γ.

For this, we need to recall the notion of orbit equivalence for measure-preserving group actions on probability spaces. Let Γ_1 and Γ_2 be two countable groups acting respectively on standard Borel spaces $(X_1, \mathcal{B}_1)$ and $(X_2, \mathcal{B}_2)$ equipped with nonatomic probability measures μ_1 and μ_2 which are invariant and ergodic. The actions $\Gamma_1 \curvearrowright (X_1, \mathcal{B}_1, \mu_1)$ and $\Gamma_2 \curvearrowright (X_2, \mathcal{B}_2, \mu_2)$ are **orbit equivalent** if there exists a Borel isomorphism $\Phi : X_1 \to X_2$ such that $\Phi_*(\mu_1) = \mu_2$ and such that

$$\Phi(\Gamma_1 \cdot x) = \Gamma_2 \cdot \Phi(x)$$

for μ_1-almost all $x \in X_1$.

The following theorem, due to Dye and Ornstein–Weiss [**Dye–59, OrWe–80**], is a fundamental result in the theory of orbit equivalence of group actions.

THEOREM 5.F.1 (**Dye and Ornstein–Weiss**). *Let Γ_1 and Γ_2 be two amenable countable groups.*

Any two nonatomic probability measure-preserving ergodic actions $\Gamma_1 \curvearrowright (X_1, \mathcal{B}_1, \mu_1)$ and $\Gamma_2 \curvearrowright (X_2, \mathcal{B}_2, \mu_2)$ are orbit equivalent.

Theorem 5.F.1 will be the main ingredient in the proof of the following result from [**Schm–77b**].

THEOREM 5.F.2 (**K. Schmidt**). *Let Γ be an amenable countable group, $(X, \mathcal{B})$ a standard Borel space, μ a nonatomic probability measure on $(X, \mathcal{B})$, and $\Gamma \curvearrowright (X, \mathcal{B})$ a measurable action. Assume that μ is Γ-invariant and ergodic.*

Then there exist uncountably many mutually nonequivalent σ-finite and nonatomic infinite measures on $(X, \mathcal{B})$ which are Γ-invariant and ergodic.

PROOF. Set $\Lambda := \Gamma \times \Gamma$ and $Y := X \times X$. Consider the product action of Λ on $(Y, \mathcal{B} \otimes \mathcal{B})$:

$$(\gamma_1, \gamma_2)(x_1, x_2) = (\gamma_1 \cdot x_1, \gamma_2 \cdot x_2) \quad \text{for all } (\gamma_1, \gamma_2) \in \Lambda, \ (x_1, x_2) \in Y.$$

This action preserves the probability measure $\nu := \mu \otimes \mu$ on $(Y, \mathcal{B} \otimes \mathcal{B})$ and is obviously ergodic.

Since μ is nonatomic and ergodic, it is well known that μ-almost all $x \in X$ has an infinite orbit (see Proposition A.5.10). Fix $x_0 \in X$ with an infinite Γ-orbit. Set

$$Y_{x_0} := \{(\gamma \cdot x_0, x) \in X \times X \mid \gamma \in \Gamma, x \in X\} = \bigsqcup_{t \in \mathcal{O}} \{t\} \times X,$$

where $\mathcal{O}$ is the Γ-orbit of x_0. Observe that Y_{x_0} is a Λ-invariant Borel subset of Y. Define a Borel measure ν_{x_0} on Y by

$$\int_Y f d\nu_{x_0} = \sum_{t \in \mathcal{O}} \int_X f(t, x) d\mu(x)$$

for every positive Borel function f on Y. Then:

- $\nu_{x_0}(Y \smallsetminus Y_{x_0}) = 0$.
- ν_{x_0} is infinite, since
$$\int_Y \mathbf{1}_Y d\mu_{x_0} = \sum_{t \in \mathcal{O}} \int_X \mathbf{1}_X(x) d\mu(x) = \sum_{t \in \mathcal{O}} 1 = \infty.$$
- ν_{x_0} is σ-finite. Indeed, $\mathcal{O}$ is countable,
$$Y = \Big(Y \smallsetminus Y_{x_0}\Big) \sqcup \Big(\bigsqcup_{t \in \mathcal{O}} \{t\} \times X\Big),$$
we have $\nu_{x_0}(Y \smallsetminus Y_{x_0}) = 0$, and $\nu_{x_0}(\{t\} \times X) = 1$ for every $t \in \mathcal{O}$.
- ν_{x_0} is nonatomic, since μ is nonatomic.
- ν_{x_0} is Λ-invariant. Indeed, let f be a positive Borel function on Y and (γ_1, γ_2) in Λ. For the translate ${}_{(\gamma_1,\gamma_2)}f$ of f by (γ_1, γ_2), we have, by Γ-invariance of μ,

$$\begin{aligned}\int_Y {}_{(\gamma_1,\gamma_2)}f d\nu_{x_0} &= \sum_{t \in \mathcal{O}} \int_X {}_{(\gamma_1,\gamma_2)}f(t, x) d\mu(x) = \sum_{t \in \mathcal{O}} \int_X f(\gamma_1^{-1} \cdot t, \gamma_2^{-1} \cdot x) d\mu(x) \\ &= \sum_{t \in \mathcal{O}} \int_X f(\gamma_1^{-1} \cdot t, x) d\mu(x) = \sum_{t \in \mathcal{O}} \int_X f(t, x) d\mu(x) = \int_Y f d\nu_{x_0}.\end{aligned}$$

- ν_{x_0} is ergodic. Indeed, let B be a Λ-invariant measurable subset of Y. For every $t \in \mathcal{O}$, the t-section

$$B_t := \{x \in X \mid (t, x) \in B\}$$

is a Γ-invariant measurable subset of X, since B is Λ-invariant. By ergodicity of μ, it follows that either $\mu(B_t) = 1$ or $\mu(B_t) = 0$ for every $t \in \mathcal{O}$. Moreover, for every $\gamma \in \Gamma$, we have $B_{\gamma \cdot t} = ((\gamma, e)^{-1}B)_t$, and hence

$$B_{\gamma \cdot t} = B_t,$$

by Λ-invariance of B. It follows that we have either $\mu(B_t) = 1$ for all $t \in \mathcal{O}$ or $\mu(B_t) = 0$ for all $t \in \mathcal{O}$. Since $B = \bigsqcup_{t \in \mathcal{O}} \{t\} \times B_t$, we have in the first case

$$\nu_{x_0}(Y \smallsetminus B) = \sum_{t \in \mathcal{O}} \mu(X \smallsetminus B_t) = 0$$

and in the second case

$$\nu_{x_0}(B) = \sum_{t \in \mathcal{O}} \mu(B_t) = 0.$$

By Theorem 5.F.1, there exists a Borel isomorphism $\Phi : Y \to X$ such that $\Phi_*(\nu) = \mu$ and

$$\Phi(\Lambda \cdot y) = \Gamma \cdot \Phi(y)$$

for ν-almost all $y \in Y$. Let $y = (x_0, x) \in Y$ be such that $\Phi(\Lambda \cdot y) = \Gamma \cdot \Phi(y)$. The image

$$\mu_{x_0} := \Phi_*(\nu_{x_0})$$

of ν_{x_0} is a measure on $(X, \mathcal{B})$, which is clearly infinite, σ-finite and nonatomic, since ν_{x_0} has these properties. Moreover, the following hold:

- μ_{x_0} is Γ-invariant. Indeed, let A be a Borel subset of X and $\gamma \in \Gamma$. Then $\mu_{x_0}(A) = \nu_{x_0}(B)$, where $B := \Phi^{-1}(A)$. Since

$$\Phi(\Lambda \cdot B) = \Gamma \cdot \Phi(B) = \Gamma \cdot A,$$

there exists a partition $B = \bigsqcup_i B_i$ in measurable subsets B_i such that

$$\Phi^{-1}(\gamma \cdot A) = \bigsqcup_i \lambda_i B_i$$

for some $\lambda_i \in \Lambda$. Then, using the Λ-invariance of ν_{x_0}, we have

$$\begin{aligned}\mu_{x_0}(\gamma A) &= \nu_{x_0}(\Phi^{-1}(\gamma A)) = \sum_i \nu_{x_0}(\lambda_i B_i) \\ &= \sum_i \nu_{x_0}(B_i) = \nu_{x_0}(B) = \mu_{x_0}(A).\end{aligned}$$

- μ_{x_0} is ergodic. Indeed let A be a Γ-invariant Borel subset of X. Then $B = \Phi^{-1}(A)$ is Λ-invariant, since

$$\Lambda \cdot B = \Phi^{-1}(\Gamma \cdot A) = \Phi^{-1}(A) = B.$$

Therefore $\nu_{x_0}(Y \smallsetminus B) = 0$ or $\nu_{x_0}(B) = 0$, that is, $\mu_{x_0}(X \smallsetminus A) = 0$ or $\mu_{x_0}(A) = 0$.

Finally, let $x_0, x_0' \in X$ with distinct Γ-orbits. Then, with the notation as above, we have

$$\nu_{x_0}(Y_{x_0'}) = 0 \quad \text{and} \quad \nu_{x_0'}(Y \smallsetminus Y_{x_0'}) = 0,$$

so that the two measures ν_{x_0} and $\nu_{x_0'}$ are mutually singular. It follows that μ_{x_0} and $\mu_{x_0'}$ are mutually singular. Since there are uncountably many Γ-orbits in X, we have constructed uncountably many nonequivalent measures on X with the required properties. □

In the situation of Theorem 5.F.2, one is sometimes interested in the existence of an uncountable family of *probability* measures on X with similar properties (see Remark 5.E.2). As we now show, this remains true at the cost of replacing the invariance of the measures by their quasi-invariance. Observe that no better result can be expected, as there might be a *unique* probability measure on X which is invariant under the Γ-action; these are called uniquely ergodic actions. For example, the action of $\mathbf{Z}$ given by an irrational rotation on the circle is uniquely ergodic (see [**BeMa–00**, Example 3.10]).

COROLLARY 5.F.3. *Let* Γ *and* $(X, \mathcal{B}, \mu)$ *be as in Theorem* 5.F.2.
Then there exist uncountably many mutually nonequivalent nonatomic proba-bility *measures on* $(X, \mathcal{B})$ *which are* Γ*-quasi-invariant and ergodic.*

PROOF. By Theorem 5.F.2, there exists an uncountable family $(\mu_\iota)_{\iota \in I}$ of non-equivalent σ-finite and nonatomic infinite measures on $(X, \mathcal{B})$ which are Γ-invariant and ergodic. By Proposition A.3.5, for every $\iota \in I$, there exists a probability measure ν_ι on $(X, \mathcal{B})$ which is equivalent to μ_ι. It is clear that every ν_ι is quasi-invariant and ergodic by Γ and that the ν_ι's are mutually nonequivalent. □

REMARK 5.F.4. For various strengthenings of Theorem 5.F.2 and other results around the existence of infinite σ-finite invariant measures, see the lecture notes [**Schm–77a**].

CHAPTER 6

Types for representations, quasi-duals, groups of type I

Let π be a representation of a second-countable LC group G in a separable Hilbert space. As seen in Section 1.G, *π can always be decomposed into a direct integral of irreducible representations, but two such decompositions of π may be completely different (Theorem* 1.G.11*). In this chapter we will define the class of type I representations of G, for which a direct integral decomposition is essentially unique.*

Following Mackey (see [**Mack–76**] *and* [**Mack–78**]*), we define in Section* 6.A *type I representations by means of the notion of subordination of representations, which is weaker than containment, and the corresponding relation of quasi-equivalence. One should think of quasi-equivalent representations as representations which have the same "kinds" of subrepresentations. Representations with only one "kind" of subrepresentation are called factor representations; more precisely, a representation is factorial if it cannot be decomposed as a direct sum of two disjoint representations. Next, we introduce multiplicity-free representations: a representation π is multiplicity-free if, for every decomposition $\pi = \pi_1 \oplus \pi_2$, the subrepresentations π_1 and π_2 are disjoint. Type I representations are defined as the representations which are quasi-equivalent to a multiplicity-free representation.*

For a representation π of a topological group G on a Hilbert space $\mathcal{H}$, the commutant $\pi(G)'$ and the bicommutant $\pi(G)''$ of $\pi(G)$ in $\mathcal{L}(\mathcal{H})$ are von Neumann subalgebras of $\mathcal{L}(\mathcal{H})$. This allows us to reformulate in Section 6.B *all the notions defined in Section* 6.A *in terms of von Neumann algebras. When G is locally compact and second-countable, every multiplicity-free representation of G has an essentially unique decomposition as a direct integral of irreducible representations; moreover, every type I representation of G has a canonical decomposition as the sum of multiplicity-free representations. We also discuss multiplicity-free representations of abelian LC groups, and multiplicity-free representations associated with Gel'fand pairs.*

In Section 6.C *we introduce the quasi-dual* $\mathrm{QD}(G)$ *of a topological group G as the space of quasi-equivalence classes of factor representations of G; it has a natural Borel structure, called the Mackey–Borel structure of* $\mathrm{QD}(G)$*. Quasi-equivalence classes of factor representations of type I constitute a subset of the quasi-dual which can be identified with the dual, so that $\widehat{G}$ can be seen as a subspace of* $\mathrm{QD}(G)$*. When G is locally compact and second-countable, every representation of G has a canonical decomposition (the so-called central decomposition) as a direct integral of factor representations, to which is associated a class of measures on* $\mathrm{QD}(G)$*.*

A topological (not necessarily LC) group G is defined to be of type I if all representations of G are of type I. Assume that G is a second-countable locally *compact group. Then G is of type I if and only if every factor representation of G is a multiple of an irreducible one (Theorem 6.D.4). When this is the case, every representation of G has a canonical decomposition as a direct integral of irreducible representations, given by a family of mutually singular measures on $\widehat{G}$ and a multiplicity function (Theorem 6.D.7).*

In Section 6.E we introduce the class of GCR (or postliminal) groups and show that every such group is of type I (there is a converse; see Glimm's Theorem in Chapter 8). We discuss in detail the important subclass of CCR (or liminal) groups. We give a complete proof of the fact that $\mathrm{SL}_2(\mathbf{R})$ belongs to this subclass and we indicate how these arguments can be extended to general semisimple connected real Lie groups.

6.A. Comparing representations, quasi-equivalence

Let G be a topological group.

6.A.a. Disjointness and quasi-equivalence. For a representation π of G and a cardinal n, recall that $n\pi$ denotes the direct sum of n representations equivalent to π. When G is second-countable, it suffices to consider countable cardinals $n \in \{1, 2, 3, \dots, \infty\}$, where ∞ stands for $\aleph_0$. For larger groups, larger cardinals are necessary; nevertheless, we will use the ambiguous notation "∞" below.

DEFINITION 6.A.1. Let π_1, π_2 be two representations of G in Hilbert spaces of positive dimensions.

Recall from Section 1.A that π_1, π_2 are **disjoint** if there does not exist a pair of nontrivial subrepresentations $\rho_1 \le \pi_1$, $\rho_2 \le \pi_2$ such that ρ_1 and ρ_2 are equivalent.

The representation π_1 is **subordinate** to π_2 if every nontrivial subrepresentation of π_1 contains a nontrivial subrepresentation equivalent to a subrepresentation of π_2, i.e., if no nontrivial subrepresentation of π_1 is disjoint from π_2.

The representations π_1 and π_2 are **quasi-equivalent**, and we write $\pi_1 \approx \pi_2$ if each one is subordinate to the other.

Here are a few straightforward immediate consequences of the definitions.

PROPOSITION 6.A.2. *Let π_1, π_2 be two representations of G.*

(1) *If π_1 is contained in π_2, then π_1 is subordinate to π_2. If π_1 and π_2 are equivalent, then they are quasi-equivalent.*
(2) *If π_1 is irreducible, then π_1 is subordinate to π_2 if and only if it is contained in π_2.*
(3) *If π_1 and π_2 are irreducible, then they are either equivalent or disjoint. Two irreducible representations are quasi-equivalent if and only if they are equivalent.*

REMARK 6.A.3. Beware that there are three distinct equivalence relations defined for representations. Two representations $\pi_j : G \to \mathcal{U}(\mathcal{H}_j)$, $j = 1, 2$, are defined to be

- equivalent, written $\pi_1 \simeq \pi_2$, if there exists a unitary operator V from $\mathcal{H}_1$ onto $\mathcal{H}_2$ such that $\pi_2(g) = V\pi_1(g)V^*$ for all $g \in G$, as defined in Section 1.A;
- weakly equivalent, written $\pi_1 \sim \pi_2$, if $\pi_1 \preceq \pi_2$ and $\pi_2 \preceq \pi_1$, as defined in Section 1.C; equivalently if $\mathrm{C^*ker}(\pi_1) = \mathrm{C^*ker}(\pi_2)$ (see Section 8.B);
- quasi-equivalent, written $\pi_1 \approx \pi_2$, as defined above.

Each of these three notions has the appropriate **Schröder–Bernstein property**. More precisely, for equivalence, if π_1 is a subrepresentation of π_2 and π_2 a subrepresentation of π_1, then π_1 and π_2 are equivalent, by Proposition 1.A.8. Similarly, for weak equivalence: if $\pi_1 \preceq \pi_2$ and $\pi_2 \preceq \pi_1$, then $\pi_1 \sim \pi_2$, by definition. Similarly, for quasi-equivalence: if π_1 is subordinate to π_2 and π_2 subordinate to π_1, then $\pi_1 \approx \pi_2$, by definition.

Let us compare equivalence and quasi-equivalence, as we did in Remark 1.C.2 for equivalence and weak equivalence. We do this first in a particularly simple situation, and then more generally in Proposition 6.A.4.

Let π, ρ be two representations of G. Assume that π and ρ can be written as direct sums of irreducible subrepresentations

$$\pi = \bigoplus_{i \in I} \pi_i \quad \text{and} \quad \rho = \bigoplus_{j \in J} \rho_j$$

(this is always possible in case G is compact). The representations π and ρ are disjoint if and only if π_i and ρ_j are not equivalent for all $(i, j) \in I \times J$. They are quasi-equivalent if and only if every π_i is equivalent to some ρ_j and every ρ_j is equivalent to some π_i.

PROPOSITION 6.A.4. *Let π_1, π_2, σ_1 be three representations of G.*

(1) *σ_1 is subordinate to π_1 if and only there exists a cardinal n such that σ_1 is equivalent to a subrepresentation of $n\pi_1$.*
(2) *π_1 and π_2 are quasi-equivalent if and only if there exists a cardinal n such that $n\pi_1$ and $n\pi_2$ are equivalent.*

PROOF. We only write the proof for representations on separable Hilbert spaces, and we will therefore only deal with countable cardinals. The nonseparable case is similar, upon using come cardinal arithmetic.

To show Claim (1), assume first that σ_1 is equivalent to a subrepresentation of $n\pi_1$. Write

$$n\pi_1 = \bigoplus_i \rho_i,$$

where the ρ_i's are representations equivalent to π_1. Let P_i be the orthogonal projection of the space of $n\pi_1$ onto the space of ρ_i. Let τ be a subrepresentation of σ_1. By hypothesis, there exists $T \neq 0$ in $\mathrm{Hom}_G(\tau, n\pi_1)$. Therefore there exists i such that $P_i T \neq 0$, so that $\mathrm{Hom}_G(\tau, \rho_i) \neq \{0\}$, and therefore $\mathrm{Hom}_G(\tau, \pi_1) \neq \{0\}$, i.e., τ and π_1 are not disjoint (Proposition 1.A.6). It follows that σ_1 is subordinate to π_1.

Assume now that σ_1 is subordinate to π_1. Denote by $\mathcal{H}$ the Hilbert space of σ_1. Consider the family $\mathcal{F}$ of G-invariant closed subspaces of $\mathcal{H}$ such that the corresponding subrepresentation of σ_1 is equivalent to a subrepresentation of π_1. By Zorn's lemma, there exists a maximal subfamily $(\mathcal{K}_j)_j$ of mutually orthogonal subspaces $\mathcal{K}_j \in \mathcal{F}$. Set $\mathcal{K} = \bigoplus_j \mathcal{K}_j$.

We claim that $\mathcal{K} = \mathcal{H}$. Indeed, assume by contradiction that this is not the case. Then the orthogonal complement $\mathcal{K}^\perp$ of $\mathcal{K}$ is nonzero and G-invariant. The subrepresentation of σ_1 defined by $\mathcal{K}^\perp$ and the representation π_1 are disjoint. This contradicts the fact that σ_1 is subordinate to π_1.

From the equality $\mathcal{K} = \mathcal{H}$, it follows that σ_1 is equivalent to a subrepresentation of $\infty\pi_1$.

To show Claim (2), assume first that $\infty\pi_1$ and $\infty\pi_2$ are equivalent. Then, by Claim (1), π_1 is subordinate to π_2 and π_2 is subordinate to π_1. This means that π_1 and π_2 are quasi-equivalent.

Assume now that π_1 and π_2 are quasi-equivalent. By Claim (1), π_1 is equivalent to a subrepresentation of $\infty\pi_2$, hence $\infty\pi_1$ is equivalent to a subrepresentation of $\infty\pi_2 \simeq \infty(\infty\pi_2)$; similarly, $\infty\pi_2$ is equivalent to a subrepresentation of $\infty\pi_1$. It follows from Proposition 1.A.8, that $\infty\pi_1$ and $\infty\pi_2$ are equivalent. □

As a consequence of Proposition 6.A.4 and Remark 1.C.2, quasi-equivalence and weak-equivalence compare as follows.

COROLLARY 6.A.5. *Let π_1, π_2 be two representations of G.*

If π_1 is subordinate to π_2, then π_1 is weakly contained in π_2. In particular, if π_1 and π_2 are quasi-equivalent, then they are weakly equivalent.

REMARK 6.A.6. The converse statement of the last claim of Corollary 6.A.5 fails in general; that is, weakly equivalent representations need not be quasi-equivalent. Indeed, consider for example the additive group $\mathbf{R}$, its regular representation $\lambda_{\mathbf{R}}$, and the direct sum

$$\pi = \bigoplus_{q \in Q} \chi_q,$$

where Q is a countable dense subset of $\mathbf{R}$, and χ_q denotes the unitary character $\chi_q : \mathbf{R} \to \mathbf{T}$, $x \mapsto e^{iqx}$. The representations $\lambda_{\mathbf{R}}$ and π are weakly equivalent (more on this in, e.g., [**BeHV–08**, F.2.7 & F.6.4]), but they are not quasi-equivalent. Indeed, otherwise χ_q would be equivalent to a subrepresentation of $\lambda_{\mathbf{R}}$, and this is absurd because $\lambda_{\mathbf{R}}$ does not have any irreducible subrepresentation.

The following proposition relates subrepresentations of quasi-equivalent representations with each other.

PROPOSITION 6.A.7. *Let π_1, π_2, σ_1 be three representations of G.*

If σ_1 is quasi-equivalent to a subrepresentation of π_1 and π_1 is quasi-equivalent to π_2, then there exists a subrepresentation σ_2 of π_2 such that σ_1 and σ_2 are quasi-equivalent.

PROOF. It suffices to consider the situation in which σ_1 is a subrepresentation of π_1. Let $\mathcal{H}_1, \mathcal{H}_2$ be the Hilbert spaces of π_1, π_2 and let $\mathcal{K}_1$ be the subspace of $\mathcal{H}_1$ defining σ_1. Let $\mathcal{L}$ be the closed subspace of $\mathcal{K}_1$ generated by the family of G-invariant closed subspaces of $\mathcal{K}_1$ such that the corresponding subrepresentation of σ_1 is equivalent to a subrepresentation of π_2. We claim that $\mathcal{L} = \mathcal{K}_1$. Indeed, assume by contradiction that this is not the case. Then the orthogonal complement $\mathcal{L}^\perp$ of $\mathcal{L}$ in $\mathcal{K}_1$ is nonzero and G-invariant. The subrepresentation of σ_1 defined by $\mathcal{L}^\perp$ is disjoint from π_2. This contradicts the fact that π_1 is subordinate to π_2.

Now let $\mathcal{K}_2$ be the closed subspace of $\mathcal{H}_2$ generated by the family of G-invariant closed subspaces of $\mathcal{H}_2$ such that the corresponding subrepresentation of π_2 is equivalent to a subrepresentation of σ_1. Let σ_2 be the subrepresentation of π_2 defined by $\mathcal{K}_2$. It is clear that σ_1 is subordinate to σ_2 and that σ_2 is subordinate to σ_1. □

6.A.b. Factor representations. We now introduce a class of representations which are called factorial, for a reason which will become clear in the next section (see Proposition 6.B.6). Another terminology for the same notion is "primary representation"; it is used, for example, in [**Mack–76**].

DEFINITION 6.A.8. A representation π of G is a **factor representation**, or is **factorial**, if, for every nontrivial decomposition $\pi = \pi_1 \oplus \pi_2$, the representations π_1 and π_2 are not disjoint.

(As a consequence of Proposition 6.B.8 below, π is factorial if and only if, for every nontrivial decomposition $\pi = \pi_1 \oplus \pi_2$, the representations π_1 and π_2 are quasi-equivalent.)

EXAMPLE 6.A.9. It follows from the definition that irreducible representations are factorial.

More generally, let π' be an irreducible representation of G and let $n \geq 1$ be a cardinal. Then $\pi := n\pi'$ is factorial. Indeed, let $\pi = \pi_1 \oplus \pi_2$ be a nontrivial direct sum decomposition. Then, for $i = 1$ and $i = 2$, we have $\mathrm{Hom}_G(\pi_i, \pi') \neq \{0\}$ by Corollary 1.A.7; since π' is irreducible, it follows from Lemma 1.A.5 that π_1 and π_2 contain subrepresentations which are equivalent to π'. It follows that π_1 and π_2 are not disjoint.

The following proposition characterizes the factor representations appearing in Example 6.A.9.

PROPOSITION 6.A.10. *Let π be representation of G. The following properties are equivalent:*

(i) *π is a factor representation which contains an irreducible representation of G;*
(ii) *π is a multiple of an irreducible representation of G.*

There is one more equivalent condition in Proposition 6.B.14.

PROOF. The fact that (ii) implies (i) was shown in Example 6.A.9. To show the converse, assume that π contains an irreducible representation π'.

Let $\mathcal{H}$ be the Hilbert space of π. There exists a maximal family $(\mathcal{K}_j)_{j \in J}$ of G-invariant and mutually orthogonal closed subspaces $\mathcal{K}_i$ of $\mathcal{H}$ such that, for every $j \in J$, the subrepresentation π_j of π defined by $\mathcal{K}_j$ is equivalent to π' (compare with the proof of Proposition 6.A.4). Set $\mathcal{K} = \bigoplus_{j \in J} \mathcal{K}_j$. Then the subrepresentation of π defined by $\mathcal{K}^\perp$ is disjoint from every π_j, hence disjoint from $\bigoplus_{j \in J} \pi_j$, by Corollary 1.A.7. Since π is factorial, we must have $\mathcal{K}^\perp = \{0\}$, that is, $\mathcal{K} = \mathcal{H}$. This shows that π is a multiple of π'. □

For examples of factor representations which are not multiples of irreducible representations, see Section 7.A and Chapter 14 below.

The following result shows that factoriality is inherited by subrepresentations.

PROPOSITION 6.A.11. *Let π be a factor representation of G, and let π_1, π_2 be nontrivial subrepresentations of π acting on mutually orthogonal subspaces.*

Then π_1 and π_2 are not disjoint. In particular, every nontrivial subrepresentation of π is factorial.

PROOF. Let $\mathcal{H}$ be the Hilbert space of π and $\mathcal{H}_1, \mathcal{H}_2$ the mutually orthogonal subspaces of $\mathcal{H}$ defining π_1 and π_2.

Assume, by contradiction, that π_1 and π_2 are disjoint. Set

$$\mathcal{H}' = \mathcal{H}_1 \oplus \mathcal{H}_2.$$

There exists a maximal family $(\mathcal{K}_j)_{j\in J}$ of G-invariant and mutually orthogonal closed subspaces $\mathcal{K}_j$ of $\mathcal{H}'^{\perp}$ such that, for every $j \in J$, the subrepresentation π_j of π defined by $\mathcal{K}_j$ is disjoint from π_1 (compare with the proof of Proposition 6.A.4).

Let $\mathcal{H}_0$ be the orthogonal complement of $\bigoplus_{j\in J}\mathcal{K}_j$ in $\mathcal{H}'^{\perp}$ and denote by π_0 the subrepresentation of π defined by $\mathcal{H}_0$. Then we have

$$\pi = (\pi_1 \oplus \pi_2) \oplus (\pi_0 \oplus \bigoplus_{j\in J}\pi_j) = (\pi_1 \oplus \pi_0) \oplus (\pi_2 \oplus \bigoplus_{j\in J}\pi_j).$$

On the one hand, $\pi_2 \oplus \bigoplus_{j\in J}\pi_j$ is disjoint from π_1 by Corollary 1.A.7. On the other hand, no subrepresentation of π_0 is disjoint from π_1, by maximality of the family $(\mathcal{K}_j)_{j\in J}$; it follows that $\pi_2 \oplus \bigoplus_{j\in J}\pi_i$ is disjoint from π_0. Therefore $\pi_2 \oplus \bigoplus_{j\in J}\pi_j$ is disjoint from $\pi_1 \oplus \pi_0$. This contradicts the fact that π is factorial. □

The next proposition shows how two factor representations can be compared. We will see later (Corollary 6.B.8) that any two nondisjoint factor representations are quasi-equivalent.

PROPOSITION 6.A.12. *Let π and π' be factor representations of G.*

Then one of the following relations holds: π and π' are disjoint, or π is contained in π', or π' is contained in π.

PROOF. Assume that π and π' are not disjoint. We have to show that either π is contained in π' or π' is contained in π.

Let $\mathcal{H}$ and $\mathcal{H}'$ be respectively the Hilbert spaces of π and π'. There exists a family $(\mathcal{K}_i, \mathcal{K}'_i)_{i\in I}$ of pairs of G-invariant closed subspaces $\mathcal{K}_i$ and $\mathcal{K}'_i$ of $\mathcal{H}$ and $\mathcal{H}'$ respectively with the following three properties:

- the $\mathcal{K}_i$'s are mutually orthogonal and the $\mathcal{K}'_i$'s are mutually orthogonal;
- for every $i \in I$, the subrepresentation π_i of π defined by $\mathcal{K}_i$ is equivalent to the subrepresentation π'_i of π' defined by $\mathcal{K}'_i$;
- $(\bigoplus_{i\in I}\mathcal{K}_i, \bigoplus_{i\in I}\mathcal{K}'_i)$ is a maximal pair of subspaces of $(\mathcal{H}, \mathcal{H}')$ with the two previous properties.

Let $\mathcal{H}_0$ and $\mathcal{H}'_0$ be the orthogonal complements of $\bigoplus_{i\in I}\mathcal{K}_i$ and $\bigoplus_{i\in I}\mathcal{K}'_i$. Denote by π_0 and π'_0 the subrepresentations of π and π' defined by $\mathcal{H}_0$ and $\mathcal{H}'_0$. We have

$$\pi = \pi_0 \oplus \bigoplus_{i\in I}\pi_i \quad \text{and} \quad \pi' = \pi'_0 \oplus \bigoplus_{i\in I}\pi'_i.$$

Moreover, $\bigoplus_{i\in I}\pi_i$ is equivalent to $\bigoplus_{i\in I}\pi'_i$ and, by maximality of the family $(\mathcal{K}_i, \mathcal{K}'_i)_{i\in I}$, the representations π_0 and π'_0 are disjoint.

Two cases may occur.

- The subrepresentation π_0 is trivial. Then π is contained in π'.
- The subrepresentation π_0 is nontrivial. We claim that π'_0 is trivial, that is, π' is contained in π.

Indeed, assume, by contradiction that π'_0 is nontrivial. Since π is factorial, π_0 and $\bigoplus_{i\in I}\pi_i$ are not disjoint. Therefore π_0 and $\bigoplus_{i\in I}\pi'_i$ are not disjoint; so, $\bigoplus_{i\in I}\pi'_i$ contains a nontrivial subrepresentation σ' which is equivalent to a subrepresentation of π_0. By Proposition 6.A.11, σ' and π'_0 are not disjoint, since π'_0 is nontrivial. It follows that π_0 and π'_0 are not disjoint, and this is a contradiction. The claim is established. □

6.A.c. Multiplicity-free and type I representations. In this subsection, we introduce type I representations and their building blocks, the multiplicity-free representations.

DEFINITION 6.A.13. A representation π of G is **multiplicity-free** if, for every nontrivial decomposition $\pi = \pi_1 \oplus \pi_2$, the representations π_1 and π_2 are disjoint.

The representation π of G is **type I**, or of **discrete type**, if it is quasi-equivalent to a multiplicity-free representation.

It follows from the definitions that the type I property is stable by quasi-equivalence: if π is a representation of G of type I, then so is every representation which is quasi-equivalent to π.

We give a few straightforward examples of multiplicity-free and type I representations. For more interesting examples, see Subsection 6.B.d.

EXAMPLE 6.A.14. (1) Let $(\pi)_{i\in I}$ be a family of mutually nonequivalent irreducible representations of G. Then $\pi := \bigoplus_{i\in I} \pi_i$ is a multiplicity-free representation.

Indeed, since the π_i's are mutually nonequivalent and irreducible, we have by Proposition 1.A.6, Schur's lemma, and Lemma 6.B.2 below,

$$\mathrm{Hom}_G(\pi, \pi) = \bigoplus_{i\in I} \mathrm{Hom}_G(\pi_i, \pi_i) = \bigoplus_{i\in I} \mathbf{C}\mathrm{Id}_{\mathcal{H}_i},$$

where $\mathcal{H}_i$ denotes the Hilbert space of π_i. In particular, $\mathrm{Hom}_G(\pi, \pi)$ is an abelian algebra and the claim follows from Proposition 6.B.10 below.

(2) Let $(n_i)_{i\in I}$ be a family of cardinals and $(\pi)_{i\in I}$ a family of mutually nonequivalent irreducible representations of G. Then $\pi := \bigoplus_{i\in I} n_i\pi_i$ is a type I representation.

Indeed, $\pi' := \bigoplus_{i\in I} \pi_i$ is multiplicity-free by (1), and π is quasi-equivalent to π', by Proposition 6.A.4.

(3) Let G be a compact group. Then every representation of G is of type I.

Indeed, as is well known, every representation of such a group is a direct sum of irreducible representations (see, for instance, [**Robe–83**, Chap. 5]) and hence the claim follows from (2).

It is obvious that a factor representation of G is multiplicity-free if and only if it is irreducible. Below we will characterize factor representations which are of type I (Proposition 6.B.14).

The following proposition shows that the property of being multiplicity-free or of type I is inherited by subrepresentations.

PROPOSITION 6.A.15. *Let π be representation and σ a subrepresentation of π.*

(1) *If π is multiplicity-free, then σ is multiplicity-free.*
(2) *If π is of type I, then σ is of type I.*

PROOF. Assume that π is multiplicity-free and let $\sigma = \sigma_1 \oplus \sigma_2$ be a nontrivial decomposition of σ. Then, we have a nontrivial decomposition

$$\pi = \sigma_1 \oplus (\sigma_2 \oplus \sigma'),$$

where σ' is the subrepresentation of π on the orthogonal complement of the subspace corresponding to σ. Therefore σ_1 and $\sigma_2 \oplus \sigma'$ are disjoint. This implies that σ_1 and σ_2 are disjoint. So, σ is multiplicity-free and (1) is proved.

Assume that π is of type I. So, π is quasi-equivalent to a multiplicity-free representation π'. By Proposition 6.A.7, σ is quasi-equivalent to a subrepresentation σ' of π'. By (1), σ' is multiplicity-free. Hence, σ is of type I and (2) is proved. □

We now show that a given quasi-equivalence class contains at most one multiplicity-free representation, up to equivalence.

PROPOSITION 6.A.16. *Let π_1, π_2 be two representations of G. Assume that π_1 and π_2 are multiplicity-free and quasi-equivalent.*
Then π_1 and π_2 are equivalent.

PROOF. In view of Proposition 1.A.8, it suffices to show that π_1 is equivalent to a subrepresentation of π_2.

As π_1 is subordinate to π_2, by the proof of Proposition 6.A.4, we have an orthogonal decomposition

$$\mathcal{H} = \bigoplus_{i \in I} \mathcal{H}_i$$

of the Hilbert space $\mathcal{H}$ of π_1 into G-invariant closed subspaces $\mathcal{H}_i$ such that the subrepresentation σ_i of π_1 corresponding to $\mathcal{H}_i$ is equivalent to a subrepresentation ρ_i of π_2; so, for every i, there exists a bijective isometric intertwiner $T_i \in \mathrm{Hom}_G(\sigma_i, \rho_i)$.

Denote by P_i the orthogonal projection from the Hilbert space $\mathcal{K}$ of π_2 onto the subspace $\mathcal{K}_i$ of ρ_i. On the one hand, we have for all i, j

$$T_j^{-1} P_j T_i \in \mathrm{Hom}_G(\sigma_i, \sigma_j).$$

On the other hand, since π_1 is multiplicity-free, the σ_i's are mutually disjoint. If $i \neq j$, it follows from Proposition 1.A.6 that $T_j^{-1} P_j T_i = 0$, hence $P_j = 0$ on the range of T_i. This shows that the subspaces $\mathcal{K}_i$ are mutually orthogonal.

Let $T : \mathcal{H} \to \mathcal{K}$ be defined by

$$T|_{\mathcal{H}_i} = T_i \qquad \text{for every} \quad i.$$

It is clear that T is an isometric G-equivariant embedding, showing that π_1 is equivalent to a subrepresentation of π_2. □

6.B. Group representations and von Neumann algebras

As in the previous section, G denotes a topological group.

The following proposition is a basic tool for the reformulation of the notions defined in Section 6.A in terms of von Neumann algebras.

PROPOSITION 6.B.1. *Let $(\pi, \mathcal{H})$ be a representation of G.*

(1) *The commutant $\pi(G)' = \mathrm{Hom}_G(\pi, \pi)$ and the bicommutant $\pi(G)''$ are von Neumann subalgebras of $\mathcal{L}(\mathcal{H})$. Moreover, $\pi(G)''$ is the von Neumann algebra generated by $\pi(G)$.*
(2) *Let $\mathcal{K}$ be a closed linear subspace of $\mathcal{H}$, and $P \in \mathcal{L}(\mathcal{H})$ the orthogonal projection on $\mathcal{K}$. Then $\mathcal{K}$ is G-invariant if and only if $P \in \pi(G)'$.*

PROOF. Since the representation π is unitary, $\pi(G)^* = \pi(G)$, and Claim (1) follows from the von Neumann bicommutant theorem; see Appendix A.11.

For (2), it is clear that $\mathcal{K}$ is G-invariant if $P \in \pi(G)'$. Conversely, assume that $\mathcal{K}$ is G-invariant. Then $\pi(g)P = P\pi(g)P$ for all $g \in G$. Therefore

$$P\pi(g) = (\pi(g^{-1})P)^* = (P\pi(g^{-1})P)^* = P\pi(g)P,$$

and hence $\pi(g)P = P\pi(g)$ for all $g \in G$. □

Let $\mathcal{H}$ be a Hilbert space which is a Hilbert direct sum of a family $(\mathcal{H})_{i\in I}$ of closed subspaces indexed by a set I. For each $i \in I$, denote by $P_i : \mathcal{H} \twoheadrightarrow \mathcal{H}_i$ the orthogonal projection. An operator $T \in \mathcal{L}(\mathcal{H})$ can be described by a matrix $(T_{i,j})_{i,j\in I}$, where $T_{i,j} \in \mathcal{L}(\mathcal{H}_j, \mathcal{H}_i)$ is defined by

$$T_{i,j}(\xi) := P_i T(\xi) \quad \text{for all } i, j \in I \text{ and } \xi \in \mathcal{H}_j,$$

and

$$T\Big((\xi_i)_{i\in I}\Big) = \Big(\sum_{j\in I} P_i T \xi_j\Big)_{i\in I} \quad \text{for all } (\xi_i)_{i\in I} \in \mathcal{H} = \bigoplus_{i\in I} \mathcal{H}_i.$$

The proof of the following lemma is straightforward and left to the reader.

LEMMA 6.B.2. *Let $(\pi_i, \mathcal{H}_i)_{i\in I}$ be a family of representations of G, and let $\pi = \bigoplus_{i\in I} \pi_i$ be their direct sum, in the Hilbert space $\mathcal{H} = \bigoplus_{i\in I} \mathcal{H}_i$. Let $T \in \mathcal{L}(\mathcal{H})$ and let $(T_{i,j})_{i,j\in I}$ be the matrix of T associated with the decomposition $\mathcal{H} = \bigoplus_{i\in I} \mathcal{H}_i$. The following properties are equivalent:*

(i) $T \in \pi(G)' = \mathrm{Hom}_G(\pi, \pi)$;
(ii) $T_{i,j} \in \mathrm{Hom}_G(\pi_j, \pi_i)$ *for every* $i, j \in I$.

Let $\mathcal{H}$ be a Hilbert space, $\mathcal{K}$ a closed subspace, and P the orthogonal projection of $\mathcal{H}$ onto $\mathcal{K}$. Let $\mathcal{M}$ be a von Neumann subalgebra of $\mathcal{L}(\mathcal{H})$. In case $P \in \mathcal{M}'$ or $P \in \mathcal{M}$, we denote by $\mathcal{M}_P$ the von Neumann subalgebra $\{PT|_\mathcal{K} \mid T \in \mathcal{M}\}$ of $\mathcal{L}(\mathcal{K})$. Observe that $\mathcal{M}_P$ is isomorphic to the subalgebra $P\mathcal{M}P$ of $\mathcal{L}(\mathcal{H})$.

When $P \in \mathcal{M}'$, the subspaces $\mathcal{K}$ and $\mathcal{K}^\perp$ are invariant under $\mathcal{M}$, and $\mathcal{M}_P$ is the induced von Neumann algebra as in Appendix A.11.

We record the following useful consequence of Lemma 6.B.2, for I consisting of two elements.

PROPOSITION 6.B.3. *Let $(\pi, \mathcal{H})$ be a representation of G and set $\mathcal{M} := \pi(G)''$. Let $P \in \mathcal{M}'$ be a projection. Then*

$$(\mathcal{M}_P)' = (\mathcal{M}')_P.$$

PROOF. Denote by $\mathcal{K}$ the image of P.

For $T \in (\mathcal{M}_P)'$, let $\widetilde{T}$ be the bounded operator on $\mathcal{H}$ with matrix $(\widetilde{T}_{i,j})_{1\le i,j\le 2}$ associated with the decomposition $\mathcal{H} = \mathcal{K} \oplus \mathcal{K}^\perp$ and given by $\widetilde{T}_{1,1} = T$ and $T_{i,j} = 0$ otherwise. Lemma 6.B.2 shows that $\widetilde{T} \in \mathcal{M}'$. Therefore $T \in (\mathcal{M}')_P$.

Conversely, let $T \in (\mathcal{M}')_P$. Then there exists $\widetilde{T} \in \mathcal{M}'$ with matrix $(\widetilde{T}_{i,j})_{1\le i,j\le 2}$ associated with the decomposition $\mathcal{H} = \mathcal{K} \oplus \mathcal{K}^\perp$ such that $\widetilde{T}_{1,1} = P\widetilde{T}P = T$. Let $S \in \mathcal{M}$. Using the fact that $P\widetilde{T}P \in \mathcal{M}'$, we have

$$TPSP = (P\widetilde{T}P)SP = P(P\widetilde{T}P)SP = PS(P\widetilde{T}P)P = PS(P\widetilde{T}P) = PSPT.$$

It follows that $T \in (\mathcal{M}_P)'$. □

6.B.a. Disjointness and quasi-equivalence, again. Let $(\pi, \mathcal{H})$ be a representation of G and $\mathcal{K}_1$ a G-invariant closed subspace of $\mathcal{H}$. Denote by π_1 and π_2 the corresponding subrepresentations of π defined by $\mathcal{K}_1$ and $\mathcal{K}_2 := \mathcal{K}_1^\perp$. Then $\pi(G)$ is contained in $\pi_1(G) \oplus \pi_2(G)$, which is itself contained in $\mathcal{L}(\mathcal{K}_1) \oplus \mathcal{L}(\mathcal{K}_2) \subset \mathcal{L}(\mathcal{H})$. This implies the inclusions

$$\pi_1(G)' \oplus \pi_2(G)' \subset \pi(G)' \tag{a.1}$$

and

$$\pi(G)'' \subset \pi_1(G)'' \oplus \pi_2(G)''. \tag{a.2}$$

We now show that equality in the inclusions (a.1) and (a.2) corresponds to disjointness of π_1 and π_2.

PROPOSITION 6.B.4. *Let $(\pi, \mathcal{H})$ be a representation of G which is the direct sum of two subrepresentations π_1, π_2 on orthogonal subspaces $\mathcal{K}_1, \mathcal{K}_2$ of $\mathcal{H}$, such that $\mathcal{H} = \mathcal{K}_1 \oplus \mathcal{K}_2$. Let P_1 be the orthogonal projection from $\mathcal{H}$ onto $\mathcal{K}_1$. The following properties are equivalent:*

(i) *π_1 and π_2 are disjoint;*
(ii) *the von Neumann algebras $\pi(G)'$ and $\pi_1(G)' \oplus \pi_2(G)'$ coincide;*
(iii) *the von Neumann algebras $\pi(G)''$ and $\pi_1(G)'' \oplus \pi_2(G)''$ coincide;*
(iv) *P_1 belongs to the centre of $\pi(G)'$, that is, belongs to $\pi(G)' \cap \pi(G)''$.*

PROOF. Assume that π_1 and π_2 are disjoint. Let $T \in \pi(G)'$. Then, by Lemma 6.B.2 and Proposition 1.A.6, the matrix $(T_{i,j})_{1 \le i,j \le 2}$ of T associated with the decomposition $\mathcal{H} = \mathcal{K}_1 \oplus \mathcal{K}_2$ is diagonal; and $T_{i,i} \in \pi_i(G)'$ for $i = 1, 2$. Therefore $\pi(G)' \subset \pi_1(G)' \oplus \pi_2(G)'$. By (a.1), $\pi(G)' = \pi_1(G)' \oplus \pi_2(G)'$. This shows that (i) implies (ii).

Assume that $\pi_1(G)' \oplus \pi_2(G)' = \pi(G)'$. Then $\pi_1(G)'' \oplus \pi_2(G)'' \subset \pi(G)''$, hence, $\pi(G)'' = \pi_1(G)'' \oplus \pi_2(G)''$ by (a.2). This shows that (ii) implies (iii).

Assume that $\pi_1(G)'' \oplus \pi_2(G)'' = \pi(G)''$. For the matrix $(T_{i,j})_{1 \le i,j \le 2}$ of P_1 associated with the decomposition $\mathcal{H} = \mathcal{K}_1 \oplus \mathcal{K}_2$, we have $T_{1,1} = \mathrm{Id}_{\mathcal{K}_1} \in \pi_1(G)''$ and $T_{i,j} = 0$ otherwise. Therefore $P_1 \in \pi(G)''$. Since $P_1 \in \pi(G)'$, it follows that P_1 belongs to the centre of $\pi(G)'$. This shows that (iii) implies (iv).

Assume that π_1 and π_2 are not disjoint. Then, by Proposition 1.A.6, there exists a nonzero element S in $\mathrm{Hom}_G(\pi_1, \pi_2)$. Let $T \in \mathcal{L}(\mathcal{H})$ be the operator with matrix $(T_{i,j})_{1 \le i,j \le 2}$ associated with the decomposition $\mathcal{H} = \mathcal{K}_1 \oplus \mathcal{K}_2$ defined by $T_{2,1} = S$ and $T_{i,j} = 0$ otherwise. Then $T \in \pi(G)'$, by Lemma 6.B.2. It is clear that $TP_1 \neq P_1 T$. So, P_1 does not belong to the centre of $\pi(G)'$. It follows that (iv) implies (i). □

The following proposition gives a characterization of quasi-equivalence of representations in terms of von Neumann algebras. For a reminder on induced von Neumann algebras such as $\pi_2(G)''_E$ below, see Appendix A.11.

PROPOSITION 6.B.5. *Let π_1 and π_2 be representations of G. The following properties are equivalent:*

(i) *π_1 is subordinate to π_2;*
(ii) *there exist a central projection $E \in \pi_2(G)''$ and an isomorphism*

$$\Phi : \pi_2(G)''_E \to \pi_1(G)''$$

such that $\Phi(E\pi_2(g)E) = \pi_1(g)$ for all $g \in G$.

In particular, π_1 and π_2 are quasi-equivalent if and only if there exists an isomorphism $\Phi : \pi_2(G)'' \to \pi_1(G)''$ such that $\Phi(\pi_2(g)) = \pi_1(g)$ for all $g \in G$.

PROOF. (i) $\Rightarrow$ (ii) Assume that π_1 is subordinate to π_2. Then π_1 is equivalent to a subrepresentation of $n\pi_2$ for some cardinal n, by Proposition 6.A.4. Denote by $\mathcal{K}$ a Hilbert space of dimension n.

Let $\mathcal{H}_1$ and $\mathcal{H}_2$ be the Hilbert spaces of π_1 and π_2. We realize $\pi := n\pi_2$ on the Hilbert space $\mathcal{H} := \mathcal{H}_2 \otimes \mathcal{K}$ by

$$\pi(g) = \pi_2(g) \otimes \mathrm{Id}_{\mathcal{K}} \quad \text{for all } g \in G.$$

Observe that

$$\pi_2(G)'' \to \pi(G)'', \quad T \mapsto T \otimes \mathrm{Id}_{\mathcal{K}}$$

is an isomorphism of von Neumann algebras mapping $\pi_2(g)$ to $\pi(g)$ for all $g \in G$. Therefore, it suffices to show the following claim: there exists a central projection $E \in \pi(G)''$ and an isomorphism $\Phi : \pi(G)''_E \to \pi_1(G)''$ such that $\Phi(E\pi(g)E) = \pi_1(g)$ for all $g \in G$.

We view π_1 as a subrepresentation of π. Let $P \in \pi(G)'$ be the orthogonal projection onto the subspace defining π_1 and let E be the central support of P (for the notion of central support, see Definition A.11.6). Since $\pi(G)''_P = \pi_1(G)''$ and since the induction map $\pi(G)''_E \to \pi(G)''_P$ is an isomorphism (Proposition A.11.7 applied to the von Neumann algebra $\mathcal{M} = \pi(G)''_E)$), the claim follows.

(ii) $\Rightarrow$ (i) Assume now that there exist a central projection $E \in \pi_2(G)''$ and an isomorphism $\Phi : E\pi_2(G)''E \to \pi_1(G)''$ such that $\Phi(E\pi_2(g)E) = \pi_1(g)$ for all $g \in G$. Upon replacing π_2 by its subrepresentation $g \mapsto E\pi_2(g)$ defined on the range of E, we may assume that E is the identity operator on $\mathcal{H}_2$.

By the structure of isomorphisms of von Neumann algebras (Proposition A.11.8), there exist

- a Hilbert space $\mathcal{K}$,
- a projection $P \in (\pi_2(G) \otimes \mathrm{Id}_{\mathcal{K}})' \subset \mathcal{L}(\mathcal{H}_2 \otimes \mathcal{K})$, with range $\mathcal{H}'_2 \subset \mathcal{H}_2 \otimes \mathcal{K}$,
- and an isomorphism of Hilbert spaces $U : \mathcal{H}'_2 \to \mathcal{H}_1$, such that

$$\Phi(T) = U \circ (T \otimes \mathrm{Id}_{\mathcal{K}})_P \circ U^{-1} \quad \text{for every } T \in \pi_2(G)''.$$

(Observe that $\mathcal{H}'_2$ is invariant under $\pi_2(G) \otimes \mathrm{Id}_{\mathcal{K}}$, since $P \in (\pi_2(G) \otimes \mathrm{Id}_{\mathcal{K}})'$.) Therefore, we have

$$\pi_1(g) = \Phi(\pi_2(g)) = U \circ (\pi_2(g) \otimes \mathrm{Id}_{\mathcal{K}}) \circ U^{-1} \quad \text{for every } g \in G.$$

This shows that π_1 is equivalent to a subrepresentation of $n\pi_2$, where n is the dimension of $\mathcal{K}$. Therefore π_1 is subordinate to π_2, by Proposition 6.A.4.

It remains to prove the last statement in Proposition 6.B.5. Assume that there exists an isomorphism $\Phi : \pi_2(G)'' \to \pi_1(G)''$ such that $\Phi(\pi_2(g)) = \pi_1(g)$ for all $g \in G$; then, by what we have just seen, π_1 and π_2 are quasi-equivalent.

Conversely, assume that π_1 and π_2 are quasi-equivalent. Then there exist a central projection $E \in \pi_2(G)''$ and an isomorphism $\Phi : \pi_2(G)''_E \to \pi_1(G)''$ such that $\Phi(E\pi_2(g)E) = \pi_1(g)$ for all $g \in G$. We claim that $E = I$. Indeed, let π', π'' be the subrepresentations of π_2 defined by E and by $I - E$. On the one hand, π' is quasi-equivalent to π_1, by what we proved above. On the other hand, π' and π'' are disjoint, by Proposition 6.B.4. Therefore π_1 and π'' are disjoint. Since $\pi_2 = \pi' \oplus \pi''$ is quasi-equivalent to π_1, it follows that π'' is trivial, that is, $E = I$. □

6.B.b. Factor representations, again. Factor representations can be characterized in terms of von Neumann algebras:

PROPOSITION 6.B.6. *For a representation π of G, the following properties are equivalent:*

(i) *π is a factor representation;*
(ii) *$\pi(G)''$ is a factor.*

PROOF. Let $\mathcal{H}$ be the Hilbert space of π. Assume that π is a factor representation and let P a nonzero projection in the centre $\mathcal{Z}$ of $\pi(G)''$. By Proposition 6.B.1, π is the direct sum of the subrepresentations π_1 and π_2 defined on the ranges of P and $\mathrm{Id}_{\mathcal{H}} - P$. By Proposition 6.B.4, the subrepresentations π_1 and π_2 are disjoint, hence π_2 is trivial, that is, $P = \mathrm{Id}_{\mathcal{H}}$. Therefore 0 and $\mathrm{Id}_{\mathcal{H}}$ are the only projections in $\mathcal{Z}$. It follows from Proposition A.11.4 that $\pi(G)''$ is a factor.

Conversely, assume that $\pi(G)''$ is a factor and let $\pi = \pi_1 \oplus \pi_2$ be a direct sum decomposition in two nontrivial subrepresentations. Since the centre of $\pi(G)'$ consists only of the scalar multiples of $\mathrm{Id}_{\mathcal{H}}$, the orthogonal projection onto the subspace defining π_1 does not belong to the centre of $\pi(G)'$. It follows from Proposition 6.B.4 that π_1 and π_2 are not disjoint. Therefore π is a factor representation. □

As a first consequence of Proposition 6.B.6, we show that factoriality of a representation is a property of its quasi-equivalence class.

COROLLARY 6.B.7. *Let π be a factor representation of G.*

Then every representation which is quasi-equivalent to π is factorial. In particular, every multiple $n\pi$ of π is factorial.

PROOF. Let σ be a representation which is quasi-equivalent to π. By Proposition 6.B.5, the von Neumann algebras $\sigma(G)''$ and $\pi(G)''$ are isomorphic. The claim now follows from Proposition 6.B.6. □

A second consequence of Proposition 6.B.6 is that subrepresentations of a factor representation belong to the same quasi-equivalence class.

COROLLARY 6.B.8. *Let π be a factor representation of G.*

Then every nontrivial subrepresentation of π is quasi-equivalent to π.

PROOF. Let $\mathcal{H}$ be the Hilbert space of π and let σ be a nontrivial subrepresentation of π. By Proposition 6.B.6, the only nonzero projection in the centre of $\pi(G)''$ is $\mathrm{Id}_{\mathcal{H}}$. Since σ is subordinate to π, it follows from Proposition 6.B.5 that there exists an isomorphism $\Phi : \pi(G)'' \to \sigma(G)''$ such that $\Phi(\pi(g)) = \sigma(g)$ for all $g \in G$, hence that σ is quasi-equivalent to π. □

The following consequence of Proposition 6.B.6 and Corollary 6.B.8 should be compared with the analogous statement for irreducible representations in Proposition 1.A.13.

COROLLARY 6.B.9. *Let X be a dense subset of G of cardinal $\aleph$.*

Every factor representation of G is quasi-equivalent to a factor representation on a Hilbert space of dimension at most $\aleph$.

In particular, if G is separable, every factor representation of G is quasi-equivalent to a factor representation on a separable Hilbert space.

PROOF. Let $(\pi, \mathcal{H})$ be a factor representation of G. Let ξ be a nonzero vector in $\mathcal{H}$ and let $\mathcal{K}$ be the closed subspace of $\mathcal{H}$ generated by $\pi(G)\xi$. Since $\pi(X)\xi$ is dense in $\pi(G)\xi$ and since X has cardinal $\aleph$, we have $\dim \mathcal{K} \leq \aleph$. Moreover, $\mathcal{K}$ is G-invariant and so defines a subrepresentation σ of π. By Corollary 6.B.8, σ is a factor representation and is quasi-equivalent to π. □

6.B.c. Multiplicity-free and type I representations, again. Multiplicity-free representations can also be characterized in terms of von Neumann algebras.

PROPOSITION 6.B.10. *For a representation π of G, the following properties are equivalent:*

(i) *π is multiplicity-free;*
(ii) *$\pi(G)'$ is an abelian von Neumann algebra.*

PROOF. Assume that π is multiplicity-free. Then, by Proposition 6.B.4, every projection contained in $\pi(G)'$ belongs to the centre of $\pi(G)'$, In view of Proposition A.11.4, it follows that $\pi(G)'$ is abelian.

Conversely, assume that $\pi(G)'$ is abelian. Let $\pi = \pi_1 \oplus \pi_2$ be a decomposition of π into subrepresentations π_1 and π_2. Since the orthogonal projection on the subspace defining π_1 belongs to $\pi(G)'$, it follows from Proposition 6.B.4 that π_1 and π_2 are disjoint. □

The direct sum of multiplicity-free representations is not multiplicity-free in general; for instance, $\pi \oplus \pi$ is not multiplicity free for any representation π. However, the following corollary of Proposition 6.B.10 holds.

COROLLARY 6.B.11. *Let $(\pi_i)_{i\in I}$ be a family of* pairwise disjoint *multiplicity-free representations of G.*

Then $\bigoplus_{i\in I} \pi_i$ is multiplicity-free.

PROOF. Set $\pi := \bigoplus_{i\in I} \pi_i$. Since the π_i's are pairwise disjoint, we have

$$\pi(G)' = \bigoplus_{i\in I} \pi_i(G)'$$

(see the proof of Proposition 6.B.4). Therefore $\pi(G)'$ is abelian, and π is multiplicity-free by Proposition 6.B.10. □

The following characterization of type I representations in terms of von Neumann algebras is a direct consequence of Proposition 6.B.5 and Proposition 6.B.10.

PROPOSITION 6.B.12. *For a representation π of G, the following properties are equivalent:*

(i) *π is of type I;*
(ii) *there exist a representation σ of G with abelian commutant $\sigma(G)'$ and an isomorphism $\Phi : \pi(G)'' \to \sigma(G)''$ such that $\Phi(\pi(g)) = \sigma(g)$ for all $g \in G$.*

Next, we show that the property of being type I is inherited by direct sums of representations.

PROPOSITION 6.B.13. (1) *Let $(\pi, \mathcal{H})$ be a representation of G and P a projection in $\pi(G)'$ with central support E. Assume that the subrepresentation of π defined on the range of P is of type I.*

Then the subrepresentation of π defined on the range of E is of type I.

(2) *Let $(\pi_i)_{i\in I}$ be a family of type I representations of G.*

Then $\bigoplus_{i\in I} \pi_i$ is a type I representation.

PROOF. (1) By Proposition 6.B.12, there exist a representation σ of G with abelian commutant $\sigma(G)'$ and an isomorphism $\Phi : \pi(G)''_P \to \sigma(G)''$ such that

$\Phi(P\pi(g)P) = \sigma(g)$ for all $g \in G$. The induction map $\Psi : \pi(G)''_E \to \pi(G)''_P$ is an isomorphism. So,

$$\Phi \circ \Psi : \pi(G)''_E \to \sigma(G)''$$

is an isomorphism and we have

$$\Phi \circ \Psi(E\pi(g)E) = \Phi(P\pi(g)P) = \sigma(g) \quad \text{for all } g \in G.$$

Therefore, again by Proposition 6.B.12, the subrepresentation of π defined on the range of E is of type I.

(2) Set

$$\mathcal{H} := \bigoplus \mathcal{H}_i \quad \text{and} \quad \pi := \bigoplus_{i \in I} \pi_i,$$

where $\mathcal{H}_i$ is the Hilbert space of π_i. For $i \in I$, denote by $P_i \in \pi(G)'$ the orthogonal projection from $\mathcal{H}$ onto $\mathcal{H}_i$ and by E_i the central support of P_i.

By Zorn's lemma, we can find a maximal family $(F_j)_{j \in J}$ of *mutually orthogonal* projections F_j in the centre $\mathcal{Z}$ of $\pi(G)''$ with the property that the subrepresentation of π defined on the range of F_j is of type I. Set

$$F := \bigoplus_{j \in J} F_j.$$

Then F is a projection in $\mathcal{Z}$. We claim that $F = \mathrm{Id}_{\mathcal{H}}$.

Indeed, assume, by contradiction, that $\mathrm{Id}_{\mathcal{H}} - F \neq 0$. Since $\bigoplus_{i \in I} P_i = \mathrm{Id}_{\mathcal{H}}$, there exists $i \in I$ such that $(\mathrm{Id}_{\mathcal{H}} - F)P_i \neq 0$ and therefore $(\mathrm{Id}_{\mathcal{H}} - F)E_i \neq 0$. The range $\mathcal{K}'$ of $(\mathrm{Id}_{\mathcal{H}} - F)E_i$ is invariant under $\pi(G)$ and $\pi(G)'$, since F and E_i belong to $\mathcal{Z}$. Therefore the orthogonal projection F' on $\mathcal{K}'$ is a nonzero projection in $\mathcal{Z}$. Let π' be the subrepresentation of π defined on the range of F'.

On the one hand, $\mathcal{K}'$ is contained in the range of E_i; so, π' is contained in the subrepresentation π_i' of π defined on the range of E_i. Since π_i is of type I, it follows from (1) that π_i' is of type I. Therefore π' is of type I, by Proposition 6.A.15. On the other hand, $\mathcal{K}'$ is contained in the range of $\mathrm{Id}_{\mathcal{H}} - F$ and so F' is orthogonal to F_j for every $j \in J$. This is a contradiction with the maximality of the family $(F_j)_{j \in J}$. Therefore, we have $\bigoplus_{j \in J} F_j = \mathrm{Id}_{\mathcal{H}}$, as claimed.

For every $j \in J$, denote by $\widetilde{\pi}_j$ the subrepresentation of π defined on the range of F_j. Then $\pi = \bigoplus_{j \in J} \widetilde{\pi}_j$. Moreover, since the F_j's are mutually orthogonal projections in the centre of $\pi(G)''$, the representations $\widetilde{\pi}_j$'s are pairwise disjoint and we have

$$\pi(G)'' = \bigoplus_{j \in J} \widetilde{\pi}_j(G)''$$

(compare with the proof of Proposition 6.B.4).

For every $j \in J$, there exist a representation σ_j of G with abelian commutant $\sigma_j(G)'$ and an isomorphism $\Phi_j : \widetilde{\pi}_j(G)'' \to \sigma_j(G)''$ such that $\Phi_j(\widetilde{\pi}_j(g)) = \sigma_j(g)$ for all $g \in G$ (Proposition 6.B.12). Then

$$\Phi := \bigoplus_{j \in J} \Phi_j : \pi(G)'' \to \bigoplus_{j \in J} \sigma_j(G)''$$

is an isomorphism of von Neumann algebras such that $\Phi_j(\pi(g)) = \bigoplus_{j \in J} \sigma_j(g)$ for all $g \in G$. Therefore (Proposition 6.B.5), π is quasi-equivalent to $\sigma := \bigoplus_{j \in J} \sigma_j$. The σ_j's are pairwise disjoint, since σ_j is quasi-equivalent to $\widetilde{\pi}_j$ and since the $\widetilde{\pi}_j$'s are pairwise disjoint. Therefore σ is multiplicity-free, by Corollary 6.B.11. Therefore, π is of type I. □

We now show that the representations appearing in Proposition 6.A.10 are exactly the factor representations of type *I*.

PROPOSITION 6.B.14. *Let π be a factor representation of G. The following properties are equivalent:*

(i) *π is of type I;*
(ii) *π is a multiple of an irreducible representation of G.*

When π has these properties, there exists an irreducible representation σ, unique up to equivalence, such that every representation of G which is quasi-equivalent to π is equivalent to a multiple of σ.

PROOF. That (ii) implies (i) is a particular case of Example 6.A.14(2).

For the converse, assume that π is type I, i.e., that π is quasi-equivalent to a multiplicity-free representation σ of G. By Corollary 6.B.7, σ is a factor representation. Proposition 6.B.10 implies that $\sigma(G)'$ is one dimensional; this means that σ is an irreducible representation. Therefore, σ is equivalent to a subrepresentation of π, by Proposition 6.A.2. Therefore π is a multiple of σ, by Proposition 6.A.10.

To show the last claim, let π' be a representation of G which is quasi-equivalent to π. Then π' is a factor representation of type I and the equivalence of (i) and (ii), applied to π', shows that π' is a multiple of an irreducible representation σ'. Since σ and σ' are quasi-equivalent and irreducible, σ and σ' are equivalent. □

It is worth summarizing the results we obtained so far regarding the quasi-equivalence class of a factor representation (Proposition 6.A.12, Corollary 6.B.7, Corollary 6.B.8, and Proposition 6.B.14).

THEOREM 6.B.15. *Let π be a factor representation of a topological group G.*

(1) *Every representation which is quasi-equivalent to π is factorial.*
(2) *Every nontrivial subrepresentation of π is factorial and quasi-equivalent to π.*
(3) *Let π_1 be a representation in the quasi-equivalence class of π; then, either π_1 is a subrepresentation of π or π is a subrepresentation of π_1.*
(4) *Let σ be a factor representation which is not quasi-equivalent to π; then σ and π are disjoint.*
(5) *The representation π is of type I if and only if π is a multiple of an irreducible representation σ of G; when π is of type I, σ is uniquely defined up to equivalence by π.*

REMARK 6.B.16. (1) In view of Claim (5) of Theorem 6.B.15, we may refine the type I classification of factor representations as follows: given a cardinal $n \geq 1$, a factor representation is said to be of type I_n if π is quasi-equivalent to an irreducible representation of dimension n.

(2) As we will see later (see Section 7.A and Chapter 14) there are factor representations which are not of type I. They are classified in various other types, via the Murray–von Neumann classification of factors in types (see Section 7.C). It is worth mentioning that, starting from Theorem 6.B.15, one may independently develop a "multiplicity theory" for factor representations and recover the Murray–von Neumann type classification, as shown in [**Mack–76**, Chap. 1].

Given a cardinal $n \geq 1$, say that a representation π of a topological group G has **multiplicity** n if π is equivalent to $n\sigma$ for a multiplicity-free representation σ of G. For a characterization of such representations in terms of von Neumann algebras, see [**Dixm–C***, 5.4.7].

The following result shows that a type I representation of G has a unique direct sum decomposition into representations of well-defined multiplicities. For the particular case of a second-countable locally compact abelian group, see the proof of Theorem 2.B.9. For the general case, we quote [**Dixm–vN**, Chap. III, § 3, no 1] (where $\mathcal{H}$ need not be separable, and the set I of cardinals need not be inside $\overline{\mathbf{N}^*}$); see also [**Dixm–C***, 5.4.9].

THEOREM 6.B.17. *Let π be a type I representation of a topological group G in a separable Hilbert space $\mathcal{H}$.*

There exist a set $I \subset \overline{\mathbf{N}^*}$ of extended positive integers and an orthogonal decomposition $\mathcal{H} = \bigoplus_{m \in I} \mathcal{H}_m$ into G-invariant closed subspaces $\mathcal{H}_m$ with the following properties:

1. *for $m, n \in I$ with $m \neq n$, the subrepresentations of π defined in $\mathcal{H}_m$ and $\mathcal{H}_n$ are disjoint;*
2. *for every $m \in I$, the subrepresentation of π defined in $\mathcal{H}_m$ has multiplicity m.*

Moreover, the orthogonal projection on $\mathcal{H}_m$ belongs to the centre of $\pi(G)'$ for every $m \in I$, and this decomposition is unique in the following sense: if we have $\mathcal{H} = \bigoplus_{n \in J} \mathcal{K}_n$ for a set J of extended integers and for G-invariant closed subspaces $\mathcal{K}_n$ with properties (1) and (2), then $I = J$ and $\mathcal{H}_m = \mathcal{K}_m$ for all $m \in I$.

Recall (Theorem 1.G.9) that every representation π of a second-countable locally LC group in a separable Hilbert space admits a direct integral decomposition into irreducible representations. When π is multiplicity-free, the following result shows that such a decomposition is essentially unique.

THEOREM 6.B.18. *Let G be a second-countable locally compact group and π a multiplicity-free representation of G in a separable Hilbert space $\mathcal{H}$. Let*

$$\pi \simeq \int_X^{\oplus} \pi_x d\mu(x) \simeq \int_Y^{\oplus} \sigma_y d\nu(y)$$

be two integral decompositions of π over standard Borel spaces X, Y equipped with σ-finite measures μ, ν, and measurable fields $x \mapsto \pi_x$, $y \mapsto \sigma_x$ of irreducible representations.

Then (after appropriate modifications on subsets of measure zero) there exists a Borel isomorphism $f : X \to Y$ such that $f_*(\mu)$ is equivalent to ν and such that $\sigma_{f(x)}$ and π_x are equivalent for μ-almost every $x \in X$.

PROOF. Let $x \mapsto \mathcal{H}_x$ and $y \mapsto \mathcal{K}_y$ be the measurable fields of Hilbert spaces defining $\int_X^{\oplus} \pi_x d\mu(x)$ and $\int_Y^{\oplus} \sigma_y d\nu(y)$. We can and will identify $\mathcal{H}$ with $\int_X^{\oplus} \mathcal{H}_x d\mu(x)$ and π with $\int_X^{\oplus} \pi_x d\mu(x)$. Set

$$\mathcal{K} := \int_Y^{\oplus} \mathcal{K}_y d\nu(y) \text{ and } \sigma := \int_Y^{\oplus} \sigma_y d\nu(y)$$

and let $U : \mathcal{H} \to \mathcal{K}$ be an isomorphism of Hilbert spaces such that $U\pi(g) = \sigma(g)U$ for every $g \in G$.

Let $\mathcal{A}$ and $\mathcal{B}$ be the algebras of diagonalisable operators in $\mathcal{L}(\mathcal{H})$ and $\mathcal{L}(\mathcal{K})$ respectively. Recall that we have isomorphisms $L^\infty(X, \mu) \to \mathcal{A}$, $\varphi \mapsto m(\varphi)$ and $L^\infty(Y, \nu) \to \mathcal{B}$, $\psi \mapsto m(\psi)$. Since the π_x's and the σ_y's are irreducible, $\mathcal{A}$ and $\mathcal{B}$ are maximal abelian subalgebras of $\pi(G)'$ and $\sigma(G)'$, by Proposition 1.G.8. As π and σ are multiplicity-free, $\pi(G)'$ and $\sigma(G)' = U\pi(G)'U^{-1}$ are abelian and hence coincide with $\mathcal{A}$ and $\mathcal{B}$, respectively. Therefore

$$m(\varphi) \mapsto Um(\varphi)U^{-1}$$

is an isomorphism from $\mathcal{A}$ onto $\mathcal{B}$. In this way, we obtain an isomorphism of involutive algebras

$$F: L^\infty(X, \mu) \to L^\infty(Y, \nu)$$

such that $Um(\varphi)U^{-1} = m(F(\varphi))$. By Theorem A.3.7, upon modifying X and Y on subsets of measure zero, we can assume that there exists a Borel isomorphism $f : X \to Y$ such that $f_*(\mu)$ is equivalent to ν and such that $F(\varphi)(f(x)) = \varphi(x)$ for all $\varphi \in L^\infty(X, \mu)$ and $x \in X$. It follows that, without loss of generality, we can identify the measure space (Y, ν) with (X, μ) and f with Id_X; so, $F = \mathrm{Id}_{L^\infty(X,\mu)}$ and we have

$$Um(\varphi)U^{-1} = m(\varphi) \quad \text{for every } \varphi \in L^\infty(X, \mu).$$

It follows from Proposition 1.H.2 that U is a decomposable operator; so, there exists a measurable field $x \to U_x$ of Hilbert space isomorphisms $U_x : \mathcal{H}_x \to \mathcal{K}_x$ such that

$$(U\xi)_x = U_x(\xi_x) \qquad \text{for } \xi \in \mathcal{H} \text{ and } x \in X.$$

Since $U\pi(g) = \sigma(g)U$ and since G contains a countable dense subset, it follows that $U_x\pi_x(g) = \sigma_x(g)U_x$ for every $g \in G$ and μ-almost every $x \in X$. □

REMARK 6.B.19. Let π be a representation of a second-countable locally LC group in a separable Hilbert space. Assume that π is of type I. Combining Theorems 6.B.17, 1.G.9, and 6.B.18, we obtain a direct integral decomposition of π into irreducible representations which is moreover unique in an appropriate sense; we will come back to this theme in Subsection 6.D.b.

6.B.d. Examples of multiplicity-free representations of abelian groups.

EXAMPLE 6.B.20. The left regular representation λ_G of LC group G is multiplicity-free if and only if the group G is abelian.

To show this, it suffices by Proposition 6.B.10 to show that $\lambda_G(G)'$ is abelian if and only if G is abelian. Assume first that $\lambda_G(G)'$ is abelian. Since $\rho_G(G)$ is obviously contained in $\lambda_G(G)'$ and since ρ_G is faithful, it follows that G is abelian. For the converse, we use the general fact that the algebra $\lambda_G(G)'$ coincides with the von Neumann algebra $\rho_G(G)''$ generated by $\rho_G(G)$; see Theorem 10.A.10 and Remark 10.A.11. When G is abelian, this implies that $\lambda_G(G)'$ is abelian.

Let G be a second-countable LC abelian group. Recall from Construction 2.B.1 that, given a probability measure μ on $\widehat{G}$, an extended positive integer $n \in \overline{\mathbf{N}^*}$, and a Hilbert space $\mathcal{K}$ of dimension n, we associate the representation $\pi_\mu^{(n)}$ of G defined on $L^2(\widehat{G}, \mu, \mathcal{K})$ by

$$(\pi_\mu^{(n)}(g)F)(\chi) = \chi(g)F(\chi) \quad \text{for } g \in G,\ F \in L^2(\widehat{G}, \mu, \mathcal{K}), \text{ and } \chi \in \widehat{G}.$$

When $n = 1$, the representation $\pi_\mu^{(1)}$ is the **canonical representation** π_μ of Construction 2.A.1 associated with μ.

THEOREM 6.B.21. *Let G be a second-countable locally compact abelian group and π a representation of G on a separable Hilbert space. The following properties are equivalent:*

(i) *π is multiplicity-free;*
(ii) *π is equivalent to a canonical representation π_μ for some probability measure on the Borel subsets of $\widehat{G}$.*

PROOF. We know that the commutant $\pi_\mu(G)'$ is abelian (Proposition 2.B.3 and Remark 2.B.4), hence π_μ is multiplicity-free, by Proposition 6.B.10. So, (ii) implies (i).

For the converse, consider a representation π of G in a separable Hilbert space. By Theorem 2.B.9, we may assume that

$$\pi = \bigoplus_{n\in I} \pi_{\mu_n}^{(n)},$$

where I is a set of extended positive integers, $(\mu_n)_{n\in I}$ a sequence of probability measures on $\widehat{G}$, and $\pi_{\mu_n}^{(m)}$ for each $n \in I$ the representation on $L^2(\widehat{G}, \mu_n, \mathcal{K}_n)$ defined above.

On the one hand, by Proposition 2.B.3, the commutant $\pi_{\mu_n}^{(n)}(G)'$ of $\pi_{\mu_n}^{(n)}(G)$ inside $\mathcal{L}(L^2(\widehat{G}, \mu_n, \mathcal{K}_n))$ consists of all decomposable operators in $\mathcal{L}(L^2(\widehat{G}, \mu_n, \mathcal{K}_n))$. On the other hand, observe that there is an obvious injection of $\pi_{\mu_n}^{(n)}(G)'$ into $\pi(G)'$.

Assume now that π is multiplicity-free, so that $\pi(G)'$ is abelian by Proposition 6.B.10, and therefore $\pi_{\mu_n}^{(n)}(G)'$ is abelian; this is possible only if $n = 1$. It follows that $I = \{1\}$, that is, $\pi = \pi_{\mu_1}$. So, (i) implies (ii). □

The following corollary is a direct consequence of Theorem 2.B.9 and Theorem 6.B.21.

COROLLARY 6.B.22. *Let G be a second-countable locally compact abelian group. Every representation π of G in a separable Hilbert space is of type I.*

In a terminology to be introduced in Section 6.D, Corollary 6.B.22 states that a second-countable LC abelian group is of type I. This result extends to every topological abelian group, as we will mention again below (Theorem 6.E.20).

6.B.e. Examples of multiplicity-free representations: quasi-regular representations of Gel'fand pairs. Let G be an LC group and K a compact subgroup. Let $C^c(K\backslash G/K)$ be the space of complex-valued continuous functions of compact support on G which are left- and right-invariant under translation by elements of K, that is,

$$C^c(K\backslash G/K) = \{\varphi \in C^c(G) \mid \varphi(kgk') = \varphi(g) \text{ for all } g \in G,\ k, k' \in K\}.$$

Recall that the convolution of two functions $f_1, f_2 \in C^c(G)$ is the function $f_1 * f_2 \in C^c(G)$ defined by

$$f_1 * f_2(g) = \int_G f_1(x) f_2(x^{-1}g) d\mu_G(x) = \int_G f_1(gx) f_2(x^{-1}) d\mu_G(x) \quad \text{for } g \in G,$$

where μ_G is a left Haar measure on G. (Note that this product $*$ does depend on the choice of the Haar measure μ_G.) Then $C^c(G)$ is an algebra for the convolution product and $C^c(K\backslash G/K)$ is a subalgebra of $C^c(G)$.

REMARK 6.B.23. We consider below as well as in Section 6.E convolutions of functions on G with functions and measures on the compact subgroup K. The definition goes as follows.

Recall that $L^1(G, \mu_G)$ can be identified with an ideal of the convolution algebra $M^b(G)$ of complex measures on G through the map $f \mapsto f\mu_G$. We can identify $M^b(K)$ with the subalgebra of $M^b(G)$ of measures on G with support in K. Let μ_K be the normalized Haar measure on K. For $\psi \in L^1(K, \mu_K)$ and $f \in L^1(G, \mu_G)$, the convolution products $\psi * f$ and $f * \psi$ are then the functions in $L^1(G, \mu_G)$ given by

$$\psi * f(g) = \int_K \psi(k)f(k^{-1}g)d\mu_K(k) \text{ and } f * \psi(g) = \int_K f(gk^{-1})\psi(k)d\mu_K(k).$$

For a measure $\nu \in M^b(K)$, viewed as a bounded measure on G with support inside K, the convolution products $\nu * f$ and $f * \nu$ are similarly in $L^1(G, \mu_G)$. For all this see Appendix A.6.

There is a linear projection $\Phi : C^c(G) \to C^c(K\backslash G/K)$ given by

$$\Phi(f) = \mu_K * f * \mu_K \qquad \text{for } f \in C^c(G),$$

that is,

$$\Phi(f)(g) = \int_K \int_K f(k_1 g k_2)\, d\mu_K(k_1)\, d\mu_K(k_2) \qquad \text{for } f \in C^c(G),\ g \in G.$$

DEFINITION 6.B.24. A **Gel'fand pair** is a pair (G, K), where G is a locally compact group and K a compact subgroup, such that the convolution algebra $C^c(K\backslash G/K) = \mu_K * C^c(G) * \mu_K$ is commutative.

Here are two straightforward examples of Gel'fand pairs.

EXAMPLE 6.B.25. (1) Let G be an LC abelian group. It is obvious that $(G, \{e\})$ is a Gel'fand pair.

(2) Let G be a **motion group**, that is, $G = K \ltimes A$ is the semi-direct product of a compact subgroup K and a closed abelian normal subgroup A. Then (G, K) is a Gel'fand pair. Indeed, the algebra $C^c(K\backslash G/K)$ is canonically isomorphic to the convolution algebra $C^c(A)$, which is obviously commutative.

More interesting examples of Gel'fand pairs are provided by the following result from [**Gelf–50**].

In the sequel, for a function f on a group G, we denote by $\check{f}$ the function on G defined by $\check{f}(g) = f(g^{-1})$ for $g \in G$.

THEOREM 6.B.26 (**Gel'fand**). *Let G be a locally compact group and K a compact subgroup. Assume that there exists a continuous involutive automorphism θ of G with the following property:*

$\theta(g^{-1})$ belongs to the double coset KgK, for every $g \in G$.

Then (G, K) is a Gel'fand pair.

PROOF. The image $\theta_*(\mu_G)$ of the left Haar measure μ_G by θ is obviously a left-invariant Radon measure on G. Therefore $\theta_*(\mu_G) = \lambda\mu_G$ for some $\lambda > 0$. Since $\theta^2 = I$, we have $\lambda = 1$, that is, μ_G is invariant under θ. It follows that θ defines an automorphism $f \mapsto f^\theta$ of the convolution algebra $C^c(G)$, given by

$$f^\theta(g) = f(\theta(g)) \qquad \text{for all } f \in C^c(G),\ g \in G.$$

Let $f_1, f_2 \in C^c(G)$. Then, on the one hand, we have

$$(*) \qquad f_1^\theta * f_2^\theta = (f_1 * f_2)^\theta.$$

On the other hand, we have

$$\begin{aligned}(\check{f}_1 * \check{f}_2)(g) &= \int_G f_1(x^{-1}) f_2(g^{-1}x) d\mu_G(x) \\ &= \int_G f_2(y) f_1(y^{-1}g^{-1}) d\mu_G(y) = f_2 * f_1(g^{-1})\end{aligned}$$

for every $g \in G$, that is,

$$(**) \qquad \check{f}_1 * \check{f}_2 = (f_2 * f_1)\check{}.$$

Now, let $f \in C^c(K\backslash G/K)$; it follows from the assumption that

$$f^\theta(g) = f(\theta(g)) = f(g^{-1}) \quad \text{for all } g \in G,$$

that is,

$$(***) \qquad f^\theta = \check{f}.$$

Relations $(*), (**)$, and $(***)$ imply that

$$f_1 * f_2 \overset{(**)}{=} (\check{f}_2 * \check{f}_1)\check{} \overset{(***)}{=} (f_2^\theta * f_1^\theta)^\theta \overset{(*)}{=} f_2 * f_1 \quad \text{for all } f_1, f_2 \in C^c(K\backslash G/K),$$

and this concludes the proof. □

EXAMPLE 6.B.27. (1) Let G be an LC group. Assume that there exists an action of G by isometries on a metric space (X, d) with the following properties:

- the stabilizer K in G of a point $x_0 \in X$ is compact;
- the action of K on the sphere $S(x_0, r) = \{x \in X \mid d(x_0, x) = r\}$ is transitive, for every $r > 0$.

Then (G, K) is a Gel'fand pair. Indeed, let us check that $\theta = \mathrm{Id}_G$ has the property stated in Theorem 6.B.26. Let $g \in G$. Then

$$d(x_0, gx_0) = d(g^{-1}x_0, x_0) = d(x_0, g^{-1}x_0),$$

and so there exists $k \in K$ with $g^{-1}x_0 = kgx_0$, and therefore $g^{-1} \in KgK$.

(2) Let $G \curvearrowright (X, d)$ be a *transitive* action of a group G by isometries on a metric space (X, d) and K the stabilizer of a point $x_0 \in X$. The action of $K \curvearrowright S(x_0, r)$ is transitive for every $r > 0$ if and only if the action G on equidistant pairs of points of X is transitive (that is, for every $(x, y), (x', y')$ in $X \times X$ such that $d(x, y) = d(x', y')$, there exists $g \in G$ such that $x' = gx$ and $y' = gy$).

Indeed, one implication being obvious, assume that $K \curvearrowright S(x_0, r)$ is transitive for every $r > 0$. Let $(x, y), (x', y')$ in $X \times X$ with $d(x, y) = d(x', y')$. Since the action of G on X is transitive, there exist $g, h \in G$ such that $x = gx_0$ and $x' = hx_0$. Then $d(x_0, g^{-1}y) = d(x_0, h^{-1}y')$, and therefore there exists $k \in K$ such that $kg^{-1}y = h^{-1}y'$. It is clear that $hkg^{-1}x = x'$ and $hkg^{-1}y = y'$.

(3) The following examples are classical. Let n be a positive integer. The Gel'fand pair $(\mathrm{SO}(n), \mathrm{SO}(n-1))$ arises from the natural action of $\mathrm{SO}(n)$ on the sphere $\mathbf{S}^{n-1}$. The Gel'fand pair $(\mathrm{SO}(n) \ltimes \mathbf{R}^n, \mathrm{SO}(n))$ arises from the natural action of $\mathrm{SO}(n) \ltimes \mathbf{R}^n$ on the Euclidean space $\mathbf{R}^n$. For $n \geq 2$, the Gel'fand pair $(\mathrm{SO}(n, 1), \mathrm{SO}(n))$ arises from the natural action of $\mathrm{SO}(n, 1)$ on the hyperbolic space $\mathbf{H}^n(\mathbf{R})$.

In all theses cases, it is easily checked that the action of the stabilizer of a point x_0 is transitive on every sphere $S(x_0, r)$.

(4) Let G be a noncompact semisimple connected real Lie group with finite centre and let K be a maximal compact subgroup of G. There exists an involutive automorphism θ of G, called the *Cartan involution*, such that K is the set of θ-fixed elements in G. Let $\mathfrak{g}$ be the Lie algebra of G and let

$$\mathfrak{p} = \{X \in \mathfrak{g} \mid d\theta_e(X) = -X\},$$

where $d\theta_e : \mathfrak{g} \to \mathfrak{g}$ is the differential of θ at the group unit. We have the so-called **Cartan decomposition**

$$G = K \exp \mathfrak{p}$$

of G (see [**Knap–02**, Theorem 6.31], and our Lemma 6.E.15 for the special case $G = \mathrm{SL}_2(\mathbf{R})$).

We claim that (G, K) is a Gel'fand pair. To show this, let us check that θ has the property stated in Theorem 6.B.26. Let $g = k \exp(X)$ with $k \in K$ and $X \in \mathfrak{p}$. Then $g^{-1} = \exp(-X)k^{-1}$, and so

$$\begin{aligned} \theta(g^{-1}) &= \theta(\exp(-X))k^{-1} = \exp(-d\theta_e(X))k^{-1} = \exp(X)k^{-1} \\ &= k^{-1}gk^{-1} \in KgK. \end{aligned}$$

An example of such a pair (G, K) is the pair $(SL_n(\mathbf{R}), \mathrm{SO}(n))$, for $n \geq 2$.

(5) Let T_d be a regular tree of some some degree $d \geq 3$. Let $\mathrm{Aut}(T_d)$ be its automorphism group, with the topology of the pointwise convergence, which makes it a locally compact group. Let K be the stabilizer of some vertex x_0 of T_d. Then K is compact and the action of K is transitive on every sphere $S(x_0, r)$. It follows that $(\mathrm{Aut}(T_d), K)$ is a Gel'fand pair.

(6) Let p be a prime. We are going to show that $(\mathrm{SL}_2(\mathbf{Q}_p), \mathrm{SL}_2(\mathbf{Z}_p))$ is a Gel'fand pair.

For this, recall that the so-called Bruhat–Tits tree T is the graph defined as follows (see Chapter II in [**Serr–77**]):

- the vertices of T are the equivalence classes $v = [L]$ of lattices L of $\mathbf{Q}_p^2$ for the relation $L \sim L'$ if $L' = \lambda L$ for some $\lambda \in \mathbf{Q}_p^\times$ (a lattice of $\mathbf{Q}_p^2$ is a finitely generated $\mathbf{Z}_p$-module which generates the $\mathbf{Q}_p$-vector space $\mathbf{Q}_p^2$);
- there is an edge between v_1 and v_2 if there exist representatives L_1 and L_2 of v_1 and v_2 such that $pL_1 \subset L_2 \subset L_1$.

The graph T is a regular tree of degree $p+1$, the natural action of the group $\mathrm{GL}_2(\mathbf{Q}_p)$ on T is a transitive action by isometries, and the stabilizer of $[L_0]$ is the compact subgroup $\mathrm{GL}_2(\mathbf{Z}_p)$, where L_0 is the standard lattice $\mathbf{Z}_p^2$.

We claim that $\mathrm{SL}_2(\mathbf{Z}_p)$ acts transitively on the sphere $S([L_0], k)$ for $k \in \mathbf{N}$. Indeed, let L, L' be lattices with $d([L_0], [L]) = d([L_0], [L']) = k$. Then, upon passing to equivalent lattices, we can find two $\mathbf{Z}_p$-bases $\{e_1, e_2\}$ and $\{e_1', e_2'\}$ of L_0 such that $\{e_1, p^k e_2\}$ and $\{e_1', p^k e_2'\}$ are bases of L and of L', respectively (see [**Serr–77**, Chap. II, § 1.1]). Let $g \in \mathrm{GL}_2(\mathbf{Z}_p)$ be defined by $g(e_i) = e_i'$ for $i = 1, 2$. Upon replacing e_1 by αe_1 for a suitable $\alpha \in \mathbf{Z}_p^\times$, we can assume that $g \in \mathrm{SL}_2(\mathbf{Z}_p)$. Since $g(e_1) = e_1'$ and $g(p^k e_2) = p^k g(e_2) = p^k e_2'$, we see that $g([L]) = [L']$ and the claim is proved. It follows from (1) that $(\mathrm{SL}_2(\mathbf{Q}_p), \mathrm{SL}_2(\mathbf{Z}_p))$ is a Gel'fand pair.

(7) Example (6) admits the following extension. Let $\mathbf{K}$ be a non-Archimedean local field, $\mathbb{G}$ be an algebraic group defined over $\mathbf{K}$, and $G = \mathbb{G}(\mathbf{K})$ the group of $\mathbf{K}$-points of $\mathbb{G}$, which is a locally compact group. Assume that $\mathbb{G}$ is reductive. Then G has maximal compact subgroups which are good, in the sense of Bruhat–Tits; let K be one of them. Then (G, K) is a Gel'fand pair [**BrTi–72**, 4.4.9].

For example, $(\mathrm{SL}_d(\mathbf{Q}_p), \mathrm{SL}_d(\mathbf{Z}_p))$ is a Gel'fand pair for all primes p and all integers $d \geq 2$.

REMARK 6.B.28. In the case of $(G, K) = (\mathrm{SL}_2(\mathbf{R}), \mathrm{SO}(2))$, we will see in Proposition 6.E.16 below that $C^c(K\backslash G/K)$ is part of a whole family of commutative algebras attached to (G, K).

Let G be an LC group with left Haar measure μ_G. Observe that the modular function Δ_G of G is determined by the equality

$$\int_G f(x^{-1}) d\mu_G(x) = \int_G \Delta_G(x^{-1}) f(x) d\mu_G(x) \quad \text{for all } f \in C_c(G)$$

(see [**BeHV–08**, Lemma A.3.4]).

LEMMA 6.B.29. *Let G be a locally compact group containing a compact subgroup K such that (G, K) is a Gel'fand pair.*

Then G is unimodular.

PROOF. It suffices to show that

$$(+) \qquad \int_G f(x^{-1}) d\mu_G(x) = \int_G f(x) d\mu_G(x) \quad \text{for all } f \in C_c(G).$$

We claim first that Equality (+) holds for $f \in C^c(K\backslash G/K)$. Indeed, choose $h \in C^c(K\backslash G/K)$ with $h(x) = h(x^{-1}) = 1$ for every $x \in \mathrm{supp}(f)$. Then, since $C^c(K\backslash G/K)$ is commutative, we have $f * h = h * f$ and hence

$$\begin{aligned} \int_G f(x^{-1}) d\mu_G(x) &= \int_G f(x^{-1}) h(x) d\mu_G(x) = (h * f)(e) = (f * h)(e) \\ &= \int_G h(x^{-1}) f(x) d\mu_G(x = \int_G f(x) d\mu_G(x). \end{aligned}$$

Next, we prove Equality (+) for arbitrary $f \in C_c(G)$.

Since K is compact and since $\Delta_G : G \to \mathbf{R}_+^*$ is a continuous homomorphism, we have $\Delta_G|_K = 1$. Therefore μ_G is right-invariant by K. It follows that, for every $f \in C_c(G)$, we have

$$\begin{aligned} \int_G (\mu_K * f * \mu_K)(x) \mu_G(x) &= \int_G \left(\int_K \int_K f(k_1 x k_2) d\mu_K(k_1) d\mu_K(k_2) \right) d\mu_G(x) \\ &= \int_G f(x) d\mu_G(x), \end{aligned}$$

where μ_K is the normalized Haar measure on K. Since $\mu_K * f * \mu_K$ belongs to $C^c(K\backslash G/K)$ and since $\mu_K * \check{f} * \mu_K = (\mu_K * f * \mu_K)\check{}$, Equality (+) holds for f. □

DEFINITION 6.B.30. Let G be a unimodular LC group and K a compact subgroup. Then G/K carries a G-invariant Borel measure $\mu_{G/K}$, which is unique up to a scalar multiple (see [**BeHV–08**, Corollary B.1.7]). We denote by $\lambda_{G/K}$ the **quasi-regular representation** $\lambda_{G/K}$ of G on $L^2(G/K, \mu_{G/K})$, that is, the natural representation of G given on $L^2(G/K, \mu_{G/K})$ by left translations (compare with Definition 1.F.5).

Let $(\pi, \mathcal{H})$ be a representation of an LC group G. Then there is an associated representation $C^c(G) \to \mathcal{L}(\mathcal{H})$, of the convolution algebra $C^c(G)$, denoted by π again, defined by

$$\pi(f) = \int_G f(g)\pi(g)d\mu_G(g). \quad \text{for all } f \in C^c(G)$$

More generally, let $M^b(G)$ be the Banach involutive algebra of complex Radon measures on G (see Appendix A.4). Then a $*$-representation $M^b(G) \to \mathcal{L}(\mathcal{H})$, again denoted by π, is defined by

$$\pi(\mu) = \int_G \pi(g)d\mu(g) \quad \text{for all } \mu \in M^b(G);$$

see Section 8.B below. The von Neumann algebras generated by $\pi(G)$, by $\pi(C^c(G))$, and by $\pi(M^b(G))$ all coincide (see Proposition 8.G.1).

The following criterion for Gel'fand pairs is well known (see [**Wolf–07**, Theorem 9.7.1]).

THEOREM 6.B.31. *Let G be locally compact group and K a compact subgroup. The following properties are equivalent:*

(i) *(G, K) is a Gel'fand pair;*
(ii) *the quasi-regular representation $\lambda_{G/K}$ of G on $L^2(G/K))$ is multiplicity-free.*

PROOF. We view $L^2(G/K)$ as the closed subspace of $L^2(G)$ consisting of the functions of $L^2(G)$ which are invariant under right translation by elements in K. Observe that $L^2(G/K)$ is invariant under the regular representation λ_G and that the quasi-regular representation $\lambda_{G/K}$ coincides with the subrepresentation of λ_G defined by $L^2(G/K)$.

The orthogonal projection P from $L^2(G)$ onto $L^2(G/K)$ is given $P = \lambda_G(\mu_K)$, that is,

$$P(f)(gK) = \int_K f(gk)d\mu_K(k),$$

for $f \in L^2(G)$ and μ_G-almost every $g \in G$, where μ_G is a Haar measure on G and μ_K the normalized Haar measure on K.

Let $\mathcal{M}$ be the von Neumann algebra generated by $\lambda_{G/K}(C^c(G))$ in $\mathcal{L}(L^2(G/K))$. On the one hand, we have

$$\mathcal{M} = P\mathcal{L}(G)P,$$

where $\mathcal{L}(G) = \lambda_G(C^c(G))'' \subset \mathcal{L}(L^2(G))$ is the von Neumann algebra of G. On the other hand, by Lemma 6.B.29, G is unimodular, and so

$$\mathcal{L}(G)' = \mathcal{R}(G),$$

where $\mathcal{R}(G) = \rho_G(G)''$ is the von Neumann algebra generated by the right regular representation ρ_G of G (see Theorem 10.A.10 below). By Proposition 6.B.3, we have

$$\mathcal{M}' = P\mathcal{R}(G)P,$$

and it follows that $P\rho_G(C^c(G))P$ is dense in $\mathcal{M}'$ for the strong operator topology. Since

$$\{P\rho_G(f)P \mid f \in C^c(G)\} = \{\rho_G(\mu_K * f * \mu_K) \mid f \in C^c(G)\} = \rho_G(C^c(K\backslash G/K))$$

and since ρ_G is faithful on $C^c(G)$, we see that $C^c(K\backslash G/K)$ is commutative if and only if $\mathcal{M}'$ is commutative, hence by Proposition 6.B.10 if and only if $\lambda_{G/K}$ is multiplicity-free. □

6.C. The quasi-dual of a topological group

In this section, we introduce the quasi-dual of a topological group G. We will see that this new dual space as well as the unitary dual of G carry natural Borel structures. In particular, these Borel structures will allow us to perform direct integrals decompositions of representations over the quasi-dual or the dual of G.

6.C.a. The quasi-dual. Let G be a topological group.

DEFINITION 6.C.1. The **quasi-dual** $\mathrm{QD}(G)$ of G is the space of quasi-equivalence classes of factor representations of G.

DEFINITION 6.C.2. Every irreducible representation of G is factorial, indeed factorial of type I, and two irreducible representations of G are quasi-equivalent if and only if they are equivalent (Proposition 6.A.2). Consequently, there is a well-defined injection $\kappa_{\mathrm{qd}}^{\mathrm{d}}$ from the dual $\widehat{G}$ into the quasi-dual $\mathrm{QD}(G)$, mapping the equivalence class of an irreducible representation of G to its quasi-equivalence class.

The image of the map $\kappa_{\mathrm{qd}}^{\mathrm{d}}$, which is the subspace of $\mathrm{QD}(G)$ consisting of the quasi-equivalence classes of factor representations of type I, is called the **type I part of the quasi-dual** and is denoted by $\mathrm{QD}(G)_{\mathrm{I}}$.

The map $\kappa_{\mathrm{qd}}^{\mathrm{d}}$ can be written as the composition of a bijection and an inclusion:

$$(\kappa 2) \qquad \kappa_{\mathrm{qd}}^{\mathrm{d}} : \widehat{G} \xrightarrow{\approx} \mathrm{QD}(G)_{\mathrm{I}} \subset \mathrm{QD}(G).$$

As will be seen in Section 7.A, the map $\kappa_{\mathrm{qd}}^{\mathrm{d}}$ is not surjective in general.

The quasi-dual $\mathrm{QD}(G)$ of G has a Borel structure that we now define (see also [**Dixm–C***, 3.8, 18.1]). We need two preliminary steps.

First step. The first step is to view $\mathrm{QD}(G)$ and G as quotient sets, as follows.

Let $\aleph$ be a cardinal. For every cardinal $n \le \aleph$, let $\mathcal{H}_n$ be a fixed Hilbert space of dimension n. Let $\mathcal{R}_n$ be a set of representations of G in $\mathcal{H}_n$; the main examples of interest of sets $\mathcal{R}_n$ are

- the set $\mathrm{Rep}_n(G)$ of all representations of G in $\mathcal{H}_n$,
- the subset $\mathrm{Fac}_n(G)$ of all factor representations of G in $\mathcal{H}_n$,
- and the subset $\mathrm{Irr}_n(G)$ of all irreducible representations of G in $\mathcal{H}_n$.

Let

$$\mathrm{Rep}_{\le\aleph} := \bigsqcup_{n\le\aleph} \mathrm{Rep}_n \qquad \mathrm{Fac}_{\le\aleph} := \bigsqcup_{n\le\aleph} \mathrm{Fac}_n \qquad \mathrm{Irr}_{\le\aleph} := \bigsqcup_{n\le\aleph} \mathrm{Irr}_n$$

be the corresponding disjoint unions.

Let $\aleph$ be a cardinal such that every factor representation of G is quasi-equivalent to a representation in $\mathrm{Fac}_{\le\aleph}(G)$ (see Corollary 6.B.9). Then

the quotient set $\mathrm{Fac}(G)_{\le\aleph}/\!\approx$ *can be identified with the quasi-dual* $\mathrm{QD}(G)$

(recall that $\approx$ denotes quasi-equivalence of representations).

Observe that every irreducible representation of G is equivalent to a representation in $\mathrm{Irr}_{\leq\aleph}(G)$. Then

the quotient set $\mathrm{Irr}_{\leq\aleph}(G)/\simeq$ *can be identified with the dual* $\widehat{G}$

(recall that $\simeq$ denotes equivalence of representations).

Since irreducible representations are equivalent if and only if they are quasi-equivalent, the identifications above show again that $\widehat{G}$ is naturally a subspace of $\mathrm{QD}(G)$.

Second step. We now introduce a Borel structure on the sets of representations defined above.

Let n be a cardinal and let $\mathcal{R}_n$ be as above. We first furnish $\mathcal{R}_n$ with the topology of weak uniform convergence on compact subsets, for which a net $(\pi_\nu)_\nu$ converges to π if, for every $\xi, \eta \in \mathcal{H}_n$, the net $(\langle \pi_\nu(g)\xi \mid \eta\rangle)_\nu$ converges to $\langle \pi(g)\xi \mid \eta\rangle$ uniformly on every compact subset of G. Then, we consider $\mathcal{R}_n$ with the associated Borel structure.

For a cardinal $\aleph$, the disjoint union $\mathcal{R}_{\leq\aleph} := \bigsqcup_{n\leq\aleph} \mathcal{R}_n$ is given the sum Borel structure, for which a subset B is Borel if and only if $B \cap \mathcal{R}_n$ is Borel for all $n \leq \aleph$.

DEFINITION 6.C.3. Let G be a topological group. Let $\aleph$ be a cardinal such that every factor representation of G is quasi-equivalent to a representation in $\mathrm{Fac}_{\leq\aleph}(G)$. Let $\mathrm{Fac}_{\leq\aleph}(G)$ and $\mathrm{Irr}_{\leq\aleph}(G)$ be equipped with the Borel structure introduced above.

The **Mackey–Borel structure** on the quasi-dual $\mathrm{QD}(G)$ of G is by definition the quotient Borel structure on $\mathrm{Fac}_{\leq\aleph}(G)/\approx$.

The **Mackey–Borel structure** on the dual $\widehat{G}$ of G is by definition the quotient Borel structure on $\mathrm{Irr}_{\leq\aleph}(G)/\simeq$.

REMARK 6.C.4. (1) The Mackey–Borel structure on $\widehat{G}$ is always finer than the Borel structure associated with the Fell topology (see Definition 1.C.5). Indeed, let U be an open subset of $\widehat{G}$ and let $n \leq \aleph$ be a cardinal. By definition of the Fell topology on $\widehat{G}$ and of the topology on $\mathrm{Irr}_n(G)$ introduced above, it is clear that the map $\mathrm{Irr}_n(G) \to \widehat{G}$ is continuous. This implies that the inverse image of U in $\mathrm{Irr}_n(G)$ is an open subset and hence a Borel subset of $\mathrm{Irr}_n(G)$. So, the inverse image of U in $\mathrm{Irr}_{\leq\aleph}(G)$ is a Borel subset.

However, the two Borel structures on $\widehat{G}$ considered above need not coincide, as will be seen below (Glimm's Theorem 8.F.3).

The following elementary example shows an analogous phenomenon. Define on the real line $\mathbf{R}$ an equivalence relation $\simeq$ by $x \simeq y$ if $y - x \in \mathbf{Q}$. On the quotient $\mathbf{R}/\simeq$, the quotient topology is trivial, with exactly two open subsets $\emptyset$ and $\mathbf{R}/\simeq$, and consequently the associated Borel structure is trivial. However, the quotient Borel structure is nontrivial: every countable subset B of $\mathbf{R}/\simeq$ has an inverse image in $\mathbf{R}$ which is countable, in particular Borel, and therefore B is Borel.

(2) Assume now that G is a *second-countable locally compact group.* Then we may choose for $\aleph$ the cardinality $\aleph_0$ of countable infinite sets. It is known that every point of $\mathrm{QD}(G)$ is a Borel subset of $\mathrm{QD}(G)$; see [**Dixm–C***, 7.2.4]. Moreover, $\mathrm{QD}(G)_{\mathrm{I}}$, the type I part of the quasi-dual, is a Borel subspace of $\mathrm{QD}(G)$ which is Borel isomorphic to $\widehat{G}$ [**Dixm–C***, 7.3.6].

6.C.b. Decomposition into factor representations. Let G be a second-countable locally compact group.

Recall that G may have representations with two direct integral decompositions involving *irreducible* representations belonging to two disjoint parts of $\widehat{G}$ (Theorem 1.G.11).

Given a representation π of G, one may consider direct integral decompositions of π into *factor* representations. Of course, uniqueness of such decompositions may also fail, as irreducible representations are factorial. However, as we now show, one can distinguish among such decompositions a *canonical* one.

For the notion of a direct integral of von Neumann algebras which is involved in the statement of the following result, see Section 1.I.

THEOREM 6.C.5. *Let G be a second-countable locally compact group and let π be a representation of G on a separable Hilbert space $\mathcal{H}$. Let $\mathcal{Z}$ be the centre of $\pi(G)'$.*

There exist a standard Borel space X, a σ-finite measure μ on X, a measurable field of Hilbert spaces $x \mapsto \mathcal{H}_x$ over X, a measurable field of representations $x \mapsto \pi_x$ of G in the $\mathcal{H}_x$'s over X, and an isomorphism of Hilbert spaces

$$U : \mathcal{H} \to \int_X^{\oplus} \mathcal{H}_x d\mu(x),$$

with the following properties:

(1) $U\pi(g)U^{-1} = \int_X^{\oplus} \pi_x(g)d\mu(x)$ *for all* $g \in G$;
(2) $U\mathcal{Z}U^{-1}$ *coincides with the algebra of diagonalisable operators on* $\int_X^{\oplus} \mathcal{H}_x d\mu(x)$;
(3) π_x *is factorial for* μ*-almost every* $x \in X$;
(4) *there exists a measurable subset N of X with $\mu(N) = 0$, such that π_x and π_y are disjoint for every $x, y \in X \smallsetminus N$ with $x \neq y$.*

Moreover, we have

$$U\pi(G)''U^{-1} = \int_X^{\oplus} \pi_x(G)''d\mu(x) \quad \textit{and} \quad U\pi(G)'U^{-1} = \int_X^{\oplus} \pi_x(G)'d\mu(x).$$

PROOF. The existence of (X, μ), $\int_X^{\oplus} \mathcal{H}_x d\mu(x)$, $\int_X^{\oplus} \pi_x d\mu(x)$, and U, as stated with properties (1) and (2), follows from Theorem 1.G.7. So, we may assume that $\mathcal{H} = \int_X^{\oplus} \mathcal{H}_x d\mu(x)$, $\pi = \int_X^{\oplus} \pi_x d\mu(x)$, and $\mathcal{Z}$ is the algebra of diagonalizable operators on $\mathcal{H}$.

The fields $x \mapsto \pi_x(G)''$ and $x \mapsto \pi_x(G)'$ of von Neumann algebras are measurable (see Example 1.I.2 and Proposition 1.I.3). Moreover, by Corollary 1.I.8, we have

$$\pi(G)'' = \int_X^{\oplus} \pi_x(G)''d\mu(x) \quad \text{and} \quad \pi(G)' = \int_X^{\oplus} \pi_x(G)'d\mu(x).$$

So, the last statement in the theorem holds. It remains to show that properties (3) and (4) are satisfied.

We claim that the field $x \mapsto \pi_x(G)'' \cap \pi_x(G)'$ of von Neumann algebras is measurable. Indeed, by Proposition 1.I.3, it suffices to show that the field

$$x \mapsto (\pi_x(G)'' \cap \pi_x(G)')'$$

is measurable. This is the case, since the von Neumann algebra $(\pi_x(G)'' \cap \pi_x(G)')'$ is generated by $\pi_x(G)' \cup \pi_x(G)''$ for every $x \in X$ and the fields $x \mapsto \pi_x(G)'$ and

$x \mapsto \pi_x(G)''$ are measurable. It follows that

$$\pi(G)'' \cap \pi(G)' = \int_X^{\oplus} (\pi_x(G)'' \cap \pi_x(G)')d\mu(x).$$

Since $\pi(G)'' \cap \pi(G)'$ is the centre $\mathcal{Z}$ of $\pi(G)''$ and since $\mathcal{Z}$ coincides with the algebra of diagonalisable operators on $\mathcal{H}$, it follows that $\pi_x(G)'' \cap \pi_x(G)'' = \mathbf{C}\mathrm{I}_{\mathcal{H}_x}$ for μ-almost every $x \in X$. So, property (3) holds.

To show that property (4) holds, let $T = \int_X^{\oplus} T_x d\mu(x) \in \pi(G)''$. Since T is a strong limit of a sequence in $\pi(C^c(G))$, there exists a set $N(T)$ with $\mu(N(T)) = 0$ and a sequence $(f_n)_{n\geq 1}$ in $C^c(G)$ such that $(\pi_x(f_n))_{n\geq 1}$ converges strongly to T_x for every $x \in X \smallsetminus N(T)$ (see Chap. II, § 2, Proposition 4 in [**Dixm–vN**]). Let $x, y \in X \smallsetminus N(T)$ and $R \in \mathrm{Hom}_G(\pi_x, \pi_y)$. Then

$$R\pi_x(f_n) = \pi_y(f_n)R \quad \text{for all} \quad n \geq 1$$

and it follows that $RT_x = T_y R$.

Now let $(X_n)_{n\geq 1}$ be a separating sequence of Borel subsets of X. For every $n \geq 1$, let $P^{(n)} \in \mathcal{Z}$ be the diagonalizable operator defined by $\mathbf{1}_{X_n}$, that is,

$$P^{(n)} = \int_X^{\oplus} \mathbf{1}_{X_n}(x)\mathrm{I}_{\mathcal{H}_x} d\mu(x).$$

For $N := \bigcup_{n\geq 1} N(P^{(n)})$, we have $\mu(N) = 0$, where $N(P^{(n)})$ is defined as above.

Let $x, y \in X \smallsetminus N$ with $x \neq y$ and let $R \in \mathrm{Hom}_G(\pi_x, \pi_y)$. Then, by what was seen above, we have

$$RP_x^{(n)} = P_y^{(n)}R \quad \text{for every} \quad n \geq 1.$$

Since $(X_n)_{n\geq 1}$ is separating, there exists $n \geq 1$ such that $\mathbf{1}_{X_n}(x) = 1$ and $\mathbf{1}_{X_n}(y) = 0$ and hence $P_x^{(n)} = \mathrm{I}_{\mathcal{H}_x}$ and $P_y^{(n)} = 0$. This implies that $R = 0$, so that π_x and π_y are disjoint. $\square$

Let π be a representation of the second countable LC group G and $\int_X^{\oplus} \pi_x(g)d\mu(x)$ a direct integral over a standard Borel space X as in Theorem 6.C.5. Denoting by $[\pi_x]$ the quasi-equivalence class of π_x, the map $\Phi : X \to \mathrm{QD}(G),\ \ x \mapsto [\pi_x]$ is injective; it can be seen that Φ is a Borel isomorphism between X and a Borel subset of $\mathrm{QD}(G)$. In this way, we obtain a direct integral decomposition of π over $\mathrm{QD}(G)$, as in the following theorem; for the details of the proof, we refer to [**Dixm–C***, 8.4.2, 18.7.6].

For the notion of a standard measure on a Borel space which appears in the following result, see Appendix A.3.

THEOREM 6.C.6. *Let G be a second-countable locally compact group and π a representation of G on a separable Hilbert space $\mathcal{H}$. Let $\mathcal{Z}$ be the centre of the von Neumann algebra generated by $\pi(G)$.*

There exist a standard measure μ on $\mathrm{QD}(G)$, a measurable field of Hilbert spaces $s \mapsto \mathcal{H}_s$ over $\mathrm{QD}(G)$, a measurable field of factor representations $s \mapsto (\rho_s, \mathcal{H}_s)$ over $\mathrm{QD}(G)$ such that the quasi-equivalence class of ρ_s is s for all $s \in \mathrm{QD}(G)$, and an isomorphism of Hilbert spaces $U : \mathcal{H} \to \int_{\mathrm{QD}(G)}^{\oplus} \mathcal{H}_s d\mu(s)$ with the following properties:

(1) *$U\pi(g)U^{-1} = \int_{\mathrm{QD}(G)}^{\oplus} \rho_s(g)d\mu(s)$ for all $g \in G$;*

(2) *$U\mathcal{Z}U^{-1}$ coincides with the von Neumann algebra of diagonalisable operators on $\int_{\mathrm{QD}(G)}^{\oplus} \mathcal{H}_s d\mu(s)$.*

Moreover, this decomposition is unique in the following sense: if π is equivalent to a direct integral of factor representations $\int_{\mathrm{QD}(G)}^{\oplus} \rho_s d\nu(s)$ as above with Properties (1) and (2), then μ and ν are equivalent, and (after appropriate modifications on subsets of measure zero) ρ_s and σ_s are equivalent for all $s \in \mathrm{QD}(G)$.

DEFINITION 6.C.7. Let G be a second-countable locally compact group and π a representation of G on a separable Hilbert space $\mathcal{H}$. A decomposition

$$\pi \simeq \int_{\mathrm{QD}(G)}^{\oplus} \rho_s d\mu(s)$$

as in Theorem 6.C.6 is called a **central direct integral decomposition** of π and μ is the **measure associated** to this decomposition. The class of the measure μ on $\mathrm{QD}(G)$ is uniquely determined by π.

The relation of subordination of representations admits a neat characterization in terms of measures on $\mathrm{QD}(G)$; see [**Dixm–C***, 8.4.4, 8.4.5].

PROPOSITION 6.C.8. *Let G be a second-countable locally compact group. Let π and ρ be representations of G on separable Hilbert spaces. Let μ and ν be the measures on $\mathrm{QD}(G)$ associated with central direct integral decompositions of π and ρ. The following properties are equivalent:*

(i) *π is quasi-equivalent to a subrepresentation of ρ;*
(ii) *μ is absolutely continuous with respect to ν.*

The following corollary is an immediate consequence of Proposition 6.C.8.

COROLLARY 6.C.9. *Let G, π, ρ and μ, ν be as in Proposition 6.C.8.*

(1) *The representations π are ρ are quasi-equivalent if and only if the measures μ and ν are equivalent.*
(2) *The representations π are ρ are disjoint if and only if the measures μ and ν are mutually singular.*

6.D. Groups of type I

We are now ready to introduce the central notion of groups of type I.

DEFINITION 6.D.1. A topological group G is **type I** if all its representations are type I.

The following result, which is a consequence of Proposition 6.B.13, implies that a separable group is of type I if all its representations on separable Hilbert spaces are type I.

PROPOSITION 6.D.2. *Let G be a topological group and X a dense subset of G of cardinal $\aleph$. Assume that every representation of G on a Hilbert space of dimension at most $\aleph$ is of type I.*

Then G is of type I.

PROOF. Let π be a representation of G on a Hilbert space $\mathcal{H}$. Then we can decompose π as a direct sum $\bigoplus_{i \in I} \pi_i$ of subrepresentations π_i in Hilbert spaces of dimension at most $\aleph$ (see the proof of Proposition 1.A.13 or that of Corollary 6.B.9). Therefore the claim follows from Proposition 6.B.13. $\square$

REMARK 6.D.3. Let G be a topological group of type I. The natural map $\kappa_{\rm qd}^{\rm d}$ of Definition 6.C.2 is a bijection from the dual $\widehat{G}$ onto the quasidual $\mathrm{QD}(G)$.

6.D.a. Factor representations of type I groups. The following theorem shows that in Definition 6.D.1 it suffices to require all factor representations of G to be of type I, at least when G is LC. Recall that a factor representation is of type I if and only if it is a multiple of an irreducible representation (Proposition 6.B.14).

THEOREM 6.D.4. *Let G be a second-countable locally compact group. The following properties are equivalent:*

(i) *G is of type I;*
(ii) *every factor representation of G is of type I.*

PROOF. We only have to show that (ii) implies (i).

Assume that every factor representation of G is of type I. Let π be a representation of G; we have to show that π is of type I. Since G is second-countable, and therefore separable, it suffices by Proposition 6.D.2 to consider the case of a representation π in a separable Hilbert space $\mathcal{H}$.

By Theorem 6.C.5, we may assume that $\mathcal{H} = \int_X^\oplus \mathcal{H}_x d\mu(x)$ and that $\pi = \int_X^\oplus \pi_x d\mu(x)$ for a measurable field $x \mapsto \mathcal{H}_x$ of Hilbert spaces and a measurable field $x \mapsto \pi_x$ of representations over a standard Borel space X with the following properties:

(i) π_x is factorial for μ-almost every $x \in X$;
(ii) the centre $\mathcal{Z}$ of $\pi(G)''$ is the algebra of diagonalizable operators on $\mathcal{H}$.

For every $p = 1, 2, \dots, \infty$, let

$$X_p = \{x \in X \mid \dim \mathcal{H}_x = p\},$$

and let $\mathcal{K}_p$ be a fixed Hilbert space of dimension p. By Proposition 1.G.2, we have

$$\mathcal{H} = \bigoplus_p \int_{X_p}^\oplus \mathcal{H}_x d\mu(x) \approx \bigoplus_p L^2(X_p, \mu, \mathcal{K}_p).$$

So, π is equivalent to a direct sum $\bigoplus_p \pi^{(p)}$, where $\pi^{(p)}$ is a representation on $L^2(X_p, \mu, \mathcal{K}_p)$ which is a direct integral $\int_{X_p} \pi_x^{(p)} d\mu(x)$ over X_p of representations in $\mathcal{K}_p$. By Proposition 6.B.13, we may therefore assume that $\mathcal{H} = L^2(X, \mu, \mathcal{K})$ and that π is a direct integral $\int_X \pi_x d\mu(x)$ over X of representations π_x in a fixed separable Hilbert space $\mathcal{K}$, with the properties (i) and (ii) above.

Recall from Theorem 6.C.5 that $x \mapsto \pi_x(G)''$ and $x \mapsto \pi_x(G)'$ are measurable fields of von Neumann algebras and that

$$\pi(G)'' = \int_X^\oplus \pi_x(G)'' d\mu(x) \ \text{ and } \ \pi(G)' = \int_X^\oplus \pi_x(G)' d\mu(x).$$

Let $(T^{(n)})_{n\geq 1}$ and $(S^{(n)})_{n\geq 1}$ be sequences respectively in $\pi(G)''$ and $\pi(G)'$ such that $(T_x^{(n)})_{n\geq 1}$ and $(S_x^{(n)})_{n\geq 1}$ generate $\pi_x(G)''$ and $\pi_x(G)'$ for μ-almost every x in X.

We may of course assume that $T^{(n)}$ (respectively $S^{(n)}$) has norm 1 and that ${T^{(n)}}^* \in \{T^{(m)} \mid m \geq 1\}$ (respectively, ${S^{(n)}}^* \in \{S^{(m)} \mid m \geq 1\}$) for every $n \geq 1$.

Since π_x is factorial, π_x is a multiple of an irreducible representation of G (Proposition 6.B.14), for μ-almost every $x \in X$; in particular, there exists a non-zero projection $P_x \in \pi_x(G)'$ such that the reduced algebra $P_x\pi_x(G)'P_x$ is abelian.

We are going to show that we can choose the P_x's so that the field $x \mapsto P_x$ is measurable.

The closed unit ball $\mathcal{L}_1$ of $\mathcal{L}(\mathcal{K})$, equipped with the weak operator topology, is metrizable and compact and hence a standard Borel space. Let X_1 be a Borel subset of X with $\mu(X \setminus X_1) = 0$ and such that π_x is factorial for every $x \in X_1$.

Let Ω be the set of pairs $(x, S) \in X_1 \times \mathcal{L}_1$ which satisfies the following conditions:

(1) $ST_x^{(n)} = T_x^{(n)}S$ for every $n \geq 1$,
(2) $S^2 = S = S^*$ (that is, S is a projection),
(3) $(SS_x^{(n)}S)(SS_x^{(m)}S) = (SS_x^{(m)}S)(SS_x^{(n)}S)$ for all $m, n \geq 1$,
(4) $S \neq 0$.

It is obvious that every one of these properties defines a Borel subset of $X_1 \times \mathcal{L}_1$; hence Ω is a Borel subset of $X_1 \times \mathcal{L}_1$. Moreover, for every $x \in X_1$, the set

$$\{S \in \mathcal{L}_1 \mid (x, S) \in \Omega\}$$

is not empty, by what we have seen above. It follows from von Neumann selection Theorem A.3.6 that there exists a Borel subset X_2 of X_1 with $\mu(X \setminus X_2) = 0$ and a Borel map

$$X_2 \to \mathcal{L}_1, \quad x \mapsto P_x$$

such that $(x, P_x) \in \Omega$ for every $x \in X_2$.

Let $x \in X_2$. Conditions (1), (2), (3), and (4) imply, respectively, that P_x belongs to $\pi_x(G)'$, that P_x is a projection, that $P_x\pi_x(G)'P_x$ is abelian, and that $P_x \neq 0$.

Set $P_x = 0$ for $x \in X \setminus X_2$ and let

$$P := \int_X^{\oplus} P_x d\mu(x).$$

Then P is a projection in $\pi(G)'$ such that $P\pi(G)'P$ is abelian.

We claim that the central support of P is $\mathrm{Id}_{\mathcal{H}}$. Indeed, let E be a central projection in $\pi(G)''$ with $P \leq E$. Since the centre of $\pi(G)''$ is the algebra of diagonalizable operators on $\mathcal{H}$, we have $E = \int_Y \mathbf{1}_x \mathrm{Id}_{\mathcal{K}}$ for a Borel subset Y of X. As $P_x \neq 0$ for every $x \in X_2$ and $\mu(X \setminus X_2) = 0$, it follows that $\mu(X \setminus Y) = 0$, that is, $E = \mathrm{Id}_{\mathcal{H}}$.

So, the induction map $\pi(G)'' \to P\pi(G)''P$ is an isomorphism of von Neumann algebras (see Proposition A.11.7). The commutant of $P\pi(G)''P$ coincides with $P\pi(G)'P$ (see Proposition 6.B.3); since $P\pi(G)'P$ is abelian, it follows from Proposition 6.B.12 that π is of type I. □

6.D.b. Decomposition of representations of type I groups. Recall that every representation of a second-countable LC group on a separable Hilbert space can be written, in general in several different ways, as a direct integral of irreducible representations (Theorem 1.G.11).

Let G be a second-countable LC group of type I. It follows from Theorem 6.B.17 that every representation of G is, in a canonical way, a direct sum of multiplicity-free representations. The following result gives a description of the multiplicity-free representations of G.

PROPOSITION 6.D.5. *Let G be a second-countable locally compact group of type I and π a multiplicity-free representation of G on a separable Hilbert space.*

There exists a positive Borel measure μ on $\widehat{G}$ such that π is equivalent to the direct integral $\int_{\widehat{G}}^{\oplus} \sigma d\mu(\sigma)$. Moreover, the measure μ is unique up to equivalence.

PROOF. Since G is of type I, we can identify $\mathrm{QD}(G)$ with $\widehat{G}$. Using the central decomposition of π as in Theorem 6.C.6, we can assume that

$$\pi = \int_{\widehat{G}}^{\oplus} \pi_\sigma d\mu(\sigma),$$

where $\pi_\sigma = n(\sigma)\sigma$ for some extended integer $n(\sigma)$.

Let $\mathcal{A}$ be the subalgebra of $\mathcal{L}(\mathcal{H})$ consisting of the decomposable operators of the form $T = \int_{\widehat{G}}^{\oplus} T_\sigma d\mu(\sigma)$, where $T_\sigma \in \pi_\sigma(G)'$ for every σ. It is clear that $\mathcal{A}$ is contained in $\pi(G)'$. Since π is multiplicity-free, $\pi(G)'$ and hence $\mathcal{A}$ is abelian. It follows that $\pi_\sigma(G)'$ is abelian and hence that $n(\sigma) = 1$ for μ-almost every σ. So, $\pi_\sigma = \sigma$ for μ-almost every σ.

The uniqueness of the class of μ follows from Corollary 6.C.9. □

REMARK 6.D.6. (1) The converse statement of Proposition 6.D.5 is also true: when G is a second-countable LC group of type I, every representation of the form $\int_{\widehat{G}}^{\oplus} \sigma d\mu(\sigma)$ for a positive Borel measure μ on $\widehat{G}$ is multiplicity-free (see [**Dixm–C***, 8.6.4]); so, the map

$$\mu \mapsto \int_{\widehat{G}}^{\oplus} \sigma d\mu(\sigma)$$

induces a bijection between equivalence classes of Borel measures on $\widehat{G}$ and equivalence classes of multiplicity-free representations of G.

(2) Even when G is not of type I, recall (Theorem 6.B.18) that a multiplicity-free representation of G admits an essentially unique integral decomposition into irreducible representations over a standard Borel space. The proof of Proposition 6.D.5 could have been based on this fact.

The following result is a refinement of Theorem 1.G.9 for LC groups of type I.

THEOREM 6.D.7 (**Canonical decomposition into irreducible representations**). *Let G be a second-countable locally compact group and π a representation of G in a separable Hilbert space $\mathcal{H}$. Assume that G is of type I.*

There exist a set $I \subset \overline{\mathbf{N}^*}$ of extended positive integers and a sequence $(\mu_m)_{m\in I}$ of positive measures on $\widehat{G}$ with the following properties:

(1) *the measures μ_m are singular with each other;*
(2) *the representation π is equivalent to*

$$\bigoplus_{m\in I} m \int_{\widehat{G}}^{\oplus} \sigma d\mu_m(\sigma).$$

Moreover, this decomposition is unique in the following sense: if π is equivalent to $\bigoplus_{n\in J} n \int_{\widehat{G}}^{\oplus} \sigma d\nu_n(\sigma)$ for a set J of extended integers and for mutually singular measures ν_n on $\widehat{G}$, then $I = J$ and μ_m and ν_m are equivalent for all $m \in I$.

PROOF. By Theorem 6.B.17, there exist a set $I \subset \overline{\mathbf{N}^*}$ of extended positive integers and an orthogonal decomposition $\pi = \bigoplus_{m\in I} \pi^{(m)}$ with these properties:

- for every $m \in I$, we have $\pi^{(m)} = m\sigma^{(m)}$ for a multiplicity-free representation $\sigma^{(m)}$ of G;
- $\pi^{(m)}$ and $\pi^{(n)}$ are disjoint for $m \neq n$.

By Proposition 6.D.5, for every $m \in I$, there exists a positive Borel measure μ on $\widehat{G}$ such that $\sigma^{(m)}$ is equivalent to the direct integral $\int_{\widehat{G}}^{\oplus} \sigma d\mu_m(\sigma)$ Therefore π is equivalent to

$$\bigoplus_{m \in I} m \int_{\widehat{G}}^{\oplus} \sigma d\mu_m(\sigma).$$

Moreover, μ_m and μ_n are disjoint for $m \neq n$ by Corollary 6.C.9.

The uniqueness statement follows from the uniqueness of the decomposition $\pi = \bigoplus_{m \in I} \pi^{(m)}$ of Theorem 6.B.17. □

REMARK 6.D.8. (1) In the special case of an abelian second-countable LC group, Theorem 6.D.7 boils down to Theorem 2.B.9.

(2) Theorem 6.D.7 fails for *every* LC group which is not of type I, as noted in Theorem 1.G.11.

6.E. A class of groups of type I

We introduce the important class of the so-called GCR groups. This is a class of type I locally compact groups; in fact, it will turn out later (Section 8.F) that every second-countable LC group of type I is a GCR group.

6.E.a. On the algebra of compact operators. Let $\mathcal{H}$ a Hilbert space. The subset $\mathcal{K}(\mathcal{H})$ of $\mathcal{L}(\mathcal{H})$ consisting of compact operators is a selfadjoint norm-closed two-sided ideal of $\mathcal{L}(\mathcal{H})$ and in particular a C*-algebra (for more details on C*-algebras and their representations, see Section 8.A). Observe that the identity representation $\mathrm{Id} : \mathcal{K}(\mathcal{H}) \to \mathcal{L}(\mathcal{H})$ is irreducible.

We recall a few standard facts about the representation theory of $\mathcal{K}(\mathcal{H})$; for a proof, we refer to [**Dixm–C***, 4.1]. For the definition of "nondegenerate", see Section 8.A.

PROPOSITION 6.E.1. *Consider the C*-algebra $\mathcal{K}(\mathcal{H})$ of compact operators on a Hilbert space $\mathcal{H}$.*

(1) *Every nondegenerate representation of $\mathcal{K}(\mathcal{H})$ is equivalent to a multiple of the identity representation* $\mathrm{Id} : \mathcal{K}(\mathcal{H}) \to \mathcal{L}(\mathcal{H})$;
(2) *$\mathcal{K}(\mathcal{H})$ is simple: the only closed two-sided ideals of $\mathcal{K}(\mathcal{H})$ are $\{0\}$ and $\mathcal{K}(\mathcal{H})$;*
(3) *let A be a C*-subalgebra of $\mathcal{K}(\mathcal{H})$ such that* $\mathrm{Id} : A \to \mathcal{L}(\mathcal{H})$ *is irreducible; then $A = \mathcal{K}(\mathcal{H})$.*

6.E.b. GCR groups. Let G be locally compact group, with left Haar measure μ_G. Let $(\pi, \mathcal{H})$ be a representation of G. Denote again by π the associated representation $C^c(G) \to \mathcal{L}(\mathcal{H})$ of the convolution algebra $C^c(G)$; this has already appeared just after Definition 6.B.30, and will appear again in Section 8.B below.

DEFINITION 6.E.2. A locally compact group G is a **GCR group**, or **postliminal group**, if, for every irreducible representation $(\pi, \mathcal{H})$ of G, the norm closure of $\pi(C^c(G))$ in $\mathcal{L}(\mathcal{H})$ contains a *nonzero* compact operator.

REMARK 6.E.3. Let G be a locally compact group.

(1) Assume that G is a GCR group and let $(\pi, \mathcal{H})$ be an irreducible representation of G. Then $\mathcal{K}(\mathcal{H})$ is contained in the norm closure of $\pi(C^c(G))$.

Indeed, denote by A the norm closure of $\pi(C^c(G))$ and set $J = A \cap \mathcal{K}(\mathcal{H})$. Then J is a closed two-sided ideal of the C*-subalgebra A of $\mathcal{L}(\mathcal{H})$, and $J \neq \{0\}$ because G is a GCR group. As $\mathrm{Id} : A \to \mathcal{L}(\mathcal{H})$ is irreducible, it follows easily that $\mathrm{Id} : J \to \mathcal{L}(\mathcal{H})$ is irreducible [**Dixm–C***, 2.11.3]. Therefore we have $J = \mathcal{K}(\mathcal{H})$, by Proposition 6.E.1.

(2) We introduce below the class of GCR C*-algebras. It turns out that G is a GCR group if and only if its maximal C*-algebra as defined in Section 8.B is GCR, at least when G is second-countable; see Section 8.F.

(3) Section 10.D we will define the notion of a **normal** representation of G. Then G is a CCR group if and only if every irreducible representation of G is normal (Proposition 10.D.3).

In several places we will need the following characterization of weak containment of representations of G in terms of operator norms. It will be discussed later: see Proposition 8.B.4 and Remark 8.B.6.

PROPOSITION 6.E.4. *Let G be a locally compact group and π_1, π_2 two representations of G. The following properties are equivalent:*

(i) *π_1 is weakly contained in π_2;*
(ii) *for every $f \in C^c(G)$, we have $\Vert\pi_1(f)\Vert \leq \Vert\pi_2(f)\Vert$.*

The importance of the class of GCR groups lies in the following result.

THEOREM 6.E.5. *Let G be a second-countable locally compact GCR group. Then G is of type I.*

PROOF. In view of Theorem 6.D.4, it suffices to show that every factor representation of G is of type I.

Let $(\pi, \mathcal{H}_\pi)$ be a factor representation of G on a Hilbert space $\mathcal{H}_\pi$. Since G is second-countable, π is weakly equivalent to an *irreducible* representation $(\sigma, \mathcal{H}_\sigma)$ of G, as we will show later (Corollary 7.F.2). Let A be the norm closure of $\sigma(C^c(G))$ in $\mathcal{L}(\mathcal{H}_\sigma)$. Since G is a GCR group, we have

$$\mathcal{K}(\mathcal{H}_\sigma) \subset A,$$

by Remark 6.E.3(1). As π is weakly equivalent to σ, we have

$$\Vert\pi(f)\Vert = \Vert\sigma(f)\Vert \quad \text{for every} \quad f \in C^c(G)$$

(see Proposition 6.E.4). This implies that the map

$$\sigma(C^c(G)) \to \pi(C^c(G)), \quad \sigma(f) \mapsto \pi(f)$$

is well-defined and extends to an isometric $*$-homomorphism $\Phi : A \to \mathcal{L}(\mathcal{H}_\pi)$, that is, to a faithful representation Φ of A on $\mathcal{H}_\pi$.

By Proposition 6.E.1, the restriction of Φ to $\mathcal{K}(\mathcal{H}_\sigma)$ is a multiple of the identity representation of $\mathcal{K}(\mathcal{H}_\sigma)$ on $\mathcal{H}_\sigma$; this means that there exist a cardinal $n \geq 1$ and a Hilbert space isomorphism $U : \mathcal{H}_\pi \to \mathcal{H}_\sigma \otimes \mathcal{H}$ such that

$$U\Phi(T)U^{-1} = T \otimes \mathrm{Id}_{\mathcal{H}} \quad \text{for every} \quad T \in \mathcal{K}(\mathcal{H}_\sigma),$$

where $\mathcal{H}$ is a Hilbert space of dimension n.

Observe that $\mathcal{K}(\mathcal{H}_\sigma)$ is dense in $\mathcal{L}(\mathcal{H}_\sigma)$ and hence in A, for the strong operator topology. It follows that we have

$$U\Phi(T)U^{-1} = T \otimes \mathrm{Id}_{\mathcal{H}} \quad \text{for every} \quad T \in A,$$

and therefore

$$U\pi(f)U^{-1} = \sigma(f) \otimes \mathrm{Id}_{\mathcal{H}} \text{ for every } f \in C^c(G).$$

Let μ_G be a Haar measure on G and $(f_n)_n$ be an approximate identity for $L^1(G, \mu_G)$ consisting of functions $f_n \in C^c(G)$ (see Section A.10). Then, for every $g \in G$, we have

$$\lim_n \pi({}_g f_n) = \pi(g) \quad \text{and} \quad \lim_n \sigma({}_g f_n) = \sigma(g)$$

in the strong operator topology, where ${}_g f_n$ is defined by ${}_g f_n(x) = f_n(g^{-1}x)$. It follows that

$$U\pi(g)U^{-1} = \sigma(g) \otimes \mathrm{Id}_{\mathcal{H}} \text{ for every } g \in G.$$

This shows that π is equivalent to $n\sigma$. Therefore π is of type I. □

REMARK 6.E.6. Theorem 6.E.5 admits a converse:

- every second-countable locally compact group of type I is a GCR group.

This is a deep result which is valid more generally in the context of separable C*-algebras and which is part of Glimm's Theorem, to be discussed in Section 8.F

6.E.c. CCR groups. We introduce an important subclass of the class of GCR groups.

DEFINITION 6.E.7. Let G be a locally compact group.

A representation $(\pi, \mathcal{H})$ of G is said to be a **CCR representation** if the algebra $\pi(C^c(G))$ is contained in $\mathcal{K}(\mathcal{H})$.

The group G is a **CCR group**, or **liminal group**, if every irreducible representation of G is a CCR representation.

Concerning the meaning of the acronyms GCR and CCR, see Remark 8.F.2.

REMARK 6.E.8. Let G be a locally compact group.

(1) Every finite-dimensional representation of G is CCR.

(2) In the case of a discrete group Γ, an infinite-dimensional representation π of Γ is not necessarily CCR. Indeed, $\pi(e) = I$ is not a compact operator.

(3) If G is a CCR group, then obviously G is a GCR group.

(4) Let $(\pi, \mathcal{H})$ be an *irreducible* CCR representation of G. Then $\pi(C^c(G))$ is dense in $\mathcal{K}(\mathcal{H})$ in the operator norm. This follows from Remark 6.E.3(1).

(5) Let $(\pi, \mathcal{H})$ be an *irreducible* CCR representation of G. Then $\{\pi\}$ is a closed point of the dual space $\widehat{G}$, equipped with the Fell topology as defined in Section 1.A (see Corollary 9.A.4 below).

As the next example shows, the class of GCR groups is strictly larger than the class of CCR groups.

EXAMPLE 6.E.9. Let $G = \mathrm{Aff}(\mathbf{R})$ be the affine group over the real line. We view G as the space $\mathbf{R}^\times \times \mathbf{R}$, equipped with the group law given by $(a, b)(a', b') = (aa', ab'+b)$, and with the natural topology. It is a second-countable locally compact group, which has a left Haar measure $dadb/a^2$. Recall from Remark 3.C.7 that, up to equivalence, G has a unique infinite-dimensional irreducible representation π defined on $L^2(\mathbf{R}^\times, dt/|t|)$ by

$$\pi(a, b)\xi(t) = e^{-2\pi i b t}\xi(at) \quad \text{for } (a, b) \in G,\ \xi \in L^2(\mathbf{R}^\times, dt/|t|),\ t \in \mathbf{R}^\times.$$

Let $f \in C^c(G)$. It can be checked (see [**Khal–74**, Corollaire p. 164]) that $\pi(f)$ is a compact operator if and only if

$$\int_{\mathbf{R}} f(a,b)db = 0 \text{ for every } a \in \mathbf{R}^\times.$$

In particular, G is a GCR group but not a CCR group. Since $\{\pi\}$ is not closed in $\widehat{G}$ as mentioned in Remark 3.C.7, the fact that G is not a CCR group is also an immediate consequence of Corollary 9.A.4 below.

Before we proceed with examples of CCR groups, we recall the following facts about the representation theory of compact groups. Let K be a compact group.

- Every irreducible representation of K is finite-dimensional.
- For $\sigma \in \widehat{K}$, let χ_σ be the character of σ, that is, the continuous function

  $$K \to \mathbf{C}, \quad x \mapsto \mathrm{Tr}(\sigma(x)).$$

  For $\sigma, \sigma' \in \widehat{K}$, we have the following "orthogonality relations":

  $$\chi_\sigma * \chi_{\sigma'} = 0 \quad \text{if } \sigma \neq \sigma' \quad \text{and} \quad \chi_\sigma * \chi_\sigma = d_\sigma^{-1}\chi_\sigma,$$

  where $*$ denotes the convolution product and d_σ the dimension of σ.
- Every representation π of K decomposes as direct sum $\pi = \bigoplus_{\sigma \in \widehat{K}} n_\sigma \sigma$; more precisely, if $(\pi, \mathcal{H})$ is a representation of K, we have a decomposition

  $$\mathcal{H} = \bigoplus_{\sigma \in \widehat{K}} \mathcal{H}_\sigma$$

  into isotypical components ($\mathcal{H}_\sigma$ is the sum of irreducible subspaces on which π is equivalent to σ) and the orthogonal projection $\mathcal{H} \to \mathcal{H}_\sigma$ is the operator $\pi(d_\sigma \overline{\chi_\sigma})$.

For all this, see for instance [**Dixm–C***, Chap. 15] or [**Foll–16**, Chap. 5].

Important examples of CCR groups are groups which contain large compact subgroups in the sense of the following definition which appears in [**Mack–63**, p. 641] (see also [**Warn–72**, Chap. 4], where a stronger notion is considered).

Definition 6.E.10. Let G be a topological group. A compact subgroup K is a **large compact subgroup** of G if every $\pi \in \widehat{G}$ has finite K-multiplicities, that is, in the decomposition $\pi|_K = \bigoplus_{\sigma \in \widehat{K}} n_\sigma \sigma$, we have $n_\sigma < \infty$ for every $\sigma \in \widehat{K}$.

Proposition 6.E.11. *Let G be a locally compact group which contains a large compact subgroup K.*

Then G is a CCR group. In particular, G is of type I.

Proof. Let $(\pi, \mathcal{H})$ be an irreducible representation of G. Let

$$\mathcal{H} = \bigoplus_{\sigma \in \widehat{K}} \mathcal{H}_\sigma$$

be the decomposition of $\mathcal{H}$ into K-isotypical components. Since K is a large compact subgroup, the restriction of π to K has finite multiplicities, and this means that $\mathcal{H}_\sigma$ is finite-dimensional for every $\sigma \in \widehat{K}$; equivalently, the operator $\pi(d_\sigma \overline{\chi_\sigma})$, which is the orthogonal projection on $\mathcal{H}_\sigma$, is of finite rank for every $\sigma \in \widehat{K}$.

Let μ_G be a Haar measure on G. Let $f \in C^c(G)$. Viewing f as a vector in $L^2(G, \mu_G)$ and considering the restriction to K of the left regular representation λ_G of G, we have

$$f = \sum_{\sigma \in \widehat{K}} \lambda_G(d_\sigma \overline{\chi_\sigma})(f) = \sum_{\sigma \in \widehat{K}} d_\sigma \overline{\chi_\sigma} * f,$$

where the sum converges in $L^2(G)$. Since the support of $\overline{\chi_\sigma} * f$ is contained in the compact set $K\mathrm{supp}(f)$, the sum $f = \sum_{\sigma \in \widehat{K}} d_\sigma \overline{\chi_\sigma} * f$ also converges in $L^1(G, \mu_G)$. Therefore, we have

$$\pi(f) = \pi\left(\sum_{\sigma \in \widehat{K}} d_\sigma \overline{\chi_\sigma} * f\right) = \sum_{\sigma \in \widehat{K}} \pi(d_\sigma \overline{\chi_\sigma})\pi(f),$$

where the last sum converges in the operator norm. Since $\pi(d_\sigma \overline{\chi_\sigma})$ is of finite rank, it follows that $\pi(f)$ is a compact operator. □

Here is an important class of LC groups with a large compact subgroup.

THEOREM 6.E.12. *Let G be a reductive linear algebraic group over a local field. Let K be a maximal compact subgroup of G.*

Then K is a large compact subgroup and so G is a CCR group.

For a proof of Theorem 6.E.12, see [**Gode–52**, Theorem 2] or [**Dixm–C***, 15.5.6] in the real case (see also [**Hari–53**, Page 230]), and [**Bern–74**, Theorem 1] in the non-Archimedean case. We will give below a complete proof of Theorem 6.E.12 in the special case $G = \mathrm{SL}_2(\mathbf{R})$. In the case of a linear semisimple real Lie group, the proof follows a similar strategy (see Remark 6.E.18), but is more involved, because K is not necessarily abelian.

EXAMPLE 6.E.13. (1) Another class of examples of groups containing a large compact subgroup is the motion groups (see Example 6.B.25). More generally, assume that the LC compact group G can be written as the product $G = KA$ of a compact subgroup K and a closed abelian subgroup A.

Then K is a large compact subgroup and so G is a CCR group (see [**Gode–52**, Theorem 5]).

(2) Every connected nilpotent Lie group is a CCR group (see [**Dixm–59**, **Kiri–62**]). For a characterization of connected and simply connected solvable Lie groups that are CCR, see Chapter V in [**AuMo–66**].

6.E.d. $\mathrm{SL}_2(\mathbf{R})$ is a CCR group. Let $G = \mathrm{SL}_2(\mathbf{R})$. We are going to show that the maximal compact subgroup $K = \mathrm{SO}(2)$ is a large subgroup of G, and therefore that G is a CCR group (see Proposition 6.E.11).

Recall that $K = \mathrm{SO}(2)$ is the group of matrices of the form

$$r(\theta) = \begin{pmatrix} \cos\theta & \sin\theta \\ -\sin\theta & \cos\theta \end{pmatrix}$$

for $\theta \in \mathbf{R}$. The dual group of the abelian group K is

$$\widehat{K} = \{\chi_n \mid n \in \mathbf{Z}\} \cong \mathbf{Z},$$

where χ_n is defined by

$$\chi_n(r(\theta)) = e^{in\theta} \qquad \text{for every } \theta \in \mathbf{R}.$$

For $n \in \mathbf{Z}$, set

$$C^c(G)_{\chi_n} = \chi_{-n} * C^c(G) * \chi_{-n} = \{\chi_{-n} * f * \chi_{-n} \mid f \in C^c(G)\}.$$

(For the convolutions $\chi_{-n} * f$ and $*f * \chi_{-n}$, see Remark 6.B.23.)

For a function $f : G \to \mathbf{C}$, define f^* by $f^*(x) = \overline{f(x^{-1})}$.

LEMMA 6.E.14. *For every $n \in \mathbf{Z}$, the following holds:*

(1) *$C^c(G)_{\chi_n}$ is a subalgebra of the convolution algebra $C^c(G)$ and is invariant under the involution $f \mapsto f^*$;*

(2) *$C^c(G)_{\chi_n}$ is the linear subspace of $C^c(G)$ consisting of the functions $f \in C^c(G)$ satisfying the equation*

$$f(r(\theta_1)xr(\theta_2)) = \chi_{-n}(r(\theta_1))f(x)\chi_{-n}(r(\theta_2)) = e^{-in\theta_1}f(x)e^{-in\theta_2}$$

for all $\theta_1, \theta_2 \in \mathbf{R}$ and $x \in G$;

PROOF. Item (1) follows from the relations

$$\chi_{-n} * \chi_{-n} = \chi_{-n} \text{ and } \chi_{-n}^* = \chi_{-n}.$$

Item (2) can be checked by a direct computation. □

Observe that, when $n = 0$, the algebra $C^c(G)_{\chi_n}$ is the algebra $C^c(K\backslash G/K)$ of K-invariant functions in $C^c(G)$ considered in Subsection 6.B.e.

The Cartan decomposition of G (see Example 6.B.27) can be further refined, as follows.

LEMMA 6.E.15. *Let A be the subgroup of diagonal matrices with positive diagonal entries in $G = \mathrm{SL}_2(\mathbf{R})$.*

Then $G = KAK$.

PROOF. Let $x \in G$. Denote by x^t the transpose of the matrix x. Then xx^t is a symmetric positive definite matrix. So, there exists a symmetric positive definite matrix y with $y^2 = x^t x$. Then $xy^{-1} \in K$, since

$$(xy^{-1})^t xy^{-1} = (y^{-1})^t x^t xy^{-1} = y^{-1}y^2y^{-1} = I.$$

So, $x = k_1 y$ for some $k_1 \in K$. (This is the polar decomposition of x as in Appendix A.8.) Moreover, since y is symmetric and positive definite, there exists $k_2 \in K$ such that $a := k_2^{-1}yk_2$ belongs to A. Therefore $x = (k_1k_2)a(k_2^{-1})$. □

There is a corresponding Cartan decomposition $G = KAK$ for any semisimple connected real Lie group G (see for example [**Knap–02**, Theorem 7.39]).

Since (G, K) is a Gel'fand pair (Example 6.B.27), the algebra $C^c(K\backslash G/K)$ is commutative. We now generalize this property to all algebras $C^c(G)_{\chi_n}$.

PROPOSITION 6.E.16. *The algebra $C^c(G)_{\chi_n}$ is commutative for every $n \in \mathbf{Z}$.*

PROOF. The proof bears similarities with the proof of Gel'fand's Theorem 6.B.26. Following the proof of Theorem 1 in [**Lang–85**, Chap. II, § 1], we consider

- the anti-automorphism τ of G of order 2 given by $x \mapsto x^t$ and
- the automorphism σ of G of order 2 given by $x \mapsto x_0 x x_0^{-1}$, where

$$x_0 = \begin{pmatrix} 1 & 0 \\ 0 & -1 \end{pmatrix}.$$

Observe that τ and σ both act as the identity on the subgroup A of diagonal matrices with positive diagonal entries. Moreover, τ and σ coincide on K: for $k = r(\theta) \in K$, we have

$$\tau(k) = k^t = k^{-1}$$

and

$$\sigma(k) = x_0 r(\theta) x_0^{-1} = r(-\theta) = k^{-1}.$$

For $f \in C^c(G)$, define $f^\tau \in C^c(G)$ by

$$f^\tau(x) = f(\tau(x)) \quad \text{for } x \in G,$$

and similarly define $f^\sigma \in C^c(G)$.

The arguments of the proof of Theorem 6.B.26 show that τ and σ preserve the Haar measure μ_G on G (recall that G is unimodular). These arguments also show that, for every $f_1, f_2 \in C^c(G)$, we have

$$(f_1 * f_2)^\tau = f_2^\tau * f_1^\tau$$

(since τ is an anti-automorphism of G) and

$$(f_1 * f_2)^\sigma = f_1^\sigma * f_2^\sigma$$

(since σ is an automorphism of G).

Let $n \in \mathbf{Z}$ and $f \in C^c(G)_{\chi_n}$. We claim that $f^\tau = f^\sigma$. Indeed, let $x \in G$; write $x = k_1 a k_2$ with $k_1, k_2 \in K$ and $a \in A$ (see Lemma 6.E.15). On the one hand, we have

$$f^\tau(x) = f(k_2^t a^t k_1^t) = f(k_2^{-1} a k_1^{-1}) = \chi_n(k_2)\chi_n(k_1) f(a);$$

on the other hand, we have

$$f^\sigma(x) = f(\sigma(k_1)\sigma(a)\sigma(k_2)) = f(k_1^{-1} a k_2^{-1}) = \chi_n(k_1)\chi_n(k_2) f(a).$$

Now, let $f_1, f_2 \in C^c(G)_{\chi_n}$. Then $(f_1 * f_2)^\tau = (f_1 * f_2)^\sigma$ and hence

$$f_2^\tau * f_1^\tau = f_1^\sigma * f_2^\sigma,$$

that is,

$$f_2^\sigma * f_1^\sigma = f_1^\sigma * f_2^\sigma,$$

Therefore,

$$f_2 * f_1 = (f_2^\sigma * f_1^\sigma)^\sigma = (f_1^\sigma * f_2^\sigma)^\sigma = f_1 * f_2,$$

and this concludes the proof. □

We can now state the main result of this subsection.

THEOREM 6.E.17. *The compact subgroup* $\mathrm{SO}(2)$ *is a large subgroup of* $\mathrm{SL}_2(\mathbf{R})$. *Consequently,* $G = \mathrm{SL}_2(\mathbf{R})$ *is a CCR group and is therefore of type I.*

PROOF. We write G for $\mathrm{SL}_2(\mathbf{R})$ and K for $\mathrm{SO}(2)$. Let $(\pi, \mathcal{H})$ be an irreducible representation of G. Let

$$\mathcal{H} = \bigoplus_{n \in \mathbf{Z}} \mathcal{H}_n$$

be the decomposition of $\mathcal{H}$ into K-isotypical components $\mathcal{H}_n := \mathcal{H}_{\chi_n}$. We will show that $\mathcal{H}_n$ is finite-dimensional, in fact at most one dimensional, for every $n \in \mathbf{Z}$.

Let $n \in \mathbf{Z}$ be fixed in the sequel.

- *First step.* We claim that $\mathcal{H}_n$ is invariant under $\pi(C^c(G)_{\chi_n})$.

Indeed, recall that the orthogonal projection $P_n : \mathcal{H} \to \mathcal{H}_n$ is the operator $\pi(\overline{\chi_n}) = \pi(\chi_{-n})$. Let $f \in C^c(G)_{\chi_n}$. Then $f = \chi_{-n} * f * \chi_{-n}$ and hence

$$\pi(f) = \pi(\chi_{-n} * f * \chi_{-n}) = \pi(\chi_{-n})\pi(f)\pi(\chi_{-n}) = P_n\pi(f)P_n.$$

This shows that $\mathcal{H}_n$ is invariant under $\pi(f)$.

As a result, we have a representation

$$\pi_n : C^c(G)_{\chi_n} \to \mathcal{L}(\mathcal{H}_n), \quad f \mapsto \pi(f)|_{\mathcal{H}_n}.$$

Recall that $C^c(G)_{\chi_n}$ is invariant under the involution $f \mapsto f^*$; therefore, π_n is a $*$-homomorphism.

• *Second step.* We claim that the commutant $\pi_n(C^c(G)_{\chi_n})'$ of the selfadjoint subalgebra $\pi_n(C^c(G)_{\chi_n})$ of $\mathcal{L}(\mathcal{H}_n)$ consists only of scalars operators.

Indeed, it suffices to show that the von Neumann algebra $\pi_n(C^c(G)_{\chi_n})''$ generated by $\pi_n(C^c(G)_{\chi_n})$ coincides with $\mathcal{L}(\mathcal{H}_n)$. Let $T \in \mathcal{L}(\mathcal{H}_n)$ and set

$$\widetilde{T} := TP_n \in \mathcal{L}(\mathcal{H}).$$

Since π is irreducible, the von Neumann algebra generated by $\pi(C^c(G))$ coincides with $\mathcal{L}(\mathcal{H})$. Therefore there exists a net $(f_i)_{i\in I}$ in $C^c(G)$ such that $\lim_i \pi(f_i) = \widetilde{T}$ in the strong (or weak) operator topology. Then $\chi_{-n} * f_i * \chi_{-n} \in C^c(G)_{\chi_n}$ and

$$\lim_i \pi(\chi_{-n} * f_i * \chi_{-n}) = \lim_i (P_n\pi(f_i)P_n) = P_n \lim_i \pi(f_i)P_n = P_n\widetilde{T}P_n = P_nTP_n.$$

Therefore

$$\lim_i \pi_n(\chi_{-n} * f_i * \chi_{-n}) = T$$

and the claim is proved.

• *Third step.* We claim that π_n is an irreducible representation of $C^c(G)_{\chi_n}$.

Indeed, let $\mathcal{K}$ be a $\pi_n(C^c(G)_{\chi_n})$-invariant closed subspace of $\mathcal{H}_n$ and let

$$P : \mathcal{H}_n \to \mathcal{K}$$

be the orthogonal projection. Let $f \in C^c(G)_{\chi_n}$. Then, $P\pi_n(f)P = \pi_n(f)P$ and therefore

$$P\pi_n(f) = (\pi_n(f^*)P)^* = (P\pi_n(f^*)P)^* = P\pi_n(f)P,$$

that is, $\pi_n(f)P = P\pi_n(f)$. So, $P \in \pi_n(C^c(G)_{\chi_n})'$. It follows from the second step that P is a scalar operator, that is, $\mathcal{K} = \{0\}$ or $\mathcal{K} = \mathcal{H}_n$.

• *Fourth step.* The dimension of $\mathcal{H}_n$ is at most one.

Indeed, $\pi_n(C^c(G)_{\chi_n})'$ consists only of scalar operators, by the second step. Since π_n is irreducible (third step) and $C^c(G)_{\chi_n}$ is commutative (Proposition 6.E.16), it follows that $\dim \mathcal{H}_n \leq 1$. □

REMARK 6.E.18. Parts of the proof of Theorem 6.E.17 are valid in a much greater generality.

Indeed, let G be a unimodular LC compact and K a compact subgroup of G. For $\sigma \in \widehat{K}$, let $\psi_\sigma = d_\sigma\overline{\chi_\sigma}$, where χ_σ is the character of σ and d_σ its dimension. Define

$$C^c(G)_\sigma = \psi_\sigma * C^c(G) * \psi_\sigma.$$

Then $C^c(G)_\sigma$ is a subalgebra of the convolution algebra $C^c(G)$ and is invariant under the involution $f \mapsto f^*$.

Let $(\pi, \mathcal{H})$ be an irreducible representation of G and let

$$\mathcal{H} = \bigoplus_{\sigma \in \widehat{K},} \mathcal{H}_\sigma$$

be the decomposition of $\mathcal{H}$ into K-isotypical components.

The arguments of the three first steps in the proof of Theorem 6.E.17 carry over, with the obvious changes, and show that the following hold:

- $\mathcal{H}_\sigma$ is invariant under $\pi(C^c(G)_\sigma)$ and so defines a $*$-representation π_σ of $C^c(G)_\sigma$;
- the commutant of $\pi_\sigma(C^c(G)_\sigma)$ in $\mathcal{L}(\mathcal{H}_\sigma)$ consists only of scalar operators;
- $(\pi_\sigma, \mathcal{H}_\sigma)$ is an irreducible representation of $C^c(G)_\sigma$.

Assume now that G is a linear semisimple real Lie group and K a maximal compact subgroup of G. Using the fact that G has a decomposition $G = KS$, where S is a closed solvable subgroup ("Iwasawa decomposition"), one shows (see [**Dixm–C***, 15.5.5] or [**Gode–52**, Theorem 2]) that the last step in the proof of Theorem 6.E.17 remains valid in the following form:

- the dimension of $\mathcal{H}_\sigma$ is at most d_σ.

On the class of groups of type I. We give a nonexhaustive list of topological groups which are known to be of type I.

THEOREM 6.E.19 (Some locally compact groups of type I). *The following locally compact groups are of type I:*

(1) *Second-countable locally compact abelian groups.*
(2) *Compact groups.*
(3) *Nilpotent connected locally compact groups, and more generally connected-by-compact locally compact groups whose solvable radical is actually nilpotent.*
(4) *Reductive algebraic groups over a local field.*
(5) *Linear algebraic groups over a local field of characteristic* 0.
(6) *Adelic reductive groups of the form* $G = \mathbf{G}(\mathbf{A})$, *where* $\mathbf{G}$ *is a reductive algebraic group defined over* $\mathbf{Q}$ *and* $\mathbf{A}$ *the adele ring of the field* $\mathbf{Q}$.
(7) *Some solvable locally compact groups, e.g., all solvable connected real Lie groups that are of dimensions at most 4, and all solvable connected real Lie groups for which the exponential map is surjective.*
(8) *Full groups of automorphisms of Bruhat–Tits trees.*

REFERENCES FOR PROOFS AND COMMENTS. Claim (1) follows from Corollary 6.B.22 and Proposition 6.D.2; Claim (2) is straightforward.

For (3) see [**Dixm–59**] and [**Lips–72**]. For (4), see Theorem 6.E.12.

For Claim (5), see [**Dixm–57**] in the case where the local field is $\mathbf{R}$ and [**BeEc–20**] in general.

Claim (6) follows from [**GGPS–69**, Chap. III, § 3, no. 3], combined with our (4). See also [**Moor–65**, Section 6].

In contrast to this and in relation with (3), it is shown in [**Moor–65**, beginning of Section 7] that, for a nilpotent algebraic group $\mathbf{G}$ defined over $\mathbf{Q}$, the adelic group $\mathbf{G}(\mathbf{A})$ is type I if and only if $\mathbf{G}$ is abelian.

Concerning (7), see [**Take–57**] for groups of which the exponential map is surjective. A necessary and sufficient condition for a solvable connected simply connected Lie group to be of type I is established in [**AuKo–71**], in terms of coadjoint orbits.

For (8), see [**Ol's–77**] and [**Ol's–80**]. For other groups of automorphisms of trees, see [**HoRa–19**]. □

THEOREM 6.E.20 (Some more topological groups of type I). *The following topological groups are type I:*

1. *Abelian topological groups.*
2. *The groups* $\mathcal{U}_\infty(\mathcal{H})$ *and* $\mathcal{U}(\mathcal{H})_{\mathrm{str}}$*, as defined above in Example* 3.G.1.
3. *The group* $\mathrm{Sym}(\mathbf{N})$ *of permutations of* $\mathbf{N}$ *with the topology of simple convergence.*
4. *The group* $\mathrm{Aut}(\mathbf{Q}, \le)$ *of automorphisms of the ordered set of the rationals with the topology of simple convergence.*
5. *The group* $\mathrm{Homeo}(2^{\mathbf{N}})$ *of homeomorphisms of the Cantor space with the compact-open topology.*

REFERENCES FOR PROOFS AND COMMENTS. Claim (1) is a consequence of the (nonobvious) fact that abelian von Neumann algebras are of type I (see Proposition 7.C.1).

For Claim (2), see [**Kiri–73**], [**Ol's–78**], and 3.G above. For a discussion of the two latter articles and further results on representations of unitary groups, see [**Neeb–14**].

For (3), see [**Lieb–72**, Theorems 3 & 4]. More generally, consider a countable infinite set X, the Polish group $\mathrm{Sym}(X)$ of permutations of X, and a subgroup G of $\mathrm{Sym}(X)$; assume that G is oligomorphic, that is, for all $n \ge 1$ the diagonal action of G on X^n has a finite number of orbits. Then G has countably many equivalence classes of irreducible representations, and every representation of G is a direct sum of irreducible representations (see [**Tsan–12**]). In particular, G is of type I. This applies to the groups $\mathrm{Aut}(\mathbf{Q}, \le)$ and $\mathrm{Homeo}(2^{\mathbf{N}})$ of (4) and (5). □

We mention a list of operations under which the class of type I groups is stable.

PROPOSITION 6.E.21 (Stability properties of type I groups).

1. *Open subgroups of type I second-countable locally compact groups are type I. For a second-countable locally compact group G and a closed subgroup H of finite index, G is type I if and only if H is type I.*
2. *Quotients of type I groups are type I: if G is a type I topological group and N a closed normal subgroup, then G/N is type I.*
3. *A Cartesian product of a finite number of type I LC groups is type I.*
4. *Let $G = \varprojlim G_\alpha$ be a projective limit of locally compact groups $G_\alpha = G/K_\alpha$, with K_α a compact normal subgroup of G for all α; then the locally compact group G is type I if and only if each G_α is type I.*
5. *Let G be a topological group and H a closed subgroup such that G/H is locally compact and possesses a nonzero, finite, G-invariant Radon measure; if H is type I, G is type I.*

   In particular, (type I)-by-compact topological groups are type I, and a fortiori abelian-by-compact topological groups are type I.

REFERENCES FOR PROOFS. For (1), see Proposition 2.4 and Corollary 2.5 in [**Kall–73**]. Fact (2) is a straightforward consequence of the definitions. For (3), see, e.g., [**Mack–76**, Theorem 3.2]. Claim (4) follows from the definitions and from Proposition 8.D.3 below (see also the remark following Proposition 2.2 in [**Moor–72**]). Claim (5) is Theorem 1 in [**Kall–70**]. □

REMARK 6.E.22. The class of type I groups is not stable under central extensions, not even under finite central extensions.

Indeed, here is an example in the class of countable discrete groups. Let p be an odd prime, $\mathbf{F}_p$ the field of order p, and let $H_\infty(\mathbf{F}_p)$ be the "Heisenberg group" as defined in Example 7.D.7 below. Then $H_\infty(\mathbf{F}_p)$ is a finite central cover of an abelian group. As $H_\infty(\mathbf{F}_p)$ is not virtually abelian, $H_\infty(\mathbf{F}_p)$ is not of type I, by Thoma's Theorem 7.D.1 below.

For an example of a connected solvable Lie group of type I whose universal covering group is not of type I, see [**Dixm–61**].

CHAPTER 7

Non type I groups

There are large classes of groups which are not of type I. Following Murray and von Neumann, we first show in Section 7.A *that the regular representation of a discrete icc group is a factor representation which is not of type I.*

Recall from Proposition 6.B.12 *that a type I representation* π *of a group* G *is characterized by a property of the von Neumann algebra* $\pi(G)''$. *Von Neumann algebras are subdivided into various types, the so-called types* I_{f}, I_∞, II_1, II_∞, *and III. We recall this type of classification in Section* 7.B, *after a short reminder on traces on operator algebras. This leads to Section* 7.C, *where we discuss the definition of group representations of corresponding types. Historically, types were first defined by Murray and von Neumann in 1936, before being imported in the theory of unitary group representations, notably by Mautner, Godement, Segal, and Mackey (see* [**Maut–50a**], [**Maut–50b**], [**Gode–51a**], [**Sega–51**], [**Mack–53**]*).*

We discuss several groups that are not of type I. In particular, we state Thoma's characterization of discrete groups of type I (Theorem 7.D.1*); we give a complete proof of this theorem in the case of a finitely generated group. The regular representation* λ_Γ *of a discrete group* Γ *generates a von Neumann algebra of finite type. It turns out that* Γ *is of type I if and only if* λ_Γ *is of type I (Theorem* 7.D.3*). We indicate a class of examples of groups* Γ *for which* λ_Γ *has both a type I component and a type II component (Theorem* 7.D.6*). We mention in Section* 7.E *results about the type of the regular representation for classes of nondiscrete locally compact groups.*

In Section 7.F, *we discuss the direct integral decompositions of a factor representation* π *of a second-countable LC group* G *into irreducible representations. We will see that almost every irreducible representation involved in such a direct integral decomposition is weakly equivalent to* π. *It follows that every factor representation of* G *is weakly equivalent to some irreducible representation of* G, *and that the quotient map* $\widehat{G} \twoheadrightarrow \mathrm{Prim}(G)$ *has a natural extension to a map* $\mathrm{QD}(G) \twoheadrightarrow \mathrm{Prim}(G)$; *when* G *is not of type I, many fibers of the map* $\widehat{G} \twoheadrightarrow \mathrm{Prim}(G)$ *are uncountable.*

7.A. A class of non type I groups

As we will see later (Theorem 7.D.3), most discrete groups are not of type I. For the moment, we single out the following venerable result of Murray and von Neumann [**MuvN–43**, Lemma 5.3.4]; see also Chap. III, §7, no. 6 in [**Dixm–vN**]. For the definition and examples of icc groups, see Appendix A.5. For the definition of factors of type II_1, see Appendix A.11.

Recall that $(\delta_\gamma)_{\gamma\in\Gamma}$ denotes the standard basis of the Hilbert space $\ell^2(\Gamma)$, where δ_γ is a Dirac function at $\gamma \in \Gamma$.

PROPOSITION 7.A.1. *Let Γ be a discrete group, $\Gamma \neq \{e\}$, and $\mathcal{L}(\Gamma) := \lambda_\Gamma(\Gamma)''$ the von Neumann algebra of the left regular representation of Γ. The following properties are equivalent:*

(i) *Γ is an icc group;*
(ii) *$\mathcal{L}(\Gamma)$ is a factor.*

NOTE. Proposition 7.C.5 will add a third property equivalent to (i) and (ii): $\mathcal{L}(\Gamma)$ is a factor of type II_1.

PROOF. Let ρ_Γ be the right regular representation of Γ on $\ell^2(\Gamma)$. Observe that $\lambda_\Gamma(\gamma)\rho_\Gamma(\gamma') = \rho_\Gamma(\gamma')\lambda_\Gamma(\gamma)$ for all $\gamma, \gamma' \in \Gamma$.
Let $T \in \mathcal{L}(\Gamma)$. Then

$$\rho_\Gamma(\gamma)T = T\rho_\Gamma(\gamma) \quad \text{for all } \gamma \in \Gamma.$$

Suppose moreover that T is in the centre of $\mathcal{L}(\Gamma)$. Then

$$\lambda_\Gamma(\gamma)T = T\lambda_\Gamma(\gamma) \quad \text{for all } \gamma \in \Gamma.$$

Set

$$f := T\delta_e \in \ell^2(\Gamma).$$

Then, for every $\gamma \in \Gamma$, we have $\lambda_\Gamma(\gamma)\rho_\Gamma(\gamma)\delta_e = \delta_e$ and

$$\lambda_\Gamma(\gamma)\rho_\Gamma(\gamma)f = (\lambda_\Gamma(\gamma)\rho_\Gamma(\gamma)T)\delta_e = (T\lambda_\Gamma(\gamma)\rho_\Gamma(\gamma))\delta_e = T\delta_e = f.$$

Therefore f is a function in $\ell^2(\Gamma)$ which is central. It follows that the support of f consists of elements with finite conjugacy class.

(i) $\Longrightarrow$ (ii) Assume that Γ is icc. Then f is supported by $\{e\}$, that is, $T\delta_e = f = c\delta_e$ for a scalar $c \in \mathbf{C}$. For every $\gamma \in \Gamma$, we have

$$T\delta_\gamma = T\lambda_\Gamma(\gamma)\delta_e = \lambda_\Gamma(\gamma)T\delta_e = c\lambda_\Gamma(\gamma)\delta_e = c\delta_\gamma.$$

Since the span of $\{\delta_\gamma \mid \gamma \in \Gamma\}$ is dense in $\ell^2(\Gamma)$, it follows that $T = cI$. This shows that $\lambda_\Gamma(\Gamma)''$ is a factor.

(ii) $\Longrightarrow$ (i) Assume now that Γ is not icc, that is, there exists a finite conjugacy class $C \subset \Gamma$ with $C \neq \{e\}$. Consider the operator

$$T := \lambda_\Gamma(\mathbf{1}_C) = \sum_{\gamma \in C} \lambda_\Gamma(\gamma),$$

which belongs to $\lambda_\Gamma(\Gamma)''$. Then $T \in \lambda_\Gamma(\Gamma)'$; indeed, for every $x \in \Gamma$, we have

$$\begin{aligned} T\lambda_\Gamma(x) &= \sum_{\gamma \in C} \lambda_\Gamma(\gamma)\lambda_\Gamma(x) = \lambda_\Gamma(x)\sum_{\gamma \in C} \lambda_\Gamma(x^{-1}\gamma x) \\ &= \lambda_\Gamma(x)\sum_{\gamma \in C} \lambda_\Gamma(\gamma) = \lambda_\Gamma(x)T. \end{aligned}$$

Therefore T belongs to the centre of $\lambda_\Gamma(\Gamma)''$. Moreover, T is not a multiple of the identity, since $T\delta_e = \mathbf{1}_C$. So, $\mathcal{L}(\Gamma)$ is not a factor. $\square$

DEFINITION 7.A.2. The **von Neumann algebra of a locally compact group** G is the von Neumann algebra $\mathcal{L}(G) := \lambda_\Gamma(G)''$ of its left regular representation $\lambda_G : G \to \mathcal{U}(L^2(G, \mu_G))$.

A remarkable property of von Neumann algebras of discrete groups, also observed by Murray and von Neumann, is that they admit finite faithful normal traces, as shown by the following proposition (these terms are defined in Section 7.B). We denote as in Appendix A.10 by $\mathcal{L}(\Gamma)_+$ the cone of positive operators in $\mathcal{L}(\Gamma)$.

PROPOSITION 7.A.3. *Let Γ be a discrete group. The linear functional*

$$\tau : \mathcal{L}(\Gamma) \to \mathbf{C}, \quad T \mapsto \langle T\delta_e \mid \delta_e \rangle$$

has the following properties:

(1) *τ is continuous for the weak operator topology on $\mathcal{L}(\Gamma)$;*
(2) *$\tau(\mathrm{Id}_{\ell^2(\Gamma)}) = 1$;*
(3) *$\tau(ST) = \tau(TS)$ for all $S, T \in \mathcal{L}(\Gamma)$;*
(4) *τ is faithful, that is, $\tau(T) > 0$ for every $T \in \mathcal{L}(\Gamma)_+$, $T \neq 0$.*

PROOF. Properties (1) and (2) are obvious.

For $\gamma \in \Gamma$, we have $\tau(\lambda_\Gamma(\gamma)) = 1$ if $\gamma = e$ and $\tau(\lambda_\Gamma(\gamma)) = 0$ otherwise. This implies that

$$\tau(\lambda_\Gamma(\gamma_1)\lambda_\Gamma(\gamma_2)) = \tau(\lambda_\Gamma(\gamma_1\gamma_2)) = \tau(\lambda_\Gamma(\gamma_2\gamma_1)) = \tau(\lambda_\Gamma(\gamma_2)\lambda_\Gamma(\gamma_1))$$

for all $\gamma_1, \gamma_2 \in \Gamma$. By linearity and density, Property (3) follows.

To show Property (4), let $T \in \mathcal{L}(\Gamma)_+$ be such that $\tau(T) = 0$; we have to show that $T = 0$. Write $T = S^*S$ for $S \in \mathcal{L}(\Gamma)$. Then

$$0 = \tau(T) = \tau(S^*S) = \langle S^*S\delta_e \mid \delta_e \rangle = \|S\delta_e\|^2,$$

that is, $S\delta_e = 0$. Now, δ_e is a cyclic vector for the right regular representation ρ_Γ and $S \in \mathcal{L}(\Gamma) \subset \rho_\Gamma(\Gamma)'$; hence $S\xi = 0$ for every $\xi \in \ell^2(\Gamma)$. It follows that $S = 0$, that is, $T = 0$. □

We will show that, for an icc group Γ, the regular representation λ_Γ is not of type I. Our proof will depend on the following fact, which is of independent interest.

PROPOSITION 7.A.4. *Let Γ be a discrete group. Assume that λ_Γ contains an irreducible subrepresentation.*

Then Γ is finite.

PROOF. Assume that λ_Γ contains an irreducible subrepresentation π. Let $\mathcal{H}$ be the Γ-invariant closed subspace of $\ell^2(\Gamma)$ defining π.

Let $P \in \mathcal{L}(\Gamma)'$ be the orthogonal projection from $\ell^2(\Gamma)$ onto $\mathcal{H}$. The induced von Neumann algebra $\mathcal{L}(\Gamma)_P = P\mathcal{L}(\Gamma)P$, viewed as a subalgebra of $\mathcal{L}(\mathcal{H})$, coincides with $\pi(\Gamma)''$ and hence with $\mathcal{L}(\mathcal{H})$, since π is irreducible. Let $E \in \mathcal{L}(\Gamma) \cap \mathcal{L}(\Gamma)'$ be the central support of P. Then the induction map

$$\Phi : \mathcal{L}(\Gamma)_E \to \mathcal{L}(\Gamma)_P$$

is an isomorphism (see Proposition A.11.7).

Consider the linear functional $\varphi : \mathcal{L}(\mathcal{H}) \to \mathbf{C}$ defined by

$$\varphi(T) = \langle \Phi^{-1}(T)\delta_0 \mid \delta_0 \rangle \quad \text{for } T \in \mathcal{L}(\mathcal{H}) = \mathcal{L}(\Gamma)_P.$$

Then, by Proposition 7.A.3, φ has the following properties:

(a) φ is continuous for the weak operator topology on $\mathcal{L}(\mathcal{H})$;
(b) $\varphi(ST) = \varphi(TS)$ for all $S, T \in \mathcal{L}(\mathcal{H})$;
(c) $\varphi(\mathrm{Id}_\mathcal{H}) = 1$.

It follows that $\mathcal{H}$ is finite-dimensional.

Indeed, by Property (b), we have $\varphi(p) = \varphi(q)$ for any two projections $p, q \in \mathcal{L}(\mathcal{H})$ with the same dimension, as p and q are conjugate under the unitary group of $\mathcal{H}$. Since $\mathrm{Id}_{\mathcal{H}} = \sum_i p_i$ in the weak operator topology for one-dimensional projections p_i, it follows from Properties (a) and (c) that $\mathcal{H}$ is finite-dimensional.

Since π is finite-dimensional, the trivial representation 1_Γ is contained in the tensor product $\pi \otimes \overline{\pi}$, where $\overline{\pi}$ is the conjugate representation of π [**BeHV–08**, A.1.13]. Therefore 1_Γ is contained in $\lambda_\Gamma \otimes \overline{\lambda_\Gamma}$, which is equivalent to λ_Γ (see [**BeHV–08**, Appendix E]). So, $\ell^2(\Gamma)$ contains a nonzero invariant vector; this is only possible if Γ is finite. □

The analogue for LC groups of Proposition 7.A.4 does not hold. Indeed, there are second-countable LC groups G which are Fell groups, i.e., which are such that the regular representation λ_G is a direct sum of irreducible representations, and which are noncompact. An example of such a group is the group M_p from Example 1.D.9. For other examples, see [**Robe–78**].

The following corollary is an immediate consequence of Propositions 7.A.1, 7.A.4, and 6.B.14.

COROLLARY 7.A.5. *Let Γ be a discrete icc group with more than one element. Then Γ is not of type I.*

7.B. Operator algebras, traces, and types

The aim of the present section is to recall some of the terminology from the theory of operator algebras relevant for the type classification of representations.

Traces on C*-algebras. Traces are important for both von Neumann algebras and C*-algebras, and also because characters of group representations are defined in terms of traces (see Chapter 10). First consider a C*-algebra A. The cone of positive elements is $A_+ := \{x \in A \mid x = y^*y \text{ for some } y \in A\}$. For $x, y \in A_+$, the order $x \leq y$ is defined by $y - x \in A_+$. More on A_+ in Appendix A.10.

DEFINITION 7.B.1. A **trace** on a C*-algebra A is a function $t : A_+ \to [0, \infty]$ such that

(1) $t(x + y) = t(x) + t(y)$ for all $x, y \in A_+$;
(2) $t(\lambda x) = \lambda t(x)$ for all $x, y \in A_+$ and $\lambda \in [0, \infty[$ (it is agreed that the product 0∞ is 0);
(3) $t(x^*x) = t(xx^*)$ for all $x \in A_+$.

Such a trace is

(4) **semi-finite** if $t(x) = \sup\{t(y) \mid y \in A_+,\ y \leq x,\ t(y) < \infty\}$ for all $x \in A_+$;
(5) **finite** if $t(x) < \infty$ for all $x \in A_+$;
(6) **lower semi-continuous** if, for all $x \in A_+$ and $\lambda \in \mathbf{R}$ such that $\lambda < t(x)$, there exists $\varepsilon > 0$ such that $\lambda < t(y)$ for all $y \in A_+$ such that $\|y - x\| < \varepsilon$;
(7) **faithful** if, for $x \in A_+$, the equality $t(x) = 0$ implies $x = 0$.

Let t_1, t_2 be two traces on A; then t_1 **dominates** t_2 if $t_1(x) \geq t_2(x)$ for all $x \in A_+$.

EXAMPLE 7.B.2. (1) Let A be a C*-algebra, not $\{0\}$. The function t defined on A_+ by $t(0) = 0$ and $t(x) = \infty$ for all $x \neq 0$ in A_+ is a lower semi-continuous faithful trace that is not semi-finite.

(2) Let $c^0(\mathbf{N})$ be the C*-algebra of sequences $(z_i)_{i\geq 1}$ of complex numbers such that $\lim_{i\to\infty} z_i = 0$, equipped with the norm $\|(z_i)_{i\geq 1}\|_\infty := \sup_{i\geq 1} |z_i|$. Its positive cone $c^0(\mathbf{N})_+$ is the subset of sequences in $c^0(\mathbf{N})$ of nonnegative real numbers. Let

$\mu = (\mu_i)_{i \geq 1}$ be a sequence of nonnegative real numbers; a trace t_μ on $c^0(\mathbf{N})$ is defined by

$$t_\mu : c^0(\mathbf{N})_+ \to [0, \infty], \qquad (z_i)_{i \geq 1} \mapsto \sum_{i=1}^{\infty} \mu_i z_i.$$

This trace is clearly semi-finite and it is an exercise in calculus to check that it is lower semi-continuous. The trace t_μ is finite if and only if $\sum_{i=1}^{\infty} \mu_i < \infty$, and faithful if and only if $\mu_i > 0$ for all $i \geq 1$.

It is easy to show that every semi-finite and semi-continuous trace on $c^0(\mathbf{N})$ is of the form t_μ for a sequence μ as above.

For a generalization of this example, see Proposition 7.B.3.

(3) Let t be the trace on $C^0(\mathbf{R})$ defined on $C^0(\mathbf{R})_+$ by $t(f) = 0$ if the support of f is compact and $t(f) = \infty$ otherwise. Then t is neither lower semi-continuous nor semi-finite.

(4) Let $\mathcal{K}(\mathcal{H})$ be the C*-algebra of compact operators on a separable infinite-dimensional Hilbert space $\mathcal{H}$. There exist so-called "Dixmier traces" t on $\mathcal{K}(\mathcal{H})$, not the zero trace, which vanish on the usual trace-class operators [**Dixm–66**]. They can be shown to be semi-finite, and not lower semi-continuous.

(5) Let G be a unimodular LC group, $C^*_\lambda(G)$ the reduced C*-algebra of G (see Chapter 8), and $C^c(G)$ the involutive subalgebra of continuous functions on G with compact supports. A semi-finite faithful trace t is defined on $C^*_\lambda(G)$ by

$$t(f * f^*) = (f * f^*)(e) = \|f\|_2^2 \quad \text{for all } f \in C^c(G),$$

and suitably extended on elements of $C^*_\lambda(G)_+$; see Example 10.A.9(3).

The trace t is finite if and only if G is discrete. In this case, we write Γ rather than G for the group, and $\mathbf{C}[\Gamma]$ rather than $C^c(\Gamma)$ for the group algebra. (See also Example 7.B.6(2).)

(6) Let G be a locally compact group. There exists a nonzero finite trace on $C^*_\lambda(G)$ if and only if the amenable radical of G is open [**FoSW, KeRa–17**].

The following proposition gives a complete description of lower semi-continuous and semi-finite traces on a commutative C*-algebra.

PROPOSITION 7.B.3. *Let X be a second countable locally compact space. Let $C^0(X)$ be the abelian C*-algebra of continuous functions on X which vanish at infinity. Let μ be a Radon measure on an open subset U of X. Define $t_\mu : C^0(X)_+ \to [0, \infty]$ by*

$$t_\mu(f) = \int_U f(x) d\mu(x) \quad \textit{for every } f \in C^0(X)_+.$$

(1) *t_μ is a lower semi-continuous and semi-finite trace on $C^0(X)$;*
(2) *t_μ is finite if and only if μ is finite;*
(3) *t_μ is faithful if and only if $\mu(V \cap U) \neq 0$ for every nonempty open subset V of X.*
(4) *Let t be a lower semi-continuous and semi-finite trace on $C^0(X)$. Then $t = t_\mu$ for a Radon measure μ on an open subset U of X.*

PROOF. (1) Let $(f_n)_{n\geq 1}$ be a sequence in $C^0(X)_+$ and $f \in C^0(X)_+$ be such that $\lim_n f_n = f$. Then, by Fatou's lemma, we have

$$t_\mu(f) = \int_U \liminf f_n(x)d\mu(x) \leq \liminf \int_U f_n(x)d\mu(x) = \liminf t_\mu(f_n),$$

and the lower continuity of t_μ follows.

Choose an increasing sequence $(\varphi_n)_{n\geq 1}$ of functions $\varphi_n \in C^c(X)_+$ with supports contained in U and such that $\lim_n \varphi_n(x) = 1$ for every $x \in U$. Let $f \in C^0(X)_+$. We have $\varphi_n f \leq f$; moreover, for every $x \in U$, the sequence $(\varphi_n(x)f(x))_{n\geq 1}$ is increasing and converges to $f(x)$. Hence, by the Lebesgue monotone convergence theorem, we have

$$\lim_n t_\mu(\varphi_n f) = \lim_n \int_U \varphi_n(x)f(x)d\mu(x) = \int_U f(x)d\mu(x) = t_\mu(f).$$

Since $\varphi_n f$ has compact support contained in U and since μ is a Radon measure on U, we have $t_\mu(\varphi_n f) < +\infty$ for every $n \geq 1$. This shows that t_μ is semi-finite.

Item (2) is obvious.

(3) Assume that t_μ is faithful and let V be a nonempty open subset of X. Let $f \in C^c(X)_+, f \neq 0$, with support contained in V. Then $t_\mu(f) \neq 0$ and it follows that $\mu(V \cap U) \neq 0$.

Assume that $\mu(V \cap U) \neq 0$ for every nonempty open subset V of X. Let $f \in C^0(X)_+$ with $f \neq 0$. The set $V_f = \{x \in X \mid f(x) > 0\}$ is open and nonempty; so, $\mu(V_f \cap U) > 0$ and therefore

$$t_\mu(f) = \int_{V_f \cap U} f(x)d\mu(x) > 0.$$

This shows that t_μ is faithful.

(4) Let $\mathfrak{m}_t$ be the ideal of definition of t, that is, the linear span of

$$\mathfrak{m}_{t,+} = \{f \in C^0(X)_+ \mid t(f) < +\infty\}$$

(see Section 10.A). The closure J of $\mathfrak{m}_t$ is a closed ideal in $C^0(X)$. Therefore, there exists a closed subset A of X such that

$$J = \{f \in C^0(X) \mid f|_A = 0\}.$$

Set $U := X \smallsetminus A$. We identify J with $C_0(U)$ and so $\mathfrak{m}_t$ with a dense ideal of $C^0(U)$.

By Lemma 7.B.5 below, $\mathfrak{m}_t$ contains $C^c(U)$. Hence, $t|_{C^c(U)}$ is a positive linear form on $C^c(U)$ and thus there exists a Radon measure μ on U such that

$$t(f) = \int_U f(x)d\mu(x) \quad \text{for all } f \in C^c(U).$$

We claim that $t = t_\mu$. Indeed, let $(\varphi_n)_{n\geq 1}$ be an increasing sequence in $C^c(U)_+$ such that $\lim_n \varphi_n = 1$ on U.

Let $f \in C^0(X)_+$. Then $\varphi_n f \in C^c(U)_+$ and $\varphi_n f \leq f$ for every $n \geq 1$; moreover we have $\lim_n \varphi_n f(x) = f(x)$ for every $x \in U$. So, by the Lebesgue monotone convergence theorem, we have

$$\lim_n t(\varphi_n f) = \lim_n \int_U \varphi_n(x)f(x)d\mu(x) = \int_U f(x)d\mu(x) = t_\mu(f).$$

It therefore suffices to show that $\lim_n t(\varphi_n f) = t(f)$ for every $f \in C^0(X)_+$.

Assume first that $f \in C^0(U)_+$. Then $\lim_n \varphi_n f = f$ and hence, by the lower semi-continuity of t, we have

$$\liminf t(\varphi_n f) \geq t(f);$$

so, $\lim_n t(\varphi_n f) = t(f)$, since $t(\varphi_n f) \leq t(f)$.

Next, let $f \in C^0(X)_+$ such that $f \notin C^0(U)_+$. Then $t(f) = +\infty$. Let $C > 0$. Since t is semi-finite, there exists $g \in \mathfrak{m}_{t,+} \subset C^0(U)_+$ such that

$$g \leq f \quad \text{and} \quad C < t(g).$$

By what we have seen above, we have $\lim_n t(\varphi_n g) = t(g)$. Hence, there exists n_0 such that

$$C < t(\varphi_n g) \quad \text{for all } n \geq n_0.$$

Since $\varphi_n g \leq \varphi_n f$, we therefore have

$$C < t(\varphi_n f) \quad \text{for all } n \geq n_0,$$

and so

$$\lim_n t(\varphi_n f) = +\infty = t(f).$$

□

REMARK 7.B.4. The result in Proposition 7.B.3 admits an appropriate generalization which holds for any C*-algebra of type I; see Corollaire 2 of Théorème 2 in [**Dixm–63**].

Every C*-algebra $\mathcal{A}$ has a unique minimal dense ideal, called the Pedersen ideal of $\mathcal{A}$ (see [**Pede–79**, § 5.6]). The following elementary lemma identifies this ideal in the case where $\mathcal{A}$ is commutative. The lemma has been used in the proof of Proposition 7.B.3.

LEMMA 7.B.5. *Let X be a locally compact topological space. Every dense ideal of $C^0(X)$ contains $C^c(X)$.*

PROOF. Let J be a dense ideal in $C^0(X)$. Let $f \in C^c(X)$; denote by K the support of f. Let $x \in K$. There exists $g_x \in J$ such that $g_x(x) \neq 0$ and we can therefore find a neighbourhood U_x of x such that $g_x \neq 0$ on U_x. By the compactness of K, there exists $x_1, \ldots, x_n \in K$ such that

$$K \subset \bigcup_{i=1}^{n} U_{x_i}.$$

Then $g := \sum_{i=1}^n |g_{x_i}|^2$ belongs to J and $g > 0$ on K. Define $h : X \to \mathbf{C}$ by

$$h(x) = \begin{cases} \dfrac{f(x)}{g(x)} & \text{for } x \in K, \\ 0 & \text{for } x \notin K. \end{cases}$$

Then h is continuous on X with compact support and therefore $f = hg$ belongs to J. □

Traces on C*-algebras are addressed again below, in Section 10.A.

Traces on von Neumann algebras. Now let $\mathcal{M}$ be a von Neumann algebra; in particular, it is a C*-algebra, and the definitions above apply. However, we will usually denote traces on von Neumann algebras by τ instead of t.

A trace τ on $\mathcal{M}$ is **normal** if, for every increasing filtering subset $\mathcal{F} \subset \mathcal{M}_+$ with least upper bound $x \in \mathcal{M}_+$, the value $\tau(x)$ is the least upper bound of the set $\{\tau(y) \in [0, \infty] \mid y \in \mathcal{F}\}$.

A trace on $\mathcal{M}$ is normal if and only if it is lower semi-continuous with respect to the weak topology [**Dixm–vN**, Chap. I, § 6, exercice 1].

EXAMPLE 7.B.6. (1) If $\mathcal{H}$ is a separable Hilbert space, the standard trace $\mathrm{T}r$ on $\mathcal{L}(\mathcal{H})$ is semi-finite, faithful, and normal. Recall that the standard trace is defined by $\mathrm{T}r(x) = \sum_{i \geq 1} \langle x e_i \mid e_i \rangle \in [0, \infty]$ for $x \in \mathcal{L}(\mathcal{H})_+$, where $(e_i)_{i \geq 1}$ is an orthonormal basis of $\mathcal{H}$; the sum is independent of the choice of the basis.

Every semi-finite normal trace on $\mathcal{L}(\mathcal{H})$ is proportional to $\mathrm{T}r$; see [**Dixm–vN**, Chap. I, § 6, no 6].

The standard trace on $\mathcal{L}(\mathcal{H})$ is finite if and only if $\mathcal{H}$ is finite-dimensional.

If $\mathcal{H}$ is infinite-dimensional, the construction referred to in Example 7.B.2(4) provides "Dixmier traces" on $\mathcal{L}(\mathcal{H})$ that are semi-finite and not normal [**Dixm–66**].

(2) Let Γ be a discrete group and λ_Γ the left regular representation of Γ on $\ell^2(\Gamma)$. The map

$$\tau : T \mapsto \langle T\delta_e \mid \delta_e \rangle$$

is a faithful finite normal trace on the von Neumann algebra $\mathcal{L}(\Gamma) \subset \mathcal{L}(\ell^2(\Gamma))$ generated by $\lambda_\Gamma(\Gamma)$, as shown in Proposition 7.A.3.

Observe that τ coincides on $C^*_\lambda(\Gamma)$ with the trace of Example 7.B.2(5).

(3) On an abelian von Neumann algebra, finite traces need not be normal. Indeed, by the Hahn–Banach Theorem, there exist nonzero positive linear functionals on $L^\infty([0,1], dt)$ which vanish on the closed subspace $C([0,1])$ of continuous functions. Such a functional is a finite trace on $L^\infty([0,1], dt)$; it is not normal since $C([0,1])$ is weakly dense in $L^\infty([0,1], dt)$.

Types of von Neumann algebras.

DEFINITION 7.B.7. A von Neumann algebra $\mathcal{M}$ is

(A) **type I** or **discrete** if it is isomorphic to a von Neumann algebra with an abelian commutant;
(B) **continuous** if there does not exist any projection $p \neq 0$ in the centre $\mathcal{M} \cap \mathcal{M}'$ such that $p\mathcal{M}p$ is discrete;
(C) **finite** [respectively **semi-finite**] if, for every $x \in \mathcal{M}_+$, $x \neq 0$, there exists a normal trace τ on $\mathcal{M}$ that is finite [respectively semi-finite] and such that $\tau(x) > 0$;
(D) **properly infinite** [respectively **purely infinite**] if there does not exist any nonzero finite normal trace on $\mathcal{M}$ [respectively any semi-finite normal trace on $\mathcal{M}$];
(E) **type II$_1$** if it is continuous and finite;
(F) **type II$_\infty$** if it is continuous, semi-finite, and properly infinite;
(G) **type III** if it is purely infinite, i.e., if the only normal semi-finite trace on $\mathcal{M}_+$ is the zero trace.

For factors (see Section 6.B.b), the typology can be refined: a factor of type I is of **type I$_n$** if it is isomorphic to the algebra $\mathrm{M}_n(\mathbf{C})$ of n-by-n complex matrices, for some integer $n \geq 1$, and of **type $\mathbf{I}_\infty$** if it is infinite-dimensional, isomorphic to the algebra $\mathcal{L}(\mathcal{H})$ for some infinite-dimensional Hilbert space $\mathcal{H}$. Moreover, type III can be subdivided in types III_λ, for $\lambda \in [0,1]$; this will appear briefly in Chapters 13 and 14.

REMARK 7.B.8. (1) Type I von Neumann algebras can also be characterized as follows: a von Neumann algebra $\mathcal{M}$ is of type I if and only if there exists a projection $p \in \mathcal{M}$ with central support I such that $p\mathcal{M}p$ is abelian [**Dixm–vN**, Chap. I, § 8, Théorème 1].

(2) Finiteness and semi-finiteness have other characterizations in terms of traces: A von Neumann algebra $\mathcal{M}$ is semi-finite if and only if it has a *faithful* semi-finite normal trace. When $\mathcal{M}$ is countably decomposable, $\mathcal{M}$ is finite if and only if it has a faithful finite normal trace [**Dixm–vN**, Chap. I, § 6, no 7]. (A von Neumann algebra $\mathcal{M}$ is countably decomposable if every collection of pairwise orthogonal nonzero projections in $\mathcal{M}$ is countable. For example, a von Neumann algebra that can act faithfully on a separable Hilbert space is countably decomposable.)

(3) The type of a von Neumann algebra $\mathcal{M}$ inside some algebra $\mathcal{L}(\mathcal{H})$ is closely related to the type of its commutant $\mathcal{M}'$:
- $\mathcal{M}$ is of type I if and only if $\mathcal{M}'$ is of type I [**Dixm–vN**, Chap. I, § 8, no 2];
- $\mathcal{M}$ is of type II if and only if $\mathcal{M}'$ is of type II (in [**Dixm–vN**, Chap. I], see §6, no. 8, cor. 1 of prop. 13, and §8, no. 2, cor. 1 of th. 1);
- $\mathcal{M}$ is of type III if and only if $\mathcal{M}'$ is of type III [**Dixm–vN**, Chap. I, § 6, no 8].

The following proposition reduces the study of general von Neumann algebras to that of von Neumann algebras of one of the types introduced above.

PROPOSITION 7.B.9. *Every von Neumann algebra $\mathcal{M}$ is canonically a product*

$$\mathcal{M}_{\mathrm{I_f}} \times \mathcal{M}_{\mathrm{I}_\infty} \times \mathcal{M}_{\mathrm{II}_1} \times \mathcal{M}_{\mathrm{II}_\infty} \times \mathcal{M}_{\mathrm{III}}$$

where the five terms in the product are respectively of type I and finite, of type I and infinite, of type II_1, of type II_∞, and of type III (and each can be absent).

REFERENCE FOR THE PROOF. In [**Dixm–vN**, Chap. I], see §6, no. 7, cor. 1 of prop. 8 and §8, no. 1, cor. 1 of prop. 1. □

7.C. Types of group representations

Let G be a topological group. Recall from Section 6.B that, for every representation $(\pi, \mathcal{H})$ of G on a Hilbert space $\mathcal{H}_\pi$, there are two associated von Neumann subalgebras of $\mathcal{L}(\mathcal{H}_\pi)$, which are commutant of each other: $\pi(G)''$ and $\pi(G)'$.

We now relate type I representations introduced in Section 6.A to type I von Neumann algebras.

PROPOSITION 7.C.1. *For a representation π of G, the following properties are equivalent:*

(i) *π is of type I;*
(ii) *the von Neumann algebra $\pi(G)''$ is of type I;*
(iii) *the von Neumann algebra $\pi(G)'$ is of type I.*

PROOF. The fact that (i) implies (ii) follows from Proposition 6.B.12 and the definition of a type I von Neumann algebra. The fact that a von Neumann algebra and its commutant are together of type I or not (see Remark 7.B.8) shows that (ii) and (iii) are equivalent.

It remains to show that (iii) implies (i). Assume that $\pi(G)'$ is of type I. Then there exists a projection $p \in \pi(G)'$ with central support I such that $p\pi(G)'p$ is abelian (see Remark 7.B.8). Let σ be the subrepresentation of π on the range of p. Since $\sigma(G)' \cong p\pi(G)'p$ (see Proposition 6.B.3), the representation σ is multiplicity-free (see Proposition 6.B.10). As the central support of p is I, it follows from Proposition 6.B.13 that the representation π is of type I. □

We can now define various types of group representations.

DEFINITION 7.C.2. Let G be a topological group. The representation π of G is said to be of type I, II, II_1, II_∞, or III if the von Neumann algebra $\pi(G)''$ is of type I, II, II_1, II_∞, or III.

The representation π of G is said to be of **discrete, continuous, semi-finite, properly infinite**, or **purely infinite type** if the von Neumann algebra $\pi(G)''$ is of discrete, continuous, semi-infinite, properly infinite, or purely infinite type.

The representation π is said to be of **weak finite type**, respectively of **finite type**, if $\pi(G)''$ is finite, respectively if $\pi(G)$" has a faithful finite normal trace. These two last notions coincide in most cases (see Remark 7.B.8); while representations of finite type play a central role in Chapters 4, 11, 12, and 15, representations of weak finite type which are not of finite type never show up in this book.

REMARK 7.C.3. Let π be a representation of G.

(1) It follows from the definitions and from Proposition 6.B.5 that every representation quasi-equivalent to π is of the same type as π.

(2) Our notion of representation of finite type does **not** agree with the notion of "finite representation" from [**Mack–76**, Page 30], which is not invariant by quasi-equivalence.

(3) π is of continuous type if and only if, for every nonzero subrepresentation ρ of π, there are two equivalent subrepresentations ρ', ρ'' of ρ in orthogonal invariant subspaces, such that $\rho' \oplus \rho'' = \rho$; for this, see [**Dixm–vN**, Chap. III, § 1, cor. 3].

(4) If π is a representation of type III on a separable Hilbert space, then every representation that is quasi-equivalent to π is equivalent to π [**Dixm–C***, 5.6.6].

As for von Neumann algebras (Proposition 7.B.9), the study of general representations can be reduced to that of representations of specific types.

PROPOSITION 7.C.4. *Every representation π of a topological group is canonically a direct sum*

$$\pi_{\text{I}_\text{f}} \oplus \pi_{\text{I}_\infty} \oplus \pi_{\text{II}_1} \oplus \pi_{\text{II}_\infty} \oplus \pi_{\text{III}}$$

where the representations in the direct sum are respectively of type I and finite, of type I and infinite, of type II_1, of type II_∞, and of type III (some of these can be in spaces of dimension 0).

With the notions we have introduced, we can rephrase parts of Propositions 7.A.1 and 7.A.3 in the following way.

PROPOSITION 7.C.5. *Let Γ be a discrete group.*

(1) *The regular representation λ_Γ is a representation of finite type.*

(2) *If $\Gamma \neq \{e\}$, the following three conditions are equivalent:*
— Γ *is icc,*
— λ_Γ *is a factor representation,*
— λ_Γ *is a factor representation of type II_1.*

7.D. Types of representations of discrete groups

All groups appearing in this section will be discrete groups.

Thoma's Theorem. We have seen before (Corollary 7.A.5) that an icc discrete group is not of type I. The following result gives a complete characterization of type I discrete groups.

For a group Γ, we denote by $FC(\Gamma)$ its FC-centre, i.e., the normal subgroup of elements in Γ with a finite conjugacy class (see Appendix A.5).

THEOREM 7.D.1 (**Thoma**). *Let Γ be a discrete group. The following properties are equivalent:*

(i) Γ *is type I;*
(ii) Γ *is virtually abelian, that is, Γ contains an abelian subgroup of finite index;*
(iii) Γ *is a Moore group, that is, $\widehat{\Gamma} = \widehat{\Gamma}_{\mathrm{fd}}$.*

PROOF. The deep implication is (i) $\Longrightarrow$ (ii); we will give a proof of this implication under the assumption that Γ is *finitely generated;* for the proof in the case of arbitrary discrete groups, we refer to the original article [**Thom–68**] (see also [**Thom–64a**] for the case of countable groups).

Assume that Γ is finitely generated and not virtually abelian. Observe that this implies that $FC(\Gamma)$ has infinite index in Γ, by a classical fact (see Proposition A.5.1). We are going to show that Γ is not of type I. In fact, we will prove the following more general result: if Γ is a countable group such that $[\Gamma : FC(\Gamma)] = \infty$, then Γ is not of type I. This result appears in [**Maut–50c**], and the proof below follows in part the arguments given there.

We consider the central direct integral decomposition of the left regular representation λ_Γ as in Theorem 6.C.5: there exists a measurable field of *factorial* representations $(\pi_x, \mathcal{H}_x)_{x\in X}$ over a standard Borel space X with a σ-finite measure μ and an isomorphism of Hilbert spaces $U : \ell^2(\Gamma) \to \int_X^\oplus \mathcal{H}_x d\mu(x)$, such that

- $U\lambda_\Gamma(\gamma)U^{-1} = \int_X^\oplus \pi_x(\gamma)d\mu(x)$ for all $\gamma \in \Gamma$;
- $U\mathcal{Z}U^{-1}$ coincides with the algebra of diagonalisable operators on $\int_X^\oplus \mathcal{H}_x d\mu(x)$, where $\mathcal{Z}$ is the centre of the von Neumann algebra $\mathcal{L}(\Gamma)$ of Γ;
- $U\lambda_\Gamma(\Gamma)'U^{-1} = \int_X^\oplus \pi_x(\Gamma)'d\mu(x)$.

Let ρ_Γ be the right regular representation of Γ. Since $\rho_\Gamma(\Gamma) \subset \lambda_\Gamma(\Gamma)'$, there exists a measurable field $x \mapsto \sigma_x$ of representations of Γ in $\int_X^\oplus \mathcal{H}_x d\mu(x)$ such that

$$U\rho_\Gamma(\gamma)U^{-1} = \int_X^\oplus \sigma_x(\gamma)d\mu(x) \qquad \text{for all } \gamma \in \Gamma.$$

Set

$$(\xi_x)_{x\in X} := U\delta_e \in \int_X^\oplus \mathcal{H}_x d\mu(x).$$

We claim that ξ_x is a cyclic vector for both π_x and σ_x, for μ-almost every $x \in X$. Indeed, δ_e is a cyclic vector for both λ_Γ and ρ_Γ; hence, $\{(\pi_x(\gamma)\xi_x)_{x\in X} \mid \gamma \in \Gamma\}$ as well as $\{(\sigma_x(\gamma)\xi_x)_{x\in X} \mid \gamma \in \Gamma\}$ are countable total subsets of $\int_X^{\oplus} \mathcal{H}_x d\mu(x)$. It follows from a general fact about direct integrals of Hilbert spaces (see Proposition 8 in Chap. II, §1 of [**Dixm–vN**]) that $\{\pi_x(\gamma)\xi_x \mid \gamma \in \Gamma\}$ as well as $\{\sigma_x(\gamma)\xi_x \mid \gamma \in \Gamma\}$ are total subsets of $\mathcal{H}_x$ for μ-almost every $x \in X$, and this proves the claim.

Since $\lambda_\Gamma(\gamma)\delta_e = \rho_\Gamma(\gamma^{-1})\delta_e$ for every $\gamma \in \Gamma$ and since Γ is countable, there exists a Borel subset X' of X with $\mu(X \smallsetminus X') = 0$ with the following properties:

(1) $\pi_x(\gamma)\xi_x = \sigma_x(\gamma^{-1})\xi_x$,
(2) $\pi_x(\gamma)\sigma_x(\gamma') = \sigma_x(\gamma')\pi_x(\gamma)$, and
(3) ξ_x is a cyclic vector for both π_x and σ_x,

for all $x \in X'$ and all $\gamma, \gamma' \in \Gamma$.

For $x \in X'$, let φ_x be the function of positive type on Γ defined by

$$\varphi_x(\gamma) = \langle \pi_x(\gamma)\xi_x \mid \xi_x\rangle \quad \text{for every} \quad \gamma \in \Gamma.$$

We claim that φ_x is conjugation-invariant, so that φ_x is a trace on Γ in the sense of a notion to be introduced in Section 11.B.

Indeed, using (1) and (2), we have, for every $\gamma_1, \gamma_2 \in \Gamma$,

$$\begin{aligned}
\varphi_x(\gamma_2\gamma_1\gamma_2^{-1}) &= \langle \pi_x(\gamma_2\gamma_1\gamma_2^{-1})\xi_x \mid \xi_x\rangle \\
&= \langle \pi_x(\gamma_2\gamma_1)\sigma_x(\gamma_2)\xi_x \mid \xi_x\rangle \\
&= \langle \sigma_x(\gamma_2)\pi_x(\gamma_2)\pi_x(\gamma_1)\xi_x \mid \xi_x\rangle \\
&= \langle \pi_x(\gamma_1)\xi_x \mid \pi_x(\gamma_2^{-1})\sigma_x(\gamma_2^{-1})\xi_x\rangle \\
&= \langle \pi_x(\gamma_1)\xi_x \mid \xi_x\rangle \\
&= \varphi_x(\gamma_1).
\end{aligned}$$

For every Borel subset E of X, the operator $P_E := U^{-1}m(\mathbf{1}_E)U$ is a projection in $\mathcal{Z}$ and we have

$$\begin{aligned}
\langle \lambda_\Gamma(\gamma)P_E\delta_e \mid \delta_e\rangle &= \langle \lambda_\Gamma(\gamma)P_E\delta_e \mid P_E\delta_e\rangle \\
&= \int_E \langle \pi_x(\gamma)\xi_x \mid \xi_x\rangle d\mu(x) \\
&= \int_E \varphi_x(\gamma) d\mu(x).
\end{aligned}$$

Let $\gamma \notin \mathrm{FC}(\Gamma)$. The proof of Proposition 7.A.1 shows that $T\delta_e(\gamma) = 0$ for every $T \in \mathcal{Z}$. It follows that

$$\int_E \varphi_x(\gamma) d\mu(x) = 0$$

for every Borel subset E of X. This clearly implies that $\varphi_x(\gamma) = 0$ for μ-almost every $x \in X$.

Since Γ is countable, it follows that there exists a Borel subset X'' of X' with $\mu(X \smallsetminus X'') = 0$ such that

(4) $\varphi_x(\gamma) = 0$ for every $x \in X''$ and every $\gamma \in \Gamma \smallsetminus \mathrm{FC}(\Gamma)$.

Let $x \in X''$. Recall that the GNS representation corresponding to φ_x is π_x and has ξ_x as a cyclic vector. We claim that the Neumann algebra $\mathcal{M}_x := \pi_x(\Gamma)''$ is a factor of type II_1.

Indeed, $\mathcal{M}_x$ is a factor since π_x is factorial. Moreover, define a positive linear functional $\tau_x : \mathcal{M}_x \to \mathbf{C}$ by $\tau_x(T) = \langle T\xi_x \mid \xi_x\rangle$ for $T \in \mathcal{M}_x$.

We show that τ_x is a (finite) trace on $\mathcal{M}_x$ by proceeding as in the proof of Proposition 7.A.3: the invariance of φ_x under conjugation implies that $\tau_x(TS) = \tau_x(ST)$ for every $T, S \in \mathcal{M}_x$; the fact that ξ_x is cyclic for σ_x (by (3) above), together with the fact that $\sigma_x(\Gamma) \subset \pi_x(\Gamma)'$, implies that τ_x is faithful.

Denote by ψ_x the restriction of φ_x to $\mathrm{FC}(\Gamma)$. By (4) above, φ_x is the trivial extension of ψ_x to Γ (see Definition 1.B.8). Hence (see Proposition 1.F.10), π_x is equivalent to an induced representation $\mathrm{Ind}_{\mathrm{FC}(\Gamma)}^{\Gamma}\rho_x$ for a representation ρ_x of $\mathrm{FC}(\Gamma)$. Since $[\Gamma : \mathrm{FC}(\Gamma)]$ is infinite, it follows that π_x is infinite-dimensional (see Remark 1.F.2 (5)). As π_x is a cyclic representation, this implies that $\mathcal{M}_x = \pi_x(\Gamma)''$ is infinite-dimensional.

As a result, $\pi_x(\Gamma)''$ is a finite factor of infinite-dimension. Hence, π_x is of type II_1 and so Γ is not of type I.

Let us prove the other implications of Theorem 7.D.1.

If (iii) holds, then Γ is a CCR group and so (i) holds, by Theorem 6.E.5 (in the case where Γ is not countable, see Comment (11) on Theorem 8.F.3 below).

It remains to show that (ii) $\Longrightarrow$ (iii). Assume that Γ contains an abelian normal subgroup N of finite index in Γ and let $(\pi, \mathcal{H})$ be an irreducible representation of Γ.

Let E be the projection-valued measure on $\widehat{N}$ associated with $\pi|_N$ and S the support of E (see Section 2.C). By Proposition 2.F.1(2), S is the closure of a Γ/N-orbit in $\widehat{N}$. Since Γ/N is finite, it follows that S is the Γ/N-orbit of some element in $\widehat{N}$. We have $\mathcal{H} = \bigoplus_{\chi \in S} \mathcal{H}^\chi$, where

$$\mathcal{H}^\chi := \{\xi \in \mathcal{H} \mid \pi(n)\xi = \chi(n)\xi \ \text{ for all } \ n \in N\};$$

moreover, since N is a normal subgroup, we have $\pi(\gamma)\mathcal{H}^\chi = \mathcal{H}^{\chi^{\gamma^{-1}}}$ for every $\chi \in S$ and every $\gamma \in \Gamma$.

Fix some element $\chi \in S$. Let $H := \Gamma_\chi$ be the stabilizer of χ and let T be a right transversal for H in Γ. Then $\mathcal{H}^\chi$ is invariant under $\pi(H)$ and, since S is the Γ-orbit of χ, we have

$$\mathcal{H} = \bigoplus_{t \in T} \mathcal{H}^{\chi^{t^{-1}}} = \bigoplus_{t \in T} \pi(t)\mathcal{H}^\chi.$$

This shows that π is equivalent to the induced representation $\mathrm{Ind}_H^\Gamma\sigma$, where σ is the subrepresentation of $\pi|_H$ defined on $\mathcal{H}^\chi$ (see Definition 1.F.1).

We claim that $\mathrm{Ind}_H^\Gamma\sigma$ is contained in $\mathrm{Ind}_N^\Gamma\chi$. Once proved, it will follow that $\dim \pi = \dim \mathrm{Ind}_H^\Gamma\sigma$ is finite, since N has finite index and so $\dim \mathrm{Ind}_N^\Gamma\chi$ is finite.

To prove the claim, observe that $\sigma|_N$ is a multiple $n\chi$ of χ for some cardinal n. Recall (see Proposition 1.F.6(4)) that $\mathrm{Ind}_N^H(\sigma|_N)$ is equivalent to the tensor product representation $\sigma \otimes \lambda_{H/N}$, where $\lambda_{H/N}$ is the quasi-regular representation. Since H/N is finite, $\lambda_{H/N}$ contains 1_H. It follows that $\mathrm{Ind}_N^H(\sigma|_N)$ contains σ. Now, $\mathrm{Ind}_N^H(\sigma|_N)$ is equivalent to $n\mathrm{Ind}_N^\Gamma\chi$ and σ is irreducible. This implies that $\mathrm{Ind}_N^H\chi$ contains σ. So, $\mathrm{Ind}_H^\Gamma(\mathrm{Ind}_N^H\chi)$, which (by induction in stages) is equivalent to $\mathrm{Ind}_N^\Gamma\chi$, contains $\mathrm{Ind}_H^\Gamma\sigma$ and the claim is proved. □

REMARK 7.D.2. (1) As mentioned in the proof above, the implication (i) $\Longrightarrow$ (ii) in Theorem 7.D.1 in the case of finitely generated groups is a direct consequence of [**Maut–50c**]. In fact, the proof shows that if Γ is a countable group such that $[\Gamma : \mathrm{FC}(\Gamma)] = \infty$, then λ_Γ is a representation of type II.

(2) Let Γ be a virtually abelian group. The proof of (ii) $\Longrightarrow$ (iii) in Theorem 7.D.1 shows that the dimensions of the irreducible representations of Γ are uniformly bounded: if N is an abelian normal subgroup of finite index in Γ, then $\dim \pi \le [\Gamma : N]$ for every irreducible representation π of Γ.

Type of the regular representation. In complement to Theorem 7.D.1, here are other results about the type of the regular representation λ_Γ of a discrete group Γ. We denote by $\mathrm{D}(\Gamma)$ the commutator group of Γ.

THEOREM 7.D.3. *Let Γ be a group.*

(1) *Γ is of type I if and only if λ_Γ is of type I.*
(2) *λ_Γ is of type II if and only if either $[\Gamma : \mathrm{FC}(\Gamma)] = \infty$, or $[\Gamma : \mathrm{FC}(\Gamma)] < \infty$ and $\mathrm{D}(\mathrm{FC}(\Gamma))$ is infinite.*
(3) *In particular, if Γ is not virtually abelian, $\mathrm{FC}(\Gamma)$ of finite index in Γ, and $\mathrm{D}(\mathrm{FC}(\Gamma))$ is a finite group, then λ_Γ has both a type I component and a type II component.*
(4) *If Γ is finitely generated, then λ_Γ is either of type I or of type II.*

NOTE. Since λ_Γ is finite, one can replace "type II" by "type II_1" in (2) to (4).

REFERENCES FOR PROOFS. Parts (1) and (2) are Sätze 2 and 1 in [**Kani–69**]. Observe that the fact that λ_Γ is of type II if $[\Gamma : \mathrm{FC}(\Gamma)] = \infty$ was already shown in [**Maut–50c**] (see Remark 7.D.2). Claim (3) follows from Parts (1) and (2).

Claim (4) follows from Theorem 7.D.1 & (2), and the fact, already mentioned above, that a finitely generated group Γ with $\mathrm{FC}(\Gamma)$ of finite index is virtually abelian, and therefore of type I. □

REMARK 7.D.4. (1) Let Γ be a countable group which is not virtually abelian. Then Γ has factor representations of type II_1, and of type III; see Theorem 8.F.4 below.

(2) Claim (1) of the theorem *does not* carry over to LC groups: there are groups with left regular representation of type I that are not of type I (see Section 7.E).

EXAMPLE 7.D.5. Except for the infinite dihedral group, our main examples of Chapter 3 are not virtually abelian, and therefore not of type I. We point out below some explicit examples of factor representations of type II_1 of these groups.

(1) The Heisenberg group $H(\mathbf{Z})$ of Section 3.B is not of type I. For factor representations of $H(\mathbf{Z})$ of types II and III, we refer to Example 14.B.2.

(2) Let Γ be an icc group, so that the left regular representation of Γ is factorial of type II_1. Assume moreover that Γ is amenable (for example solvable). Then $\mathcal{L}(\Gamma)$ is *the* hyperfinite factor of type II_1, by the famous result of Connes on the uniqueness of the hyperfinite factor of type II_1 [**Conn–76**, see in particular Corollary 7.2]. This applies to

- $\mathrm{Aff}(\mathbf{K})$, the affine group over an infinite field $\mathbf{K}$, as in Section 3.C,
- the Baumslag–Solitar group $\mathrm{BS}(1, n)$ for every integer $n \ge 2$, as in Section 3.D,
- the restricted wreath product $H \wr A$ whenever H and A are nontrivial abelian countable groups, for example the lamplighter group $\mathbf{Z} \wr (\mathbf{Z}/2\mathbf{Z})$ of Section 3.E.

(3) For an infinite field $\mathbf{K}$ and an integer $n \ge 2$, the general linear group $\Gamma = \mathrm{GL}_n(\mathbf{K})$ of Section 3.F is not type I. The quotient group $\mathrm{PGL}_n(\mathbf{K}) = \mathrm{GL}_n(\mathbf{K})/\mathbf{K}^\times$

can be shown to be icc, for example because it is Zariski dense in an appropriate algebraic group with trivial centre (compare [**BeHa–94**, Proposition 2]), so that its left regular representation is of type II_1. It follows that the quasi-regular representation $\lambda_{\mathrm{GL}_n(\mathbf{K})/\mathbf{K}^\times}$ of $\mathrm{GL}_n(\mathbf{K})$ in $\ell^2(\mathrm{PGL}_n(\mathbf{K}))$ is factorial of type II_1.

(4) The question to know whether appropriate quasi-regular representations provide examples of factor representation of types II and III has attracted some attention. Here is one example:

For the integral translation subgroup of the rational affine group

$$T = \begin{pmatrix} 1 & \mathbf{Z} \\ 0 & 1 \end{pmatrix} \subset \mathrm{Aff}(\mathbf{Q}) = \begin{pmatrix} \mathbf{Q}^\times & \mathbf{Q} \\ 0 & 1 \end{pmatrix},$$

the quasi-regular representation $\lambda_{\mathrm{Aff}(\mathbf{Q})/T}$ is of type III [**Bind–93**, Proposition 3.6].

Groups with regular representation of mixed type. Let Γ be a discrete group which is not virtually abelian; assume that $[\Gamma : \mathrm{FC}(\Gamma)] < \infty$ and that $\mathrm{D}(\mathrm{FC}(\Gamma))$ is finite. It follows from Theorem 7.D.3 that the regular representation of Γ must involve representations of different types. The following result of Kaplansky [**Kapl–51a**, Theorem 2] gives the explicit type decomposition of λ_Γ for a class of examples. Recall that $\mathcal{L}(\Gamma)$ denotes the von Neumann algebra generated by λ_Γ.

THEOREM 7.D.6. *Let Γ be an infinite discrete group with centre Z. Assume that Z is finite of prime order p and that $\mathrm{D}(\Gamma) = Z$.*
Then

$$\mathcal{L}(\Gamma) = \mathcal{A} \oplus \bigoplus_{i=1}^{p-1} \mathcal{M}_i,$$

where $\mathcal{A}$ is an abelian von Neumann algebra and the $\mathcal{M}_i$'s are factors of type II_1.

PROOF. Let z_0 be a generator for the cyclic group Z of order p. The unitary dual $\widehat{Z}$ can be identified with the group C_p of p-th roots of unity in $\mathbf{C}$: to every $\omega \in C_p$ corresponds the unitary character $\chi_\omega : Z \to \mathbf{T}$ defined by $\chi_\omega(z_0^j) = \omega^j$ for $j \in \{0, 1, \dots, p-1\}$.

We have an orthogonal decomposition of $\ell^2(\Gamma)$ under the action of $\lambda_\Gamma(z_0)$, given by

$$\ell^2(\Gamma) = \bigoplus_{\omega \in C_p} \mathcal{H}_\omega,$$

where

$$\mathcal{H}_\omega = \{f \in \ell^2(\Gamma) \mid f(z_0 x) = \omega f(x) \text{ for every } x \in \Gamma\}.$$

Let $\omega \in C_p$. Observe that $\mathcal{H}_\omega$ is invariant under $\lambda_\Gamma(\Gamma)$ as well as under $\rho_\Gamma(\Gamma)$. The corresponding orthogonal projection $P_\omega : \ell^2(\Gamma) \to \mathcal{H}_\omega$ belongs to the centre of the von Neumann algebra $\mathcal{L}(\Gamma)$ and is given by

$$P_\omega(f)(x) = \frac{1}{p}\sum_{i=0}^{p-1} \omega^{-i} f(z_0^i x) \quad \text{for } f \in \ell^2(\Gamma) \text{ and } x \in \Gamma.$$

Denote by π_ω the restriction of λ_Γ to $\mathcal{H}_\omega$. The von Neumann algebra $\pi_\omega(\Gamma)''$ can be identified with the induced von Neumann algebra $\mathcal{L}(\Gamma)_{P_\omega}$. It is clear that

$$\pi = \bigoplus_{\omega \in C_p} \pi_\omega$$

and that

$$\mathcal{L}(\Gamma) = \bigoplus_{\omega \in C_p} \pi_\omega(\Gamma)''.$$

• *First case.* Assume that $\omega = 1$. We claim that $\pi_1(\Gamma)''$ is abelian.

Indeed, $\mathcal{H}_1$ can be identified with $\ell^2(\Gamma/Z)$ and π_1 with $\lambda_{\Gamma/Z}$. The claim follows, since Γ/Z is abelian.

• *Second case.* Assume that $\omega \neq 1$. We claim that π_ω is factorial of type II_1.

Indeed, set

$$f_\omega := P_\omega(\delta_e) \in \mathcal{H}_\omega.$$

Then f_ω is a cyclic vector for the restriction of ρ_Γ to $\mathcal{H}_\omega$, since δ_e is a cyclic vector for ρ_Γ and since $P_\omega \in \rho_\Gamma(\Gamma)'$. Therefore, the linear map

$$\pi_\omega(\Gamma)'' \to \mathcal{H}_\omega, \qquad T \mapsto T(f_\omega)$$

is injective.

We first prove that $\pi_\omega(\Gamma)''$ is a factor. Let $T \in \mathcal{L}(\mathcal{H}_\omega)$ be in the centre of $\pi_\omega(\Gamma)''$. In order to show that T is a scalar multiple of the identity, it suffices check that the function

$$f := T(f_\omega) \in \mathcal{H}_\omega$$

is a scalar multiple of f_ω.

As in the proof of Proposition 7.A.1, f is invariant under conjugation by elements of Γ. We claim that $f(x) = 0$ for every $x \in \Gamma \smallsetminus Z$.

Indeed, assume that $f(x) \neq 0$ for some $x \in \Gamma \smallsetminus Z$. Then there exists $y \in \Gamma$ with $yxy^{-1} \neq x$. Since $\mathrm{D}(\Gamma) = Z$, we therefore have $yxy^{-1} = z_0^i x$ for some $i \in \{1, \dots, p-1\}$. It follows that

$$f(x) = f(yxy^{-1}) = f(z_0^i x) = \omega^i f(x),$$

and this is a contradiction, since $\omega^i \neq 1$.

So, $f = 0$ on $\Gamma \smallsetminus Z$. Since $f \in \mathcal{H}_\omega$, we have, on the one hand,

$$f(z_0^i) = \omega^i f(z_0) \quad \text{for every } i \in \{0, \dots, p-1\};$$

on the other hand, a straightforward computation shows that

$$f_\omega(x) = \begin{cases} \dfrac{\omega^i}{p} & \text{if } x = z_0^i \text{ for } i \in \{0, \dots, p-1\}, \\ 0 & \text{otherwise.} \end{cases}$$

Therefore f is a scalar multiple of f_ω. So, $\pi_\omega(\Gamma)''$ is a factor.

Next, we show that the factor $\pi_\omega(\Gamma)''$ is of finite type. The canonical trace of $\mathcal{L}(\Gamma)$, as in Proposition 7.A.3, defines by restriction a faithful trace on

$$\pi_\omega(\Gamma)'' \cong P_\omega \mathcal{L}(\mathcal{H}) P_\omega.$$

Therefore $\pi_\omega(\Gamma)''$ is a finite factor.

It remains to show that $\pi_\omega(\Gamma)''$ is infinite-dimensional. For this, it suffices to check that the image of $\pi_\omega(\Gamma)''$ under the map

$$\pi_\omega(\Gamma)'' \to \ell^2(\Gamma), \ \ T \mapsto T(f_\omega)$$

is infinite-dimensional. This is indeed the case: since Γ is infinite and Z is finite, we can find an infinite sequence $(x_n)_{n \geq 1}$ of elements $x_n \in \Gamma$ with $x_m Z \neq x_n Z$ for all m, n with $m \neq n$. Set

$$f_n := \pi_\omega(x_n)(f_\omega) \quad \text{for } n \geq 1.$$

One checks that $\mathrm{supp}(f_n) = x_n Z$. Therefore the family $(f_n)_{n\geq 1}$ is linearly independent. □

EXAMPLE 7.D.7. A specific example as in Theorem 7.D.6. is the "Heisenberg group" $H_\infty(\mathbf{F}_p)$ defined as follows. Let p be an odd prime. Denote by $\mathbf{F}_p$ the field of order p, by $V = \mathbf{F}_p^{(\mathbf{N})}$ a vector space over $\mathbf{F}_p$ of countable infinite dimension, by $(x_i)_{i\in\mathbf{N}}$ a typical element of V, with $x_i \in \mathbf{F}_p$ and $x_i = 0$ for almost all $i \in \mathbf{N}$, and by ω the symplectic form on $V \oplus V$, defined by

$$\omega((x,y),(x',y')) = \sum_{i\in\mathbf{N}} (x_i y_i' - y_i x_i') \quad \text{for } (x,y),(x',y') \in V \oplus V.$$

Then $H_\infty(\mathbf{F}_p)$ is the group with underlying set $V \oplus V \oplus \mathbf{F}_p$ and with multiplication defined by

$$(x,y,z)(x',y',z') = (x+x', y+y', z+z' + \omega((x,y),(x',y')))$$

for $(x,y,z),(x',y',z') \in H_\infty(\mathbf{F}_p)$. The additive group of $\mathbf{F}_p$ is identified with the subgroup of $H_\infty(\mathbf{F}_p)$ of elements of the form $(0,0,z)$. It is straightforward to check (using that p is odd) that

(1) the commutator group and the centre of $H_\infty(\mathbf{F}_p)$ both coincide with $\mathbf{F}_p$,
(2) all conjugacy classes of $H_\infty(\mathbf{F}_p)$ are finite, i.e., $\mathrm{FC}(\Gamma) = \Gamma$,
(3) $H_\infty(\mathbf{F}_p)$ is not virtually abelian.

It follows from Theorem 7.D.6 that $\lambda(H_\infty(\mathbf{F}_p))''$ is a direct sum of an abelian von Neumann algebra and $p-1$ factors of type II_1.

7.E. On types of representations of locally compact groups

We quote below some known results about the possible types of the regular representation for some classes of locally compact groups G.

(1) If G is unimodular, then λ_G has no type III part [**Sega–50b**]. Also, if G is connected, then λ_G has no type III part; this is a fact established in [**Dixm–69**], building up on previous results for second-countable groups and solvable Lie groups from [**Gode–51b**], [**Sega–50a**], [**Maut–50b**, Lemma 7.4], and Pukánszky ([**Puka–71**, Section IV] and a previous announcement).

If G is connected, then G does not have any representation of type II_1 [**KaSi–52**] (see Corollary 15.D.3).

(2) If G is a LC group, λ_G is finite if and only if G is a SIN group (see Theorem 15.C.3).

(3) More generally, for G locally compact, λ_G has a nontrivial finite part if and only if there exists at least one invariant compact neighbourhood of e. Moreover, when this holds, there exists a compact normal subgroup K of G, say with quotient $Q := G/K$, such that the finite part of $\lambda_G(G)''$ is isomorphic to $\lambda_Q(Q)''$ [**Tayl–76**, Section 4]. There are characterizations of locally compact groups G such that $\lambda_G(G)''$ has a nonzero type I finite part [**Tayl–76**, Section 5].

(4) A tensor product $\mathcal{M} \otimes \mathcal{N}$ is a factor of type II_∞ whenever $\mathcal{M}$ is a factor of type II_1 and $\mathcal{N}$ a factor of type I_∞. As we have already noted in Remark 3.C.7, the left regular representation of the affine group $\mathrm{Aff}(\mathbf{R})$ is a factorial representation of type I_∞. From this, Proposition 7.C.5, and the previous observation, it follows

that, for any icc discrete group Γ, the left regular representation of the Cartesian product $\Gamma \times \mathrm{Aff}(\mathbf{R})$ is of type II_∞.

(5) Nilpotent connected LC groups are of type I, as mentioned in Theorem 6.E.19 but connected solvable LC groups need not be. Consider for example the Mautner group, that is, the semi-direct product $M := \mathbf{R} \ltimes \mathbf{C}^2$ for the action of $\mathbf{R}$ on $\mathbf{C}^2$ given by $t(z, z') = (e^{2\pi i t}z, e^{2\pi i \lambda t}z')$ for all $t \in \mathbf{R}$ and $(z, z') \in \mathbf{C}^2$, for some irrational $\lambda \in \mathbf{R}$. This group is not of type I, see [**AuMo–66**, Page 138, and Theorem 9 Page 110]; indeed its regular representation is of type II_∞ [**Kaji–83**, Proposition 4.1]. Observe that the Mautner group is unimodular. As much as we know, Mautner himself has not published anything more on this group than some allusions in [**Maut–50b**].

(6) Mackey has found an LC group that has the following properties: it is solvable, simply connected, not unimodular, not type I, but its regular representation is type I [**Mack–61**]; see also [**KlLi–73**, Example 3.e].

(7) For examples of groups with λ_G of type III, see Section 14.D below.

7.F. Non type I factor representations and irreducible representations

Let G be a second-countable LC group and π a representation of G on a Hilbert space. Recall from Section 1.G that π can be written as a direct integral of irreducible representations $\pi \simeq \int_X^\oplus \pi_x d\mu(x)$ and that such a decomposition is usually not unique. However, if π is a factor representation, the following result shows that almost all π_x's are weakly equivalent to π. The proof we give, which apparently is due to Fell, is patterned after the one sketched in [**Dixm–60a**, Remarque on page 100].

Proposition 7.F.1. *Let G be a second-countable locally compact group and $x \mapsto \pi_x$ a measurable field of representations of G over a standard Borel space X equipped with σ-finite measure μ. Assume that the direct integral $\pi := \int_X^\oplus \pi_x d\mu(x)$ is a factor representation.*

Then there exists a Borel subset N of X with $\mu(N) = 0$ such that π_x is weakly equivalent to π, for every $x \in X \smallsetminus N$.

Proof. Let $E \subset X$ be a Borel subset with $\mu(E) > 0$. Let

$$\pi_E := \int_E \pi_x d\mu(x)$$

be the direct integral of the π_x's over E. It is clear that π_E is a nontrivial subrepresentation of π. Since π is factorial, π_E is quasi-equivalent to π by Corollary 6.B.8. It follows that π_E is weakly equivalent to π (Corollary 6.A.5). Therefore, on the one hand, we have, by Proposition 6.E.4,

$$\|\pi_E(f)\| = \|\pi(f)\| \text{ for every } f \in C^c(G).$$

On the other hand, since $\pi_E(f)$ is a decomposable operator, $\|\pi_E(f)\|$ is the μ-essential supremum over E of the function $x \mapsto \|\pi_x(f)\|$ (see Section 1.H).

We have shown that $\|\pi(f)\|$ coincides with the μ-essential supremum over E of $x \mapsto \|\pi_x(f)\|$ for every $f \in C^c(G)$ and every Borel subset E of X with $\mu(E) > 0$.

Since G is second-countable, we can find a sequence $(f_n)_{n\geq 1}$ in $C^c(G)$ which is dense for the $L^1(G)$-norm. For every pair of integers $n, m \geq 1$, set

$$E_{n,m} := \left\{ x \in X \mid \|\pi_x(f_n)\| \leq \|\pi(f_n)\| - \frac{1}{m} \right\}.$$

Then $E_{n,m}$ is a measurable subset of X and it follows from what we have just shown that $\mu(E_{n,m}) = 0$. Set

$$N = \Big(\bigcup_{n,m} E_{m,n} \Big) \cup \Big(\bigcup_{n} \{ x \in X \mid \|\pi(f_n)\| < \|\pi_x(f_n)\| \} \Big).$$

Then $\mu(N) = 0$, and for $x \in X \smallsetminus N$, we have

$$\|\pi_x(f_n)\| = \|\pi(f_n)\| \quad \text{for every} \quad n \geq 1.$$

Therefore π_x is weakly equivalent to π for $x \in X \smallsetminus N$, by Proposition 6.E.4 again. □

The following corollary of Proposition 7.F.1 establishes an important link between the quasi-dual $\mathrm{QD}(G)$ and the primitive dual $\mathrm{Prim}(G)$ of G, introduced respectively in Sections 6.C and 1.E. Recall that $\mathrm{Prim}(G)$ is in a canonical way a quotient of $\widehat{G}$.

COROLLARY 7.F.2. *Let G be a second-countable locally compact group and π a factor representation of G.*

Then there exists an irreducible *representation of G which is weakly equivalent to π.*

The map $\mathrm{QD}(G) \twoheadrightarrow \mathrm{Prim}(G)$, *which to π associates the weak equivalence class of any $\sigma \in \widehat{G}$ weakly equivalent to π, extends the natural map $\kappa^{\mathrm{d}}_{\mathrm{prim}} : \widehat{G} \twoheadrightarrow \mathrm{Prim}(G)$ of Definition* 1.E.1.

PROOF. By Corollary 6.B.9 and Corollary 6.A.5, we can assume that the Hilbert space of π is separable. Then π admits a direct integral decomposition

$$\pi \simeq \int_X^{\oplus} \pi_x d\mu(x)$$

into irreducible representations (see Theorem 1.G.9) and the claim follows from Proposition 7.F.1. □

REMARK 7.F.3. (1) Corollary 7.F.2 follows also from Corollary 8.A.21 below.

(2) Corollary 7.F.2 holds more generally for a σ-compact locally compact group G (see Proposition 8.D.3).

In the case of non type I factor representations, we can prove the following stronger version of Corollary 7.F.2.

COROLLARY 7.F.4. *Let G be a second-countable locally compact group and π a factor representation of G. Assume that π is not of type I.*

Then there exist uncountably many pairwise nonequivalent irreducible representations of G which are weakly equivalent to π.

PROOF. As in the proof of Corollary 7.F.2, we can assume that the Hilbert space of π is separable. Let

$$\pi \simeq \int_X^{\oplus} \pi_x d\mu(x)$$

be a direct integral decomposition into irreducible representations over a standard Borel space X defined by a measurable field $x \mapsto \mathcal{H}_x$ of Hilbert spaces. By Proposition 7.F.1, we can assume that the π_x's are all weakly equivalent to π.

Let σ be an irreducible representation of G. Set

$$E_\sigma := \{x \in X \mid \pi_x \simeq \sigma\}$$

(recall that $\simeq$ denotes equivalence of representations).

We claim that E_σ is a Borel subset of X. Indeed, let $p \in \{1, 2, \ldots, +\infty\}$ be the dimension of σ, let

$$X_p = \{x \in X \mid \dim \mathcal{H}_x = p\},$$

and let $\mathcal{K}$ be a fixed Hilbert space of dimension p. By Proposition 1.G.2, X_p is a Borel subset of X and the subrepresentation of π defined on

$$\int_{X_p}^{\oplus} \mathcal{H}_x d\mu(x) \approx L^2(X_p, \mu, \mathcal{K})$$

is equivalent to a representation on $L^2(X_p, \mu, \mathcal{K})$ which is a direct integral $\int_{X_p} \pi_x^{(p)} d\mu(x)$ over X_p of representations $\pi_x^{(p)}$ in $\mathcal{K}$. Let $\mathrm{Irr}_p(G)$ be the set of irreducible representations of G in $\mathcal{K}$, equipped with its Mackey–Borel structure as in Section 6.C. The set

$$\Omega_\sigma = \{\rho \in \mathrm{Irr}_p(G) \mid \rho \simeq \sigma\}$$

consists of the representations $\rho \in \mathrm{Irr}_p(G)$ for which $\mathrm{Hom}_G(\rho, \sigma)$ has dimension 1. By [**Dixm–C***, Lemma 3.7.3], Ω_σ is a Borel subset of $\mathrm{Irr}_p(G)$. Since

$$X_p \to \mathrm{Irr}_p(G), \quad x \mapsto \pi_x^{(p)}$$

is measurable and since

$$E_\sigma = \{x \in X_p \mid \pi_x^{(p)} \in \Omega_\sigma\},$$

E_σ is a Borel subset of X.

We claim that $\mu(E_\sigma) = 0$. Once proved, this will imply the claim, since it will follow that $\mu(\bigcup_{\sigma \in C} E_\sigma) = 0$ for every countable subset C of $\widehat{G}$.

Assume by contradiction $\mu(E_\sigma) > 0$ for some $\sigma \in \widehat{G}$, with Hilbert space $\mathcal{H}_\sigma$. Consider the representation $(\pi_{E_\sigma}, \mathcal{H}_{E_\sigma})$, where

$$\pi_{E_\sigma} := \int_{E_\sigma}^{\oplus} \pi_x d\mu(x) \text{ and } \mathcal{H}_{E_\sigma} := \int_{E_\sigma}^{\oplus} \mathcal{H}_x d\mu(x).$$

For every $x \in E_\sigma$, there exists an isomorphism of Hilbert spaces $U_x : \mathcal{H}_\sigma \to \mathcal{H}_x$ intertwining σ and π_x. It follows that there exists an isomorphism of Hilbert spaces

$$\mathcal{H}_\sigma \otimes L^2(E_\sigma, \mu) \to \mathcal{H}_{E_\sigma}$$

intertwining a multiple of σ and π_{E_σ} (see Théorème 2 in [**Dixm–vN**, Chap. II, § 2]). So, π_{E_σ} is equivalent to a multiple of σ and is therefore of type I. However, Corollary 6.B.8 shows that π_{E_σ} is quasi-equivalent to π, since π is a factor representation and since π_{E_σ} is a nontrivial subrepresentation of π. It follows that π is of type I and this is a contradiction. □

CHAPTER 8

Representations of C*-algebras and of LC groups, the Glimm Theorem

Let G be a second-countable LC group and $\widehat{G}$ its dual, equipped with the Mackey–Borel structure, as defined in Section 6.C*. On the one hand, as we will see in this chapter (Glimm's Theorem* 8.F.3*), $\widehat{G}$ is not countably separated, unless G is of type I, as defined in Section* 6.D*. On the other hand, a result of Effros (Theorem* 8.A.25*) shows that the primitive dual* $\mathrm{Prim}(G)$ *is always a standard Borel space. This shows that* $\mathrm{Prim}(G)$ *is much better behaved than $\widehat{G}$, and one may hope to classify* $\mathrm{Prim}(G)$ *for some specific non type I groups. This will be done in Chapter* 9 *for most of our test examples.*

Given a representation π of a locally compact group G, it is useful to "extend" π to group algebras associated with G. Such a group algebra, which is universal in an appropriate sense, is the maximal C^*-algebra $C^*_{\max}(G)$ of G, which is defined in Section 8.B*. The representations of the group G correspond precisely to the nondegenerate representations of the C^*-algebra $C^*_{\max}(G)$. The spectrum (or dual space) $\widehat{A}$ and the primitive dual (or primitive ideal space)* $\mathrm{Prim}(A)$ *are defined for a general C^*-algebra A, and agree with $\widehat{G}$ and* $\mathrm{Prim}(G)$ *in case $A = C^*_{\max}(G)$.*

Section 8.A *is devoted to the study of* $\mathrm{Prim}(A)$ *for a separable C^*-algebra A, where* $\mathrm{Prim}(A)$ *is equipped with the Borel structure induced by the Jacobson topology. In particular, we give a complete proof of Effros' Theorem:* $\mathrm{Prim}(A)$ *is a standard Borel space.*

In Section 8.C*, we discuss in some length the functorial properties of maximal C^*-algebras and reduced C^*-algebras of locally compact groups. The assignment $G \rightsquigarrow C^*_{\max}(G)$ for a general locally compact group G does not have good functorial properties, with respect to arbitrary continuous homomorphisms. However, as we will see, $\Gamma \rightsquigarrow C^*_{\max}(\Gamma)$ is a covariant functor from the category of discrete groups to the category of C^*-algebras. In contrast, this fails for the assignment $\Gamma \rightsquigarrow C^*_{\lambda}(\Gamma)$, where $C^*_{\lambda}(\Gamma)$ denotes the reduced C^*-algebra of the group Γ.*

Several results about a second-countable LC group G use the fact that $C^*_{\max}(G)$ is a separable C^*-algebra. In Section 8.D*, we show how some of these results can be extended to σ-compact LC groups, using as one tool the Kakutani–Kodaira Theorem* 8.D.1*. In particular, the fact that a factor representation of G is weakly equivalent to an irreducible one (which was previously proved in case G is second-countable, see Corollary* 7.F.2*) is valid for any σ-compact LC group.*

In Section 8.E*, we come back to the notion of a central character of a group representation, that has already appeared in Section* 1.A*, and we show that it can be associated with a primitive ideal.*

Section 8.F *contains a statement of Glimm's Theorem, which characterizes type I for second-countable LC groups G. In particular, this class of groups coincides* *with the class of GCR groups from Section* 6.E.b. *Actually, Glimm's Theorem characterizes separable C*-algebras of type I (see Remark* 8.F.2 *for this notion). We prefer to state a formulation in terms of second-countable groups, and to show how it extends to the case of σ-compact LC groups.*

In Section 8.G, *we discuss alternative definitions of the von Neumann algebra $\pi(G)''$ associated with a representation π of a locally compact group G (see Section* 6.B*). In Section* 8.H, *we briefly discuss one way to replace the maximal C*-algebra of a locally compact group G by other group algebras.*

8.A. Spectrum and primitive ideal space of a C*-algebra

Our goal in this section is to prove Effros' Theorem 8.A.25 about primitive duals of second-countable LC groups, which is a special case of a more general result (Theorem 8.A.24) about separable C*-algebras. For this, we need to relate the primitive ideal space of a C*-algebra to its space of pure states.

8.A.a. Spectrum and primitive ideal space. We first discuss basic facts about representations of complex involutive algebras and C*-algebras; for a reminder on theses algebras, see A.10.

Let A be a complex involutive algebra. A **representation** of A on a Hilbert space $\mathcal{H}$ is a morphism of involutive algebras $\pi : A \to \mathcal{L}(\mathcal{H})$. For a family $(\pi_\iota : A \to \mathcal{L}(\mathcal{H}_\iota))_{\iota \in I}$, there is a natural notion of direct sum $\bigoplus_{\iota\in I} \pi_\iota : A \to \mathcal{L}\big(\bigoplus_{\iota \in I} \mathcal{H}_\iota\big)$. Two representations π_1, π_2 of A on Hilbert spaces $\mathcal{H}_1, \mathcal{H}_2$ are **equivalent** if there exists a surjective linear isometry $U : \mathcal{H}_1 \to \mathcal{H}_2$ such that $U\pi_1(x) = \pi_2(x)U$ for all $x \in A$.

A representation π of a complex involutive algebra A on a Hilbert space $\mathcal{H}$ is **trivial** if $\pi(A) = \{0\}$. It is **nondegenerate** if the closed linear span $\overline{\pi(A)\mathcal{H}}$ of

$$\{\pi(a)\xi \mid a \in A, \ \xi \in \mathcal{H}\}$$

is $\mathcal{H}$ itself. Every representation $\pi : A \to \mathcal{L}(\mathcal{H})$ is a direct sum

$$\pi \ = \ \pi_0 \oplus \pi_1$$

of a trivial representation π_0 and a nondegenerate representation π_1. Indeed,

$$\mathcal{H}_0 \ = \ \{\xi \in \mathcal{H} \mid \pi(A)\xi = \{0\}\} \qquad \text{and} \qquad \mathcal{H}_1 = \overline{\pi(A)\mathcal{H}}$$

are $\pi(A)$-invariant closed subspaces of $\mathcal{H}$, the corresponding subrepresentations π_0 and π_1 are respectively trivial and nondegenerate, and $\mathcal{H} = \mathcal{H}_0 \oplus \mathcal{H}_1$.

A representation π of A in $\mathcal{H}$ is **irreducible** if there are exactly two $\pi(A)$-invariant closed subspaces of $\mathcal{H}$, namely $\{0\}$ and $\mathcal{H}$. Note that an irreducible representation is either trivial and in $\mathcal{H} \approx \mathbf{C}$, or nondegenerate.

Assume now that A is a C*-algebra. Every representation $\pi : A \to \mathcal{L}(\mathcal{H})$ of A is automatically norm decreasing: $\Vert \pi(x) \Vert \le \Vert x \Vert$ for all $x \in A$ [**Dixm–C***, 1.3.7]; moreover, when π is faithful, i.e., when π is injective, then $\Vert \pi(x) \Vert = \Vert x \Vert$ for all $x \in A$ [**Dixm–C***, 1.8.1]. A representation $\pi : A \to \mathcal{L}(\mathcal{H})$ of A is irreducible, as defined above, if and only if there are exactly two $\pi(A)$-invariant subspaces of $\mathcal{H}$ (closed or not), namely $\{0\}$ and $\mathcal{H}$ [**Dixm–C***, 2.8.4].

We introduce two dual spaces of a C*-algebra, which are analogues of the unitary dual and the primitive dual of a group (for more on this analogy, see Section 8.B). There is also an analogue of the quasi-dual of a group (see Remark 8.A.22).

DEFINITION 8.A.1. Let A be a C*-algebra. The **spectrum** $\widehat{A}$ of A is the space of equivalence classes of nontrivial irreducible representations of A.

A two-sided ideal I in A is **primitive** if there exists a nontrivial irreducible representation $\pi : A \to \mathcal{L}(\mathcal{H}_\pi)$ such that $I = \ker(\pi)$. The **primitive ideal space** $\mathrm{Prim}(A)$ of A is the set of primitive ideals of A.

Observe that a primitive ideal is closed and proper, by definition. The canonical map

$$(\kappa 1_A) \qquad \kappa^{\mathrm{d}}_{\mathrm{prim}} : \widehat{A} \twoheadrightarrow \mathrm{Prim}(A), \qquad \pi \mapsto \ker(\pi)$$

is surjective, by definition (compare with the map ($\kappa 1$) in Section 1.E).

REMARK 8.A.2. (1) The notion of primitive ideal space is defined more generally for associative algebras. Consider a complex associative algebra $\mathcal{A}$. A two-sided ideal I of $\mathcal{A}$ is **primitive** if there exists a complex vector space V and an irreducible representation $\pi_{\mathrm{alg}} : \mathcal{A} \to \mathrm{End}(V)$, in the sense of associative algebras, such that $I = \ker(\pi_{\mathrm{alg}})$.

(2) For C*-algebras, primitive ideals as defined in (1) are the same as primitive ideals as in Definition 8.A.1. More precisely:

Let A be a C*-algebra. Let π_{alg} be an irreducible representation of A on a complex vector space, in the sense of associative algebras. Then there exists a C*-algebra representation π_* of A in some Hilbert space that is equivalent to π_{alg}. Moreover, if π_*, π'_* are two irreducible C*-algebra representations of A, they are algebraically equivalent (i.e., there exists an invertible linear operator T from the space of π_* to that of π'_* such that $\pi'_*(x) = T\pi_*(x)T^{-1}$ for all $x \in A$) if and only if they are equivalent in the sense of C*-algebras (i.e., there exists a surjective linear isometry V from the space of π_* to that of π'_* such that $\pi'_*(x) = V\pi_*(x)V^{-1}$ for all $x \in A$) [**Dixm–C***, 2.9.6].

We now introduce a topology on the primitive ideal space of a C*-algebra.

DEFINITION 8.A.3. Let A be a C*-algebra. The space $\mathrm{Prim}(A)$ is furnished with the **Jacobson topology**, also called the **hull-kernel topology**, defined as: the closure of a subset Σ of $\mathrm{Prim}(A)$ is the set of all $J \in \mathrm{Prim}(A)$ such that

$$\bigcap_{I \in \Sigma} I \subset J.$$

Here is an important result about the separation properties of the topological space $\mathrm{Prim}(A)$. We will simply say **maximal ideal** of A for an ideal which is maximal among the closed and proper two-sided ideals of A.

PROPOSITION 8.A.4. *Let A be a C*-algebra.*

(1) $\mathrm{Prim}(A)$ *is a T_0-space.*
(2) *Let $I \in \mathrm{Prim}(A)$. Then I is a closed point in* $\mathrm{Prim}(A)$ *if and only if I is a maximal ideal of A. In particular,* $\mathrm{Prim}(A)$ *is a T_1-space if and only if every primitive ideal of A is a maximal ideal.*

PROOF. (1) Let I_1, I_2 be two distinct primitive ideals in A. We have therefore, say, $I_1 \not\subset I_2$. By definition of the Jacobson topology, the closure of I_1 is

$$X = \{I \in \mathrm{Prim}(A) \mid I_1 \subset I\}.$$

Therefore, $\mathrm{Prim}(A) \smallsetminus X$ is a neighbourhood of I_2 which does not contains I_1.

(2) It is clear that $\{I\}$ is closed in $\mathrm{Prim}(A)$ if and only if I is maximal among the primitive ideals of A (see the proof of (1)). We claim that if I is maximal among the primitive ideals of A, then I is maximal among the closed and proper two-sided ideal of A.

Indeed, let J be a closed and proper two-sided ideal of A with $I \subset J$. Let π be an irreducible representation of the quotient C*-algebra A/J. Then π lifts to an irreducible representation, again denoted by π, of A such that $J \subset \ker(\pi)$. Since $\ker(\pi) \in \mathrm{Prim}(A)$, it follows that $I = \ker(\pi)$ and hence $I = J$. □

REMARK 8.A.5. (1) The fact that $\mathrm{Prim}(A)$ is a T_0-space should be compared with the definition of $\mathrm{Prim}(G)$ in Reformulation 1.E.2 together with Corollary 8.B.5.

(2) The space $\mathrm{Prim}(A)$ is a Baire space (see Proposition 8.A.15 below).

(3) When A is separable, $\mathrm{Prim}(A)$ is second-countable (see Corollary 8.A.14 below).

We can now define a topology and the induced Borel structure on the spectrum of A.

DEFINITION 8.A.6. The **topology on the spectrum** $\widehat{A}$ of a C*-algebra A is defined as follows: a subset of $\widehat{A}$ is open if and only if it is the inverse image under the map

$$\kappa_{\mathrm{prim}}^{\mathrm{d}} : \widehat{A} \twoheadrightarrow \mathrm{Prim}(A)$$

of an open subset of $\mathrm{Prim}(A)$. This topology induces the **Borel structure on the spectrum** $\widehat{A}$.

REMARK 8.A.7. (1) There are several equivalent ways to define the topology on $\widehat{A}$; see [**Dixm–C***, § 3].

(2) It is obvious that the map $\kappa_{\mathrm{prim}}^{\mathrm{d}} : \widehat{A} \twoheadrightarrow \mathrm{Prim}(A)$ is continuous and open.

(3) Using Proposition 8.A.4, it is an easy exercise to show that $\widehat{A}$ is a T_0-space if and only if the map $\kappa_{\mathrm{prim}}^{\mathrm{d}}$ is a homeomorphism.

(4) In general, the map $\kappa_{\mathrm{prim}}^{\mathrm{d}}$ is far from being a bijection; see 9.G below.

8.A.b. Pure states and primitive ideal space. Let A be a C*-algebra. Let A_+^* denote the set of positive linear functionals on A; recall that these functionals are continuous, i.e., that A_+^* is a subset of the Banach space dual A^* of A [**Dixm–C***, 2.1.8]. Denote by $A_{+,\leq 1}^*$, respectively $A_{+,1}^*$, the subset of A_+^* of forms of norm at most 1, respectively equal 1; a **state** on A is an element of $A_{+,1}^*$.

When A has a unit, it is easy to check that $\|\varphi\| = \varphi(1)$ for all $\varphi \in A_+^*$ [**Dixm–C***, 2.1.4], so that $A_{+,1}^* = \{\varphi \in A_+^* \mid \varphi(1) = 1\}$.

As for functions of positive type on a group (see Section 1.B), there is a GNS construction for positive linear functionals on a C*-algebra.

CONSTRUCTION 8.A.8. The **Gel'fand–Naimark–Segal construction** associates to every $\varphi \in A_+^* \smallsetminus \{0\}$ a GNS triple $(\pi_\varphi, \mathcal{H}_\varphi, \xi_\varphi)$, consisting of a representation $(\pi_\varphi, \mathcal{H}_\varphi)$ of A and a cyclic vector $\xi_\varphi \in \mathcal{H}_\varphi$ such that $\varphi(x) = \langle \pi_\varphi(x)\xi_\varphi \mid \xi_\varphi \rangle$, in the following way.

The set $J = \{x \in A \mid \varphi(x^*x) = 0\}$ is a closed left ideal of A. A scalar product is defined on A/J by

$$\langle x + J \mid y + J \rangle = \varphi(y^*x) \qquad \text{for all } x, y \in A.$$

Let $\mathcal{H}_\varphi$ be the Hilbert space completion of A/J. Then a representation π_φ of A on $\mathcal{H}_\varphi$ is defined on A/J by

$$\pi_\varphi(a)(x+J) = ax+J \qquad \text{for all } \ a, x \in A.$$

Moreover, $x+J \mapsto \varphi(x)$ is well-defined on A/J and extends to a continuous linear functional on the Hilbert space $\mathcal{H}_\varphi$, again denoted by φ. By Riesz' theorem, there exists a unique vector $\xi_\varphi \in \mathcal{H}_\varphi$ such that

$$\varphi(x) = \langle x+J \mid \xi_\varphi \rangle \qquad \text{for all } \ x \in A.$$

(When A is unital ξ_φ is simply the image of 1 in $\mathcal{H}_\varphi$.) We have

$$\begin{aligned} \langle y+J \mid \pi_\varphi(x)\xi_\varphi \rangle &= \langle \pi_\varphi(x^*)(y+J) \mid \xi_\varphi \rangle = \langle x^*y+J \mid \xi_\varphi \rangle \\ &= \varphi(x^*y) = \langle y+J \mid x+J \rangle \end{aligned}$$

for all $x, y \in A$, and hence $\pi_\varphi(x)\xi_\varphi = x+J$ for every $x \in A$. This implies that ξ_φ is cyclic and $\varphi(x) = \langle \pi_\varphi(x)\xi_\varphi \mid \xi_\varphi \rangle$ for every $x \in A$.

The GNS triple $(\pi_\varphi, \mathcal{H}_\varphi, \xi_\varphi)$ associated as above to φ is unique up to unitary equivalence: if $(\pi, \mathcal{H})$ is a representation of A with cyclic vector ξ such that $\|\xi\| = \|\xi_\varphi\|$ and $\varphi(x) = \langle \pi(x)\xi \mid \xi \rangle$ for all $x \in A$, then there exists a unitary Hilbert space isomorphism $S : \mathcal{H} \to \mathcal{H}_\varphi$ intertwining π and π_φ and such that $S(\xi) = \xi_\varphi$.

EXAMPLE 8.A.9. (1) Consider a compact space X and the C*-algebra $C(X)$. The space of states $C(X)^*_{+,1}$ can be identified with the space of probability Radon measures on X; this is one form of the Riesz representation theorem, see Theorem A.4.3. For such a measure μ, there are identifications of the Hilbert space $\mathcal{H}_\mu$ with $L^2(X, \mu)$, of the cyclic vector ξ_μ with the constant function of value 1 on X, and of the operator $\pi_\mu(f)$ with the operator $m(f)$ of multiplication by f on $L^2(X, \mu)$ for all $f \in C(X)$. The representation π_μ is irreducible if and only if it is of dimension 1, if and only if μ is a Dirac measure.

(2) In anticipation of Section 8.B, consider a locally compact group G with left Haar measure μ_G, its group algebra $L^1(G, \mu_G)$, and its maximal C*-algebra $C^*_{\text{max}}(G)$. Every normalized function of positive type $\varphi \in P_1(G)$ defines a normalized positive linear form ω^1_φ on $L^1(G, \mu_G)$ by $\omega^1_\varphi(f) = \int_G \varphi(g)f(g)d\mu_G(g)$, which extends canonically to a state $\omega = \omega^*_\varphi \in C^*_{\text{max}}(G)^*_{+,1}$ [**Dixm–C***, 13.4.5, 2.7.5]. The GNS representation π_ω of $C^*_{\text{max}}(G)$ is the extension of the GNS representation π_φ of G, in the sense of Section 8.B.

Let A be a C*-algebra. The subsets $A^*_{+,\leq 1}$ and $A^*_{+,1}$ of the unit ball of the dual Banach space A^* are convex.

We denote the corresponding sets of indecomposable elements by $\text{Extr}(A^*_{+,\leq 1})$ and $\text{P}(A)$; the set $\text{P}(A)$ is the set of **pure states** of A. The set $A^*_{+,\leq 1}$ is compact in the unit ball of A^* for the weak*-topology, and we have $\text{Extr}(A^*_{+,\leq 1}) = \{0\} \sqcup \text{P}(A)$ [**Dixm–C***, 2.5.5]. Note that $P(A)$ is closed in $\text{Extr}(A^*_{+,\leq 1})$ when A has a unit, and that 0 is in its closure when A has no unit [**Dixm–C***, 2.12.13].

For $\varphi \in A^*_{+,\leq 1} \smallsetminus \{0\}$, the associated GNS representation $(\pi_\varphi, \mathcal{H}_\varphi)$ is irreducible if and only if $\varphi \in \text{P}(A)$ [**Dixm–C***, 2.5.4]. We therefore have a map

$$\text{P}(A) \twoheadrightarrow \widehat{A}, \qquad \varphi \mapsto \pi_\varphi,$$

where $\widehat{A}$ is the spectrum of A (as defined in Subsection 8.A.a). This map is surjective: the inverse image of $\pi \in \widehat{A}$ consists of the states of the form $x \mapsto \langle \pi(x)\xi \mid \xi \rangle$, where ξ is a unit vector in the Hilbert space of π. Composition of this map with the canonical surjection

$$\kappa^{\mathrm{d}}_{\mathrm{prim}} : \widehat{A} \twoheadrightarrow \mathrm{Prim}(A), \quad \pi \mapsto \ker(\pi)$$

of Definition 8.A.1 yields a surjective map

$$\Phi : \mathrm{P}(A) \twoheadrightarrow \mathrm{Prim}(A), \quad \varphi \mapsto \ker(\pi_\varphi).$$

PROPOSITION 8.A.10. *Let A be a C^*-algebra. Let $\mathrm{P}(A)$ be equipped with the weak*-topology and $\mathrm{Prim}(A)$ with the Jacobson topology.*

The maps $\Phi : \mathrm{P}(A) \twoheadrightarrow \mathrm{Prim}(A)$ and $\mathrm{P}(A) \twoheadrightarrow \widehat{A}$ defined above are continuous and open.

PROOF. The topology on $\mathrm{Prim}(A)$ may be described in terms of pure states as follows (see [**Dixm–C***, 3.4.10]). Let $\Sigma \subset \mathrm{Prim}(A)$ and $J \in \mathrm{Prim}(A)$; set

$$Q(\Sigma) := \Phi^{-1}(\Sigma) \subset \mathrm{P}(A).$$

Then $J \in \overline{\Sigma}$ if and only if $\Phi^{-1}(J) \in \overline{Q(\Sigma)}$. This shows that Φ is continuous.

Let U be an open subset of $\mathrm{P}(A)$; we claim that $V := \Phi(U)$ is open in $\mathrm{Prim}(A)$. Indeed, assume, by contradiction, that V is not open. Then, some $J \in V$ is in the closure of

$$\Sigma := \mathrm{Prim}(A) \smallsetminus V.$$

Let $\varphi \in U$ be such that $\Phi(\varphi) = J$. Then φ belongs to the closure of the set

$$Q(\Sigma) = \mathrm{P}(A) \smallsetminus \Phi^{-1}(V).$$

Since $\mathrm{P}(A) \smallsetminus \Phi^{-1}(V) \subset \mathrm{P}(A) \smallsetminus U$ and since $\mathrm{P}(A) \smallsetminus U$ is closed, it follows that $\varphi \in \mathrm{P}(A) \smallsetminus U$ and this is a contradiction.

The claim for $\mathrm{P}(A) \twoheadrightarrow \widehat{A}$ follows. □

It was shown in Propositions 1.D.1 and 1.D.10 that the dual space $\widehat{G}$ of an LC compact group G is locally quasi-compact and that $\widehat{G}$ is quasi-compact when G is discrete. We extend this result to the spectrum of a C*-algebra.

PROPOSITION 8.A.11. *Let A be a C^*-algebra.*

Then $\widehat{A}$ and $\mathrm{Prim}(A)$ are locally quasi-compact topological spaces. When A is unital, these spaces are quasi-compact.

PROOF. We adapt the arguments of the proof of Proposition 1.D.1 to the context of C*-algebras.

Let $\pi \in \widehat{A}$ and let $U \subset \widehat{A}$ be an open neighborhood of π. Fix $\varphi_0 \in P(A)$ such that $\Phi(\varphi_0) = \pi$, where Φ is the open and continuous surjective map $\Phi : \mathrm{P}(A) \twoheadrightarrow \widehat{A}$ as in Proposition 8.A.10

Since $U^c := \widehat{A} \smallsetminus U$ is a closed subset in $\widehat{A}$ and $\pi \notin U^c$, there exists an element $a_0 \in A$ such that $\rho(a_0) = 0$ for every $\rho \in U^c$ and $\varphi_0(a_0) = 1$. Then

$$V := \{\varphi \in P(A) \mid \mathrm{Re}\,\varphi(a_0) \geq 1/2\}$$

is a neighbourhoud of φ_0 in $P(A)$ and $K := \Phi(V)$ is a neighbourhood of π in $\widehat{A}$. Since $\psi(a_0) = 0$ for every $\psi \in P(A)$ with $\Phi(\psi) \in U^c$, we have $K \subset U$.

Let $(\pi_\lambda)_{\lambda \in \Lambda}$ be a net in K. For every $\lambda \in \Lambda$, let $\varphi_\lambda \in V$ be such that $\Phi(\varphi_\lambda) = \pi_\lambda$. Since $A^*_{+,\leq 1}$ is compact, upon passing to a subnet, we can assume that $(\varphi_\lambda)_{\lambda \in \Lambda}$ is weak*-convergent to some functional $\varphi_1 \in A^*_{+,\leq 1}$. Then $\mathrm{Re}\,\varphi_1(a_0) \geq 1/2$ and, in particular, $\varphi_1 \neq 0$.

Let π_1 be the representation of A associated to φ_1 by the GNS construction and set $I = \ker \pi_1$. We view $(A/I)^*_{+,\leq 1}$ and $P(A/I)$ as subsets of $A^*_{+,\leq 1}$ and $P(A)$ in the natural way (see [**Dixm–C***, 2.11.8]) and consider the continuous affine map

$$\Psi : (A/I)^*_{+,\leq 1} \to \mathbf{R}, \qquad \varphi \mapsto \operatorname{Re} \varphi(a_0).$$

Let $b = \max \Psi\left((A/I)^*_{+,\leq 1}\right)$. Since $\varphi_1 \in (A/I)^*_{+,\leq 1}$ and $\Psi(\varphi_1) \geq 1/2$, we have $b \geq 1/2$. As in the proof of Proposition 1.D.1, we can find $\varphi \in P(A/I) \subset P(A)$ such that $\Psi(\varphi) = b$. Since $\operatorname{Re} \varphi(a_0) = \Psi(\varphi) = b \geq 1/2$, we have $\varphi \in V$.

Set $\sigma := \Phi(\varphi) \in K$. Let $(\pi_{\lambda'})_{\lambda' \in \Lambda'}$ be a subnet of $(\pi_\lambda)_{\lambda \in \Lambda}$. Since $(\varphi_{\lambda'})_{\lambda' \in \Lambda'}$ converges to φ_1, the ideal $\bigcap_{\lambda' \in \Lambda'} \ker \pi_{\lambda'}$ is contained in I and hence in $\ker \sigma$. This shows that σ is in the closure of $\{\pi_{\lambda'} \mid \lambda' \in \Lambda'\}$. It follows that $(\pi_\lambda)_{\lambda \in \Lambda}$ converges to σ.

We have proved that $\widehat{A}$ is locally quasi-compact. The claim for $\operatorname{Prim}(A)$ follows immediately.

Assume now that A has a unit 1. Let $\pi \in \widehat{A}$ be arbitrary. Set $U = \widehat{A}$ and take $a_0 = 1$ in the proof above. Then $V = P(A)$ and the previous arguments show that $\widehat{A} = \Phi(V)$ is quasi-compact. The claim for $\operatorname{Prim}(A)$ follows immediately. □

REMARK 8.A.12. For a slightly different proof of Proposition 8.A.11, see [**Dixm–C***, 3.3.8, 3.1.8].

Observe that a nonunital C*-algebra may have a quasi-compact (even compact) spectrum: the C*-algebra of compact operators on a separable infinite-dimensional Hilbert space has no unit, and its spectrum consists of one point (see Proposition 6.E.1).

The following proposition will play a crucial role in the sequel.

PROPOSITION 8.A.13. *Let A be a separable C*-algebra.*

The space $\mathrm{P}(A)$, *equipped with the weak*-topology, is a Polish space.*

PROOF. Since A is separable, the unit ball $A^*_{\leq 1}$ of A^*, equipped with the weak*-topology, is a metrizable compact space and hence a Polish space. Therefore, the same holds for the closed subspace $X := A^*_{+,\leq 1}$ of $A^*_{\leq 1}$.

Let d be a metric on X defining its topology; set $X_0 = \{0\}$ and, for $n \geq 1$, let X_n be the set of $\varphi \in X$ such that there exist $\varphi_1, \varphi_2 \in X$ with $\varphi = \frac{1}{2}(\varphi_1 + \varphi_2)$ and $d(\varphi_1, \varphi_2) \geq 1/n$. Then every X_n is closed in X and

$$\mathrm{P}(A) = X \smallsetminus \bigcup_{n \geq 0} X_n = \bigcap_{n \geq 0} (X \smallsetminus X_n)$$

is a G_δ set in the Polish space X. Therefore, $\mathrm{P}(A)$ is a Polish space (see Proposition A.1.2). □

Here is a first consequence of Proposition 8.A.13.

COROLLARY 8.A.14. *Let A be a separable C*-algebra.*

The primitive ideal space $\operatorname{Prim}(A)$ *and the spectrum* $\widehat{A}$ *are second-countable topological spaces.*

PROOF. By Proposition 8.A.13, there exists a sequence $(U_n)_n$ of open sets generating the topology of $\mathrm{P}(A)$. Since $\Phi : \mathrm{P}(A) \twoheadrightarrow \operatorname{Prim}(A)$ is open and continuous (Proposition 8.A.10), the sequence of open sets $(\Phi(U_n))_n$ generates the topology of $\operatorname{Prim}(A)$. So, $\operatorname{Prim}(A)$ is second-countable.

The claim for $\widehat{A}$ follows. □

Here is a second consequence of Proposition 8.A.13.

PROPOSITION 8.A.15. *Let A be a C*-algebra.*
The primitive ideal space $\mathrm{Prim}(A)$ *and the spectrum* $\widehat{A}$ *are Baire spaces.*

PROOF. We provide a proof for the particular case of a separable C*-algebra A, and refer to [**Dixm–C***, 3.4.13] for the general case.

Let $(V_n)_n$ be a sequence of open dense subsets of $\mathrm{Prim}(A)$. Since Φ is continuous and open, every $U_n := \Phi^{-1}(V_n)$ is open and dense in $\mathrm{P}(A)$. However, the space $\mathrm{P}(A)$, being a Polish space (Proposition 8.A.13), is a Baire space. Therefore $\bigcap_n U_n$ is dense in $\mathrm{P}(A)$; since Φ is continuous, it follows that

$$\bigcap_n V_n = \Phi\left(\bigcap_n U_n\right)$$

is dense in $\mathrm{Prim}(A)$. So, $\mathrm{Prim}(A)$ is a Baire space.

The claim for $\widehat{A}$ follows. □

8.A.c. A primitive ideal space is a standard Borel space. Let A be a C*-algebra. We relate primitive ideals of A to C*-semi-norms on A.

A **C*-semi-norm** on A is a semi-norm $N : A \to \mathbf{R}_+$ such that

(1) $N(x) \leq \|x\|$,
(2) $N(xy) \leq N(x)N(y)$,
(3) $N(x^*x) = N(x)^2$,

for all $x, y \in A$.

Denote by $\mathcal{J}(A)$ the set of closed two-sided ideals of A and by $\mathcal{N}(A)$ the set of C*-semi-norms on A.

EXAMPLE 8.A.16. (1) Let $J \in \mathcal{J}(A)$. Then A/J, equipped with the quotient norm

$$x + J \mapsto \|x + J\| := \inf_{y \in J} \|x + y\|,$$

is a C*-algebra and the map

$$N_J : A \to \mathbf{R}_+ \quad \text{defined by} \quad N_J(x) = \|x + J\| \quad \text{for all } x \in A$$

is a C*-semi-norm on A. Observe that $\{x \in A \mid N_J(x) = 0\} = J$.

(2) The maximum of two C*-semi-norms is a C*-semi-norm (but the minimum is not).

The next lemma shows that every C*-semi-norm on A is of the form N_J for some closed two-sided ideal J, as in Example 8.A.16(1).

LEMMA 8.A.17. *We keep the notation of Example* 8.A.16 *(1).*

(1) *The map $J \mapsto N_J$ is a one-to-one correspondence from $\mathcal{J}(A)$ onto $\mathcal{N}(A)$.*
(2) *For every $J_1, J_2 \in \mathcal{J}(A)$, we have* $N_{J_1 \cap J_2} = \max\{N_{J_1}, N_{J_2}\}$.

PROOF. (1) Let $N \in \mathcal{N}(A)$. Set $J := \{x \in A \mid N(x) = 0\}$. Then $J \in \mathcal{J}(A)$. Moreover, the map $\overline{N} : x + J \mapsto N(x)$ is a norm on the quotient A/J. The completion $\overline{A}$ of A/J with respect to $\overline{N}$ is a C*-algebra. The inclusion map from A/J, equipped with the quotient norm of A, to $\overline{A}$, is an injective homomorphism between C*-algebras and is therefore an isometry. So, $N = N_J$ and this shows that $J \mapsto N_J$ is surjective.

Let $J_1, J_2 \in \mathcal{J}(A)$ be such that $N_{J_1} = N_{J_2}$. Then

$$J_1 = \{x \in A \mid N_{J_1}(x) = 0\} = \{x \in A \mid N_{J_2}(x) = 0\} = J_2$$

and this shows that $J \mapsto N_J$ is injective.

(2) The involutive algebra $A/J_1 \oplus A/J_2$, equipped with the norm

$$\|(x + J_1, y + J_2)\| = \max\{\|x + J_1\|, \|y + J_2\|\},$$

is a C*-algebra. The map

$$A/(J_1 \cap J_2) \to A/J_1 \oplus A/J_2, \qquad x + J_1 \cap J_2 \mapsto (x + J_1, x + J_2)$$

is an injective homomorphism of C*-algebras and is hence an isometry. Therefore, we have

$$N_{J_1 \cap J_2}(x) = \|x + J_1 \cap J_2\| = \max\{\|x + J_1\|, \|x + J_2\|\} = \max\{N_{J_1}(x), N_{J_2}(x)\},$$

for all $x \in A$. □

Recall that a two-sided ideal J in A is **prime** if $J \neq A$ and if, for $J_1, J_2 \in \mathcal{J}(A)$ such that $J_1 J_2 \subset J$, at least one of the inclusions $J_1 \subset J$, $J_2 \subset J$ holds. Here, we are only interested in closed prime ideals. For examples of closed prime ideals, see Proposition 8.A.19.

Observe that, for $J_1, J_2 \in \mathcal{J}(A)$, we have $J_1 J_2 = J_1 \cap J_2$. Indeed, it is obvious that $J_1 J_2 \subset J_1 \cap J_2$. To show the reverse inclusion, it suffices to show that the positive part $(J_1 \cap J_2)_+$ of $J_1 \cap J_2$ is contained in $J_1 J_2$, since the C*-algebra $J_1 \cap J_2$ is the linear span of $(J_1 \cap J_2)_+$. Let $x \in (J_1 \cap J_2)_+$. Then $x^{1/2} \in J_1 \cap J_2$ and hence $x = x^{1/2} x^{1/2} \in J_1 J_2$.

A C*-semi-norm $N \in \mathcal{N}(A)$ is **extremal** if, whenever $N_1, N_2 \in \mathcal{N}(A)$ are such that $N = \max\{N_1, N_2\}$, then $N_1 = N$ or $N_2 = N$. We denote by $\mathcal{E}(A)$ the set of extremal C*-semi-norms on A that are different from 0.

The following lemma shows that prime ideals correspond to extremal C*-norms under the map $J \mapsto N_J$.

LEMMA 8.A.18. *Let $J \in \mathcal{J}(A)$. The following properties are equivalent:*

(i) *J is a closed prime ideal;*
(ii) *whenever $J = J_1 \cap J_2$ for $J_1, J_2 \in \mathcal{J}(A)$ holds, then $J_1 = J$ or $J_2 = J$;*
(iii) *$N_J \in \mathcal{E}(A)$.*

PROOF. Assume that J is prime. If $J = J_1 \cap J_2$, then $J = J_1 J_2$ and hence $J_1 = J$ or $J_2 = J$. This shows that (i) implies (ii).

Assume that (ii) holds and let $J_1, J_2 \in \mathcal{J}(A)$ be such that $J_1 \cap J_2 \subset J$. Set $I_1 = \overline{J + J_1}$ and $I_2 = \overline{J + J_2}$. Then $I_1, I_2 \in \mathcal{J}(A)$ and $I_1 I_2 \subset J$; hence, we have

$$I_1 \cap I_2 = I_1 I_2 \subset J \subset I_1 \cap I_2,$$

that is, $J = I_1 \cap I_2$. Therefore, $I_1 = J$ or $I_2 = J$, that is, $J_1 \subset J$ or $J_2 \subset J$. So, J is a prime ideal and (i) holds. This shows that (i) and (ii) are equivalent.

By Lemma 8.A.17, the C*-semi-norm N_J is extremal if and only if, whenever $N_J = \max\{N_{J_1}, N_{J_2}\}$ for $J_1, J_2 \in \mathcal{J}(A)$ holds, then $J_1 = J$ or $J_2 = J$. Moreover, for $J_1, J_2 \in \mathcal{J}(A)$, we have $\max\{N_{J_1}, N_{J_2}\} = N_{J_1 \cap J_2}$, again by Lemma 8.A.17. This shows that (iii) and (ii) are equivalent. □

Observe that, for a subset Σ of $\mathrm{Prim}(A)$, we have $\bigcap_{J \in \Sigma} J = \{0\}$ if and only if Σ is dense in $\mathrm{Prim}(A)$. In particular, if Σ is a closed subset of $\mathrm{Prim}(A)$ such that $\bigcap_{J \in \Sigma} J = \{0\}$, then $\Sigma = \mathrm{Prim}(A)$.

PROPOSITION 8.A.19. *Let A be a C*-algebra and J a closed two-sided ideal in A.*

(1) *If $J \in \mathrm{Prim}(A)$, then J is a closed prime ideal.*
(2) *Assume that A is separable. If J is a closed prime ideal, then $J \in \mathrm{Prim}(A)$.*

PROOF. (1) Let $(\pi, \mathcal{H})$ be an irreducible representation of A such that $J = \ker(\pi)$. Let $J_1, J_2 \in \mathcal{J}(A)$ be such that $J_1 J_2 \subset J$. Assume that J_2 is not contained in J, that is, $\pi(J_2) \neq \{0\}$. Then $\overline{\pi(J_2)\mathcal{H}}$ is a nonzero closed $\pi(A)$-invariant subspace and therefore $\overline{\pi(J_2)\mathcal{H}} = \mathcal{H}$, by irreducibility of π.

Since

$$\pi(J_1)\overline{\pi(J_2)\mathcal{H}} \subset \overline{\pi(J_1)\pi(J_2)\mathcal{H}} = \overline{\pi(J_1J_2)\mathcal{H}} \subset \overline{\pi(J)\mathcal{H}} = \{0\},$$

it follows that $\pi(J_1) = \{0\}$, that is, $J_1 \subset J$.

(2) It suffices to show that, if $\{0\}$ is a prime ideal of A, then $\{0\}$ is a primitive. Indeed, assume this is proved and let J be a closed prime ideal of A. Then $\{0\}$ is a prime ideal of the separable C*-algebra A/J and hence $\{0\} \in \mathrm{Prim}(A/J)$; this means that A/J has a faithful irreducible representation π. Then $\pi \circ p \in \widehat{A}$, where $p : A \twoheadrightarrow A/J$ is the canonical projection, and hence $J = \ker(\pi \circ p) \in \mathrm{Prim}(A)$.

Assume that $\{0\}$ is a prime ideal of A. Since $\mathrm{Prim}(A)$ is second-countable (Corollary 8.A.14), there exists a sequence $(V_n)_n$ of nonempty open subsets generating the topology of $\mathrm{Prim}(A)$.

For n, set

$$I_n = \bigcap_{J \in V_n} J \qquad \text{and} \qquad J_n = \bigcap_{J \in \mathrm{Prim}(A) \smallsetminus V_n} J.$$

Then I_n and J_n are closed two-sided ideals of A and $I_n \cap J_n = \{0\}$. Moreover, $J_n \neq \{0\}$, since $\mathrm{Prim}(A) \smallsetminus V_n$ is a closed proper subset of $\mathrm{Prim}(A)$. The ideal $\{0\}$ being prime, it follows that $I_n = 0$. This means that V_n is a dense subset of $\mathrm{Prim}(A)$. Since $\mathrm{Prim}(A)$ is a Baire space (Corollary 8.A.14), $\bigcap_n V_n$ is therefore dense in $\mathrm{Prim}(A)$. In particular, $\bigcap_n V_n$ is not empty.

Let $J \in \bigcap_n V_n$. Then $\bigcap_n V_n = \{J\}$, since the sequence $(V_n)_n$ generates the topology of $\mathrm{Prim}(A)$. So, $\{J\}$ is dense in $\mathrm{Prim}(A)$. Therefore, $J \subset I$ for every $I \in \mathrm{Prim}(A)$ and hence $J = \{0\}$. This shows that $\{0\}$ is a primitive ideal. □

REMARK 8.A.20. There are examples of nonseparable C*-algebras in which the closed ideal $\{0\}$ is prime and not primitive [**Weav–03**].

As for group representations (see Proposition 6.B.6), a representation π of the C*-algebra A is said to be **factorial** (or a **factor representation**) if the von Neumann algebra $\pi(A)''$ generated by $\pi(A)$ is a factor.

As a consequence of Proposition 8.A.19, we now show that kernels of factor representations of a separable C*-algebra are primitive ideals. A first proof of this result was already given in Corollary 7.F.2, when A is the maximal C*-algebra of a second-countable LC group (see Section 8.B). For another proof, see [**Dixm–60a**, Corollaire 3].

COROLLARY 8.A.21. *Let A be separable C*-algebra and π a factor representation of A.*

Then $\ker \pi \in \mathrm{Prim}(A)$.

PROOF. Let $\mathcal{H}$ be the Hilbert space of π and set $J := \ker \pi$. By Proposition 8.A.19(2) it suffices to show that J is a prime ideal.

Let $J_1, J_2 \in \mathcal{J}(A)$ be such that $J_1 J_2 \subset J$. Assume that J_2 is not contained in J, that is, $\pi(J_2) \neq \{0\}$.

Then $\overline{\pi(J_2)\mathcal{H}}$ is a nonzero closed subspace of $\mathcal{H}$ which is invariant under both $\pi(A)$ and its commutant $\pi(A)'$. It follows that the orthogonal projection

$$P : \mathcal{H} \to \overline{\pi(J_2)\mathcal{H}}$$

is a nonzero projection in the centre of $\pi(A)''$. Since π is factorial, we have $P = \mathrm{Id}_{\mathcal{H}}$, that is, $\overline{\pi(J_2)\mathcal{H}} = \mathcal{H}$. Therefore,

$$\pi(J_1)\mathcal{H} = \pi(J_1)\overline{\pi(J_2)\mathcal{H}} \subset \overline{\pi(J_1)\pi(J_2)\mathcal{H}} = \overline{\pi(J_1 J_2)\mathcal{H}} \subset \overline{\pi(J)\mathcal{H}} = \{0\},$$

and so $\pi(J_1) = \{0\}$, that is, $J_1 \subset J$. □

REMARK 8.A.22. In analogy with the group case (see Proposition 6.B.5), two factor representations π_1 and π_2 of A are said to be **quasi-equivalent** if there exists an isomorphism $\Phi : \pi_1(A)'' \to \pi_2(A)''$ such that $\Phi(\pi_1(a)) = \pi_2(a)$ for every $a \in A$. The **quasi-spectrum** $\mathrm{QS}(A)$ of A is the space of quasi-equivalence classes of factor representations of A. Observe that two quasi-equivalent representations of A have the same kernels.

When A is separable, Corollary 8.A.21 shows that the map $\pi \mapsto \ker \pi$ induces a map $\mathrm{QS}(A) \twoheadrightarrow \mathrm{Prim}(A)$. Observe that this map is an extension to $\mathrm{QS}(A)$ of the canonical map $\widehat{A} \twoheadrightarrow \mathrm{Prim}(A)$.

LEMMA 8.A.23. *Let A be a separable C^*-algebra. Let $\mathcal{N}(A)$ be equipped with the topology of pointwise convergence on A.*

(1) *$\mathcal{N}(A)$ is a compact metrizable space; in particular, $\mathcal{N}(A)$ is a Polish space.*
(2) *$\mathcal{E}(A)$ is a G_δ set in the Polish space $\mathcal{N}(A)$; in particular, $\mathcal{E}(A)$ is a standard Borel space.*

PROOF. Let $(x_n)_{n\in\mathbf{N}}$ be a dense sequence in A.

(1) The product space $X = \prod_{n\in\mathbf{N}}[-\|x_n\|, \|x_n\|]$, equipped with the product topology, is a compact metrizable space. Let $N \in \mathcal{N}(A)$. Since $N(x) \leq \|x\|$ for every $x \in A$, we have $(N(x_n))_{n\in\mathbf{N}} \in X$. The map

$$\psi : \mathcal{N}(A) \to X, \qquad N \mapsto (N(x_n))_{n\in\mathbf{N}}$$

is easily seen to be a homeomorphism between $\mathcal{N}(A)$ and its image $\psi(\mathcal{N}(A))$. Moreover, it is clear that $\psi(\mathcal{N}(A))$ is closed in X.

(2) For $i, j \in \mathbf{N}$ with $i \neq j$ and $k \geq 1$, set

$$U_{i,j}^{(k)} := \left\{(N, N') \in \mathcal{N}(A) \times \mathcal{N}(A) \middle| N(x_i) + \frac{1}{k} \leq N'(x_i) \text{ and } N'(x_j) + \frac{1}{k} \leq N(x_j)\right\}.$$

Then $U_{i,j}^{(k)}$ is a closed and hence compact subset of $\mathcal{N}(A) \times \mathcal{N}(A)$.

The map

$$f : \mathcal{N}(A) \times \mathcal{N}(A) \to \mathcal{N}(A), \qquad (N, N') \mapsto \max\{N, N'\}$$

is continuous and we have

$$f\left(\bigcup_{i,j\in\mathbf{N}, i\neq j, k\geq 1} U_{i,j}^{(k)}\right) \cup \{0\} = \mathcal{N}(A) \smallsetminus \mathcal{E}(A).$$

Since $\{0\}$ is closed and each $f(U_{i,j}^{(k)})$ is compact, it follows that $\mathcal{N}(A) \setminus \mathcal{E}(A)$ is a countable union of closed subsets of $\mathcal{N}(A)$; therefore, $\mathcal{E}(A)$ is a G_δ set in $\mathcal{N}(A)$. □

We are now ready to prove the main result of this section, which is Theorem 2.1 in [**Effr–63**].

THEOREM 8.A.24 (**Effros**). *Let A be separable C*-algebra. The primitive ideal space* $\mathrm{Prim}(A)$ *of A is a standard Borel space.*

PROOF. Consider the injective map

$$F : \mathrm{Prim}(A) \hookrightarrow \mathcal{N}(A), \qquad J \mapsto N_J.$$

By Lemma 8.A.18 and Proposition 8.A.19, we have $F(\mathrm{Prim}(A)) = \mathcal{E}(A)$. In view of Lemma 8.A.23, it suffices to show that $F : \mathrm{Prim}(A) \to \mathcal{E}(A)$ is an isomorphism of Borel spaces.

Let us show that F is a Borel map. By the definition of the topology of $\mathcal{N}(A)$, it suffices to show that $J \mapsto N_J(x)$ is a Borel function for every $x \in A$. We are going to prove that $J \mapsto N_J(x)$ is even lower semi-continuous.

To show this, let $\alpha \in \mathbf{R}$. Let $J_0 \in \mathrm{Prim}(A)$ be in the closure of the set

$$X_\alpha := \{J \in \mathrm{Prim}(A) \mid N_J(x) \leq \alpha\}.$$

Set $I_\alpha = \bigcap_{J \in X_\alpha} J$; then $I_\alpha \subset J_0$ and hence A/J_0 is a quotient of the C*-algebra A/I_α. Therefore, we have

$$N_{J_0}(x) = \|x + J_0\| \leq \|x + I_\alpha\| = \sup_{J \in X_\alpha} \|x + J\| = \sup_{J \in X_\alpha} N_J(x)$$

and hence $N_{J_0}(x) \leq \alpha$, that is, $J_0 \in X_\alpha$. This shows that X_α is closed for every $\alpha \in \mathbf{R}$; therefore $J \mapsto N_J(x)$ is lower semi-continuous for every $x \in A$.

Next, we show that F^{-1} is even a continuous map. Let Σ be a closed subset of $\mathrm{Prim}(A)$. Set $I = \bigcap_{J \in \Sigma} J$. Since Σ is closed, I is the intersection of the primitive ideals that contain I. Therefore

$$F(\Sigma) = \{N_J \mid J \in \Sigma\} = \{N_J \mid I \subset J\} = \{N \in \mathcal{E}(A) \mid N(I) = 0\}.$$

The set

$$\{N \in \mathcal{E}(A) \mid N(I) = 0\} = \bigcap_{x \in I} \{N \in \mathcal{E}(A) \mid N(x) = 0\}$$

is closed in $\mathcal{E}(A)$. Therefore F^{-1} is continuous. □

Anticipating on the next section, and applying Theorem 8.A.24 to LC groups, we obtain the following result.

THEOREM 8.A.25. *Let G be a second-countable locally compact group. The primitive dual* $\mathrm{Prim}(G)$ *is a standard Borel space.*

8.B. C*-algebras and representations of LC groups

Let G be a locally compact group, with left-invariant Haar measure μ_G. Let $L^1(G, \mu_G)$, or for short $L^1(G)$, denote the Banach involutive algebra of integrable functions $G \to \mathbf{C}$ (up to equality μ_G-almost everywhere), with the convolution product and the usual involution; see Appendix A.6.

To a representation $\pi : G \to \mathcal{U}(\mathcal{H}_\pi)$, there is associated an algebra homomorphism $L^1(G) \to \mathcal{L}(\mathcal{H}_\pi)$, denoted by π again, defined by

$$\pi(f) = \int_G f(g)\pi(g)d\mu_G(g) \qquad \text{for all } f \in L^1(G).$$

Moreover, since π is a unitary representation of G, we have

$$\begin{aligned}\pi(f^*) &= \int_G f^*(g)\pi(g)d\mu_G(g) = \int_G \Delta(g^{-1})\overline{f(g^{-1})}\pi(g)d\mu_G(g) \\ &= \int_G \overline{f(g)}\pi(g^{-1})d\mu_G(g) = \left(\int_G f(g)\pi(g)d\mu_G(g)\right)^* = \pi(f)^*\end{aligned}$$

for all $f \in L^1(G)$, so that

$$\pi : L^1(G) \to \mathcal{L}(\mathcal{H}_\pi)$$

is a $*$-representation of the involutive algebra $L^1(G)$. This has an extension to a $*$-representation $M^b(G) \to \mathcal{L}(\mathcal{H})$ of the convolution algebra of complex Radon measures on G as defined in Section A.4, defined by

$$\pi(\mu) = \int_G \pi(g)d\mu(g) \qquad \text{for all } \mu \in M^b(G).$$

We denote by $C^*_\pi(G)$ the C*-subalgebra of $\mathcal{L}(\mathcal{H}_\pi)$ generated by the set of operators $\pi(L^1(G))$, that is, the closure of $\pi(L^1(G))$ in $\mathcal{L}(\mathcal{H}_\pi)$ for the operator norm.

There are two particularly important cases.

Definition 8.B.1. Let G be an LC group.

Let λ be left regular representation of G on $L^2(G, \mu_G)$. The algebra $C^*_\lambda(G)$ is called the **reduced C*-algebra** of G.

Let $\pi_{\max}$ be the direct sum of all equivalence classes of cyclic representations of G. The algebra $C^*_{\pi_{\max}}(G)$ is called the **maximal C*-algebra** of G and is denoted by $C^*_{\max}(G)$.

Remark 8.B.2. The maximal C*-algebra $C^*_{\max}(G)$ is the completion of $L^1(G)$ with respect to the norm

$$\|f\|_{\max} := \sup_\pi \|\pi(f)\|_{\mathcal{L}(\mathcal{H}_\pi)},$$

where the supremum is taken over all equivalence classes of cyclic representations π of G. The condition of cyclicity ensures that these classes of representations constitute a set. For the definition of $C^*_{\max}(G)$, one could just as well consider the set of all equivalence classes of *irreducible* representations of G; see [**Dixm–C***, § 2.7].

Let π be a representation of G. The associated $*$-representation $L^1(G) \to \mathcal{L}(\mathcal{H}_\pi)$ extends to a nondegenerate C*-representation $C^*_{\max}(G) \to \mathcal{L}(\mathcal{H}_\pi)$, often denoted by π again. Conversely, every nondegenerate C*-representation of $C^*_{\max}(G)$ is associated in this way to exactly one representation of G. Moreover, a closed subspace of $\mathcal{H}_\pi$ is $\pi(G)$-invariant if and only if it is $\pi(C^*_{\max}(G))$-invariant. For all this, see [**Dixm–C***, 13.3.4, 13.9.3] or [**Foll–16**, Sections 3.2 and 7.2].

We record in the following proposition the main points of the preceding discussion.

Proposition 8.B.3. *For a locally compact group G, the construction above provides a bijection between group representations of G and nondegenerate C*-representations of $C^*_{\max}(G)$.*

This bijection respects equivalence and irreducibility. In particular, it induces a bijection

$$\widehat{G} \xrightarrow{\approx} \widehat{C^*_{\max}(G)}$$

from the dual $\widehat{G}$ *of the group* G *onto the spectrum of the* C^**-algebra* $C^*_{\max}(G)$.

For $\pi : G \to \mathcal{U}(\mathcal{H}_\pi)$ as above, the image of $\pi : C^*_{\max}(G) \to \mathcal{L}(\mathcal{H}_\pi)$ is dense in $C^*_\pi(G)$ (because the image of $L^1(G)$ is already dense in $C^*_\pi(G)$, by definition), and therefore is $C^*_\pi(G)$ itself (because images of morphisms of C*-algebras are closed [**Dixm–C***, 1.8.3]), so that the representation π of G gives rise to a surjective morphism of C*-algebras

$$C^*_{\max}(G) \twoheadrightarrow C^*_\pi(G).$$

The kernel of this morphism, denoted by $\mathrm{C^*ker}(\pi)$, is the **C*-kernel** of π.

C*-kernels provide for representations of LC groups an equivalent definition of weak containment and weak equivalence; for the proof of the following proposition, see [**Dixm–C***, 3.4.4 & 18.1.4].

PROPOSITION 8.B.4. *Let* G *be a locally compact group. Let* π_1, π_2 *be two representations of* G.

The representation π_1 *is weakly contained in* π_2 *if and only if* $\mathrm{C^*ker}(\pi_2) \subset \mathrm{C^*ker}(\pi_1)$, *and therefore* π_1 *and* π_2 *are weakly equivalent if and only if* $\mathrm{C^*ker}(\pi_1) = \mathrm{C^*ker}(\pi_2)$.

Let $\widehat{G}$ be equipped with the Fell topology as in Section 1.C. The primitive dual $\mathrm{Prim}(G)$ of G, which is the quotient of $\widehat{G}$ by the relation of weak equivalence, is equipped with the quotient topology as in Section 1.E.

The following corollary is an immediate consequence of Proposition 8.B.4 and of the characterization of the Fell topology on $\widehat{G}$ in terms of weak containment (Proposition 1.C.8).

COROLLARY 8.B.5. *Let* G *be a locally compact group.*
The bijection

$$\widehat{G} \xrightarrow{\approx} \widehat{C^*_{\max}(G)}$$

of Proposition 8.B.3 *is a homeomorphism and induces a homeomorphism*

$$\mathrm{Prim}(G) \xrightarrow{\approx} \mathrm{Prim}(C^*_{\max}(G)).$$

REMARK 8.B.6. (1) The canonical surjective map (κ1) of Section 1.E can now be written

$$(\kappa 1) \qquad \kappa^{\mathrm{d}}_{\mathrm{prim}} : \widehat{G} \twoheadrightarrow \mathrm{Prim}(G) = \mathrm{Prim}(C^*_{\max}(G)), \qquad \pi \mapsto \mathrm{C^*ker}(\pi).$$

(2) Let π_1, π_2 be two representations of an LC group G and let B be a dense subspace of $C^*_{\max}(G)$, for example $B = C^c(G)$ or $B = L^1(G)$. The following properties are equivalent:

(i) π_1 is weakly contained in π_2;
(ii) $\Vert \pi_1(x) \Vert \le \Vert \pi_2(x) \Vert$ for every $x \in B$;
(iii) $\Vert \pi_1(\mu) \Vert \le \Vert \pi_2(\mu) \Vert$ for every $\mu \in M^b(G)$.

Indeed, (ii) is equivalent to $\Vert \pi_1(x) \Vert \le \Vert \pi_2(x) \Vert$ for every $x \in C^*_{\max}(G)$, by the density of B; this last property is equivalent to $\mathrm{C^*ker}(\pi_2) \subset \mathrm{C^*ker}(\pi_1)$ and hence to (i), by Proposition 8.B.4. The equivalence of (iii) and (ii) follows from the fact that $L^1(G)$ is dense in $M^b(G)$ for the weak*-topology.

REMARK 8.B.7. (1) The Fell topology on the dual $\widehat{G}$ of an LC group G can be defined in at least two ways: in terms of functions of positive type, as in Section 1.A, or in terms of the Jacobson topology on $\mathrm{Prim}(G) = \mathrm{Prim}(C^*_{\max}(G))$ and the map (κ1). For more on the equivalence of several definitions of this topology, see [**Dixm–C***, § 3].

(2) The map of (κ1) is injective (and therefore bijective) in important cases, see Glimm's Theorem 8.F.3, but not in general, see Section 9.G.

(3) The C*-kernel of π should not be confused with the kernel of the group homomorphism $\pi : G \to \mathcal{U}(\mathcal{H}_\pi)$. For example, if λ is the left regular representation of G as above, so that $\lambda : G \to \mathcal{U}(L^2(G))$ is injective in all cases, we have $\mathrm{C^*ker}(\lambda) = \{0\}$ if and only if G is amenable (Theorem 1.C.3).

(4) The bijection of Proposition 8.B.3 also respects quasi-equivalence, as introduced in Remark 8.A.22.

(5) There are some topological groups G that are not locally compact for which a C*-algebra $C^*_{\max}(G)$ can nevertheless be defined, and be of some use. See [**StVo–75**] for $\mathcal{U}(\infty)$, an appropriate inductive limit of the compact Lie groups $\mathcal{U}(n)$. C*-algebras of groups that are not locally compact will not be mentioned again in the present book.

EXAMPLE 8.B.8. Let G be a locally compact group that is amenable. We repeat that the left regular representation λ of G establishes an isomorphism from $C^*_{\max}(G)$ onto $C^*_\lambda(G)$; in other words, every representation of G is weakly contained in the regular representation.

(1) In case G is an abelian LC group, the Fourier transform $\mathcal{F} : L^2(G) \to L^2(\widehat{G})$ is a unitary isomorphism of Hilbert spaces (the L^2 spaces are defined in terms of Haar measures on G and $\widehat{G}$ that are associated with each other), and $\mathcal{F}C^*_\lambda(G)\mathcal{F}^{-1} = C^0(\widehat{G})$; recall that $C^0(\widehat{G})$ stands for the algebra (for the pointwise product) of complex-valued continuous functions on $\widehat{G}$ vanishing at infinity. In particular, the C*-algebras $C^*_{\max}(G)$ and $C^0(\widehat{G})$ are isomorphic.

(2) When G is compact, its C*-algebra $C^*_{\max}(G)$ is isomorphic to the restricted product $\prod^{c_0}_{\pi \in \widehat{G}} \mathrm{End}(\mathcal{H}_\pi)$ of the family of full matrix algebras, indexed by the dual of G. Recall that $1 \le \dim \mathcal{H}_\pi < \infty$ for all $\pi \in \widehat{G}$, so that $\mathrm{End}(\mathcal{H}_\pi)$ is isomorphic to the algebra $\mathrm{M}_n(\mathbf{C})$ of complex n-by-n matrices for $n = \dim \mathcal{H}_\pi$; see for instance Proposition 3.4 in [**Will–07**].

Recall also that, given a family $(A_\iota)_{\iota \in I}$ of C*-algebras, their Cartesian product $\prod_{\iota \in I} A_\iota$ is the C*-algebra constituted of $(x_\iota)_{\iota \in I}$ with $x_\iota \in A_\iota$ for all $\iota \in I$ and $\sup_{\iota \in I} \|x_\iota\| < \infty$, and their **restricted product** $\prod^{c_0}_{\iota \in I} A_\iota$ is the C*-algebra constituted of $(x_\iota)_{\iota \in I}$ such that, for all $\varepsilon > 0$, there is a cofinite subset J_ε in I such that $\|x_\iota\| < \varepsilon$ for all $\iota \in J_\varepsilon$.

REMARK 8.B.9. (1) Let G be a discrete group. Then $C^*_\pi(G)$ is a C*-algebra with unit for every representation π of G. There is some kind of converse: an LC group G is discrete if and only if its reduced C*-algebra $C^*_\lambda(G)$ has a unit, if and only if its maximal C*-algebra $C^*_{\max}(G)$ has a unit [**Miln–71**].

(2) Specific descriptions of maximal C*-algebras have been worked out for a few groups, often as spaces of sections of bundles or continuous fields. Besides abelian

and compact groups, here is a small sample of groups for which such descriptions are known, with some of the relevant references:

- the special linear groups $\mathrm{SL}_2(\mathbf{C})$ [**Fell–61**] and $\mathrm{SL}_2(\mathbf{R})$ [**Mili–71, Vale–84**];
- the group Spin(4, 1), universal cover of the de Sitter group $\mathrm{SO}(4,1)_0$ [**BoMa–77**];
- the motion group $\mathbf{R}^n \rtimes \mathrm{SO}(n)$ [**AbEL–11**];
- the Heisenberg group $H(\mathbf{R})$, as in Remark 3.B.19; see [**LuTu–11**], and also [**Gorb–80**];
- the affine group $\begin{pmatrix} \mathbf{K}^\times & \mathbf{K} \\ 0 & 1 \end{pmatrix}$ over a nondiscrete locally compact field $\mathbf{K}$, for example over $\mathbf{Q}_p$ [**Rose–76**].

8.C. Functorial properties of group C*-algebras

For a locally compact group G, the assignments $G \rightsquigarrow C^*_{\max}(G)$ and $G \rightsquigarrow C^*_\lambda(G)$ do not have good functorial properties with respect to general continuous group homomorphisms. Here is an example which illustrates this fact.

EXAMPLE 8.C.1. Let G be a nondiscrete LC group and let G_{disc} be the group G equipped with the discrete topology. The identity map $G_{\mathrm{disc}} \to G$ is a continuous isomorphism. However, $C^*_{\max}(G)$ is not isomorphic to a quotient of $C^*_{\max}(G_{\mathrm{disc}})$. Indeed, $C^*_{\max}(G_{\mathrm{disc}})$ is a unital C*-algebra, whereas, as mentioned above (Remark 8.B.9), $C^*_{\max}(G)$ has no unit.

The reason for the lack of functoriality of $G \rightsquigarrow C^*_{\max}(G)$ lies in the bad functorial properties of the assignment $G \rightsquigarrow L^1(G)$; more precisely, given a continuous group homomorphism $G_1 \to G_2$, there is in general no naturally associated homomorphism between the convolution algebras $L^1(G_1)$ and $L^1(G_2)$. The same holds for $C^c(G_1)$ and $C^c(G_2)$.

We will see that $G \rightsquigarrow C^*_{\max}(G)$ and, to a lesser extent, $G \rightsquigarrow C^*_\lambda(G)$, do have satisfactory functorial properties with respect to some special continuous group homomorphisms.

8.C.a. On the functor $G \rightsquigarrow M^b(G)$. Let G be an LC group. Recall from Section 8.B that the measure algebra $M^b(G)$ of complex Radon measures on G is a Banach involutive algebra and that every representation π of G induces a $*$-representation $\pi : M^b(G) \to \mathcal{L}(\mathcal{H}_\pi)$.

As a preliminary step, we show that the assignment $G \rightsquigarrow M^b(G)$ has optimal functorial properties.

PROPOSITION 8.C.2. *Let G_1, G_2 be locally compact groups and $\theta : G_1 \to G_2$ a continuous homomorphism. For $\mu \in M^b(G_1)$, denote by $\theta_*(\mu) \in M^b(G_2)$ the image of μ under θ.*

The map $\theta_* : M^b(G_1) \to M^b(G_2)$ is a continuous homomorphism of involutive algebras such that $\|\theta_*\| \le 1$.

PROOF. Observe that, for $\mu \in M^b(G_1)$, the measure $\theta_*(\mu) \in M^b(G_2)$ is determined by

$$\int_{G_2} \varphi(y) d(\theta_*(\mu))(y) = \int_{G_1} \varphi(\theta(x)) d\mu(x) \qquad \text{for all } \varphi \in C^b(G_2).$$

It is clear that the map θ_* is a continuous linear map and that $\|\theta_*\| \le 1$.

For $\mu_1, \mu_2 \in M^b(G_1)$ and $\varphi \in C^b(G_2)$, we have

$$\begin{aligned}\int_{G_2} \varphi(y) d(\theta_*(\mu_1 * \mu_2))(y) &= \int_{G_1} \varphi(\theta(x)) d(\mu_1 * \mu_2)(x) \\ &= \int_{G_1}\int_{G_1} \varphi(\theta(xz)) d\mu_1(x) d\mu_2(z) \\ &= \int_{G_1}\int_{G_1} \varphi(\theta(x)\theta(z)) d\mu_1(x) d\mu_2(z) \\ &= \int_{G_2}\int_{G_2} \varphi(uv) d(\theta_*(\mu_1))(u) d(\theta_*(\mu_2))(v) \\ &= \int_{G_2} \varphi(y) d(\theta_*(\mu_1) * \theta_*(\mu_2))(y).\end{aligned}$$

Therefore $\theta_*(\mu_1 * \mu_2) = \theta_*(\mu_1) * \theta_*(\mu_2)$.

For $\mu \in M^b(G_1)$ and $\varphi \in C^b(G_2)$, we have

$$\begin{aligned}\int_{G_2} \varphi(y) d(\theta_*(\mu^*))(y) &= \int_{G_1} \varphi(\theta(x)) d\mu^*(x) = \int_{G_1} \overline{\varphi(\theta(x^{-1}))} d\mu(x) \\ &= \int_{G_2} \overline{\varphi(z^{-1})} d\theta_*(\mu)(z) = \int_{G_2} \varphi(y) d(\theta_*(\mu))^*(y).\end{aligned}$$

Therefore $\theta_*(\mu^*) = (\theta_*(\mu))^*$. This proves the claim. □

REMARK 8.C.3. Proposition 8.C.2 shows that the assignment $G \rightsquigarrow M^b(G)$ is a functor from the category of LC groups to the category of Banach involutive algebras. Since $(\beta \circ \alpha)_* = \beta_* \circ \alpha_*$, when $\alpha : G_1 \to G_2$ and $\beta : G_2 \to G_3$ are continuous homomorphisms between LC groups G_1, G_2, G_3, this functor is covariant.

We need to study the action of representations of G_2 on measures in $\theta_*(M_b(G_1))$.

PROPOSITION 8.C.4. *Let $\theta : G_1 \to G_2$ and $\theta_* : M^b(G_1) \to M^b(G_2)$ be as in Proposition* 8.C.2.

For every $\mu \in M^b(G_1)$ and every representation π of G_2, we have

$$\pi(\theta_*(\mu)) = (\pi \circ \theta)_*(\mu).$$

PROOF. Let π be a representation of G_2 and $\mu \in M^b(G_1)$. Then

$$\pi(\theta_*(\mu)) = \int_{G_2} \pi(y) d\theta_*(\mu)(y) = \int_{G_1} \pi(\theta(x)) d\mu(x) = (\pi \circ \theta)_*(\mu).$$

This proves the claim. □

For our study of the assignment $G \rightsquigarrow C^*_{\max}(G)$, it will be helpful to introduce C*-algebras associated with the Banach involutive algebra $M^b(G)$, as we did for the Banach involutive algebra $L^1(G)$ in Section 8.B.

DEFINITION 8.C.5. Let G be an LC group. The **enveloping C*-algebra** of $M^b(G)$, denoted by $M^*_{\max}(G)$, is the completion of $M^b(G)$ for the norm

$$\|\mu\|_{\max} := \sup_\pi \|\pi(\mu)\|_{\mathcal{L}(\mathcal{H}_\pi)},$$

where the supremum is taken over all equivalence classes of cyclic (or, equivalently, irreducible) representations π of G.

The **reduced enveloping C*-algebra** of $M^b(G)$, denoted by $M^*_\lambda(G)$, is the completion of $M^b(G)$ for the norm

$$\|\mu\|_\lambda := \|\lambda(\mu)\|.$$

Equivalently, $M^*_\lambda(G)$ could be defined as the norm closure of $\lambda(M^b(G))$ in $\mathcal{L}(L^2(G))$.

The next proposition shows in particular that the assignment $G \rightsquigarrow M^*_{\max}(G)$ is a covariant functor from the category of LC groups to the category of C*-algebras.

THEOREM 8.C.6. *Let G_1, G_2 be locally compact groups and $\theta : G_1 \to G_2$ a continuous homomorphism.*

The map $\theta_* : M^b(G_1) \to M^b(G_2)$ of Proposition 8.C.2 extends to a morphism $\theta_* : M^*_{\max}(G_1) \to M^*_{\max}(G_2)$ of C-algebras.*

PROOF. In the sequel, $\| \cdot \|_{\max}$ will denote the norm on $M^*_{\max}(G_1)$ as well as the norm on $M^*_{\max}(G_2)$.

Let $\mu \in M^b(G_1)$ and π a representation of G_2. By Proposition 8.C.4, we have $\pi(\theta_*(\mu)) = (\pi \circ \theta)_*(\mu)$. It follows from the definition of the norm $\| \cdot \|_{\max}$ that

$$\|\pi(\theta_*(\mu))\| = \|(\pi \circ \theta)_*(\mu)\| \le \|\mu\|_{\max}.$$

Therefore we have

$$\|\theta_*(\mu)\|_{\max} \le \|\mu\|_{\max} \qquad \text{for every } \mu \in M^b(G_1).$$

So, the map $\theta_* : M^b(G_1) \to M^b(G_2)$, which is a homomorphism of involutive algebras, extends to a morphism of C*-algebras $M^*_{\max}(G_1) \to M^*_{\max}(G_2)$. $\square$

8.C.b. On the assignment $G \rightsquigarrow C^*_{\max}(G)$. Let G be an LC group. Recall that $C^c(G)$ is an involutive algebra for the convolution product and the involution given by $f^*(x) = \Delta(x^{-1})\overline{f(x^{-1})}$ for $f \in C^c(G)$. Recall also that $C^c(G)$ may be viewed as a $*$-subalgebra of the Banach involutive algebra $M^b(G)$ by means of the map

$$f \mapsto f(x)d\mu_G(x) \qquad \text{for } f \in C^c(G).$$

Let G_1, G_2 be LC groups and $\theta : G_1 \to G_2$ a continuous group homomorphism. Let $\theta_* : M^b(G_1) \to M^b(G_2)$ be the associated map as in Proposition 8.C.2. Assume that θ is an *open* homomorphism, that is, $\theta(U)$ is open in G_2 for every open subset U of G_1. In Theorem 8.C.12, we will show that, in this case, $\theta_*(C^c(G_1))$ is contained in $C^c(G_2)$ and that $\theta_* : C^c(G_1) \to C^c(G_2)$ extends to a morphism $C^*_{\max}(G_1) \to C^*_{\max}(G_2)$. To this end, we need to separately treat three types of homomorphisms: topological isomorphisms, quotient maps, and open injective homomorphisms.

PROPOSITION 8.C.7. *Let G_1, G_2 be LC groups with Haar measures μ_{G_1}, μ_{G_2}, and let $\theta : G_1 \to G_2$ be an isomorphism of topological groups.*

(1) *We have $\theta_*(C^c(G_1)) = C^c(G_2)$ and, for an appropriate normalization of μ_{G_1} and μ_{G_2}, the map $\theta_* : C^c(G_1) \to C^c(G_2)$ is an isomorphism of involutive algebras.*
(2) *The map $\theta_* : C^c(G_1) \to C^c(G_2)$ extends to an isomorphism $C^*_{\max}(G_1) \to C^*_{\max}(G_2)$ of C*-algebras.*

PROOF. Since θ is an isomorphism of topological groups, it follows from Proposition 8.C.2 that $\theta_* : M^b(G_1) \to M^b(G_2)$ is an isomorphism of involutive algebras. It is obvious that $\theta_*(\mu_{G_1})$ is a multiple of μ_{G_2}, that $\theta_*(C^c(G_1)) = C^c(G_2)$, and that the restriction of θ_* is a linear bijection $C^c(G_1) \to C^c(G_2)$. We may normalize μ_{G_1} and μ_{G_2} so that $\mu_{G_2} = \theta_*(\mu_{G_1})$. With this normalization, $\theta_* : C^c(G_1) \to C^c(G_2)$ is a morphism of involutive algebras. This proves Item (1).

Let $f \in C^c(G_1)$. Denoting by $\|\cdot\|_{\max}$ the norms on $M^*_{\max}(G_1)$ and $M^*_{\max}(G_2)$, we have, by Theorem 8.C.6,

$$\|\theta_*(f)\|_{\max} \leq \|f\|_{\max}.$$

Since the same arguments also apply to θ^{-1}, it follows that

$$\|\theta_*(f)\|_{\max} = \|f\|_{\max}, \qquad \text{for every } f \in C^c(G_1).$$

Therefore the $*$-isomorphism $\theta_* : C^c(G_1) \to C^c(G_2)$ extends to an isomorphism

$$C^*_{\max}(G_1) \to C^*_{\max}(G_2)$$

of C*-algebras. □

Next, we examine the case of quotient maps.

PROPOSITION 8.C.8. *Let G be a locally compact group, N a closed normal subgroup of G, and $\theta : G \to G/N$ the canonical epimorphism.*

(1) *For a proper normalization of the Haar measures μ_G, μ_N. and $\mu_{G/N}$, we have $\theta_*(C^c(G)) = C^c(G/N)$ and $\theta_* : C^c(G) \to C^c(G/N)$ is a surjective homomorphism of involutive algebras.*
(2) *The map $\theta_* : C^c(G) \twoheadrightarrow C^c(G/N)$ extends to a surjective morphism $C^*_{\max}(G) \twoheadrightarrow C^*_{\max}(G/N)$ of C*-algebras.*

PROOF. Fix a Haar measure μ_N on N and let $P : C^c(G) \to C^c(G/N)$ be the linear map defined by

$$P(f)(xN) = \int_N f(xy)d\mu_N(y) \qquad \text{for } f \in C^c(G), x \in G.$$

It is a surjective linear map [**BeHV–08**, Lemma B.1.2]. For a proper normalization of the Haar measures μ_G and $\mu_{G/N}$ of G and G/N, we have the so-called Weil's formula

$$\int_G f(x)d\mu_G(x) = \int_{G/N} P(f)(xN)d\mu_{G/N}(xN) \qquad \text{for all } f \in C^c(G)$$

(see [**Weil–40**, Chap. II, § 9]).

Let $f \in C^c(G)$ and $\varphi \in C^c(G/N)$. Since $P((\varphi \circ \theta)f) = \varphi P(f)$, we have

$$\int_{G/N} \varphi(xN)P(f)(xN)d\mu_{G/N}(xN) = \int_G \varphi(\theta(x))f(x)d\mu_G(x),$$

by Weil's formula. This shows that

$$\theta_*(f) = P(f) \qquad \text{for all } f \in C^c(G),$$

that is, the restriction of θ_* to $C^c(G)$ coincides with P. This implies that

$$\theta_*(C^c(G)) = P(C^c(G)) = C^c(G/N).$$

Moreover, $\theta_* : C^c(G) \to M^b(G/N)$ is a homomorphism of involutive algebras, by Proposition 8.C.2. Therefore $\theta_* : C^c(G) \to C^c(G/N)$ is a surjective homomorphism of involutive algebras. This proves Item (1).

Let $f \in C^c(G)$. By Theorem 8.C.6, we have

$$\|\theta_*(f)\|_{\max} \leq \|f\|_{\max},$$

where, as before, $\|\cdot\|_{\max}$ denotes the norm on the enveloping C*-algebras of the measure algebras. Therefore $\theta_* : C^c(G) \to C^c(G/N)$ extends to a morphism $C^*_{\max}(G) \to C^*_{\max}(G/H)$ of C*-algebras; this morphism is surjective, since $\theta_*(C^c(G))$ coincides with $C^c(G/N)$, and since $C^c(G/N)$ is dense in $C^*_{\max}(G/H)$. □

Finally, we treat the inclusion of open subgroups. We first give a preliminary result on the inclusion homomorphism θ of a closed, not necessarily open, subgroup H of an LC group G. For $f \in C_c(H)$, observe that $\theta_*(f)$ is the measure $f(x)d\mu_H(x)$ on G. Observe also that, for every representation π of G, we have

$$\pi(\theta_*(\mu)) = (\pi|_H)(\mu) \quad \text{for all } \mu \in M^b(H).$$

PROPOSITION 8.C.9. *Let G be an LC group, H a closed subgroup of G, and $\theta : H \to G$ the canonical inclusion.*

The injective $*$-homomorphism $\theta_* : C^c(H) \to M^b(G)$ extends to a morphism $\theta_* : C^*_{\max}(H) \to M^*_{\max}(G)$ of C-algebras. Moreover, the following properties are equivalent:*

(i) *$\theta_* : C^*_{\max}(H) \to M^*_{\max}(G)$ is injective;*
(ii) *for every representation σ of H, there exists a representation π of G such that σ is weakly contained in $\pi|_H$.*

PROOF. By Theorem 8.C.6, the map $\theta_* : C^c(H) \to M^b(G)$ extends to a morphism $\theta_* : M^*_{\max}(H) \to M^*_{\max}(G)$ of C*-algebras. Since obviously $C^*_{\max}(H)$ may be viewed as a C*-subalgebra of $M^*_{\max}(H)$, we obtain by restriction a morphism $\theta_* : C^*_{\max}(H) \to M^*_{\max}(G)$ of C*-algebras.

Assume that (i) holds. Then θ_* is isometric, that is,

$$(*) \qquad \|\theta_*(f)\|_{\max} = \|f\|_{\max} \quad \text{for all } f \in C^c(H).$$

Let σ be a representation of H. Set $\pi_{max} := \bigoplus_{\pi \in \widehat{G}} \pi$. Since

$$\|\mu\|_{\max} = \|\pi_{max}(\mu)\| \quad \text{for all } \mu \in M^b(G),$$

it follows from $(*)$ that

$$\|\sigma(f)\| \le \|\pi_{max}(\theta_*(f))\| = \|(\pi_{max}|_H)(f)\| \quad \text{for all } f \in C^c(H);$$

this shows that σ is weakly contained in $\pi_{max}|_H$.

Conversely, assume that (ii) holds. Let

$$\sigma_{max} := \bigoplus_{\sigma \in \widehat{H}} \sigma.$$

There exists a representation π of G such that σ_{max} is weakly contained in $\pi|_H$. Therefore, we have

$$\|f\|_{\max} = \|\sigma_{max}(f)\| \le \|(\pi|_H)(f)\| = \|\pi(\theta_*(f))\| \le \|\theta_*(f)\|_{\max},$$

and hence $\|f\|_{\max} = \|\theta_*(f)\|_{\max}$ for all $f \in C^c(H)$. So, $\theta_* : C^*_{\max}(H) \to M^*_{\max}(G)$ is injective. □

We indicate two classes of subgroups H for which the map $\theta_* : C^*_{\max}(H) \to M^*_{\max}(G)$ of Proposition 8.C.9 is injective.

COROLLARY 8.C.10. *Let G be an LC group, H a closed subgroup of G, and $\theta : H \to G$ the canonical injection.*

(1) *Assume either that H is open or that H is amenable. Then the morphism $\theta_* : C^*_{\max}(H) \to M^*_{\max}(G)$ is injective.*

(2) *Assume that H is open. For an appropriate normalization of μ_G and μ_H, the image of $C^*_{\max}(H)$ under θ_* is a subalgebra of $C^*_{\max}(G)$.*

PROOF. To show Item (1), it suffices to check that Condition (ii) of Proposition 8.C.9 is satisfied, when H is either open or amenable.

• *First case.* Assume that H is open. Let σ be a representation of H. Consider the induced representation $\pi := \mathrm{Ind}_H^G\sigma$. Then σ is contained in $\pi|_H$ (see Definition 1.F.1).

• *Second case.* Assume that H is amenable. Let σ be a representation of H. Then σ is weakly contained in λ_H. Since $\lambda_G|_H$ is weakly equivalent to λ_H, it follows that σ is weakly contained in $\lambda_G|_H$.

To show Item (2), note first that $C^c(H)$ is naturally a subspace of $C^c(G)$, since H is open in G. Choose μ_G and μ_H such that $\mu_G|_H = \mu_H$. Then, for every $f \in C^c(H)$, we have

$$\theta_*(f) = f(x)d\mu_H(x) = f(x)d\mu_G(x),$$

and hence θ_* is an inclusion of $C^c(H)$ as a subalgebra of $C^c(G)$. Since the closure of $C^c(G)$ in $M^*_{\max}(G)$ coincides with $C^*_{\max}(G)$, the claim follows. □

REMARK 8.C.11. (1) Condition (ii) of Proposition 8.C.9 was introduced in [**Fell–64**, Page 442] as "Property (WP3)" and studied as a weak version of Frobenius reciprocity for finite groups. It also appeared in [**Rief–74**, page 209] and [**BeVa–95**] in a context similar to ours, with the multiplier C*-algebra $M(C^*_{\max}(G))$ of $C^*_{\max}(G)$ replacing the algebra $M^*_{\max}(G)$.

(2) Condition (ii) from Proposition 8.C.9 does not always hold. Indeed, this condition fails for $G = \mathrm{SL}_3(\mathbf{C})$ and a subgroup H isomorphic to $\mathrm{SL}_2(\mathbf{C})$, as shown in [**Fell–64**, Theorem 6.1], with an incomplete proof which was corrected in [**BeLS–92**, Remark 1.13 (i)]. The failure of Condition (ii) was established in [**BeVa–95**] for any lattice H in a higher rank simple Lie group G as well as for many lattices in a rank one simple Lie group.

We are now ready to state the final result concerning the assignment $G \rightsquigarrow C^*_{\max}(G)$.

THEOREM 8.C.12. *Let G_1, G_2 be locally compact groups with Haar measures μ_{G_1}, μ_{G_2} and let $\theta : G_1 \to G_2$ be an open continuous homomorphism.*

(1) *We have $\theta_*(C^c(G_1)) \subset C^c(G_2)$ and, for appropriate normalizations of μ_{G_1} and μ_{G_2}, the map $\theta_* : C^c(G_1) \to C^c(G_2)$ is a morphism of involutive algebras.*
(2) *This map $\theta_* : C^c(G_1) \to C^c(G_2)$ extends to a morphism of C*-algebras $\theta_* : C^*_{\max}(G_1) \to C^*_{\max}(G_2)$.*
(3) *$\theta_* : C^*_{\max}(G_1) \to C^*_{\max}(G_2)$ is surjective if and only if $\theta : G_1 \to G_2$ is surjective.*
(4) *$\theta_* : C^*_{\max}(G_1) \to C^*_{\max}(G_2)$ is injective if and only if $\theta : G_1 \to G_2$ is injective*

PROOF. Let $N \subset G_1$ be the kernel and $H \subset G_2$ the image of θ. We can write

$$\theta = \theta_3 \circ \theta_2 \circ \theta_1,$$

where

• $\theta_1 : G_1 \to G_1/N$ is the canonical epimorphism,

- $\theta_2 : G_1/N \to H$ is the group isomorphism induced by θ,
- $\theta_3 : H \to G_2$ is the canonical injection.

Since θ is an open map, H is an open subgroup of G_2 and $\theta_2 : G_1/N \to H$ is an isomorphism of topological groups. Therefore Items (1) and (2) follow from Proposition 8.C.7, Proposition 8.C.8, and Corollary 8.C.10.

Items (3) and (4) are immediate consequences of the following observations:

- the map $C^*_{\max}(G_1) \to C^*_{\max}(G_1/N)$ induced by θ_1 is surjective;
- the maps $C^*_{\max}(G_1/N) \to C^*_{\max}(H)$ and $C^*_{\max}(H) \to C^*_{\max}(G)$ induced by θ_2 and θ_3 are injective;
- the map $C^*_{\max}(G_1) \to C^*_{\max}(G_1/N)$ is injective if and only if $N = \{e\}$, that is, if and only if θ is injective;
- the map $C^*_{\max}(H) \to C^*_{\max}(G_2)$ is surjective if and only if $H = G_2$, that is, if and only if θ is surjective. □

If Γ is a discrete group, observe that $C^c(\Gamma)$ coincides with the group algebra $\mathbf{C}[\Gamma]$. We will identify Γ in the obvious way as a subset of $\mathbf{C}[\Gamma]$. Under this identification, if $\theta : \Gamma_1 \to \Gamma_2$ is a homomorphism between discrete groups Γ_1 and Γ_2, then $\theta_* : \mathbf{C}[\Gamma_1] \to \mathbf{C}[\Gamma_2]$ coincides with the linear extension of θ to $\mathbf{C}[\Gamma_1]$.

The following immediate consequence of Theorem 8.C.12 shows in particular that the assignment $\Gamma \rightsquigarrow C^*_{\max}(\Gamma)$ is a covariant functor from the category of discrete groups to the category of C*-algebras.

COROLLARY 8.C.13. *Let Γ_1, Γ_2 be discrete groups.*

Every group homomorphism $\theta : \Gamma_1 \to \Gamma_2$ extends to a morphism of C-algebras $\theta_* : C^*_{\max}(\Gamma_1) \to C^*_{\max}(\Gamma_2)$.*

8.C.c. On the assignment $G \rightsquigarrow C^*_\lambda(G)$. As we now discuss, reduced group C*-algebras have less functorial properties than maximal group C*-algebras.

Let G_1, G_2 be LC groups and $\theta : G_1 \to G_2$ an open continuous homomorphism. Recall from Theorem 8.C.12 that θ induces a homomorphism $\theta_* : C^c(G_1) \to C^c(G_2)$ of involutive algebras, upon an appropriate normalization of μ_{G_1} and μ_{G_2}. The following result gives a characterization of the homomorphisms θ inducing a morphism between the respective reduced C*-algebras.

THEOREM 8.C.14. *Let G_1, G_2 be LC groups and $\theta : G_1 \to G_2$ a continuous* open *homomorphism. Assume that the Haar measures μ_{G_1}, μ_{G_2} are normalized as in Theorem* 8.C.12(1). *The following properties are equivalent:*

(i) *the induced homomorphism $\theta_* : C^c(G_1) \to C^c(G_2)$ extends to a morphism $C^*_\lambda(G_1) \to C^*_\lambda(G_2)$ of C*-algebras;*
(ii) *the kernel of θ is an amenable subgroup of G_1.*

PROOF. Recall that λ_G denotes the left regular representation of an LC group G.

• *First step.* We claim that $\theta_* : C^c(G_1) \to C^c(G_2)$ extends to $C^*_\lambda(G_1)$ if and only if the representation $\lambda_{G_2} \circ \theta$ of G_1 is weakly contained in λ_{G_1}.

Indeed, let $f \in C^c(G_1)$. By Proposition 8.C.4, we have

$$\lambda_{G_2}(\theta_*(f)) = (\lambda_{G_2} \circ \theta)(f).$$

This shows that the inequality

$$\|\lambda_{G_2}(\theta_*(f))\| \leq \|\lambda_{G_1}(f)\|$$

is equivalent to the inequality

$$\|(\lambda_{G_2} \circ \theta)(f)\| \leq \|\lambda_{G_1}(f)\|.$$

Therefore θ_* extends from $C^c(G_1)$ to $C^*_\lambda(G_1)$ if and only if

$$\|(\lambda_{G_2} \circ \theta)(f)\| \leq \|\lambda_{G_1}(f)\| \quad \text{for every } f \in C^c(G_1),$$

that is, if and only if $\lambda_{G_2} \circ \theta$ is weakly contained in λ_{G_1}.

• *Second step.* Assume that $\theta : G_1 \to G_2$ is an isomorphism. We claim that $\lambda_{G_2} \circ \theta$ equivalent to λ_{G_1}.

Indeed, define a linear map $U : L^2(G_1) \to L^2(G_2)$ by

$$(U\xi)(y) = \xi(\theta^{-1}(y)) \quad \text{for } \xi \in L^2(G_1),\ y \in G_2.$$

By the normalization of μ_{G_1} and μ_{G_2}, we have $\theta_*(\mu_{G_1}) = \mu_{G_2}$ (see the proof of Proposition 8.C.7). It follows, as is easily checked, that U is an isomorphism of Hilbert spaces. Moreover, we have

$$\begin{aligned}(\lambda_{G_2}(\theta(g))U\xi)(y) &= (U\xi)(\theta(g^{-1})y) = \xi(g^{-1}\theta^{-1}(y)) \\ &= (\lambda_{G_1}(g)\xi)(\theta^{-1}(y)) = (U\lambda_{G_1}(g)\xi)(y),\end{aligned}$$

for every $\xi \in L^2(G_1), g \in G_1$, and $y \in G_2$. So, U intertwines $\lambda_{G_2} \circ \theta$ and λ_{G_1}.

• *Third step.* Assume that G_1 is an open subgroup of G_2 and that $\theta : G_1 \to G_2$ is the canonical injection. We claim that $\lambda_{G_2} \circ \theta$ is weakly equivalent to λ_{G_1}.

Indeed, this holds since we have $\lambda_{G_2} \circ \theta = \lambda_{G_2}|_{G_1}$ in this case.

• *Fourth step.* Assume that $G_2 = G_1/N$ for a closed normal subgroup N of G_1 and that $\theta : G_1 \to G_1/N$ is the canonical epimorphism. We claim that $\lambda_{G_2} \circ \theta$ is weakly contained in λ_{G_1} if and only if N is amenable.

Indeed, observe first that $\lambda_{G_2} \circ \theta$ is equivalent to the induced representation $\mathrm{Ind}_N^{G_1} 1_N$.

Assume that N is amenable. Then 1_N is weakly contained in λ_N and hence, by continuity of induction, $\mathrm{Ind}_N^{G_1} 1_N$ is weakly contained in $\mathrm{Ind}_N^{G_1} \lambda_N$, which is equivalent to λ_{G_1}. Therefore $\lambda_{G_2} \circ \theta$ is weakly contained in λ_{G_1}.

Conversely, assume that $\lambda_{G_2} \circ \theta$ is weakly contained in λ_{G_1}. Then, by continuity of restriction, the representation $(\lambda_{G_2} \circ \theta)|_N$ is weakly contained in $\lambda_{G_1}|_N$. Observe that $(\lambda_{G_2} \circ \theta)|_N$ is equivalent to a multiple of the trivial representation 1_N. Since $\lambda_{G_1}|_N$ is weakly equivalent to λ_N, it follows that 1_N is weakly contained in λ_N. Therefore N is amenable.

• *Fifth step.* Let $\theta : G_1 \to G_2$ be an arbitrary open homomorphism. Denote by N the kernel of θ and by H the image of θ. As in the proof of Theorem 8.C.12, we can write

$$\theta = \theta_3 \circ \theta_2 \circ \theta_1,$$

where $\theta_1 : G_1 \to G_1/N$ is the canonical epimorphism, $\theta_2 : G_1/N \to H$ is the group isomorphism induced by θ, and $\theta_3 : H \to G_2$ is the canonical injection.

By the three first steps, θ_2 and θ_3 induce morphisms $C^*_\lambda(G_1/N) \to C^*_\lambda(H)$ and $C^*_\lambda(H) \to C^*_\lambda(G_2)$. The fourth step shows that θ_1 induces a morphism $C^*_\lambda(G_1) \to C^*_\lambda(G_1/N)$ if and only if N is amenable. □

The following corollary is an immediate consequence of Theorem 8.C.14; it should be compared with Corollary 8.C.13, in which we established that

$$\Gamma \rightsquigarrow C^*_{\max}(\Gamma)$$

is a functor on the category of discrete groups to that of C*-algebras. Recall that we view a discrete group Γ as a subset of its group algebra $\mathbf{C}[\Gamma]$, which is itself a subalgebra of $C^*_\lambda(\Gamma)$.

COROLLARY 8.C.15. *Let Γ_1, Γ_2 be discrete groups and $\theta : \Gamma_1 \to \Gamma_2$ a homomorphism. The following properties are equivalent:*

(i) *θ extends to a morphism $\theta_* : C^*_\lambda(\Gamma_1) \to C^*_\lambda(\Gamma_2)$ of C^*-algebras;*
(ii) *the kernel of θ is an amenable subgroup of Γ_1.*

8.D. Second-countable and σ-compact LC groups

Some properties of representations of σ-compact locally compact groups are well-documented in the literature for second-countable LC groups only. For example, the fact that C*-kernels of factor representations of σ-compact LC groups are primitive is apparently known to experts (see Line 5 of [**Pogu–83**]); however, the only references of which we are aware deal with second-countable LC groups (see [**Dixm–60a**, Corollaire 3], and also Corollaries 8.A.21 and 7.F.2). Observe that a σ-compact LC group need not be second-countable. For example, an uncountable Cartesian product of nontrivial compact groups is compact and is not second-countable.

We indicate in this section how, in several situations, the case of a σ-compact LC group can be reduced to that of second-countable LC groups. The first tool for this reduction is the Kakutani–Kodaira Theorem; for a proof, see the original article [**KaKo–44**], or [**CoHa–16**, 2.B.6].

THEOREM 8.D.1 (Kakutani–Kodaira). *Let G be a σ-compact locally compact group.*

For every neighbourhood U of e in G, there exists a compact normal subgroup K of G such that $K \subset U$ and G/K is second-countable.

Using the notion of projective limits of topological groups (see [**BTG1–4**, Chap. III, § 7]), Theorem 8.D.1 shows that every σ-compact LC group is the projective limit of second-countable LC groups.

The second tool for the reduction is the following crucial lemma; a similar result appears in [**Moor–72**, Proposition 2.2], under more restrictive assumptions and with a different proof.

LEMMA 8.D.2. *Let G be a topological group and $\mathcal{K}$ a family of compact normal subgroups of G with the following property: for every neighbourhood U of e in G, there exists $K \in \mathcal{K}$ with $K \subset U$. Let π be a factor representation of G.*

Then π factorizes by a quotient G/K for some $K \in \mathcal{K}$.

PROOF. Let ξ be a unit vector in $\mathcal{H}_\pi$. Since π is continuous, there exists a neighbourhood U of e in G such that

$$\|\pi(u)\xi - \xi\| < \sqrt{2} \quad \text{for all } u \in U.$$

Therefore there exists $K \in \mathcal{K}$ such that

$$\|\pi(g)\xi - \xi\| < \sqrt{2} \quad \text{for all } g \in K.$$

Since K is a group, this implies that the space $\mathcal{H}_\pi^K$ of $\pi(K)$-invariant vectors in $\mathcal{H}_\pi$ is nonzero (see [**BeHV–08**, Proposition 1.1.5]). Since K is normal in G, the subspace $\mathcal{H}_\pi^K$ is $\pi(G)$-invariant. Moreover, $\mathcal{H}_\pi^K$ is obviously invariant under the commutant $\pi(G)'$ of $\pi(G)$. Therefore the orthogonal projection $p \colon \mathcal{H}_\pi \to \mathcal{H}_\pi^K$ belongs to $\pi(G)' \cap \pi(G)''$, that is, to the centre of the von Neumann algebra $\pi(G)''$. Since π is factorial and $\mathcal{H}_\pi^K \neq \{0\}$, we have $p = \mathrm{id}_{\mathcal{H}_\pi}$, that is, $\mathcal{H}_\pi = \mathcal{H}_\pi^K$; this means that π factorizes by G/K. □

Lemma 8.D.2, combined with Theorem 8.D.1, allows us to relate the dual and the quasi-dual a σ-compact LC group to the duals and quasi-duals of appropriate second-countable LC groups.

PROPOSITION 8.D.3. *Let G be a σ-compact locally compact group, and let $\mathcal{K}_G$ denote the set of compact normal subgroups K in G such that G/K is second-countable, ordered by inclusion.*

(1) *For every factor representation π of G, there exists a compact normal subgroup $K \in \mathcal{K}_G$ such that π factorizes by G/K.*
(2) *For every factor representation π of G, the C*-kernel $\mathrm{C^*ker}(\pi)$ is a primitive ideal of $C^*_{\mathrm{max}}(G)$.*
(3) *The dual $\widehat{G}$ is homeomorphic to the inductive limit*

$$\widehat{G} = \varinjlim_{K \in \mathcal{K}_G} \widehat{G/K}.$$

(4) *Similarly, the quasi-dual $\mathrm{QD}(G)$ is Borel-isomorphic to the inductive limit*

$$\mathrm{QD}(G) = \varinjlim_{K \in \mathcal{K}_G} \mathrm{QD}(G/K).$$

PROOF. Claim (1) follows from Lemma 8.D.2 and Theorem 8.D.1.

(2) Let π be a factor representation of G. By (1), there exists $K \in \mathcal{K}_G$ and a factor representation π' of G/K such that $\pi = \pi' \circ p$, where $p \colon G \to G/K$ stands for the canonical epimorphism. Let $p_* : C^*_{\mathrm{max}}(G) \twoheadrightarrow C^*_{\mathrm{max}}(G/K)$ be the morphism induced on the maximal C*-algebras (see Proposition 8.C.8). By Corollary 7.F.2 and Proposition 8.B.4, there exists an irreducible representation ρ' of G/K such that $\mathrm{C^*ker}(\rho') = \mathrm{C^*ker}(\pi')$. Consider the irreducible representation ρ of G given by the composition $\rho : C^*_{\mathrm{max}}(G) \overset{p_*}{\twoheadrightarrow} C^*_{\mathrm{max}}(G/K) \overset{\rho'}{\to} \mathcal{L}(\mathcal{H}_{\rho'})$. Then

$$\mathrm{C^*ker}(\pi) = (p_*)^{-1}(\mathrm{C^*ker}(\pi')) = (p_*)^{-1}(\mathrm{C^*ker}(\rho')) = \mathrm{C^*ker}(\rho)$$

and the ideal $\mathrm{C^*ker}(\pi)$ of $C^*_{\mathrm{max}}(G)$ is primitive.

(3) & (4) These claims follow from (1). Let us spell out the homeomorphism. For $K \in \mathcal{K}_G$, the space $\widehat{G/K}$ with its Fell topology can be seen as a closed subspace of $\widehat{G}$. The Fell topology on $\widehat{G}$ coincides with the inductive limit topology, that is, the one for which a subset A in $\widehat{G}$ is open if and only if $A \cap \widehat{G/K}$ is open for all $K \in \mathcal{K}_G$.

Similarly, for each $K \in \mathcal{K}_G$, let the quasi-dual of G/K be given its Mackey–Borel structure. For its own Mackey–Borel structure, a subset A in $\mathrm{QD}(G)$ is Borel if and only if the intersection $A \cap \mathrm{QD}(G/K)$ is Borel for all $K \in \mathcal{K}_G$. □

As a corollary, we show that, for a σ-compact locally compact group G, the quotient map $\widehat{G} \twoheadrightarrow \mathrm{Prim}(G)$ has a natural extension to the quasi-dual $\mathrm{QD}(G)$.

COROLLARY 8.D.4. *For a σ-compact locally compact group G, the map*

$$(\kappa 3) \qquad \kappa^{\mathrm{qd}}_{\mathrm{prim}} : \mathrm{QD}(G) \twoheadrightarrow \mathrm{Prim}(G), \qquad \rho \mapsto \mathrm{C^*ker}(\rho)$$

is an extension of the canonical map $\kappa^{\mathrm{d}}_{\mathrm{prim}} : \widehat{G} \twoheadrightarrow \mathrm{Prim}(G)$ of Section 1.E.

PROOF. Let ρ be a factor representation of G. On the one hand, every factor representation of G quasi-equivalent to ρ has the same C*-kernel as ρ (Corollary 6.A.5 and Proposition 8.B.4); on the other hand, $\mathrm{C^*ker}(\rho)$ is a primitive ideal (Proposition 8.D.3(2)). □

8.E. The central character of a representation, of a primitive ideal

Let G be a topological group. The **projective kernel** of a representation $(\pi, \mathcal{H})$ of G is the set $\mathrm{Pker}\pi$ defined by

$$\mathrm{Pker}\pi \;=\; \{g \in G \mid \pi(g) = \chi_\pi(g)\mathrm{Id}_\mathcal{H} \;\text{ for some number }\; \chi_\pi(g) \in \mathbf{T}\}.$$

Observe that $\mathrm{Pker}\pi$ is a normal closed subgroup of G containing $\ker \pi$, that the function

$$\chi_\pi \,:\, \mathrm{Pker}\pi \to \mathbf{T}, \qquad g \mapsto \chi_\pi(g)$$

is a unitary character, and that

$$\ker(\pi : G \to \mathcal{U}(\mathcal{H})) \;=\; \ker(\chi_\pi \,:\, \mathrm{Pker}\pi \to \mathbf{T}).$$

For the next proposition, recall that central characters have been defined above (Definition 1.A.12).

PROPOSITION 8.E.1. *Let G be a topological group and π a factor representation of G. Let $\chi_\pi : \mathrm{Pker}\pi \to \mathbf{T}$ be as above.*

(1) *The centre Z of G is contained in $\mathrm{Pker}\pi$, so that the factor representation π admits a central character, say $\psi_\pi \in \widehat{Z}$, which is the restriction of χ_π to Z.*

(2) *Let H be a closed subgroup which contains the commutator subgroup of G (and therefore which is normal in G) and $p : G \twoheadrightarrow G/(\ker\pi \cap H)$ the canonical projection.*
Then $\mathrm{Pker}\pi = p^{-1}(\overline{Z})$, where $\overline{Z}$ is the centre of $G/(\ker\pi \cap H)$.

(3) *Let ρ be any representation of G which is weakly equivalent to π.*
Then $\mathrm{Pker}\rho = \mathrm{Pker}\pi$ and $\chi_\rho = \chi_\pi$.

PROOF. (1) For $g \in Z$, the operator $\pi(g)$ belongs to the centre of the factor $\pi(G)''$ and hence is a scalar operator.

(2) Let $g \in G$. Then $\pi(g)$ commutes with $\pi(g')$ for all $g' \in G$ if and only if $\pi(g)$ belongs to the centre of the factor $\pi(G)''$, that is, if and only if $\pi(g)$ is a scalar operator. Therefore, we have $g \in \mathrm{Pker}\pi$ if and only if

$$\pi([g, g']) \;=\; [\pi(g), \pi(g')] \;=\; \mathrm{Id}_\mathcal{H} \qquad \text{for all } \; g' \in G,$$

that is, if and only if $[g, g'] \in \ker\pi$ for all $g' \in G$.

Since H contains $[G, G]$, the last condition holds if and only if $[g, g'] \in \ker\pi \cap H$ for all $g' \in G$, i.e., if and only if $g \in p^{-1}(\overline{Z})$.

(3) The restriction $\rho|_{\mathrm{Pker}\pi}$ is weakly contained in $\pi|_{\mathrm{Pker}\pi}$, and therefore in χ_π. This implies that every normalized function of positive type associated with $\rho|_{\mathrm{Pker}\pi}$ coincides with χ_π, and therefore that

$$\rho(g) \;=\; \chi_\pi(g)\mathrm{Id}_\mathcal{H} \qquad \text{for all } \; g \in \mathrm{Pker}\pi.$$

It follows that $\mathrm{Pker}\pi \subset \mathrm{Pker}\rho$ and $\psi_\rho|_{\mathrm{Pker}\pi} = \chi_\pi$. Similarly, we have $\mathrm{Pker}\rho \subset \mathrm{Pker}\pi$ and $\chi_\pi|_{\mathrm{Pker}\rho} = \chi_\rho$. □

REMARK 8.E.2. By (1) of Proposition 8.E.1, there are two maps

$$\widehat{G} \to \widehat{Z} \quad \text{and} \quad \mathrm{QD}(G) \to \widehat{Z}$$

which associate to $\pi \in \widehat{G}$ or $\pi \in \mathrm{QD}(G)$ the central character $\psi_\pi \in \widehat{Z}$. By (3) of Proposition 8.E.1, the map $\widehat{G} \to \widehat{Z}$ factorizes by the map

$$\kappa^{\mathrm{d}}_{\mathrm{prim}} : \widehat{G} \twoheadrightarrow \mathrm{Prim}(G)$$

of (κ1) in Section 8.B. Similarly, when G is a σ-compact LC group, the map $\mathrm{QD}(G) \to \widehat{Z}$ factorizes by the map

$$\kappa^{\mathrm{qd}}_{\mathrm{prim}} : \mathrm{QD}(G) \twoheadrightarrow \mathrm{Prim}(G),$$

of Corollary 8.D.4. When $\mathrm{Prim}(G)$ is viewed as the space of primitive ideals in the C*-algebra $C^*_{\mathrm{max}}(G)$, this shows that there is a well-defined central character attached to such primitive ideals.

8.F. Characterization of type I groups: The Glimm Theorem

This section is devoted to the statement of Glimm's Theorem, which gives several characterizations of LC groups of type I.

8.F.a. CCR and GCR, or liminal and postliminal C*-algebras. We introduce a class of C*-algebras which play a central role in Glimm's Theorem.

Given a Hilbert space $\mathcal{H}$, recall that $\mathcal{K}(\mathcal{H})$ denotes the ideal of compact operators in $\mathcal{L}(\mathcal{H})$.

Let A be a C*-algebra. For a representation π of A, consider the two-sided ideal

$$\pi^{-1}(\mathcal{K}(\mathcal{H}_\pi)) = \{a \in A \mid \pi(a) \text{ is compact}\}$$

of A; note that it depends on the equivalence class of π only. Define

$$\mathrm{CCR}(A) = \bigcap_{\pi \in \widehat{A}} (\pi^{-1}(\mathcal{K}(\mathcal{H}_\pi));$$

so, $\mathrm{CCR}(A)$ is the ideal of A consisting of the elements a for which $\pi(a)$ is compact for every irreducible representation π of A.

DEFINITION 8.F.1. Let A be a C*-algebra A.

A is defined to be **CCR** or **liminal** if, for every irreducible representation $\pi : A \to \mathcal{L}(\mathcal{H}_\pi)$, the image $\pi(A)$ consists of compact operators on $\mathcal{H}_\pi$.

A is defined to be **GCR** or **postliminal** if $\mathrm{CCR}(A/J) \neq \{0\}$ for every closed two-sided ideal J in A.

REMARK 8.F.2. (1) The acronym CCR refers to "completely continuous representations". The term "liminal" indicates that these algebras are the first to consider; etymologically, "liminal" comes from the Latin "limen", meaning threshold. The acronym GCR refers to "generalized CCR algebra".

(2) We comment on the relationship between the notions of GCR groups and CCR groups, as defined in Section 6.E, and the notions of CCR and GCR C*-algebras we have just introduced.

- It is clear that an LC group G is a CCR group if and only if its maximal C*-algebra $C^*_{\mathrm{max}}(G)$ is CCR;
- it is also clear that, if $C^*_{\mathrm{max}}(G)$ is GCR, then G is a GCR group;

- conversely, it is true that $C^*_{\max}(G)$ is GCR if G is a GCR group; however, this is a deep fact which is part of Glimm's Theorem 8.F.3 below (see also Comment (11) after this theorem).

(3) The notion of a type I group from Section 6.D makes sense in the context of C*-algebras: a C*-algebra A is of type I if, for every representation π of A, the von Neumann algebra $\pi(A)''$ is of type I. Glimm's Theorem 8.F.3, in its general formulation, shows that the class of separable C*-algebras coincides with the class of separable GCR C*-algebras.

8.F.b. Glimm's Theorem. In the fundamental 1961 article of Glimm, only one out of nine theorems is formulated for second-countable locally compact groups, and the other theorems are formulated for separable C*-algebras. Here, we choose a formulation for groups, more precisely for σ-compact LC groups.

THEOREM 8.F.3. *For a σ-compact locally compact group G, the following conditions are equivalent:*

(i) *G is type I, as defined in Section* 6.D*;*
(ii) *G is a GCR group, as defined in Section* 6.E*;*
(iii) *$C^*_{\max}(G)$ is a GCR C*-algebra;*
(iv) *the map $\kappa^{\mathrm{d}}_{\mathrm{prim}} : \widehat{G} \twoheadrightarrow \mathrm{Prim}(G)$ of (κ1) which appears in Definition* 1.E.1 *and Remark* 8.B.6 *is a homeomorphism;*
(v) *the Mackey–Borel structure on $\widehat{G}$ is countably separated;*
(vi) *the Mackey–Borel structure on $\widehat{G}$ is standard;*
(vii) *the Mackey–Borel structure on $\widehat{G}$ is the same as the Borel structure defined by the Fell topology.*

Below we give several comments on Glimm's Theorem; for the complete proof, we refer to the original article [**Glim–61a**] and [**Dixm–C***, Chap. 9].

Comments on Theorem 8.F.3.

(1) The equivalence between the properties (i) to (vi) listed above is proved in [**Glim–61a**] in the context of separable C*-algebras; applied to the C*-algebra $C^*_{\max}(G)$, this implies Theorem 8.F.3 for a second-countable LC group G. Concerning the equivalence of these properties with property (vii), see Comment (8) below.

(2) The equivalence of all properties, except (iii), carries over from second-countable LC groups to the more general case of a σ-compact LC group G, by Proposition 8.D.3. For the equivalence of (iii) with the other properties, see Comment (11) below.

(3) The equivalence (i) $\Longleftrightarrow$ (vi) was conjectured by Mackey [**Mack–57**, Page 163].

(4) Property (i) is equivalent to the fact that every factor representation of G is a multiple of an irreducible representation. See Theorem 6.D.4 and Proposition 6.B.14.

(5) The implication (ii) $\Longrightarrow$ (i) is proved in Theorem 6.E.5.

(6) The implication (iv) $\Longrightarrow$ (i) is a consequence of Corollary 7.F.4;

(7) The implications (iii) $\Longrightarrow$ (ii) and (vi) $\Longrightarrow$ (v) are obvious.

(8) The proofs in [**Glim–61a**] are independent of previous work, mainly by Dixmier, Fell, and Kaplansky, in which the equivalence of several of the properties listed above were established, in the context of separable C*-algebras:
the implication (iii) $\Longrightarrow$ (i) is an early result from [**Kapl–51b**];
the equivalence (iii) $\Longleftrightarrow$ (iv) was proved in [**Dixm–60a**];
the equivalence (iii) $\Longleftrightarrow$ (v) $\Longleftrightarrow$ (vi) $\Longleftrightarrow$ (vii) was proved in [**Dixm–60c**];
the implication (iv) $\Longrightarrow$ (vi) was established in [**Fell–60b**, Theorem 4.1].

(9) In view of the previous comments, the main novelty in [**Glim–61a**] is the proof of the implication (i) $\Longrightarrow$ (iii), apart from the independent proofs it offers for all the other implications.

(10) A conceptually new proof of (i) $\Longrightarrow$ (vi) is given in [**Effr–65**].

(11) Let G be a locally compact group, σ-compact or not. It is known that Properties (i), (ii), and (iii) of Theorem 8.F.3 are equivalent; see [**Saka–66**], [**Saka–67**].

(12) Recall from Theorem 1.D.13 (1) that the conditions of Theorem 8.F.3 are moreover equivalent to the fact that the dual $\widehat{G}$ is a T_0 space.

(13) Considering that groups of type I enjoy many good properties which other groups lack, Kirillov has suggested that a second-countable locally compact group should be called **tame** if it is type I and wild otherwise [**Kiri–76**, see Section 8.4].

Let G be a locally compact group which is not type I. Then G has a factor representation of type II or a factor representation of type III. In fact, when G is second-countable, it was shown in [**Glim–61a**] that G has factor representations of both types.

THEOREM 8.F.4. *Let G be a locally compact group which is not of type I.*

(1) *The group G has factor representations of type III.*
(2) *If G is σ-compact, then G has factor representations of type II_∞.*
(3) *Let Γ be a countable group that is not of type I. Then Γ has a factor representation of type II_1.*

ON THE PROOF. For (1), see [**Glim–61a**] in case G is second-countable and [**Saka–66**] in general.

Claim (2) follows from [**Glim–61a**] and [**Mare–75**], by translating from separable C*-algebras to σ-compact groups.

For (3), it follows from Theorem 7.D.3 that the regular representation of Γ has a part of type II, say λ_Γ^{II}. In the central decomposition

$$\lambda_\Gamma^{II} = \int_{\mathrm{QD}(\Gamma)}^{\oplus} \pi_\omega d\mu(\omega)$$

(see Theorem 6.C.6), π_ω is a factor representation of type II_1 for almost every $\omega \in \mathrm{QD}(\Gamma)$ [**Dixm–vN**, Chap. II, § 3, no 5]. □

8.G. The von Neumann algebra of a group representation

In Section 6.B, we have associated two von Neumann algebras, $\pi(G)''$ and $\pi(G)'$, to any representation π of a topological group G. Let us record another way to define these when G is locally compact.

Let G be a locally compact group and π a representation of G on a Hilbert space $\mathcal{H}_\pi$. Recall that $C^c(G)$ stands for the convolution algebra of complex-valued continuous functions on G with compact support. We denote again by π each of the corresponding morphisms from $C^c(G)$, $L^1(G)$, $M^b(G)$, and $C^*_{\max}(G)$ to $\mathcal{L}(\mathcal{H}_\pi)$. Since $C^c(G)$, $L^1(G)$, $C^*_{\max}(G)$ have approximate units, the following is elementary to check; see [**Dixm–C***, 13.3.5].

PROPOSITION 8.G.1. *Let π be a representation of a locally compact group G. With the notation as above, $\pi(G)$, $\pi(C^c(G))$, $\pi(L^1(G))$, and $\pi(C^*_{\max}(G))$ all generate the same von Neumann algebra:*

$$\pi(G)'' = \pi(C^c(G))'' = \pi(L^1(G))'' = \pi(M^b(G))'' = \pi(C^*_{\max}(G))''.$$

Given a second-countable locally compact group G, the second dual of $C^*_{\max}(G)$, say $\mathrm{vN}_{\max}(G)$, has a natural structure of von Neumann algebra, of which the normal $*$-representations are in natural one-to-one correspondence with the unitary representations of G. The algebra $\mathrm{vN}_{\max}(G)$ is the enveloping von Neumann algebra of $C^*_{\max}(G)$ [**Dixm–C***, § 12.1]. On this subject, we only quote two articles by Ernest [**Erne–64, Erne–65**].

REMARK 8.G.2. In contrast to the equality $\pi(C^*_{\max}(G))'' = \pi(G)''$ of Proposition 8.G.1, $C^*_\pi(G)$ *need not* coincide with the C*-subalgebra $C^*(\pi(G))$ of $\mathcal{L}(\mathcal{H}_\pi)$ generated by $\pi(G)$, as the following examples illustrate.

(1) Note that $C^*(\lambda(G))$ is always a C*-algebra with unit, whereas, when G is not discrete, the reduced C*-algebra $C^*_\lambda(G)$ does not have a unit (see Remark 8.B.9). More information on $C^*(\lambda(G))$ can be found in [**Bedo–94**].

(2) Let G be a locally compact abelian group. We give a precise description of $C^*_\lambda(G)$ and $C^*(\lambda(G))$ and compare them to each other.

Plancherel's theorem shows that $C^*_\lambda(G)$ is isomorphic to the algebra $C^0(\widehat{G})$. Observe that, if G is not discrete, $\widehat{G}$ is not compact and so $C^0(\widehat{G})$ does not have a unit.

Plancherel's theorem also shows that $C^*(\lambda(G))$ is isomorphic to the algebra $AP(\widehat{G})$ of almost periodic functions on $\widehat{G}$, which is the closure of the algebra $\mathrm{Trig}(\widehat{G})$ of trigonometric polynomials for the topology of uniform convergence on $\widehat{G}$ (see Section A.7).

Let $\mathrm{Bohr}(\widehat{G}) = \widehat{G_{\mathrm{disc}}}$ be the Bohr compactification of $\widehat{G}$ (see Proposition 4.C.4). Every function in $AP(\widehat{G})$ extends uniquely to a continuous function on $\mathrm{Bohr}(\widehat{G})$ and the so-defined map $AP(\widehat{G}) \to C(\mathrm{Bohr}(\widehat{G}))$ is an algebra isomorphism. It follows that $C^*(\lambda(G))$ is isomorphic to $C(\mathrm{Bohr}(\widehat{G}))$. For more details on this example, see [**KoKa–43**].

8.H. Variants

Given an LC group G, there are other group algebras than $C^*_{\max}(G)$ and other primitive duals than $\mathrm{Prim}(G)$ that can be attached to G. One is the space $\mathrm{Prim}_*(L^1(G))$ consisting of kernels of irreducible $*$-representations of $L^1(G)$ associated with irreducible representations of G. Another one is the space $\mathrm{Prim}(L^1(G))$ of all primitive ideals of $L^1(G)$, i.e., of all kernels of simple $L^1(G)$-modules (in the algebraic sense). There are also spaces of maximal ideals, $\mathrm{Max}(C^*_{\max}(G))$ and

$\mathrm{Max}(L^1(G))$. There is a natural injection of $\mathrm{Max}(C^*_{\max}(G))$ into $\mathrm{Prim}(G)$, and this is a bijection if and only if $\mathrm{Prim}(G)$ is a T_1 space.

The only variant about which we add a comment is the space $\mathrm{Prim}_*(L^1(G))$. There is a natural map

$$\Phi : \mathrm{Prim}(G) \to \mathrm{Prim}_*(L^1(G)), \quad \mathrm{C^*ker}(\pi) \mapsto L^1\ker(\pi) = \mathrm{C^*ker}(\pi) \cap L^1(G)$$

that is surjective, and continuous for the Jacobson topologies. In [**Dixm–60b**, Théorème 4], it is shown that this map is injective (and therefore a homeomorphism) when G satisfies some "Property (P)", in particular when G is a connected Lie group that is either nilpotent or semisimple. It is also known that Φ is a homeomorphism when G is an LC-group with polynomial growth [**BLSV–78**, Satz 2].

We end this section by an example showing that Φ need not be injective.

PROPOSITION 8.H.1. *Let Γ be a lattice in $G = \mathrm{PSL}_2(\mathbf{R})$ and let Φ be as above. Then $\Phi : Prim(\Gamma) \to Prim_*(\ell^1(\Gamma))$ is not injective.*

PROOF. Let π_1 and π_2 be nonequivalent representations of G in the complementary series. Then the restrictions $\pi_1|_\Gamma$ and $\pi_2|_\Gamma$ to Γ are irreducible, by [**CoSt–91**, Proposition 2.5]. By [**BeHa–94**, Théorème 2], we have on the one hand

$$\mathrm{C^*ker}\,(\pi_1|_\Gamma) \neq \mathrm{C^*ker}\,(\pi_2|_\Gamma)\,,$$

and on the other hand

$$\mathrm{C^*ker}\,(\pi_1|_\Gamma) \subset \mathrm{C^*ker}(\lambda_\Gamma) \quad \text{and} \quad \mathrm{C^*ker}\,(\pi_2|_\Gamma) \subset \mathrm{C^*ker}(\lambda_\Gamma),$$

where λ_Γ stands for the left regular representations of Γ. It follows that

$$L^1\ker(\pi_1|_\Gamma) \subset L^1\ker(\lambda_\Gamma) \quad \text{and} \quad L^1\ker(\pi_2|_\Gamma) \subset L^1\ker(\lambda_\Gamma).$$

Now λ_Γ is faithful on $\ell^1(\Gamma)$, i.e., $L^1\ker(\lambda_\Gamma) = \{0\}$. Therefore

$$L^1\ker(\pi_1|_\Gamma) = L^1\ker(\pi_2|_\Gamma) = \{0\},$$

and that concludes the proof. □

CHAPTER 9

Examples of primitive duals

In this chapter, the primitive duals are determined for the examples of groups considered in Chapter 3*, that is, for D_∞, for two-step nilpotent discrete groups and in particular the Heisenberg groups $H(\mathbf{K})$ and $H(\mathbf{Z})$, for* $\mathrm{Aff}(\mathbf{K})$*, for* $\mathrm{BS}(1,p)$*, for the lamplighter group, and for $\mathrm{GL}_n(\mathbf{K})$ when $\mathbf{K}$ is an infinite algebraic extension of a finite field and $n \geq 2$. For such a field $\mathbf{K}$, we will also determine the primitive dual of $\mathrm{SL}_n(\mathbf{K})$ for $n \geq 3$ and deduce the remarkable fact that* $\mathrm{Prim}(\mathrm{PSL}_n(\mathbf{K}))$ *consists of only two points.*

The main tool we will use, Proposition 9.A.1*, is a weak containment result for induced representations from an abelian normal subgroup.*

Given a second-countable LC group G which is not *of type I, the map $\kappa^{\mathrm{d}}_{\mathrm{prim}} : \widehat{G} \twoheadrightarrow$* $\mathrm{Prim}(G)$ *of Sections* 1.E *and* 8.B *has fibers of uncountable cardinality, as a consequence of Corollary* 7.F.4*). In Section* 9.G*, we will check this fact directly with specific representations, for our main examples of discrete groups.*

Let Γ be a countable group which is not abelian-by-finite. Then the dual space $\widehat{\Gamma}$ is not a standard Borel space, by Thoma's Theorem 7.D.1 *and Glimm's Theorem* 8.F.3*. A natural problem is to compare the "complexity" of the Borel spaces $\widehat{\Gamma}_1$ and $\widehat{\Gamma}_2$ for two such groups Γ_1 and Γ_2. We address this question in Section* 9.H*. The main result (Theorem* 9.H.9*) shows that Γ_1 and Γ_2 are always Borel isomorphic if Γ_1 and Γ_2 are moreover* amenable.

The discussion for the first of our examples is straightforward: the infinite dihedral group D_∞ is abelian-by-finite, and is therefore a type I group. In particular, the canonical projection from its dual to its primitive ideal space is a homeomorphism. The dual of D_∞ has been described in Section 3.A.

9.A. A weak containment result for induced representations

Proposition 9.A.1 is implicitly contained in [**Guic–63**, Chap. II, § 2, Lemme 1] and [**Guic–65**, § 3, Lemme 2]. It is a particular case of the "generalized Effros–Hahn conjecture", a result about the structure of the primitive ideal space of crossed product C*-algebras established in [**GoRo–79**].

PROPOSITION 9.A.1. *Let Γ be a discrete group, $(\pi, \mathcal{H})$ a representation of Γ, and N an* abelian *normal subgroup of Γ. Assume that there exists $\chi_0 \in \widehat{N}$ which is weakly contained in $\pi|_N$ and such that $\chi_0^\gamma \neq \chi_0$ for every $\gamma \in \Gamma \smallsetminus N$.*

Then the induced representation $\mathrm{Ind}_N^\Gamma \chi_0$ is weakly contained in π.

PROOF. Set $\sigma := \mathrm{Ind}_N^\Gamma \chi_0$. Let $\widetilde{\chi}_0$ be the trivial extension of χ_0, equal to χ_0 on N and to 0 on $\Gamma \smallsetminus N$; recall from Proposition 1.F.10 that the representation σ of Γ has a cyclic vector, say ξ_σ, and that $\widetilde{\chi}_0$ is the corresponding function of positive type:

$$\widetilde{\chi}_0(\gamma) = \langle \sigma(\gamma)\xi_\sigma \mid \xi_\sigma \rangle \quad \text{for all } \gamma \in \Gamma.$$

(Even if it is not relevant to the present proof, recall moreover that σ is irreducible; see Corollary 1.F.16.)

Let $\varepsilon > 0$, and let F be a finite subset of Γ. By Proposition 1.C.4 it is enough to show that there exists a function of positive type φ associated with π such that $\sup_{\gamma \in F} |\varphi(\gamma) - \widetilde{\chi_0}(\gamma)| < \varepsilon$. We write F_1 for $F \cap N$ and F_2 for $F \cap (\Gamma \smallsetminus N)$, so that $F = F_1 \sqcup F_2$.

Let $E : \mathcal{B}(\widehat{N}) \twoheadrightarrow \mathrm{Proj}(\mathcal{H})$ be the projection-valued measure associated with the representation $(\pi|_N, \mathcal{H})$ of the abelian group N; we have

$$\pi(\gamma) = \int_{\widehat{N}} \chi(\gamma) dE(\chi) \qquad \text{for all } \gamma \in N.$$

By assumption, $\chi_0^\gamma \neq \chi_0$ for every $\gamma \in F_2$. Therefore by continuity of the action $\widehat{N} \curvearrowleft \Gamma$, there exists a neighbourhood U of χ_0 in $\widehat{N}$ such that

$$(*) \qquad U^\gamma \cap U = \emptyset \qquad \text{for all } \gamma \in F_2.$$

Upon replacing U by a smaller neighbourhood of χ_0 if necessary, we can assume that, moreover, we have

$$(**) \qquad |\chi(\gamma) - \chi_0(\gamma)| < \varepsilon \qquad \text{for all } \chi \in U \text{ and } \gamma \in F_1.$$

Since χ_0 is weakly contained in $\pi|_N$, we have $E(U) \neq 0$, by Proposition 2.D.1. Let $\xi \in \mathcal{H}$ be a unit vector in the range of the projection $E(U)$. We claim that

$$(\sharp) \qquad |\langle \pi(\gamma)\xi \mid \xi \rangle - \widetilde{\chi_0}(\gamma)| < \varepsilon \qquad \text{for all } \gamma \in F.$$

Indeed, we have

$$\langle \pi(\gamma)\xi \mid \xi \rangle = \int_{\widehat{N}} \chi(\gamma) d\mu_\xi(\chi) \qquad \text{for all } \gamma \in N,$$

where μ_ξ is the probability measure on $\widehat{N}$ associated with E and ξ. Since

$$E(B)\xi = E(B)E(U)\xi = 0$$

for every $B \in \mathcal{B}(\widehat{N})$ with $B \cap U = \emptyset$, the support of μ_ξ is contained in U and therefore

$$\langle \pi(\gamma)\xi \mid \xi \rangle = \int_U \chi(\gamma) d\mu_\xi(\chi) \qquad \text{for all } \gamma \in N.$$

It follows from $(**)$ that we have

$$(\sharp_1) \qquad \begin{aligned} |\langle \pi(\gamma)\xi \mid \xi \rangle - \widetilde{\chi_0}(\gamma)| &= \left| \int_U (\chi(\gamma) - \chi_0(\gamma)) d\mu_\xi(\chi) \right| \\ &\leq \int_U |\chi(\gamma) - \chi_0(\gamma)| d\mu_\xi(\chi) < \varepsilon \qquad \text{for all } \gamma \in F_1. \end{aligned}$$

Now, let $\gamma \in F_2$. We have

$$\pi(\gamma)E(B)\pi(\gamma)^{-1} = E\left(B^{\gamma^{-1}}\right) \qquad \text{for all } B \in \mathcal{B}(\widehat{N})$$

(see Proposition 2.F.1). Since $\xi = E(U)\xi$, it follows that

$$\pi(\gamma)\xi = \pi(\gamma)E(U)\xi = \pi(\gamma)E(U)\pi(\gamma)^{-1}\pi(\gamma)\xi = E(U^{\gamma^{-1}})\pi(\gamma)\xi$$

and hence, using $(*)$,

$$\langle \pi(\gamma)\xi \mid \xi \rangle = \langle E(U^{\gamma^{-1}})\pi(\gamma)\xi \mid E(U)\xi \rangle = \langle E(U)E(U^{\gamma^{-1}})\pi(\gamma)\xi \mid \xi \rangle = 0.$$

Therefore, we have

$$(\sharp_2) \qquad \langle \pi(\gamma)\xi \mid \xi \rangle = \widetilde{\chi_0}(\gamma) \qquad \text{for all } \gamma \in F_2.$$

Since $(\sharp_1)$ and $(\sharp_2)$ imply $(\sharp)$, this concludes the proof. □

REMARK 9.A.2. Under the assumption of Proposition 9.A.1, the induced representation $\mathrm{Ind}_N^\Gamma \chi_0^\gamma$ is weakly contained in π for every $\gamma \in \Gamma$. Indeed, $\mathrm{Ind}_N^\Gamma \chi_0^\gamma$ is equivalent to $\mathrm{Ind}_N^\Gamma \chi_0$ (Proposition 1.F.9).

We need at several places the following general result on CCR representations, as defined in Definition 6.E.7.

PROPOSITION 9.A.3. *Let G be a locally compact group and $(\pi, \mathcal{H})$ an irreducible CCR representation of G.*

(1) *Let ρ be a representation of G which is weakly contained in π. Then ρ is equivalent to a multiple of π.*
(2) *Assume that G is discrete; let H be a subgroup of G and σ a representation of H. If π is contained in the induced representation $\mathrm{Ind}_H^G \sigma$, then H has finite index in G.*

PROOF. (1) On the one hand, since ρ is weakly contained in π, the representation $\rho : C^*_{\max}(G) \to \mathcal{L}(\mathcal{H})$ factorizes by the quotient algebra $C^*_{\max}(G)/\mathrm{C^*ker}(\pi)$.

On the other hand, since π is a CCR representation and is irreducible, the C*-algebra $C^*_{\max}(G)/\mathrm{C^*ker}(\pi)$ is isomorphic to the algebra $\mathcal{K}(\mathcal{H})$ of compact operators on $\mathcal{H}$. Since every representation of $\mathcal{K}(\mathcal{H})$ is equivalent to a multiple of the identity representation on $\mathcal{H}$ (Proposition 6.E.1), this implies that ρ is equivalent to a multiple of π.

(2) Since G is discrete and π is a CCR representation, the identity operator $\pi(e)$ is compact. Therefore π is finite-dimensional.

Assume that π is contained in the induced representation $\lambda := \mathrm{Ind}_H^G \sigma$. Since π is finite-dimensional, 1_G is contained in the tensor product $\pi \otimes \overline{\pi}$, where $\overline{\pi}$ is the conjugate representation of π [**BeHV–08**, A.1.13]. Therefore 1_G is contained in $\lambda \otimes \overline{\lambda}$. However, $\lambda \otimes \overline{\lambda}$ is equivalent to the induced representation $\mathrm{Ind}_H^G(\sigma \otimes (\overline{\lambda}|_H))$; see [**BeHV–08**, E.2.5]. This implies that H has finite index in G [**BeHV–08**, E.3.1]. □

Note that, as a particular case of (1), if ρ, π are irreducible representations of G and π is finite-dimensional, then ρ is weakly contained in π if and only if ρ is contained in π.

The following corollary is an immediate consequence of Proposition 9.A.3(1).

COROLLARY 9.A.4. *Let G be a locally compact group and $(\pi, \mathcal{H})$ an irreducible CCR representation of G.*

Then the weak equivalence class of π in $\widehat{G}$ consists of π itself; in other words, $\{\pi\}$ is a closed point in $\widehat{G}$.

In particular, if $\widehat{G}_{\mathrm{fd}}$ is the subset of $\widehat{G}$ of finite-dimensional irreducible representations of G, as in Remark 1.C.14, the canonical projection of $\widehat{G}$ onto $\mathrm{Prim}(G)$ restricts to an injection $\widehat{G}_{\mathrm{fd}} \to \mathrm{Prim}(G)$.

9.B. Two-step nilpotent groups

Let Γ be a two-step nilpotent discrete group, with centre Z. (Recall that, by our definition of two-step nilpotent groups, Γ could also be an abelian group.) In Sections 3.B and 4.B, we have constructed families of nonequivalent irreducible representations of Γ. In this section, we describe $\mathrm{Prim}(\Gamma)$, using Proposition 9.A.1 and some facts established in Section 3.B.

For $\psi \in \widehat{Z}$, denote as in Section 3.B by $\overline{\Gamma}$ the quotient $\Gamma / \ker \psi$, by $\overline{Z}$ its centre, and by $p : \Gamma \twoheadrightarrow \overline{\Gamma}$ the canonical projection. Set

$$Z_\psi := p^{-1}(\overline{Z}),$$

which is a normal subgroup of Γ containing Z. Denote by p' the restriction $Z_\psi \twoheadrightarrow \overline{Z}$ of p. The map $\overline{\chi} \mapsto \overline{\chi} \circ p'$ is a homeomorphism from the dual group $\widehat{\overline{Z}}$ of $\overline{Z}$ onto the subspace $\{\chi \in \widehat{Z_\psi} \mid \chi|_{\ker \psi} = 1\}$ of $\widehat{Z_\psi}$. In particular, the set

$$\{\chi \in \widehat{Z_\psi} \mid \chi|_Z = \psi\}$$

consists of unitary characters of Z_ψ. (Recall that Z_ψ need not be abelian, so that the sign $\widehat{\ }$ in $\widehat{Z_\psi}$ indicates a dual, not a Pontryagin dual.)

Let π be a factor representation of Γ. The projective kernel $\mathrm{Pker}\pi$, the associated unitary character $\chi_\pi \in \widehat{\mathrm{Pker}\pi}$, and the central character $\psi_\pi = \chi_\pi|_Z \in \widehat{Z}$ of π have been defined in Section 8.E. Since $\ker \pi \cap Z = \ker \psi_\pi$ and since $[\Gamma, \Gamma] \subset Z$, we have

$$\mathrm{Pker}\pi = Z_{\psi_\pi},$$

by Proposition 8.E.1(2).

THEOREM 9.B.1. *Let Γ be a two-step nilpotent group with centre Z. For every $\psi \in \widehat{Z}$, let Z_ψ be the normal subgroup of Γ defined as above.*
The map

$$\Phi : \begin{cases} \bigsqcup_{\psi \in \widehat{Z}} \left(\{\psi\} \times \left\{ \chi \in \widehat{Z_\psi} \mid \chi|_Z = \psi \right\} \right) & \to \quad \mathrm{Prim}(\Gamma) \\ (\psi, \chi) & \mapsto \quad \mathrm{C^*ker}(\mathrm{Ind}_{Z_\psi}^\Gamma \chi) \end{cases}$$

is a bijection.

The inverse map Φ^{-1} is the following. Let J be a primitive ideal in $C^*_{\max}(\Gamma)$. Choose an irreducible representation π of Γ such that $\mathrm{C^*ker}(\pi) = J$. Let $\chi_\pi \in \widehat{\mathrm{Pker}\pi}$ and $\psi_\pi = \chi_\pi|_Z$ be as above. Then $\Phi^{-1}(J) = (\psi_\chi, \chi_\pi)$.

PROOF. • *First step.* We show that every primitive ideal in $C^*_{\max}(\Gamma)$ is the image of a pair (ψ, χ) in the domain of Φ. Such a primitive ideal can be written as $\mathrm{C^*ker}(\pi)$ for some irreducible representation π of Γ, say with central character $\psi = \psi_\pi \in \widehat{Z}$ and unitary character $\chi = \chi_\pi \in \widehat{\mathrm{Pker}\pi} = \widehat{Z_\psi}$. It suffices to show that

$$\mathrm{C^*ker}(\pi) = \mathrm{C^*ker}(\mathrm{Ind}_{Z_\psi}^\Gamma \chi),$$

i.e., that π is weakly equivalent to $\mathrm{Ind}_{Z_\psi}^\Gamma \chi$.

∘ *First case.* Assume that ψ is faithful. Then $Z_\psi = Z$ and we have to show that π is weakly equivalent to $\mathrm{Ind}_Z^\Gamma \psi$.

Let N be a maximal abelian subgroup of Γ. Let $\chi_0 \in \widehat{N}$ be such that χ_0 is weakly contained in $\pi|_N$. By continuity of the restriction process, $\chi_0|_Z$ is weakly contained in $\pi|_Z$ and hence $\chi_0|_Z = \psi$, that is,

$$\begin{aligned} \chi_0 \in \widehat{N}(\psi) &= \{\sigma \in \widehat{N} \mid \text{the restriction of } \sigma \text{ to } Z \text{ is } \psi\} \\ &= \left\{ \sigma \in \widehat{N} \mid \sigma = \chi_0 \rho \ \text{ for some } \rho \in \widehat{N/Z} \right\} \end{aligned}$$

($\widehat{N}(\psi)$ is here as in Lemma 3.B.3).

By Lemma 3.B.3(3), the group Γ/N acts freely on $\widehat{N}(\psi)$. By Proposition 9.A.1 and Remark 9.A.2, it follows that $\mathrm{Ind}_N^\Gamma \chi_0^\gamma$ is weakly contained in π for every $\gamma \in \Gamma$.

By Lemma 3.B.3(4), the set $\{\chi_0^\gamma \mid \gamma \in \Gamma\}$ is dense in $\widehat{N}(\psi)$. It follows that $\mathrm{Ind}_N^\Gamma \sigma$ is weakly contained in π for every $\sigma \in \widehat{N}(\psi)$, and therefore that $\bigoplus_{\sigma \in \widehat{N}(\psi)} \mathrm{Ind}_N^\Gamma \sigma$ is weakly contained in π.

On the one hand, by Proposition 1.F.21(2), we have a weak equivalence

$$\mathrm{Ind}_Z^N \psi \sim \bigoplus_{\rho \in \widehat{N/Z}} \chi_0 \rho = \bigoplus_{\sigma \in \widehat{N}(\psi)} \sigma.$$

By continuity of the induction process (see [**BeHV–08**, F.3.5]), it follows that the representation $\mathrm{Ind}_Z^\Gamma \psi \simeq \mathrm{Ind}_N^\Gamma(\mathrm{Ind}_Z^N \psi)$ is weakly equivalent to

$$\mathrm{Ind}_N^\Gamma \Bigg(\bigoplus_{\sigma \in \widehat{N}(\psi)} \sigma \Bigg) \simeq \bigoplus_{\sigma \in \widehat{N}(\psi)} \mathrm{Ind}_N^\Gamma \sigma,$$

and therefore that $\mathrm{Ind}_Z^\Gamma \psi$ is weakly contained in π.

On the other hand, by Proposition 1.F.21(1) and since Γ/Z is amenable, π is weakly contained in $\mathrm{Ind}_Z^\Gamma(\pi|_Z)$. Since $\mathrm{Ind}_Z^\Gamma(\pi|_Z)$ is weakly equivalent to $\mathrm{Ind}_Z^\Gamma \psi$, it follows that π is weakly contained in $\mathrm{Ind}_Z^\Gamma \psi$.

Therefore, π is weakly equivalent to $\mathrm{Ind}_Z^\Gamma \psi$, and the claim is proved in case ψ is faithful.

∘ *Second case.* Assume now that ψ is not faithful. We have

$$\mathrm{Pker}\pi = Z_\psi = p^{-1}(\overline{Z}),$$

by Proposition 8.E.1.

The representation π factorizes into a representation $\overline{\pi}$ of $\overline{\Gamma}$ such that $\pi = \overline{\pi} \circ p$. The unitary character χ of $\mathrm{Pker}\pi$ factorizes into a faithful unitary character $\overline{\chi}$ of $\overline{Z}$, which is also the central character of $\overline{\pi}$.

By what was proved in the first case, $\overline{\pi}$ is weakly equivalent to $\mathrm{Ind}_{\overline{Z}}^{\overline{\Gamma}} \overline{\chi}$. Since

$$\mathrm{Ind}_{Z_\psi}^\Gamma \chi = \left(\mathrm{Ind}_{\overline{Z}}^{\overline{\Gamma}} \overline{\chi} \right) \circ p,$$

it follows that $\pi = \overline{\pi} \circ p$ is weakly equivalent to $\mathrm{Ind}_{Z_\psi}^\Gamma \chi$.

• *Second step.* We claim that every ideal $\mathrm{C^*ker}(\mathrm{Ind}_{Z_\psi}^\Gamma \chi)$ in the image Φ is a primitive ideal of $C^*_{\mathrm{max}}(\Gamma)$. More precisely, given $\psi \in \widehat{Z}$ and $\chi \in \widehat{Z_\psi}$ extending χ, we claim that there exists $\pi \in \widehat{\Gamma}$ such that

$$\mathrm{C^*ker}(\pi) = \mathrm{C^*ker}(\mathrm{Ind}_{Z_\psi}^\Gamma \chi)$$

and that, moreover, ψ coincides with the central character ψ_π of π and χ with the unitary character χ_π of $\widehat{\mathrm{Pker}\pi} = \widehat{Z_\psi}$.

The unitary character χ factorizes into a unitary character of $\overline{Z}$, say $\overline{\chi}$. Let $\overline{\pi}$ be an irreducible representation of $\overline{\Gamma}$ which is weakly contained in $\mathrm{Ind}_{\overline{Z}}^{\overline{\Gamma}} \overline{\chi}$, i.e., which is in the support of $\mathrm{Ind}_{\overline{Z}}^{\overline{\Gamma}} \overline{\chi}$ (see Remark 1.C.14). Then $\overline{\pi}$ has $\overline{\chi}$ as central character, since the restriction of $\mathrm{Ind}_{\overline{Z}}^{\overline{\Gamma}} \overline{\chi}$ to $\overline{Z}$ is a multiple of $\overline{\chi}$ (see Proposition 1.F.9(2)). Therefore, by the first step,

$$\mathrm{C^*ker}(\overline{\pi}) = \mathrm{C^*ker}(\mathrm{Ind}_{\overline{Z}}^{\overline{\Gamma}} \overline{\chi});$$

lifting $\overline{\pi}$ to a representation π of Γ, we have

$$\mathrm{C^*ker}(\pi) = \mathrm{C^*ker}(\mathrm{Ind}_{Z_\psi}^\Gamma \chi).$$

Observe that we also have $\chi_\pi = \chi$ and $\psi_\pi = \psi$, as claimed.

• *Third step.* We claim that the map Φ is injective.

Indeed, for $i = 1, 2$, let $\psi_i \in \widehat{Z}$ and $\chi_i \in \widehat{Z_{\psi_i}}$ extending ψ_i. By the second step, there exists $\pi_i \in \widehat{\Gamma}$ with the following properties: $\Phi(\psi_i, \chi_i) = \mathrm{C^*ker}(\pi_i)$, the central character of π_i is ψ_i, and the unitary character of $\mathrm{Pker}\pi_i = Z_{\psi_i}$ is χ_i.

Assume now that $\Phi(\psi_1, \chi_1) = \Phi(\psi_2, \chi_2)$. Then π_1 and π_2 are weakly equivalent. Therefore $\mathrm{Pker}\pi_1 = \mathrm{Pker}\pi_2$ and $\chi_1 = \chi_2$, by Proposition 8.E.1(3). It is then obvious that $\psi_1 = \psi_2$. □

REMARK 9.B.2. (1) The description in Theorem 9.B.1 of $\mathrm{Prim}(\Gamma)$ for a two-step nilpotent group Γ appears in [**Kani–82**, Lemma 2], which follows ideas from [**Howe–77b**, Proposition 5]. The work [**Howe–77b**] provides a description of $\mathrm{Prim}(\Gamma)$ for Γ a finitely generated, torsion-free, discrete nilpotent group. For an extension of this description to a large class of nilpotent LC groups, see [**EcKl–12**].

(2) As we will see later (Section 12.B), every representation $\mathrm{Ind}_{Z_\psi}^{\Gamma}\chi$ of Γ appearing in Theorem 9.B.1 is a **factor** representation of finite type. Except in some very particular cases, $\mathrm{Ind}_{Z_\psi}^{\Gamma}\chi$ is not irreducible.

We are going to apply Theorem 9.B.1 to Heisenberg groups.

Heisenberg group over a commutative ring. Let R be a unital commutative ring. As in Section 3.B, consider the Heisenberg group $H(R)$, with centre $Z \approx R$. Recall that, for $\psi \in \widehat{Z}$, we have

- the ideal $I_\psi = \{a \in R \mid aR \subset \ker \psi\}$ of R;
- the normal subgroup Z_ψ of $H(R)$ defined earlier in this section, which is $Z_\psi = \{(a, b, c) \in \Gamma \mid a, b \in I_\psi,\ c \in R\}$ by Lemma 3.B.7.

For every unitary character χ of Z_ψ such that $\chi|_Z = \psi$, there is a pair (α, β) in $\widehat{I_\psi} \times \widehat{I_\psi}$ such that χ is equal to

$$\chi_{\alpha,\beta,\psi} : Z_\psi \to \mathbf{T} \ \ (a, b, c) \mapsto \alpha(a)\beta(b)\psi(c).$$

The following result is a direct consequence of Theorem 9.B.1.

COROLLARY 9.B.3. *Let $\Gamma = H(R)$ for a unital commutative ring R. For every $\psi \in \widehat{Z}$, let I_ψ and Z_ψ be as above.*

The map

$$\Phi : \begin{cases} \bigsqcup_{\psi \in \widehat{Z}} \{\psi\} \times \widehat{I_\psi^{\,2}} & \to \quad \mathrm{Prim}(\Gamma) \\ (\psi, (\alpha, \beta)) & \mapsto \quad \mathrm{C^*ker}(\mathrm{Ind}_{Z_\psi}^{\Gamma}\chi_{\alpha,\beta,\psi}) \end{cases}$$

is a bijection

We consider the particular case $\Gamma = H(\mathbf{K})$, where $R = \mathbf{K}$ is a field. Recall that $Z \approx \mathbf{K}$, and $\Gamma/Z \approx \mathbf{K}^2$. For $\psi \in \widehat{\mathbf{K}} \smallsetminus \{1_{\mathbf{K}}\}$, we have $I_\psi = \{0\}$ and hence $Z_\psi = Z$. For $\psi = 1_{\mathbf{K}}$, we have $I_\psi = \mathbf{K}$ and hence $Z_\psi = H(\mathbf{K})$. We therefore obtain the following consequence of Corollary 9.B.3.

COROLLARY 9.B.4. *Let $\mathbf{K}$ be a field, $\Gamma = H(\mathbf{K})$ the Heisenberg group over $\mathbf{K}$, Z the centre of $H(\mathbf{K})$, and $p : \Gamma \twoheadrightarrow \Gamma/Z$ the canonical epimorphism.*

The map

$$\Phi : \left(\widehat{Z} \smallsetminus \{1_Z\}\right) \sqcup \widehat{\Gamma/Z} \to \mathrm{Prim}(\Gamma)$$

defined by

$$\Phi(\psi) = \text{C*ker}(\text{Ind}_Z^\Gamma \psi) \qquad \textit{for} \ \ \psi \in \widehat{Z} \smallsetminus \{1_Z\} \approx \widehat{\mathbf{K}} \smallsetminus \{1_{\mathbf{K}}\},$$
$$\Phi(\chi) = \text{C*ker}(\chi \circ p) \qquad \textit{for} \ \ \chi \in \widehat{\Gamma/Z} \approx \widehat{\mathbf{K}^2},$$

is a bijection.

We will see later (Corollary 12.A.6) that $\text{Ind}_Z^\Gamma \psi$ is a **factor** representation of $\Gamma = H(\mathbf{K})$, which moreover is **not** irreducible.

As a consequence of Corollary 9.B.4, we obtain an explicit description of Prim(Γ) in terms of weak equivalence classes of irreducible representations.

COROLLARY 9.B.5. *Let* $\mathbf{K}$ *be a field and* $\Gamma = H(\mathbf{K})$ *the Heisenberg group over* $\mathbf{K}$. *Let* Z *be the centre* $\{(0,0,c)\} \mid c \in \mathbf{K}\}$ *of* Γ *and* N *the maximal abelian subgroup* $\{(0,b,c) \mid b,c \in \mathbf{K}\}$ *of* Γ, *as in Definition* 3.C.1.

(1) *Let* $\psi \in \widehat{Z} \smallsetminus \{1_Z\}$*; denote by* χ_ψ *the unitary character of* N *defined by* $\chi_\psi(0,b,c) = \psi(c)$ *for* $b,c \in \mathbf{K}$. *The representation* $\text{Ind}_N^\Gamma \chi_\psi$ *is irreducible; its weak equivalence class in* $\widehat{\Gamma}$ *consists of all irreducible representations of* Γ *with central character* ψ.
(2) *For* $\chi \in \widehat{\Gamma/Z}$, *the weak equivalence class of* $\chi \circ p$ *in* $\widehat{\Gamma}$ *consists of* $\chi \circ p$ *itself.*

PROOF. (1) By Corollary 3.B.11, the representation $\text{Ind}_N^\Gamma \chi_\psi$ is irreducible and its central character is ψ. By Proposition 8.E.1, every factor representation of Γ which is weakly equivalent to $\text{Ind}_N^\Gamma \chi_\psi$ has central character ψ. Let π be an irreducible representation of Γ which has central character ψ; by Corollary 9.B.4, there exists $\psi' \in \widehat{Z} \smallsetminus \{1_Z\}$ such that π is weakly equivalent to $\text{Ind}_Z^\Gamma \psi'$; by Proposition 8.E.1 again, $\psi' = \psi$ and π is therefore weakly equivalent to $\text{Ind}_N^\Gamma \chi_\psi$.

(2) Let π be an irreducible representation of Γ which is weakly equivalent to $\chi \circ p$. By Proposition 8.E.1, the central character of π is 1_Z, so that π can be seen as an irreducible representation of the abelian group Γ/Z; hence π is of dimension 1 and is equivalent to $\chi \circ p$ (see Remark 1.C.2(3)). □

REMARK 9.B.6. Corollary 9.B.5 shows that the map

$$\text{Prim}(H(\mathbf{K})) \to \widehat{\mathbf{Z}} \approx \widehat{\mathbf{K}},$$

that associates to a weak equivalence class its central character (see Section 8.E) is a surjective map with the property that the inverse image of an element in $\widehat{\mathbf{K}} \smallsetminus \{1_{\mathbf{K}}\}$ is one point and the inverse image of $\{1_{\mathbf{K}}\}$ is a copy of $\widehat{\mathbf{K}}^2$.

Next, we consider the case where R is the ring $\mathbf{Z}$ of rational integers.

Recall that

$$\widehat{Z} \approx \widehat{\mathbf{Z}} = \{\psi_\theta \mid \theta \in [0,1[\},$$

with $\psi_\theta(c) = e^{2\pi i \theta c}$ for all $c \in \mathbf{Z}$. We denote by $\widehat{Z}_\infty$ the elements of infinite order in $\widehat{Z}$ and by $\widehat{Z}_n$ the elements of order n for any finite integer $n \geq 1$.

Let $\psi = \psi_\theta \in \widehat{Z}$. Two cases can occur:

- $\psi \in \widehat{Z}_\infty$; then $I_\psi = \{0\}$ and hence $Z_\psi = Z$.
- $\psi \in \widehat{Z}_n$ for some $n \geq 1$. When $n = 1$ then $\theta = 0$ and $\psi_\theta = 1$, so that $I_\psi = \mathbf{Z}$ and $Z_\psi = H(\mathbf{Z})$. When $n \geq 2$ then $\theta = p/n$ for some integer $p \in \{1, \ldots, n-1\}$ such that p and n are coprime, so that $I_\psi = n\mathbf{Z}$ and Z_ψ coincides with the normal subgroup

$$\Lambda(n) = \{(a, b, c) \in \Gamma \mid a \in n\mathbf{Z},\ b \in n\mathbf{Z},\ c \in \mathbf{Z}\},$$

which only depends on n.

Fix $\psi = \psi_\theta \in \widehat{Z}_n$. Every unitary character of $\Lambda(n)$ which coincides with ψ on Z is of the form $\chi_{\alpha,\beta,\psi}$, where

$$\chi_{\alpha,\beta,\psi}(a, b, c) = e^{2\pi i(\alpha a + \beta b)}\psi(c) \quad \text{for } a, b \in n\mathbf{Z},\ c \in \mathbf{Z}$$

for a uniquely determined pair $(\alpha, \beta) \in [0, 1/n[^2$.

In view of these remarks, the following result is an immediate consequence of Corollary 9.B.3.

COROLLARY 9.B.7. *Let $\Gamma = H(\mathbf{Z})$ be the Heisenberg group over the integers and Z its centre.*

The map

$$\Phi : \widehat{Z}_\infty \sqcup \Big(\bigsqcup_{n \geq 1} \bigsqcup_{\psi \in \widehat{Z}_n} \{\psi\} \times [0, 1/n[^2 \Big) \to \mathrm{Prim}(\Gamma)$$

defined by

$$\begin{aligned} \Phi(\psi) &= \mathrm{C^*ker}(\mathrm{Ind}_Z^\Gamma \psi) && \textit{for } \psi \in \widehat{Z}_\infty, \\ \Phi(\psi, \alpha, \beta) &= \mathrm{C^*ker}(\mathrm{Ind}_{\Lambda(n)}^\Gamma \chi_{\alpha,\beta,\psi}) && \textit{for } (\psi, \alpha, \beta) \in \widehat{Z}_n \times [0, 1/n[^2 \end{aligned}$$

is a bijection.

Note that the complement of $\widehat{Z}_\infty$ in the domain of Φ appears already in Corollary 4.B.5 as the space X which parametrizes $\widehat{\Gamma}_{\mathrm{fd}}$. This illustrates the fact that the restriction of the canonical map $\widehat{\Gamma} \to \mathrm{Prim}(\Gamma)$ to $\widehat{\Gamma}_{\mathrm{fd}}$ is a bijection onto its image; see Remark 1.E.3(2).

As in the case of the Heisenberg group over a field, we give an explicit parametrization of $\mathrm{Prim}(\Gamma)$ in terms of weak equivalence classes of irreducible representations.

For $\psi \in \widehat{Z}_\infty$, denote as before by χ_ψ the unitary character of the maximal abelian subgroup $N = \{(0, b, c) \mid b, c \in \mathbf{Z}\}$ defined by $\chi_\psi(0, b, c) = \psi(c)$ for $b, c \in \mathbf{Z}$. For an integer $n \geq 1$, denote by $\Gamma(n)$ the normal subgroup

$$\Gamma(n) = \{(a, b, c) \in \Gamma \mid a \in n\mathbf{Z},\ b, c \in \mathbf{Z}\}$$

of Γ (which appeared already in Corollaries 3.B.17 and 4.B.5). For $\psi \in \widehat{Z}_n$ and $(\alpha, \beta) \in [0, 1/n[^2$, the previously defined unitary character $\chi_{\alpha,\beta,\psi}$ of $\Lambda(n)$ extends to a unitary character of $\Gamma(n)$, again denoted by $\chi_{\alpha,\beta,\psi}$ and defined by the same formula:

$$\chi_{\alpha,\beta,\psi}(a, b, c) = e^{2\pi i(\alpha a + \beta b)}\psi(c) \quad \text{for } a \in n\mathbf{Z},\ b \in \mathbf{Z},\ c \in \mathbf{Z}.$$

COROLLARY 9.B.8. *Let $\Gamma = H(\mathbf{Z})$ and Z be its centre.*

(1) *For* $\psi \in \widehat{Z}_\infty$, *the representation* $\mathrm{Ind}_N^\Gamma \chi_\psi$ *is irreducible; its weak equivalence class consists of all irreducible representations of* Γ *with central character* ψ.

(2) *For every* $n \geq 1$ *and* $(\psi, \alpha, \beta) \in \widehat{Z}_n \times [0, 1/n[^2$, *the representation* $\pi_{\alpha,\beta,\psi} = \mathrm{Ind}_{\Gamma(n)}^\Gamma \chi_{\alpha,\beta,\psi}$ *is irreducible. Moreover, every irreducible representation* π *of* Γ *with central character* $\psi \in \widehat{Z}_n$ *is weakly equivalent to* $\pi_{\alpha,\beta,\psi}$ *for the unique pair* $(\alpha, \beta) \in [0, 1/n[^2$ *such that* $\pi|_{\Lambda(n)}$ *is a multiple of* $\chi_{\alpha,\beta,\psi}$.

PROOF. This is a direct consequence of Corollaries 3.B.14, 3.B.17, 9.B.7, and Proposition 8.E.1. Compare with the proof of Corollary 9.B.5. □

REMARK 9.B.9. For $\Gamma = H(\mathbf{Z})$, consider the surjective map

$$\mathrm{Prim}(\Gamma) \twoheadrightarrow \widehat{Z} \approx \mathbf{T},$$

which associates to a weak equivalence class of irreducible representations of Γ its central character. By Corollary 9.B.8, the inverse image of every element of infinite order in $\widehat{Z}$ consists of one point, and the inverse image of every element of finite order is a 2-torus.

Consider now the canonical surjective map

$$\kappa_{\mathrm{prim}}^{\mathrm{d}} : \widehat{\Gamma} \twoheadrightarrow \mathrm{Prim}(\Gamma).$$

First let $\psi \in \widehat{Z}$ be an element of infinite order, say $\psi(c) = e^{2\pi i\theta c}$ for all $c \in Z \approx \mathbf{Z}$ and some $\theta \in [0, 1[$, $\theta \notin \mathbf{Q}$. Denote by $\widehat{\Gamma}(\psi)$ the subset of the dual $\widehat{\Gamma}$ of classes of irreducible representations with central character ψ. On the one hand, all irreducible representations in $\widehat{\Gamma}(\psi)$ correspond to the same primitive ideal of Γ, that is, the image of $\widehat{\Gamma}(\psi)$ by $\kappa_{\mathrm{prim}}^{\mathrm{d}}$ consists of a unique point (see Corollary 9.B.7). On the other hand, $\widehat{\Gamma}(\psi)$ contains uncountably many elements (see Corollary 3.B.14 and Remark 3.B.15).

For every $\pi \in \widehat{\Gamma}(\psi)$ the C*-subalgebra $\pi(C^*_{\max}(\Gamma))$ of $\mathcal{L}(\mathcal{H}_\pi)$ is isomorphic to the C*-algebra A_θ of the irrational rotation of angle $2\pi\theta$ [**Rief–81**]. Therefore, $\widehat{\Gamma}(\psi)$ is in natural bijection with the spectrum $\widehat{A_\theta}$ of A_θ. For more on the representation theory of A_θ, see [**O'Do–81**].

Now let π be an irreducible representation of Γ which has a central character of finite order. Then π is finite-dimensional, and any finite-dimensional irreducible representation of Γ weakly equivalent to π is equivalent to π (Corollary 9.A.4). In other words, the inverse image by $\kappa_{\mathrm{prim}}^{d}$ consists of a single point.

There are descriptions of the topology of $\mathrm{Prim}(\Gamma)$. We refer to [**Howe–77a**], [**BaPa–94**], and [**KaTa–13**, Example 5.71].

9.C. Affine groups of infinite fields

Let $\mathbf{K}$ be an *infinite* field and $\mathrm{Aff}(\mathbf{K})$ the group of affine transformations of $\mathbf{K}$. Let N be its derived group, so that $N \approx \mathbf{K}$ and $\mathrm{Aff}(\mathbf{K})/N \approx \mathbf{K}^\times$, as in Section 3.C.

THEOREM 9.C.1. *For* $\Gamma = \mathrm{Aff}(\mathbf{K})$ *and* N *as above, we have*

$$\mathrm{Prim}(\Gamma) = \{0\} \sqcup \left\{ \mathrm{C^*ker}(\chi \circ p) \mid \chi \in \widehat{\Gamma/N} \right\} \approx \{0\} \sqcup \widehat{\Gamma/N}$$

where p *is the quotient map* $\Gamma \twoheadrightarrow \Gamma/N$.

PROOF. Let π be an irreducible representation of Γ. There are two cases to consider.

Suppose first that π is trivial on N. Then $\pi = \chi \circ p$ for some unitary character $\chi \in \widehat{\Gamma/N}$ and hence $\mathrm{C^*ker}(\pi) = \mathrm{C^*ker}(\chi \circ p)$.

Suppose now that π is not trivial on N. We claim that $\mathrm{C^*ker}(\pi) = \{0\}$.

Indeed, in this case, there exists $\chi_0 \in \widehat{N}$ with $\chi_0 \neq 1_N$ which is weakly contained in $\pi|_N$. Since Γ/N acts freely on $\widehat{N} \smallsetminus \{1_N\}$ (Lemma 3.C.2), $\mathrm{Ind}_N^\Gamma \chi_0^\gamma$ is weakly contained in π for every $\gamma \in \Gamma$ (Proposition 9.A.1 and Remark 9.A.2). As $\{\chi_0^\gamma \mid \gamma \in \Gamma\}$ is dense in $\widehat{N}$ (Lemma 3.C.2), it follows that $\mathrm{Ind}_N^\Gamma \chi$ is weakly contained in π for every $\chi \in \widehat{N}$, hence the direct sum $\bigoplus_{\chi \in \widehat{N}} \mathrm{Ind}_N^\Gamma \chi$ is weakly contained in π.

On the one hand, since N is amenable, $\bigoplus_{\chi \in \widehat{N}} \chi$ is weakly equivalent to the regular representation λ_N of N, hence $\bigoplus_{\chi \in \widehat{N}} \mathrm{Ind}_N^\Gamma \chi$ is weakly equivalent to the regular representation $\lambda_\Gamma = \mathrm{Ind}_N^\Gamma \lambda_N$ of Γ, by continuity of induction (Proposition 1.F.6(5)). It follows that λ_Γ is weakly contained in π. On the other hand, since Γ is amenable, π is weakly contained in λ_Γ.

Therefore π is weakly equivalent to λ_Γ, that is,

$$\mathrm{C^*ker}(\pi) \,=\, \mathrm{C^*ker}(\lambda_\Gamma) \,=\, \{0\},$$

as claimed.

This shows that $\mathrm{Prim}(\Gamma)$ is contained in the right-hand side of the equality of the theorem.

Conversely, we clearly have $\mathrm{C^*ker}(\chi \circ p) \in \mathrm{Prim}(\Gamma)$ for every $\chi \in \widehat{\Gamma/N}$. Also, there exists an irreducible representation of Γ which is nontrivial on N (see Section 3.C), say π, and we have shown above that $\mathrm{C^*ker}(\pi) = \{0\}$, hence $\{0\} \in \mathrm{Prim}(\Gamma)$. □

REMARK 9.C.2. Let $\Gamma = \mathrm{Aff}(\mathbf{K})$ and $N \approx \mathbf{K}$ be as above. As in the case of the Heisenberg groups, we can give an explicit parametrization of $\mathrm{Prim}(\Gamma)$ in terms of weak equivalence classes of irreducible representations. Indeed, choose any $\psi \in \widehat{\mathbf{K}} \smallsetminus \{1_{\mathbf{K}}\}$ and recall from Theorem 3.C.3 that $\pi_\psi = \mathrm{Ind}_N^\Gamma \psi$ is irreducible.

For every $\chi \in \widehat{\Gamma/N}$, the weak equivalence class of the unitary character $\chi \circ p$ contains only $\chi \circ p$ itself. By contrast, the weak equivalence class of π_ψ contains all other irreducible representations of Γ, which all have C*-kernel $\{0\}$.

9.D. Solvable Baumslag–Solitar groups

Let p be a prime. Recall from Section 3.D that the Baumslag–Solitar group

$$\Gamma \,=\, \mathrm{BS}(1,p) \,=\, \begin{pmatrix} p^{\mathbf{Z}} & \mathbf{Z}[1/p] \\ 0 & 1 \end{pmatrix} \,=\, \left\{ \begin{pmatrix} p^n & b \\ 0 & 1 \end{pmatrix} \in \mathrm{GL}_2(\mathbf{Z}[1/p]) \,\Big|\, n \in \mathbf{Z}, b \in \mathbf{Z}[1/p] \right\}$$

is the semi-direct product $\Gamma = A \ltimes N \approx \mathbf{Z} \ltimes \mathbf{Z}[1/p]$, where

$$A \,=\, \left\{ \begin{pmatrix} p^k & 0 \\ 0 & 1 \end{pmatrix} \mid k \in \mathbf{Z} \right\} \,\approx\, \mathbf{Z} \quad \text{and} \quad N \,=\, \left\{ \begin{pmatrix} 1 & b \\ 0 & 1 \end{pmatrix} \mid b \in \mathbf{Z}[1/p] \right\} \,\approx\, \mathbf{Z}[1/p],$$

and the generator $\begin{pmatrix} p & 0 \\ 0 & 1 \end{pmatrix}$ of A acts on N by multiplication by p. We often write (k, b) for $\begin{pmatrix} p^k & b \\ 0 & 1 \end{pmatrix}$.

We proceed to describe Prim(Γ). The result (Theorem 9.D.1) can be derived from [**Guic–65**], where a more general situation is considered and where the case $p = 2$ is treated with details our exposition is in many respects different from the one in [**Guic–65**].

Recall from Section 3.D that we can identify $\widehat{N}$ with the p-adic solenoid

$$\mathbf{Sol}_p = (\mathbf{Q}_p \times \mathbf{R})/\Delta, \qquad \text{where } \Delta = \{(a, -a) \in \mathbf{Q}_p \times \mathbf{R} \mid a \in \mathbf{Z}[1/p]\},$$

and that we write $[(x, y)]$ for the class in $\mathbf{Sol}_p$ of (x, y) in $\mathbf{Q}_p \times \mathbf{R}$. The action of A on $\widehat{N}$ corresponds to the action of $\mathbf{Z}$ on $\mathbf{Sol}_p$ for which the generator $1 \in \mathbf{Z}$ acts by the map

$$T_p : \mathbf{Sol}_p \to \mathbf{Sol}_p, \ \ [(x, y)] \mapsto [px, py]$$

induced by the multiplication by p. We denote by χ_s the element in $\widehat{N}$ corresponding to $s \in \mathbf{Sol}_p$. For $n \in \mathbf{N}^*$, let $\mathbf{Sol}_p(n)$ be the set of elements of $\mathbf{Sol}_p$ with period n. Let

$$\mathrm{Per}(T_p) = \bigsqcup_{n \geq 1} \mathbf{Sol}_p(n)$$

denote the set of periodic elements, and

$$\mathbf{Sol}_p(\infty) := \mathbf{Sol}_p \smallsetminus \mathrm{Per}(T_p)$$

the set of elements with an infinite T_p-orbit.

We have determined in Corollary 4.B.7 the finite-dimensional representations of Γ which are associated with T_p-periodic elements. More precisely, given $n \geq 1$, $s \in \mathbf{Sol}_p(n)$, and $\theta \in [0, 1/n[$, the induced representation $\pi_{s,\theta} = \mathrm{Ind}_{\Gamma(n)}^{\Gamma} \chi_{s,\theta}$ is an irreducible finite-dimensional representation, where $\chi_{s,\theta}$ is a unitary character of the finite index normal subgroup $\Gamma(n) = \{(k, b) \in \Gamma \mid k \in n\mathbf{Z},\ b \in \mathbf{Z}[1/p]\}$, with $\chi_{s,\theta} = \chi_s$ on N.

As in Corollary 4.B.7, we set

$$X_{\mathrm{fd}} = \bigsqcup_{n \geq 1} \left(\mathbf{Sol}_p(n) \times [0, 1/n[\right),$$

and define

$$\widetilde{T_p} : X_{\mathrm{fd}} \sqcup \mathbf{Sol}_p(\infty) \to X_{\mathrm{fd}} \sqcup \mathbf{Sol}_p(\infty)$$

by $\widetilde{T_p}(s, \theta) = (T_p s, \theta)$ for $(s, \theta) \in X_{\mathrm{fd}}$ and $\widetilde{T_p}(s) = T_p(s)$ for $s \in \mathbf{Sol}_p(\infty)$. We will need to consider the quasi-orbit space of the $\widetilde{T_p}$-action on $X_{\mathrm{fd}} \sqcup \mathbf{Sol}_p(\infty)$.

Given an action of a group G on a topological X, the corresponding **quasi-orbit space** is the quotient space $X/\sim$ for the equivalence relation $\sim$ defined on X by

$$x \sim x' \quad \text{if} \quad \overline{Gx} = \overline{Gx'};$$

see Appendix A.1.

THEOREM 9.D.1. *Let $\Gamma = \mathrm{BS}(1, p) = A \ltimes N$ be the Baumslag–Solitar group; let X_{fd} and $\mathbf{Sol}_p(\infty)$ be as above.*

The map

$$\Phi : X_{\mathrm{fd}} \sqcup \mathbf{Sol}_p(\infty) \to \mathrm{Prim}(\Gamma),$$

defined by

$$\Phi(s) = \mathrm{C^*ker}(\mathrm{Ind}_{\Gamma(n)}^{\Gamma} \chi_{s,\theta}) \quad \textit{for } n \geq 1,\ (s, \theta) \in \mathbf{Sol}_p(n) \times [0, 1/n[\subset X_{\mathrm{fd}}$$

and

$$\Phi(s) = \mathrm{C^*ker}(\mathrm{Ind}_N^{\Gamma} \chi_s) \quad \textit{for } s \in \mathbf{Sol}_p(\infty),$$

factorizes into a bijection from the quasi-orbit space of the $\widetilde{T}_p$-action on $X_{\rm fd} \sqcup \mathbf{Sol}_p(\infty)$ *onto* $\mathrm{Prim}(\Gamma)$.

PROOF. • *First step.* We first prove that every primitive ideal is contained in the range of Φ.

Let π be an irreducible representation of Γ. Let $E : \mathcal{B}(\widehat{N}) \to \mathrm{Proj}(\mathcal{H})$ be the projection-valued measure associated with the restriction $\pi|_N$ of π to N and let S be the support of E.

By Proposition 2.F.1(2), there exists $s \in \mathbf{Sol}_p$ such that S is the closure of the Γ-orbit $\{\chi_s^\gamma \mid \gamma \in A\}$. Two cases may occur: $s \in \mathbf{Sol}_p(\infty)$ or $s \in \mathrm{Per}(T_p)$.

Assume first that $s \in \mathbf{Sol}_p(\infty)$. We claim that

$$\mathrm{C^*ker}(\pi) \ = \ \mathrm{C^*ker}(\mathrm{Ind}_N^\Gamma \chi_s).$$

Indeed, since A acts freely on the orbit of χ_s, the induced representation $\mathrm{Ind}_N^\Gamma \chi_s^\gamma$ is weakly contained in π for every $\gamma \in \Gamma$, by Proposition 9.A.1. It follows that $\mathrm{Ind}_N^\Gamma \chi$ is weakly contained in π for every $\chi \in S$. Therefore the representation

$$\rho := \mathrm{Ind}_N^\Gamma \Big(\bigoplus_{\chi \in S} \chi\Big) \ \simeq \ \bigoplus_{\chi \in S} (\mathrm{Ind}_N^\Gamma \chi)$$

is weakly contained in π.

On the one hand, observe that $\mathrm{Ind}_N^\Gamma(\pi|_N)$ is weakly equivalent to ρ. Moreover, since $\mathrm{Ind}_N^\Gamma(\pi|_N) \simeq \pi \otimes \mathrm{Ind}_N^\Gamma 1_N$ and since Γ/N is amenable, π is weakly contained in $\mathrm{Ind}_N^\Gamma(\pi|_N)$; compare with the proofs of Theorems 9.B.1 and 9.C.1. Therefore, π is weakly contained in ρ, hence π is weakly equivalent to ρ.

On the other hand, since $\mathrm{Ind}_N^\Gamma \chi_s^\gamma$ is equivalent to $\mathrm{Ind}_N^\Gamma \chi_s$ for every $\gamma \in A$ and since S is the closure of the A-orbit of χ_s, continuity of induction shows that ρ is weakly equivalent to $\mathrm{Ind}_N^\Gamma \chi_s$. Therefore π is weakly equivalent to $\mathrm{Ind}_N^\Gamma \chi_s$, as claimed.

Assume now that s is a periodic point for T_p, that is, $s \in \mathbf{Sol}_p(n)$ for some $n \geq 1$. We claim that π is equivalent to $\mathrm{Ind}_{\Gamma(n)}^\Gamma \chi_{s,\theta}$ for some $\theta \in [0, 1/n[$.

Indeed, the support S of E consists of the finite A-orbit of χ_s. Therefore, by Corollary 2.D.2, we have $\mathcal{H}^\chi \neq \{0\}$ for every $\chi \in S$ and $\mathcal{H} = \bigoplus_{\chi \in S} \mathcal{H}^\chi$, where

$$\mathcal{H}^\chi \ = \ \{\xi \in \mathcal{H} \mid \pi(n)\xi = \chi(n)\xi \quad \text{for all } \ n \in N\}.$$

We now proceed as in proof of the Third Step of Theorem 4.A.3. Since N is a normal subgroup of Γ, we have $\pi(\gamma^{-1})\mathcal{H}^\chi = \mathcal{H}^{\chi^\gamma}$ for every $\gamma \in \Gamma$ and $\chi \in S$. In particular, we have $\pi(\gamma)\mathcal{H}^{\chi_s} = \mathcal{H}^{\chi_s}$ for every γ in the stabilizer $\Gamma(n)$ of χ_s in Γ. It follows that π is equivalent to the induced representation $\mathrm{Ind}_{\Gamma(n)}^\Gamma \sigma$, where σ is the subrepresentation of $\pi|_{\Gamma(n)}$ defined on the $\Gamma(n)$-invariant subspace $\mathcal{H}^{\chi_s}$.

The representation σ is an irreducible representation of $\Gamma(n) = A(n) \ltimes N$ which is trivial on the commutator subgroup of $\Gamma(n)$. It follows that σ is a unitary character of $\Gamma(n)$. Therefore, π is equivalent to $\mathrm{Ind}_{\Gamma(n)}^\Gamma \chi_{s,\theta}$.

• *Second step.* We claim that every ideal in the image Φ is a primitive ideal.

Indeed, let $s \in \mathbf{Sol}_p(\infty)$. The representation $\mathrm{Ind}_N^\Gamma \chi_s$ is irreducible (Proposition 3.D.6). So, $\Phi(s) \in \mathrm{Prim}(\Gamma)$.

Similarly, for $n \geq 1$ and $(s, \theta) \in \mathbf{Sol}_p(n)$, the induced representation $\mathrm{Ind}_{\Gamma(n)}^\Gamma \chi_{s,\theta}$ is irreducible (Proposition 3.D.8 or Corollary 4.B.7), and so $\Phi(s, \theta) \in \mathrm{Prim}(\Gamma)$.

• *Third step.* We claim that Φ factorizes into a map from the quasi-orbit space of $X_{\rm fd} \sqcup \mathbf{Sol}_p(\infty)$ to $\mathrm{Prim}(\Gamma)$.

Indeed, let $s, s' \in \mathbf{Sol}_p(\infty)$ be two points with the same quasi-orbit. Then, the first step applied to $\pi = \mathrm{Ind}_N^\Gamma \chi_{s'}$ shows that

$$\mathrm{C^*ker}(\mathrm{Ind}_N^\Gamma \chi_{s'}) \; = \; \mathrm{C^*ker}(\mathrm{Ind}_N^\Gamma \chi_s).$$

If $(s, \theta), (s', \theta') \in X_{\mathrm{fd}}$ are in the same quasi-orbit, then $\theta = \theta'$ and s, s' belong to the same T_p-orbit. Therefore, $\mathrm{Ind}_{\Gamma(n)}^\Gamma \chi_{s,\theta}$ and $\mathrm{Ind}_{\Gamma(n)}^\Gamma \chi_{s',\theta}$ are equivalent.

• *Fourth step.* We claim that the map induced by Φ between the quasi-orbit space of $X_{\mathrm{fd}} \sqcup \mathbf{Sol}_p(\infty)$ and $\mathrm{Prim}(\Gamma)$ is injective.

Indeed, let $s \in \mathbf{Sol}_p(\infty)$ and set $\pi = \mathrm{Ind}_N^\Gamma \chi_s$. Then the support of the projection-valued measure associated with $\pi|_N$ is the closure of the A-orbit of χ_s. Let $s' \in \mathbf{Sol}_p(\infty)$ be such that $\pi' = \mathrm{Ind}_N^\Gamma \chi_{s'}$ is weakly equivalent to π. Then $\pi'|_N$ is weakly equivalent to $\pi'|_N$. Therefore, the closures of the orbits of s' and s coincide, that is, s' and s belong to the same quasi-orbit.

Let $s \in \mathbf{Sol}_p(n)$ and $\theta \in [0, 1/n[$; set $\pi = \mathrm{Ind}_{\Gamma(n)}^\Gamma \chi_{s,\theta}$. Let $s' \in \mathbf{Sol}_p(m)$ and $\theta' \in [0, 1/m[$ be such that $\pi' = \mathrm{Ind}_{\Gamma(m)}^\Gamma \chi_{s',\theta'}$ is weakly equivalent to π. Then the irreducible representations π and π' are equivalent, since they are finite-dimensional (see Proposition 9.A.3). Therefore, by Corollary 4.B.7, (s, θ) and (s', θ') belong to the same $\widetilde{T}_p$-orbit in X_{fd}. □

REMARK 9.D.2. Let μ be the normalized Haar measure on $\mathbf{Sol}_p$. As was observed in Remark 3.D.7, the action of T_p on $(\mathbf{Sol}_p, \mu)$ is ergodic; it follows that the T_p-orbit of μ-almost every element in $\mathbf{Sol}_p$ is dense (see [**BeMa–00**, Chap IV, Proposition 1.5]). However, besides the periodic points, there are points in $\mathbf{Sol}_p$ with a nondense orbit (see Lemma 14.E.5).

9.E. Lamplighter group

Recall from Section 3.E that the lamplighter group is the semi-direct product $\Gamma = A \ltimes N$ of $A = \mathbf{Z}$ with $N = \bigoplus_{k\in\mathbf{Z}} \mathbf{Z}/2\mathbf{Z}$, where the action of $\mathbf{Z}$ on $\bigoplus_{k\in\mathbf{Z}} \mathbf{Z}/2\mathbf{Z}$ is given by shifting the coordinates. Recall also that $\widehat{N}$ can be identified with $X = \prod_{k\in\mathbf{Z}}\{0, 1\}$, the dual action $\mathbf{Z}$ on $\widehat{N}$ being given by the shift transformation T on X.

As in Section 3.E, $\mathrm{Per}(T)$ denotes the set of T-periodic points and $X(n)$ the set of points in X with T-period $n \geq 1$.

For $x \in X$, we denote by χ_x the corresponding element in $\widehat{N}$. Recall that, for $x \in X(n)$ and $\theta \in [0, 1/n[$, the induced representation $\pi_{x,\theta} = \mathrm{Ind}_{\Gamma(n)}^\Gamma \chi_{x,\theta}$ is an irreducible finite-dimensional representation of Γ, where $\chi_{x,\theta}$ is a unitary character of the finite index normal subgroup $\Gamma(n) = n\mathbf{Z} \ltimes N$ with $\chi_{x,\theta} = \chi_x$ on N.

As in Proposition 3.E.3, we set

$$X_{\mathrm{fd}} \; = \; \bigsqcup_{n\geq 1} \; (X(n) \times [0, 1/n[),$$

and $T : X_{\mathrm{fd}} \to X_{\mathrm{fd}}$ is defined by $T(x, \theta) = (Tx, \theta)$. We also set

$$X(\infty) \; = \; X \smallsetminus \mathrm{Per}(T).$$

The proof of Theorem 9.D.1 carries over *mutatis mutandis* and yields the following result.

THEOREM 9.E.1. *Let $\Gamma = A \ltimes N$ be the lamplighter group; we keep the notation above.*

The map

$$\Phi : X_{\text{fd}} \sqcup X(\infty) \to \text{Prim}(\Gamma),$$

defined by

$$\Phi(x) = \text{C*ker}(\text{Ind}_{\Gamma(n)}^{\Gamma} \chi_{x,\theta}) \quad \textit{for } n \geq 1,\ (x,\theta) \in X(n) \times [0, 1/n[$$

and

$$\Phi(x) = \text{C*ker}(\text{Ind}_N^{\Gamma} \chi_x) \quad \textit{for } x \in X(\infty),$$

factorizes into a bijection from the quasi-orbit space of the T-action on $X_{\text{fd}} \sqcup X(\infty)$ to Prim(Γ).

9.F. General and special linear groups

Let $\mathbf{K}$ be an infinite field. Following [**HoRo–89**], we will determine the maximal ideals in the C*-algebra $C^*_{\text{max}}(\Gamma)$ for the general linear group $\Gamma = \text{GL}_n(\mathbf{K})$ when $n \geq 2$ as well as for the special linear group $\Gamma = \text{SL}_n(\mathbf{K})$ when $n \geq 3$, viewed as discrete groups. When Γ is amenable (equivalently, when $\mathbf{K}$ is an algebraic extension of a finite field; see Proposition 9.F.8), we will describe the primitive dual Prim(Γ).

9.F.a. General linear groups. We first treat the case of general linear groups. In this subsection, we set

$$\Gamma := \text{GL}_n(\mathbf{K}),$$

for an integer $n \geq 1$. The group Γ acts naturally on $\mathbf{K}^n$ (from the left) and hence on $\widehat{\mathbf{K}}^n$ by duality (from the right), through the formula

$$\chi^{\gamma}(x) = \chi(\gamma x) \quad \text{for } \gamma \in \text{GL}_n(\mathbf{K}),\ \chi \in \widehat{\mathbf{K}}^n,\ x \in \mathbf{K}^n.$$

We will need several preliminary lemmas.

LEMMA 9.F.1. *We keep the notation above and assume that $\mathbf{K}$ is* infinite.

(1) *The Γ-orbit of every $\chi \in \widehat{\mathbf{K}}^n \smallsetminus \{1\}$ is dense in $\widehat{\mathbf{K}}^n$.*
(2) *Assume moreover that $\mathbf{K}$ is countable. The set of $\chi \in \widehat{\mathbf{K}}^n$ with trivial stabilizer in Γ is a dense G_δ set in $\widehat{\mathbf{K}}^n$. In particular, this set is not empty.*

PROOF. (1) Let $\chi = (\chi_1, \ldots, \chi_n) \in \widehat{\mathbf{K}}^n \smallsetminus \{1\}$. We first claim that there exists $\chi' = (\chi'_1, \ldots, \chi'_n)$ in the Γ-orbit of χ with $\chi'_i \neq 1$ for every $i \in \{1, \ldots, n\}$.

Indeed, choose $i_0 \in \{1, \ldots, n\}$ such that $\chi_{i_0} \neq 1$, and let $i \in \{1, \ldots, n\}$ be such that $\chi_i = 1$. Let γ be the elementary matrix $E_{i,i_0}(1)$. Then $\chi^{\gamma} = (\chi'_1, \ldots, \chi'_n)$, where $\chi'_i = \chi_{i_0}$ and $\chi'_j = \chi_j$ for $j \neq i$. Therefore $\chi'_i \neq 1$. The claim follows by iteration of this procedure.

So, it suffices to prove (1) in case $\chi_i \neq 1$ for every $i \in \{1, \ldots, n\}$. For a diagonal matrix γ with diagonal entries $a_1, \ldots, a_n \in \mathbf{K}^{\times}$, we have

$$\chi^{\gamma} = (\chi_1^{a_1}, \ldots, \chi_n^{a_n}),$$

where $(a, \chi_i) \to \chi_i^a$ denotes the action of $\mathbf{K}^{\times} = \text{GL}_1(\mathbf{K})$ on $\widehat{\mathbf{K}}$. Since $\chi_i \neq 1$, the $\mathbf{K}^{\times}$-orbit of χ_i is dense in $\widehat{\mathbf{K}}$ for every $i \in \{1, \ldots, n\}$, by Lemma 3.C.2. This proves the claim.

(2) Let $\gamma \in \Gamma$ be distinct from the identity matrix I_n. Denote by Fix(γ) the set of fixed points of γ in $\widehat{\mathbf{K}}^n$.

For $\chi \in \widehat{\mathbf{K}}^n$, we have $\chi \in \mathrm{Fix}(\gamma)$ if and only if the subspace $V := (\gamma - I_n)\mathbf{K}^n$ of $\mathbf{K}^n$ is contained in $\ker \chi$. Therefore, $\mathrm{Fix}(\gamma)$ coincides with the annihilator $V^\perp$ of V in $\widehat{\mathbf{K}}^n$. In particular, $\mathrm{Fix}(\gamma)$ is a closed subgroup of $\widehat{\mathbf{K}}^n$. Observe that $V \neq \{0\}$, since $\gamma \neq I_n$.

We claim that $\mathrm{Fix}(\gamma)$ has empty interior. Indeed, assume by contradiction that this is not the case. Then $\mathrm{Fix}(\gamma)$ is an open subgroup of $\widehat{\mathbf{K}}^n$. Since $\widehat{\mathbf{K}}^n$ is compact, it follows that $\mathrm{Fix}(\gamma)$ has finite index in $\widehat{\mathbf{K}}^n$. However, by Pontryagin duality, the dual group $\widehat{V}$ of the subgroup V of $\mathbf{K}^n$ can be identified with $\widehat{\mathbf{K}}^n/V^\perp$. Since $\mathrm{Fix}(\gamma) = V^\perp$, it follows that $\widehat{V}$ and hence V are finite. This contradicts the fact that $V \neq \{0\}$ and that $\mathbf{K}$ is infinite. Therefore $\mathrm{Fix}(\gamma)$ has empty interior for every $\gamma \in \Gamma$ with $\gamma \neq I_n$.

Since $\mathbf{K}$ is countable, Γ is countable and it follows from the Baire Category Theorem that the F_σ subset

$$X := \bigcup_{\gamma \in \Gamma, \gamma \neq I_n} \mathrm{Fix}(\gamma)$$

of $\widehat{\mathbf{K}}^n$ has empty interior. The set of $\chi \in \widehat{\mathbf{K}}^n$ with trivial stabilizer in Γ coincides with the complement of X in $\widehat{\mathbf{K}}^n$ and is therefore a dense G_δ set. □

As in Section 3.F, we denote by A the subgroup of diagonal matrices and by B the subgroup of upper triangular matrices in Γ. Let $W \approx \mathcal{S}_n$ be the subgroup of the permutation matrices in Γ, that is, the matrices with exactly one 1 in each row and column and with 0's elsewhere. Observe that W acts on $A \approx (\mathbf{K}^\times)^n$ by permuting coordinates. Recall the **Bruhat decomposition** of Γ (see for instance [**AlBe–95**, Chap. 2]):

$$\mathrm{GL}_n(\mathbf{K}) = \bigsqcup_{w \in W} BwB.$$

LEMMA 9.F.2. *Let $(\pi, \mathcal{H})$ be a cyclic representation of Γ, with cyclic unit vector $\xi \in \mathcal{H}$. Assume that there exists a unitary character $\chi = (\chi_1, \ldots, \chi_n)$ of B with $\chi_1, \ldots, \chi_n$ pairwise distinct such that*

$$\langle \pi(b)\xi \mid \xi \rangle = \chi(b) \quad \textit{for every } b \in B.$$

Then π is equivalent to the principal series representation $\mathrm{Ind}_B^\Gamma \chi$.

PROOF. Let T be a set of representatives for the classes in $B\backslash\Gamma$ with $e \in T$. Then, with the notation as in Section 1.F, $\rho = \mathrm{Ind}_B^\Gamma \chi$ is realized on $\ell^2(T)$ by

$$(\rho(\gamma)f)(t) = \chi(\alpha(t, \gamma))f(t \cdot \gamma) \quad \text{for all } \gamma \in \Gamma,\ f \in \ell^2(T), \text{ and } t \in T.$$

The delta function δ_e is a cyclic vector for ρ. To show that π are ρ are equivalent, it suffices, by Proposition 1.B.10, to prove that the two functions of positive type

$$\varphi_\pi : \gamma \mapsto \langle \pi(\gamma)\xi \mid \xi \rangle$$

and

$$\varphi_\rho : \gamma \mapsto \langle \rho(\gamma)\delta_e \mid \delta_e \rangle$$

coincide on Γ.

We have

$$\varphi_\rho(b) = \langle \rho(b)\delta_e \mid \delta_e \rangle = \chi(b) \quad \text{for every } b \in B$$

and $\varphi_\rho(\gamma) = 0$ for $\gamma \in \Gamma \smallsetminus B$. This means that φ_ρ is the trivial extension $\widetilde{\psi}$ of ψ to Γ (see also Proposition 1.F.10). It follows from the assumption that $\varphi_\pi = \varphi_\rho$ on B; it therefore remains to show that $\varphi_\pi = 0$ on $\Gamma \smallsetminus B$.

Let $b \in B$. We have, by assumption, $\langle \pi(b)\xi \mid \xi \rangle = \chi(b)$. It follows from the equality case of the Cauchy–Schwarz inequality that $\pi(b)\xi = \chi(b)\xi$.

Let $w \in W$ and $b_1, b_2 \in B$. Then, we have

$$\begin{aligned} \varphi_\pi(b_1 w b_2) &= \langle \pi(b_1 w b_2)\xi \mid \xi \rangle = \langle \pi(b_1)\pi(w)\pi(b_2)\xi \mid \xi \rangle \\ &= \langle \pi(w)\pi(b_2)\xi \mid \pi(b_1^{-1})\xi \rangle = \chi(b_2)\chi(b_1)\langle \pi(w)\xi \mid \xi \rangle \\ &= \chi(b_2)\chi(b_1)\varphi_\pi(w). \end{aligned}$$

In view of the Bruhat decomposition of Γ, it therefore suffices to prove that $\varphi_\pi(w) = 0$ for every $w \in W \smallsetminus \{e\}$.

Assume by contradiction that $\varphi_\pi(w) \neq 0$ for some $w \in W \smallsetminus \{e\}$. Then, for every $a \in A$ we have $waw^{-1} \in A \subset B$ and hence

$$\begin{aligned} \chi(a)\varphi_\pi(w) &= \langle \pi(wa)\xi \mid \xi \rangle \\ &= \langle \pi(waw^{-1})\pi(w)\xi \mid \xi \rangle \\ &= \langle \pi(w)\xi \mid \pi(wa^{-1}w^{-1})\xi \rangle \\ &= \overline{\chi(wa^{-1}w^{-1})}\langle \pi(w)\xi \mid \xi \rangle \\ &= \chi(waw^{-1})\varphi_\pi(w). \end{aligned}$$

Since $\varphi_\pi(w) \neq 0$, it follows that $\chi(a) = \chi(waw^{-1})$ for every $a \in A$. Let $\sigma \in \mathcal{S}$ be the permutation corresponding to w, so that

$$waw^{-1} = (a_{\sigma(1)}, \dots, a_{\sigma(n)}) \quad \text{for all } a = (a_1, \dots, a_n) \in A.$$

Then, for every $a = (a_1, \dots, a_n) \in A$, we have

$$\begin{aligned} \chi_1(a_1) \cdots \chi_n(a_n) &= \chi(a) = \chi(waw^{-1}) \\ &= \chi_1(a_{\sigma(1)}) \cdots \chi_n(a_{\sigma(n)}) \end{aligned}$$

and hence $\chi_{\sigma(i)} = \chi_i$ for all $i \in \{1, \dots, n\}$. Since $w \neq e$, there exists $i \in \{1, \dots, n\}$ with $\sigma(i) \neq i$ and this is a contradiction to the assumption that the χ_j's are all pairwise distinct. □

Recall that the centre Z of Γ is the subgroup of scalar matrices.

LEMMA 9.F.3. *Let χ be a unitary character of B and set $\psi := \chi|_Z \in \widehat{Z}$. The principal series representation $\mathrm{Ind}_B^\Gamma \chi$ is weakly equivalent to $\lambda_\psi := \mathrm{Ind}_Z^\Gamma \psi$.*

PROOF. • *First step.* We claim that λ_ψ is weakly contained in $\rho := \mathrm{Ind}_B^\Gamma \chi$.

Let $\widetilde{\psi}$ be the trivial extension of ψ to Γ. Since δ_e is a cyclic vector for λ_ψ and since $\widetilde{\psi}$ is the function of positive type associated with λ_ψ and δ_e (see proof of Lemma 9.F.2), it suffices to show that $\widetilde{\psi}$ is the uniform limit over finite subsets of normalized functions of positive type associated with ρ.

Since $Z \subset B$ and since $\psi = \chi|_Z$, the induced representation $\mathrm{Ind}_B^\Gamma \chi$ has ψ as central character (see Proposition 1.F.7).

Let F be a finite subset of Γ and write $F = F_1 \cup F_2$ with $F_1 \subset Z$ and $F_2 \subset \Gamma \smallsetminus Z$. Let $\gamma \in F_2$. Set

$$X^\gamma := \{x \in \Gamma \mid Bx\gamma = Bx\} = \{x \in \Gamma \mid x\gamma x^{-1} \in B\}.$$

Since B is an algebraic subvariety of $\Gamma = \mathrm{GL}_n(\mathbf{K})$, we see that X^γ is an algebraic subvariety of Γ. Observe that

$$\bigcap_{x\in\Gamma} x^{-1}Bx = Z.$$

Therefore X^γ has positive codimension in $\mathrm{GL}_n(\mathbf{K})$, because $\gamma \notin Z$. Since $\mathbf{K}$ is infinite, it follows that $\bigcup_{\gamma\in F_2} X^\gamma$ is a proper subset of Γ. Therefore there exists $x \in \Gamma$ such that $x \notin X^\gamma$, that is, such that $Bx\gamma \neq Bx$ for every $\gamma \in F_2$.

Let T be a set of representatives for $B\backslash G$ with $x \in T$. For all $\gamma \in F_2$, we have $x\cdot\gamma \neq x$ and hence

$$\begin{aligned}\langle \rho(\gamma)\delta_x \mid \delta_x\rangle &= \langle (\mathrm{Ind}_B^\Gamma \chi)(\gamma)\delta_x \mid \delta_x\rangle \\ &= \chi(\alpha(x,\gamma))\langle \delta_{x\cdot\gamma} \mid \delta_x\rangle \\ &= 0.\end{aligned}$$

Moreover, for $\gamma \in F_1$ we have

$$\begin{aligned}\langle \rho(\gamma)\delta_x \mid \delta_x\rangle &= \psi(\gamma)\langle \delta_x \mid \delta_x\rangle \\ &= \psi(\gamma),\end{aligned}$$

since $F_1 \subset Z$. The claim is therefore proved.

• *Second step.* We claim that $\rho = \mathrm{Ind}_B^\Gamma\chi$ is weakly contained in λ_ψ.

Indeed, B is solvable and hence amenable. Therefore, 1_B is weakly contained in the quasi-regular representation $\lambda_{B/Z} = \mathrm{Ind}_Z^B 1_Z$. Using Proposition 1.F.6(4), we have

$$\mathrm{Ind}_Z^B\psi = \mathrm{Ind}_Z^B(\chi|_Z) \simeq \chi\otimes\mathrm{Ind}_Z^B 1_Z = \chi\otimes\lambda_{B/Z},$$

and it follows that $\chi \simeq \chi\otimes 1_B$ is weakly contained in $\chi\otimes\lambda_{B/Z} \simeq \mathrm{Ind}_Z^B\psi$. Therefore $\rho = \mathrm{Ind}_B^\Gamma\chi$ is weakly contained in $\mathrm{Ind}_B^\Gamma(\mathrm{Ind}_Z^B\psi) \simeq \mathrm{Ind}_Z^\Gamma\psi = \lambda_\psi$, as was to be shown. □

REMARK 9.F.4. Let $\mathrm{Ind}_B^\Gamma\chi_1$ and $\mathrm{Ind}_B^\Gamma\chi_2$ be two principal series representations of Γ with the same central character, that is, such that $\chi_1|_Z = \chi_2|_Z$. It follows from Lemma 9.F.3 that $\mathrm{Ind}_B^\Gamma\chi_1$ and $\mathrm{Ind}_B^\Gamma\chi_2$ are weakly equivalent. However, observe that $\mathrm{Ind}_B^\Gamma\chi_1$ and $\mathrm{Ind}_B^\Gamma\chi_2$ are not equivalent if $\chi_1 \neq \chi_2$, by Theorem 3.F.2.

Recall that the commutator subgroup of $\Gamma = \mathrm{GL}_n(\mathbf{K})$ is $\mathrm{SL}_n(\mathbf{K})$; hence, the one-dimensional representations of Γ correspond to the unitary characters of the abelianized group

$$\Gamma_{\mathrm{ab}} = \mathrm{GL}_n(\mathbf{K})/\mathrm{SL}_n(\mathbf{K}) \approx \mathbf{K}^\times.$$

One checks that $\psi \in \widehat{Z} \approx \mathbf{K}^\times$ arises as the central character of a one-dimensional representation of Γ if and only if $\psi = 1$ on the group $\mu_n^{\mathbf{K}}$ of the n-th roots of unity in $\mathbf{K}$.

Here is the first main result of this subsection.

THEOREM 9.F.5. *Let $\mathbf{K}$ be an infinite field and $n \geq 2$ an integer. Let π be a representation of $\Gamma = \mathrm{GL}_n(\mathbf{K})$ which is not trivial on $\mathrm{SL}_n(\mathbf{K})$. Assume that π admits a central character $\psi \in \widehat{Z}$.*

Then $\lambda_\psi = \mathrm{Ind}_Z^\Gamma\psi$ is weakly contained in π.

PROOF. • *First step.* We claim that it suffices to prove the theorem in the case of a countable infinite field $\mathbf{K}$.

Indeed, assume that the theorem has been proved in the case of a countable field. Since δ_e is a cyclic vector for λ_ψ, we have to show that the trivial extension $\widetilde{\psi}$ of ψ to Γ is the uniform limit over finite subsets of normalized functions of positive type associated with π (see the proof of Lemma 9.F.3).

Let F be a finite subset of Γ. Since π is not trivial on $\mathrm{SL}_n(\mathbf{K})$, upon adding to F a suitable element of $\mathrm{SL}_n(\mathbf{K})$, we can assume that $\pi(\gamma)$ is not the identity for some $\gamma \in F$. Let $\mathbf{L}$ be the subfield of $\mathbf{K}$ generated by the coefficients of all matrices in F (in case $\mathbf{L}$ is finite, we replace $\mathbf{L}$ by a countable infinite subfield of $\mathbf{K}$ containing $\mathbf{L}$). Then $\mathbf{L}$ is countable and π is not trivial on $\mathrm{SL}_n(\mathbf{L})$. Set $\Gamma' = \mathrm{GL}_n(\mathbf{L})$. Then $\mathrm{Ind}_{Z'}^{\Gamma'}\psi'$ is weakly contained in $\pi|_{\Gamma'}$, where $Z' = Z \cap \Gamma$ is the centre of Γ' and $\psi' = \psi|_{Z'}$. Since $F \subset \Gamma'$, it follows that $\widetilde{\psi'}$ and hence $\widetilde{\psi}$ is the uniform limit over F of normalized functions of positive type associated with π.

The claim is proved and so we may assume from now on that $\mathbf{K}$ is countable.

Next, we will need to consider the subgroup

$$Q = H \ltimes U \approx \mathrm{GL}_{n-1}(\mathbf{K}) \ltimes \mathbf{K}^{n-1}$$

of Γ, where

$$H = \left\{ \begin{pmatrix} g & 0 \\ 0 & 1 \end{pmatrix} \mid g \in \mathrm{GL}_{n-1}(\mathbf{K}) \right\} \approx \mathrm{GL}_{n-1}(\mathbf{K})$$

and

$$U = \left\{ \begin{pmatrix} I_{n-1} & x \\ 0 & 1 \end{pmatrix} \mid x \in \mathbf{K}^{n-1} \right\} \approx \mathbf{K}^{n-1}.$$

Observe that the action of H on U corresponds to the natural action of $\mathrm{GL}_{n-1}(\mathbf{K})$ on $\mathbf{K}^{n-1}$.

• *Second step.* We claim that the restriction $\pi|_U$ of π to U is not trivial.

Indeed, assume that $\pi|_U$ is trivial, that is, $U \subset \ker\pi$. Every elementary matrix $E_{i,j}(\alpha)$ for $\alpha \in \mathbf{K}$ and $i, j \in \{1, \dots, n\}, i \neq j$, is conjugate to an elementary matrix contained in U. It follows that $\ker\pi$ contains all elementary matrices $E_{i,j}(\alpha)$ and hence $\mathrm{SL}_n(\mathbf{K})$, since $\mathrm{SL}_n(\mathbf{K})$ is generated by these matrices. This is a contradiction, as π is not trivial on $\mathrm{SL}_n(\mathbf{K})$.

• *Third step.* We claim that the regular representation λ_Q of Q is weakly contained in $\pi|_Q$.

Indeed, by the second step, there exists $\chi \in \widehat{U} \approx \widehat{\mathbf{K}}^n$ with $\chi \neq 1$ which is weakly contained in $\pi|_U$. Then χ^γ is weakly contained in $\pi|_U$ for every $\gamma \in H$ (see Remark 9.A.2). It follows from Lemma 9.F.1 (1) that every $\chi \in \widehat{U}$ is weakly contained $\pi|_U$. By Lemma 9.F.1(2) (here we use the countability of $\mathbf{K}$), there exists $\chi_0 \in \widehat{U}$ with a trivial stabilizer in H which is weakly contained in $\pi|_U$. Proposition 9.A.1 therefore implies that $\mathrm{Ind}_U^Q\chi_0^\gamma$ is weakly contained in $\pi|_Q$ for every $\gamma \in H$. Again by Lemma 9.F.1 (1), the set $\{\chi_0^\gamma \mid \gamma \in H\}$ is dense in $\widehat{U}$. It follows that $\mathrm{Ind}_U^Q\chi$ is weakly contained in $\pi|_Q$ for every $\chi \in \widehat{U}$. Since the direct sum $\bigoplus_{\chi\in\widehat{U}} \chi$ is weakly equivalent to the regular representation λ_U, this implies that $\lambda_Q = \mathrm{Ind}_U^Q\lambda_U$ is weakly contained $\pi|_Q$.

• *Fourth step.* We claim that $\pi|_B$ weakly contains a unitary character $\chi = (\chi_1, \dots, \chi_n)$ of B with $\chi|_Z = \psi$ and such that the χ_i's are pairwise distinct.

Indeed, let $B(n-1)$ be the subgroup of the upper-triangular matrices in $H \approx \mathrm{GL}_{n-1}(\mathbf{K})$ and consider the subgroup $S = B(n-1) \ltimes U$ of Q. By the third step, λ_Q is weakly contained in $\pi|_Q$. Therefore λ_S is weakly contained in $\pi|_S$. Since S is amenable, it follows that every one-dimensional representation of S is weakly contained in $\pi|_S$.

Observe that B is the Cartesian product of the subgroup S with the centre Z. Since ψ is the central character of π, it follows that every unitary character χ of B with $\chi|_Z = \psi$ is weakly contained in $\pi|_B$.

It remains to show that there exists a unitary character $\chi = (\chi_1, \ldots, \chi_n) \in \widehat{(\mathbf{K}^\times)^n}$ of B with $\chi|_Z = \psi$ and all χ_i pairwise distinct.

We proceed by induction on n. For $n \geq 2$, assume, by induction hypothesis, that there exist pairwise distinct $\delta_1, \ldots, \delta_{n-1} \in \widehat{\mathbf{K}^\times}$ such that $\delta|_Z = \psi$ for $\delta = (\delta_1, \ldots, \delta_{n-1})$, that is, such that $\delta_1 \cdots \delta_{n-1} = \psi$.

Observe that, since $\mathbf{K}$ is infinite, the subgroup $K_n = \{x^n \mid x \in \mathbf{K}^\times\}$ of n-th powers in $\mathbf{K}^\times$ is infinite. Hence, the dual group of K_n is infinite, that is, the annihilator $K_n^\perp$ of K_n has infinite index in $\widehat{\mathbf{K}^\times}$. Since

$$K_n^\perp = \{\delta \in \widehat{\mathbf{K}^\times} \mid \delta^n = 1_{\mathbf{K}^\times}\},$$

this implies that $\{\delta^n \mid \delta \in \widehat{\mathbf{K}^\times}\}$ is an infinite subgroup of $\widehat{\mathbf{K}^\times}$. We can therefore find $\delta \in \widehat{\mathbf{K}^\times}$ such that $\delta^n \neq \delta_i^{-1}$ for every $i \in \{1, \ldots, n-1\}$.

We set

$$\chi_1 := \delta\delta_1, \chi_2 := \delta\delta_2, \ldots, \chi_{n-1} := \delta\delta_{n-1} \text{ and } \chi_n := (\delta^{n-1})^{-1}.$$

It is clear that the χ_i's are pairwise distinct and that $\chi|_Z = \psi$ for $\chi = (\chi_1, \ldots, \chi_n)$.

• *Fifth step.* We claim that π weakly contains a principal series representation $\mathrm{Ind}_B^\Gamma \chi$ for a unitary character χ of B with $\chi|_Z = \psi$.

Indeed, let $\chi = (\chi_1, \ldots, \chi_n)$ be a unitary character of B as in the fourth step. Since χ is weakly contained in $\pi|_B$, there exists a sequence $(\varphi_n)_{n\geq 1}$ of normalized functions of positive type associated with π such that

$$\lim_n \varphi_n(b) = \chi(b) \qquad \text{for all } b \in B.$$

By compactness of the set of normalized functions of positive type on Γ for the topology of pointwise convergence, upon passing to a subsequence, we can assume that $(\varphi_n)_{n\geq 1}$ converges to a normalized function of positive type φ on Γ. Let $(\rho, \mathcal{H}, \xi)$ be GNS a triple associated to φ (see Construction 1.B.6). Since $\varphi|_B = \chi$ and since the χ_i's are pairwise distinct, Lemma 9.F.2 shows that ρ is equivalent to the principal series representations $\mathrm{Ind}_B^\Gamma \chi$. As $\lim_n \varphi_n = \varphi$, the representation ρ and hence $\mathrm{Ind}_B^\Gamma \chi$ is weakly contained in π.

• *Sixth step.* We claim that π weakly contains the representation $\lambda_\psi = \mathrm{Ind}_Z^\Gamma \psi$. Indeed, this follows from the fifth step and from Lemma 9.F.3. □

As a consequence of Theorem 9.F.5, we obtain a description of the maximal ideals in the C*-algebra $C^*_{\max}(\mathrm{GL}_n(\mathbf{K}))$ for an arbitrary infinite field $\mathbf{K}$.

Corollary 9.F.6. *Let $\mathbf{K}$ be an infinite field and $\Gamma = \mathrm{GL}_n(\mathbf{K})$ for $n \geq 2$. The maximal two-sided ideals of $C^*_{\max}(\Gamma)$ are:*

- *either C*-kernels of one-dimensional representations of Γ, that is, ideals of the form $\mathrm{C^*ker}(\chi)$ for a unitary character χ of $\Gamma_{\mathrm{ab}} = \mathrm{GL}_n(\mathbf{K})/\mathrm{SL}_n(\mathbf{K})$;*
- *or ideals of the form $\mathrm{C^*ker}(\mathrm{Ind}_Z^\Gamma \psi)$, for some $\psi \in \widehat{Z}$.*

PROOF. Let J be a maximal ideal of $C^*_{\max}(\Gamma)$; then J is a primitive ideal of Γ and hence $J = \mathrm{C^*ker}(\pi)$ for an irreducible representation π of Γ.

If π is one dimensional, then $J = \mathrm{C^*ker}(\chi)$ for a unitary character χ of Γ_{ab}.

Assume now that π is not one dimensional. Then, by Theorem 9.F.5, $\mathrm{Ind}_Z^\Gamma \psi$ is weakly contained in π, that is,

$$J = \mathrm{C^*ker}(\pi) \subset \mathrm{C^*ker}(\mathrm{Ind}_Z^\Gamma \psi).$$

Therefore $J = \mathrm{C^*ker}(\mathrm{Ind}_Z^\Gamma \psi)$, by maximality of J. □

REMARK 9.F.7. In Corollary 9.F.6, it may happen that an ideal of the form $\mathrm{C^*ker}(\mathrm{Ind}_Z^\Gamma \psi)$ for $\psi \in \widehat{Z}$ is not maximal. Indeed, assume that Γ is amenable (equivalently, that $\mathbf{K}$ is an infinite algebraic extension of a finite field; see Proposition 9.F.8 below). Then the trivial representation 1_Γ is weakly contained in $\mathrm{Ind}_Z^\Gamma 1_Z$, that is, $\mathrm{C^*ker}(\mathrm{Ind}_Z^\Gamma 1_Z) \subset \mathrm{C^*ker}(1_\Gamma)$. However, $\mathrm{Ind}_Z^\Gamma 1_Z$ is obviously not weakly contained in 1_Γ. Therefore $\mathrm{C^*ker}(\mathrm{Ind}_Z^\Gamma 1_Z)$ is not a maximal ideal.

When $\mathrm{GL}_n(\mathbf{K})$ is amenable, Theorem 9.F.5 enables us to give a complete description of the primitive dual of $\mathrm{GL}_n(\mathbf{K})$. The fields $\mathbf{K}$ for which $\mathrm{GL}_n(\mathbf{K})$ (or $\mathrm{SL}_n(\mathbf{K})$) is amenable can easily be characterized.

PROPOSITION 9.F.8. *Let* $\mathbf{K}$ *be a field and* $n \geq 2$. *The following properties are equivalent:*

(i) *the group* $\mathrm{GL}_n(\mathbf{K})$ *is amenable;*
(ii) *the group* $\mathrm{SL}_n(\mathbf{K})$ *is amenable;*
(iii) *the field* $\mathbf{K}$ *is an algebraic extension of a finite field.*

PROOF. Since amenability is inherited by subgroups, it is obvious that (i) implies (ii).

Assume that $\mathbf{K}$ is an algebraic extension of a finite field $\mathbf{F}_q$. Then $\mathbf{K} = \bigcup_m \mathbf{K}_m$ for an increasing family of finite extensions $\mathbf{K}_m$ of $\mathbf{F}_q$, and therefore $\mathrm{GL}_n(\mathbf{K}) = \bigcup_m \mathrm{GL}_n(\mathbf{K}_m)$ is the inductive limit of the finite and hence amenable groups $\mathrm{GL}_n(\mathbf{K}_m)$. It follows that $\mathrm{GL}_n(\mathbf{K})$ is amenable. So, (iii) implies (i).

Assume that $\mathbf{K}$ is not an algebraic extension of a finite field. If the characteristic of $\mathbf{K}$ is 0, then $\mathbf{K}$ contains $\mathbf{Q}$; if $\mathbf{K}$ is of characteristic p, then $\mathbf{K}$ contains the purely transcendental extension $\mathbf{F}_p(T)$ of $\mathbf{F}_p$. Therefore $\mathrm{SL}_n(\mathbf{K})$ contains a copy of $\mathrm{SL}_2(\mathbf{Q})$ or a copy of $\mathrm{SL}_2(\mathbf{F}_p(T))$.

As is well known, each of the groups $\mathrm{SL}_2(\mathbf{Q})$ or $\mathrm{SL}_2(\mathbf{F}_p(T))$ contains a free group on two generators.

Indeed, denote by K either $\mathbf{Q}$ or $\mathbf{F}_p(T)$. Let $|\cdot|$ be an absolute value on K with respect to which the completion k of K is a local field. In the case $K = \mathbf{Q}$, we could take for $|\cdot|$ the usual Archimedean absolute value for which $k = \mathbf{R}$; however, in the case $K = \mathbf{F}_p(T)$, there is essentially only one choice for $|\cdot|$ and then k is the field of Laurent series $\mathbf{F}_p((T^{-1}))$ over $\mathbf{F}_p$.

Choose $\lambda \in K^*$ with $|\lambda| \neq 1$ and $a \in K \smallsetminus \{0, 1\}$. Consider the following matrices $A, B \in \mathrm{SL}_2(K)$:

$$A = \begin{pmatrix} \lambda & 0 \\ 0 & \lambda^{-1} \end{pmatrix} \quad \text{and} \quad B = CAC^{-1}, \quad \text{where} \quad C = \begin{pmatrix} 1 & a \\ 1 & 1 \end{pmatrix},$$

and their action on the projective line $\mathbf{P}(k) = k \cup \{\infty\}$. The set of fixed points of A is $\{0, \infty\}$ and is disjoint from the set $\{a, 1\}$ of fixed points of B. Observe that $\lim_{n \to +\infty} A^{\pm n} x \in \{0, \infty\}$ and $\lim_{n \to +\infty} B^{\pm n} x \in \{a, 1\}$ for every $x \in \mathbf{P}(k)$. Choose compact and disjoint subsets X_1 and X_2 of $\mathbf{P}(k)$ containing respectively $\{0, \infty\}$ and $\{a, 1\}$. There exists $N \geq 1$ such that we have

$$A^{\pm m}(X_2) \subset X_1 \quad \text{and} \quad B^{\pm m}(X_1) \subset X_2,$$

for every $m \geq N$. It follows from the classical ping-pong lemma that A^N and B^N generate a free subgroup of $\mathrm{SL}_2(K)$. Consequently, $\mathrm{SL}_n(\mathbf{K})$ is not amenable and this shows that (ii) implies (iii). □

We describe the primitive dual of $\mathrm{GL}_n(\mathbf{K})$, when $\mathrm{GL}_n(\mathbf{K})$ is amenable.

THEOREM 9.F.9. *Let* $\Gamma = \mathrm{GL}_n(\mathbf{K})$, *where* $n \geq 2$ *and* $\mathbf{K}$ *is an infinite algebraic extension of a finite field.*

The map

$$\Phi : \widehat{\Gamma_{\mathrm{ab}}} \sqcup \widehat{Z} \to \mathrm{Prim}(\Gamma),$$

defined by

$$\Phi(\chi) = \mathrm{C^*ker}(\chi) \quad \textit{for } \chi \in \widehat{\Gamma}_{\mathrm{ab}}$$

and

$$\Phi(\psi) = \mathrm{C^*ker}(\mathrm{Ind}_Z^\Gamma \psi) \quad \textit{for } \psi \in \widehat{Z} \approx \widehat{\mathbf{K}^\times},$$

is a bijection.

PROOF. • *First step.* We claim that the image of the map Φ is indeed contained in $\mathrm{Prim}(\Gamma)$.

Observe first that, for every $\chi \in \widehat{\Gamma_{\mathrm{ab}}}$, the ideal $\mathrm{C^*ker}(\chi)$ is of course a primitive ideal.

Let $\psi \in \widehat{Z}$ and set $J = \mathrm{C^*ker}(\mathrm{Ind}_Z^\Gamma \psi)$. We can choose a unitary character $\chi = (\chi_1, \ldots, \chi_n) \in \widehat{\mathbf{K}^\times}^n$ of B such that $\chi|_Z = \psi$ (for instance, by taking $\chi_1 = \psi$ and $\chi_i = 1$ for all $i \neq 1$). Let $\pi = \mathrm{Ind}_B^\Gamma \chi$ be the corresponding principal series representation. Then π is irreducible (Theorem 3.F.2) and the central character of π is ψ. On the one hand, since π is not one dimensional, $\mathrm{Ind}_Z^\Gamma \psi$ is weakly contained in π, by Theorem 9.F.5. On the other hand, since Γ is amenable and since $\mathrm{Ind}_Z^\Gamma(\pi|_Z)$ and $\pi \otimes \mathrm{Ind}_Z^\Gamma 1_Z$ are equivalent, π is weakly contained in $\mathrm{Ind}_Z^G(\pi|_Z)$ which is weakly equivalent to $\mathrm{Ind}_Z^\Gamma \psi$. Therefore π and $\mathrm{Ind}_Z^\Gamma \psi$ are weakly equivalent. This shows that $J = \mathrm{C^*ker}(\pi)$ and hence that J is a primitive ideal. So, the image of the map Φ is indeed contained in $\mathrm{Prim}(\Gamma)$.

• *Second step.* We claim that the map Φ is surjective.

Indeed, let π be an irreducible representation of Γ, with central character ψ. Assume that π is not one dimensional. On the one hand, $\mathrm{Ind}_Z^\Gamma \psi$ is weakly contained in π, by Theorem 9.F.5. On the other hand, as in the first step, π is weakly contained in $\mathrm{Ind}_Z^\Gamma \psi$, since Γ is amenable. This shows that

$$\mathrm{C^*ker}(\pi) = \mathrm{C^*ker}(\mathrm{Ind}_Z^\Gamma \psi).$$

• *Third step.* We claim that the map Φ is injective.

Indeed, let $\chi_1, \chi_2 \in \widehat{\Gamma_{\mathrm{ab}}}$ be such that $\Phi(\chi_1) = \Phi(\chi_2)$, that is, such that χ_1 and χ_2 are weakly equivalent. Then $\chi_1 = \chi_2$.

Next, let $\psi_1, \psi_2 \in \widehat{Z}$ be such that $\Phi(\psi_1) = \Phi(\psi_2)$, that is, such that $\lambda_{\psi_1} = \mathrm{Ind}_Z^\Gamma \psi_1$ and $\lambda_{\psi_2} = \mathrm{Ind}_Z^\Gamma \psi_2$ are weakly equivalent. Then $\psi_1 = \psi_2$, since the central characters of λ_{ψ_1} and λ_{ψ_2} are respectively ψ_1 and ψ_2.

Finally, let $\chi \in \widehat{\Gamma_{\mathrm{ab}}}$ and $\psi \in \widehat{Z}$. As in the first step, there exists a principal series representation π of Γ such that π is weakly equivalent to $\mathrm{Ind}_Z^\Gamma \psi$. Since π is an infinite-dimensional irreducible representation, π cannot be weakly equivalent to the finite-dimensional representation χ (see Proposition 9.A.3(1)) and hence $\Phi(\chi) \neq \Phi(\psi)$. □

Theorem 9.F.9 takes a striking form, when we go over to the projective linear group $\mathrm{PGL}_n(\mathbf{K}) = \mathrm{GL}_n(\mathbf{K})/Z$.

COROLLARY 9.F.10. *Let* $\Gamma = \mathrm{PGL}_n(\mathbf{K})$, *where* $n \geq 2$ *and* $\mathbf{K}$ *is an infinite algebraic extension of a finite field. Then*

$$\mathrm{Prim}(\Gamma) = \{0\} \cup \{C^*\mathrm{ker}(\chi) \mid \chi \in \widehat{\Gamma_{\mathrm{ab}}}\}.$$

PROOF. Let $G = \mathrm{GL}_n(\mathbf{K})$. For $\psi \in \widehat{Z}$, the representation $\mathrm{Ind}_Z^G \psi$ factorizes through Γ if and only if $\psi = 1_Z$. Moreover, $\mathrm{Ind}_Z^G 1_Z$ factorizes into the regular representation λ_Γ of Γ. Since Γ is amenable, $C^*\mathrm{ker}(\lambda_\Gamma) = \{0\}$. Therefore, the result is a consequence of Theorem 9.F.9. □

9.F.b. Special linear groups. We now treat the case of the special linear group $\mathrm{SL}_n(\mathbf{K})$ for an infinite field $\mathbf{K}$ and an integer $n \geq 2$.

We will need to replace subgroup Q of $\mathrm{GL}_n(\mathbf{K})$ appearing in the proof of Theorem 9.F.5 by the subgroup

$$Q' = \left\{ \begin{pmatrix} g & x \\ 0 & (\det g)^{-1} \end{pmatrix} \;\middle|\; g \in \mathrm{GL}_{n-1}(\mathbf{K}),\ x \in \mathbf{K}^{n-1} \right\}$$

of $\mathrm{SL}_n(\mathbf{K})$. Observe that $Q' \approx \mathrm{GL}_{n-1}(\mathbf{K}) \ltimes \mathbf{K}^{n-1}$, where the action of $\mathrm{GL}_{n-1}(\mathbf{K})$ on $\mathbf{K}^{n-1}$ is now given by

$$g \cdot x = (\det g) gx \qquad \text{for } g \in \mathrm{GL}_{n-1}(\mathbf{K}),\ x \in \mathbf{K}^{n-1}.$$

It is clear that $g \in \mathrm{GL}_{n-1}(\mathbf{K})$ acts trivially on $\mathbf{K}^{n-1}$ if and only if $g = \lambda I_{n-1}$ for $\lambda \in \mu_n^{\mathbf{K}}$, where $\mu_n^{\mathbf{K}}$ denotes the set of n-th roots of unity in $\mathbf{K}$.

We need to prove the analogue of Lemma 9.F.1 for the new action introduced above.

LEMMA 9.F.11. *Let* $(\gamma, x) \mapsto \gamma \cdot x$ *denote the (left) action of* $\mathrm{GL}_n(\mathbf{K})$ *on* $\mathbf{K}^n$ *given by* $\gamma \cdot x = (\det \gamma)\gamma x$ *and* $(\gamma, \chi) \mapsto \gamma \cdot \chi$ *the corresponding dual (right) action on* $\widehat{\mathbf{K}}^n$, *defined by* $\gamma \cdot \chi(x) = \chi(\gamma \cdot x)$.

(1) *Assume that* $n \geq 2$. *For every* $\chi \in \widehat{\mathbf{K}}^n \smallsetminus \{1\}$, *the* $\mathrm{GL}_n(\mathbf{K})$*-orbit of* χ *is dense in* $\widehat{\mathbf{K}}^n$.
(2) *Assume that* $n \geq 1$ *and that* $\mathbf{K}$ *is countable. The set of* $\chi \in \widehat{\mathbf{K}}^n$ *for which the stabilizer in* $\mathrm{GL}_n(\mathbf{K})$ *is* $\{\lambda I_n \mid \lambda \in \mu_{n+1}^{\mathbf{K}}\}$ *is a dense* G_δ *set in* $\widehat{\mathbf{K}}^n$. *In particular, this set is not empty.*

PROOF. (1) Let $\chi = (\chi_1, \ldots, \chi_n) \in \widehat{\mathbf{K}}^n \smallsetminus \{1\}$. Let $i_0 \in \{1, \ldots, n\}$ be such that $\chi_{i_0} \neq 1$. Since $n \geq 2$, we can choose $i \in \{1, \ldots, n\}$ with $i \neq i_0$. For $t \in \mathbf{K}^\times$, let γ be the elementary matrix $E_{i,i_0}(t)$. Observe that $\gamma \in \mathrm{SL}_n(\mathbf{K})$ and $\gamma \cdot \chi = (\chi_1', \ldots, \chi_n')$, where

$$\chi_i' = \chi_{i_0}^t \chi_i \qquad \text{and} \qquad \chi_j' = \chi_j \ \text{ for every } j \neq i.$$

Since the $\mathbf{K}^\times$-orbit of χ_{i_0} is dense in $\widehat{\mathbf{K}}$ (Lemma 3.C.2), it follows that the closure of the $\mathrm{SL}_n(\mathbf{K})$-orbit of χ contains every character $(\chi_1', \ldots, \chi_n') \in \widehat{\mathbf{K}}^n$ with $\chi_{i_0}' = \chi_{i_0}$. Replacing i_0 by any other $i \neq i_0$, the same argument implies that the $\mathrm{SL}_n(\mathbf{K})$-orbit of χ is dense in $\widehat{\mathbf{K}}^n$.

(2) Let $\gamma \in \mathrm{GL}_n(\mathbf{K})$ with $\gamma \notin \{\lambda I_n \mid \lambda \in \mu_{n+1}^{\mathbf{K}}\}$. The set $\mathrm{Fix}(\gamma)$ of fixed points of γ in $\widehat{\mathbf{K}}^n$ is the annihilator $V^\perp$ in $\widehat{\mathbf{K}}^n$ of the subspace $V := ((\det\gamma)\gamma - I_n)\mathbf{K}^n$ of $\mathbf{K}^n$. Since $(\det\gamma)\gamma \neq I_n$, we have $V \neq \{0\}$. As in the proof of Lemma 9.F.1(2), this implies that $\mathrm{Fix}(\gamma)$ has empty interior and, since $\mathbf{K}$ is countable, the claim follows from the Baire Category Theorem. □

REMARK 9.F.12. In contrast to Lemma 9.F.1 for the natural action of $\mathrm{GL}_n(\mathbf{K})$ on $\mathbf{K}^n$, Lemma 9.F.11(1) fails in general when $n = 1$. Indeed, let $\mathbf{K} = \mathbf{F}_2(T)$ be the purely transcendental extension of $\mathbf{F}_2$. The set $\mathbf{K}^2 = \{x^2 \mid x \in \mathbf{K}\}$ of squares in $\mathbf{K}$ is a *proper* subgroup of $\mathbf{K}$. Let χ be a nontrivial character of $\mathbf{K}$ such that $\chi = 1$ on $\mathbf{K}^2$. For every $t \in \mathbf{K}^\times$ and $x \in \mathbf{K}$, we have $t \cdot \chi(x) = \chi(t^2 x)$ and therefore $t \cdot \chi = 1$ on $\mathbf{K}^2$. In particular, $\{t \cdot \chi \mid t \in \mathbf{K}^\times\}$ is not dense in $\widehat{\mathbf{K}}$.

We denote by A and B respectively the subgroups of diagonal and upper triangular matrices in $\Gamma := \mathrm{SL}_n(\mathbf{K})$.

The **Weyl group** of Γ is the quotient $N_\Gamma(A)/A$, where $N_\Gamma(A)$ is the normalizer of A in Γ. For every permutation matrix P_σ, with $\sigma \in \mathcal{S}_n$, let $T_\sigma = I_n$ if σ is even and let T_σ be the diagonal matrix $\mathrm{diag}(-1, 1, \ldots, 1)$ if σ is odd. Then $P_\sigma T_\sigma \in \mathrm{SL}_n(\mathbf{K})$ and the set

$$W := \{P_\sigma T_\sigma \mid \sigma \in \mathcal{S}_n\}$$

is a transversal for A in $N_\Gamma(A)$; moreover, W acts on A by permuting coordinates:

$$waw^{-1} = \mathrm{diag}(a_{\sigma(1)}, \ldots, a_{\sigma(n)}) \quad \text{for all } a = \mathrm{diag}(a_1, \ldots, a_n) \in A,\ w \in W.$$

In particular, the Weyl group of Γ is isomorphic to $\mathcal{S}_n$. As in the case of $\mathrm{GL}_n(\mathbf{K})$, we have a **Bruhat decomposition** of Γ :

$$\mathrm{SL}_n(\mathbf{K}) = \bigsqcup_{w \in W} BwB.$$

The following analogue of Lemma 9.F.2 holds for $\mathrm{SL}_n(\mathbf{K})$.

LEMMA 9.F.13. *Let $(\pi, \mathcal{H})$ be a cyclic representation of Γ, with cyclic unit vector $\xi \in \mathcal{H}$. Assume that there exists a unitary character χ of B with $\chi^w \neq \chi$ for every $w \in W \smallsetminus \{e\}$ such that*

$$\langle \pi(b)\xi \mid \xi \rangle = \chi(b) \quad \textit{for every } b \in B.$$

Then π is equivalent to the principal series representation $\mathrm{Ind}_B^\Gamma \chi$ of Γ.

PROOF. As in the proof of Lemma 9.F.2, it suffices to show that $\varphi_\pi(w) = 0$ for every $w \in W \smallsetminus \{e\}$, where φ_π is the function of positive type associated to (π, ξ).

Again as in the proof of Lemma 9.F.2, we have $\chi(a)\varphi_\pi(w) = \chi(waw^{-1})\varphi_\pi(w)$ for every $a \in A$. Since $\chi^w \neq \chi$, the claim follows. □

Recall that the centre Z of Γ is the subgroup of scalar matrices $\{\lambda I_n \mid \lambda \in \mu_n^{\mathbf{K}}\}$.

LEMMA 9.F.14. *Let χ be a unitary character of B, and set $\psi := \chi|_Z \in \widehat{Z}$. The principal series representation $\mathrm{Ind}_B^\Gamma \chi$ is weakly equivalent to $\mathrm{Ind}_Z^\Gamma \psi$.*

PROOF. The proof of the corresponding Lemma 9.F.3 for the case of $\mathrm{GL}_n(\mathbf{K})$ carries over without change. □

The analogue of Theorem 9.F.5 holds for $\mathrm{SL}_n(\mathbf{K})$ when $n \geq 3$.

THEOREM 9.F.15. *Let $\mathbf{K}$ be an infinite field and $n \geq 3$ an integer. Let π be a nontrivial representation of $\Gamma = \mathrm{SL}_n(\mathbf{K})$. Assume that π admits a central character $\psi \in \widehat{Z}$.*

Then $\lambda_\psi = \mathrm{Ind}_Z^\Gamma \psi$ is weakly contained in π.

PROOF. We check that the steps in the proof of Theorem 9.F.5 carry over for $\Gamma = \mathrm{SL}_n(\mathbf{K})$, after suitable modifications.

• *First step.* We may assume that $\mathbf{K}$ is countable. Indeed, this follows with the same arguments as in the first step of the proof of Theorem 9.F.5.

We consider the subgroup $Q = H \ltimes U \approx \mathrm{GL}_{n-1}(\mathbf{K}) \ltimes \mathbf{K}^{n-1}$ of Γ, where

$$H = \left\{ \begin{pmatrix} g & 0 \\ 0 & (\det g)^{-1} \end{pmatrix} \,\middle|\, g \in \mathrm{GL}_{n-1}(\mathbf{K}) \right\} \approx \mathrm{GL}_{n-1}(\mathbf{K})$$

and

$$U = \left\{ \begin{pmatrix} I_{n-1} & x \\ 0 & 1 \end{pmatrix} \,\middle|\, x \in \mathbf{K}^{n-1} \right\} \approx \mathbf{K}^{n-1}.$$

The action of H on U corresponds to the action $(g, x) \mapsto (\det g)gx$ of $\mathrm{GL}_{n-1}(\mathbf{K})$ on $\mathbf{K}^{n-1}$, introduced above.

Observe that Z is a subgroup of Q.

• *Second step.* The restriction $\pi|_U$ of π to U is not trivial. Indeed, the arguments in the second step of the proof of Theorem 9.F.5 apply here without change.

• *Third step.* The induced representation $\mathrm{Ind}_Z^Q \psi$ of Q is weakly contained in $\pi|_Q$.

Indeed, by the second step, there exists $\chi \in \widehat{U} \approx \widehat{\mathbf{K}}^n$ with $\chi \neq 1$ which is weakly contained in $\pi|_U$. It follows from Lemma 9.F.11(1) that every $\chi \in \widehat{U}$ is weakly contained $\pi|_U$ (here we use the assumption that $n \geq 3$). Hence, by Lemma 9.F.11 (2), $\pi|_U$ weakly contains some unitary character $\chi_0 \in \widehat{U}$ with stabilizer in H equal to $Z = \{\lambda I_n \mid \lambda \in \mu_n^{\mathbf{K}}\}$.

Observe that ZU is a normal subgroup of Q and that ZU is isomorphic to the direct product $Z \times U$. The stabilizer of the unitary character $\psi \otimes \chi_0$ of ZU in Q is ZU. It therefore follows from Proposition 9.A.1 that $\mathrm{Ind}_{ZU}^Q (\psi \otimes \chi_0^\gamma)$ is weakly contained in $\pi|_Q$ for every $\gamma \in H$. Again using Lemma 9.F.1 (1), this implies that $\mathrm{Ind}_{ZU}^Q(\psi \otimes \chi)$ is weakly contained in $\pi|_Q$ for every $\chi \in \widehat{U}$. Since U is abelian, it follows that $\mathrm{Ind}_{ZU}^Q(\psi \otimes \lambda_U)$ is weakly contained in $\pi|_Q$, where λ_U is the regular representation of U.

Now, $\mathrm{Ind}_Z^Q \psi$ is equivalent to $\mathrm{Ind}_{ZU}^Q(\mathrm{Ind}_Z^{ZU} \psi)$, which in turn is equivalent to $\mathrm{Ind}_{ZU}^Q(\psi \otimes \lambda_U)$. Therefore, $\mathrm{Ind}_Z^Q \psi$ is weakly contained in $\pi|_Q$.

• *Fourth step.* We claim that $\pi|_B$ weakly contains a unitary character χ of B with $\chi|_Z = \psi$ and such that $\chi^w \neq \chi$ for every $w \in W \smallsetminus \{e\}$.

Indeed, it follows from the third step that $(\mathrm{Ind}_Z^Q \psi)|_B$ is weakly contained in $\pi|_B$. Now, $\mathrm{Ind}_Z^Q \psi$ is equivalent to $\mathrm{Ind}_B^Q \rho_\psi$, where $\rho_\psi = \mathrm{Ind}_Z^B \psi$. Since ρ_ψ is contained in $(\mathrm{Ind}_B^Q \rho_\psi)|_B$, this implies that ρ_ψ is weakly contained in $\pi|_B$.

Let $\widetilde{\psi} \in \widehat{A}$ be an extension of ψ to A. Viewing $\widetilde{\psi}$ as a unitary character of B, we have

$$\rho_\psi = \mathrm{Ind}_Z^B \psi = \mathrm{Ind}_Z^B (\widetilde{\psi}|_Z) \simeq \widetilde{\psi} \otimes \mathrm{Ind}_Z^B 1_Z.$$

As B/Z is amenable, the support of $\mathrm{Ind}_Z^B 1_Z$ consists of the irreducible representations of B/Z lifted to B. The support of $\rho_\psi \simeq \widetilde{\psi} \otimes \mathrm{Ind}_Z^B 1_Z$ therefore consists of the irreducible representations of B with ψ as central character. In particular, ρ_ψ weakly contains every unitary character χ of B with $\chi|_Z = \psi$.

Choose $\chi_1 \in \widehat{\mathbf{K}^\times}$ with $\chi_1|_{\mu_n^{\mathbf{K}}} = \psi$. As $\mathbf{K}$ is infinite, we can find pairwise distinct $\chi_2, \ldots, \chi_{n-1} \in \widehat{\mathbf{K}^\times}$ with $\chi_i|_{\mu_n^{\mathbf{K}}} = 1$ for all $i \in \{2, \ldots, n-1\}$. Let $\chi \in \widehat{A}$ be defined by

$$\chi(\mathrm{diag}(a_1, a_2, \ldots, a_n)) = \chi_1(a_1)\chi_2(a_2)\cdots\chi_{n-1}(a_{n-1}).$$

Then $\chi|_Z = \psi$ and, as easily checked, we have $\chi^w \neq \chi$ for every $w \in W \smallsetminus \{e\}$.

• *Fifth step.* We claim that π weakly contains a principal series representation $\mathrm{Ind}_B^\Gamma \chi$ for a unitary character χ of B with $\chi|_Z = \psi$.

Indeed, let χ be a unitary character of B as in the fourth step. The same arguments as in the fifth step of the proof of Theorem 9.F.5 in combination with Lemma 9.F.13 show that $\mathrm{Ind}_B^\Gamma \chi$ is weakly contained in π.

• *Sixth step.* It follows from the fifth step and from Lemma 9.F.14 that π weakly contains the representation $\lambda_\psi = \mathrm{Ind}_Z^\Gamma \psi$. □

As in the case of GL_n (Corollary 9.F.6), we obtain a description of the maximal ideals in the C*-algebra $C^*_{\max}(\mathrm{SL}_n(\mathbf{K}))$ for an arbitrary infinite field $\mathbf{K}$ and for $n \geq 3$. The proof is a direct consequence of Theorem 9.F.15.

COROLLARY 9.F.16. *Let $\mathbf{K}$ be an infinite field and $\Gamma = \mathrm{SL}_n(\mathbf{K})$ for $n \geq 3$.*

Apart from the ideal $\mathrm{C^*ker}(1_\Gamma)$ of codimension 1, every maximal two-sided ideal of $C^*_{\max}(\Gamma)$ is of the form $\mathrm{C^*ker}(\mathrm{Ind}_Z^\Gamma \psi)$, for some $\psi \in \widehat{Z}$.

REMARK 9.F.17. It may happen that $\mathrm{C^*ker}(1_\Gamma)$ is the only maximal ideal in $C^*_{\max}(\Gamma)$; see Theorem 9.F.18 below and compare with Remark 9.F.7.

We are now ready to describe $\mathrm{Prim}(\Gamma)$ when Γ is amenable and $n \geq 3$.

THEOREM 9.F.18. *Let $\Gamma = \mathrm{SL}_n(\mathbf{K})$, where $n \geq 3$ and $\mathbf{K}$ is an infinite algebraic extension of a finite field.*

The map

$$\Phi : \{1_\Gamma\} \sqcup \widehat{Z} \to \mathrm{Prim}(\Gamma),$$

defined by $\Phi(1_\Gamma) = \mathrm{C^*ker}(1_\Gamma)$ and

$$\Phi(\psi) = \mathrm{C^*ker}(\mathrm{Ind}_Z^\Gamma \psi) \quad \textit{for } \psi \in \widehat{Z} \approx \widehat{\mu_n^{\mathbf{K}}},$$

is a bijection.

PROOF. We claim that the image of the map Φ is indeed contained in $\mathrm{Prim}(\Gamma)$. Indeed, observe first that the ideal $\mathrm{C^*ker}(1_\Gamma)$ is of course a primitive ideal.

Let $\psi \in \widehat{Z}$ and set $J = \mathrm{C^*ker}(\mathrm{Ind}_Z^\Gamma \psi)$. Choose a unitary character χ of B such that $\chi|_Z = \psi$. Let $\pi = \mathrm{Ind}_B^\Gamma \chi$ be the corresponding principal series representation. Since B is its own commensurator in Γ (the proof given in Lemma 3.F.1 is also valid for $\mathrm{SL}_n(\mathbf{K})$), the representation π is irreducible; moreover, the central character of π is ψ. On the one hand, since π is not trivial, $\mathrm{Ind}_Z^\Gamma \psi$ is weakly contained in π, by Theorem 9.F.15. On the other hand, since Γ is amenable, π is weakly contained in $\mathrm{Ind}_Z^\Gamma \psi$ (see the first step of the proof of Theorem 9.F.9). Therefore π and $\mathrm{Ind}_Z^\Gamma \psi$ are weakly equivalent. Therefore, $J = \mathrm{C^*ker}(\pi)$ and this shows that $J \in \mathrm{Prim}(\Gamma)$. So, the image of the map Φ is indeed contained in $\mathrm{Prim}(\Gamma)$.

We claim that the map Φ is bijective. Indeed, the arguments given for the second and third steps of the proof of Theorem 9.F.9 carry over without change in the case of $\mathrm{SL}_n(\mathbf{K})$. □

A consequence of Theorem 9.F.18 is that $\mathrm{Prim}(\mathrm{SL}_n(\mathbf{K}))$ is finite when $n \geq 3$ and $\mathbf{K}$ is an infinite algebraic extension of a finite field. The result for $\Gamma = \mathrm{PSL}_n(\mathbf{K})$ is even more striking.

COROLLARY 9.F.19. *Let* $\Gamma = \mathrm{PSL}_n(\mathbf{K})$, *where* $n \geq 3$ *and* $\mathbf{K}$ *is an infinite algebraic extension of a finite field.*

The primitive dual $\mathrm{Prim}(\Gamma)$ *consists of exactly two ideals:* $\{0\}$ *and* $\mathrm{C^*ker}(1_\Gamma)$.

REMARK 9.F.20. (1) Let $\Gamma = \mathrm{PSL}_n(\mathbf{K})$, where $\mathbf{K}$ is an infinite algebraic extension of a finite field and $n \geq 3$. It is worth rephrasing Corollary 9.F.19 in two ways:

- In terms of weak containment: every irreducible representation of Γ which is not the trivial representation 1_Γ is weakly equivalent to the regular representation λ_Γ.
- In terms of the ideal structure of the C*-algebra of Γ: the only nontrivial quotient of $C^*_{\max}(\Gamma) = C^*_\lambda(\Gamma)$ is the copy of $\mathbf{C}$ given by the trivial representation 1_Γ.

(2) Let Γ be a group, with $\Gamma \neq \{e\}$. Then $\mathrm{Prim}(\Gamma)$ has at least two elements: indeed, $\mathrm{C^*ker}(1_\Gamma)$ is a closed point in $\mathrm{Prim}(\Gamma)$; moreover, there exists an irreducible representation π of Γ which is distinct from 1_Γ; then $\mathrm{C^*ker}(\pi) \in \mathrm{Prim}(\Gamma)$ and $\mathrm{C^*ker}(\pi) \neq \mathrm{C^*ker}(1_\Gamma)$. The groups $\mathrm{PSL}_n(\mathbf{K})$ as in Corollary 9.F.19 or $\mathrm{SL}_n(\mathbf{K})$ as in Remark 9.F.21 are the only infinite groups Γ for which we know that $\mathrm{Prim}(\Gamma)$ consists of exactly two points.

(3) QUESTION. Does Corollary 9.F.19 also hold for $n = 2$?

REMARK 9.F.21. Note that there are cases for which $\Gamma = \mathrm{SL}_n(\mathbf{K})$ already has a trivial centre for $n \geq 3$; in this case, the same conclusion as in Corollary 9.F.19 holds: the primitive dual $\mathrm{Prim}(\Gamma)$ consists of exactly the two ideals $\{0\}$ and $\mathrm{C^*ker}(1_\Gamma)$. This is true in particular for $\mathrm{SL}_3(\overline{\mathbf{F}_3})$, since $\overline{\mathbf{F}_3}$ has no nontrivial cubic root of 1, and more generally for $\mathrm{SL}_p(\mathbf{K})$, whenever p is an odd prime and $\mathbf{K}$ an infinite algebraic extension of $\mathbf{F}_p$.

9.G. On the noninjectivity of the map from the dual to the primitive dual

Let G be a second-countable LC group which *is not* of type I. Then Corollary 7.F.4 shows that the map $\kappa^{\mathrm{d}}_{\mathrm{prim}} : \widehat{G} \twoheadrightarrow \mathrm{Prim}(G)$ of Sections 1.E and 8.B has fibers of uncountable cardinality. For several of the discrete groups Γ we have met so far, this can be checked with specific representations, as shown by the next examples.

EXAMPLE 9.G.1. (1) For the Heisenberg group $\Gamma = H(\mathbf{Z})$ over the integers, this was observed in Remark 9.B.9.

(2) For the affine group $\mathrm{Aff}(\mathbf{K})$ of an infinite field $\mathbf{K}$, see Remark 9.C.2.

(3) Let p be a prime and $\Gamma = \mathrm{BS}(1,p)$ be the Baumslag–Solitar group, as in Scetions 3.D and 9.D. Proposition 3.D.6 shows that there is an injective map Φ from the set of nonperiodic T_p-orbits in the solenoid $\mathbf{Sol}_p$ to $\widehat{\Gamma}$, where T_p is the transformation given by multiplication by p. There are uncountably many dense T_p-orbits (see the Remark 3.D.7). By Theorem 9.D.1 (see First Step of the proof),

every irreducible representation of Γ corresponding to a dense T_p-orbit maps under $\kappa^{\mathrm{d}}_{\mathrm{prim}}$ to the primitive ideal $\{0\}$.

(4) Let $\Gamma = \mathbf{Z} \ltimes \bigoplus_{k\in\mathbf{Z}} \mathbf{Z}/2\mathbf{Z}$ be the lamplighter group, as in Sections 3.E and 9.E. There are uncountably many dense T-orbits for the shift map T on $X = \prod_{k\in\mathbf{Z}} \mathbf{Z}/2\mathbf{Z}$. As in the case of the Baumslag–Solitar group, every irreducible representation of Γ corresponding to a dense T-orbit maps under $\kappa^{\mathrm{d}}_{\mathrm{prim}}$ to the primitive ideal $\{0\}$, by Theorem 9.E.1.

(5) Let $\Gamma = \mathrm{GL}_n(\mathbf{K})$ for an infinite field $\mathbf{K}$ and $n \geq 2$. Let ψ be a unitary character of the centre Z of Γ. Denote by X_ψ the set of unitary characters χ of the subgroup B of upper triangular matrices in Γ such that $\chi|_Z = \psi$. Observe that X_ψ is uncountable.

On the one hand, by Theorem 3.F.2, the principal series representations $\mathrm{Ind}_B^\Gamma \chi$ are all irreducible and mutually not equivalent. On the other hand, by Lemma 9.F.3 (see also Remark 9.F.4), the $\mathrm{Ind}_B^\Gamma \chi$'s are mutually weakly equivalent for $\chi \in X_\psi$. So, the $\mathrm{Ind}_B^\Gamma \chi$'s for $\chi \in X_\psi$ map to the same point in $\mathrm{Prim}(\Gamma)$.

EXAMPLE 9.G.2. Let $\Gamma = \mathrm{PSL}_2(\mathbf{Z})$ and $H \approx \mathbf{Z}$ be the subgroup which is the image in Γ of the subgroup $\begin{pmatrix} 1 & \mathbf{Z} \\ 0 & 1 \end{pmatrix}$ of $\mathrm{SL}_2(\mathbf{Z})$. For $\theta \in [0, 1[$, denote by $\chi_\theta \in \widehat{H}$ the unitary character defined by $\chi_\theta(n) = e^{2\pi i k\theta}$ for all $k \in \mathbf{Z}$, and set $\pi_\theta = \mathrm{Ind}_H^\Gamma \chi_\theta$. We claim that

(1) the regular representation λ of Γ is a direct integral $\lambda = \int_{\widehat{H}}^\oplus \pi_\theta d\theta$;
(2) all π_θ are irreducible and pairwise not equivalent to each other;
(3) all π_θ are weakly equivalent to each other and to λ as well; in particular, all π_θ have the same image in $\mathrm{Prim}(\Gamma)$.

To show (1), observe that, by induction in stages, $\lambda = \mathrm{Ind}_H^\Gamma \lambda_H$ for the regular representation λ_H of H. Let $d\theta$ denote the normalized Haar measure on $\widehat{H} \approx \mathbf{T}$; by the Plancherel Theorem A.7.6 and Example 1.G.10, λ_H is equivalent to the direct integral $\int_{\widehat{H}}^\oplus \chi_\theta d\theta$. It follows that

$$\lambda = \mathrm{Ind}_H^\Gamma \left(\int_{\widehat{H}}^\oplus \chi_\theta d\theta \right) = \int_{\widehat{H}}^\oplus \left(\mathrm{Ind}_H^\Gamma \chi_\theta \right) d\theta = \int_{\widehat{H}}^\oplus \pi_\theta d\theta.$$

To show (2), it suffices to prove that H is equal to its commensurator $\mathrm{Comm}_\Gamma(H)$ in Γ, by the Mackey–Shoda irreducibility and equivalence criteria (Theorems 1.F.12 and 1.F.17).

Let X be the set of equivalence classes of primitive vectors in $\mathbf{Z}^2$, where two primitive vectors x, y are equivalent if $y = \pm x$; the group Γ acts transitively on X, and H is the stabilizer of the class x_0 of $\begin{pmatrix} 1 \\ 0 \end{pmatrix}$. Moreover, $\{x_0\}$ is the unique finite orbit of H in X; hence, $H = \mathrm{Comm}_\Gamma(H)$.

Finally, since H is amenable, χ_θ is weakly contained in λ_H for every $\theta \in \widehat{H}$ by Theorem 1.C.3, hence $\pi_\theta = \mathrm{Ind}_H^\Gamma \chi_\theta$ is weakly contained in $\lambda = \int_{\widehat{H}}^\oplus \pi_\theta d\theta$ [**BeHV–08**, F.3.5]. As Γ is C*-simple (that is, the reduced C*-algebra $C^*_\lambda(\Gamma)$ is simple [**Harp–07**]), it follows that π_θ is weakly equivalent to λ for every $\theta \in \widehat{H}$ and Item (3) is proved.

EXAMPLE 9.G.3. Let G be a connected real Lie group that is simple, non-compact, with centre $\{1\}$, and let Γ be a lattice in G. Let π, π' be nonequivalent irreducible representations of G that are in the principal series of G; denote by $\pi|_\Gamma, \pi'|_\Gamma$ their restrictions to Γ. Then:

$\pi|_\Gamma$ and $\pi'|_\Gamma$ are irreducible and nonequivalent [**CoSt–91**];

$\pi|_\Gamma$ and $\pi'|_\Gamma$ are weakly equivalent to each other [**BeHa–94**]. (This was already cited for the particular case of $G = \mathrm{SL}_2(\mathbf{R})$ in the proof of Proposition 8.H.1.) In particular, there are fibers of uncountable cardinality in the projection $\widehat{\Gamma} \twoheadrightarrow \mathrm{Prim}(\Gamma)$.

EXAMPLE 9.G.4. Let Γ be a weakly branch group acting on the d-regular rooted tree, for some $d \geq 2$; see [**Grig–11**].

Assume that Γ is subexponentially bounded. Then the projection $\widehat{\Gamma} \twoheadrightarrow \mathrm{Prim}(\Gamma)$ has uncountable fibres. This covers several of the basic examples of groups acting on rooted trees, such as the first Grigorchuk group, Gupta–Sidki p-groups, and the Basilica group. See [**DuGr–17a**], and Corollary 1 in [**DuGr–17b**].

9.H. Borel comparison for duals of groups

It is a natural question to ask what are the possible spaces that can arise as duals of groups. We now address this question for duals of countable discrete groups, viewed as Borel spaces.

PROPOSITION 9.H.1. *The dual of a countable infinite discrete group is an uncountable Borel space.*

PROOF. Let Γ be a countable discrete group.

Assume first that Γ is not of type I. Then Γ has a factor representation π which is not of type I (Theorem 6.D.4). By Corollary 7.F.4, there are uncountably many pairwise nonequivalent irreducible representations of Γ which are weakly equivalent to π. In particular, $\widehat{\Gamma}$ is uncountable.

Now let Γ be of type I. Assume that $\widehat{\Gamma}$ is at most countable; so $\widehat{\Gamma} = \{\pi_k \mid k \in I\}$ for a subset I of $\mathbf{N}$. By Theorem 6.D.7, there exists a sequence of extended non-negative integers $(m_k)_{k \in I}$ such that the left regular representation λ_Γ is equivalent to a direct sum

$$\bigoplus_{k \in I} m_k \pi_k.$$

In particular, λ_Γ contains an irreducible subrepresentation of Γ. It follows from Proposition 7.A.4 that Γ is finite. $\square$

The easiest case to deal with is that of the abelian-by-finite groups. As a consequence of Proposition 9.H.1 and Theorem 8.F.3(vi), we have:

COROLLARY 9.H.2. *Let Γ be a countable infinite abelian-by-finite group.*

The dual $\widehat{\Gamma}$ is an uncountable standard Borel space, and in particular is Borel isomorphic to the dual $\widehat{\mathbf{Z}} = \mathbf{T}$.

Let X be a standard Borel space. An equivalence relation $\mathcal{R}$ on X is **Borel** if R is a Borel subset of $X \times X$, and **smooth** if there exists a Borel map $\varphi : X \to \mathbf{R}$ such that, for $(x, x') \in X \times X$, we have $(x, x') \in \mathcal{R}$ if and only if $\varphi(x) = \varphi(x')$.

Consider a countable infinite group Γ. Let $\mathcal{H}$ be an infinite-dimensional separable Hilbert space, and $\mathcal{U}(\mathcal{H})$ its unitary group, with the strong topology; recall that $\mathcal{U}(\mathcal{H})$ is a Polish group.

Denote by $\mathrm{Rep}_\infty(\Gamma)$ the space of representations of Γ in $\mathcal{H}$, equipped with the coarsest Borel structure such that the maps $\pi \mapsto \langle \pi(\gamma)\xi \mid \eta \rangle$ are Borel measurable, for all $\gamma \in \Gamma$ and $\xi, \eta \in \mathcal{H}_\pi$ (see Section 6.C).

Let $\mathrm{Irr}_\infty(\Gamma)$ be the subspace of $\mathrm{Rep}_\infty(\Gamma)$ consisting of the irreducible representations. In $\mathrm{Irr}_\infty(\Gamma)$, equivalence of representations is a Borel equivalence relation; we denote it here by $\simeq_\Gamma$. Observe that the quotient space $\mathrm{Irr}_\infty(\Gamma)/\simeq_\Gamma$ is the complement of the finite-dimensional part $\widehat{\Gamma}_{\mathrm{fd}}$ of the dual of Γ, viewed here with its Mackey–Borel structure.

The following result is a consequence of Glimm's Theorem 8.F.3 and Thoma's Theorem 7.D.1.

THEOREM 9.H.3. *For a group Γ, the following conditions are equivalent:*

(i) *Γ is not abelian-by-finite,*
(ii) *Γ has an infinite-dimensional irreducible representation, i.e., $\mathrm{Irr}_\infty(\Gamma) \neq \emptyset$,*
(iii) *the dual $\widehat{\Gamma}$ is not countably separated,*
(iv) *the equivalence relation $\simeq_\Gamma$ on the space $\mathrm{Irr}_\infty(\Gamma)$ is not smooth.*

Let $\mathcal{R}, \mathcal{S}$ be two Borel equivalence relations on Polish spaces X, Y respectively. Then $\mathcal{R}$ is **Borel reducible to $\mathcal{S}$**, and this is denoted by $\mathcal{R} \leq_B \mathcal{S}$, if there exists a Borel map $\varphi : X \to Y$ such that $(x, x') \in \mathcal{R}$ if and only if $(\varphi(x), \varphi(x')) \in \mathcal{S}$; intuitively, this means that $\mathcal{S}$ is not less complicated than $\mathcal{R}$. For example, a Borel equivalence relation $\mathcal{R}$ is smooth if and only if it is Borel reducible to the equality equivalence relation $\{(x, y) \in \mathbf{R} \times \mathbf{R} \mid x = y\}$ on $\mathbf{R}$. The relation $\mathcal{R}$ is **Borel equivalent to $\mathcal{S}$**, denoted by $\mathcal{R} \sim_B \mathcal{S}$, if $\mathcal{R} \leq_B \mathcal{S}$ and $\mathcal{S} \leq_B \mathcal{R}$.

Consider the equivalence relation E_0 on $2^{\mathbf{N}}$ defined for $\omega, \omega' \in 2^{\mathbf{N}}$ by $(\omega, \omega') \in E_0$ if $\omega(n) = \omega'(n)$ for all $n \in \mathbf{N}$ large enough; this relation E_0 is not smooth. For a Borel equivalence relation $\mathcal{R}$ on a Polish space X, it is a fundamental result of [**HaKL–90**], building up on previous results by Glimm [**Glim–61a, Glim–61b**] and Effros, that

- either $\mathcal{R}$ is smooth,
- or $E_0 \leq_B \mathcal{R}$.

In particular, $E_0 \leq_B \simeq_\Gamma$ for any countable group Γ that is not abelian-by-finite.

As mentioned in [**Thos–15**, Corollary 1.7], the following fact is a direct consequence of [**Mott–12**, Theorem 2.5], which is a more general result about equivalence relations induced by actions of Polish groups.

PROPOSITION 9.H.4. *Let Γ, Δ be two countable groups which are not abelian-by-finite.*

Then $\simeq_\Gamma \sim_B \simeq_\Delta$ if and only if the duals $\widehat{\Gamma}$ and $\widehat{\Delta}$ are Borel isomorphic.

Let us turn to *amenable* non type I groups. We will show that the unitary dual spaces of any two such groups are Borel isomorphic. This remarkable result (Theorem 9.H.9) is due to Thomas [**Thos–15**]. There are two main ingredients in its proof:

- the first one is a result of Sutherland (Theorem 9.H.7) showing that the dual space of every amenable non type I countable group is Borel reducible to the dual space of the lamplighter group; the main property of amenable groups used here is Dye's Theorem 5.F.1 about orbit equivalence of actions of such groups;

- the second ingredient is a result of Elliott (Theorem 9.H.8) showing that the dual space of a locally finite group is Borel reducible to the dual space of any non type I countable group.

We will need results from Section 5.D about cocycles for actions on probability spaces and the associated representations.

Let $(X, \mu) \curvearrowleft \Gamma$ be a measure class preserving action of a countable group Γ on a probability space (X, μ).

For a Hilbert space $\mathcal{K}$, we denote by $Z^1_\mu(X \times \Gamma, \mathcal{U}(\mathcal{K}))$ the space of cocycles $\alpha : X \times \Gamma \to \mathcal{U}(\mathcal{K})$ over (X, μ) with values in the unitary group of $\mathcal{K}$, where two such cocycles are identified if they agree on a set $X' \times \Gamma$ with $\mu(X') = 1$ (see Section 5.C). We will equip $Z^1_\mu(X \times \Gamma, \mathcal{U}(\mathcal{K}))$ with the coarsest Borel structure for which the maps

$$\alpha \to \int_X \langle \alpha(x, \gamma)\xi \mid \eta\rangle d\mu(x)$$

are measurable for all $\gamma \in \Gamma$ and $\xi, \eta \in \mathcal{K}$.

In what follows, a crucial role will be played by the fact that $Z^1_\mu(X \times \Gamma, \mathcal{U}(\mathcal{K}))$ only depends on the orbit equivalence class of the action $(X, \mu) \curvearrowleft \Gamma$ (see Section 5.F for this notion). For the notion of irreducible cocycles, see Definition 5.C.4.

PROPOSITION 9.H.5 (**Feldman–Moore**). *Let $(X, \mu) \curvearrowleft \Gamma_1$ and $(Y, \nu) \curvearrowleft \Gamma_2$ be orbit equivalent ergodic and essentially free actions of countable groups Γ_1 and Γ_2 by measure-preserving transformations on nonatomic probability spaces.*

For every Hilbert space $\mathcal{K}$, there exists a (natural) Borel isomorphism

$$f : Z^1_\mu(X \times \Gamma_1, \mathcal{U}(\mathcal{K})) \to Z^1_\nu(Y \times \Gamma_2, \mathcal{U}(\mathcal{K})),$$

which maps cohomologous cocycles to cohomologous ones and irreducible cocycles to irreducible ones.

For the proof, we refer to [**FeMo–77**, Theorem 5].

From now on, we will denote by $\mathcal{K}$ the standard Hilbert space of infinite countable dimension.

PROPOSITION 9.H.6. *Let Γ, G be countable groups, $(X, \mu) \curvearrowleft \Gamma$ a probability measure-preserving ergodic action of a countable group Γ, and $\varphi : \Gamma \to G$ a surjective homomorphism. Assume that the restriction of the action $(X, \mu) \curvearrowleft \Gamma$ to $\ker \varphi$ is ergodic. Let $(\pi, \mathcal{K})$ be a representation of G. Define a map*

$$\mathrm{Rep}_\infty(G) \to Z^1_\mu(X \times \Gamma, \mathcal{U}(\mathcal{K})), \qquad \pi \mapsto \alpha_\pi$$

by

$$\alpha_\pi(x, \gamma) = \pi(\varphi(\gamma)) \quad \textit{for} \quad x \in X, \ \gamma \in \Gamma$$

for $\pi \in \mathrm{Rep}_\infty(G)$.

(1) *Let $\pi \in \mathrm{Irr}_\infty(G)$; then α_π is an irreducible cocycle.*
(2) *Let $\pi_1, \pi_2 \in \mathrm{Irr}_\infty(G)$; then the cocycles α_{π_1} and α_{π_2} are cohomologous if and only if the representations π_1 and π_2 are equivalent.*

PROOF. The fact that α_π is a cocycle over (X, μ) with values in $\mathcal{U}(\mathcal{K})$ for every representation $(\pi, \mathcal{K})$ of G is immediate (compare also Example 5.B.3(4)).

Let $(\pi_1, \mathcal{K})$ and $(\pi_2, \mathcal{K})$ be representations of G. Let

$$X \to \mathcal{U}(\mathcal{K}), \qquad x \mapsto T(x)$$

be an intertwiner between α_{π_1} and α_{π_2}; we then have

$$\alpha_{\pi_1}(x,\gamma)T(x\gamma) = T(x)\alpha_{\pi_2}(x,\gamma),$$

that is,

$$(*) \qquad \pi_2(\varphi(\gamma))T(x\gamma) = T(x)\pi_1(\varphi(\gamma)),$$

for all $\gamma \in \Gamma$ and μ-almost all $x \in X$. In particular, we have $T(x\gamma) = T(x)$ for all $\gamma \in \ker\varphi$ and μ-almost all $x \in X$.

Let $(\xi_n)_n$ be a Hilbert basis of $\mathcal{K}$. Then, for any n and m, the measurable function

$$x \mapsto \langle T(x)\xi_n \mid \xi_m\rangle$$

is invariant under $\ker\varphi$ and hence μ-almost constant, since the action of $\ker\varphi$ on (X,μ) is ergodic. It follows that $x \mapsto T(x)$ is μ-almost constant.

To show (1), assume that π_1 is irreducible. Then, taking $\pi_2 = \pi_1$, it follows from Relation $(*)$ that $T(x)$ is a multiple of the identity for μ-almost all $x \in X$, since φ is surjective.

To show (2), assume that π_1 and π_2 are irreducible and not equivalent. Since φ is surjective, it follows again from Relation $(*)$ that $T(x) = 0$ for μ-almost all $x \in X$. So, α_{π_1} and α_{π_2} are not cohomologous.

The proof that, conversely, α_{π_1} and α_{π_2} are cohomologous if π_1 and π_2 are equivalent is immediate and left to the reader. □

In the sequel, Γ_0 will be the lamplighter group $\mathbf{Z} \wr (\mathbf{Z}/2\mathbf{Z})$. Recall that

$$\Gamma_0 = A \ltimes N \quad \text{with} \quad A = \mathbf{Z} \quad \text{and} \quad N = C_2^{(\mathbf{Z})},$$

where $C_2^{(\mathbf{N})}$ is the restricted product indexed by $\mathbf{Z}$ of copies of the cyclic group C_2 of order 2. Recall also that $\widehat{N}$ can be identified with $X = \{0,1\}^{\mathbf{Z}}$, the dual action $\mathbf{Z}$ on $\widehat{N}$ being given by the shift transformation T on X. Moreover, the normalized Haar measure on $\widehat{N} = X$ is the measure $\mu = \bigotimes^{\mathbf{Z}} p$, where p is the uniform probability measure on $\{0,1\}$. Observe that the action of A on (X,μ) is essentially free and ergodic. The following theorem is from [**Suth–83**].

THEOREM 9.H.7 (**Sutherland**). *Let Γ be a countable infinite amenable group which is not abelian-by-finite.*

There exists a Borel map

$$\Phi : \mathrm{Rep}_\infty(\Gamma) \to Z^1_\mu(X \times A, \mathcal{U}(\mathcal{K}))$$

with the following properties:

(1) *for every $\pi \in \mathrm{Irr}_\infty(\Gamma)$, the cocycle $\Phi(\pi)$ is irreducible;*
(2) *for $\pi_1, \pi_2 \in \mathrm{Irr}_\infty(\Gamma)$, the cocycles $\Phi(\pi_1)$ and $\Phi(\pi_2)$ are cohomologous if and only if the representations π_1 and π_2 are equivalent.*

In particular, we have

$$\simeq_\Gamma \le_B \simeq_{\Gamma_0} .$$

PROOF. Let $(Z,m) \curvearrowleft \Gamma$ be a measure-preserving ergodic and essentially free action of Γ on a nonatomic probability measure space (Z,m); an example of such an action is the Bernoulli shift on $Z = \{0,1\}^\Gamma$ equipped with the measure $m = \otimes^\Gamma p$, with p as above.

Set

$$(Y, \nu) := \prod_{k \in \mathbf{Z}} (Z, m)$$

and let Γ act diagonally on (Y, ν); also let $\mathbf{Z}$ act on (Y, ν) by shifting the coordinates. In this way, we obtain an essentially free ergodic action of $A \times \mathbf{Z} \times \Gamma$ on $(X \times Y, \mu \otimes \nu)$, which remains ergodic when restricted to $A \times \mathbf{Z}$.

Since the groups $A \times \mathbf{Z} \times \Gamma$ and A are amenable, their actions on $(X \times Y, \mu \otimes \nu)$ and (X, μ) are orbit equivalent, by Dye's Theorem 5.F.1. By Proposition 9.H.5, there is a Borel isomorphism

$$f : Z^1_{\mu \times \nu}((X \times Y) \times (A \times \mathbf{Z} \times \Gamma), \mathcal{U}(\mathcal{K})) \to Z^1_\mu(X \times A, \mathcal{U}(\mathcal{K})),$$

which preserves the equivalence and the irreducibility of cocycles. Now define a Borel map

$$\Phi : \operatorname{Rep}_\infty(\Gamma) \to Z^1_\mu(X \times A, \mathcal{U}(\mathcal{K})), \qquad \pi \mapsto f(\alpha_\pi),$$

with $\alpha_\pi \in Z^1_{\mu \times \nu}((X \times Y) \times (A \times \mathbf{Z} \times \Gamma), \mathcal{U}(\mathcal{K}))$ defined as in Proposition 9.H.6.

Let $\pi \in \operatorname{Irr}_\infty(\Gamma)$. Then, by Proposition 9.H.6, α_π is an irreducible cocycle over $(X \times Y, \mu \otimes \nu)$, since $A \times \mathbf{Z}$ acts ergodically on $(X \times Y, \mu \otimes \nu)$. Hence, $f(\alpha_\pi)$ is an irreducible cocycle over (X, μ), by Proposition 9.H.5. This establishes Property (1).

Let $\pi_1, \pi_2 \in \operatorname{Irr}_\infty(\Gamma)$. By Proposition 9.H.6, the cocycles $\Phi(\pi_1)$ and $\Phi(\pi_2)$ are cohomologous if and only if π_1 and π_2 are equivalent. This establishes Property (2).

To show the last claim, let $\pi \in \operatorname{Irr}_\infty(\Gamma)$. By Theorem 5.D.4, the representation $\Psi(\pi)$ of $\Gamma_0 = A \ltimes N$ on $L^2(X, \mu, \mathcal{K})$ associated to the cocycle $\Phi(\pi)$ is irreducible. Again by Theorem 5.D.4, the resulting Borel map

$$\Psi : \operatorname{Irr}_\infty(\Gamma) \to \operatorname{Irr}_\infty(\Gamma_0)$$

is such that $\Psi(\pi_1)$ and $\Psi(\pi_2)$ are equivalent if and only if π_1 and π_2 are equivalent. So, $\simeq_\Gamma \leq_B \simeq_{\Gamma_0}$. □

A C*-algebra $\mathcal{A}$ is said to be **approximately finite-dimensional** if $\mathcal{A}$ is the closure of the union $\bigcup_n \mathcal{A}_n$ of an increasing sequence of finite-dimensional C*-subalgebras $\mathcal{A}_n$. The next result is from [**Elli–77**].

THEOREM 9.H.8 (**Elliott**). *Let Γ_1 be a countable locally finite group and Γ a countable non type I group. Then*

$$\simeq_{\Gamma_1} \leq_B \simeq_\Gamma .$$

PROOF. Since Γ_1 is a countable locally finite group, we can write Γ_1 as a union

$$\Gamma_1 = \bigcup_n H_n$$

for an increasing sequence of finite subgroups H_n. For the group algebra $\mathbf{C}[\Gamma_1]$, we have

$$\mathbf{C}[\Gamma_1] = \bigcup_n \mathbf{C}[H_n].$$

For every n, we can view the C*-algebra $C^*_{\max}(H_n) = \mathbf{C}[H_n]$ of H_n as a C*-subalgebra of the C*-algebra $C^*_{\max}(\Gamma_1)$ of Γ_1, by Theorem 8.C.12. So, $\bigcup_n C^*_{\max}(H_n)$ is dense in $C^*_{\max}(\Gamma_1)$. This shows that $C^*_{\max}(\Gamma_1)$ is an approximately finite-dimensional C*-algebra.

Since $C^*_{\max}(\Gamma)$ is a separable non type I C*-algebra, the claim follows from a general result [**Elli–77**] about the spectrum of approximately finite-dimensional C*-algebras. □

We are now ready to give the answer [**Thos–15**, Corollary 6.2] to the question about the relative Borel complexity of dual spaces of amenable groups.

THEOREM 9.H.9 (**Thomas**). *Let* Γ *and* Λ *be countable amenable groups, both not abelian-by-finite.*

Then we have

$$\simeq_\Gamma \ \sim_B \ \simeq_\Lambda .$$

Equivalently, the dual space $\widehat{\Gamma}$ *and* $\widehat{\Lambda}$ *are Borel isomorphic.*

PROOF. Let

$$\Gamma_1 \ = \ \mathrm{Sym}(3)^{(\mathbf{N})}$$

be the restricted product of a countable infinite number of copies of the symmetric group of order 6. Since Γ_1 is a countable locally finite group, Theorem 9.H.8 shows that $\simeq_{\Gamma_1} \ \leq_B \ \simeq_\Gamma$ for every group Γ which is not of type I.

Recall that $\Gamma_0 = A \ltimes N$ is the lamplighter group, with $A = \mathbf{Z}$ and $N = C_2^{(\mathbf{Z})}$. In view of Theorem 9.H.7, it suffices to show that there is a Borel map

$$\Psi : \ Z^1_\mu(X \times A, \mathcal{U}(\mathcal{K})) \to \mathrm{Rep}_\infty(\Gamma_1)$$

with the following properties:

- if $\alpha \in Z^1_\mu(X \times A, \mathcal{U}(\mathcal{K}))$ is an irreducible cocycle, then $\Psi(\alpha)$ is an irreducible representation of Γ_1;
- if $\alpha_1, \alpha_2 \in Z^1_\mu(X \times \Gamma_0, \mathcal{U}(\mathcal{K}))$ are irreducible cocycles, then the representations $\Psi(\alpha_1)$ and $\Psi(\alpha_2)$ are equivalent if and only if the cocycles α_1 and α_2 are cohomologous.

Since $\mathrm{Sym}(3) \simeq C_2 \ltimes C_3$, for copies of the cyclic groups C_2 and C_3 of order 2 and 3, we can write Γ_1 as a semi-direct product

$$\Gamma_1 \ = \ H \ltimes M, \qquad \text{for} \qquad H = C_2^{(\mathbf{N})} \quad \text{and} \qquad M = C_3^{(\mathbf{N})}.$$

The dual group $\widehat{M}$ of M can be identified with the direct product $Z = C_3^{\mathbf{N}}$. For a generator s of C_3, let

$$Y \ := \ \{s, s^2\}^{\mathbf{N}} \subset Z$$

be equipped with the probability measure $\nu \ = \ \bigotimes^{\mathbf{N}} p$, where p is the uniform measure on $\{s, s^2\}$. Then Y is H-invariant and the measure-preserving action $(Y, \nu) \curvearrowleft H$ is essentially free and ergodic. The actions $(Y, \nu) \curvearrowleft H$ and $(X, \mu) \curvearrowleft A$ are orbit equivalent and hence, by Proposition 9.H.5, there exists a Borel isomorphism

$$f : \ Z^1_\mu(X \times A, \mathcal{U}(\mathcal{K})) \to Z^1_\nu(Y \times H, \mathcal{U}(\mathcal{K})),$$

which preserves equivalence and irreducibility of cocycles.

For $\alpha \in Z^1_\mu(X \times A, \mathcal{U}(\mathcal{K}))$, denote by $\Psi(\alpha)$ the representation of $\Gamma_1 = H \ltimes M$ associated to the cocycle $f(\alpha)$ as in Section 5.C. By Theorem 5.D.4, the map

$$\Psi : \ Z^1_\mu(X \times A, \mathcal{U}(\mathcal{K})) \to \mathrm{Rep}_\infty(\Gamma_1)$$

has all the required properties. □

Little is known about the Borel complexity of duals of nonamenable groups. Following [**Thos–15**], define a countable group Γ to be **representation universal** if $\simeq_\Lambda$ is Borel reducible to $\simeq_\Gamma$, for every countable group Λ.

Let F_∞ be the free group on a countable infinite set of generators. It is a simple observation that F_∞ is representation universal; indeed, if $p : F_\infty \to \Lambda$ is a surjective group homomorphism, then the natural map $\pi \mapsto \pi \circ p$ from $\mathrm{Irr}_\infty(\Lambda)$ to $\mathrm{Irr}_\infty(F_\infty)$ is a Borel reduction from $\simeq_\Lambda$ to $\simeq_{F_\infty}$. As was shown in [**Thos–15**, Theorem 1.10], the free group F_2 group on 2 generators is also representation universal.

The following question is open.

QUESTION 9.H.10. Does there exist a countable infinite group Γ which is not abelian-by-finite and such that $\simeq_\Gamma \not\sim_B \simeq_{F_\infty}$? Equivalently, which is not abelian-by-finite and such that $\widehat{\Gamma}$ is not Borel isomorphic to $\widehat{F_\infty}$?

CHAPTER 10

Normal quasi-dual and characters

Let G be a second-countable LC group. In general, the primitive dual $\mathrm{Prim}(G)$ *is a far more accessible object than $\widehat{G}$. Indeed,* $\mathrm{Prim}(G)$ *is a standard Borel space (Theorem* 8.A.25*), whereas the dual $\widehat{G}$ is standard if and only if G is of type I (Theorem* 8.F.3*).*

With the ultimate goal to parametrize the primitive dual $\mathrm{Prim}(G)$*, we introduce the notions of a traceable representation (Section* 10.C*) and of a normal factor representation of G (Section* 10.D*). The normal quasi-dual* $\mathrm{QD}(G)_{\mathrm{norm}}$ *of G, which is defined as the set of quasi-equivalence classes of normal factor representations, is a standard Borel subspace of the quasi-dual* $\mathrm{QD}(G)$*; see Theorem* 10.D.7*. We will see in Section* 11.D *that, for some classes of groups G, the primitive dual* $\mathrm{Prim}(G)$ *is naturally in bijection with* $\mathrm{QD}(G)_{\mathrm{norm}}$.

Normal factor representations of G can be described in terms of characters. Characters of G are (possibly infinite) traces on the maximal C^*-algebra $C^*_{\mathrm{max}}(G)$ of G, with an appropriate extremality property. We will show in Section 10.D *that there is a natural bijective map from* $\mathrm{QD}(G)_{\mathrm{norm}}$ *onto the space* $\mathrm{Char}(G)$ *of equivalence classes of characters of G, where two characters are equivalent if they are proportional to each other.*

A detailed account on Hilbert algebras is given in Section 10.A*. Such an algebra is associated with every lower semi-continuous trace on a C^*-algebra (Proposition* 10.A.12*), a fact which is crucial in order to show the surjectivity of the map* $\mathrm{QD}(G)_{\mathrm{norm}} \to \mathrm{Char}(G)$*. Hilbert algebras also appear in the construction of the standard representation of a semi-finite von Neumann algebra (Section* 10.B*), as well as in connection with traces on general topological groups (Section* 11.B*).*

10.A. Traces and Hilbert algebras

Let A be a C*-algebra. Recall that traces on A have been introduced in Section 7.B. The traces which appear in this chapter will be assumed to be semi-finite; moreover, depending on whether A is a general C*-algebra or a von Neumann algebra, we will require such traces to be lower semi-continuous or normal. We will usually denote by $t, t', \dots$ traces on a general C*-algebra and by $\tau, \tau', \dots$ traces on a von Neumann algebra.

DEFINITION 10.A.1. Let A be a C*-algebra and $t : A_+ \to [0, \infty]$ a trace on A. The set

$$\mathfrak{n}_t := \{x \in A \mid t(x^*x) < \infty\}$$

is a selfadjoint two-sided ideal of A. The self-adjoint two-sided ideal

$$\mathfrak{m}_t = \mathfrak{n}_t^2$$

is the set of linear combinations of elements in $\mathfrak{m}_{t,+} = \mathfrak{m}_t \cap A_+$, and

$$\mathfrak{m}_{t,+} = \{x \in A_+ \mid t(x) < \infty\}.$$

The two-sided ideal $\mathfrak{m}_t$ is the **ideal of definition** of t. The trace t extends from $\mathfrak{m}_{t,+}$ to a linear functional $\mathfrak{m}_t \to \mathbf{C}$, denoted by t again, and

$$\begin{aligned} t(x^*) &= \overline{t(x)} && \text{for all } x \in \mathfrak{m}_t, \\ t(xy) &= t(yx) && \text{for all } x \in \mathfrak{m}_t \text{ and } y \in A, \\ t(xy) &= t(yx) && \text{for all } x, y \in \mathfrak{n}_t. \end{aligned}$$

(See [**Dixm–C***, 6.1.1].)

There is a construction that associates to every lower semi-continuous semi-finite trace t on A a representation λ_t of A such that the generated von Neumann algebra $\lambda_t(A)''$ is semi-finite; see Proposition 10.A.12. The construction involves basic objects that are Hilbert algebras. We give a complete account on two main facts concerning these algebras, i.e., the Commutation Theorem and the existence of a canonical trace, following [**Dixm–vN**, Chap. I, § 5]; see Theorems 10.A.4 and 10.A.8.

Definition 10.A.2. A **Hilbert algebra** is a complex involutive algebra $\mathcal{A}$, equipped with a scalar product $\langle \cdot \mid \cdot \rangle$, with the following properties:

(1) $\langle x \mid y \rangle = \langle y^* \mid x^* \rangle$ for all $x, y \in \mathcal{A}$;
(2) $\langle xy \mid z \rangle = \langle y \mid x^* z \rangle$ for all $x, y, z \in \mathcal{A}$;
(3) for every $x \in \mathcal{A}$, the map $\mathcal{A} \to \mathcal{A}$, $y \mapsto xy$ is continuous;
(4) the set $\{xy \mid x, y \in \mathcal{A}\}$ is total in $\mathcal{A}$.

Let $\mathcal{H}$ be the Hilbert space completion of $\mathcal{A}$. It follows from (1) to (4) that:

(5) the map $x \mapsto x^*$ from $\mathcal{A}$ to $\mathcal{A}$ extends to an involutive conjugate linear isometry $J : \mathcal{H} \to \mathcal{H}$;
(6) for every $x \in \mathcal{A}$, the maps defined on $\mathcal{A}$ by $y \mapsto xy$ and $y \mapsto yx$ extend to bounded linear operators $\lambda(x)$ and $\rho(x)$ on $\mathcal{H}$, and $x \mapsto \lambda(x)$ [respectively $x \mapsto \rho(x)$] is a nondegenerate $*$-representation of the involutive algebra $\mathcal{A}$ [respectively of the opposite algebra of $\mathcal{A}$];
(7) $\lambda(x)\rho(y) = \rho(y)\lambda(x)$ for all $x, y \in \mathcal{A}$;
(8) $J\lambda(x)J = \rho(x^*)$, and $J\rho(x)J = \lambda(x^*)$ for all $x \in \mathcal{A}$.

Let $\lambda(\mathcal{A})''$ and $\rho(\mathcal{A})''$ be the von Neumann subalgebras of $\mathcal{L}(\mathcal{H})$ generated by $\{\lambda(x) \mid x \in \mathcal{A}\}$ and $\{\rho(x) \mid x \in \mathcal{A}\}$ respectively. It is clear that $\lambda(\mathcal{A})'' \subset \rho(\mathcal{A})'$ and $\rho(\mathcal{A})'' \subset \lambda(\mathcal{A})'$. Theorem 10.A.4 below shows that these von Neumann algebras are in fact each other's commutants. The proof is based on properties of the so-called bounded elements in the Hilbert space $\mathcal{H}$, that we introduce now.

An element $a \in \mathcal{H}$ is said to be **bounded** if the map $\mathcal{A} \to \mathcal{H}$, $x \mapsto \rho(x)a$ extends from $\mathcal{A}$ to a bounded operator on $\mathcal{H}$; when this is the case, this extension is a uniquely determined operator, denoted by $\lambda(a)$. Denote by $\mathcal{A}_b$ the set of bounded elements in $\mathcal{H}$; it is clear that $\mathcal{A}_b$ is a linear subspace of $\mathcal{H}$. Observe that $\mathcal{A} \subset \mathcal{A}_b$ and that the map $\lambda : \mathcal{A}_b \to \mathcal{L}(\mathcal{H})$ coincides on $\mathcal{A}$ with the representation λ of $\mathcal{A}$ introduced above. For more on $\mathcal{A}_b$, see Remark 10.A.5.

Lemma 10.A.3. *Let $\mathcal{A}$ be a Hilbert algebra, and let $\mathcal{H}, J, \lambda, \rho, \mathcal{A}_b$ be as above.*

(1) *The identity operator $I \in \mathcal{L}(\mathcal{H})$ is in the closure of both the involutive algebras $\lambda(\mathcal{A})$ and $\rho(\mathcal{A})$, for the strong operator topology;*

(2) *the linear map* $a \mapsto \lambda(a)$ *is injective on* $\mathcal{A}_b$;
(3) *for* $a \in \mathcal{A}_b$, *we have* $Ja \in \mathcal{A}_b$ *and* $\lambda(Ja) = \lambda(a)^*$;
(4) *for* $a \in \mathcal{A}_b$, *there exists a unique bounded operator on* $\mathcal{H}$, *denoted* $\rho(a)$, *which extends the map* $x \mapsto \lambda(x)a$ *from* $\mathcal{A}$ *to* $\mathcal{H}$; *moreover,* $\rho(a) = J\lambda(a)^*J$;
(5) *the set* $\lambda(\mathcal{A}_b)$ *[respectively* $\rho(\mathcal{A}_b)$*] is a two-sided ideal of* $\rho(\mathcal{A})'$ *[respectively* $\lambda(\mathcal{A})'$*]; more precisely, for* $a \in \mathcal{A}_b$ *and* $T \in \rho(\mathcal{A})'$ *[respectively* $\lambda(\mathcal{A})'$*] we have* $Ta \in \mathcal{A}_b$, $JT^*Ja \in \mathcal{A}_b$, *and*

$$T\lambda(a) = \lambda(Ta) \quad \text{and} \quad \lambda(a)T = \lambda(JT^*Ja)$$

[respectively, $T\rho(a) = \rho(Ta)$ *and* $\rho(a)T = \rho(JT^*Ja)$*].*

PROOF. (1) By Property (4) in the definition of a Hilbert algebra, the linear span of $\lambda(\mathcal{A})\mathcal{H}$ is dense in $\mathcal{H}$. By von Neumann's bicommutant theorem (Theorem A.11.1), $\lambda(\mathcal{A})''$ is the closure of $\lambda(\mathcal{A})$ for the strong operator topology; since $I \in \lambda(\mathcal{A})''$, this proves the claim for $\lambda(\mathcal{A})$. The proof for $\rho(\mathcal{A})$ is similar.

(2) Let $a \in \mathcal{A}_b$ be such that $\lambda(a) = 0$. Then $\rho(x)a = 0$ for all $x \in \mathcal{A}$ and it follows from (1) that $a = 0$.

(3) For $a \in \mathcal{A}_b$ and $x, y \in \mathcal{A}$, we have

$$\begin{aligned}\langle \lambda(a)x \mid y\rangle &= \langle \rho(x)a \mid y\rangle = \langle a \mid \rho(x)^*y\rangle \\ &= \langle a \mid yx^*\rangle = \langle xy^* \mid Ja\rangle \\ &= \langle \rho(y^*)x \mid Ja\rangle,\end{aligned}$$

and hence $\lambda(a)^*y = \rho(y)Ja$, showing that $Ja \in \mathcal{A}_b$ and $\lambda(Ja) = \lambda(a)^*$.

(4) Using (3), we have, for $a \in \mathcal{A}_b$ and $x \in \mathcal{A}$,

$$\lambda(x)a = J\rho(x^*)Ja = J\lambda(Ja)x^* = J\lambda(a)^*x^* = J\lambda(a)^*Jx,$$

showing that $J\lambda(a)^*J$ coincides with $x \mapsto \lambda(x)a$ on $\mathcal{A}$.

(5) For $a \in \mathcal{A}_b$ and $x, y \in \mathcal{A}$, we have

$$\lambda(a)\rho(y)x = \lambda(a)(yx) = \rho(yx)a = \rho(y)\rho(x)a = \rho(y)\lambda(a)x,$$

showing that $\lambda(\mathcal{A}_b) \subset \rho(\mathcal{A})'$.

Let $T \in \rho(\mathcal{A})'$, $a \in \mathcal{A}_b$. For $x \in \mathcal{A}$, we have

$$\rho(x)Ta = T\rho(x)a = T\lambda(a)x;$$

this shows that $Ta \in \mathcal{A}_b$ and that $\lambda(Ta) = T\lambda(a)$. Since $T^* \in \rho(\mathcal{A})'$, using (3) twice, we therefore have

$$\lambda(a)T = (T^*\lambda(a)^*)^* = (T^*\lambda(Ja))^* = \lambda(T^*Ja)^* = \lambda(JT^*Ja).$$

The claim for $\rho(\mathcal{A}_b)$ is proved in a similar way. □

We are ready to prove the Commutation Theorem for a Hilbert algebra.

THEOREM 10.A.4 (**Commutation Theorem**). *Let* $\mathcal{A}$ *be a Hilbert algebra,* $\mathcal{H}$ *its completion, and* $\lambda(\mathcal{A})''$ *and* $\rho(\mathcal{A})''$ *the von Neumann subalgebras of* $\mathcal{L}(\mathcal{H})$ *generated by the left and right multiplications by elements of* $\mathcal{A}$. *We have*

$$\lambda(\mathcal{A})' = \rho(\mathcal{A})'' \quad \text{and} \quad \rho(\mathcal{A})' = \lambda(\mathcal{A})''.$$

PROOF. • *First step.* We claim that $\lambda(\mathcal{A}_b)$ is dense in $\rho(\mathcal{A})'$ and that $\rho(\mathcal{A}_b)$ is dense in $\lambda(\mathcal{A})'$ for the strong operator topology.

Indeed, let $T \in \rho(\mathcal{A})'$. Since I is in the strong closure of $\lambda(\mathcal{A})'$ by Lemma 10.A.3(1), there exists a sequence $(T_n)_{n\geq 1}$ in $\lambda(\mathcal{A})'$ converging strongly to I; then the sequence $(TT_n)_{n\geq 1}$ converges strongly to T. By Lemma 10.A.3 (5), $TT_n \in \lambda(\mathcal{A}_b)$ for all $n \geq 1$, and it follows that T is in the strong closure of $\lambda(\mathcal{A}_b)$.

The claim for $\rho(\mathcal{A}_b)$ is proved similarly.

• *Second step.* We claim that, for $a, b \in \mathcal{A}_b$, we have $\lambda(a)b = \rho(b)a$. Indeed, for $x \in \mathcal{A}$, we have, using Lemma 10.A.3(3) and (4),

$$\begin{aligned}\langle \lambda(a)b \mid x\rangle &= \langle b \mid \lambda(a)^*x\rangle = \langle b \mid \lambda(Ja)x\rangle = \langle b \mid \rho(x)Ja\rangle \\ &= \langle \rho(x^*)b \mid Ja\rangle = \langle J\lambda(x)Jb \mid Ja\rangle = \langle a \mid \lambda(x)Jb\rangle \\ &= \langle a \mid \lambda(x)b^*\rangle = \langle a \mid \rho(b)^*x\rangle = \langle \rho(b)a \mid x\rangle.\end{aligned}$$

• *Third step.* We claim that $\lambda(\mathcal{A})' = \rho(\mathcal{A})''$ and $\rho(\mathcal{A})' = \lambda(\mathcal{A})''$.

As mentioned earlier, we have $\rho(\mathcal{A})'' \subset \lambda(\mathcal{A})'$. In order to show that $\lambda(\mathcal{A})' \subset \rho(\mathcal{A})''$, we have to prove that $ST = TS$ for $S \in \lambda(\mathcal{A})'$ and $T \in \rho(\mathcal{A})'$. By the first step, it suffices to show that $ST = TS$ for $S = \rho(a)$ and $T = \lambda(b)$, with $a, b \in \mathcal{A}_b$. Using Lemma 10.A.3(5) and the second step, we have for every $x \in \mathcal{A}$,

$$STx = \rho(a)\lambda(b)x = \rho(a)\rho(x)b = \rho(\rho(a)x)b = \lambda(b)\rho(a)x = TSx.$$

The proof that $\rho(\mathcal{A})' = \lambda(\mathcal{A})''$ is similar. □

REMARK 10.A.5. The set $\mathcal{A}_b$ of bounded elements of a Hilbert algebra $\mathcal{A}$ is a linear subspace of the Hilbert space completion $\mathcal{H}$ of $\mathcal{A}$; moreover, for $a, b \in \mathcal{A}_b$, we have $\lambda(a)b \in \mathcal{A}_b$ and $Ja \in \mathcal{A}_b$ (see Lemma 10.A.3). We define maps

$$\mathcal{A}_b \times \mathcal{A}_b \to \mathcal{A}_b, \ \ (a, b) \mapsto ab \quad \text{and} \quad \mathcal{A}_b \to \mathcal{A}_b, \ \ a \mapsto a^*$$

by $ab := \lambda(a)b$ and $a^* := Ja$. These maps extend the multiplication law and the involution of $\mathcal{A}$ and turn $\mathcal{A}_b$ into a complex involutive algebra. Indeed, $(a, b) \mapsto ab$ is clearly bilinear and $a \mapsto a^*$ is an anti-linear involution; moreover, we have $\lambda(a)b = \rho(b)a$ for $a, b \in \mathcal{A}_b$ (see the second step of the proof of Theorem 10.A.4). Therefore for $a, b, c \in \mathcal{A}_b$, we have

$$a(bc) = \lambda(a)(\rho(c)b) = \rho(c)(\lambda(a)b) = (ab)c$$

and

$$a^*b^* = \lambda(Ja)(Jb) = J(J\lambda(Ja)J)(b) = J\rho(a)(b) = J(ba) = (ba)^*.$$

One checks that $\mathcal{A}_b$, endowed with the scalar product induced from $\mathcal{H}$, is a Hilbert algebra which contains $\mathcal{A}$ as a $*$-subalgebra. Every element in $\mathcal{H}$ which is bounded with respect to the Hilbert algebra $\mathcal{A}_b$ is bounded with respect to the Hilbert algebra $\mathcal{A}$ and therefore belongs to $\mathcal{A}_b$, that is, we have $(\mathcal{A}_b)_b = \mathcal{A}_b$. A Hilbert algebra with this property is called a **full Hilbert algebra**.

The next theorem establishes that the von Neumann algebra $\lambda(\mathcal{A})''$ associated with a Hilbert algebra $\mathcal{A}$ carries a canonical trace, called the natural trace associated with $\mathcal{A}$; and similarly for $\rho(\mathcal{A})''$. For a proof, we will need the following two elementary lemmas.

LEMMA 10.A.6. *Let $\mathcal{M} \subset \mathcal{L}(\mathcal{H})$ be a von Neumann algebra and $S, T \in \mathcal{M}_+$ with $S \leq T$.*

There exists a unique operator $X \in \mathcal{M}$ with $X|_{T(\mathcal{H})^\perp} = 0$ such that $S^{1/2} = XT^{1/2}$. Moreover, we have $\|X\| \leq 1$.

PROOF. For every $\xi \in \mathcal{H}$, we have

$$\|S^{1/2}\xi\|^2 = \langle S\xi \mid \xi\rangle \leq \langle T\xi \mid \xi\rangle = \|T^{1/2}\xi\|^2;$$

the map $T^{1/2}(\mathcal{H}) \to \mathcal{H}$, $T^{1/2}\xi \to S^{1/2}\xi$, is therefore well defined and continuous, of norm at most 1. It has a linear continuous extension $Y : \overline{T^{1/2}(\mathcal{H})} \to \mathcal{H}$ and we have $S^{1/2} = YT^{1/2}$.

Observe that we have $\ker T = \ker T^{1/2}$, and therefore

$$\overline{T(\mathcal{H})} = (\ker T)^\perp = (\ker T^{1/2})^\perp = \overline{T^{1/2}(\mathcal{H})}.$$

Define $X \in \mathcal{L}(\mathcal{H})$ by $X = Y$ on $\overline{T(\mathcal{H})}$ and $X = 0$ on $T(\mathcal{H})^\perp$. Then $S^{1/2} = XT^{1/2}$.

The uniqueness of $X \in \mathcal{L}(\mathcal{H})$ with the previous properties is clear; it follows that $UXU^{-1} = X$ for every unitary operator $U \in \mathcal{M}'$. Therefore $X \in \mathcal{M}'' = \mathcal{M}$. Since $\|Y\| \leq 1$, it is clear that $\|X\| \leq 1$. □

LEMMA 10.A.7. *Let $(T_i)_i$ in $\mathcal{L}(\mathcal{H})_+$ be a net of positive operators converging to $T \in \mathcal{L}(\mathcal{H})_+$ in the strong operator topology. Assume that $\sup_i \|T_i\| \leq \|T\|$.*

Then $(T_i^{1/2})_i$ converges to $T^{1/2}$ in the strong operator topology.

PROOF. We claim first that $(P(T_i))_i$ converges to $P(T)$ in the strong operator topology for every polynomial $P \in \mathbf{C}[X]$. To show this, it suffices to prove that $(T_i^n)_i$ converges to T^n for every integer $n \geq 0$.

We proceed by induction on n; the case $n = 0$ is trivial and the case $n = 1$ is true by assumption. Assume that the claim is true for some $n \geq 1$. Let $\xi \in \mathcal{H}$. Then $\lim_i T_i^n\xi = T^n\xi$ and $\lim_i T_i(T^n\xi) = T(T^n\xi)$. Since

$$\begin{aligned}\|T_i^{n+1}\xi - T^{n+1}\xi\| &\leq \|T_i(T_i^n\xi - T^n\xi)\| + \|T_i(T^n\xi) - T(T^n\xi)\| \\ &\leq \|T\|\|T_i^n\xi - T^n\xi\| + \|T_i(T^n\xi) - T(T^n\xi)\|,\end{aligned}$$

the claim is also true for $n + 1$.

Observe that, for every $S \in \mathcal{L}(\mathcal{H})_+$ with $\|S\| \leq \|T\|$, the C*-algebra homomorphism

$$C([0, \|T\|]) \to \mathcal{L}(\mathcal{H}), \qquad f \mapsto f(S)$$

defined by functional calculus [**Dixm–C***, § 1.5] is norm decreasing.

Fix $\varepsilon > 0$. Let $P \in \mathbf{C}[X]$ be such that $\sup_{t\in[0,\|T\|]} |P(t) - t^{1/2}| \leq \varepsilon$; then

$$\|P(T) - T^{1/2}\| \leq \varepsilon \quad \text{and} \quad \|P(T_i) - T_i^{1/2}\| \leq \varepsilon \quad \text{for all } i.$$

Let $\xi \in \mathcal{H}$. Since $\lim_i P(T_i)\xi = P(T)\xi$, we have $\|T_i^{1/2}\xi - T^{1/2}\xi\| \leq 3\varepsilon$, for i large enough. □

We are ready to prove the second fundamental result about Hilbert algebras. It contains a definition of the **natural trace** τ associated to $\mathcal{A}$.

THEOREM 10.A.8 (**Natural trace associated with a Hilbert algebra**). *Let $\mathcal{A}$ be a Hilbert algebra, $\mathcal{H}$ its completion, $\mathcal{A}_b$ the set of its bounded elements, and $\lambda(\mathcal{A})''$ and $\rho(\mathcal{A})''$ the von Neumann subalgebras of $\mathcal{L}(\mathcal{H})$ generated by the left and right multiplications by elements of $\mathcal{A}$. For $T \in \lambda(\mathcal{A})''_+$, set*

$$\begin{aligned}\tau(T) &= \langle a \mid a\rangle \ \textit{if}\ T^{1/2} = \lambda(a) \quad \textit{for some } a \in \mathcal{A}_b, \\ \tau(T) &= +\infty \quad \textit{otherwise}.\end{aligned}$$

Then τ is a semi-finite faithful normal trace on $\lambda(\mathcal{A})''$. Moreover, we have

$$\mathfrak{n}_\tau = \{\lambda(a) \mid a \in \mathcal{A}_b\}$$

and

$$\tau(\lambda(b)^*\lambda(a)) = \langle a \mid b\rangle \quad \textit{for all} \quad a, b \in \mathcal{A}_b.$$

A similar trace is defined on $\rho(\mathcal{A})'' = \lambda(\mathcal{A})'$.

In particular, the von Neumann algebras $\lambda(\mathcal{A})''$ and $\rho(\mathcal{A})''$ are semi-finite.

PROOF. • *First step.* It is clear that τ takes its values in $[0, \infty]$ and that $\tau(\alpha T) = \alpha\tau(T)$ for all $T \in \lambda(\mathcal{A})''_+$ and $\alpha \geq 0$.

• *Second step.* Let $S, T \in \lambda(\mathcal{A})''_+$ be such that $S \leq T$. We claim that $\tau(S) \leq \tau(T)$.

Indeed, since this is obvious if $\tau(T) = +\infty$, we can assume that $T^{1/2} = \lambda(a)$ for some $a \in \mathcal{A}_b$. By Lemma 10.A.6, there exists $A \in \lambda(\mathcal{A})''$ with $\|A\| \leq 1$ such that $S^{1/2} = AT^{1/2}$. It follows from Lemma 10.A.3(5) that $Aa \in \mathcal{A}_b$ and $S^{1/2} = \lambda(Aa)$. We then have

$$\tau(S) = \|Aa\|^2 \leq \|a\|^2 = \tau(T).$$

• *Third step.* Let $T_1, T_2 \in \lambda(\mathcal{A})''_+$; we claim that $\tau(T_1 + T_2) = \tau(T_1) + \tau(T_2)$.

Indeed, set $T := T_1 + T_2$. By Lemma 10.A.6, for each $j \in \{1, 2\}$, there exists $A_j \in \lambda(\mathcal{A})''$ such that $T_j^{1/2} = A_j T^{1/2}$ and such that $A_j = 0$ on $T(\mathcal{H})^\perp$. Set $S := A_1^*A_1 + A_2^*A_2$. We have

$$T = T_1^{1/2}T_1^{1/2} + T_2^{1/2}T_2^{1/2} = T^{1/2}A_1^*A_1T^{1/2} + T^{1/2}A_2^*A_2T^{1/2} = T^{1/2}ST^{1/2}$$

and hence

$$\langle ST^{1/2}\xi \mid T^{1/2}\xi\rangle = \langle T^{1/2}\xi \mid T^{1/2}\xi\rangle \qquad \text{for all } \xi \in \mathcal{H};$$

it follows that $\langle S\eta \mid \eta\rangle = \langle \eta \mid \eta\rangle$ for all $\eta \in \overline{T(\mathcal{H})}$. Since $S = 0$ on $T(\mathcal{H})^\perp$, we see that S is the orthogonal projection on $\overline{T(\mathcal{H})} = \overline{T^{1/2}(\mathcal{H})}$, and therefore $T^{1/2} = ST^{1/2}$.

If $\tau(T_1) = +\infty$ or $\tau(T_2) = +\infty$, then $\tau(T) = +\infty$, by the second step. So, we can assume that $T_1^{1/2} = \lambda(a_1)$ and $T_2^{1/2} = \lambda(a_2)$ for $a_1, a_2 \in \mathcal{A}_b$. Then

$$T^{1/2} = ST^{1/2} = A_1^*A_1T^{1/2} + A_2^*A_2T^{1/2} = A_1^*T_1^{1/2} + A_2^*T_2^{1/2}$$

and hence $T^{1/2} = \lambda(a)$ for $a = A_1^*a_1 + A_2^*a_2 \in \mathcal{A}_b$.

Observe that $A_j a = a_j$, since

$$\lambda(a_j) = T_j^{1/2} = A_jT^{1/2} = A_j\lambda(a) = \lambda(A_ja).$$

We claim that $a \in \overline{T(\mathcal{H})} = \overline{T^{1/2}(\mathcal{H})}$; indeed, by Lemma 10.A.3(1), there exists $x_i \in \mathcal{A}$ with $\lim_i \rho(x_i) = I$ in the strong operator topology and then

$$a = \lim_i \rho(x_i)a = \lim_i \lambda(a)x_i = \lim_i T^{1/2}x_i.$$

It follows that $Sa = a$ and therefore

$$\begin{aligned}\tau(T) &= \langle a \mid a\rangle = \langle Sa \mid a\rangle \\ &= \langle A_1a \mid A_1a\rangle + \langle A_2a \mid A_2a\rangle \\ &= \langle a_1 \mid a_1\rangle + \langle a_2 \mid a_2\rangle = \tau(T_1) + \tau(T_2).\end{aligned}$$

• *Fourth step.* Let $T \in \lambda(\mathcal{A})''$; we claim that $\tau(TT^*) = \tau(T^*T)$.

Let $T = W|T|$ be the polar decomposition of T (see Section A.8). Recall $|T| = (T^*T)^{1/2}$ and that W is a partial isometry with initial space $(\ker T)^\perp = \overline{|T|(\mathcal{H})}$ and final space $\overline{T(\mathcal{H})}$; recall also that W and $|T|$ belong to the von Neumann algebra generated by T and hence to $\lambda(\mathcal{A})''$.

Assume that $\tau(T^*T) < +\infty$. Then $|T| = \lambda(a)$ for some $a \in \mathcal{A}$ and

$$\tau(T^*T) \,=\, \|a\|^2.$$

Since $a \in \overline{|T|(\mathcal{H})}$ (see the third step), we have, on the one hand,

$$\|Wa\| \,=\, \|a\|.$$

On the other hand, using Lemma 10.A.3(3) and (4), we have

$$T^* \,=\, (W\lambda(a))^* \,=\, \lambda(a)^*W^* \,=\, \lambda(Ja)W^* \,=\, \lambda(JWa),$$

and it follows (by an argument from the third step applied to JWa) that $JWa \in \overline{T^*(\mathcal{H})} = \overline{|T|(\mathcal{H})}$; therefore

$$\|WJWa\| \,=\, \|JWa\| \,=\, \|Wa\| \,=\, \|a\|.$$

As $|T^*| = WT^* = \lambda(WJWa)$, it follows that

$$\tau(TT^*) \,=\, \|WJWa\|^2 \,=\, \|a\|^2 \,=\, \tau(T^*T).$$

We have shown, for any $T \in \lambda(\mathcal{A})''$, that, if $\tau(TT^*) < +\infty$, then $\tau(T^*T) < +\infty$ and $\tau(T^*T) = \tau(TT^*)$; it follows that, for any $T \in \lambda(\mathcal{A})''$, we have $\tau(T^*T) < +\infty$ if and only if $\tau(TT^*) < +\infty$ and then $\tau(T^*T) = \tau(TT^*)$. This shows that $\tau(T^*T) = \tau(TT^*)$ for every $T \in \lambda(\mathcal{A})''$.

• *Fifth step.* The preceding steps show that τ is a trace on $\lambda(\mathcal{A})''$. We claim that τ is faithful.

Indeed, let $T \in \lambda(\mathcal{A})''_+$ be such that $\tau(T) = 0$. Then $T^{1/2} = \lambda(a)$ for some $a \in \mathcal{A}_b$ and we have $\tau(T) = \|a\|^2 = 0$, that is, $a = 0$; hence $T^{1/2} = 0$ and $T = 0$.

• *Sixth step.* We claim that τ is normal.

Indeed, let $T \in \lambda(\mathcal{A})''_+$ and let $(T_i)_i$ be an increasing filtering subset of $\lambda(\mathcal{A})''_+$ with $T = \sup_i T_i$. By the second step, we have $\sup_i \tau(T_i) \le \tau(T)$. So, it suffices to show that $\tau(T) \le \sup_i \tau(T_i)$. Since this inequality is obvious if $\sup_i \tau(T_i) = +\infty$, we may assume that $\alpha := \sup_i \tau(T_i) < +\infty$.

We have $T_i = \lambda(a_i)$ for some $a_i \in \mathcal{A}_b$ and

$$\langle a_i \mid a_i \rangle \,=\, \tau(T_i) \le \alpha \qquad \text{for all } i.$$

Since $\lim_i T_i = T$ in the strong operator topology and since $T_i \le T$, we have $\lim T_i^{1/2} = T^{1/2}$ in the strong operator topology (Lemma 10.A.7). Therefore we have, for $x, y \in \mathcal{A}$,

$$\langle T^{1/2}y \mid x\rangle \,=\, \lim_i \langle T_i^{1/2}y \mid x\rangle \,=\, \lim_i \langle \lambda(a_i)y \mid x\rangle \,=\, \lim_i \langle \rho(y)a_i \mid x\rangle \,=\, \lim_i \langle a_i \mid xy^*\rangle.$$

Since the set $\{xy^* \mid x, y \in \mathcal{A}\}$ is total in $\mathcal{H}$ and since $\sup_i \|a_i\| \le \alpha$, the limit $\lim_i a_i = a \in \mathcal{H}$ exists in the weak topology on $\mathcal{H}$; we then have $\|a\| \le \alpha$ and

$$\langle T^{1/2}y \mid x\rangle \,=\, \langle a \mid xy^*\rangle \,=\, \langle \rho(y)a \mid x\rangle$$

for all $x, y \in \mathcal{A}$. It follows that $a \in \mathcal{A}_b$, that $T^{1/2} = \lambda(a)$, and that

$$\tau(T) \,=\, \langle a \mid a\rangle \,\le\, \alpha \,=\, \sup_i \tau(T_i).$$

• *Seventh step.* We claim that $\mathfrak{n}_\tau = \lambda(\mathcal{A}_b)$ and that $\tau(\lambda(b)^*\lambda(a)) = \langle a \mid b\rangle$ for all $a, b \in \mathcal{A}_b$.

Let $T \in \lambda(\mathcal{A})''_+$. We have to show $\tau(T^*T) < +\infty$ if and only if $T \in \lambda(\mathcal{A}_b)$. Let $T = W|T|$ be the polar decomposition of T. Assume that $\tau(T^*T) = \tau(|T|^2) < +\infty$;

then $|T| \in \lambda(\mathcal{A}_b)$, by definition of τ. Since $\lambda(\mathcal{A}_b)$ is an ideal of $\lambda(\mathcal{A})'' = \rho(\mathcal{A})'$, it follows that $T = W|T| \in \lambda(\mathcal{A}_b)$.

Conversely, assume that $T \in \lambda(\mathcal{A}_b)$. Then $|T| = W^*T \in \lambda(\mathcal{A}_b)$, that is, $|T| = \lambda(a)$ for some $a \in \mathcal{A}_b$. It follows that

$$\tau(T^*T) = \tau(|T|^2) = \langle a \mid a \rangle < +\infty.$$

Observe that $T = \lambda(Wa)$ and, since $\|Wa\| = \|a\|$ (see fourth step), we also have $\tau(T^*T) = \langle Wa \mid Wa \rangle$. It follows by polarization that

$$\tau(\lambda(b)^*\lambda(a)) = \langle a \mid b \rangle \qquad \text{for all} \quad a, b \in \mathcal{A}_b.$$

- *Eighth step.* We claim that τ is semi-finite.

Since τ is normal, it suffices to prove the following claim (see Corollaire 3 in [**Dixm–vN**, Chap. I, § 6]). Let $T \in \lambda(\mathcal{A})''_+$ with $T \neq 0$. Then there exists $S \in \lambda(\mathcal{A})''_+$ with $0 \neq S \leq T$ and $\tau(S) < +\infty$.

Let $a \in \mathcal{A}_b$. Since $\lambda(\mathcal{A}_b)$ is a right ideal in $\lambda(\mathcal{A})''$, we have $\lambda(a)T^{1/2} \in \lambda(\mathcal{A}_b)$ and hence, by the seventh step,

$$\tau(T^{1/2}\lambda(a)^*\lambda(a)T^{1/2}) < +\infty.$$

Observe that, for every $\xi \in \mathcal{H}$, we have

$$\begin{aligned}\langle T^{1/2}\lambda(a)^*\lambda(a)T^{1/2}\xi \mid \xi\rangle &= \|\lambda(a)T^{1/2}\xi\|^2 \\ &\leq \|\lambda(a)\|^2\|T^{1/2}\xi\|^2 = \|\lambda(a)\|^2\langle T\xi \mid \xi\rangle,\end{aligned}$$

so that

$$T^{1/2}\lambda(a)^*\lambda(a)T^{1/2} \leq \|\lambda(a)\|^2 T.$$

We claim that there exists $a \in \mathcal{A}_b$ with $\lambda(a)T^{1/2} \neq 0$.

Indeed, otherwise we would have $T^{1/2} = 0$ and hence $T = 0$, since I belongs to the closure of $\lambda(\mathcal{A})$ in the strong operator topology.

Let $a \in \mathcal{A}_b$ be such that $\lambda(a)T^{1/2} \neq 0$. Then $T^{1/2}\lambda(a)^*\lambda(a)T^{1/2} \neq 0$. Set

$$S := T^{1/2}\lambda(b)^*\lambda(b)T^{1/2} \qquad \text{for } b = \frac{1}{\|\lambda(a)\|}a \in \mathcal{A}_b.$$

We have

$$S \neq 0, \qquad \tau(S) < +\infty, \qquad \text{and} \qquad S \leq T.$$

This concludes the proof. □

EXAMPLE 10.A.9. (1) Let (X, μ) be a measure space. Then

$$\mathcal{A} = L^\infty(X,\mu) \cap L^2(X,\mu)$$

is a commutative Hilbert algebra with Hilbert space completion $L^2(X,\mu)$. The two associated von Neumann algebras $\lambda(\mathcal{A})''$ and $\rho(\mathcal{A})''$ are the same and coincide with the algebra

$$\{m(\varphi) \mid \varphi \in L^\infty(X,\mu)\} \approx L^\infty(X,\mu)$$

of multiplication operators by essentially bounded functions. The space of bounded elements $\mathcal{A}_b$ coincides with $\mathcal{A} = L^\infty(X,\mu) \cap L^2(X,\mu)$, and the natural trace τ on $L^\infty(X,\mu)$ has $L^\infty(X,\mu) \cap L^1(X,\mu)$ as an ideal of definition and is given by integration against μ.

(2) Let $\mathcal{H}$ be a separable Hilbert space; the involutive algebra $\mathrm{TC}(\mathcal{H})$ of trace-class operators on $\mathcal{H}$ is a Hilbert algebra for the scalar product

$$\langle S \mid T \rangle = \mathrm{Tr}(T^* S);$$

the completion of $\mathrm{TC}(\mathcal{H})$ is the Hilbert space $\mathrm{HS}(\mathcal{H})$ of Hilbert-Schmidt operators on $\mathcal{H}$. Every element in $\mathrm{HS}(\mathcal{H})$ is bounded. The associated von Neumann algebras are the isomorphic images of $\mathcal{L}(\mathcal{H})$ acting on $\mathrm{HS}(\mathcal{H})$ by left and right composition respectively. If these algebras are identified with $\mathcal{L}(\mathcal{H})$, the associated natural trace τ is the standard trace Tr on $\mathcal{L}(\mathcal{H})$.

(3) Let G be a unimodular LC group, with Haar measure μ_G. The convolution algebra $C^c(G)$, equipped with the involution $f^*(g) = \overline{f(g^{-1})}$ and the scalar product

$$\langle f_1 \mid f_2 \rangle = \int_G f_1(g)\overline{f_2(g)} d\mu_G(g),$$

is a Hilbert algebra; the completion of $C^c(G)$ is $L^2(G)$. A function $\xi \in L^2(G)$ is a bounded element of this Hilbert algebra if and only if $\xi * \eta \in L^2(G)$ for every $\eta \in L^2(G)$; in this case $\lambda(\xi)$ is defined by $\lambda(\xi)(\eta) = \xi * \eta$ for all $\eta \in L^2(G)$. The associated von Neumann algebras $\lambda(C^c(G))''$ and $\rho(C^c(G))''$ are the von Neumann algebras generated by the left and the right regular representations of G. The associated natural trace τ is given by

$$\tau(f^* * f) = \|f\|_2^2 \quad \text{for all } f \in C^c(G).$$

In particular, when G is compact, $L^2(G)$ is itself a Hilbert algebra.

(4) Let $\mathcal{M}$ be a von Neumann algebra, τ a semi-finite faithful normal trace on $\mathcal{M}$, and $\mathfrak{n}_\tau$ the corresponding two-sided ideal, as in Definition 10.A.1. Then $\mathfrak{n}_\tau$, with the scalar product defined by $\langle x \mid y \rangle = \tau(xy^*)$, is a Hilbert algebra. This is a special case of the more general construction of Proposition 10.A.12 below. We come back to this example in Section 10.B.

Note that (2) above is a particular case of (4).

(5) Every dense $*$-subalgebra of a Hilbert algebra is a Hilbert algebra. For example, in the situation of (2), the algebra of operators of finite rank on $\mathcal{H}$ is a Hilbert algebra.

(6) Proposition 11.B.5 below shows that there is a Hilbert algebra associated with any nonzero trace on a topological group.

THEOREM 10.A.10. *Let G be a unimodular locally compact group. Let $\mathcal{L}(G) := \lambda(C^c(G))''$ and $\mathcal{R}(G) := \rho(C^c(G))''$ be the von Neumann algebras generated by the left and the right regular representations of G.*

Then $\mathcal{L}(G)$ and $\mathcal{R}(G)$ are semi-finite von Neumann algebras and we have

$$\mathcal{L}(G)' = \mathcal{R}(G) \quad \text{and} \quad \mathcal{R}(G)' = \mathcal{L}(G).$$

Moreover, there exists a normal, faithful, semi-finite trace τ on $\mathcal{L}(G)$ such that

$$\tau(\lambda(f)\lambda(f)^*) = \|f\|_2^2 \quad \text{for all } f \in L^1(G) \cap L^2(G).$$

PROOF. It suffices to particularize to the Hilbert algebra $\mathcal{A} = C^c(G)$ of Example 10.A.9(3) Theorem 10.A.4 and Theorem 10.A.8. Observe that, in this case, $\mathcal{A}_b = \{\xi \in L^2(G) \mid \xi * \eta \in L^2(G) \text{ for all } \eta \in L^2(G)\}$ contains $L^1(G) \cap L^2(G)$. □

REMARK 10.A.11. Let G be a *nonunimodular* LC group, and let $\mathcal{L}(G)$ and $\mathcal{R}(G)$ be as above. Then it is still true that $\mathcal{L}(G)' = \mathcal{R}(G)$; see [**Dixm–vN**, Chap. I, § 5, Th. 1 and Exercice 5]. However, $\mathcal{L}(G)'$ need no longer be semi-finite; indeed, we will give in Section 14.D below examples of LC groups G for which $\mathcal{L}(G)'$ is a von Neumann algebra of type III.

The following proposition shows that a lower semi-continuous semi-finite trace on a C*-algebra gives rise to a Hilbert algebra.

PROPOSITION 10.A.12. *Let A be a C^*-algebra, t a lower semi-continuous semi-finite trace on A, $t \neq 0$, and $\mathfrak{n}_t$, $\mathfrak{m}_t$ the corresponding two-sided ideals, as in Definition* 10.A.1.

(a) *The map $(x, y) \mapsto t(y^*x)$ is a positive Hermitian form on $\mathfrak{n}_t$.*
(b) *The set $N_t = \{x \in \mathfrak{n}_t \mid t(x^*x) = 0\}$ is a selfadjoint two-sided ideal of A; moreover*

$$N_t = \{x \in \mathfrak{n}_t \mid t(y^*x) = t(xy^*) = 0 \quad \text{for all } y \in \mathfrak{n}_t\}.$$

The map $(x+N_t, y+N_t) \mapsto \langle x+N_t \mid y+N_t\rangle := t(y^*x)$ is a scalar product on $\mathfrak{n}_t/N_t$.
(c) *The algebra $\mathcal{A}_t := \mathfrak{n}_t/N_t$, equipped with the scalar product defined in* (b) *and the involution defined by $(x + N_t)^* = x^* + N_t$ for all $x \in \mathfrak{n}_t$, is a* Hilbert algebra.
(d) *Let λ_t and ρ_t be the representations of A on the Hilbert space completion $\mathcal{H}_t$ of $\mathcal{A}_t$, defined by $\lambda_t(a) : x+N_t \mapsto ax+N_t$ and $\rho_t(a) : x+N_t \mapsto xa^*+N_t$ for $a \in A$. Then the von Neumann subalgebras $\lambda_t(A)''$ and $\rho_t(A)''$ of $\mathcal{L}(\mathcal{H}_t)$ coincide with $\lambda(\mathcal{A}_t)''$ and $\rho(\mathcal{A}_t)''$, respectively, and are each other's commutants:*

$$\lambda_t(A)' = \rho_t(A)'' \quad \text{and} \quad \rho_t(A)' = \lambda_t(A)''.$$

(e) *Let τ_t denote the natural trace on $\lambda_t(A)''$; then*

$$\tau_t(\lambda_t(x)) = t(x) \quad \text{for every } x \in \mathfrak{m}_{t,+}.$$

PROOF. (a) It is straightforward to check that $(x, y) \mapsto t(y^*x)$ is a positive Hermitian form on $\mathfrak{n}_t$.

(b) Let $x \in N_t$, that is, $t(x^*x) = 0$. Cauchy–Schwarz inequality shows that $t(y^*x) = 0$ for every $y \in \mathfrak{n}_t$. Therefore, N_t is a left ideal of $\mathfrak{n}_t$. Since t is a trace, we have $t(yx^*) = t(x^*y)$ and hence, by sesquilinearity,

$$t(yx^*) = t((y^*x)^*) = \overline{t(y^*x)} = 0$$

for every $y \in \mathfrak{n}_t$. This shows that N_t is a selfadjoint left ideal and hence a two-sided ideal of $\mathfrak{n}_t$.

(c) Properties (1) and (2) in the definition of a Hilbert algebra are straightforward to check.

Let us show Property (3). The selfadjoint ideal $\mathfrak{m}_t = \mathfrak{n}_t^2$ has an approximate identity (see Section A.10) and, for every $y \in \mathfrak{n}_t$, the map $x \mapsto t(y^*xy)$ is a continuous positive linear functional on $\mathfrak{m}_t$; hence

$$t(y^*x^*xy) \leq \|x^*x\| t(y^*y) \quad \text{for all } x, y \in \mathfrak{n}_t.$$

See [**Dixm–C***, Proposition 2.1.5]. It follows that, for all $x, y \in \mathfrak{n}_t$, we have

$$\begin{aligned} \|xy + N_t\|^2 &= t((xy)^*yx) = t(y^*x^*xy) \\ &\leq \|x^*x\| t(y^*y) = \|x^*x\| \|y + N_t\|^2, \end{aligned}$$

and this shows that $y + N_t \mapsto xy + N_t$ is continuous for every $x \in \mathfrak{n}_t$.

To show Property (4), let $(u_\iota)_\iota$ be an increasing approximate identity for $\mathfrak{n}_t$ in $\mathfrak{n}_{t,+}$ (see Section A.10). Let $x \in \mathfrak{n}_t$. Then

$$\begin{aligned} \|(u_\iota x - x) + N_t\|^2 &= t((u_\iota x - x)^*(u_\iota x - x)) \\ &= t(x^*u_\iota^2 x) - 2t(x^*u_\iota x) + t(x^*x) \\ &\leq \|u_\iota^2\| t(x^*x) - 2t(x^*u_\iota x) + t(x^*x) \\ &\leq 2\left(t(x^*x) - t(x^*u_\iota x)\right). \end{aligned}$$

Since $(x^*u_\iota x)_\iota$ is an increasing family with $\lim_\iota x^*u_\iota x = x^*x$ and since t is lower semi-continuous, we have $\lim_\iota t(x^*u_\iota x) = t(x^*x)$. This shows that

$$\lim_\iota \|(u_\iota x - x) + N_t\| = 0.$$

(d) This is a direct consequence of Theorem 10.A.4.

(e) Let $x \in \mathfrak{m}_{t,+}$. Then $a := x^{1/2} \in \mathfrak{n}_{t,+}$ and $a + N_t \in \mathcal{A}_t$, hence

$$\tau_t(\lambda_t(x)) = \langle a + N_t \mid a + N_t \rangle = t(a^*a) = t(x),$$

and the proof is complete. □

10.B. The standard representation of a semi-finite von Neumann algebra

We will need to consider the so-called standard representation of a semi-finite von Neumann algebra. It appears below in Lemma 10.C.5 and Proposition 11.B.7. Our exposition rests on [**Dixm–vN**, Chap. I, § 6, no 2].

Let $\mathcal{M}$ be a von Neumann algebra, equipped with a faithful semi-finite normal trace τ. Let $\mathfrak{m} = \mathfrak{m}_\tau$ be the ideal of definition of τ, and $\mathfrak{n} = \mathfrak{n}_\tau$ the ideal of all $x \in \mathcal{M}$ with $\tau(x^*x) < \infty$; recall that $\mathfrak{m} = \mathfrak{n}^2$.

Since τ is normal, τ is lower-continuous for the weak topology and hence for the norm topology. By Proposition 10.A.12, the space $\mathfrak{n}$, equipped with the scalar product $(x, y) \mapsto \tau(y^*x)$, is therefore a Hilbert algebra.

Let $L^2(\mathfrak{n}, \tau)$ denote the Hilbert space completion of $\mathfrak{n}$, with respect to this scalar product. Instead of λ, we denote by $x \mapsto \widehat{x}$ the associated representation of $\mathcal{M}$ on $L^2(\mathfrak{n}, \tau)$, where $\widehat{x}$ is the continuous extension to $L^2(\mathfrak{n}, \tau)$ of the linear map

$$\mathfrak{n} \to \mathfrak{n}, \qquad y \mapsto xy.$$

Moreover, if $\widehat{\tau}$ denotes the natural trace on $\mathfrak{n}$, then

$$\widehat{\tau}(\widehat{x}) = \tau(x) \quad \text{for all } x \in \mathcal{M}.$$

PROPOSITION 10.B.1. *The algebra*

$$\widehat{\mathcal{M}} := \{\widehat{x} \mid x \in \mathcal{M}\}$$

is a von Neumann subalgebra of $\mathcal{L}(L^2(\mathcal{M}, \tau))$.

PROOF. Since the map

$$\mathcal{M} \to \mathcal{L}(L^2(\mathcal{M},\tau)), \qquad x \mapsto \widehat{x}$$

is a unital $*$-algebra homomorphism, it suffices to show that this map is normal (see Corollaire 2 of [**Dixm–vN**, Chap. I, §4, 39]).

Let $(x_i)_i$ be an increasing family in $\mathcal{M}_+$ with $x = \sup_i x_i \in \mathcal{M}_+$. Then $(\widehat{x_i})_i$ is an increasing family in $\mathcal{L}(L^2(\mathcal{M},\tau))_+$ with $\widehat{x_i} \leq \widehat{x}$ for all i and hence $\sup_i \widehat{x_i} \leq \widehat{x}$.

Let $y \in \mathfrak{n}$. Then

$$\langle \widehat{x_i} y \mid y \rangle = \langle x_i y \mid y \rangle = \tau(y^* x_i y) \qquad \text{for all } i.$$

Since $\sup_i y^* x y_i = y^* x y$ and since τ is normal, we have $\sup_i \tau(y^* x y_i) = \tau(y^* x y)$ and hence

$$\sup_i \langle \widehat{x_i} y \mid y \rangle = \tau(y^* x y) = \langle \widehat{x} y \mid y \rangle \qquad \text{for all } y \in \mathfrak{n}.$$

As $\mathfrak{n}$ is dense in $L^2(\mathcal{M},\tau)$, it follows that $\sup_i \widehat{x_i} = \widehat{x}$. □

It follows from the previous lemma that $\widehat{\mathcal{M}}$ coincides with the von Neumann algebra $\lambda(\mathfrak{n})''$ associated with the Hilbert algebra $\mathfrak{n}$ as in Section 10.A.

The representation

$$\mathcal{M} \to \widehat{\mathcal{M}}, \qquad x \mapsto \widehat{x}$$

is called the **standard representation** of the pair $(\mathcal{M},\tau)$. Observe that the standard representation is an isomorphism between $\mathcal{M}$ and $\widehat{\mathcal{M}}$; indeed, if $\widehat{x} = 0$ for $x \in \mathcal{M}$, then $x \in \mathfrak{n}$ and $\tau(x^* x) = 0$ and hence $x = 0$, since τ is faithful. The von Neumann algebra $\widehat{\mathcal{M}}$ is often called the **standard form** of $\mathcal{M}$.

EXAMPLE 10.B.2. (1) Let $\mathcal{H}$ be a separable Hilbert space and $\mathcal{M} = \mathcal{L}(\mathcal{H})$. The standard representation of the von Neumann algebra $\mathcal{L}(\mathcal{H})$ is the representation $T \mapsto \widehat{T}$ of $\mathcal{L}(\mathcal{H})$ on the Hilbert space $\mathrm{HS}(\mathcal{H})$ of Hilbert-Schmidt operators on $\mathcal{H}$ given by

$$\widehat{T}(S) = TS, \qquad \text{for all } S \in \mathrm{HS}(\mathcal{H}).$$

(2) Let G be a unimodular LC group. The standard representation of the von Neumann algebra $\lambda_G(G)''$ of the left regular representation of G coincides with $\lambda_G(G)''$ acting on $L^2(G)$.

10.C. Group representations associated with traces

Let G be an LC group. Let t be a nonzero lower semi-continuous semi-finite trace on the maximal C*-algebra $C^*_{\mathrm{max}}(G)$ of G. Let λ_t be the representation of $C^*_{\mathrm{max}}(G)$ associated with t, as in Section 10.A; then λ_t is a traceable representation of G, in the following sense.

DEFINITION 10.C.1. A **trace representation** of an LC group G is a pair (π, τ), with π a representation of G and τ a normal faithful semi-finite trace on the von Neumann algebra $\pi(G)''$ such that the following condition holds:

(*) the set of all $\pi(x)$, for $x \in C^*_{\mathrm{max}}(G)$ such that $0 < \tau(\pi(x^* x)) < +\infty$, is weakly dense in $\pi(G)''$.

(Recall from Proposition 8.G.1 that $\pi(G)'' = \pi(C^*_{\mathrm{max}}(G))''$.)

Note that the existence of a normal faithful semi-finite trace on $\pi(G)''$ *does not* imply Condition (*), even when π is factorial; see Examples 10.D.5(4) and (5) below.

We say that two trace representations (π_1, τ_1) and (π_1, τ_1) are **quasi-equivalent** if there exists an isomorphism $\Phi : \pi_1(G)'' \to \pi_2(G)''$ such that $\Phi(\pi_1(g)) = \pi_2(g)$ for all $g \in G$, and such that $\tau_1 = \tau_2 \circ \Phi$.

A representation π of G is **traceable** if there exists a normal faithful semi-finite trace τ on the von Neumann algebra $\pi(G)''$ such that the pair (π, τ) is a trace representation.

Traces and traceable representations appear for C*-algebras in [**Dixm–C***, 6.6 and 6.7], and for LC groups in [**Dixm–C***, 17.1].

Being traceable only depends on the quasi-equivalence class of the representation:

PROPOSITION 10.C.2. *Let π and ρ be quasi-equivalent representations of G. If π is traceable, then ρ is traceable.*

PROOF. Let τ be a normal faithful semi-finite trace on $\pi(G)''$ such that (π, τ) is a trace representation and let $\Phi : \pi(G)'' \to \rho(G)''$ be an isomorphism with $\Phi(\pi(g)) = \rho(g)$ for all $g \in G$. Since Φ is an isomorphism of von Neumann algebras, Φ and Φ^{-1} are normal maps; see [**Dixm–vN**, Chap. I, § 4, Corollaire 1]. Therefore $\tau' := \tau \circ \Phi^{-1}$ is a normal faithful semi-finite trace on $\rho(G)''$ and we have $\Phi(\pi(x)) = \rho(x)$ for all $x \in C^*_{\max}(G)$. It follows that Condition (*) of Definition 10.C.1 is satisfied for τ' and hence (ρ, τ') is a trace representation of G. □

EXAMPLE 10.C.3. (1) Let $\pi : G \to \mathcal{U}(n)$ be a finite-dimensional representation of an LC group G. Then (π, τ) is a trace representation, where τ is the restriction of the standard trace on $M_n(\mathbf{C})$ to the algebra $\pi(G)''$, which is the linear span of $\pi(G)$. Assume that π is not a factor representation, that is, π is not a multiple of an irreducible representation. Then there are more faithful traces on $\pi(G)''$ than positive multiples of τ. For instance, assume that $\pi = \pi_1 \oplus \pi_2$ is the direct sum of two nonequivalent irreducible representations π_1 and π_2 on subspaces V_1 and V_2 of $\mathbf{C}^n$. Then, for every $a > 0, b > 0$, a faithful trace $\tau_{a,b}$ is defined on

$$\pi(G)'' = \pi_1(G)'' \oplus \pi_2(G)'' \approx \mathcal{L}(V_1) \oplus \mathcal{L}(V_2)$$

by

$$\tau_{a,b}(T \oplus S) = a\tau_1(T) + b\tau_2(S)$$

where τ_i is the standard trace on $\mathcal{L}(V_i)$.

(2) Examples of traceable representations of a unimodular LC group G arise often as follows. Let ν be complex-valued Radon measure (that is, a linear combination of positive Radon measures) on G. Assume that ν is of positive type, that is,

$$\int_G (f^* * f)(x) d\nu(x) \geq 0 \qquad \text{for all } f \in C^c(G),$$

where f^* is defined by $f^*(x) = \overline{f(x^{-1})}$ (because G is unimodular). Assume also that ν is central, that is,

$$\int_G (f_1 * f_2)(x) d\nu(x) = \int_G (f_2 * f_1)(x) d\nu(x) \qquad \text{for all } f_1, f_2 \in C^c(G).$$

Then ν defines a lower semi-continuous, semi-finite trace t_ν on $C^*_{\max}(G)$, with the ideal of definition containing $C^c(G)$, and given there by

$$t_\nu(f * f^*) = \int_G (f * f^*)(x) d\nu(x) \qquad \text{for all } f \in C^c(G).$$

The representation λ_ν associated with t_ν as in Proposition 10.A.12 is traceable.

For instance, when $\nu = \delta_e$ is the Dirac measure at the group unit, t_ν is the semi-finite trace given by

$$t_\nu(f * f^*) = (f * f^*)(e) = \|f\|_2^2 \qquad \text{for all } f \in C^c(G).$$

and λ_ν is the left regular representation (see also Example 10.A.9 (3)). The trace t_ν is finite only when G is discrete. For all this, see [**Gode–54**].

(3) When G is a real Lie group, the construction of (2) can be — and has to be — extended to central distributions of positive type on G. The resulting lower semi-continuous, semi-finite traces on $C^*_{\max}(G)$ have ideals of definition which contain the space of smooth functions with compact support on G.

As shown in Proposition 10.A.12, for every lower semi-continuous semi-finite trace t on $C^*_{\max}(G)$, there exists a representation λ_t of G (equivalently of $C^*_{\max}(G)$) and a trace τ_t on $\lambda_t(G)''$ such that (λ_t, τ_t) is a trace representation of G, with $t = \tau_t \circ \lambda_t$.

Proposition 10.C.6 will show that every trace representation of G is of this kind. It is convenient to state two preliminary lemmas.

LEMMA 10.C.4. *Let (π, τ) be a trace representation of G.*
Then $\tau \circ \pi$ is a lower semi-continuous semi-finite trace on $C^*_{\max}(G)$.

PROOF. It is clear that $t := \tau \circ \pi$ is a trace on $C^*_{\max}(G)$. Since (π, τ) is a trace representation, $\pi(\mathfrak{m}_t)$ is weakly dense in $\pi(G)''$ and hence the restriction of π to the ideal $\overline{\mathfrak{m}_t}$ is nondegenerate. Let $(u_\iota)_\iota$ be an increasing approximate identity for $\overline{\mathfrak{m}_t}$ with $u_\iota \in \mathfrak{m}_+$. Then $(\pi(u_\iota))_\iota$ converges to I for the strong operator topology.

Let $x \in C^*_{\max}(G)_+$. Since $(\pi(x^{1/2}u_\iota x^{1/2}))_\iota$ is an increasing family in $\pi(G)''_+$ with

$$\lim_\iota \pi(x^{1/2}u_\iota x^{1/2}) = \pi(x)$$

and since τ is normal, we have

$$\lim_\iota \tau(\pi(x^{1/2}u_\iota x^{1/2})) = \tau(\pi(x)).$$

This shows that $\lim_\iota t(x^{1/2}u_\iota x^{1/2}) = t(x)$. Since $x^{1/2}u_\iota x^{1/2} \in \mathfrak{m}_t$, we have

$$t(x^{1/2}u_\iota x^{1/2}) < +\infty$$

and hence t is semi-finite. Moreover, being normal, τ is lower semi-continuous for the weak topology and therefore for the norm topology on $\pi(G)''$. It follows that $t = \tau \circ \pi$ is lower semi-continuous. □

LEMMA 10.C.5. *Let (π_1, τ_1) and (π_2, τ_2) be trace representations of G such that $\tau_1 \circ \pi_1 = \tau_2 \circ \pi_2$.*
Then (π_1, τ_1) and (π_2, τ_2) are quasi-equivalent.

PROOF. For $j = 1, 2$, set

$$\mathcal{M}_j = \pi_j(G)'' = \pi_j(C^*_{\max}(G))'' \quad \text{and} \quad \mathfrak{n}_j = \mathfrak{n}_{\tau_j}.$$

Recall that $\pi_j(C^*_{\max}(G)) \cap \mathfrak{n}_j$ is weakly dense in $\mathcal{M}_j$, since π_j is a trace representation.

We consider the standard representation $x \mapsto \widehat{x}$ of $\mathcal{M}_j$ on $L^2(\mathfrak{n}_j, \tau_j)$. The map $\Phi_j : x \mapsto \widehat{x}$ is an isomorphism between the von Neumann algebras $\mathcal{M}_j$ and

$$\widehat{\mathcal{M}_j} := \{\widehat{x} \mid x \in \mathcal{M}_j\}$$

(see Section 10.B). Moreover, $\widehat{\tau_j} \circ \Phi_j = \tau_j$, where $\widehat{\tau_j}$ is the natural trace on $\widehat{\mathcal{M}_j}$.

Denote by $\widehat{\pi_j}$ the representation of G on $L^2(\mathfrak{n}_j, \tau_j)$ defined by $\widehat{\pi_j} = \Phi_j \circ \pi_j$, that is,

$$\widehat{\pi_j}(g) = \widehat{\pi_j(g)} \quad \text{for all } g \in G.$$

We have, for every $a \in C^*_{\max}(G)$,

$$\widehat{\tau_j}(\widehat{\pi_j(a)}) = \tau_j(\pi_j(a)),$$

and hence $\widehat{\tau_j}(\widehat{\pi_j(a)}) < +\infty$ if and only if $\tau_j(\pi_j(a)) < +\infty$. Since $\Phi_j : \mathcal{M}_j \to \widehat{\mathcal{M}_j}$ is an isomorphism, this shows that $(\widehat{\pi_j}, \widehat{\tau_j})$ is a trace representation and that $(\widehat{\pi_j}, \widehat{\tau_j})$ is quasi-equivalent to the trace representation (π_j, τ_j).

As a consequence, it suffices to show that the trace representations $(\widehat{\pi_1}, \widehat{\tau_1})$ and $(\widehat{\pi_2}, \widehat{\tau_2})$ are quasi-equivalent.

Set

$$\mathcal{A}_j := \widehat{\pi_j}(C^*_{\max}(G)) \cap \mathfrak{n}_{\widehat{\tau_j}};$$

then $\mathcal{A}_j$ is a Hilbert subalgebra of the Hilbert algebra $\mathfrak{n}_{\widehat{\tau_j}}$ and, since $\mathcal{A}_j$ is weakly dense in $\widehat{\mathcal{M}_j}$, the associated left von Neumann algebra coincides with $\widehat{\mathcal{M}_j}$ (see Proposition 1 in [**Dixm–vN**, Chap. I, § 5, 2]). Therefore $\mathcal{A}_j$ is dense in the Hilbert space completion of $\mathfrak{n}_{\widehat{\tau_j}}$, which is $L^2(\mathfrak{n}_j, \tau_j)$.

Observe that, by assumption, we have

$$\widehat{\tau_1}(\widehat{\pi_1(a)}) = \tau_1(\pi_1(a)) = \tau_2(\pi_2(a)) = \widehat{\tau_2}(\widehat{\pi_2(a)}),$$

for all $a \in C^*_{\max}(G)$. In particular, we have $\widehat{\pi_1}^{-1}(\mathcal{A}_1) = \widehat{\pi_2}^{-1}(\mathcal{A}_2)$. Since $\widehat{\tau_1}$ and $\widehat{\tau_2}$ are faithful, it follows that the map

$$\mathcal{A}_1 \to \mathcal{A}_2, \qquad \widehat{\pi_1}(a) \mapsto \widehat{\pi_2}(a), \qquad \text{for all } a \in \widehat{\pi_1}^{-1}(\mathcal{A}_1)$$

is well-defined and extends to a Hilbert space isomorphism

$$U : L^2(\mathcal{M}_1, \tau_1) \to L^2(\mathcal{M}_2, \tau_2).$$

Let

$$\Phi : \mathcal{L}(L^2(\mathcal{M}_1, \tau_1)) \to \mathcal{L}(L^2(\mathcal{M}_2, \tau_2)), \qquad T \mapsto UTU^{-1}.$$

Then Φ restricts to an isomorphism from $\widehat{\mathcal{M}_1}$ to $\widehat{\mathcal{M}_2}$, such that $\Phi(\widehat{\pi_1}(a)) = \widehat{\pi_2}(a)$ for all $a \in C^*_{\max}(G)$. It follows that $\Phi(\widehat{\pi_1}(g)) = \widehat{\pi_2}(g)$ for all $g \in G$. Moreover, we have $\widehat{\tau_1} = \widehat{\tau_2} \circ \Phi$. Therefore $(\widehat{\pi_1}, \widehat{\tau_1})$ and $(\widehat{\pi_2}, \widehat{\tau_2})$ are quasi-equivalent. □

Let t be a lower semi-continuous semi-finite trace on $C^*_{\max}(G)$, $t \neq 0$; recall from Proposition 10.A.12 that there is an associated trace representation (λ_t, τ_t) such that $\tau_t \circ \lambda_t = t$. Conversely, let (π, τ) be a trace representation of G; by Lemma 10.C.4, $\tau \circ \pi$ is a lower semi-continuous semi-finite trace on $C^*_{\max}(G)$.

PROPOSITION 10.C.6. *Let G be a locally compact group. For a lower semi-continuous semi-finite trace $t \neq 0$ on the C^*-algebra $C^*_{\max}(G)$, let (λ_t, τ_t) be as in Proposition* 10.A.12 *the corresponding trace representation. For a trace representation (π, τ) of G, consider as in Lemma* 10.C.4 *the lower semi-continuous semi-finite trace $\tau \circ \pi$ on $C^*_{\max}(G)$.*

Then the maps $t \mapsto$ (class of (λ_t, τ_t)) and (class of (π, τ)) $\mapsto \tau \circ \pi$ induce bijections inverse to each other between

(i) *the set of lower semi-continuous semi-finite traces t on the maximal C^*-algebra $C^*_{\max}(G)$, and*
(ii) *the set of trace representations (π, τ) of G, up to quasi-equivalence.*

PROOF. We know from Lemma 10.C.5 that $\tau \circ \pi$ only depends on the quasi-equivalence class of (π, τ).

We have to check that the composition of these two maps, in each order, is the appropriate identity.

For t as in (i), we have $\tau_t \circ \lambda_t = t$ by Proposition 10.A.12(e).

For (π, τ) and $t = \tau \circ \pi$ as in (ii), the trace representations (π, τ) and (λ_t, τ_t) are quasi-equivalent by Lemma 10.C.5. □

10.D. Normal factor representations and characters

In this section, we will be mostly interested in traceable representations which are *factorial*.

DEFINITION 10.D.1. A factor representation of a locally compact group is **normal** if it is traceable.

Normal factor representations were introduced in [**Gode–54**] and further studied and extended to the context of Banach involutive algebras in [**Guic–63**]. We continue to follow [**Dixm–C***, Chap.7] for our exposition.

PROPOSITION 10.D.2. *For a factor representation π of a locally compact group G, the following properties are equivalent:*

(i) *π is a normal factor representation;*
(ii) *the factor $\pi(G)''$ is semi-finite (that is, of type I or II) and there exists an element $x \in C^*_{\max}(G)$ such that $0 < \tau(\pi(x^*x)) < +\infty$, where τ is a faithful normal trace on $\pi(G)''$.*

PROOF. It is clear that (i) implies (ii). Assume that (ii) holds, and let τ be a semi-finite faithful normal trace on $\pi(G)''$. Then

$$J := \mathfrak{n}_\tau \cap \pi(C^*_{\max}(G))$$

is a nonzero two-sided ideal of $\pi(C^*_{\max}(G))$. Therefore the weak closure B of J is a nonzero weakly closed two-sided ideal of the von Neumann algebra $\pi(C^*_{\max}(G))'' = \pi(G)''$. Since $\pi(G)''$ is a factor, we have $B = \pi(G)''$; see [**Dixm–vN**, Chap. I, § 4, no 4, Corollaire 3]. Therefore π is a normal factor representation. □

We first characterize irreducible representations which are normal.

PROPOSITION 10.D.3. *Let G be a locally compact group.*

An irreducible representation π of G is normal if and only if $\pi(C^*_{\max}(G))$ contains the space $\mathcal{K}(\mathcal{H}_\pi)$ of compact operators on $\mathcal{H}_\pi$.

PROOF. Recall that the standard trace is the unique normal semi-finite trace on $\mathcal{L}(\mathcal{H}_\pi)$, up to scalar multiples. It is obvious that π is normal if $\pi(C^*_{\max}(G))$ contains $\mathcal{K}(\mathcal{H}_\pi)$. Conversely, assume that π is traceable. Then $\pi(C^*_{\max}(G))$ contains a nonzero trace-class (and hence compact) operator. Since π is irreducible, this implies that $\pi(C^*_{\max}(G))$ contains all compact operators (see Remark 6.E.3(1)). □

The following corollary is a consequence of Proposition 10.D.3 and Glimm's Theorem 8.F.3 (see also Comment (11) after this theorem).

COROLLARY 10.D.4. *Let G be a locally compact group. The following properties are equivalent:*

(i) *every irreducible representation of G is normal;*
(ii) *G is of type I.*

EXAMPLE 10.D.5. Let G be a locally compact group.

(1) A finite-dimensional irreducible representation of G is normal.

(2) A factor representation π of finite type is normal. Indeed, when $\pi(G)''$ is a factor that is either of type I_n for some finite integer n or of type II_1, there is a faithful finite trace on this factor.

(3) Let Γ be a discrete group which is not of type I. Denote by $\widehat{\Gamma}_\infty$ the set of equivalence classes of infinite-dimensional irreducible representations of Γ and by $\widehat{\Gamma}_{\mathrm{norm},\infty}$ the set of all representations in $\widehat{\Gamma}_\infty$ which are normal. Corollary 10.D.4 shows that $\widehat{\Gamma}_{\mathrm{norm},\infty}$ is strictly contained in $\widehat{\Gamma}_\infty$.

It may happen that $\widehat{\Gamma}_{\mathrm{norm},\infty}$ is empty; this is the case for the Heisenberg group $H(\mathbf{K})$ over an infinite field $\mathbf{K}$ (see Proposition 11.D.2). We will give in Chapter 14 examples of representations in $\widehat{\Gamma}_{\mathrm{norm},\infty}$, when Γ is $BS(1,p)$ or the lamplighter group. For examples of such representations when Γ is $\mathrm{GL}_n(\mathbf{Q})$ or $\mathrm{SL}_n(\mathbf{Z})$ for $n \geq 2$, see [**Bekk–19**].

(4) We are going to define a representation π of the free group F_2 on two generators which is infinite-dimensional and irreducible, in particular which is factorial of type I_∞, and which is not normal.

Let a, b be two free generators of F_2. Denote by C the infinite cyclic subgroup of F_2 generated by a. Let π be the quasi-regular representation of F_2 acting on $\mathcal{H}_\pi := \ell^2(F_2/C)$. The representation π is clearly infinite-dimensional. Since C is its own commensurator in F_2, the representation π is irreducible (Example 1.F.15). It remains to check that π is not normal; by Proposition 10.D.3, it suffices to check that $\pi(C^*_{\max}(F_2))$ does not contain the space of compact operators on $\mathcal{H}_\pi$.

Since C is amenable, the unit representation 1_C is weakly contained in the regular representation λ_C of C, so that $\pi = \mathrm{Ind}_C^{F_2}(1_C)$ is weakly contained in the regular representation $\lambda = \mathrm{Ind}_C^{F_2}(\lambda_C)$ of F_2, and therefore $\pi : C^*_{\max}(F_2) \to \mathcal{L}(\mathcal{H}_\pi)$ is the composition of the canonical surjection $\lambda : C^*_{\max}(F_2) \twoheadrightarrow C^*_\lambda(F_2)$ with a C*-algebra morphism $\pi' : C^*_\lambda(F_2) \to \mathcal{L}(\mathcal{H}_\pi)$. Since the reduced C*-algebra $C^*_\lambda(F_2)$ is simple [**Powe–75**], the ideal $(\pi')^{-1}(\mathcal{K}(\mathcal{H}_\pi))$ is either $\{0\}$ or $C^*_\lambda(F_2)$; but the second case is impossible, because the image by π' of the unit in $C^*_\lambda(F_2)$, which is the identity on the infinite-dimensional space $\mathcal{H}_\pi$, is not compact, hence π is not normal.

(5) A factor representation of type II_∞ need not be normal. For example, consider an infinite field $\mathbf{K}$ and the Heisenberg group $H(\mathbf{K})$, which is a nilpotent discrete group. Then $H(\mathbf{K})$ has factor representations of type II_∞ (Theorems 7.D.3(1) and 8.F.4), but every factor representation which is normal is of finite type (Proposition 11.D.2).

(6) A factor representation of type III is not normal.

(7) Assume that G is unimodular and second-countable. By Theorem 10.A.10, $\lambda_G(G)''$ carries a semi-finite faithful normal trace. In other terms, the left regular representation λ_G is traceable. It follows that G has a family of normal factor representations which separate the points of $L^1(G)$, and in particular which separate the points of G; see Corollaire to Théorème 1 in Chap. I, §3 of [**Guic–63**].

(8) There are separable C*-algebras which are purely infinite, that is, which have no nonzero traces with nonzero ideal of definition. An example is the Cuntz algebra O_2 of [**Cuntz–77**] (it appears already in [**Dixm–64a**, Example 2.1]).

For an LC group G, the maximal C*-algebra of G has at least one normal factor representation, the unit representation 1_G. It has many other ones when G is unimodular and not $\{1\}$, see (7) above, or when G is a connected real Lie group, see Theorem 11.D.1 below. The following question seems to be open.

QUESTION. Does there exist an LC group G such that 1_G is the only normal factor representation of G?

We now introduce an interesting subset of the quasi-dual of an LC group.

DEFINITION 10.D.6. Let G be a locally compact group. The **normal quasi-dual** $\mathrm{QD}(G)_{\mathrm{norm}}$ of G is the set of quasi-equivalence classes of normal factor representations of G.

The following result from [**Halp–75**] holds for general separable C*-algebras. In case of a second-countable locally compact group G which is not of type I, it shows that $\mathrm{QD}(G)_{\mathrm{norm}}$ is better behaved than the dual $\widehat{G}$ or the quasi-dual $\mathrm{QD}(G)$.

THEOREM 10.D.7. *Let G be a second-countable locally compact group.*

The space $\mathrm{QD}(G)_{\mathrm{norm}}$, equipped with the Mackey–Borel structure induced by that of $\mathrm{QD}(G)$, is a standard Borel space.

Normal factor representations of LC group G can be described in terms of characters.

DEFINITION 10.D.8. A **character** of an LC group G is a nonzero lower semi-continuous semi-finite trace t on $C^*_{\mathrm{max}}(G)$, which satisfies the following condition:

every lower semi-continuous semi-finite trace on $C^*_{\mathrm{max}}(G)$ dominated by t is proportional to t on $(\overline{\mathfrak{m}_t})_+$.

We denote by $\mathrm{Char}(G)$ the set of equivalence classes of characters of G, where two characters are equivalent if they are proportional to each other.

Recall from Lemma 10.C.4 that, if (π, τ) is a trace representation of G, then $\tau \circ \pi$ is a lower semi-continuous semi-finite trace on $C^*_{\mathrm{max}}(G)$.

Let t be a lower semi-continuous semi-finite trace on $C^*_{\mathrm{max}}(G)$, $t \neq 0$. We use the notation of Proposition 10.A.12 for the ideals $\mathfrak{n}_t$, N_t, and $\mathfrak{m}_t$, for the Hilbert algebra $\mathcal{A}_t = \mathfrak{n}_t/N_t$, and for the Hilbert space completion $\mathcal{H}_t$ of $\mathcal{A}_t$. For $x \in \mathfrak{n}_t$, we denote by $[x]$ the image of x in $\mathcal{H}_t$, so that we have

$$t(y^*x) = \langle [x] \mid [y] \rangle \quad \text{for all } x, y \in \mathfrak{n}_t.$$

Recall also that there are corresponding commuting representations λ_t and ρ_t of G on $\mathcal{H}_t$, and there exists a faithful normal trace τ_t on $\lambda_t(G)''$ such that $t = \tau_t \circ \lambda_t$.

As we now show, every semi-continuous semi-finite trace on $C^*_{\max}(G)$ dominated by t on $(\overline{\mathfrak{m}_t})_+$ defines an element in the centre of $\lambda_t(G)''$.

LEMMA 10.D.9. *Let t and t' be lower semi-continuous semi-finite traces on $C^*_{\max}(G)$ such that $t'(x) \le t(x)$ for every $x \in (\overline{\mathfrak{m}_t})_+$.*

There exists $T \in \lambda_t(G)' \cap \rho_t(G)'$ such that $0 \le T \le I$ and such that

$$t'(y^*x) = \langle T[x] \mid [y] \rangle \quad \textit{for all } x, y \in \mathfrak{n}_t.$$

PROOF. Since $t'(x) \le t(x)$ for every $x \in (\overline{\mathfrak{m}_t})_+$, the map

$$\Phi : \mathcal{A}_t \times \mathcal{A}_t \to \mathbf{C}, \ \ (x, y) \mapsto t'(y^*x)$$

is a well-defined positive Hermitian form on $\mathcal{A}_t$, which is separately continuous for the Hilbert space norm (compare with Lemma 1.B.13). Therefore there exists $T \in \mathcal{L}(\mathcal{H}_t)$ such that $0 \le T \le I$ and such that

$$\Phi(x, y) = \langle T[x] \mid [y] \rangle \quad \text{for all } x, y \in \mathfrak{n}_t.$$

Let us check that T is in the centre of $\lambda_t(G)''$. For $z \in C^*_{\max}(G), x, y \in \mathfrak{n}_t$, we have

$$\begin{aligned} \langle T\lambda_t(z)[x] \mid [y] \rangle &= \langle T[zx] \mid [y] \rangle = t'(y^*zx) = t'((z^*y)^*x) \\ &= \langle T[x] \mid [z^*y] \rangle = \langle T[x] \mid \lambda_t(z^*)[y] \rangle = \langle \lambda_t(z)T[x] \mid [y] \rangle; \end{aligned}$$

this shows that $T \in \lambda_t(C^*_{\max}(G))' = \lambda_t(G)'$. Since t' is a trace, we also have

$$\begin{aligned} \langle T\rho_t(z)[x] \mid [y] \rangle &= \langle T[xz^*] \mid [y] \rangle = t'(y^*xz^*) = t'(z^*y^*x) = t'((yz)^*x) \\ &= \langle T[x] \mid [yz] \rangle = \langle T[x] \mid \rho_t(z^*)[y] \rangle = \langle \rho_t(z)T[x] \mid [y] \rangle, \end{aligned}$$

and hence $T \in \rho_t(C^*_{\max}(G))' = \rho_t(G)'$. □

Let π_1 and π_2 be normal factor representations of G. For $j \in \{1, 2\}$, let τ_j be a semi-finite faithful normal trace on $\pi_j(G)''$. Observe that, if π_1 and π_2 are quasi-equivalent, and if $\Phi : \pi_1(G)'' \to \pi_2(G)''$ is an isomorphism such that $\pi_2(g) = \Phi(\pi_1(g))$ for all $g \in G$, then the trace $\tau_2 \circ \Phi$ is proportional to τ_1. Indeed, since $\pi_1(G)''$ is a factor, $\tau_2 \circ \Phi$ is proportional to τ_1 (see the Corollaire in [**Dixm–vN**, Chap. I, § 6, no 4]).

As a consequence, this yields a mapping $\mathrm{QD}(G)_{\mathrm{norm}} \to \mathrm{Char}(G)$ which maps a normal factor representation π (up to quasi-equivalence) to $\tau \circ \pi$ (up to a positive scalar factor) which is called **the character of the representation** π. The following proposition shows that this mapping is in fact a bijection.

PROPOSITION 10.D.10. *Let G be a locally compact group.*

The map $\pi \mapsto \tau \circ \pi$ described above induces a bijection

$$\mathrm{QD}(G)_{\mathrm{norm}} \ \xrightarrow{\approx} \ \mathrm{Char}(G),$$

with inverse bijection induced by the map $t \mapsto \lambda_t$ of Proposition 10.A.12.

PROOF. Let π be a normal factor representation of G; let τ be a semi-finite faithful normal trace on $\pi(G)''$ and $t := \tau \circ \pi$. Upon passing to a quasi-equivalent representation, we may assume by Proposition 10.C.6 that $\pi = \lambda_t$. Let t' be a lower semi-continuous semi-finite trace on $C^*_{\max}(G)$ such that $t'(x) \le t(x)$ for every

$x \in (\overline{\mathfrak{m}_t})_+$. By Lemma 10.D.9, there exists an operator T in the centre of $\lambda_t(G)''$ such that $t'(y^*x) = \langle T[x] \mid [y]\rangle$ for all $x, y \in \mathfrak{n}_t$. Since $\lambda_t(G)''$ is a factor, T is a scalar multiple of I, hence t' is proportional to t on $(\overline{\mathfrak{m}_t})_+$. Therefore, t is a character of G.

Conversely, let t be a character of G and let (λ_t, τ_t) be the associated trace representation on the Hilbert space $\mathcal{H}_t$. We claim that λ_t is factorial. Indeed, let T be an operator in the centre of $\lambda_t(G)''$ such that $0 \leq T \leq I$. The map

$$t' : x \mapsto \tau_t(T\lambda_t(x))$$

is a trace on $C^*_{\max}(G)$. By Lemma 10.C.4, t' is lower semi-continuous and semi-finite. Since $t = \tau_t \circ \lambda_t$ and $0 \leq T \leq I$, we have $t'(x) \leq t(x)$ for every $x \in (\overline{\mathfrak{m}_t})_+$. Therefore there exists $\alpha \in [0, 1]$ such that $t' = \alpha t$ on $(\overline{\mathfrak{m}_t})_+$.

Let $x \in \mathfrak{n}_t$. Then $T[x] \in \mathcal{H}_t$ is a bounded element for the Hilbert algebra associated with t, by Lemma 10.A.3(5), where $[x]$ denotes the image of x in $\mathcal{H}_t$. By Theorem 10.A.8, we have therefore

$$\langle T[x] \mid [x]\rangle = \tau_t(T\lambda_t(x^*)\lambda_t(x)) = t'(x^*x) = \alpha t(x^*x) = \alpha\langle [x], [x]\rangle.$$

It follows that $T = \alpha I$ and hence that $\lambda_t(G)''$ is a factor. □

For the subclass of normal irreducible representations (see Proposition 10.D.3), we obtain the following corollary.

COROLLARY 10.D.11. *Let G be a locally compact group.*

(1) *Let π be a normal irreducible representation of G. The character of π can be normalized to be $t_\pi := \mathrm{Tr} \circ \pi$, where Tr is the canonical trace on $\mathcal{L}(\mathcal{H}_\pi)$. The ideal of definition of t_π is the set of $x \in C^*_{\max}(G)$ for which $\pi(x)$ is a trace class operator.*

(2) *Let π and π' be normal irreducible representations of G. Then t_π and $t_{\pi'}$ are proportional if and only π and π' are equivalent.*

PROOF. Item (1) follows from Proposition 10.D.3 and from the fact that Tr is the unique normal semi-finite trace on $\mathcal{L}(\mathcal{H}_\pi)$, up to positive scalar multiples.

Item (2) is a consequence of Item (1), of Proposition 10.D.10, and of the fact that two irreducible representations are equivalent if and only if they are quasi-equivalent. □

EXAMPLE 10.D.12. (1) Let G be a compact group. Characters of G are finite. Corollary 10.D.11 establishes the classical fact that the map $\pi \mapsto \mathrm{Tr}_\pi \circ \pi$ provides a bijection from the dual $\widehat{G}$ of G to the set $\mathrm{Char}(G)$ of its irreducible characters, where Tr_π is the standard trace on $\mathcal{L}(\mathcal{H}_\pi)$.

Note that characters are defined here for factor representations only, whereas, in the setting of finite and compact groups, it is usual to define a character for every finite-dimensional representation of the group.

(2) More generally, let G be a locally compact group of type I. Every irreducible representation is normal (Corollary 10.D.4), and so every character on G is proportional to t_π for a unique irreducible representation π on G, with t_π as in Corollary 10.D.11. In this case, the map $\pi \mapsto t_\pi$ provides again a bijection $\widehat{G} \overset{\approx}{\longrightarrow} \mathrm{Char}(G)$.

(3) Let Γ be a discrete group, not the one-element group. The Dirac function δ_e at the group unit is of positive type and central. The associated representation

is the left regular representation λ_Γ of Γ in $\ell^2(\Gamma)$; it is of finite type and traceable. The group Γ is icc if and only if λ_Γ is factorial (Proposition 7.A.1). When this is the case, λ_Γ is of type II_1 and its character $C^*_{\mathrm{max}}(\Gamma) \to \mathbf{C}$ is given by $f \mapsto \langle \lambda_\Gamma(f)\delta_e \mid \delta_e \rangle$; anticipating on the notation of Section 11.A, we have $\delta_e \in E(\Gamma)$.

(4) Let G be a unimodular LC group, and ν a complex-valued Radon measure on G which is of positive type and central. As discussed in Example 10.C.3, ν defines a lower semi-continuous, semi-finite trace t_ν on $C^*_{\mathrm{max}}(G)$; the associated representation π_ν is factorial (and hence normal) if and only if t_ν is a character. This can be repeated for a unimodular real Lie group G and a distribution ν on G which is of positive type and central.

(5) Let G be a semisimple or nilpotent connected Lie group. Every irreducible representation π of G is normal (since G is of type I) and the corresponding character t_π is given by a central distribution of positive type on G. See [**Hari–54**, § 5] and [**Kiri–62**, § 7].

(6) Let Γ be a countable group and $(X, \mathcal{B}, \mu)$ a standard probability space on which Γ acts by measure preserving transformations. For every $\gamma \in \Gamma$, let

$$X^\gamma = \{x \in X \mid \gamma x = x\}$$

denote the corresponding fixed point set. As we will show below (Theorem 15.F.6), the function

$$\varphi : \Gamma \to \mathbf{R}, \quad \gamma \mapsto \mu(X^\gamma)$$

is central and of positive type, and we will describe the associated GNS representation. Note that φ defines a finite trace t_φ on $C^*_{\mathrm{max}}(\Gamma)$; see Section 11.A.

CHAPTER 11

Finite characters and Thoma's dual

In this chapter, we study the finite part $\mathrm{QD}(G)_{\mathrm{fin}}$ *of the quasi-dual* $\mathrm{QD}(G)$ *of a topological group* G*, that is, the space of quasi-equivalence classes of factor representations of* G *which are of finite type.*

When G *is locally compact,* $\mathrm{QD}(G)_{\mathrm{fin}}$ *is in a bijective correspondence with the set* $\mathrm{Char}(G)_{\mathrm{fin}}$ *of finite characters on* G*, that is, traces on the maximal C*-algebra* $C^*_{\max}(G)$ *of* G *which are everywhere finite (see Section* 11.A*).*

For an arbitrary topological group G*, let* $\mathrm{Tr}(G)$ *be the set of traces of* G*, that is, the set of continuous functions on* G *which are central and of positive type. In Section* 11.B*, we show that the GNS-construction provides a surjective map from* $\mathrm{Tr}(G)$ *to the set of quasi-equivalence classes of representations* π *of* G *such that* $\pi(G)''$ *is a finite von Neumann algebra. In this context, a crucial role is played by the fact that a Hilbert algebra is associated with every trace on* G *(see Chapter* 10.A*). As a consequence, we will see in Section* 11.C *that* $\mathrm{QD}(G)_{\mathrm{fin}}$ *is parametrized by the Thoma dual* $E(G)$*, which is the set of extreme rays of* $\mathrm{Tr}(G)$*.*

In Section 11.D*, we review for a second countable locally compact group* G *the relationship between the quasi-dual spaces* $\mathrm{QD}(G)_{\mathrm{norm}}$ *and* $\mathrm{QD}(G)_{\mathrm{fin}}$*, on the one hand, and the primitive dual* $\mathrm{Prim}(G)$*, on the other hand,*

11.A. Factor representations of finite type and finite characters

Recall that the type of a representation π of a topological group only depends on the quasi-equivalence class of π (see Remark 7.C.3).

DEFINITION 11.A.1. Let G be a topological group. The **finite part of the quasi-dual** $\mathrm{QD}(G)_{\mathrm{fin}}$ is the space of quasi-equivalence classes of factor representations π of G which are of finite type, that is, such that $\pi(G)''$ is either a finite-dimensional factor or a factor of type II_1.

Suppose now that G is locally compact; recall that the set $\mathrm{Char}(G)$ of characters of G has been introduced in Definition 10.D.8. A character $t \in \mathrm{Char}(G)$ is **finite** if t is a finite trace on the maximal C*-algebra $C^*_{\max}(G)$ of G. We denote by $\mathrm{Char}(G)_{\mathrm{fin}}$ the subset of $\mathrm{Char}(G)$ consisting of the finite characters t, normalized by $\|t\| = 1$.

The following fact is a consequence of Proposition 10.D.10.

PROPOSITION 11.A.2. *Let* G *be a locally compact group.*
The map $\pi \mapsto t \circ \pi$ *of Proposition* 10.D.10 *induces a bijection*

$$\mathrm{QD}(G)_{\mathrm{fin}} \overset{\approx}{\longrightarrow} \mathrm{Char}(G)_{\mathrm{fin}}.$$

Let μ_G be a Haar measure on the LC group G. Let t be a finite trace on $C^*_{\max}(G)$. The restriction $t|_{L^1(G,\mu_G)}$ is a positive linear functional on $L^1(G, \mu_G)$

and is therefore defined by a unique continuous function of positive type φ_t on G, such that $\varphi_t(e) = \|t\|$ (see Remark 1.B.16(2) and Example 8.A.9(2)). Since t is a trace, φ_t is central, that is, φ_t is invariant under conjugation by elements of G.

Conversely, let φ be a central continuous function of positive type on G. Then φ defines a positive linear functional on $L^1(G, \mu_G)$ which extends uniquely to a positive linear functional t_φ of $C^*_{\max}(G)$ with $\|t_\varphi\| = \varphi(e)$. Since φ is central, t_φ is a trace on $C^*_{\max}(G)$.

Observe that the maps $t \mapsto \varphi_t$ and $\varphi \mapsto t_\varphi$ are continuous affine maps which are inverse to each other.

Observe also that the set of central continuous functions of positive type φ on G with $\|\varphi\|_\infty = \varphi(e) \leq 1$ is a convex subset of $L^\infty(G, \mu_G)$, which is compact for the weak*-topology. As a consequence of the discussion above, we obtain the following result.

PROPOSITION 11.A.3. *Let G be a locally compact group.*

(1) *The map $t \mapsto \varphi_t$ described above is a bijection between the set of finite traces on $C^*_{\max}(G)$ and the set of central continuous functions of positive type on G.*
(2) *The map $t \mapsto \varphi_t$ restricted to* $\mathrm{Char}(G)_{\mathrm{fin}}$ *is a bijection between* $\mathrm{Char}(G)_{\mathrm{fin}}$ *and the set of central continuous functions of positive type φ on G such that $\varphi(e) = 1$, and which are extremal with these properties.*

When G is a locally compact group, Propositions 11.A.2 and 11.A.3 yield a parametrization of $\mathrm{QD}(G)_{\mathrm{fin}}$ in terms of central continuous functions of positive type on G.

When G is not locally compact, the space $\mathrm{Char}(G)_{\mathrm{fin}}$ is not defined. Nevertheless, as we will see in the next sections, the parametrization of $\mathrm{QD}(G)_{\mathrm{fin}}$ in terms of central continuous functions of positive type on G carries over to an arbitrary topological group G.

11.B. GNS Construction for traces on groups

Let us return to functions of positive type and the GNS construction, keeping the notation of Section 1.B.

Let G be a topological group (not necessarily a locally compact group). Recall that a function φ on G is **central** if $\varphi(gh) = \varphi(hg)$ for all $g, h \in G$.

DEFINITION 11.B.1. (1) A **trace** on G is a continuous function $\varphi : G \to \mathbf{C}$ which is central and of positive type.

Denote by $\mathrm{Tr}(G)$ the space of traces on G. Let $\mathrm{Tr}_{\leq 1}(G)$ be the subset of functions $\varphi \in \mathrm{Tr}(G)$ such that $\|\varphi\|_\infty = \varphi(e) \leq 1$, and $\mathrm{Tr}_1(G)$ be the subset of functions $\varphi \in \mathrm{Tr}_{\leq 1}(G)$ normalized by $\varphi(e) = 1$.

(2) Recall (see Definition 7.C.2) that a representation π of G is of **finite type** if the von Neumann algebra $\pi(G)''$ has a faithful finite normal trace.

The term "trace" as defined above was coined in [**CaMo–84**]. Traces are often called "characters" in the literature (see, for instance, [**DuMe–14**] and [**PeTh–16**]).

REMARK 11.B.2. (1) Let π be a representation of G. Assume that $\pi(G)''$ is countably decomposable (this is, for instance, the case if the Hilbert space of π is separable or if $\pi(G)''$ is a factor). Then $\pi(G)''$ is of finite type if and only if $\pi(G)''$ is a finite von Neumann algebra (see Remark 7.B.8).

(2) Let π be a finite type representation of G and assume that $\pi(G)''$ is a factor. Then $\pi(G)''$ has a *unique* normalized finite normal trace; in fact, there is a unique linear functional τ on $\pi(G)''$ such that $\tau(xy) = \tau(yx)$ for all $x, y \in \pi(G)''$ and $\tau(I) = 1$ (see Corollaire in [**Dixm–vN**, Chap. III, § 5, no 1]).

EXAMPLE 11.B.3. Let G be a topological group.

(1) The constant function 1_G of value 1 is a trace on G. More generally, every unitary character on G is a trace on G.

(2) The characteristic function of an open normal subgroup of G is a trace on G. In particular, when the group is discrete, the Dirac function supported at the unit element of the group is a trace on the group.

(3) (Compare with Definition 1.B.4.) Let π be a finite type representation of G and τ a faithful finite normal trace on $\pi(G)''$. Then the continuous function $\varphi_{\pi,\tau}$ defined by

$$\varphi_{\pi,\tau}(g) = \tau(\pi(g)) \qquad \text{for all } g \in G$$

is clearly central and nonzero. It is also of positive type; indeed, for $n \geq 1$, $c_1, \ldots, c_n \in \mathbf{C}$, and $g_1, \ldots, g_n \in G$, we have

$$\sum_{i,j=1}^{n} c_i \overline{c_j} \varphi_{\pi,\tau}(g_j^{-1} g_i) = \tau\Big(\big(\sum_{i=1}^{n} c_i \pi(g_i)\big)\big(\sum_{j=1}^{n} c_j \pi(g_j)\big)^*\Big) \geq 0.$$

Therefore $\varphi_{\pi,\tau} \in \mathrm{Tr}(G)$. If τ is normalized, then $\varphi_{\pi,\tau} \in \mathrm{Tr}_1(G)$.

CONSTRUCTION 11.B.4 (Compare with Construction 1.B.6). The present construction shows that every nonzero element in $\mathrm{Tr}(G)$ arises as in the previous example. More precisely, it shows that, given $\varphi \in \mathrm{Tr}(G)$, $\varphi \neq 0$, there exists a representation π_φ of G (the GNS representation of Construction 1.B.6) and a faithful finite normal trace τ_φ on $\pi_\varphi(G)''$ such that $\varphi(g) = \tau_\varphi(\pi_\varphi(g))$ for all $g \in G$.

Let $\varphi \in \mathrm{Tr}(G)$, $\varphi \neq 0$. Recall from Section 1.B how the corresponding GNS triple $(\pi_\varphi, \mathcal{H}_\varphi, \xi_\varphi)$ is constructed.

The set $\mathbf{C}[G]$ of complex-valued functions on G with finite support is an involutive algebra endowed with a positive Hermitian form Φ_φ, and the set J_φ of functions f such that $\Phi_\varphi(f, f) = 0$ is a left ideal. We denote by $\mathcal{A}_\varphi$ the quotient $\mathbf{C}[G]/J_\varphi$. We write $\delta_g \in \mathbf{C}[G]$ for the characteristic function of $\{g\}$ for each $g \in G$, and $[f]$ for the class in $\mathcal{A}_\varphi$ of a function $f \in \mathbf{C}[G]$. We denote by $\mathcal{H}_\varphi$ the Hilbert space obtained by completion of $\mathcal{A}_\varphi$, with respect to the scalar product induced on $\mathcal{A}_\varphi$ by Φ_φ, and by ξ_φ the vector $[\delta_e]$ viewed in $\mathcal{H}_\varphi$. For $g \in G$, the operator $\pi_\varphi(g)$ on $\mathcal{H}_\varphi$ is defined as in Construction 1.B.6 to be the continuous linear extension to $\mathcal{H}_\varphi$ of the left translation $[f] \mapsto [g \cdot f]$, where $(g \cdot f)(x) = f(g^{-1}x)$ for $g, x \in G$ and $f \in \mathcal{A}_\varphi$.

As φ is central, J_φ is now a *two-sided* $*$-ideal, the quotient $\mathcal{A}_\varphi = \mathbf{C}[G]/J_\varphi$ is an involutive algebra, and there is for all $g \in G$ an operator $\rho_\varphi(g)$ on $\mathcal{H}_\varphi$ which is the continuous linear extension of the *right translation* $[f] \mapsto [f \cdot g]$, where $(f \cdot g)(x) = f(xg)$ for $g, x \in G$ and $f \in \mathcal{A}_\varphi$. One checks, as for π_φ in Section 1.B, that the assignment $g \mapsto \rho_\varphi(g)$ is continuous, and is a representation of G on $\mathcal{H}_\varphi$ with cyclic vector $\xi_\varphi = [\delta_e]$. We have $\pi_\varphi(g)\xi_\varphi = [\delta_g]$ and

$$\pi_\varphi(g)\rho_\varphi(h) = \rho_\varphi(h)\pi_\varphi(g) \qquad \text{for all } g, h \in G.$$

Note: For coherence with Section 10.A, we could have written λ_φ instead of π_φ. We have chosen the latter for coherence with Section 1.B.

We are going to show that the von Neumann algebras $\pi_\varphi(G)''$ and $\rho_\varphi(G)''$ are each commutants and have a faithful finite normal trace.

PROPOSITION 11.B.5. *Let G be a topological group and $\varphi \in \mathrm{Tr}(G)$, $\varphi \neq 0$. Let $\mathcal{A}_\varphi, \Phi_\varphi, \pi_\varphi$ and ρ_φ be as in Definition* 11.B.4.

(a) *The involutive algebra $\mathcal{A}_\varphi$, equipped with the scalar product induced by Φ_φ, is a Hilbert algebra.*
(b) *We have $\pi_\varphi(G)'' = \rho_\varphi(G)'$ and $\rho_\varphi(G)'' = \pi_\varphi(G)'$. As a consequence, $\pi_\varphi(G)' \cap \rho_\varphi(G)'$ is the common centre of the von Neumann algebras $\pi_\varphi(G)''$ and $\rho_\varphi(G)''$.*
(c) *The natural trace τ_φ associated to $\mathcal{A}_\varphi$ as in Theorem* 10.A.8 *is a faithful finite normal trace on $\pi_\varphi(G)''$ and is given by*

$$\tau_\varphi : T \mapsto \langle T\xi_\varphi \mid \xi_\varphi \rangle;$$

in particular, we have $\varphi = \tau_\varphi \circ \pi_\varphi$.

PROOF. (a) Property (1) of the definition of a Hilbert algebra (Definition 10.A.2), follows from the fact that φ is central. It is straightforward to check Properties (2) and (3), and Property (4) is obvious since $\mathcal{A}_\varphi$ is unital.

Claim (b) follows from (a) and from the Commutation Theorem 10.A.4.

(c) Observe that the cyclic vector $\xi_\varphi = [\delta_e]$ belongs to the set $(\mathcal{A}_\varphi)_b$ of bounded elements of the Hilbert algebra $\mathcal{A}_\varphi$; the corresponding operator $\pi_\varphi(\xi_\varphi) \in \pi_\varphi(G)''$ is the identity operator I on $\mathcal{H}_\varphi$.

Let $T \in \pi_\varphi(G)''$. Then $T = \pi_\varphi(\xi_\varphi) \in \pi_\varphi(\mathcal{A}_\varphi)_b$ (see Lemma 10.A.3). It follows (see Theorem 10.A.8) that the natural trace τ_φ associated to $\mathcal{A}_\varphi$ has $\pi_\varphi(G)''$ as the ideal of definition and that

$$\tau_\varphi(T) = \tau_\varphi\left(\pi_\varphi(\xi_\varphi)^*\pi_\varphi(T\xi_\varphi)\right) = \langle T\xi_\varphi \mid \xi_\varphi \rangle.$$

The fact that τ_φ is a faithful finite normal trace on $\pi_\varphi(G)''$ follows from Theorem 10.A.8. This fact can also be checked directly, with arguments as in the proof of Proposition 7.A.3. □

Let π be a finite type representation of G and τ a normalized faithful finite normal trace on $\pi(G)''$. Recall from Example 11.B.3 that $\varphi_{\pi,\tau} = \tau \circ \pi$ belongs to $\mathrm{Tr}_1(G)$.

An equivalence relation was defined in Section 10.C for arbitrary trace representations in the case where G is *locally compact.* We extend this definition as follows for finite type representations of a general topological group.

DEFINITION 11.B.6. For $i = 1, 2$, let π_i be a finite type representation of the topological group G and let τ_i be a normalized faithful finite normal trace on the von Neumann algebra $\pi_i(G)''$. We say that the pairs (π_1, τ_1) and (π_2, τ_2) are **quasi-equivalent** if there exists an isomorphism $\Phi : \pi_1(G)'' \to \pi_2(G)''$ such that $\Phi(\pi_1(g)) = \pi_2(g)$ for all $g \in G$, and such that $\tau_1 = \tau_2 \circ \Phi$. It is clear in this case that the associated functions $\varphi_{\pi_1,\tau_1} = \tau_1 \circ \pi_1$ and $\varphi_{\pi_2,\tau_2} = \tau_2 \circ \pi_2$ in $\mathrm{Tr}_1(G)$ are equal.

The next proposition, which appears in [**HiHi–05**], shows that the converse statement is also true. Observe that, when G is locally compact, this proposition follows easily from Proposition 11.A.3 combined with Proposition 10.C.6.

PROPOSITION 11.B.7 (compare with Proposition 1.B.10). *Let G be a topological group.*

The assignments $\varphi \rightsquigarrow (\pi_\varphi, \tau_\varphi)$ given by the GNS construction and $(\pi, \tau) \mapsto \tau \circ \pi$ induce bijections inverse to each other between

(i) *the set $\mathrm{Tr}_1(G)$ of normalized central functions of positive type φ on G, and*
(ii) *the set of quasi-equivalence classes of pairs (π, τ), where π is a finite type representation of G and τ a normalized faithful finite normal trace on $\pi(G)''$.*

PROOF. We first recall some facts about the standard representation (see Section 10.B) in the special case of a **finite** von Neumann algebra.

Let $\mathcal{M}$ be a finite von Neumann algebra, with a faithful normalized finite normal trace τ. Equipped with the scalar product $(x, y) \mapsto \tau(y^*x)$, the space $\mathcal{M}$ is a Hilbert algebra. Denote by $L^2(\mathcal{M}, \tau)$ the Hilbert space completion of $\mathcal{M}$ and let

$$\mathcal{M} \to \widehat{\mathcal{M}}, \qquad x \mapsto \widehat{x},$$

be the standard representation of $(\mathcal{M}, \tau)$ on $L^2(\mathcal{M}, \tau)$. It is clear that the image ξ of I in $L^2(\mathcal{M}, \tau)$ is a cyclic vector for the representation $x \mapsto \widehat{x}$; the natural trace $\widehat{\tau}$ on $\widehat{\mathcal{M}}$ is given by

$$\widehat{\tau}(\widehat{x}) \,=\, \langle \widehat{x}\xi \mid \xi \rangle \,=\, \tau(x) \qquad \text{for all } x \in \mathcal{M}.$$

For $j = 1, 2$, let $\pi_j : G \to \mathcal{U}(\mathcal{H}_j)$ be a representation such that the von Neumann algebra $\mathcal{M}_j := \pi_j(G)''$ has a faithful normalized finite normal trace τ_j. Assume that $\varphi_{\pi_1, \tau_1} = \varphi_{\pi_2, \tau_2}$. We have to show that the pairs (π_1, τ_1) and (π_2, τ_2) are quasi-equivalent.

Still for $j = 1, 2$, consider the standard representation $x \mapsto \widehat{x}$ of the Neumann algebra $\mathcal{M}_j$ on $L^2(\mathcal{M}_j, \tau_j)$. The map $\Phi_j : x \mapsto \widehat{x}$ is an isomorphism between the von Neumann algebras $\mathcal{M}_j$ and $\widehat{\mathcal{M}_j} := \{\widehat{x} : x \in \mathcal{M}_|\}$, and $\widehat{\tau_j} \circ \Phi_j = \tau_j$, where $\widehat{\tau_j}$ is the natural trace on $\widehat{\mathcal{M}_j}$. Therefore the pair (π_j, τ_j) is quasi-equivalent to the pair $(\widehat{\pi_j}, \widehat{\tau_j})$, where $\widehat{\pi_j}$ is the representation of G on $L^2(\mathcal{M}_j, \tau_j)$ defined by $\widehat{\pi_j}(g) = \widehat{\pi_j(g)}$ for $g \in G$. As a consequence, it suffices to show that the pairs $(\widehat{\pi_1}, \widehat{\tau_1})$ and $(\widehat{\pi_2}, \widehat{\tau_2})$ are quasi-equivalent.

Observe that, by assumption, we have

$$\widehat{\tau_1}(g) \,=\, \varphi_{\pi_1, \tau_1}(g) \,=\, \varphi_{\pi_2, \tau_2}(g) \,=\, \widehat{\tau_2}(g),$$

that is,

$$\langle \widehat{\pi_1}(g)\xi_1 \mid \xi_1 \rangle \,=\, \langle \widehat{\pi_2}(g)\xi_2 \mid \xi_2 \rangle,$$

for all $g \in G$. Since, for all $n \geq 1$, $g_1, \ldots, g_n \in G$ and $c_1, \ldots, c_n \in \mathbf{C}$, we have

$$\Big\Vert \sum_{i=1}^{n} c_i \widehat{\pi_2}(g_i)\xi_2 \Big\Vert^2 \,=\, \sum_{i,j=1}^{n} c_i \overline{c_j} \widehat{\tau_2}(g_j^{-1} g_i) \,=\, \sum_{i,j=1}^{n} c_i \overline{c_j} \widehat{\tau_1}(g_j^{-1} g_i) \,=\, \Big\Vert \sum_{i=1}^{n} c_i \widehat{\pi_1}(g_i)\xi_1 \Big\Vert^2,$$

the map $\sum_{i=1}^n c_i \widehat{\pi_1}(g_i)\xi_1 \mapsto \sum_{i=1}^n c_i \widehat{\pi_2}(g_i)\xi_2$ is well defined on the linear span of $\widehat{\pi_1}(G)\xi_1$ and extends to a Hilbert space isomorphism $U : L^2(\mathcal{M}_1, \tau_1) \to L^2(\mathcal{M}_2, \tau_2)$. Let

$$\Phi : \mathcal{L}(L^2(\mathcal{M}_1, \tau_1)) \to \mathcal{L}(L^2(\mathcal{M}_2, \tau_2)), \qquad T \mapsto UTU^{-1}.$$

Then Φ restricts to an isomorphism from $\widehat{\mathcal{M}_1}$ to $\widehat{\mathcal{M}_2}$, such that $\Phi(\widehat{\pi_1}(g)) = \widehat{\pi_2}(g)$ for every $g \in G$. Moreover, we have $\widehat{\tau_1} = \widehat{\tau_2} \circ \Phi$. Therefore $(\widehat{\pi_1}, \widehat{\tau_1})$ and $(\widehat{\pi_2}, \widehat{\tau_2})$ are quasi-equivalent, and this concludes the proof. □

11.C. Thoma's dual

Let G be a topological group. The sets of traces $\mathrm{Tr}_{\leq 1}(G)$ and $\mathrm{Tr}_1(G)$ as in Definition 11.B.1 are convex subsets of the vector space of all complex-valued functions on G.

DEFINITION 11.C.1. Define $\mathrm{Extr}(\mathrm{Tr}_{\leq 1}(G))$ as the set of functions φ in $\mathrm{Tr}_{\leq 1}(G)$ that are **indecomposable** in the following sense: if $\varphi = t\varphi_1 + (1-t)\varphi_2$ for $\varphi_1, \varphi_2 \in \mathrm{Tr}_{\leq 1}(G)$ and $t \in \,]0,1[$, then $\varphi = \varphi_1 = \varphi_2$. Similarly define $\mathrm{Extr}(\mathrm{Tr}_1(G))$, and call it the **Thoma dual** of G. From now on, we use Thoma's notation, $E(G)$, rather than $\mathrm{Extr}(\mathrm{Tr}_1(G))$.

Thoma used properties of the space $E(G)$ in order to achieve his characterization of discrete groups which are of type I (Theorem 7.D.1) and he advocated its use as a substitute for the dual of discrete groups [**Thom–64a, Thom–64b**].

The following lemma, which relates $E(G)$ and $\mathrm{Extr}(\mathrm{Tr}_{\leq 1}(G))$, is proved exactly in the same way as Lemma 1.B.12, which relates $\mathrm{Extr}(P_1(G))$ and $\mathrm{Extr}(P_{\leq 1}(G))$.

LEMMA 11.C.2 (compare with Lemma 1.B.12). *For every topological group G, we have*

$$\mathrm{Extr}(\mathrm{Tr}_{\leq 1}(G)) \;=\; E(G) \cup \{0\}.$$

Proposition 11.C.4 provides a necessary and sufficient condition on φ for the GNS representation π_φ of Sections 1.B and 11.B to be factorial.

Let $\varphi \in \mathrm{Tr}(G)$ and let $\psi \in \mathrm{Tr}(G)$ be such that $\psi \leq \varphi$, that is, such that $\varphi - \psi \in P(G)$; it is then obvious that $\varphi - \psi$ belongs to $\mathrm{Tr}(G)$. For $T \in \pi_\varphi(G)'$ with $0 \leq T \leq I$, define $\varphi_T \in P(G)$ as in Section 1.B by

$$\varphi_T(g) \;=\; \langle \pi_\varphi(g) T\xi_\varphi \mid \xi_\varphi \rangle \;=\; \langle \pi_\varphi(g) T^{1/2}\xi_\varphi \mid T^{1/2}\xi_\varphi \rangle.$$

LEMMA 11.C.3 (compare with Lemma 1.B.13). *Let $\varphi \in \mathrm{Tr}(G)$; let π_φ and ρ_φ be as in Construction* 11.B.4.

The map $T \to \varphi_T$ is a bijective correspondence from the set of operators T in $\pi_\varphi(G)' \cap \rho_\varphi(G)'$ with $0 \leq T \leq I$ to the set of traces $\psi \in \mathrm{Tr}(G)$ with $\psi \leq \varphi$.

PROOF. For $T \in \pi_\varphi(G)' \cap \rho_\varphi(G)'$ with $0 \leq T \leq I$, the function φ_T is central; indeed, for $g, x \in G$, we have

$$\begin{aligned}
\varphi_T(x^{-1}gx) \;&=\; \langle \pi_\varphi(g)\pi_\varphi(x)T\xi_\varphi \mid \pi_\varphi(x)\xi_\varphi \rangle \;=\; \langle \pi_\varphi(g)T\pi_\varphi(x)\xi_\varphi \mid \pi_\varphi(x)\xi_\varphi \rangle \\
&=\; \langle \pi_\varphi(g)T\pi_\varphi(x)\xi_\varphi \mid \rho_\varphi(x^{-1})\xi_\varphi \rangle \;=\; \langle \pi_\varphi(g)T\pi_\varphi(x)\rho_\varphi(x)\xi_\varphi \mid \xi_\varphi \rangle \\
&=\; \langle \pi_\varphi(g)T\xi_\varphi \mid \xi_\varphi \rangle \;=\; \varphi_T(g).
\end{aligned}$$

Moreover, since $\varphi - \varphi_T = \varphi_{I-T}$, we have $\varphi_T \leq \varphi$.

It follows from Lemma 1.B.13 that the map $T \mapsto \varphi_T$ is injective.

Let $\psi \in \mathrm{Tr}(G)$ with $\psi \leq \varphi$. The proof of Lemma 1.B.13 shows that there exists $T \in \pi_\varphi(G)'$ with $0 \leq T \leq I$ such that $\psi = \varphi_T$. It remains to show that $T \in \rho_\varphi(g)'$.

We have, for $x, y, z \in G$,

$$\begin{aligned}\psi(y(z^{-1}x)y^{-1}) &= \langle \pi_\varphi(y(z^{-1}x)y^{-1})\xi_\varphi \mid T\xi_\varphi\rangle \\ &= \langle \pi_\varphi(z^{-1}x)\pi_\xi(y^{-1})\xi_\varphi \mid T\pi_\varphi(y^{-1})\xi_\varphi\rangle \\ &= \langle \pi_\varphi(z^{-1}x)\rho_\xi(y)\xi_\varphi \mid T\rho_\varphi(y)\xi_\varphi\rangle \\ &= \langle \pi_\varphi(x)\xi_\varphi \mid \rho_\xi(y^{-1})T\rho_\varphi(y)\pi_\varphi(z)\xi_\varphi\rangle.\end{aligned}$$

However, since ψ is central, we also have

$$\psi(y(z^{-1}x)y^{-1}) = \psi(z^{-1}x) = \langle \pi_\varphi(x)\xi_\varphi \mid T\pi_\varphi(z)\xi_\varphi\rangle.$$

Because ξ_φ is a cyclic vector, it follows that $\rho_\varphi(y^{-1})T\rho_\varphi(y) = T$ for all $y \in G$, that is, $T \in \rho_\varphi(G)'$. □

PROPOSITION 11.C.4 (compare with Proposition 1.B.14). *Let G be a topological group. Let $\varphi \in \mathrm{Tr}_1(G)$, and π_φ be the corresponding GNS representation, as in Construction 11.B.4 and Proposition 11.B.7.*

Then π_φ is factorial if and only if $\varphi \in E(G)$.

PROOF. Assume that π_φ is factorial; let $\varphi_1, \varphi_2 \in \mathrm{Tr}_1(G)$ and $t \in \,]0,1[$ be such that $\varphi = t\varphi_1 + (1-t)\varphi_2$. Let $j \in \{1,2\}$; since $\varphi_j \le \varphi$, there exists by Lemma 11.C.3 $T_j \in \pi_\varphi(G)' \cap \rho_\varphi(G)'$ with $0 \le T_j \le I$ such that $\varphi_j = \varphi_{T_j}$. Since π_φ is factorial, $\pi_\varphi(G)' \cap \rho_\varphi(G)' = \pi_\varphi(G)' \cap \pi_\varphi(G)'' = \mathrm{Id}_{\mathcal{H}_\varphi}$ (see Proposition 11.B.5), so that both T_1 and T_2 are scalar multiples of the identity. It follows that $\varphi_1 = \varphi_2 = \varphi$, and consequently that $\varphi \in E(G)$.

For the converse implication, we leave it to the reader to adapt the proof of Proposition 1.B.14. □

For $i = 1, 2$, let π_i be a *factor* representation of finite type of G and let τ_i be the unique normalized trace on $\pi_i(G)''$ (see Remark 11.B.2). If the representations π_1 and π_2 are quasi-equivalent, then the pairs (π_1, τ_1) and (π_2, τ_2) are quasi-equivalent. Indeed, let Φ be a quasi-equivalence between π_1 and π_2; then $\tau_2 \circ \Phi$ is a normalized finite normal trace on $\pi_1(G)''$, and hence $\tau_2 \circ \Phi = \tau_1$.

The following result is therefore a consequence of Proposition 11.C.4. Observe that Item (2) was already established in Proposition 11.A.3.

THEOREM 11.C.5. *Let G be a topological group. For $\varphi \in E(G)$, let π_φ and τ_φ be as in* 11.B.4 *and* 11.B.7.

(1) *The map*

$$E(G) \to \mathrm{QD}(G)_{\mathrm{fin}}, \qquad \varphi \mapsto \pi_\varphi$$

establishes a bijection from the Thoma dual onto the set of quasi-equivalence classes of factor representations of finite type of G.

(2) *Assume that G is locally compact. The map*

$$E(G) \to \mathrm{Char}(G)_{\mathrm{fin}}, \qquad \varphi \mapsto t_\varphi := \tau_\varphi \circ \pi_\varphi$$

is a bijection from the Thoma dual onto the set of finite normalized characters of G.

EXAMPLE 11.C.6. Let G be a topological group.

(1) When G is locally compact and abelian, the Thoma dual $E(G)$ is naturally identified with the Pontryagin dual $\widehat{G}$. When G is compact, $E(G)$ is the set of normalized traces of irreducible representations of G. These are particular cases of Theorem 11.C.5.

(2) When G is a locally compact group of type I, $E(G)$ is the set of normalized traces of finite-dimensional irreducible representations of G. See Theorem 11.C.5(1) and Corollary 10.D.4.

For example, the Thoma dual of the infinite dihedral group D_∞ consists of the normalized characters of the irreducible representations

$$\rho_{0,1},\ \rho_{0,-1},\ \rho_{\pi,1},\ \rho_{\pi,-1}, \quad \text{and } \{\rho_\theta\}_{0<\theta<\pi},$$

explicitly written in Section 3.A.

(3) When G is a connected LC group, $E(G)$ consists exactly of the normalized traces of finite-dimensional irreducible unitary representations of G (see Corollary 15.D.3 below).

(4) For some results about $E(G)$ in the case where G is a totally disconnected LC group, we refer to Section 15.D. For instance, it is shown there (Corollary 15.D.7) that if, moreover, G is compactly generated and if the only open normal subgroup of G is G itself, then $E(G) = \{1_G\}$. This applies for instance to $G = \mathrm{PSL}_n(\mathbf{Q}_p)$ or to $G = \mathrm{Aut}^+(T)$, the group of type preserving automorphisms of a locally finite regular tree T.

(5) The Thoma duals of the discrete groups appearing as the main examples of Chapter 3 will be described in Chapter 12.

EXAMPLE 11.C.7. The problem of the description of $E(\Gamma)$ for other *discrete* groups Γ has been considered by several authors, as the following list shows:

(1) For the infinite symmetric group $\Gamma = \mathrm{Sym}_{\mathrm{fin}}(\mathbf{N})$, the group of permutations with finite supports of a countable infinite set, see [**Thom–64c**] and, for other methods, [**VeKe–81**].
(2) For the group $\Gamma = \mathrm{GL}(\infty, \mathbf{F})$, where $\mathbf{F}$ is a finite field, see [**Skud–76**], and also [**GoKV–14**].
(3) For Γ a nilpotent discrete group, see among others [**Smir–66**], [**Howe–77a**], and [**CaMo–84**].
(4) For the group $\Gamma = \mathrm{SL}_n(\mathbf{Z}), n \geq 3$, see [**Bekk–07**]. For the extension to the case where Γ is a lattice in a simple connected Lie group of real rank ≥ 2, see [**Pete–15**] and [**BoHo–19**].
(5) For Γ from the Higman–Thompson families $F_{n,r}$ or $G_{n,r}$, see [**DuMe–14**].
(6) For $\Gamma = \mathbf{G}(\mathbf{k})$ the group of $\mathbf{k}$-points of a simple algebraic group $\mathbf{G}$ over an infinite field $\mathbf{k}$, see [**Bekk–20**]; for the case where $\mathbf{k} = \mathbf{Q}$ and $\mathbf{G}$ is a general algebraic group over $\mathbf{Q}$, see [**BeFr–20**].

Here are the few results concerning *non locally compact* groups G to which we are aware.

(7) There is a description of $E(G)$ for $G = \mathcal{U}(\infty)$, the inductive limit of the unitary groups $\mathcal{U}(n)$ equipped with the inductive limit topology, in [**Voic–76**], [**StVo–75**], and [**Boye–83**].

(8) $E(G)$ has been described in [**Dudk–11**] for G the full group of an ergodic hyperfinite equivalence relation; this result has been extended in [**CaLe–16**] to so-called orbit full groups.

We end this section with two elementary lemmas on central functions of positive type, to be used in the next chapter.

Let G be a topological group. Let $\varphi \in P_1(G)$ be a normalized continuous function of positive type on G. Set

$$P_\varphi = \{g \in G \mid |\varphi(g)| = 1\} \quad \text{and} \quad K_\varphi = \{g \in G \mid \varphi(g) = 1\},$$

which are closed subsets of G containing e, and $P_\varphi \supset K_\varphi$. Let $(\mathcal{H}_\varphi, \xi_\varphi, \pi_\varphi)$ be the GNS triple defined by φ, as in Construction 1.B.6. Let $\chi_\varphi = \chi_{\pi_\varphi}$ be the associated unitary character of the projective kernel $\mathrm{Pker}\pi_\varphi$ of π_φ as in Section 8.E.

Recall from Proposition 8.E.1 that every factor representation has a central character.

LEMMA 11.C.8. *Let G be a topological group, $\varphi \in P_1(G)$, and $\pi_\varphi, P_\varphi, K_\varphi$ as above.*

(1) *We have*

$$P_\varphi = \{g \in G \mid \pi_\varphi(g)\xi_\varphi = \varphi(g)\xi_\varphi\}$$

and

$$K_\varphi = \{g \in G \mid \pi_\varphi(g)\xi_\varphi = \xi_\varphi\}.$$

In particular, P_φ and K_φ are closed subgroups of G, and the restriction $\varphi|_{P_\varphi}$ is a unitary character $P_\varphi \to \mathbf{T}$.

(2) *We have $\varphi(g_1 g g_2) = \varphi(g_1)\varphi(g)\varphi(g_2)$ for all $g_1, g_2 \in P_\varphi$ and $g \in G$.*

Assume moreover that φ is central, i.e., that $\varphi \in \mathrm{Tr}_1(G)$.

(3) *We have $P_\varphi = \mathrm{Pker}\pi_\varphi$ and $K_\varphi = \ker \pi_\varphi$. Moreover, $\varphi|_{P_\varphi}$ coincides with the unitary character χ_φ of $\mathrm{Pker}\pi_\varphi$ associated with π_φ.*

Assume moreover that φ is indecomposable, i.e., that $\varphi \in E(G)$.

(4) *The restriction $\varphi|_Z$ of φ to the centre Z coincides with the central character of π_φ.*

PROOF. (1) For every $g \in G$, we have $|\varphi(g)| = |\langle \pi_\varphi(g)\xi_\varphi \mid \xi_\varphi\rangle| \leq 1$ by the Cauchy–Schwarz inequality. By the equality case, if $g \in P_\varphi$, then $\pi_\varphi(g)\xi_\varphi$ is a scalar multiple of ξ_φ, and this scalar multiple is precisely $\varphi(g)$. Consequently,

$$P_\varphi = \{g \in G \mid \ \pi_\varphi(g)\xi_\varphi = \varphi(g)\xi_\varphi\},$$

P_φ is a subgroup of G, and the map $P_\varphi \to \mathbf{T}$, $g \mapsto \varphi(g)$ is a unitary character, with kernel K_φ.

(2) For $g_1, g_2 \in P_\varphi$ and $g \in G$, it follows from (1) that

$$\begin{aligned} \varphi(g_1 g g_2) &= \langle \pi_\varphi(g_1 g g_2)\xi_\varphi \mid \xi_\varphi\rangle = \langle \pi_\varphi(g)\pi_\varphi(g_2)\xi_\varphi \mid \pi_\varphi(g_1^{-1})\xi_\varphi\rangle \\ &= \langle \pi_\varphi(g)\varphi(g_2)\xi_\varphi \mid \varphi(g_1^{-1})\xi_\varphi\rangle = \varphi(g_1)\varphi(g)\varphi(g_2). \end{aligned}$$

(3) Let $g \in G$; the triple $(\mathcal{H}_\varphi, \pi(g^{-1})\xi_\varphi, \pi_\varphi)$ is a GNS triple for φ^g, where φ is the function of positive type on G defined by $\varphi^g(x) = \varphi(gxg^{-1})$.

Assume now that φ is central and let $x \in P_\varphi$. Since $P_\varphi = P_{\varphi^g}$, it follows from (1) that

$$\pi_\varphi(x)\pi_\varphi(g^{-1})\xi_\varphi = \varphi^g(x)\pi_\varphi(g^{-1})\xi_\varphi = \varphi(x)\pi_\varphi(g^{-1})\xi_\varphi \quad \text{for all } g \in G$$

and hence that $\pi_\varphi(x) = \varphi(x)I$, since ξ_φ is a cyclic vector for π_φ. This shows that P_φ is contained in $\mathrm{Pker}\pi$. Since $\mathrm{Pker}\pi$ is obviously contained in P_φ, it follows that $P_\varphi = \mathrm{Pker}\pi$; it is obvious that $\varphi|_{P_\varphi}$ coincides with χ_φ and that $K_\varphi = \ker_{\pi_\varphi}$.

(4) Since $\varphi \in E(\Gamma)$, the representation π_φ is factorial and the claim follows from Proposition 8.E.1. □

The next elementary lemma will be the crucial tool in our study of traces on two-step nilpotent groups. Recall that $[g, h]$ denotes the commutator $g^{-1}h^{-1}gh$ of two elements $g, h \in G$.

LEMMA 11.C.9. *Let G be a topological group, $\varphi \in \mathrm{Tr}(G)$, and $g \in G$. Assume that there exists $h \in G$ such that $[g, h] \in P_\varphi \smallsetminus K_\varphi$.*

Then $\varphi(g) = 0$.

PROOF. Using the fact φ is central and Lemma 11.C.8, we have

$$\varphi(g) \;=\; \varphi(h^{-1}gh) \;=\; \varphi(g[g,h]) \;=\; \varphi(g)\varphi([g,h]),$$

and the claim follows. □

11.D. Characters and primitive duals

Let G be a second-countable LC group. As Theorem 10.D.7 shows, $\mathrm{QD}(G)_{\mathrm{norm}}$ is a standard Borel set. One may thus seek to parametrize $\mathrm{Prim}(G)$ by the elements of $\mathrm{QD}(G)_{\mathrm{norm}}$ and consider the restriction to $\mathrm{QD}(G)_{\mathrm{norm}}$ of the map (κ3) of Corollary 8.D.4:

$$(\kappa 4) \qquad \kappa_{\mathrm{prim}}^{\mathrm{norm}} : \mathrm{QD}(G)_{\mathrm{norm}} \to \mathrm{Prim}(G), \qquad \pi \mapsto \mathrm{C^*ker}(\pi).$$

In contrast to the previous maps $\kappa_{\mathrm{prim}}^{d}$ and $\kappa_{\mathrm{prim}}^{\mathrm{qd}}$, it is not known whether $\kappa_{\mathrm{prim}}^{\mathrm{norm}}$ is always surjective.

When π_1 and π_2 are *normal* irreducible representations such that $\kappa_{\mathrm{prim}}^{\mathrm{norm}}(\pi_1) = \kappa_{\mathrm{prim}}^{\mathrm{norm}}(\pi_2)$, then π_1 and π_2 are equivalent (this follows from Proposition 10.D.3 and [**Dixm–C***, 4.1.10]). So, $\kappa_{\mathrm{prim}}^{\mathrm{norm}}$ is injective on the type I part of $\mathrm{QD}(G)_{\mathrm{norm}}$. However, $\kappa_{\mathrm{prim}}^{\mathrm{norm}}$ is not injective in general on the type II part of $\mathrm{QD}(G)_{\mathrm{norm}}$.

Indeed, for a prime p, consider the Baumslag–Solitar group $\mathrm{BS}(1,p)$ as in Section 3.D. As we will see in Corollary 14.E.4, there exist uncountably many normal factor representations π_i of Γ of type II_1 such that $\kappa_{\mathrm{prim}}^{\mathrm{norm}}(\pi_i) = \kappa_{\mathrm{prim}}^{\mathrm{norm}}(\pi_j)$ for all i and j and such that π_i and π_j are not quasi-equivalent for $i \neq j$.

The map $\kappa_{\mathrm{prim}}^{\mathrm{norm}}$ also fails to be injective in the case of Γ the group $\mathrm{Sym}_{\mathrm{fin}}(\mathbf{N})$ of bijections of $\mathbf{N}$ with finite support, as shown in [**Haue–86**].

On the positive side, here is a remarkable result of Pukánszky for the class of connected Lie groups.

THEOREM 11.D.1 (**Pukánszky**). *Let G be a connected real Lie group. The map*

$$\kappa_{\mathrm{prim}}^{\mathrm{norm}} : \mathrm{QD}(G)_{\mathrm{norm}} \to \mathrm{Prim}(G)$$

is a bijection.

REFERENCES FOR THE PROOF. The result has first appeared as Theorem 1 in [**Puka–74**]. An exposition of the proof is given in Chapter III of the monograph [**Puka–99**]. Pukánszky's Theorem can now be deduced from results by Poguntke; see the Corollary to Theorem 2 in [**Pogu–83**]. □

Moreover, in the case of G a solvable connected and simply connected Lie group, [**Puka–74**] and [**Puka–99**] provide an explicit description of the (usually infinite) characters associated to the normal factor representations of G in terms of a "generalized Kirillov theory", involving orbits of the co-adjoint action of G on its Lie algebra.

There are only few nonconnected LC groups G for which it is known that the map $\kappa_{\mathrm{prim}}^{\mathrm{norm}}$ is a bijection. There is some evidence that this might be true for any nilpotent LC group. For what we know:

PROPOSITION 11.D.2. *Let Γ be a nilpotent-by-finite discrete group.*

(1) *Every character of Γ is finite, that is,* $\mathrm{Char}(\Gamma) = \mathrm{Char}(\Gamma)_{\mathrm{fin}}$*;*
(2) *the map $\kappa_{\mathrm{prim}}^{\mathrm{norm}}$ is surjective.*

NOTE. By Proposition 10.D.10 and Theorem 11.C.5, the equality in (1) could equally well be written $\mathrm{QD}(\Gamma)_{\mathrm{norm}} = \mathrm{QD}(\Gamma)_{\mathrm{fin}}$, and the sets

$$\mathrm{QD}(\Gamma)_{\mathrm{norm}}, \quad \mathrm{Char}(\Gamma), \quad \text{and} \quad E(\Gamma)$$

are in natural bijection with each other.

REFERENCES FOR THE PROOF. The proof was given in [**CaMo–84**] for nilpotent groups only; however, the arguments there carry over to nilpotent-by-finite groups.

Indeed, the proof of (1) is an easy consequence of the fact, which was mentioned in Remark 1.E.4(3), that every primitive ideal of a nilpotent group is maximal, and this is still true for a nilpotent-by-finite group (see [**Pogu–82**]). Claim (2) is a consequence of the amenability of Γ and, again, of the fact that primitive ideals of $C^*_{\mathrm{max}}(\Gamma)$ are maximal. □

For a topological group G, we have a canonical map $E(G) \to \mathrm{Prim}(G)$ which is the composition of the natural bijection $E(G) \to \mathrm{QD}(G)_{\mathrm{fin}}$ of Theorem 11.C.5, the inclusion $\mathrm{QD}(G)_{\mathrm{fin}} \subset \mathrm{QD}(G)_{\mathrm{norm}}$ (see Example 10.D.5(2)), and the canonical map $\kappa_{\mathrm{prim}}^{\mathrm{norm}} : \mathrm{QD}(G)_{\mathrm{norm}} \to \mathrm{Prim}(G)$ of the present section.

THEOREM 11.D.3. *The canonical map $E(\Gamma) \to \mathrm{Prim}(\Gamma)$ is a bijection in the following cases:*

(1) *Γ is a finitely generated group with polynomial growth;*
(2) *$\Gamma = \mathbf{G}(\mathbf{Q})$ is the group of rational points of a unipotent algebraic group over $\mathbf{Q}$.*
(3) *$\Gamma = \mathrm{GL}_n(\mathbf{K})$, where $\mathbf{K}$ is an infinite algebraic extension of a finite field.*

REFERENCES FOR THE PROOF. Result (1) was established in [**Howe–77a**] for finitely generated torsion free nilpotent groups, and extended in [**Kani–80**] to finite extensions of finitely generated nilpotent groups, which are the groups with polynomial growth by Gromov's Theorem.

Part (2) is from [**Pfef–95**]. Concerning (3), see Corollary 12.E.2 below. □

CHAPTER 12

Examples of Thoma's duals

We determine Thoma's duals $E(\Gamma)$ for our four main test classes of discrete groups Γ:

- *two-step nilpotent discrete groups, in particular the Heisenberg group over an infinite field* $\mathbf{K}$ *or over the integers (Section* 12.A*),*
- *the affine group* $\mathrm{Aff}(\mathbf{K})$ *over* $\mathbf{K}$ *(Section* 12.B*),*
- *the Baumslag–Solitar group* $\mathrm{BS}(1,p)$ *for a prime* p *(Section* 12.C*),*
- *the lamplighter group* $\mathbf{Z} \ltimes \bigoplus_{k\in\mathbf{Z}} \mathbf{Z}/2\mathbf{Z}$ *(Section* 12.D*),*
- *and the general linear group* $\mathrm{GL}_n(\mathbf{K})$ *over* $\mathbf{K}$ *(Section* 12.E*).*

The results are well known; see [**ArKa–97**, Example 2.4] *for the Heisenberg group,* [**Thom–64b**, § 1, Beispiel 2] *for the affine group,* [**Guic–63**, § 3, Proposition 1] *for* $\mathrm{BS}(1,2)$*, and* [**Kiri–65**] *for the general linear group. The proofs we give for the affine group and for the general linear group are modelled after the approach used in* [**Bekk–07**] *for the case* $\mathrm{SL}_n(\mathbf{Z}), n \geq 3$.

12.A. Two-step nilpotent groups

Let Γ be a two-step nilpotent discrete group, and Z its centre, as in Sections 3.B, 4.B, and 9.B. For $\psi \in \widehat{Z}$, we denote as before by Z_ψ the inverse image of the centre of $\Gamma/\ker\psi$ by the canonical epimorphism $\Gamma \twoheadrightarrow \Gamma/\ker\psi$. If φ is a function of positive type on a subgroup H of Γ, then its trivial extension $\widetilde{\varphi}$ to Γ is a function of positive type on Γ (Proposition 1.F.10). In the following theorem, the set X is the same as that of Theorems 9.B.1 and 12.A.2.

Recall that the projective kernel $P\ker\pi$ of a representation π of a group G and the unitary character χ_π of $P\ker\pi$ have been introduced in Section 8.E. Recall also (Proposition 8.E.1) that if π is factorial, then π admits a central character ψ_π and that ψ_π coincides with the restriction of χ_π on the centre of G.

THEOREM 12.A.1. *Let* Γ *be a two-step nilpotent discrete group, with centre* Z. *Set*

$$X := \bigsqcup_{\psi\in\widehat{Z}} \left(\{\psi\} \times \left\{\chi \in \widehat{Z_\psi} \mid \chi|_Z = \psi\right\}\right).$$

Then the map

$$\Phi : X \to E(\Gamma), \qquad (\psi,\chi) \mapsto \widetilde{\chi}$$

is a bijection. The inverse map $\Phi^{-1} : E(\Gamma) \to X$ *is given by*

$$\Phi^{-1}(\varphi) = (\psi_{\pi_\varphi}, \lambda_{\pi_\varphi}),$$

where ψ_{π_φ} *is the central character of the GNS representation* π_φ *associated with* φ *and* χ_{π_φ} *the unitary character of* $\mathrm{Pker}\pi_\varphi$ *determined by* π_φ.

PROOF. • *First step.* We claim that every element in $E(\Gamma)$ is of the form $\Phi(\psi,\chi)$ for a pair (ψ,χ) in the set X.

Let $\varphi \in E(\Gamma)$; let $\pi = \pi_\varphi$ be the associated GNS representation. The quotient group $\overline{\Gamma} = \Gamma/\ker\pi$ is two-step nilpotent, and its centre is $\mathrm{Pker}\pi/\ker\pi$ (see Proposition 8.E.1(2)).

A function $\varphi \in E(\Gamma)$ is a unitary character, say χ, if and only if $\pi_\varphi(g) = \chi(g)\mathrm{Id}_{\mathcal{H}_\varphi}$ for all $g \in G$, if and only if $\mathrm{Pker}\pi_\varphi = \Gamma$. If these conditions hold, then $\widetilde{\chi} = \chi$ is in the image of Φ. We can therefore assume from now on that $\mathrm{Pker}\varphi \neq \Gamma$.

Let $\gamma \in \Gamma$ such that $\gamma \notin \mathrm{Pker}\pi$. Then there exists $\delta \in \Gamma$ such that $[\gamma,\delta] \notin \ker\pi$. Since $\overline{\Gamma}$ is two-step nilpotent, the image of $[\gamma,\delta]$ in $\overline{\Gamma}$ belongs to the centre of $\overline{\Gamma}$, hence $[\gamma,\delta] \in \mathrm{Pker}\pi$. Therefore $\varphi(\gamma) = 0$ by Lemma 11.C.9. This shows that φ is the trivial extension to Γ of the unitary character χ_π of $\mathrm{Pker}\pi$ determined by π.

Let $\psi \in \widehat{Z}$ be the central character of the factor representation π. By Proposition 8.E.1(2), we have $\mathrm{Pker}\pi = Z_\psi$.

Since $\chi_\pi|_Z$ coincides with ψ, we have $(\psi,\chi) \in X$ and $\varphi = \Phi(\psi,\chi)$ for $\chi = \chi_\pi$.

• *Second step.* Let $\psi \in \widehat{Z}$ and $\varphi \in \mathrm{Tr}_1(\Gamma)$ be such that $\varphi|_Z = \psi$. We claim that $\varphi = 0$ on $\Gamma \smallsetminus Z_\psi$.

Indeed, since the convex hull of $E(\Gamma)$ is dense in $\mathrm{Tr}_1(\Gamma)$ for the topology of pointwise convergence, it suffices to prove the claim in the case where $\varphi = \sum_{i=1}^n t_i\varphi_i$ is a convex combination of $n \geq 2$ functions $\varphi_i \in E(\Gamma)$ with $0 < t_i < 1$ for every i.

For every $z \in Z$, we have

$$\psi(z) = \varphi(z) = \sum_{i=1}^n t_i\varphi_i(z);$$

since $|\psi(z)| = 1$ and $|\varphi_i(z)| \leq 1$, it follows that $\varphi_i|_Z = \psi$ for every $i = 1,\dots,n$. Therefore we have, by the first step, $\varphi_i = 0$ on $\Gamma \smallsetminus Z_\psi$ for every $i = 1,\dots,n$. So, $\varphi = 0$ on $\Gamma \smallsetminus Z_\psi$.

• *Third step.* We claim that every function in the image of Φ belongs to $E(\Gamma)$.

Let $(\psi,\chi) \in X$; we have to show that the trivial extension $\widetilde{\chi}$ of χ belongs to $E(\Gamma)$. Since $\widetilde{\chi}$ is in $\mathrm{Tr}_1(\Gamma)$ by Proposition 1.F.10, we have only to check that $\widetilde{\chi}$ is indecomposable.

Let $\varphi_1,\varphi_2 \in \mathrm{Tr}_1(\Gamma)$ and $t \in \,]0,1[$ be such that $\widetilde{\chi} = t\varphi_1 + (1-t)\varphi_2$. For every $\gamma \in Z_\psi$, we have

$$\chi(\gamma) = \varphi(\gamma) = t\varphi_1(\gamma) + (1-t)\varphi_2(\gamma).$$

Since $|\chi(\gamma)| = 1$ and $|\varphi_1(\gamma)| \leq 1, |\varphi_2(\gamma)| \leq 1$, it follows that $\varphi_1|_{Z_\psi} = \varphi_2|_{Z_\psi} = \chi$. In particular, $\varphi_1|_Z = \varphi_2|_Z = \chi|_Z = \psi$.

It follows from the second step that $\varphi_1 = 0$ and $\varphi_2 = 0$ on $\Gamma \smallsetminus Z_\psi$. Therefore $\varphi_1 = \varphi_2 = \widetilde{\chi}$.

• *Fourth step.* The map Φ is injective. Indeed, let $(\psi_1,\chi_1), (\psi_2,\chi_2) \in X$ be such that $\Phi(\psi_1,\chi_1) = \Phi(\psi_2,\chi_2)$, that is, $\widetilde{\chi}_1 = \widetilde{\chi}_2$. Then $Z_{\psi_1} = Z_{\psi_2}$, $\chi_1 = \chi_2$, and

$$\psi_1 = \chi_1|_Z = \chi_2|_Z = \psi_2.$$

Therefore, $(\psi_1,\chi_1) = (\psi_2,\chi_2)$. □

We rephrase Theorem 12.A.1 in terms of the space $\mathrm{QD}(\Gamma)_{\mathrm{fin}}$ of quasi-equivalence classes of factor representations of finite type, specifying moreover their possible types.

THEOREM 12.A.2. *Let Γ be a two-step nilpotent discrete group, with centre Z, and let X be as in Theorem* 12.A.1.

Then the map

$$\Phi : X \to \mathrm{QD}(\Gamma)_{\mathrm{fin}}, \qquad (\psi, \chi) \mapsto \mathrm{Ind}_{Z_\psi}^{\Gamma} \chi$$

is a bijection.

Moreover, for $(\psi, \chi) \in X$, the factor representation $\mathrm{Ind}_{Z_\psi}^{\Gamma} \chi$ is of type II_1 if Z_ψ has infinite index in Γ, and of type I_{finite} otherwise.

PROOF. The fact that Φ is a bijection between X and $\mathrm{QD}(G)_{\mathrm{fin}}$ follows from Theorem 12.A.1, in combination with Theorem 11.C.5 and Proposition 1.F.10.

Let $(\psi, \chi) \in X$. If Z_ψ has finite index in Γ, then $\mathrm{Ind}_{Z_\psi}^{\Gamma} \chi$ is finite-dimensional and hence $\mathrm{Ind}_{Z_\psi}^{\Gamma} \chi$ is of type I_n for an integer n. If Z_ψ has infinite index, then $\mathrm{Ind}_{Z_\psi}^{\Gamma} \chi$ is of type II_1; indeed, otherwise, $\mathrm{Ind}_{Z_\psi}^{\Gamma} \chi$ would be a multiple of a finite-dimensional irreducible representation, and this is impossible by Proposition 9.A.3(2). $\square$

The following corollary is an immediate consequence of Theorem 12.A.1 and the description of $\mathrm{Prim}(\Gamma)$ in Theorem 9.B.1. The canonical map $E(\Gamma) \to \mathrm{Prim}(\Gamma)$ is that of Theorem 11.D.3.

COROLLARY 12.A.3. *For a two-step nilpotent discrete group Γ, the canonical map*

$$E(\Gamma) \to \mathrm{Prim}(\Gamma)$$

is a bijection

We apply now Theorem 12.A.1 to Heisenberg groups.

Thoma's dual for Heisenberg groups. As in Section 3.B, here $H(R)$ denotes the Heisenberg group over a unital commutative ring R.

Fix a unitary character ψ of the centre Z of $H(R)$. Let I_ψ be the largest ideal of R contained in $\ker \psi$ (which already appeared in Sections 3.B, 4.B, and 9.B). By Lemma 3.B.7, we have

$$Z_\psi = \{(a, b, c) \in \Gamma \mid (a, b) \in I_\psi^2, \ c \in R\}.$$

Every unitary character of Z_ψ with $\chi|_Z = \psi$ is of the form $\chi_{\alpha,\beta,\psi}$ for a unique pair $(\alpha, \beta) \in \widehat{I_\psi}^2$, where $\chi_{\alpha,\beta,\psi} : Z_\psi \to \mathbf{T}$ is defined by the formula

$$\chi_{\alpha,\beta,\psi}(a, b, c) = \alpha(a)\beta(b)\psi(c) \qquad \text{for } \ (a, b) \in I_\psi^2, \ b, c \in R.$$

In view of these facts, the following result is a direct consequence of Theorem 12.A.1.

COROLLARY 12.A.4. *Let $\Gamma = H(R)$ be the Heisenberg group over a unital commutative ring R. Set*

$$X := \bigsqcup_{\psi \in \widehat{Z}} \{\psi\} \times \widehat{I_\psi}^2.$$

The map $\Phi : X \to E(\Gamma)$ defined by

$$\Phi(\psi, (\alpha, \beta)) = \widetilde{\chi_{\alpha,\beta,\psi}}$$

(where $\widetilde{\chi_{\alpha,\beta,\psi}}$ is the trivial extension of $\chi_{\alpha,\beta,\psi}$ to Γ) is a bijection.

We first consider the particular case where $R = \mathbf{K}$ is a field. For $\psi \in \widehat{\mathbf{K}} \smallsetminus \{1_{\mathbf{K}}\}$, we have $I_\psi = \{0\}$ and hence $Z_\psi = Z$; for $\psi = 1_{\mathbf{K}}$, we have $I_\psi = \mathbf{K}$ and hence $Z_\psi = H(\mathbf{K})$. The following result is therefore an immediate consequence of Corollary 12.A.4. The set X has already appeared in Corollary 9.B.4.

Corollary 12.A.5. *Let $\Gamma = H(\mathbf{K})$ for a field $\mathbf{K}$ and Z the centre of Γ. Let $p : \Gamma \twoheadrightarrow \Gamma/Z$ be the canonical projection. Set*

$$X = \left(\widehat{Z} \smallsetminus \{1_Z\}\right) \sqcup \widehat{\Gamma/Z} \approx \left(\widehat{\mathbf{K}} \smallsetminus \{1_{\mathbf{K}}\}\right) \sqcup \widehat{\mathbf{K}}^2.$$

The map $\Phi : X \to E(\Gamma)$, defined by

$$\Phi(\psi) = \widetilde{\psi} \quad \textit{for } \psi \in \widehat{Z} \smallsetminus \{1_Z\}$$

and

$$\Phi(\chi) = \chi \circ p \quad \textit{for } \chi \in \widehat{\Gamma/Z},$$

is a bijection.

We rephrase Corollary 12.A.5 in terms of factor representations, specifying moreover their possible types.

Corollary 12.A.6. *Let $\Gamma = H(\mathbf{K})$, $p : \Gamma \twoheadrightarrow \Gamma/Z$, and X be as in Corollary* 12.A.5.

The map Φ defined by

$$\Phi(\psi) = \mathrm{Ind}_Z^\Gamma \psi \quad \textit{for } \psi \in \widehat{Z} \smallsetminus \{1_Z\}$$

and

$$\Phi(\chi) = \chi \circ p \quad \textit{for } \chi \in \widehat{\Gamma/Z}$$

is a bijection between X and the set $\mathrm{QD}(\Gamma)_{\mathrm{fin}}$. Moreover, for $\psi \in \widehat{Z} \smallsetminus \{1_Z\}$, the factor representation $\mathrm{Ind}_Z^\Gamma \psi$ is of type II_1 if $\mathbf{K}$ is infinite and of type I_q if $\mathbf{K}$ is finite of cardinality q.

Proof. The fact that Φ is a bijection between X and $\mathrm{QD}(G)_{\mathrm{fin}}$ is a straightforward consequence of Theorem 12.A.2 and Corollary 12.A.5. That $\mathrm{Ind}_Z^\Gamma \psi$ is of type II_1 if $\mathbf{K}$ is infinite also follows from Theorem 12.A.2.

Assume that $\mathbf{K}$ is finite of cardinality q and let $\psi \in \widehat{Z} \smallsetminus \{1_Z\}$. Since $\mathrm{Ind}_Z^\Gamma \psi$ is a factor representation, it is a multiple of an irreducible representation π of Γ. Since π has ψ as a central character, π is not one dimensional and so has dimension q, by Corollary 4.B.3. It follows that $\mathrm{Ind}_Z^\Gamma \psi$ is of type I_q. □

The following corollary shows that Heisenberg groups over nondiscrete topological fields have no interesting finite factor representations.

Corollary 12.A.7. *Let $\mathbf{K}$ be a nondiscrete topological field. Now consider $H(\mathbf{K})$ as a topological group, with its topology making it homeomorphic to the topological space $\mathbf{K}^3$, and therefore $H(\mathbf{K})/Z$ topologically isomorphic to $\mathbf{K}^2$.*

Every function in $E(H(\mathbf{K}))$ is the lift of a unitary character of the quotient group $H(\mathbf{K})/Z$.

Equivalently, the factor representations of finite type of $H(\mathbf{K})$ are the multiples of one-dimensional representations.

PROOF. Denote here by $H(\mathbf{K})_{\rm disc}$ the group Γ of Corollary 12.A.5, where "disc" stands for "discrete". A function $\varphi : H(\mathbf{K}) \to \mathbf{C}$ that is in $E(H(\mathbf{K})_{\rm disc})$ is in $E(H(\mathbf{K}))$ if and only if it is continuous. In particular, a function of the type $\widetilde{\psi}$ for $\psi \in \widehat{Z}$ is not in $E(H(\mathbf{K}))$, because Z is not open in $H(\mathbf{K})$. □

Next, we consider the case where $R = \mathbf{Z}$ is the ring of rational integers.

We recall some notation introduced before Corollary 9.B.7. We denote by $\widehat{Z}_\infty$ the set of elements of infinite order and $\widehat{Z}_n$ the elements of order $n \geq 1$ in $\widehat{Z}$.

Let $\psi \in \widehat{Z}$. If $\psi \in \widehat{Z}_\infty$, then $Z_\psi = Z$. If $\psi \in \widehat{Z}_n$ for some $n \geq 1$, then $Z_\psi = \Lambda(n)$, where

$$\Lambda(n) = \{(a,b,c) \in \Gamma \mid a,b \in n\mathbf{Z},\ c \in \mathbf{Z}\}.$$

Recall that

$$\widehat{Z} \approx \widehat{\mathbf{Z}} = \{\psi_\theta \mid \theta \in [0,1[\},$$

with $\psi_\theta(n) = e^{2\pi i n\theta}$ for $n \in \mathbf{Z}$.

Fix $\psi = \psi_\theta \in \widehat{Z}_n$. For $(\alpha,\beta) \in [0,1/n[^2$, we denote by $\chi_{\alpha,\beta,\psi}$ the unitary character of $\Lambda(n)$ which coincides with ψ on Z, defined by

$$\chi_{\alpha,\beta,\psi}(a,b,c) = e^{2\pi i(\alpha a + \beta b)}\psi(c) \quad \text{for } a,b \in n\mathbf{Z},\ c \in \mathbf{Z}.$$

The following result is an immediate consequence of Corollary 12.A.4. The set X has already appeared in Corollary 9.B.7.

COROLLARY 12.A.8. *Let $\Gamma = H(\mathbf{Z})$ be the Heisenberg group over the integers and Z its centre. Set*

$$X = \widehat{Z}_\infty \sqcup \Big(\bigsqcup_{n\geq 1} \bigsqcup_{\psi \in \widehat{Z}_n} \{\psi\} \times [0,1/n[^2 \Big).$$

The map $\Phi : X \to E(\Gamma)$, defined by

$$\Phi(\psi) = \widetilde{\psi} \quad \textit{for } \psi \in \widehat{Z}_\infty$$

and

$$\Phi(\psi,\alpha,\beta) = \widetilde{\chi_{\alpha,\beta,\psi}} \quad \textit{for } (\psi,\alpha,\beta) \in \widehat{Z}_n \times [0,1/n[^2,$$

is a bijection.

We rephrase Corollary 12.A.8 in terms of factors representations, specifying moreover their possible types.

COROLLARY 12.A.9. *Let $\Gamma = H(\mathbf{Z})$ and X as in Corollary 12.A.8. The map Φ defined by*

$$\Phi(\psi) = \mathrm{Ind}_Z^\Gamma \psi \quad \textit{for } \psi \in \widehat{Z}_\infty$$

and

$$\Phi(\psi,\alpha,\beta) = \mathrm{Ind}_{\Lambda(n)}^\Gamma \chi_{\alpha,\beta,\psi} \quad \textit{for } (\psi,\alpha,\beta) \in \widehat{Z}_n \times [0,1/n[^2,$$

is a bijection between X and the set $\mathrm{QD}(\Gamma)_{\rm fin}$.

Moreover, $\mathrm{Ind}_Z^\Gamma \psi$ is of type II_1 for $\psi \in \widehat{Z}_\infty$ and $\mathrm{Ind}_{\Lambda(n)}^\Gamma \chi_{\alpha,\beta,\psi}$ is of type I_n for $(\psi,\alpha,\beta) \in \widehat{Z}_n \times [0,1/n[^2$.

PROOF. The fact that Φ is a bijection between X and $\mathrm{QD}(\Gamma)_{\mathrm{fin}}$ is a consequence of Theorem 12.A.2 as well as the fact that $\mathrm{Ind}_Z^\Gamma \psi$ is of type II_1 for $\psi \in \widehat{Z}_\infty$.

Let $(\psi, \alpha, \beta) \in \widehat{Z}_n \times [0, 1/n[^2$. The factor representation $\mathrm{Ind}_{\Lambda(n)}^\Gamma \chi_{\alpha,\beta,\psi}$ is a multiple of the irreducible representation $\mathrm{Ind}_{\Gamma(n)}^\Gamma \chi_{\alpha,\beta,\psi}$, of dimension n of Corollary 3.B.17, where

$$\Gamma(n) = \{(a, b, c) \in \Gamma \mid a \in n\mathbf{Z},\ b \in \mathbf{Z},\ c \in \mathbf{Z}\}.$$

Therefore $\mathrm{Ind}_{\Lambda(n)}^\Gamma \chi_{\alpha,\beta,\psi}$ is of type I_n. □

REMARK 12.A.10. For an odd prime p, let Γ be the "infinite-dimensional" Heisenberg group $H_\infty(\mathbf{F}_p) = V \oplus V \oplus \mathbf{F}_p$ over the field $\mathbf{F}_p$, of Example 7.D.7. Since Γ is two-step nilpotent, we can use Theorem 12.A.1 to determine $E(\Gamma)$: as every nontrivial character of the center $Z \approx \mathbf{F}_p$ is faithful, we have

$$E(\Gamma) = \{\widetilde{\psi} \mid \psi \in \widehat{Z} \smallsetminus \{1_Z\}\} \ \cup\ \{\chi \circ r \mid \in \widehat{\Gamma/Z}\},$$

where $r : \Gamma \twoheadrightarrow \Gamma/Z \approx V \oplus V$ is the canonical projection. Observe that, by Theorem 7.D.6, the infinite-dimensional factor representations of Γ are all subrepresentations of the regular representation λ_Γ.

12.B. Affine groups of infinite fields

As in Sections 3.C, 4.B, and 9.C, let $\mathbf{K}$ be an *infinite* field, $\Gamma = \mathrm{Aff}(\mathbf{K})$ the group of affine transformations of $\mathbf{K}$, and N its derived group.

We will describe below (Theorem 12.B.3) the Thoma dual of Γ. The main ingredients for the proof are the following two lemmas that deal with restrictions of characters to normal subgroups.

LEMMA 12.B.1. *Let Γ be a group and N a normal subgroup of Γ. Let $\varphi \in \mathrm{Tr}_1(\Gamma)$ be such that $\varphi|_N = \delta_e$.*

Then $\varphi(\gamma) = 0$ for every $\gamma \in \Gamma$ for which the set of commutators $\{[\gamma, x] \mid x \in N\}$ is infinite.

PROOF. By assumption, there exists a sequence $(x_k)_k$ of elements in N such that, for $y_k := [\gamma, x_k]$, we have

$$y_k \neq y_l \quad \text{for all } k \neq l.$$

Let $(\pi, \mathcal{H}, \xi)$ be a GNS triple associated with φ. Since $y_l^{-1} y_k \in N \smallsetminus \{e\}$, we have $\varphi(y_l^{-1} y_k) = 0$, that is, $\langle \pi(y_k)\xi \mid \pi(y_l)\xi \rangle = 0$ for all $k \neq l$. This means that $(\pi(y_k)\xi)_k$ is an orthonormal sequence in $\mathcal{H}$, so that $(\pi(y_k)\xi)_k$ is weakly convergent to 0 in $\mathcal{H}$.

For every k, we have

$$\varphi(\gamma) = \varphi(x_k \gamma x_k^{-1}) = \varphi(\gamma[\gamma, x_k]) = \langle \pi(y_k)\xi \mid \pi(\gamma^{-1})\xi \rangle.$$

Therefore $\varphi(\gamma) = \lim_k \langle \pi(y_k)\xi \mid \pi(\gamma^{-1})\xi \rangle = 0$. □

LEMMA 12.B.2. *Let N be an* infinite *abelian normal subgroup of the group Γ. Assume that the Γ-conjugacy classes contained in N are $\{e\}$ and $N \smallsetminus \{e\}$. Let φ be a normalized function of positive type on N which is Γ-invariant.*

Then $\varphi = t 1_N + (1 - t)\delta_e$ for some $t \in [0, 1]$.

PROOF. Since φ is Γ-invariant, we have $\varphi = t1_N + s\delta_e$ for some $s, t \in \mathbf{C}$. Moreover, $s + t = 1$, as $\varphi(e) = 1$. It remains to check that $t \in [0, 1]$

Since φ is a function of positive type on the abelian group N, by Bochner's Theorem A.7.5, φ is the Fourier–Stieltjes transform $\mathcal{F}(\mu)$ of a probability measure μ on the compact dual group $\widehat{N}$ of N. Then

$$\mu = t\delta_e + s\mu_{\widehat{N}},$$

where $\mu_{\widehat{N}}$ is the normalized Haar measure on $\widehat{N}$ and δ_e the Dirac measure at the group unit of $\widehat{N}$. As N is infinite and discrete, $\widehat{N}$ is infinite and nondiscrete. Therefore $\mu_{\widehat{N}}$ has no atoms and hence $\mu(\{e\}) = t \in [0, 1]$. $\square$

We are now in a position to describe the Thoma dual of Γ.

THEOREM 12.B.3. *Let* $\mathbf{K}$ *be an infinite field and* $\mathrm{Aff}(\mathbf{K})$ *the corresponding affine group. A function* $\varphi : \Gamma \to \mathbf{C}$ *belongs to* $E(\Gamma)$ *if and only if* φ *has one of the following forms:*

(1) *the lift to* Γ *of a unitary character of the quotient group* Γ/N *(the latter isomorphic to the discrete group* $\mathbf{K}^\times$*);*
(2) *the Dirac function* δ_e *at the group unit.*

In other terms, $E(\Gamma)$ *is in bijection with* $\widehat{\mathbf{K}^\times} \sqcup \{\delta_e\}$.

Equivalently, every factor representation of finite type of Γ *is quasi-equivalent to one of:*

(3) *the composition of the quotient map* $\Gamma \to \Gamma/N$ *with a unitary character* $\Gamma/N \to \mathbf{T}$*, which is of type* I_1*;*
(4) *the left regular representation of* Γ*, which is of type* II_1.

PROOF. We identify $\Gamma = \mathrm{Aff}(\mathbf{K})$ with $\mathbf{K}^\times \ltimes \mathbf{K}$ and write (a, s) for the matrix

$$\begin{pmatrix} a & s \\ 0 & 1 \end{pmatrix}.$$

Observe that the conjugacy class of an element $(a, s) \in \Gamma$ is

$$\{(a, t(1-a) + bs) \mid t \in \mathbf{K}, b \in \mathbf{K}^\times\};$$

in particular, the conjugacy class of every $\gamma \in \Gamma \smallsetminus \{e\}$ is infinite.

A function φ as in (1) is clearly in $E(\Gamma)$.

Let $\varphi = \delta_e$ be as in (2). Then $\varphi \in \mathrm{Tr}_1(\Gamma)$ by Proposition 1.F.10; since Γ is icc, we have $\varphi \in E(\Gamma)$ by Proposition 7.A.1.

We have to show that, conversely, every $\varphi \in E(\Gamma)$ is either as in (1) or as in (2).

Let $\varphi \in E(\Gamma)$ and let $(\pi, \mathcal{H}, \xi)$ be the GNS triple associated with φ. Consider the closed subspace $\mathcal{H}^N$ of $\pi(N)$-invariant vectors in $\mathcal{H}$, where N is the normal subgroup

$$N = \{(1, s) \mid s \in \mathbf{K}\} \approx \mathbf{K}.$$

Two cases case can occur.

∘ *First case.* Assume that $\mathcal{H}^N \neq \{0\}$. On the one hand, since N is normal in Γ, the space $\mathcal{H}^N$ is $\pi(\Gamma)$-invariant. On the other hand, $\mathcal{H}^N$ is clearly invariant under the commutant $\pi(\Gamma)'$ of $\pi(\Gamma)$ in $\mathcal{L}(\mathcal{H})$. So, the orthogonal projection $p : \mathcal{H} \twoheadrightarrow \mathcal{H}^N$ onto $\mathcal{H}^N$ belongs to $\pi(\Gamma)' \cap \pi(\Gamma)''$, which is the centre of $\pi(\Gamma)''$. Since $\pi(\Gamma)''$ is a factor and since $p \neq 0$, we have $p = \mathrm{Id}_{\mathcal{H}}$. Then $\pi(s) = \mathrm{Id}_{\mathcal{H}}$ for all $s \in N$; hence, N is in the kernel of π. So, φ is the lift to Γ of a unitary character of $\Gamma/N \approx \mathbf{K}^\times$.

∘ *Second case.* Assume that $\mathcal{H}^N = \{0\}$. Consider the restriction $\psi := \varphi|_N$ of φ to N.

Observe that the conjugacy classes in Γ contained in the normal subgroup $N = \{(1, s) \mid s \in \mathbf{K}\} \approx \mathbf{K}$ are $\{0\}$ and $N \smallsetminus \{0\}$.

Since ψ is constant on Γ-conjugacy classes in N, it follows from Lemma 12.B.2 that $\psi = t1_N + (1-t)\delta_e$ for some $t \in [0, 1]$. We claim that $t = 0$.

Indeed, otherwise, the representation of N on the closed linear subspace generated by $\pi(N)\xi$ would contain the trivial representation 1_N; this would be a contradiction, since $\mathcal{H}^N = \{0\}$.

As a result, we have $\psi = \varphi|_N = \delta_e$. We claim that $\varphi = \delta_e$.

Indeed, let $\gamma = (a, s) \in \Gamma \smallsetminus N$; then $a \neq 1$ and hence

$$\{[\gamma, x] \mid x \in N\} \;=\; \{(1, (a^{-1} - 1)t) \mid t \in \mathbf{K}\}$$

is infinite. So, $\varphi(\gamma) = 0$, by Lemma 12.B.1. This ends the proof. □

The following corollary is an immediate consequence of Theorem 12.B.3 and the description of $\mathrm{Prim}(\mathrm{Aff}(\mathbf{K}))$ from Theorem 9.C.1. The canonical map $E(\Gamma) \to \mathrm{Prim}(\Gamma)$ is that of Theorem 11.D.3.

COROLLARY 12.B.4. *For* $\Gamma = \mathrm{Aff}(\mathbf{K})$, *the canonical map*

$$E(\Gamma) \to \mathrm{Prim}(\Gamma)$$

is a bijection

Observe that, when $\mathbf{K}$ is a nondiscrete topological field, the functions φ of Theorem 12.B.3 are continuous on the topological group $\mathrm{Aff}(\mathbf{K})$ only if φ is the lift of a continuous unitary character of $\mathrm{Aff}(\mathbf{K})/N$.

COROLLARY 12.B.5. *Let* $\mathbf{K}$ *be a nondiscrete topological field. Now consider* $\mathrm{Aff}(\mathbf{K})$ *as a topological group, with its topology making it homeomorphic to* $\mathbf{K}^\times \times \mathbf{K}$, *and therefore* $\mathrm{Aff}(\mathbf{K})/N$ *topologically isomorphic to* $\mathbf{K}^\times$.

Every function in $E(\mathrm{Aff}(\mathbf{K}))$ *is the lift of a unitary character of the quotient group* $\mathrm{Aff}(\mathbf{K})/N$.

Equivalently, every factor representation of finite type of $\mathrm{Aff}(\mathbf{K})$ *is the composition of the quotient map* $\mathrm{Aff}(\mathbf{K}) \to \mathrm{Aff}(\mathbf{K})/N \approx \mathbf{K}^\times$ *with a unitary character* $\mathbf{K}^\times \to \mathbf{T}$.

12.C. Solvable Baumslag–Solitar groups

Let p be a prime. Consider the Baumslag–Solitar group $\Gamma = \mathrm{BS}(1, p) = A \ltimes N \approx \mathbf{Z} \ltimes \mathbf{Z}[1/p]$, as in Sections 3.D, 4.B, and 9.D. Recall that we can identify $\widehat{N}$ with the p-adic solenoid

$$\mathbf{Sol}_p \;=\; (\mathbf{Q}_p \times \mathbf{R})/\Delta \quad \text{for } \Delta \;=\; \{(a, -a) \mid a \in \mathbf{Z}[1/p]\},$$

and that the action of A on $\widehat{N}$ corresponds to the action of $\mathbf{Z}$ on $\mathbf{Sol}_p$ given by the transformation $T_p : \mathbf{Sol}_p \to \mathbf{Sol}_p$ induced by the multiplication by p (see Lemma 3.D.2).

Recall also that we parametrized in Corollary 4.B.7 the space $\widehat{\Gamma}_{\mathrm{fd}}$ of finite-dimensional irreducible representations of Γ by the set $X_{\mathrm{fd}}/\sim$ of T_p-orbits in X_{fd}, where

$$X_{\mathrm{fd}} \;=\; \bigsqcup_{n \geq 1} \left(\mathbf{Sol}_p(n) \times [0, 1/n[\right),$$

where $\mathbf{Sol}_p(n)$ is the set of points in $\mathbf{Sol}_p$ of T_p-period n. For $(s,\theta) \in \mathbf{Sol}_p(n) \times [0, 1/n[$, we denote by $\pi_{s,\theta} = \mathrm{Ind}_{\Gamma(n)}^{\Gamma} \chi_{s,\theta}$ the corresponding finite-dimensional irreducible representation of Γ. The associated character $\varphi_{s,\theta} \in E(\Gamma)$ of Γ is the function defined by

$$\varphi_{s,\theta}(\gamma) = \mathrm{Tr}(\pi_{s,\theta}(\gamma)) \quad \text{for } \gamma \in \Gamma,$$

where $\mathrm{Tr}(z)$ denotes the normalized trace of a matrix z.

One can give an explicit formula for $\varphi_{s,\theta}$: for $n \geq 1$, $(s,\theta) \in \mathbf{Sol}_p(n) \times [0, 1/n[$, and $(k, b) \in \Gamma$, we have

$$\varphi_{s,\theta}(k,b) = \begin{cases} \frac{1}{n} e^{2\pi i k\theta/n} \sum_{\chi \in \mathcal{O}_s} \chi(b) & \text{if } k \in n\mathbf{Z}, \\ 0 & \text{otherwise,} \end{cases}$$

where $\mathcal{O}_s$ denotes the A-orbit of $\chi_{s,\theta}$ in $\widehat{N}$.

Denote by $\mathrm{EP}(\mathbf{Sol}_p)$ the set of ergodic T_p-invariant probability measures μ on $\mathbf{Sol}_p$, which are *nonatomic* (that is, such that $\mu(\{s\}) = 0$ for every $s \in \mathbf{Sol}_p$). We view the elements of $\mathrm{EP}(\mathbf{Sol}_p)$ as ergodic A-invariant atomfree probability measures on $\widehat{N}$, under the identification $\widehat{N} \approx \mathbf{Sol}_p$.

Observe that, by Bochner's Theorem A.7.5, the (inverse) Fourier–Stieltjes transform $\overline{\mathcal{F}}(\mu)$ of a probability measure μ on $\widehat{N}$ is a function of positive type on N. The trivial extension $\widetilde{\overline{\mathcal{F}}(\mu)}$ of $\overline{\mathcal{F}}(\mu)$ to Γ is then a function of positive type on Γ (Proposition 1.F.10).

THEOREM 12.C.1. *Let* $\Gamma = A \ltimes N \approx \mathbf{Z} \ltimes \mathbf{Z}[1/p]$ *be the Baumslag–Solitar group* $\mathrm{BS}(1,p)$. *Set*

$$X = \mathrm{EP}(\mathbf{Sol}_p) \sqcup X_{\mathrm{fd}} = \mathrm{EP}(\mathbf{Sol}_p) \sqcup \bigsqcup_{n \geq 1} \left(\mathbf{Sol}_p(n) \times [0, 1/n[\right),$$

and identify $\widehat{N}$ *with* $\mathbf{Sol}_p$.

The map $\Phi : X \to E(\Gamma)$, *defined by*

$$\Phi(\mu) = \widetilde{\overline{\mathcal{F}}(\mu)} \quad \textit{for } \mu \in \mathrm{EP}(\mathbf{Sol}_p)$$

and

$$\Phi(s,\theta) = \varphi_{s,\theta} \quad \textit{for } (s,\theta) \in X_{\mathrm{fd}},$$

is a surjection.

PROOF. • *First step.* We claim that every element in $E(\Gamma)$ is in the image of Φ.

Let $\varphi \in E(\Gamma)$; let $(\pi, \mathcal{H}, \xi)$ be a GNS triple associated with φ. Consider the projection-valued measure $E : \mathcal{B}(\widehat{N})$ on $\widehat{N}$ associated with $\pi|_N$. Let $\mu = \mu_\xi$ be the probability measure on $\widehat{N}$ corresponding to (E, ξ), which is defined by

$$\mu(B) = \langle E(B)\xi \mid \xi \rangle \quad \text{for } B \in \mathcal{B}(\widehat{N}).$$

Recall that, for every $B \in \mathcal{B}(\widehat{N})$, the projection $E(B)$ belongs to the von Neumann algebra $\pi(N)''$ and hence to $\pi(\Gamma)''$.

Observe that the support of μ coincides with the support of E. Indeed, let $B \in \mathcal{B}(\widehat{N})$; if $E(B) = 0$, then obviously $\mu(B) = 0$. Conversely, assume that $\mu(B) = 0$. Then

$$\mathrm{Tr}_\varphi(E(B)) = \langle E(B)\xi \mid \xi \rangle = 0,$$

where Tr_φ is the trace on $\pi(\Gamma)''$ associated with φ as in Construction 11.B.4. Therefore $E(B) = 0$ since Tr_φ is faithful.

Since

$$(*) \qquad \pi(\gamma)E(B)\pi(\gamma)^{-1} = E(B^{\gamma^{-1}}) \quad \text{for all } B \in \mathcal{B}(\widehat{N}), \ \gamma \in \Gamma$$

(see Proposition 2.F.1), and since φ is conjugation-invariant, the measure μ is Γ-invariant.

We claim that μ is ergodic under the Γ-action. Indeed, let $B \in \mathcal{B}(\widehat{N})$ be a Γ-invariant subset with $\mu(B) > 0$. Then $E(B)$ is a nonzero projection which belongs to the von Neumann algebra $\pi(\Gamma)''$. Moreover, relation $(*)$ above shows that $E(B)$ commutes with $\pi(\gamma)$ for every $\gamma \in \Gamma$. So, $E(B)$ is in the centre of $\pi(\Gamma)''$. Therefore $E(B) = I$, since π is factorial and $E(B) \neq 0$. It follows that

$$E(\widehat{N} \smallsetminus B) = I - E(B) = 0$$

and hence $\mu(\widehat{N} \smallsetminus B) - 0$; this proves the claim.

Two cases may now occur.

◦ *First case.* μ is atomic, that is, there exists a point $s \in \widehat{N} \approx \mathbf{Sol}_p$ such that $\mu(\{s\}) > 0$. Since μ is a Γ-invariant ergodic probability measure, it follows that s is a periodic point for T_p and the support of μ, which is the support of E, is the T_p-orbit of s. Now, since π is factorial, π is weakly equivalent to an *irreducible* representation ρ of Γ (Proposition 8.D.3(2)). Then $\pi|_N$ and $\rho|_N$ are weakly equivalent and hence the projection-valued measure E' associated with $\rho|_N$ has the same support as E (see Proposition 2.D.1(ii)). So, the support of E' is a finite T_p-orbit. By Theorem 9.D.1, ρ has to be a finite-dimensional representation of the form

$$\pi_{s,\theta} = \mathrm{Ind}_{\Gamma(n)}^{\Gamma} \chi_{s,\theta}.$$

It follows that π is a multiple of ρ, by Proposition 9.A.3. Therefore φ coincides with the normalized trace $\varphi_{s,\theta}$ of $\rho = \pi_{s,\theta}$.

◦ *Second case.* The measure μ is nonatomic. We claim that $\varphi = \widetilde{\overline{\mathcal{F}}(\mu)}$.

Indeed, on the one hand, for every $n \in N$, we have by the SNAG Theorem 2.C.3,

$$\varphi(n) = \langle \pi(n)\xi \mid \xi \rangle = \int_{\widehat{N}} \chi(n) d\mu(n) = \overline{\mathcal{F}}(\mu)(n),$$

and so $\varphi = \overline{\mathcal{F}}(\mu)$ on N.

On the other hand, let $\gamma \in \Gamma \smallsetminus N$. Since $\mu(\mathrm{Per}(T_p)) = 0$, the action of the quotient group Γ/N on the measure space $(\widehat{N}, \mu)$ is essentially free (see Section 13.A for this notion). Therefore by Proposition 13.A.2 below, there exists a sequence $(B_n)_n$ in $\mathcal{B}(\widehat{N})$ with

$$\begin{aligned} \mathbf{1}_{B_n}\mathbf{1}_{B_m} &= 0 \quad \text{for } n \neq m \\ \sum_n \mathbf{1}_{B_n} &= \mathbf{1}_{\widehat{N}} \quad \text{and} \\ \mathbf{1}_{B_n}\mathbf{1}_{B_n^{\gamma^{-1}}} &= 0 \quad \text{for every } n. \end{aligned}$$

(We have used the notation $\mathbf{1}_B$ for the characteristic function of a subset B of $\widehat{N}$, defined by $\mathbf{1}_B(\chi) = 1$ if $\chi \in B$ and $\mathbf{1}_B(\chi) = 0$ if $\chi \in \widehat{N} \smallsetminus B$.) It follows from the general properties of a projection-valued measure that

$$\begin{aligned} E(B_n)E(B_m) &= 0 \quad \text{for } n \neq m \\ \sum_n E(B_n) &= I \quad \text{(in the strong operator topology) and} \\ E(B_n)E(B_n^{\gamma^{-1}}) &= 0 \quad \text{for every } n. \end{aligned}$$

Using the trace property of Tr_φ, it follows that

$$\begin{aligned} \langle \pi(\gamma)E(B_n)\xi \mid E(B_m)\xi \rangle &= \mathrm{Tr}_\varphi(E(B_m)\pi(\gamma)E(B_n)) \\ &= \mathrm{Tr}_\varphi(E(B_n)E(B_m)\pi(\gamma)) = 0, \end{aligned}$$

for $n \neq m$. Therefore we have

$$\begin{aligned} \varphi(\gamma) &= \langle \pi(\gamma)\xi \mid \xi \rangle = \langle \pi(\gamma)(\sum_n E(B_n)\xi) \mid \sum_n E(B_n)\xi \rangle \\ &= \sum_{n,m} \langle \pi(\gamma)E(B_n)\xi \mid E(B_m)\xi \rangle = \sum_n \langle \pi(\gamma)E(B_n)\xi \mid E(B_n)\xi \rangle \\ &= \sum_n \langle E(B_n^{\gamma^{-1}})\pi(\gamma)\xi \mid E(B_n)\xi \rangle = \sum_n \langle E(B_n)E(B_n^{\gamma^{-1}})\pi(\gamma)\xi \mid \xi \rangle \\ &= 0. \end{aligned}$$

Summarizing, we have shown that φ coincides with the trivial extension of $\overline{\mathcal{F}}(\mu)$ to Γ.

• *Second step.* We claim that every function in the image of Φ belongs to $E(\Gamma)$. This is clear for the functions of the form $\varphi_{s,\theta}$.

Now let μ be an ergodic Γ-invariant nonatomic probability measure on $\widehat{N}$ and set $\varphi = \widetilde{\overline{\mathcal{F}}(\mu)}$. It is clear (compare Proposition 1.F.10) that φ is in $\mathrm{Tr}_1(\Gamma)$; so, we only have to check that φ is indecomposable.

Let $\varphi_1, \varphi_2 \in \mathrm{Tr}_1(\Gamma)$ and $t \in]0,1[$ be such that $\varphi = t\varphi_1 + (1-t)\varphi_2$. As above, we have $\varphi_1|_N = \overline{\mathcal{F}}(\mu_1)$ and $\varphi_2|_N = \overline{\mathcal{F}}(\mu_1)$ for Γ-invariant probability measures μ_1 and μ_2 on $\widehat{N}$. By injectivity of the Fourier–Stieltjes transform, we then have $\mu = t\mu_1 + (1-t)\mu_2$. Since μ is ergodic, it follows that $\mu_1 = \mu_2 = \mu$. In particular, μ_1 and μ_2 are nonatomic.

As in the second case of the first step, we can use the essential freeness of the action of Γ/N on $(\widehat{N}, \mu_i) = (\widehat{N}, \mu)$ and the trace property of Tr_{φ_i}; it follows that $\varphi_i = 0$ on $\Gamma \smallsetminus N$ for $i = 1, 2$. Moreover φ_i coincides with φ on N, since $\mu_i = \mu$. Therefore $\varphi_1 = \varphi_2 = \widetilde{\overline{\mathcal{F}}(\mu)}$. □

REMARK 12.C.2. An ergodic T_p-invariant probability measure on $\mathbf{Sol}_p$ which is atomic is necessarily the uniform measure on a periodic T_p-orbit; so, there are countably many such measures.

Observe that the normalized Haar measure on $\mathbf{Sol}_p$ is a T_p-invariant ergodic nonatomic probability measure (see Remark 3.D.7); so, $\mathrm{EP}(\mathbf{Sol}_p)$ is nonempty. In fact, as we will show (see Corollary 12.D.5), the set $\mathrm{EP}(\mathbf{Sol}_p)$ is uncountable.

12.D. Lamplighter group

As in Sections 3.E, 4.B, and 9.E, the lamplighter group is the semi-direct product $\Gamma = A \ltimes N$ of $A = \mathbf{Z}$ with $N = \bigoplus_{k \in \mathbf{Z}} \mathbf{Z}/2\mathbf{Z}$, where the action of $\mathbf{Z}$ on $\bigoplus_{k \in \mathbf{Z}} \mathbf{Z}/2\mathbf{Z}$ is given by shifting the coordinates. Recall that $\widehat{N}$ can be identified with $X = \{0,1\}^{\mathbf{Z}}$, the dual action $\mathbf{Z}$ on $\widehat{N}$ being given by the shift transformation T on X.

As in Section 3.E, $X(n)$ denotes the set of points in X with T-period $n \geq 1$. Recall from Proposition 3.E.3 that the space $\widehat{\Gamma}_{\mathrm{fd}}$ of finite-dimensional irreducible representations of Γ is parametrized by the set of T-orbits in

$$\bigsqcup_{n\geq 1} (X(n) \times [0,1/n[).$$

For $(x,\theta) \in X(n) \times [0,1/n[$, we denote by $\pi_{x,\theta} = \mathrm{Ind}_{\Gamma(n)}^{\Gamma} \chi_{x,\theta}$ the corresponding finite-dimensional irreducible representation of Γ and by $\varphi_{x,\theta}$ the associated character in $E(\Gamma)$.

Denote by $\mathrm{EP}(X)$ the set of ergodic T-invariant nonatomic probability measures on X, which we view as ergodic A-invariant probability measures on $\widehat{N}$, under the identification $\widehat{N} \approx X$.

The following result can be proved in exactly the same way as Theorem 12.C.1.

THEOREM 12.D.1. *Let $\Gamma = A \ltimes N$ be the lamplighter group. Set*

$$\Omega := \mathrm{EP}(X) \sqcup \bigsqcup_{n\geq 1} (X(n) \times [0,1/n[),$$

and identify $\widehat{N}$ with X.

The map $\Phi : \Omega \to E(\Gamma)$, defined by

$$\Phi(\mu) = \widetilde{\overline{\mathcal{F}}(\mu)} \quad \textit{for } \mu \in \mathrm{EP}(X)$$

and

$$\Phi(x,\theta) = \varphi_{x,\theta} \quad \textit{for } (X,\theta) \in X(n) \times [0,1/n[,\ n \geq 1,$$

is a surjection.

REMARK 12.D.2. The atomic ergodic T-invariant probability measures on X are the uniform measures on periodic T-orbits. So, there are countably many such measures. By contrast, the set $\mathrm{EP}(X)$ is uncountable, as we now see (Proposition 12.D.3). There is no known classification of the measures in $\mathrm{EP}(X)$.

For later use (see Corollary 12.D.5), we need to work on a shift space (X,T) over possibly more than two symbols. For an integer $K \geq 2$, we consider the product space

$$X = \{0,\ldots,K-1\}^{\mathbf{Z}}$$

equipped with the product topology and the shift transformation $T : X \to X$.

PROPOSITION 12.D.3. *Let (X,T) be the two-sided shift over $\{0,\ldots,K-1\}$ as above.*

There exist uncountably many mutually singular ergodic T-invariant nonatomic probability measures on X, all with (topological) support equal to X.

PROOF. For a real number $t \in \,]0,1[$, let ν_t be a probability measure on $\{0,\ldots,K-1\}$ such that $\nu_t(\{i\}) > 0$ for every $i \in \{0,\ldots,K-1\}$; below, we will impose a further condition depending on t. We have equipped X with the product measure $\mu_t = \bigotimes^{\mathbf{Z}} \nu_t$ of copies of ν_t. Observe that μ_t is a nonatomic probability measure on X; moreover, it is a classical fact (see for instance [**Walt–82**, Theorem 1.12]) that μ_t is T-invariant and ergodic.

We claim that the topological support of μ_t is X. Indeed, let U be a nonempty open subset of X. Then U contains a cylinder of the form

$$C^{n_1,\ldots,n_k}_{a_1,\ldots,a_k} = \{(x_n)_{n\in\mathbf{Z}} \in X \mid x_{n_i} = a_i \text{ for } i = 1,\ldots,k\},$$

for integers $n_1 < \cdots < n_k$ in $\mathbf{Z}$ and elements $a_1,\ldots,a_k$ in $\{0,\ldots,K-1\}$. Since

$$\mu_t(C^{n_1,\ldots,n_k}_{a_1,\ldots,a_k}) = \prod_{i=1}^{k} \nu_t(a_i) > 0,$$

it follows that $\mu_t(U) > 0$ and the claim is proved.

We require now that the mean of ν_t is equal to t, that is,

$$\sum_{i=0}^{K-1} i\nu_t(i) = t.$$

For definiteness, we could define ν_t by

$$\nu_t(\{0\}) = 1 - \frac{2t}{K} \quad \text{and} \quad \nu_t(\{i\}) = \frac{2t}{K(K-1)} \quad \text{for } i \in \{1,\ldots,K-1\}.$$

We claim that the μ_t's are mutually singular. Indeed, let $t \in \,]0,1[$. By the law of large numbers, there exists a measurable subset A_t of X with $\mu_t(A_t) = 1$ such that

$$\lim_{N\to+\infty} \frac{1}{N}\sum_{n=1}^{N} x_n = t \quad \text{for all } x = (x_n)_{n\in\mathbf{Z}} \in A_t.$$

It follows that $A_s \cap A_t = \emptyset$ for $s,t \in \,]0,1[$ with $s \neq t$, and this ends the proof. □

Recall that every real number $x \in [0,1[$ has a *unique* K-expansion

$$x = 0.x_1x_2x_3\ldots = \sum_{i=1}^{+\infty} \frac{x_i}{K^i}$$

with $x_i \in \{0,\ldots,K-1\}$, provided we require that $x_i \neq K-1$ for infinitely many i's.

Let p be a prime, $p \geq 2$. Let $\mathbf{Sol}_p$ be the p-adic solenoid, and

$$T_p : \mathbf{Sol}_p \to \mathbf{Sol}_p$$

the multiplication by p, as in Section 12.C. The following proposition shows that $(\mathbf{Sol}_p, T_p)$ is strongly related to the two-sided shift (X,T) over $\{0,\ldots,p-1\}$, as a Borel dynamical system (that is, as a Borel space equipped with a Borel transformation).

Denote by X_1 the measurable T-invariant subset of

$$X := \{0,\ldots,p-1\}^{\mathbf{Z}}$$

consisting of sequences $(x_n)_{n\in\mathbf{Z}}$ in X which are eventually constant equal to $p-1$, that is, for which there exists an $N \geq 0$ such that $x_n = p-1$ for every $n \geq N$.

PROPOSITION 12.D.4. *There is a Borel isomorphism*

$$\Phi : X \smallsetminus X_1 \to \mathbf{Sol}_p$$

such that $\Phi(Tx) = T_p(\Phi(x))$ *for every* $X \smallsetminus X_1$.

PROOF. Recall from Section 3.D that $\mathbf{Sol}_p = (\mathbf{Q}_p \times \mathbf{R})/\mathbf{Z}[1/p]$, with $\mathbf{Z}[1/p]$ identified with the subgroup $\{(a,-a) \in \mathbf{Q}_p \times \mathbf{R} \mid a \in \mathbf{Z}[1/p]\}$.

Let $\mathbf{Z}_p$ be the ring of p-adic integers. The obvious map

$$\mathbf{Z}_p \times \mathbf{R} \to (\mathbf{Q}_p \times \mathbf{R})/\mathbf{Z}[1/p]$$

has $\mathbf{Z}$ as a kernel (identified as above with a subgroup of $\mathbf{Q}_p \times \mathbf{R}$) and is clearly surjective. It follows that $\mathbf{Sol}_p$ is isomorphic (as topological group) to $(\mathbf{Z}_p \times \mathbf{R})/\mathbf{Z}$. Therefore $\mathbf{Z}_p \times [0,1[$ is a transversal for the $\mathbf{Z}$-cosets in $(\mathbf{Z}_p \times \mathbf{R})/\mathbf{Z}$, and so we have an identification of Borel spaces

$$\mathbf{Sol}_p \cong \mathbf{Z}_p \times [0,1[.$$

Under this identification, the map T_p is given on $\mathbf{Z}_p \times [0,1[$ by

$$T_p(a,\lambda) = (pa + [p\lambda], \{p\lambda\})$$

for $a \in \mathbf{Z}_p$ and $\lambda \in [0,1[$, where $[\alpha] \in \mathbf{N}$ and $\{\alpha\} \in [0,1[$ denote the integer part and the fractional part of a real number α.

We define a measurable map

$$\Phi : X \smallsetminus X_1 \to \mathbf{Z}_p \times [0,1[$$

as follows. For $x = (x_n)_{n\in\mathbf{Z}} \in X \smallsetminus X_1$, set

$$x^+ = (x_n)_{n\in\mathbf{N}} \quad \text{and} \quad x^- = (x_{-n})_{n\in\mathbf{N}^*}.$$

Let $\lambda(x^+)$ be the real number in $[0,1[$ with p-expansion given by the sequence x^+, that is,

$$\lambda(x^+) = 0.x_0x_1x_2\cdots = \sum_{n=0}^{+\infty} \frac{x_n}{p^{n+1}},$$

and let $a(x^-)$ be the p-adic integer

$$a(x^-) = \sum_{n=1}^{+\infty} x_{-n}p^{n-1} = x_{-1} + x_{-2}p + x_{-3}p^2 + \cdots .$$

Define

$$\Phi(x) := (a(x^-), \lambda(x^+)) \quad \text{for} \quad x \in X \smallsetminus X_1.$$

Since every element $[0,1[$ has a unique p-expansion by a sequence which is not eventually constant equal to $p-1$, the map Φ is a bijection.

The inverse map of Φ is given as follows. For $(a,x) \in \mathbf{Z}_p \times [0,1[$, let

$$x = 0.x_0x_1x_2\cdots = \sum_{i=0}^{+\infty} \frac{x_i}{p^{i+1}}$$

be the unique p-expansion of x by a sequence which is not eventually constant equal to $p-1$ and let

$$a = x_{-1} + x_{-2}p + x_{-3}p^2 + \cdots$$

be the p-adic expansion of a. Then $\Phi^{-1}(a,\lambda) = (x^-, x^+)$, where

$$x^+ = (x_0, x_1, x_2, \ldots) \quad \text{and} \quad x^- = (\ldots, x_{-3}, x_{-2}, x_{-1}).$$

Since Φ^{-1} is obviously measurable, $\Phi : X \smallsetminus X_1 \to \mathbf{Z}_2 \times [0,1[$ is a Borel isomorphism.

We claim that

$$\Phi(Tx) = T_p(\Phi(x)) \quad \text{for} \quad x = (x_n)_{n\in\mathbf{Z}} \in X \smallsetminus X_1.$$

Indeed, we have

$$(Tx)^+ \;=\; (x_{n+1})_{n\in\mathbf{N}} \quad \text{and} \quad (Tx)^- \;=\; (x_{-n+1})_{n\in\mathbf{N}^*},$$

and so

$$\lambda((Tx)^+) \;=\; 0.x_1x_2x_3\cdots \quad \text{and} \quad a((Tx)^-) \;=\; x_0 + x_{-1}p + x_{-2}p^2 + \cdots .$$

Therefore

$$\lambda((Tx)^+) \;=\; \{p\lambda(x^+)\} \quad \text{and} \quad a((Tx)^-) \;=\; pa(x^-) + x_0 \;=\; pa(x^-) + [p\lambda(x^+)],$$

and it follows that

$$\Phi(Tx) \;=\; (a((Tx)^-), \lambda((Tx)^+)) \;=\; T_p(a(x^-), \lambda(x^+)) \;=\; T_p(\Phi(x)),$$

as was to be shown. □

We now use Propositions 12.D.3 and 12.D.4 in order to show that $\mathbf{Sol}_p$ has uncountably many ergodic T_p-invariant nonatomic probability measures.

COROLLARY 12.D.5. *Let $\mathbf{Sol}_p$ be the p-adic solenoid and $T_p : \mathbf{Sol}_p \to \mathbf{Sol}_p$ be as above.*

There exist uncountably many mutually singular ergodic T_p-invariant nonatomic probability measures on $\mathbf{Sol}_p$, all with (topological) support equal to $\mathbf{Sol}_p$.

PROOF. For $t \in (0, t)$, let ν_t and μ_t be the probability measures on $\{0, \ldots, p-1\}$ and $X = \{0, \ldots, p-1\}^{\mathbf{Z}}$ defined in the proof of Proposition 12.D.3.

We claim that $\mu_t(X_1) = 0$, where X_1 is the measurable subset of X as in Proposition 12.D.4. Indeed, for $N \geq 0$, let X_1^N be the set of sequences $(x_n)_{n\in\mathbf{Z}}$ in X such that $x_n = p - 1$ for every $n \geq N$. Then

$$\mu_t(X_1^N) \;=\; \lim_{n\to+\infty} (\nu_t(\{p-1\}))^{n-N} \;=\; 0$$

and hence $\mu_t(X_1) = 0$, since $X_1 = \bigcup_{N\geq 0} X_1^N$. As a result, we may view μ_t as a probability measure on $X \smallsetminus X_1$.

Let $m_t = \Phi_*(\mu_t)$ be the image of μ_t under the map

$$\Phi : X \smallsetminus X_1 \to \mathbf{Sol}_p$$

of Proposition 12.D.4. The probability measure m_t is T_p-invariant, ergodic and nonatomic, since μ_t is ergodic T-invariant and nonatomic and since Φ is an equivariant Borel isomorphism.

Moreover, the m_t's are mutually singular measures on $\mathbf{Sol}_p$, since the μ_t's are mutually singular measures on $X \smallsetminus X_1$.

We claim that $\mathrm{supp}(m_t) = \mathbf{Sol}_p$. Indeed, let U be a nonempty open subset of $\mathbf{Sol}_p$. Its inverse image V in $\mathbf{Z}_p \times \mathbf{R}$ is open. So, there exists $N \geq 2$ such that $U' := V \cap (\mathbf{Z}_p \times [0, 1[)$ contains the set $p^N\mathbf{Z}_2 \times [1/p^N, 1/p^{N-1}[$. Therefore $\Phi^{-1}(U')$ contains the cylinder

$$C_N \;=\; \left\{ (x_n)_{n\in\mathbf{Z}} \in X \smallsetminus X_1 \;\middle|\; \begin{array}{l} x_{-N} = \cdots = x_{-1} = 0 \text{ and} \\ x_0 = \cdots = x_{N-2} = 0, \;\; x_{N-1} \neq 0 \end{array} \right\}.$$

Since $\mu_t(C_N) > 0$, it follows that $m_t(U) = m_t(U') > 0$. □

12.E. General linear groups

Let $\mathbf{K}$ be an infinite field and n an integer, $n \geq 2$. Recall that the commutator group of the general linear group $\mathrm{GL}_n(\mathbf{K})$ is the special linear group $\mathrm{SL}_n(\mathbf{K})$, and the centre Z of $\mathrm{GL}_n(\mathbf{K})$ consists of the scalar matrices λI_n, for $\lambda \in \mathbf{K}^\times$. The following theorem is from [**Kiri–65**].

In the proof, we will use the notion of group representations with vanishing matrix coefficients. Let G be a locally compact group. A representation $(\pi, \mathcal{H})$ of G has **vanishing matrix coefficients** if all its matrix coefficients of π, that is, all functions

$$\varphi_{\pi,\xi,\eta} : G \to \mathbf{C}, \quad g \mapsto \langle \pi(g)\xi \mid \xi' \rangle$$

for $\xi, \eta \in \mathcal{H}$, belong to $C^0(G)$; this already holds if there exists a total subset D of $\mathcal{H}$ such that $\varphi_{\pi,\xi,\eta} \in C^0(G)$ for all $\xi, \eta \in D$.

The regular representation λ_G of G has vanishing matrix coefficients, since

$$\varphi_{\lambda_G,f,f'} \in C^c(G) \qquad \text{for } f, f' \in C^c(G).$$

If $(\pi_i)_{i\in I}$ is a family of representations of G with vanishing matrix coefficients, then their direct sum $\bigoplus_{i\in I} \pi_i$ also has vanishing matrix coefficients. A representation of a noncompact LC group which has a nonzero fixed vector does not have vanishing matrix coefficients.

Let π be a representation of G with vanishing matrix coefficients. Every representation of G which is equivalent to π has vanishing matrix coefficients. Observe also that the restriction $\pi|_H$ of π to any closed subgroup H of G has vanishing matrix coefficients; this implies that $\pi|_H$ does not contain the trivial representation 1_H if H is not compact.

THEOREM 12.E.1 (**Kirillov**). *Let* $\mathrm{GL}_n(\mathbf{K})$ *be as above. A function* $\varphi : \mathrm{GL}_n(\mathbf{K}) \to \mathbf{C}$ *belongs to* $E(\mathrm{GL}_n(\mathbf{K}))$ *if and only if* φ *has one of the following forms:*

(1) *the lift to* $\mathrm{GL}_n(\mathbf{K})$ *of a unitary character of* $\mathrm{GL}_n(\mathbf{K})/\mathrm{SL}_n(\mathbf{K}) \approx \mathbf{K}^\times$,
(2) *the trivial extension* $\widetilde{\chi}$ *of a unitary character* χ *of the centre* $Z \approx \mathbf{K}^\times$ *of* $\mathrm{GL}_n(\mathbf{K})$.

In other words, $E(\mathrm{GL}_n(\mathbf{K}))$ *is in natural bijection with* $\widehat{\mathbf{K}^\times} \sqcup \widehat{Z}$.

Equivalently, every factor representation of finite type of $\mathrm{GL}_n(\mathbf{K})$ *is quasi-equivalent to one of:*

(3) *the composition of the quotient map* $\mathrm{GL}_n(\mathbf{K}) \twoheadrightarrow \mathrm{GL}_n(\mathbf{K})/\mathrm{SL}_n(\mathbf{K}) \approx \mathbf{K}^\times$ *with a unitary character* $\mathbf{K}^\times \to \mathbf{T}$;
(4) *the induced representation* $\mathrm{Ind}_Z^{\mathrm{GL}_n(\mathbf{K})} \chi$ *for a unitary character* χ *of* Z.

Proof of Theorem 12.E.1 for $n = 2$. We write Γ for $\mathrm{GL}_2(\mathbf{K})$, and we consider the following subgroups of Γ:

$$\begin{aligned} P &= \left\{ \begin{pmatrix} a & b \\ 0 & 1 \end{pmatrix} \;\middle|\; a \in \mathbf{K}^\times,\ b \in \mathbf{K} \right\}, \\ P^t &= \left\{ \begin{pmatrix} a & 0 \\ c & 1 \end{pmatrix} \;\middle|\; a \in \mathbf{K}^\times,\ c \in \mathbf{K} \right\}, \\ N &= \left\{ \begin{pmatrix} 1 & b \\ 0 & 1 \end{pmatrix} \;\middle|\; b \in \mathbf{K} \right\}, \\ N^t &= \left\{ \begin{pmatrix} 1 & 0 \\ c & 1 \end{pmatrix} \;\middle|\; c \in \mathbf{K} \right\}, \\ A = P \cap P^t &= \left\{ \begin{pmatrix} a & 0 \\ 0 & 1 \end{pmatrix} \;\middle|\; a \in \mathbf{K}^\times \right\}. \end{aligned}$$

Observe that $P = A \ltimes N$ and $P^t = A \ltimes N^t$ are isomorphic to the group $\mathrm{Aff}(\mathbf{K})$ of affine transformations of $\mathbf{K}$.

Let $\varphi \in E(\Gamma)$. The restrictions $\varphi|_P$ and $\varphi|_{P^t}$ to P and P^t are in $\mathrm{Tr}_1(P)$ and $\mathrm{Tr}_1(P^t)$, respectively. By Lemma 12.B.2, there exist $\varphi_1 \in \mathrm{Tr}_1(P), \varphi_2 \in \mathrm{Tr}_1(P^t)$ with $\varphi_1|_N = 1, \varphi_2|_{N^t} = 1$ and real numbers $t_1, t_2 \in [0, 1]$ such that

$$\varphi|_P = t_1\varphi_1 + (1 - t_1)\delta_e \qquad \text{and} \qquad \varphi|_{P^t} = t_2\varphi_2 + (1 - t_2)\delta_e.$$

Let $(\pi, \mathcal{H}, \xi)$ be the GNS triple associated with φ. Fix $g \in \Gamma$ and let $\mathcal{H}_g$ be the closed linear span of $\pi(P)\pi(g)\xi$. It is clear that $\mathcal{H}_g$ is $\pi(P)$-invariant and that $\pi(g)\xi$ is a cyclic vector for the associated representation of P on $\mathcal{H}_g$. Observe that, for every $\gamma \in P$, we have

$$\langle \pi(\gamma)\pi(g)\xi \mid \pi(g)\xi \rangle = \varphi(g^{-1}\gamma g) = \varphi(\gamma) = t_1\varphi_1(\gamma) + (1 - t_1)\delta_e(\gamma),$$

because φ is a central function. This implies that we have an orthogonal decomposition

$$\mathcal{H}_g = \mathcal{H}_{(1,g)} \oplus \mathcal{H}^\infty_{(1,g)},$$

such that P acts on $\mathcal{H}^\infty_{(1,g)}$ as a multiple of the regular representation λ_P of P and N acts as the identity on $\mathcal{H}_{(1,g)}$. In particular, the representation of P on $\mathcal{H}^\infty_{(1,g)}$ has vanishing matrix coefficients.

Let $\mathcal{H}_{(1)}$ be the closed linear span of $\bigcup_{g \in \Gamma} \mathcal{H}_{(1,g)}$ and $\mathcal{H}^\infty_{(1)}$ the closed linear span of $\bigcup_{g \in \Gamma} \mathcal{H}^\infty_{(1,g)}$. Observe that the representation of P on $\mathcal{H}^\infty_{(1)}$ has vanishing matrix coefficients and that N acts as the identity on $\mathcal{H}_{(1)}$.

Since $\mathcal{H}$ is the closed linear span of $\bigcup_{g \in \Gamma} \mathcal{H}_g$, it is clear that $\mathcal{H}_{(1)} + \mathcal{H}^\infty_{(1)}$ is a dense subspace of $\mathcal{H}$.

We claim that $\mathcal{H}_{(1)}$ and $\mathcal{H}^\infty_{(1)}$ are orthogonal to each other. Indeed, assume by contradiction that $p|_{\mathcal{H}_{(1)}} : \mathcal{H}_{(1)} \to \mathcal{H}^\infty_{(1)}$ is nonzero, where $p : \mathcal{H} \to \mathcal{H}^\infty_{(1)}$ is the orthogonal projection. Since $p|_{\mathcal{H}_{(1)}}$ is an intertwining operator between the representations $\pi_{(1)}$ and $\pi^\infty_{(1)}$ of P defined by $\mathcal{H}_{(1)}$ and $\mathcal{H}^\infty_{(1)}$, it follows (see Lemma 1.A.5) that the representations $\pi_{(1)}$ and $\pi^\infty_{(1)}$ have equivalent subrepresentations. Now, $\pi_{(1)}|_N$ is a multiple of the trivial representation 1_N and the representation $\pi^\infty_{(1)}$ of P has vanishing matrix coefficients. Hence N is finite and this is a contradiction.

As a result, we have an orthogonal decomposition

$$\mathcal{H} = \mathcal{H}_{(1)} \oplus \mathcal{H}^\infty_{(1)}$$

into $\pi(P)$-invariant subspaces with the following properties: the representation of P on $\mathcal{H}_{(1)}^{\infty}$ has vanishing matrix coefficients and N acts as the identity on $\mathcal{H}_{(1)}$.

Similarly, we have a second orthogonal decomposition

$$\mathcal{H} = \mathcal{H}_{(2)} \oplus \mathcal{H}_{(2)}^{\infty}$$

into $\pi(P^t)$-invariant subspaces, with the same properties: the representation of P^t on $\mathcal{H}_{(2)}^{\infty}$ has vanishing matrix coefficients and N^t acts as the identity on $\mathcal{H}_{(2)}$. (Observe that some of the subspaces $\mathcal{H}_{(i)}$ or $\mathcal{H}_{(i)}^{\infty}$ above may be trivial.)

As mentioned above, $\mathcal{H}_{(1)}^{\infty}$ [respectively $\mathcal{H}_{(2)}^{\infty}$] contains no nonzero vector invariant under N [respectively N^t]. It follows that $\mathcal{H}_{(1)}$ is the space $\mathcal{H}^N$ of $\pi(N)$-invariant vectors and $\mathcal{H}_{(2)}$ is the space $\mathcal{H}^{N^t}$ of $\pi(N^t)$-invariant vectors in $\mathcal{H}$.

We claim that $\mathcal{H}^N \cap \mathcal{H}^{N^t}$ is invariant under $\pi(\Gamma)'' \cup \pi(\Gamma)'$.

Indeed, $\mathcal{H}^N \cap \mathcal{H}^{N^t}$ is a closed subspace of $\mathcal{H}$ which is invariant under $\pi(A)$, as A normalizes N and N^t. So, $\mathcal{H}^N \cap \mathcal{H}^{N^t}$ is invariant under $\pi(P)$ and $\pi(P^t)$, hence under $\pi(\Gamma)$ since $P \cup P^t$ generates $\Gamma = \mathrm{GL}_2(\mathbf{K})$, hence under $\pi(\Gamma)''$. Moreover, $\mathcal{H}^N$ and $\mathcal{H}^{N^t}$ are invariant under the commutants $\pi(N)'$ and $\pi(N^t)'$ of $\pi(N)$ and $\pi(N^t)$, respectively; therefore $\mathcal{H}^N \cap \mathcal{H}^{N^t}$ is invariant under $\pi(\Gamma)'$.

- *First case.* Assume that $\mathcal{H}^N \cap \mathcal{H}^{N^t} \neq \{0\}$. We claim that $\varphi(\gamma) = 1$ for all $\gamma \in \mathrm{SL}_2(\mathbf{K})$.

Indeed, since $\mathcal{H}^N \cap \mathcal{H}^{N^t}$ is invariant by $\pi(\Gamma)'' \cup \pi(\Gamma)'$, the orthogonal projection $q : \mathcal{H} \to \mathcal{H}^N \cap \mathcal{H}^{N^t}$ belongs to the centre of $\pi(\Gamma)''$. However, π is a factor representation, as $\varphi \in E(\Gamma)$. It follows that $q = \mathrm{Id}_{\mathcal{H}}$, that is, $\mathcal{H} = \mathcal{H}^N \cap \mathcal{H}^{N^t}$. This shows that $\varphi(\gamma) = 1$ for all γ either in N or in N^t, and therefore also for all γ in the subgroup $\mathrm{SL}_2(\mathbf{K})$ generated by $N \cup N^t$.

- *Second case.* Assume that $\mathcal{H}^N \cap \mathcal{H}^{N^t} = \{0\}$. Set

$$\mathcal{K} := \mathcal{H}_{(1)}^{\infty} + \mathcal{H}_{(2)}^{\infty}.$$

Then $\mathcal{K}$ is a dense subspace of $\mathcal{H}$, because $\mathcal{K}^{\perp} = \mathcal{H}^N \cap \mathcal{H}^{N^t} = \{0\}$.

Recall that the representation of P on $\mathcal{H}_{(1)}^{\infty}$ [respectively the representation of P^t on $\mathcal{H}_{(2)}^{\infty}$] has vanishing matrix coefficients. Since $A = P \cap P^t$, it follows that the representation $\pi|_A$ of A on $\mathcal{H}$ has vanishing matrix coefficients.

Let $\gamma \in \Gamma \smallsetminus Z$. We claim that $\varphi(\gamma) = 0$.

To show this, observe first that γ is conjugated inside Γ to a matrix of the form $\begin{pmatrix} 0 & b \\ 1 & d \end{pmatrix}$, for $b \in \mathbf{K}^{\times}$ and $d \in \mathbf{K}$. Indeed, choose $v \in \mathbf{K}^2$ such that $\gamma(v)$ is not a multiple of v (this is possible, since $\gamma \notin Z$); if $\delta \in \Gamma$ maps the canonical basis of $\mathbf{K}^2$ onto $\{v, \gamma(v)\}$, then $\delta^{-1}\gamma\delta$ is of the form $\begin{pmatrix} 0 & b \\ 1 & d \end{pmatrix}$. So, it suffices to show that $\varphi(\gamma) = 0$ for $\gamma = \begin{pmatrix} 0 & b \\ 1 & d \end{pmatrix}$. Set $\varepsilon = \begin{pmatrix} 0 & 1 \\ 1 & 0 \end{pmatrix}$.

Let $(a_k)_{k \geq 1}$ be a sequence of pairwise distinct elements in $\mathbf{K}^{\times}$; for every $k \geq 1$, set

$$\delta_k = \begin{pmatrix} a_k & 0 \\ 0 & 1 \end{pmatrix} \quad \text{and} \quad \gamma_k = \begin{pmatrix} a_k^{-1} & d \\ 0 & a_k b \end{pmatrix}.$$

Then

$$\delta_k \gamma \delta_k^{-1} = \begin{pmatrix} 0 & a_k b \\ a_k^{-1} & d \end{pmatrix} = \begin{pmatrix} 0 & 1 \\ 1 & 0 \end{pmatrix} \begin{pmatrix} a_k^{-1} & d \\ 0 & a_k b \end{pmatrix} = \varepsilon \gamma_k,$$

and we have $\varphi(\gamma) = \varphi(\varepsilon\gamma_k)$. Write $\gamma_k = z_k\alpha_k n_k$ with

$$z_k = \begin{pmatrix} a_k b & 0 \\ 0 & a_k b \end{pmatrix} \in Z, \ \alpha_k = \begin{pmatrix} a_k^{-2}b^{-1} & \\ 0 & 1 \end{pmatrix} \in A, \ \text{and } n_k = \begin{pmatrix} 1 & da_k \\ 0 & 1 \end{pmatrix} \in N.$$

Since $z_k \in Z$, we have, using Lemma 11.C.8(4),

$$\varphi(\gamma) = \varphi(\varepsilon\gamma_k) = \varphi(\varepsilon z_k\alpha_k n_k) = \varphi(z_k)\varphi(\varepsilon\alpha_k n_k).$$

Set $\eta := \pi(\varepsilon)\xi$ and write the orthogonal decompositions

$$\xi = \xi_1 + \xi_2 \ \text{ and } \ \eta = \eta_1 + \eta_2$$

of ξ and η, with $\xi_1, \eta_1 \in \mathcal{H}^N$ and $\xi_2, \eta_2 \in \mathcal{H}^\infty_{(1)}$; for every k, we have

$$\begin{aligned} \varphi(\varepsilon\alpha_k n_k) &= \langle \pi(\alpha_k n_k)\xi \mid \eta \rangle \\ &= \langle \pi(\alpha_k n_k)\xi_1 \mid \eta_1 \rangle + \langle \pi(\alpha_k n_k)\xi_2 \mid \eta_2 \rangle \\ &= \langle \pi(\alpha_k)\xi_1 \mid \eta_1 \rangle + \langle \pi(\alpha_k n_k)\xi_2 \mid \eta_2 \rangle. \end{aligned}$$

Observe that the set $\{\alpha_k \mid k \geq 1\}$ is infinite, since the a_k's are pairwise distinct. On the one hand, as $(\pi|_A, \mathcal{H})$ has vanishing matrix coefficients, we therefore have

$$\lim_k \langle \pi(\alpha_k)\xi_1, \mid \eta_1 \rangle = 0.$$

On the other hand, since the representation of P on $\mathcal{H}^\infty_{(1)}$ has vanishing matrix coefficients, we also have

$$\lim_k \langle \pi(\alpha_k n_k)\xi_2 \mid \eta_2 \rangle = 0.$$

Therefore,

$$\varphi(\gamma) = \lim_k \varphi(z_k)\big(\langle \pi(\alpha_k)\xi_1 \mid \eta_1 \rangle + \langle \pi(\alpha_k n_k)\xi_2 \mid \eta_2 \rangle\big) = 0.$$

It remains to show that, if a function φ on Γ has one of the forms stated in Theorem 12.E.1, then $\varphi \in E(\Gamma)$. This is obvious if φ is the lift of a unitary character of $\mathrm{GL}_2(\mathbf{K})/SL_2(\mathbf{K})$. Assume that $\varphi = \widetilde{\chi}$ is the trivial extension of a unitary character χ of the centre Z.

Since the function $\widetilde{\chi}$ is in $\mathrm{Tr}_1(\Gamma)$ by Proposition 1.F.10, we have to check that it is indecomposable. Let $\varphi_1, \varphi_2 \in \mathrm{Tr}_1(\Gamma)$ and $t \in]0,1[$ be such that

$$\widetilde{\chi} = t\varphi_1 + (1-t)\varphi_2.$$

By what we have just seen, for $j = 1, 2$, we can write

$$\varphi_j = t_j\psi_1^{(j)} + (1-t_j)\psi_2^{(j)},$$

where $t_j \in [0,1]$, $\psi_1^{(j)} \in \mathrm{Tr}_1(\Gamma)$ is equal to 1 on $\mathrm{SL}_2(\mathbf{K})$, and $\psi_2^{(j)} \in \mathrm{Tr}_1(\Gamma)$ is equal to 0 on $\Gamma \smallsetminus Z$. It follows that we can assume that each φ_j is either equal to 1 on $\mathrm{SL}_2(\mathbf{K})$ or equal to 0 on $\Gamma \smallsetminus Z$. Since φ vanishes outside Z, we necessarily have that both φ_1 and φ_2 are 0 on $\Gamma \smallsetminus Z$. Then, as in the proof of Theorem 12.B.3, φ_1 and φ_2 have to agree with χ on Z. So, $\varphi_1 = \varphi_2 = \varphi$ and hence $\varphi \in E(\mathrm{GL}_n(\mathbf{K}))$.

□

Proof of Theorem 12.E.1 for $n \geq 3$. We denote by $e_1, \ldots, e_n$ the vectors of the canonical basis of $\mathbf{K}^n$. For every $i \in \{1, \ldots, n\}$, we consider the copy V_i of $\mathbf{K}^{n-1}$ inside $\Gamma = \mathrm{GL}_n(\mathbf{K})$ consisting of the matrices $\gamma \in \mathrm{SL}_n(\mathbf{K})$ such that $\gamma(e_j) = e_j$ for $j \neq i$. Observe that V_i is normalized by a copy G_i of $\mathrm{GL}_{n-1}(\mathbf{K}) \ltimes \mathbf{K}^{n-1}$ inside Γ. We will also have to consider the transpose subgroups V_i^t; of course, V_i^t is normalized by G_i^t. Observe that $V_j \cap V_i^t \approx \mathbf{K}$ is a infinite subgroup of Γ for all

$i \neq j$. We will refer to subgroups of the form V_i or V_i^t as to the copies of $\mathbf{K}^{n-1}$ inside Γ.

Let $\varphi \in E(\Gamma)$. Fix a copy V of $\mathbf{K}^{n-1}$ inside Γ. Lemma 12.B.2 applied to $\mathrm{GL}_{n-1}(\mathbf{K}) \ltimes \mathbf{K}^{n-1}$ (with $N = V$) shows that

$$\varphi|_V = t 1_V + (1-t)\delta_e,$$

for some $t \in [0,1]$. Let $(\pi, \mathcal{H}, \xi)$ be the GNS triple associated with φ. The same arguments as in the case $n = 2$ above show that we have a corresponding orthogonal decomposition of $\mathcal{H}$,

$$\mathcal{H} = \mathcal{H}^V \oplus \mathcal{H}_V^\infty,$$

into $\pi(V)$-invariant closed subspaces, where the representation of V on $\mathcal{H}_V^\infty$ has vanishing matrix coefficients and where V acts trivially on $\mathcal{H}^V$.

- *First step.* Let W be another copy of $\mathbf{K}^{n-1}$ with $V \cap W \neq \{0\}$. We claim that $\mathcal{H}^V = \mathcal{H}^W$ and $\mathcal{H}_\infty^V = \mathcal{H}_\infty^W$.

Indeed, the representations on $\mathcal{H}_V^\infty$ and on $\mathcal{H}_W^\infty$ of the infinite subgroup $V \cap W$ have vanishing matrix coefficients; it follows that

$$\mathcal{H}^{V\cap W} = \mathcal{H}^V \quad \text{and} \quad \mathcal{H}^{V\cap W} = \mathcal{H}^W$$

and hence $\mathcal{H}^V = \mathcal{H}^W$.

- *Second step.* Let W be an arbitrary copy of $\mathbf{K}^{n-1}$. We claim that $\mathcal{H}^V = \mathcal{H}^W$ and $\mathcal{H}_V^\infty = \mathcal{H}_W^\infty$.

Indeed, we can find two copies W^1 and W^2 of $\mathbf{K}^{n-1}$ with

$$V \cap W^1 \neq \{0\}, \quad W^1 \cap W^2 \neq \{0\}, \quad \text{and} \quad W^2 \cap W \neq \{0\}.$$

The claim follows from the first step.

- *Third step.* Assume that $\mathcal{H}^V \neq \{0\}$. We claim that $\varphi = 1$ on $\mathrm{SL}_n(\mathbf{K})$.

Indeed, by the second step, $\mathcal{H}^V = \mathcal{H}^W$ for every copy W of $\mathbf{K}^{n-1}$ and hence $\mathcal{H}^V = \mathcal{H}^{\mathrm{SL}_n(\mathbf{K})}$, since $\mathrm{SL}_n(\mathbf{K})$ is generated by the various W's. In particular, $\mathcal{H}^V$ is invariant under $\pi(\Gamma)$, as $\mathrm{SL}_n(\mathbf{K})$ is normal in Γ. Moreover, $\mathcal{H}^V$ is also invariant under the commutant $\pi(\Gamma)'$ of $\pi(\Gamma)$. Therefore $\mathcal{H}^V = \mathcal{H}$, as π is a factor representation.

- *Fourth step.* Assume that $\mathcal{H}^V = \{0\}$. We claim that $\varphi = 0$ on $\Gamma \smallsetminus Z$.

Indeed, by the second step, $\mathcal{H} = \mathcal{H}_V^\infty$ and hence $\varphi|_V = \delta_e$. Let $G_V \approx \mathrm{GL}_{n-1}(\mathbf{K}) \ltimes \mathbf{K}^{n-1}$ be the subgroup normalizing V. Since $\{[g,v] \mid v \in V\}$ is infinite for every $g \in G_V \smallsetminus V$, it follows from Lemma 12.B.1 that $\varphi|_{G_V} = \delta_e$.

Let $\gamma \in \Gamma \smallsetminus Z$. As in the last part of the proof of the case $n = 2$, upon passing to a conjugate of γ in Γ, we can find $\varepsilon \in \Gamma$, a sequence $(z_k)_k$ in Z, and a sequence $(\gamma_k)_k$ of pairwise distinct elements of G_V such that $\gamma = z_k \varepsilon \gamma_k$. Since

$$\langle \pi(\gamma_k)\xi \mid \pi(\gamma_l)\xi \rangle = \varphi(\gamma_l^{-1}\gamma_k) = 0$$

for $k \neq l$, the sequence $(\pi(\gamma_k)\xi)_k$ is weakly convergent to 0 in $\mathcal{H}$ and therefore $\varphi(\gamma) = 0$.

Finally, one checks, as in the case $n = 2$, that every function on Γ which has one of the forms stated in Theorem 12.E.1 is in $E(\Gamma)$. □

For $\psi \in \widehat{Z}$, we have $\widetilde{\chi} \in E(\Gamma)$; so, the induced representation $\mathrm{Ind}_Z^\Gamma \psi$, which is a GNS representation associated to $\widetilde{\chi}$, is a factor representation. The following corollary is therefore an immediate consequence of Theorem 12.E.1 and Theorem 9.F.9. The canonical map $E(\Gamma) \to \mathrm{Prim}(\Gamma)$ is that of Theorem 11.D.3.

COROLLARY 12.E.2. *Let* $\mathbf{K}$ *be an infinite algebraic extension of a finite field,* $n \geq 2$, *and* $\Gamma = \mathrm{GL}_n(\mathbf{K})$. *The map*

$$E(\Gamma) \to \mathrm{Prim}(\Gamma)$$

is a bijection.

Now consider the group $\Gamma = \mathrm{SL}_n(\mathbf{K})$. Its centre Z is finite, it consists of the matrices λI_n with λ an nth root of unity in $\mathbf{K}$, that is, $Z \approx \mu_n^{\mathbf{K}} := \{\lambda \in \mathbf{K}^* \mid \lambda^n = 1\}$. The quotient $\mathrm{PSL}_n(K)$ is an infinite simple group. The proof of Theorem 12.E.1 carries over without change to $\Gamma = \mathrm{SL}_n(\mathbf{K})$ for $n \geq 3$, after replacing there the copies of $\mathrm{GL}_{n-1}(\mathbf{K}) \ltimes \mathbf{K}^{n-1}$ by copies of $\mathrm{SL}_{n-1}(\mathbf{K}) \ltimes \mathbf{K}^{n-1}$. However, for $\Gamma = \mathrm{SL}_2(\mathbf{K})$, the proof involves further arguments; see [**PeTh–16**]. Here is the final result for $n \geq 2$.

THEOREM 12.E.3. *Let* $\mathbf{K}$ *be an infinite field and* n *an integer,* $n \geq 2$.
Then

$$E(\mathrm{SL}_n(\mathbf{K})) = \{\mathbf{1}_{\mathrm{SL}_n(\mathbf{K})}\} \sqcup \{\widetilde{\chi} \mid \chi \in \widehat{Z}\},$$

where $\widetilde{\chi}$ *denotes the trivial extension of a unitary character* χ *of the centre* $Z \approx \mu_n^{\mathbf{K}}$ *of* $\mathrm{SL}_n(\mathbf{K})$.

Consequently, we have

$$E(\mathrm{PSL}_n(\mathbf{K})) = \{\mathbf{1}_{\mathrm{PSL}_n(\mathbf{K})},\ \delta_e\}.$$

CHAPTER 13

The group measure space construction

The previous chapters have shown some examples of representations of discrete groups which generate factors of various types. As a preliminary to the construction of other examples in the next chapter, here we discuss the group measure space construction of Murray and von Neumann [**MuvN–36**], *which plays a fundamental role in the theory of operator algebras. This construction associates to an action of a discrete group* Γ *on a standard Borel space* $(X, \mathcal{B})$, *equipped with a quasi-invariant measure* μ, *a von Neumann algebra* $L^\infty(X, \mu) \rtimes \Gamma$. *The group measure space construction is described in detail in Section* 13.A.

Assume that the Γ*-action on* (X, μ) *is ergodic and essentially free. We will show that* $L^\infty(X, \mu) \rtimes \Gamma$ *is a factor and that its type depends on properties of the measure* μ. *In particular, this factor is of type III when there is no* Γ*-invariant measure on* $(X, \mathcal{B})$ *equivalent to* μ. *In order to check this last condition, in Section* 13.B *we discuss a criterion based on the essential range of the action of* Γ *on* (X, μ). *This implies for example that* $L^\infty(\mathbf{P}^1(\mathbf{R}), \mu) \rtimes \Gamma$ *is a factor of type* III_1 *for any lattice* Γ *in* $\mathrm{PSL}_2(\mathbf{R})$ *and for any* Γ*-quasi-invariant measure* μ *on* $\mathbf{P}^1(\mathbf{R})$.

13.A. Construction of factors of different types

The main result of this section is Theorem 13.A.13. Our exposition follows [**Dixm–vN**, Chap. I, § 9] and [**Take–79**, Chapter V, Section 7].

13.A.a. Countable group actions on measured standard Borel spaces. Consider a standard Borel space $(X, \mathcal{B})$ equipped with a σ-finite measure μ. Recall that a Borel isomorphism θ of $(X, \mathcal{B})$ is **nonsingular** if the direct image $\theta_*(\mu)$ is measure equivalent to μ. An action of a group Γ by Borel isomorphisms $X \to X$ is nonsingular if the measure μ is equivalent with its direct image by every $\gamma \in \Gamma$.

Definition 13.A.1. Let $(X, \mathcal{B})$ be a standard Borel space, equipped with a σ-finite measure μ, and Γ a countable group. Consider a nonsingular action of Γ on $(X, \mathcal{B}, \mu)$, say from the right. Set

$$F := \{x \in X \mid x\gamma = x \text{ for some } \gamma \neq e \text{ in } \Gamma\} = \bigcup_{\gamma \in \Gamma \smallsetminus \{e\}} X^\gamma,$$

where X^γ is the set of fixed points of $\gamma \in \Gamma$ in X. The action is

- **essentially free** if $\mu(F) = 0$; equivalently, if $\mu(X^\gamma) = 0$ for every γ in $\Gamma \smallsetminus \{e\}$,
- **ergodic** if either $\mu(Y) = 0$ or $\mu(X \smallsetminus Y) = 0$ for every Γ-invariant subset $Y \in \mathcal{B}$ (ergodic actions are defined more generally in Section A.5 for actions of second-countable locally compact groups),
- **essentially transitive** if there exists an orbit $Y = x\Gamma$ in X such that $\mu(X \smallsetminus Y) = 0$.

The action of Γ on $(X, \mathcal{B}, \mu)$ provides a *left action* $\gamma \mapsto \theta_\gamma$ of Γ on $L^\infty(X, \mu)$ by automorphisms of von Neumann algebras, defined by

$$(*) \qquad \theta_\gamma(\varphi)(x) = \varphi(x\gamma) \quad \text{for all } \gamma \in \Gamma,\ \varphi \in L^\infty(X, \mu), \text{ and } x \in X.$$

It will be useful to characterize the essential freeness of an action of Γ on $(X, \mathcal{B}, \mu)$ in terms of the associated action on the von Neumann algebra $L^\infty(X, \mu)$. In this section, we will only use the equivalence of (i) and (iii); the equivalence with (ii) has been used earlier, in the proof of Theorem 12.C.1.

Recall that, for $Y \in \mathcal{B}$, we denote by $\mathbf{1}_Y$ the characteristic function of Y, defined by $\mathbf{1}_Y(x) = 1$ if $x \in Y$ and $\mathbf{1}_Y(x) = 0$ if $x \in X \smallsetminus Y$. We view $\mathbf{1}_Y$ as an element of the von Neumann algebra $L^\infty(X, \mu)$; observe that $\theta_\gamma(\mathbf{1}_Y) = \mathbf{1}_{Y\gamma^{-1}}$ for every $\gamma \in \Gamma$.

PROPOSITION 13.A.2. *Let Γ be a countable group acting by nonsingular measurable transformations on a standard Borel space $(X, \mathcal{B})$ equipped with a σ-finite measure μ. The following properties are equivalent:*

(i) *the action of Γ on $(X, \mathcal{B}, \mu)$ is essentially free;*
(ii) *for every $\gamma \in \Gamma \smallsetminus \{e\}$, there exists a sequence $(Y_n)_{n \geq 1}$ in $\mathcal{B}$ with the following properties:*

$$\begin{aligned} \mathbf{1}_{Y_n}\mathbf{1}_{Y_m} &= 0 && \textit{for } m, n \geq 1 \textit{ such that } m \neq n, \\ \sum_{n \geq 1} \mathbf{1}_{Y_n} &= \mathbf{1}_X, \\ \mathbf{1}_{Y_n}\theta_\gamma(\mathbf{1}_{Y_n}) &= 0 && \textit{for every } n \geq 1; \end{aligned}$$

(iii) *if $\varphi \in L^\infty(X, \mu)$ and $\gamma \in \Gamma \smallsetminus \{e\}$ are such that $\varphi\theta_\gamma(\psi) = \varphi\psi$ for every $\psi \in L^\infty(X, \mu)$, then $\varphi = 0$.*

PROOF. (i) $\Rightarrow$ (ii) Since X is a standard Borel space, there exists a sequence $(B_n)_{n \geq 1}$ in $\mathcal{B}$ which separates the points of X.

Fix $\gamma \in \Gamma \smallsetminus \{e\}$. Set $X'^\gamma := X \smallsetminus X^\gamma$. For every $x \in X'^\gamma$, we have $x \neq x\gamma$, and so there exists $n \geq 1$ such that $x \in B_n$ and $x\gamma \notin B_n$. Therefore

$$X'^\gamma = \bigcup_{n \geq 1} (X'^\gamma \cap (B_n \smallsetminus B_n\gamma^{-1})).$$

For every $n \geq 1$, set $Z_n := X'^\gamma \cap (B_n \smallsetminus B_n\gamma^{-1})$. We have $Z_n \cap Z_n\gamma^{-1} = \emptyset$. Define recursively a sequence of Borel sets $(Y_n)_{n \geq 1}$ by $Y_1 = Z_1$ and

$$Y_n = Z_n \smallsetminus \bigcup_{k < n} Z_k \quad \text{for } n \geq 2.$$

Then

$$(\sharp) \qquad \begin{aligned} Y_m \cap Y_n &= \emptyset && \text{for } m, n \geq 1 \text{ such that } m \neq n, \\ X'^\gamma &= \bigcup_{n \geq 0} Y_n, \\ Y_n \cap Y_n\gamma^{-1} &= \emptyset && \text{for every } n \geq 1. \end{aligned}$$

Assume now that (i) holds. Then $\mu(X \smallsetminus X'^\gamma) = \mu(X^\gamma) = 0$, so that $\mathbf{1}_{X'^\gamma} = \mathbf{1}_X \in L^\infty(X, \mu)$, and ($\sharp$) shows that (ii) holds.

(ii) $\Rightarrow$ (iii) Assume that there exists a sequence $(Y_n)_{n\geq 1}$ as in (ii). Let $\varphi \in L^\infty(X,\mu)$ and $\gamma \in \Gamma \smallsetminus \{e\}$ be such that $\varphi\theta_\gamma(\psi) = \varphi\psi$ for all $\psi \in L^\infty(X,\mu)$. Then, for all $n \geq 1$, we have $\varphi\theta_\gamma(\mathbf{1}_{Y_n}) = \varphi\mathbf{1}_{Y_n}$ and

$$0 = \varphi\left(\mathbf{1}_{Y_n}\theta_\gamma(\mathbf{1}_{Y_n})\right) = \left(\varphi\theta_\gamma(\mathbf{1}_{Y_n})\right)\mathbf{1}_{Y_n} = \left(\varphi\mathbf{1}_{Y_n}\right)\mathbf{1}_{Y_n} = \varphi\mathbf{1}_{Y_n}.$$

Therefore

$$\varphi = \sum_{n\geq 1} \varphi\mathbf{1}_{Y_n} = 0$$

and this shows that (ii) implies (iii).

(iii) $\Rightarrow$ (i) Let $\gamma \in \Gamma \smallsetminus \{e\}$. For $x \in X^\gamma$ and $\psi \in L^\infty(X,\mu)$, we have

$$\theta_\gamma(\psi)(x) = \psi(x\gamma) = \psi(x).$$

Therefore for $\varphi := \mathbf{1}_{X^\gamma}$, we have $\varphi\theta_\gamma(\psi) = \varphi\psi$. By assumption (iii), we therefore have $\varphi = 0$, that is, $\mu(X^\gamma) = 0$. This shows that (iii) implies (i). $\square$

We also translate the ergodicity of a group action on a measure space (X,μ) in terms of the associated action on $L^\infty(X,\mu)$.

PROPOSITION 13.A.3. *Let Γ be a countable group acting by nonsingular measurable transformations on a standard Borel space $(X,\mathcal{B})$ equipped with a σ-finite measure μ. The following properties are equivalent:*

(i) *the action of Γ on $(X,\mathcal{B},\mu)$ is ergodic;*
(ii) *every $\varphi \in L^\infty(X,\mu)$ such that $\theta_\gamma(\varphi) = \varphi$ for all $\gamma \in \Gamma$ is constant μ-almost everywhere.*

NOTE. In (ii), one can equivalently replace $L^\infty(X,\mu)$ by the space of complex-valued measurable functions on X up to equality μ-almost everywhere.

PROOF. Assume that (ii) holds and let $Y \in \mathcal{B}$ be Γ-invariant. Then $\theta_\gamma(\mathbf{1}_Y) = \mathbf{1}_Y$ for all $\gamma \in \Gamma$, and hence $\mathbf{1}_Y$ is constant μ-almost everywhere. This implies that $\mu(Y) = 0$ or $\mu(X \smallsetminus Y) = 0$.

Conversely, assume that the action of Γ on $(X,\mathcal{B},\mu)$ is ergodic. Let $\varphi \in L^\infty(X,\mu)$ be such that $\theta_\gamma(\varphi) = \varphi$, for every $\gamma \in \Gamma$. Upon considering the real and imaginary part of φ, we may assume that φ is real-valued.

For every $\gamma \in \Gamma$, we have $\mu(N_\gamma) = 0$, where

$$N_\gamma = \{x \in X \mid \varphi(x\gamma) \neq \varphi(x)\}.$$

Since Γ is countable, we have $\mu(N) = 0$, for $N := \bigcup_{\gamma\in\Gamma\smallsetminus\{e\}} N_\gamma$. Let $\varphi' : X \to \mathbf{R}$ be defined by $\varphi' = \varphi$ on $X \smallsetminus N$ and $\varphi' = 0$ on N; then φ' coincides with φ almost everywhere and φ' is Γ-invariant: $\varphi'(x\gamma) = \varphi'(x)$ for every $x \in X$ and $\gamma \in \Gamma$. So, me may assume that φ is Γ-invariant.

Consider the "distribution function" $F : \mathbf{R} \to [0,+\infty[\cup\{+\infty\}$ of φ, defined by

$$F(t) = \mu(\{x \in X \mid \varphi(x) \leq t\}) \quad \text{for } t \in \mathbf{R}.$$

On the one hand, F is an increasing function with

$$\lim_{t\to-\infty} F(t) = 0 \quad \text{and} \quad \lim_{t\to+\infty} F(t) = \mu(X).$$

On the other hand, since $\{x \in X \mid \varphi(x) \leq t\}$ is Γ-invariant, for every $t \in \mathbf{R}$, we have either $F(t) = 0$ or $F(t) = \mu(X)$, by ergodicity of the Γ-action. Then

$$t_0 := \inf\{t \in \mathbf{R} \mid F(t) = \mu(X)\}$$

exists in $\mathbf{R}$. For every $n \geq 1$, we have

$$\mu\left(\left\{x \in X \mid \varphi(x) \leq t_0 - \frac{1}{n}\right\}\right) = 0 \text{ and } \mu\left(\left\{x \in X \mid \varphi(x) > t_0 + \frac{1}{n}\right\}\right) = 0.$$

This implies that $\mu(\{x \in X \mid \varphi(x) = t_0\}) = \mu(X)$ and φ is constant μ-almost everywhere. □

REMARK 13.A.4. Let Γ be a countable group acting by nonsingular measurable transformations on a standard Borel space $(X, \mathcal{B})$ equipped with a σ-finite measure μ. The following properties are equivalent:

(i) the action of Γ on $(X, \mathcal{B}, \mu)$ is essentially transitive;
(ii) the action is ergodic and the measure μ is atomic.

DEFINITION 13.A.5. Let Γ be a countable group acting by nonsingular measurable transformations on a standard Borel space $(X, \mathcal{B})$ equipped with a σ-finite measure μ.

Assume first that the action is a **left** action. The associated **Koopman representation** is the representation of Γ on $L^2(X, \mu)$ which associates to $\gamma \in \Gamma$ the unitary operator u_γ defined by

$$(u_\gamma f)(x) = c(\gamma^{-1}, x)^{1/2} f(\gamma^{-1} x) \quad \text{for all } \gamma \in \Gamma,\ f \in L^2(X, \mu),\ x \in X,$$

where $c(\gamma, x) = \frac{d(\gamma^{-1})_*(\mu)}{d\mu}(x)$ denotes the Radon–Nikodym cocycle of the action $\Gamma \curvearrowright (X, \mathcal{B}, \mu)$, as in Appendix A.3 and in Definition 5.A.1.

In case of a **right** group action $(X, \mathcal{B}, \mu) \curvearrowleft \Gamma$ by nonsingular measurable transformations, the Radon–Nikodym cocycle associated to the action of an element $\gamma \in \Gamma$ is exactly the same as before; we prefer to denote it by $c(x, \gamma)$ instead of $c(\gamma, x)$. The Koopman representation $\gamma \mapsto u_\gamma$ of Γ on $L^2(X, \mu)$ is then defined by

$$(u_\gamma f)(x) = c(x, \gamma)^{1/2} f(x\gamma) \quad \text{for all } \gamma \in \Gamma,\ f \in L^2(X, \mu),\ x \in X.$$

The definition of Koopman representations can be extended to σ-compact locally compact groups acting by nonsingular automorphisms on measure spaces $(X, \mathcal{B}, \mu)$ such that $L^2(X, \mu)$ is a separable Hilbert space; these technical conditions, σ-compactness and separability, are needed for the associated representation to be continuous. See Proposition A.6.1 in [**BeHV–08**].

The terminology refers to an article by Koopman [**Koop–31**], where the author considers the time evolution on an energy level in the (p, q)-space of a Hamiltonian system and associates to this situation a "one-parameter group of unitary transformations" in the appropriate Hilbert space, i.e., a representation of $\mathbf{R}$.

13.A.b. Factors obtained by the group measure space construction. Consider a nonsingular action $(X, \mathcal{B}, \mu) \curvearrowleft \Gamma$. There are two representations on $L^2(X, \mu)$: the Koopman representation of the group Γ, as above, and the representation $\varphi \mapsto m(\varphi)$ of the abelian von Neumann algebra $\mathcal{A} := L^\infty(X, \mu)$ on $L^2(X, \mu)$ by multiplication operators, defined by

$$m(\varphi)(f)(x) = \varphi(x) f(x) \quad \text{for all } \varphi \in \mathcal{A},\ f \in L^2(X, \mu),\ x \in X.$$

The following relation holds between these two representations:

$$(**) \qquad u_\gamma m(\varphi) u_\gamma^{-1} = m(\theta_\gamma(\varphi)) \quad \text{for all } \gamma \in \Gamma,\ \varphi \in \mathcal{A}.$$

Denote by $\mathcal{L}$ the von Neumann algebra acting on the Hilbert space $L^2(X,\mu)$ and generated by

$$\{u_\gamma \in \mathcal{U}(L^2(X,\mu)) \mid \gamma \in \Gamma\} \cup \{m(\varphi) \in \mathcal{L}(L^2(X,\mu)) \mid \varphi \in \mathcal{A}\}.$$

PROPOSITION 13.A.6. *Let Γ be a countable group acting on the right by nonsingular measurable transformations on a standard Borel space $(X,\mathcal{B})$ equipped with a σ-finite measure μ; assume that this action is ergodic. Let $\mathcal{L}$ be the von Neumann algebra on $L^2(X,\mu)$ defined above.*

Then $\mathcal{L}$ is the algebra $\mathcal{L}(L^2(X,\mathcal{B},\mu))$ of all bounded operators on $L^2(X,\mathcal{B},\mu)$. In particular, $\mathcal{L}$ is a factor of type I.

PROOF. We continue to write $\mathcal{A}$ for $L^\infty(X,\mu)$. The subset

$$\{m(\varphi) \in \mathcal{L}(L^2(X,\mu)) \mid \varphi \in \mathcal{A}\}$$

is a maximal abelian subalgebra of $\mathcal{L}(L^2(X,\mathcal{B},\mu))$, by Corollary 1.H.6.

Let $T \in \mathcal{L}'$. On the one hand, T commutes with $m(\varphi)$ for all $\varphi \in \mathcal{A}$, hence $T = m(\psi)$ for some $\psi \in \mathcal{A}$. On the other hand, $T = m(\psi)$ commutes with $\{u_\gamma \in \mathcal{U}(L^2(X,\mu)) \mid \gamma \in \Gamma\}$, and it follows from relation $(**)$ that ψ is Γ-invariant. By Proposition 13.A.3, ψ is constant almost everywhere, that is, T is a scalar operator. Therefore $\mathcal{L}' = \mathbf{C}\mathrm{Id}_{L^2(X,\mu)}$, and $\mathcal{L} = \mathcal{L}(L^2(X,\mu))$. □

In order to construct factors which are not of type I, we now consider representations of Γ and $\mathcal{A} = L^\infty(X,\mu)$ on the Hilbert space $\ell^2(\Gamma, L^2(X,\mu))$, namely:

- the representation $\gamma \mapsto U_\gamma$ of the group Γ defined by
$$U_\gamma(F)(\gamma') = u_\gamma(F(\gamma'\gamma))$$
for all $\gamma, \gamma' \in \Gamma$ and $F \in \ell^2(\Gamma, L^2(X,\mu))$,
- the representation $\varphi \mapsto M(\varphi)$ of $\mathcal{A}$ defined by
$$M(\varphi)(F)(\gamma) = m(\varphi)(F(\gamma))$$
for all $\varphi \in \mathcal{A}$, $F \in \ell^2(\Gamma, L^2(X,\mu))$, and $\gamma \in \Gamma$.

The following relations hold:

$$(***) \qquad M(\varphi)^* = M(\overline{\varphi}), \qquad U_\gamma^* = U_{\gamma^{-1}}, \qquad U_\gamma M(\varphi) U_\gamma^{-1} = M(\theta_\gamma(\varphi)),$$

for all $\varphi \in \mathcal{A}$ and $\gamma \in \Gamma$.

DEFINITION 13.A.7. Let the notation be as above. The **group measure space von Neumann algebra** or the **crossed product von Neumann algebra** associated with the nonsingular action of Γ on $(X,\mathcal{B},\mu)$ is the von Neumann algebra denoted by

$$L^\infty(X,\mu) \rtimes \Gamma, \qquad \text{or} \qquad \mathcal{A} \rtimes \Gamma, \qquad \text{or below} \qquad \mathcal{M},$$

of operators on $\ell^2(\Gamma, L^2(X,\mu))$ generated by

$$\{M(\varphi) \mid \varphi \in \mathcal{A}\} \cup \{U_\gamma \mid \gamma \in \Gamma\}.$$

[We feel obliged to respect tradition and write Γ *on the right* in $L^\infty(X,\mu) \rtimes \Gamma$, even though the action of Γ on $L^\infty(X,\mu)$ is here a *left action.*]

Observe that, as relations $(***)$ show, the linear subspace

$$\mathcal{M}_0 := \Big\{ \sum_{k=1}^n M(\varphi_k) U_{\gamma_k} \;\Big|\; n \in \mathbf{N},\ \varphi_k \in \mathcal{A},\ \gamma_k \in \Gamma \Big\}$$

is an algebra, which is moreover selfadjoint and weakly dense in $\mathcal{M}$.

We identify $\ell^2(\Gamma, L^2(X,\mu))$ with the tensor product $\ell^2(\Gamma) \otimes L^2(X,\mu)$. For $T \in \mathcal{L}(\ell^2(\Gamma, L^2(X,\mu)))$, we denote by $(T_{s,t})_{s,t\in\Gamma}$ the matrix associated with T with respect to the decomposition

$$\ell^2(\Gamma) \otimes L^2(X,\mu) = \bigoplus_{s\in\Gamma} (\mathbf{C}\delta_s \otimes L^2(X,\mu)),$$

where $(\delta_s)_{s\in\Gamma}$ is the orthonormal basis of $\ell^2(\Gamma)$ constituted of Dirac delta functions. Thus, $T_{s,t} \in \mathcal{L}(L^2(X,\mu))$ is defined by

$$\langle T_{s,t} f_t \mid f_s\rangle = \langle T(\delta_t \otimes f_t) \mid \delta_s \otimes f_s\rangle \quad \text{for } s,t \in \Gamma \text{ and } f_t, f_s \in L^2(X,\mu).$$

In the following proposition, we record a few facts about the entries of the matrix of an element from $\mathcal{M}$.

PROPOSITION 13.A.8. *Let $\mathcal{M} = L^\infty(X,\mu) \rtimes \Gamma$ be as before and let $T \in \mathcal{M}$.*

(1) *There exists a family $\{\varphi_\gamma\}_{\gamma\in\Gamma}$ in $L^\infty(X,\mu)$ such that the matrix of T as above is $(m(\varphi_{st^{-1}})u_{st^{-1}})_{s,t\in\Gamma}$.*
(2) *Let $(m(\varphi_{st^{-1}})u_{st^{-1}})_{s,t\in\Gamma}$, $(m(\psi_{st^{-1}})u_{st^{-1}})_{s,t\in\Gamma}$, and $(m(\psi'_{st^{-1}})u_{st^{-1}})_{s,t\in\Gamma}$ be the matrices as in (1) of respectively T, TT^*, and T^*T. Then*

$$\psi_e = \sum_{\gamma\in\Gamma} |\varphi_\gamma|^2 \quad \text{and} \quad \psi'_e = \sum_{\gamma\in\Gamma} |\theta_\gamma(\varphi_{\gamma^{-1}})|^2,$$

where the infinite sums are weakly convergent in $L^\infty(X,\mu)$.

PROOF. (1) Let $\varphi \in L^\infty(X,\mu)$ and $\gamma \in \Gamma$. The matrix $(M_{s,t})_{s,t\in\Gamma}$ of $M(\varphi)$ is given by

$$M_{s,t} = \begin{cases} m(\varphi) & \text{for } s = t, \\ 0 & \text{otherwise,} \end{cases}$$

and the matrix $(U_{s,t})_{s,t\in\Gamma}$ of U_γ is given by

$$U_{s,t} = \begin{cases} u_\gamma & \text{if } st^{-1} = \gamma, \\ 0 & \text{otherwise.} \end{cases}$$

Assume that $T \in \mathcal{M}_0$, that is, $T = \sum_{k=1}^n M(\varphi_{\gamma_k})U_{\gamma_k} \in \mathcal{M}_0$ with $\gamma_k \neq \gamma_l$ for $k \neq l$. The matrix $(T_{s,t})_{s,t\in\Gamma}$ of T is then given by

$$T_{s,t} = \begin{cases} m(\varphi_{\gamma_k})u_{\gamma_k} & \text{if } st^{-1} = \gamma_k \in \{\gamma_1, \ldots, \gamma_n\}, \\ 0 & \text{otherwise.} \end{cases}$$

So, we see that, for every $T \in \mathcal{M}_0$, there exists a family $\{\varphi_\gamma\}_{\gamma\in\Gamma}$ in $L^\infty(X,\mu)$ such that the matrix of T is $(m(\varphi_{st^{-1}})u_{st^{-1}})_{s,t\in\Gamma}$. Since this last property is preserved by going over to weak limits, the claim is true for every $T \in \mathcal{M}$.

(2) Denoting by $(S_{s,t})_{s,t\in\Gamma}$ the matrix of $S \in \mathcal{L}(L^2(X,\mu))$, we have

$$\begin{aligned}(TT^*)_{e,e} &= \sum_{t\in\Gamma} T_{e,t}T^*_{t,e} = \sum_{t\in\Gamma} T_{e,t}(T_{e,t})^* \\ &= \sum_{t\in\Gamma}(m(\varphi_{t^{-1}})u_{t^{-1}})(u^*_{t^{-1}}m(\overline{\varphi_{t^{-1}}})) \\ &= \sum_{t\in\Gamma} m(\varphi_{t^{-1}})m(\overline{\varphi_{t^{-1}}}) \\ &= m\left(\sum_{t\in\Gamma}|\varphi_\gamma|^2\right),\end{aligned}$$

and, using $(***)$,

$$\begin{aligned}(T^*T)_{e,e} &= \sum_{t\in\Gamma} T^*_{e,t}T_{t,e} = \sum_{t\in\Gamma}(T_{t,e})^*T_{t,e} \\ &= \sum_{t\in\Gamma}(u_t^*m(\overline{\varphi_t})(m(\varphi_t)u_t)) \\ &= m\left(\sum_{t\in\Gamma}|\theta_t(\varphi_{t^{-1}})|^2\right).\end{aligned}$$

This proves the claim. □

Henceforth, for $T \in \mathcal{M}$, we will write $T = \sum_{\gamma\in\Gamma}\varphi_\gamma u_\gamma$ instead of writing its matrix $(m(\varphi_{st^{-1}})u_{st^{-1}})_{s,t\in\Gamma}$ as in Proposition 13.A.8(1).

We will also often consider $\mathcal{A}$ as a subalgebra of $\mathcal{M}$, by identifying a function $\varphi \in \mathcal{A}$ with the corresponding operator $M(\varphi) \in \mathcal{M}$.

LEMMA 13.A.9. *Let Γ be a countable group acting on the right by nonsingular measurable transformations on a standard Borel space $(X,\mathcal{B})$ equipped with a σ-finite measure μ; assume that this action is essentially free.*

Then the von Neumann subalgebra $M(\mathcal{A}) \approx \mathcal{A}$ of $\mathcal{M} = L^\infty(X,\mu)\rtimes\Gamma$ is maximal abelian.

PROOF. Let $T = \sum_{\gamma\in\Gamma}\varphi_\gamma u_\gamma \in \mathcal{M}$ be in the commutant of $M(\mathcal{A})$. For every $\varphi \in \mathcal{A}$, we have $TM(\varphi) = M(\varphi)T$ and therefore

$$\sum_{\gamma\in\Gamma}\varphi_\gamma\theta_\gamma(\varphi)u_\gamma = \sum_{\gamma\in\Gamma}\varphi\varphi_\gamma u_\gamma.$$

This means that

$$\varphi_\gamma\theta_\gamma(\varphi) = \varphi\varphi_\gamma \quad \text{for all } \varphi\in\mathcal{A} \text{ and } \gamma\in\Gamma.$$

Since the action of Γ is essentially free, it follows from Proposition 13.A.2 that $\varphi_\gamma = 0$ for every $\gamma \in \Gamma \smallsetminus \{e\}$. This shows that $T = M(\varphi_e)$ and the lemma is proved. □

We define a linear map

$$E : \mathcal{M} = L^\infty(X,\mu) \rtimes \Gamma \longrightarrow \mathcal{A} = L^\infty(X,\mu),$$

called the **conditional expectation** onto $\mathcal{A}$, by

$$E(T) = \varphi_e \quad \text{for} \quad T = \sum_{\gamma\in\Gamma}\varphi_\gamma u_\gamma \in \mathcal{M}.$$

For $T \in \mathcal{M}$, observe that $E(T) = T_{e,e}$, and so $E(T)$ is the function in $L^\infty(X, \mu)$ determined by the identity

$$\int_X E(T)(x) f_1(x) \overline{f_2(x)} d\mu(x) = \langle T(\delta_e \otimes f_1) \mid \delta_e \otimes f_2 \rangle \quad \text{for all } f_1, f_2 \in L^2(X, \mu).$$

Item (2) of the following lemma is Item (1) of Proposition 13.A.8; the other items are easily checked and left to the reader.

LEMMA 13.A.10. *Let $E : \mathcal{M} \to \mathcal{A}$ be as above.*

(1) *$E(I) = I$. For $T \in \mathcal{M}_+$, we have $E(T) \in \mathcal{A}_+$.*
(2) *For $T = \sum_{\gamma \in \Gamma} \varphi_\gamma u_\gamma \in \mathcal{M}$, we have*

$$E(TT^*) = \sum_{\gamma \in \Gamma} |\varphi_\gamma|^2 \quad \text{and} \quad E(T^*T) = \sum_{\gamma \in \Gamma} |\theta_\gamma(\varphi_{\gamma^{-1}})|^2.$$

(3) *For $T \in \mathcal{M}$, we have $E(T^*T) = 0$ if and only if $T = 0$.*
(4) *E is a normal map between the von Neumann algebras $\mathcal{M}$ and $\mathcal{A}$.*
(5) *For $\varphi, \psi \in \mathcal{A}$ and $T \in \mathcal{M}$, we have $E(\varphi T \psi) = \varphi E(T) \psi$; in particular, E is the identity map on $\mathcal{A}$.*

In the following proposition, we relate traces on the crossed product algebra $L^\infty(X, \mu) \rtimes \Gamma$ and invariant measures on X.

PROPOSITION 13.A.11. *Let $(X, \mathcal{B}, \mu) \curvearrowleft \Gamma$ be a nonsingular action of a countable group Γ on the standard Borel space $(X, \mathcal{B})$, equipped with a σ-finite measure μ. Let $\mathcal{M}, \mathcal{A}$, and E be as above.*

(1) *Assume that there exists a σ-finite Γ-invariant measure ν on $(X, \mathcal{B})$ which is equivalent to μ. Then*

$$\tau : \mathcal{M}_+ \to [0, +\infty], \quad T \mapsto \nu \circ E(T) = \int_X E(T)(x) d\nu(x)$$

is a semi-finite faithful normal trace on $\mathcal{M}$. Moreover, τ is finite if and only if the measure ν is finite.
(2) *Assume that there exists a semi-finite faithful normal trace*

$$\tau : \mathcal{M}_+ \to [0, +\infty]$$

on $\mathcal{M}$ and, moreover, that the action of Γ is ergodic and essentially free. Then there exists a σ-finite Γ-invariant measure ν on $(X, \mathcal{B})$ which is equivalent to μ and such that $\tau|_{\mathcal{A}} = \nu$, that is,

$$\tau(M(\varphi)) = \int_X \varphi(x) d\nu(x) \quad \text{for all } \varphi \in \mathcal{A}_+.$$

PROOF. (1) For $S, T \in \mathcal{M}_+$ and $\lambda \geq 0$, it is straightforward to check that

$$\tau(S + T) = \tau(S) + \tau(T) \quad \text{and} \quad \tau(\lambda S) = \lambda \tau(S).$$

For $T = \sum_{\gamma\in\Gamma} \varphi_\gamma u_\gamma$, we have, using the formulae in Lemma 13.A.10(2) and the Γ-invariance of the measure ν,

$$\begin{aligned}\tau(T^*T) &= \int_X \left(\sum_{\gamma\in\Gamma} |\theta_\gamma(\varphi_{\gamma^{-1}})(x)|^2\right) d\nu(x) = \sum_{\gamma\in\Gamma}\int_X |\theta_\gamma(\varphi_{\gamma^{-1}})(x)|^2 d\nu(x)\\ &= \sum_{\gamma\in\Gamma}\int_X |\varphi_{\gamma^{-1}}(x)|^2 d\nu(x) = \int_X \left(\sum_{\gamma\in\Gamma} |\varphi_{\gamma^{-1}}(x)|^2\right) d\nu(x)\\ &= \int_X \left(\sum_{\gamma\in\Gamma} |\varphi_\gamma(x)|^2\right) d\nu(x) = \tau(TT^*).\end{aligned}$$

So, $\tau(T^*T) = \tau(TT^*)$ for every $T \in \mathcal{M}$. The formula above also shows that τ is faithful on $\mathcal{M}_+$ since ν is faithful on $\mathcal{A}_+$. Moreover, τ is normal since E and ν are normal. It remains to show that τ is semi-finite.

Since ν is σ-finite, there exists an increasing sequence $(\varphi_n)_{n\geq 1}$ in $\mathcal{A}_+$ such that $\sup_n \varphi_n = \mathbf{1}_X$ and $\int_X \varphi_n(x)d\nu(x) < +\infty$ for all $n \geq 1$. Then

$$\tau(M(\varphi_n)) = \int_X \varphi_n(x)d\nu(x) < +\infty \qquad \text{for all } n \geq 1$$

and $\sup_n M(\varphi_n) = I$. This implies that τ is semi-finite; see for instance [**Dixm–vN**, Chap. I, § 6, Prop. 1, Cor. 2].

(2) We define $\nu : \mathcal{B} \to [0, +\infty]$ by

$$\nu(B) = \tau(M(\mathbf{1}_B)) \qquad \text{for all } B \in \mathcal{B}.$$

Since τ is normal, ν is σ-additive. Also, since τ is faithful, ν is equivalent to μ. Moreover, ν is Γ-invariant. Indeed, let $B \in \mathcal{B}$ and $\gamma \in \Gamma$; using the fact that τ is a trace, we have

$$\begin{aligned}\nu(B\gamma^{-1}) &= \tau(M(\mathbf{1}_{B\gamma^{-1}})) = \tau(M(\theta_\gamma(\mathbf{1}_B)))\\ &= \tau(U_\gamma M(\mathbf{1}_B)U_\gamma^{-1}) = \tau(M(\mathbf{1}_B)) = \nu(B).\end{aligned}$$

It remains to show that ν is σ-finite. To prove this, it suffices to show the following

Claim. There exists $B \in \mathcal{B}$ with $0 < \nu(B) < +\infty$.

Indeed, assume that the claim is proved. Since the action of Γ on X is μ-ergodic and since ν is equivalent to μ, this action is also ν-ergodic. Let $X' := \bigcup_\gamma B\gamma$; then $X' \in \mathcal{B}$, X' is Γ-invariant, and $\nu(X') > 0$. Therefore $\nu(X \smallsetminus X') = 0$. Moreover, $\nu(B\gamma) = \nu(B) < +\infty$ for every $\gamma \in \Gamma$. Since Γ is countable, this shows that ν is σ-finite.

To prove the claim, fix $T \in \mathcal{M}_+$ with $0 < \tau(T) < +\infty$. The first step is to show that $\tau(E(T)) < +\infty$.

Indeed, denote by $K(T)$ the convex hull of $\{UTU^* \mid U \in \mathcal{U}(\mathcal{A})\}$, where $\mathcal{U}(\mathcal{A})$ is the unitary group of $\mathcal{A}$. Let $K'(T)$ be the closure of $K(T)$ for the ultra-weak topology on $\mathcal{L}(\mathcal{H})$, where $\mathcal{H} = \ell^2(\Gamma, L^2(X, \mu))$. Since $K(T)$ is norm bounded, $K'(T)$ is a compact convex subset for the ultra-weak topology. (On the ultra-weak topology, see Appendix A.8.) The group $\mathcal{U}(\mathcal{A})$ acts on $K'(T)$ by the affine ultra-weakly continuous transformations given by

$$K'(T) \ni S \mapsto USU^* \in K'(T)$$

for $U \in \mathcal{U}(\mathcal{A})$. Since $\mathcal{U}(\mathcal{A})$ is abelian, it follows from the Markov–Kakutani Theorem that there exists a common fixed point $T_0 \in K'(T)$ for all $U \in \mathcal{U}(\mathcal{A})$. (For this theorem, see for example [**BEVT1–5**, Chap. IV, Appendice].) Thus, we have $UT_0 = T_0U$ for every $U \in \mathcal{U}(\mathcal{A})$. Since $\mathcal{A}$ coincides with the linear span of $\mathcal{U}(\mathcal{A})$, it follows that $ST_0 = T_0S$ for every $S \in \mathcal{A}$. As the Γ-action is essentially free, $\mathcal{A}$ is a maximal abelian subalgebra of $\mathcal{M}$ (Lemma 13.A.9) and it follows that $T_0 \in \mathcal{A}$.

Using Lemma 13.A.10(5), we have

$$E(UTU^*) \,=\, UE(T)U^* \,=\, E(T) \qquad \text{for all } U \in \mathcal{U}(\mathcal{A}).$$

Since $T_0 \in K'(T)$ and since $E : \mathcal{M} \to \mathcal{A}$ is normal and therefore continuous for the ultra-weak topology, it follows that $T_0 = E(T)$.

We have $\tau(S) = \tau(T)$ for every $S \in K(T)$, by the trace property of τ. Moreover, since τ is normal, τ is lower semi-continuous for the weak topology (as mentioned before Example 7.B.6) and hence for the ultra-weak topology. It follows that $\tau(S) \leq \tau(T)$ for every $S \in K'(T)$; in particular,

$$\tau(T_0) \,=\, \tau(E(T)) \,\leq\, \tau(T) \,<\, +\infty.$$

Now let φ be the function in $\mathcal{A} = L^\infty(X, \mu)$ such that $M(\varphi) = T_0 = E(T)$. By Items (1) and (3) of Lemma 13.A.10, we have $\varphi \neq 0$ and $\varphi \geq 0$, since $T \neq 0$ and $T \in \mathcal{M}_+$. So, $\lambda := \|\varphi\|_\infty > 0$. Therefore there exists $B \in \mathcal{B}$ with $0 < \mu(B) < +\infty$ such that $\varphi(x) \geq \lambda/2$ for every $x \in B$.

Since ν is equivalent to μ, we have $\nu(B) > 0$. Moreover, we have $\mathbf{1}_B \leq \dfrac{2}{\lambda}\varphi$ and hence

$$\nu(B) \,=\, \tau(M(\mathbf{1}_B)) \,\leq\, \frac{2}{\lambda}\tau(M(\varphi)) \,=\, \frac{2}{\lambda}\tau(T_0) \,<\, \infty.$$

The proof of the Claim is now complete. □

REMARK 13.A.12. Under the assumptions of Proposition 13.A.11(2), a stronger result is true: if τ is a semi-finite faithful normal trace on $\mathcal{M} = L^\infty(X, \mu) \rtimes \Gamma$, then τ is of the form $\nu \circ E$ for a σ-finite Γ-invariant measure ν on $(X, \mathcal{B})$.

Indeed, as we will see below (Theorem 13.A.13), the crossed product algebra $\mathcal{M}$ is a factor. On the one hand, $\tau|_\mathcal{A}$ is given by a σ-finite Γ-invariant measure ν on X. On the other hand, by Proposition 13.A.11(1), ν defines a semi-finite faithful normal trace τ' on $\mathcal{M}$ such that $\tau' = \nu' \circ E$. However, since $\mathcal{M}$ is a factor, τ and τ' have to be proportional to each other (Corollaire to Théorème 3 in [**Dixm–vN**, Chap. I, § 6]); hence, we have $\tau = \tau'$, since τ and τ' coincide on $\mathcal{A}$.

The following result is due to Murray and von Neumann; see Lemmas 12.3.4, 13.1.1, and 13.1.2 in [**MuvN–36**], and Theorem IX in [**vNeu–40**].

THEOREM 13.A.13. *Let $(X, \mathcal{B}, \mu) \curvearrowleft \Gamma$ be a nonsingular action of a countable group Γ on the measured standard Borel space $(X, \mathcal{B}, \mu)$, as above. Denote by $\mathcal{M}$ the corresponding crossed product $L^\infty(X, \mu) \rtimes \Gamma$. Assume that the action of Γ is essentially free and ergodic.*

1. *The von Neumann algebra $\mathcal{M}$ is a factor.*
2. *Assume that the action of Γ on $(X, \mathcal{B})$ is essentially transitive; then $\mathcal{M}$ is a factor of type I.*
3. *Assume that there exists a Γ-invariant nonatomic probability measure on $(X, \mathcal{B})$ which is equivalent to μ; then $\mathcal{M}$ is a factor of type II_1.*

(4) *Assume that there exists a Γ-invariant infinite σ-finite nonatomic measure on $(X, \mathcal{B})$ which is equivalent to μ; then $\mathcal{M}$ is a factor of type II_∞.*

(5) *Assume that there does not exist any Γ-invariant σ-finite measure on X which is equivalent to μ; then $\mathcal{M}$ is a factor of type III.*

PROOF. (1) Let $T \in \mathcal{M} \cap \mathcal{M}'$. In particular, $T \in M(\mathcal{A})'$. Since the action of Γ is essentially free, $M(\mathcal{A})$ is a maximal abelian von Neumann subalgebra of $\mathcal{M}$ (Lemma 13.A.9) and hence $T = M(\varphi)$ for some $\varphi \in \mathcal{A}$. As T commutes with U_γ, we then have, by relation $(***)$ above, $\theta_\gamma(\varphi) = \varphi$ for every $\gamma \in \Gamma$. By ergodicity of the Γ-action, it follows that φ is constant almost everywhere, that is, T is a multiple of the identity operator. This shows that $\mathcal{M}$ is a factor.

(2) Upon removing a set of μ-measure 0, we may assume that X is a Γ-orbit. Since Γ is countable, X is countable and $\mu(\{x\}) > 0$ for every $x \in X$. Let $x \in X$. Then $p := M(1_{\{x\}})$ is a nonzero minimal projection in the von Neumann algebra $M(\mathcal{A}) \approx \mathcal{A}$. We claim that p is a nonzero minimal projection in $\mathcal{M}$. Once proved, this will show that $\mathcal{M}$ is a factor of type I.

Let $q \in \mathcal{M}$ be a nonzero projection with $q \leq p$, that is, such that $pq = qp = q$. Then, for every $\varphi \in \mathcal{A}$, we have

$$qM(\varphi) = qpM(\varphi) = qM(1_{\{x\}}\varphi) = \varphi(x)qM(1_{\{x\}}) = \varphi(x)qp = \varphi(x)q$$

and

$$M(\varphi)q = M(\varphi)pq = M(\varphi 1_{\{x\}})q = \varphi(x)M(1_{\{x\}})q = \varphi(x)pq = \varphi(x)q,$$

showing that q commutes with $M(\mathcal{A})$. Since $M(\mathcal{A})$ is a maximal abelian subalgebra of $\mathcal{M}$ (Lemma 13.A.9), it follows that $q \in M(\mathcal{A})$ and hence $q = p$. So, p is a minimal projection in $\mathcal{M}$.

(3) and (4) Let ν be a Γ-invariant σ-finite nonatomic measure on $(X, \mathcal{B})$ which is equivalent to μ. It follows from (1) and Proposition 13.A.11(1) that $\mathcal{M}$ is a semi-finite factor, with trace τ such that

$$\tau(M(\varphi)) = \int_X \varphi(x) d\nu(x) \qquad \text{for all } \varphi \in \mathcal{A}_+.$$

Assume that ν is a probability measure. Then $\tau(I) = \tau(M(\mathbf{1}_X)) = 1$ and this shows that τ is a finite trace. Moreover, since ν is nonatomic, for every $\alpha \in [0,1]$, there exists $B \in \mathcal{B}$ such that

$$\nu(B) = \int_X \mathbf{1}_B(x) d\nu(x) = \alpha$$

and hence $\tau(M(\mathbf{1}_B)) = \alpha$. This shows that τ takes on projections in $\mathcal{M}$ for every value in $[0,1]$. Therefore, $\mathcal{M}$ is a factor of type II_1.

Assume that ν is an infinite measure. Then, similarly, τ takes on projections in $\mathcal{M}$ for every value in $[0, +\infty]$. Therefore, $\mathcal{M}$ is a factor of type II_∞.

(5) follows from Proposition 13.A.11(2). □

Here is an addendum to Theorem 13.A.13, concerning the assumptions of essential freeness and ergodicity.

PROPOSITION 13.A.14. *Let $(X, \mu) \curvearrowleft \Gamma$ and $\mathcal{M} = L^\infty(X, \mu) \rtimes \Gamma = \mathcal{A} \rtimes \Gamma$ be as in the previous theorem.*

(1) *The algebra $M(\mathcal{A}) \approx \mathcal{A}$ is a maximal abelian von Neumann subalgebra of $\mathcal{M}$ if and only the action $(X, \mu) \curvearrowleft \Gamma$ is essentially free.*

(2) *If $\mathcal{M}$ is a factor, then the action $(X, \mu) \curvearrowleft \Gamma$ is ergodic.*

PROOF. (1) We have already proved (Lemma 13.A.9) that, if $(X,\mu) \curvearrowleft \Gamma$ is essentially free, then $M(\mathcal{A})$ is a maximal abelian von Neumann subalgebra of $\mathcal{M}$.

Assume that $(X,\mu) \curvearrowleft \Gamma$ is not essentially free. Then, by Proposition 13.A.2, there exists $\varphi \in L^\infty(X,\mu)$ with $\varphi \neq 0$ and $\gamma_0 \in \Gamma \smallsetminus \{e\}$ such that $\varphi\theta_{\gamma_0}(\psi) = \psi\varphi$ for every $\psi \in L^\infty(X,\mu)$. Let $T := M(\varphi)U_{\gamma_0} \in \mathcal{M}$. We claim that $T \in M(\mathcal{A})'$.

Indeed, let $S = M(\psi) \in \mathcal{A}$. Then

$$TS = M(\varphi)U_{\gamma_0}M(\psi) = M(\varphi)M(\theta_{\gamma_0}(\psi))U_{\gamma_0} = M(\psi\varphi)U_{\gamma_0} = ST.$$

Since $\varphi \neq 0$ and $\gamma_0 \neq e$, we have $T \notin M(\mathcal{A})$, and it follows that $M(\mathcal{A})$ is not maximal abelian.

(2) Assume that $(X,\mu) \curvearrowleft \Gamma$ is not ergodic. Then, there exists (Proposition 13.A.3) a nonconstant $\varphi \in L^\infty(X,\mu)$ such that $\theta_\gamma(\varphi) = \varphi$ for all $\gamma \in \Gamma$. Let $T := M(\varphi) \in M(\mathcal{A})$. We claim that $T \in \mathcal{M}'$. Indeed, let $S = \sum_{\gamma\in\Gamma} \psi_\gamma u_\gamma \in \mathcal{M}$. Then

$$\begin{aligned} TS &= \sum_{\gamma\in\Gamma} \varphi\psi_\gamma u_\gamma = \sum_{\gamma\in\Gamma} \psi_\gamma \varphi u_\gamma = \sum_{\gamma\in\Gamma} \psi_\gamma u_\gamma \theta_{\gamma^{-1}}(\varphi) \\ &= \sum_{\gamma\in\Gamma} \psi_\gamma u_\gamma \varphi = ST. \end{aligned}$$

So, T belongs to the centre of $\mathcal{M}$ and is not a scalar multiple of the identity. Therefore $\mathcal{M}$ is not a factor. □

REMARK 13.A.15. Unlike the ergodicity assumption, the essential freeness assumption for the action $(X,\mu) \curvearrowleft \Gamma$ is not necessary for $L^\infty(X,\mu) \rtimes \Gamma$ to be a factor.

An easy example is given by the action of an icc countable group Γ on a one-point space $X = \{*\}$. In this case, $L^\infty(X,\mu) \rtimes \Gamma$ coincides with the von Neumann algebra $\mathcal{L}(\Gamma)$ of the regular representation of Γ and is therefore a factor (Proposition 7.A.1).

Using Theorem 13.A.13, we give various examples of group actions $(X,\mu) \curvearrowleft \Gamma$ on measure spaces (X,μ), for which the crossed product von Neumann algebra $L^\infty(X,\mu) \rtimes \Gamma$ is a factor of type II_1, II_∞, or III. For other examples of the use of Theorem 13.A.13, in particular in connection with the regular representation of a group, see Chapter 14.

EXAMPLE 13.A.16 (**factors of type $\mathbf{II}_1$**). Let G be an infinite metrizable compact group, with normalized Haar measure μ_G. Let Γ be a countable dense subgroup of G. The action of Γ on (G,μ_G) by right translations is measure preserving, free, and ergodic. Since G is infinite, μ_G is nonatomic. Therefore by Theorem 13.A.13, $L^\infty(G,\mu_G) \rtimes \Gamma$ is a factor of type II_1.

A first example of such a pair (G,Γ) is $(\mathbf{T},\Gamma_\theta)$ with $\Gamma_\theta = \{e^{2\pi i n\theta} \mid n \in \mathbf{Z}\}$ for an irrational number θ. A second example is the pair $(\mathrm{SO}(3),\Gamma)$, where Γ is one of the subgroups of $\mathrm{SO}(3)$ isomorphic to $(\mathbf{Z}/2\mathbf{Z}) * (\mathbf{Z}/3\mathbf{Z})$ mentioned in Remark 4.C.17(3).

EXAMPLE 13.A.17 (**factors of type $\mathbf{II}_1$**). Let G be a simple noncompact connected real Lie group with trivial centre and let Λ be a lattice in G. Let μ be the unique G-invariant probability measure on the Borel subsets of $\Lambda\backslash G$. Let Γ be a countable subgroup of G which is not relatively compact in G. The action of Γ on $(\Lambda\backslash G,\mu)$ by right translations is measure preserving. Moreover, this action is essentially free (see [**Boul–16**, Lemma 4.1]), and ergodic by Moore's ergodicity theorem (see [**Zimm–84**, Theorem 2.2.6]). Therefore, by Theorem 13.A.13, $L^\infty(\Lambda\backslash G, \mu) \rtimes \Gamma$ is a factor of type II_1.

An example of a triple (G, Λ, Γ) as above is for instance $(\mathrm{PSL}_n(\mathbf{R}), \mathrm{PSL}_n(\mathbf{Z}), \Gamma)$ for $n \geq 2$, where Γ is any countable subgroup of $\mathrm{PSL}_n(\mathbf{R})$ which is not relatively compact.

EXAMPLE 13.A.18 (**factors of type $\mathbf{II}_\infty$**). Let G be a noncompact and nondiscrete second-countable LC group, with right Haar measure μ_G. Let Γ be a countable dense subgroup of G. The action of Γ on (G, μ_G) by right translations is measure preserving, free, and ergodic. Since G is nondiscrete, μ_G is nonatomic. Therefore by Theorem 13.A.13, $L^\infty(G, \mu_G) \rtimes \Gamma$ is a factor of type II_∞.

Examples of such pairs (G, Γ) include $(\mathbf{R}, \mathbf{Q})$, and $(SL_n(\mathbf{R}), \mathrm{SL}_n(\mathbf{Q}))$ for $n \geq 2$.

EXAMPLE 13.A.19 (**factors of type III**). The group $G = \mathrm{PSL}_2(\mathbf{R})$ acts (on the right) on the real projective line $\mathbf{P}^1(\mathbf{R})$ by Möbius transformations (see the lines before Proposition 13.B.7 below). The Lebesgue measure μ on $\mathbf{P}^1(\mathbf{R}) \cong \mathbf{R} \cup \{\infty\}$ is quasi-invariant (but not invariant) by the action of G. Let Γ be a lattice in G (as, for instance, $\Gamma = \mathrm{PSL}_2(\mathbf{Z})$).

- The action $(\mathbf{P}^1(\mathbf{R}), \mu) \curvearrowleft \Gamma$ is essentially free. Indeed, every $g \in G, g \neq e$, has at most two fixed points in $\mathbf{P}^1(\mathbf{R})$.
- The action $(\mathbf{P}^1(\mathbf{R}), \mu) \curvearrowleft \Gamma$ is ergodic. Indeed, observe that G acts transitively on $\mathbf{P}^1(\mathbf{R})$ and that the stabilizer of the point ∞ is the image P in $G = \mathrm{PSL}_2(\mathbf{R})$ of the group
$$\left\{ \begin{pmatrix} a & 0 \\ c & d \end{pmatrix} \,\middle|\, a, c, d \in \mathbf{R},\ ad = 1 \right\}$$
of lower triangular matrices. So, $\mathbf{P}^1(\mathbf{R})$ can be identified, as a G-space, with $P\backslash G$. By Moore's duality theorem (see [**Zimm–84**, Corollary 2.2.3]), it suffices to show that the measure-preserving action of P on the probability space $(G/\Gamma, \nu)$ is ergodic, where ν is the unique G-invariant probability measure on the Borel subset of G/Γ. Since P is not relatively compact in G, this is the case by Moore ergodicity theorem again (see the reference in Example 13.A.17 above).
- There exists no Γ-invariant σ-finite measure on $\mathbf{P}^1(\mathbf{R})$ which is equivalent to μ. This is shown below, see Corollary 13.B.9.

Therefore, by Theorem 13.A.13, $L^\infty(\mathbf{P}^1(\mathbf{R}), \mu) \rtimes \Gamma$ is a factor of type III.

13.B. Ergodic group actions without invariant measure

Theorem 13.A.13(5) and Example 13.A.19 are motivations for the study of actions of groups on measured standard Borel spaces without invariant measures.

Let Γ be a countable group, $(X, \mathcal{B})$ a standard Borel space, μ a σ-finite measure on it, and $(X, \mu) \curvearrowleft \Gamma$ an action of Γ on (X, μ) by nonsingular measurable transformations. Recall from Section 5.A the Radon–Nikodym cocycle $c_\mu : X \times \Gamma \to \mathbf{R}_+^\times$, given by
$$c_\mu(x, \gamma) = \frac{d\gamma_*^{-1}(\mu)}{d\mu}(x) \qquad \text{for all } x \in X,\ \gamma \in \Gamma.$$

Assume that the action $(X, \mu) \curvearrowleft \Gamma$ is ergodic and μ nonatomic. Following [**Schm–77a**, § 3] and [**FeMo–77**, § 8], we define the essential range of c_μ.

DEFINITION 13.B.1. We denote by $\overline{\mathbf{R}_+^\times} = \mathbf{R}_+^\times \cup \{\infty\}$ the one-point compactification of $\mathbf{R}_+^\times$. A number $\lambda \in \mathbf{R}_+^\times$ is an **essential value** of c_μ if, for every $\varepsilon > 0$ and every measurable subset A of X with $\mu(A) > 0$, there exists a measurable subset $B \subset A$ with $\mu(B) > 0$ and $\gamma \in \Gamma$ such that $B\gamma \subset A$ and

$$|c_\mu(x,\gamma) - \lambda| \leq \varepsilon \qquad \text{for all } x \in B;$$

the point ∞ is an essential value of c_μ if, for every $n \in \mathbf{N}^*$ and every measurable subset A of X with $\mu(A) > 0$, there exists a measurable subset $B \subset A$ with $\mu(B) > 0$ and $\gamma \in \Gamma$ such that $B\gamma \subset A$ and

$$|c_\mu(x,\gamma)| \notin [1/n, n] \qquad \text{for all } x \in B.$$

The set of essential values of c_μ in $\overline{\mathbf{R}_+^\times}$ is called the **essential range** of the action $(X,\mu) \curvearrowleft \Gamma$ or of the cocycle c_μ, and will be denoted by $r(X,\Gamma,\mu)$; it is also called the asymptotic range or the ratio range of Krieger–Araki–Woods of this action.

It is easy to check that the essential range $r(X,\Gamma,\mu)$ only depends on the equivalence class of μ and on the orbit equivalence class of $(X,\mu) \curvearrowleft \Gamma$. It is obvious that 1 belongs to $r(X,\Gamma,\mu)$ and that $r(X,\Gamma,\mu)$ is a closed subset of $\overline{\mathbf{R}_+^\times}$; moreover, $r(X,\Gamma,\mu) \cap \mathbf{R}_+^\times$ is a closed subgroup of the multiplicative group $\mathbf{R}_+^\times$ (see Lemma 3.3 in [**Schm–77a**]). Therefore the possible values of the asymptotic range $r(X,\Gamma,\mu)$ are

(1) $r(X,\Gamma,\mu) = \{1\}$;
(2) $r(X,\Gamma,\mu) = \{1,\infty\}$;
(3) $r(X,\Gamma,\mu) = \overline{\mathbf{R}_+^\times}$;
(4) $r(X,\Gamma,\mu) = \lambda^{\mathbf{Z}} \cup \{\infty\}$ for some $\lambda \in \mathbf{R}_+^\times$ with $0 < \lambda < 1$.

DEFINITION 13.B.2. In the situation described above, the action $(X,\mu) \curvearrowleft \Gamma$ is said to be

of **type II** in case (1),
of **type III$_0$** in case (2),
of **type III$_1$** in case (3),
of **type III$_\lambda$** in case (4).

REMARK 13.B.3. Actually, given any Borel cocycle $c : X \times \Gamma \to G$ with values in a second-countable locally compact group G, one can define the essential range of c in exactly the same way (see [**Schm–77a**, Definition 3.1]).

Next, we show that if $(X,\mu) \curvearrowleft \Gamma$ is of type III, then there does not exist a σ-finite Γ-invariant measure on X which is equivalent to μ.

PROPOSITION 13.B.4. *Assume that there exists a Γ-invariant σ-finite measure on X which is equivalent to μ.*

Then the action $(X,\mu) \curvearrowleft \Gamma$ is of type II.

PROOF. We have to show that $r(X,\Gamma,\mu) = \{1\}$.

Replacing μ by an equivalent measure if necessary, we can assume that μ is a probability measure (Proposition A.3.5). Since X is a standard Borel space and μ is nonatomic, we can assume that $X = [0,1]$ and that μ is the Lebesgue measure (Theorem A.3.8).

It follows from the assumption that the cocycle c_μ is a coboundary, that is, we have

$$c_\mu(x,\gamma) = \frac{\varphi(x)}{\varphi(x\gamma)} \quad \text{for all } x \in X,\ \gamma \in \Gamma$$

for some measurable function $\varphi : X \to \mathbf{R}_+^\times$ (see Example 5.B.3(2)).

Fix $\varepsilon > 0$. Since φ is measurable, there exists by Lusin's Theorem (see [**Rudi–66**, 2.23]) a continuous function $\psi : X \to \mathbf{R}_+^\times$ such that the measurable subset

$$X_0 := \{x \in X \mid \varphi(x) = \psi(x)\}$$

has positive measure. By uniform continuity of ψ, there exists $N > 0$ such that, for every $i \in \{0, \dots, N-1\}$,

$$\left| \frac{\psi(x)}{\psi(y)} - 1 \right| \le \varepsilon \quad \text{for all } x, y \in \left[\frac{i}{N}, \frac{i+1}{N}\right].$$

Since $\mu(X_0) > 0$, we have $\mu(X_0 \cap \left[\frac{i}{N}, \frac{i+1}{N}\right]) > 0$ for some $i \in \{0, \dots, N-1\}$. Set

$$A := X_0 \cap \left[\frac{i}{N}, \frac{i+1}{N}\right].$$

We have

$$\left| \frac{\varphi(x)}{\varphi(y)} - 1 \right| \le \varepsilon \quad \text{for all } x, y \in A.$$

Let B be a measurable subset with $B \subset A$ and let $\gamma \in \Gamma$ be such that $B\gamma \subset A$. Let $x \in B$. Since $x \in A$ and $x\gamma \in A$, we have

$$|c_\mu(x,\gamma) - 1| = \left| \frac{\varphi(x)}{\varphi(x\gamma)} - 1 \right| \le \varepsilon \quad \text{for all } x \in B.$$

Let $\lambda \in \mathbf{R}_+^\times \smallsetminus \{1\}$. Choose $\varepsilon > 0$ such that $|\lambda - 1| > 2\varepsilon$. Then, with A and B as above, we have $|c_\mu(x,\gamma) - \lambda| > \varepsilon$ for every $x \in B$. It follows from the definition that λ is not an essential value of c_μ.

A similar argument also shows that ∞ is not an essential value of c_μ. As a result, we have $r(X, \Gamma, \mu) = \{1\}$. □

REMARK 13.B.5. (1) The converse statement in Proposition 13.B.4 is also true: if $r(X, \Gamma, \mu) = \{1\}$ (that is, if $(X, \mu) \curvearrowleft \Gamma$ is of type II), then there exists a Γ-invariant σ-finite measure on X which is equivalent to μ; see Theorem 3.9 in [**Schm–77a**].

(2) There is a subdivision of von Neumann factors of type III into further subclasses: types III_0, III_1, and III_λ for $\lambda \in (0,1)$; see Chapter 3 in [**Sund–87**]. It is known that, for a group action $(X, \mu) \curvearrowleft \Gamma$ as in Theorem 13.A.13, the crossed product $L^\infty(X, \mu) \rtimes \Gamma$ is of type III_λ if and only if $(X, \mu) \curvearrowleft \Gamma$ is of type III_λ for $\lambda \in [0, 1]$ (see Proposition 4.3.18 in [**Sund–87**]).

Let $(X, \mathcal{B}, \mu) \curvearrowleft \Gamma$ be a nonsingular action of a countable group Γ on a measure space equipped with a σ-finite measure. Consider the product space

$$\widetilde{X} := X \times \mathbf{R}$$

equipped with the product σ-algebra and with the measure

$$d\widetilde{\mu}(x,t) = d\mu(x)e^{-t}dt,$$

where dt is the Lebesgue measure on $\mathbf{R}$. We are going to associate to the Radon–Nikodym cocycle c_μ defined by the action of Γ on X a *measure-preserving* action of Γ on $(\widetilde{X}, \widetilde{\mu})$.

For $\gamma \in \Gamma$ and $(x, t) \in \widetilde{X}$, define

$$(x,t)\gamma = (x\gamma, t + \log c_\mu(x, \gamma)).$$

Using the cocycle identity, we check that this indeed defines an action $\widetilde{X} \curvearrowleft \Gamma$: for $\gamma, \delta \in \Gamma$ and $(x,t) \in \widetilde{X}$, we have

$$\begin{aligned} ((x,t)\gamma)\delta &= (x\gamma, t + \log c_\mu(x,\gamma))\delta \\ &= ((x\gamma)\delta, t + \log c_\mu(x,\gamma) + \log c_\mu(x\gamma,\delta)) \\ &= (x(\gamma\delta), t + \log c_\mu(x, \gamma\delta)) \\ &= (x,t)(\gamma\delta). \end{aligned}$$

The action $\widetilde{X} \curvearrowleft \Gamma$ preserves $\widetilde{\mu}$: for $\gamma \in \Gamma$ and a measurable function $f : \widetilde{X} \to \mathbf{R}_+$, we have

$$\begin{aligned} \int_{\widetilde{X}} f((x,t)\gamma) d\widetilde{\mu}(x,t) &= \int_X \int_{\mathbf{R}} f(x\gamma, t + \log c_\mu(x,\gamma))\, e^{-t} d\mu(x) dt \\ &= \int_X \int_{\mathbf{R}} f(x, t + \log c_\mu(x\gamma^{-1},\gamma))\, c_\mu(x,\gamma^{-1})\, e^{-t} d\mu(x) dt \\ &= \int_X \int_{\mathbf{R}} f(x,t)\, c_\mu(x,\gamma^{-1})\, e^{-t} c_\mu(x\gamma^{-1},\gamma)\, d\mu(x) dt \\ &= \int_X \int_{\mathbf{R}} f(x,t)\, e^{-t} d\mu(x) dt \\ &= \int_{\widetilde{X}} f(x,t)\, d\widetilde{\mu}(x,t). \end{aligned}$$

The action $(\widetilde{X}, \widetilde{\mu}) \curvearrowleft \Gamma$ is called the **Maharam extension**, or Maharam skew product extension, of the action $(X, \mu) \curvearrowleft \Gamma$ (see [**Aaro–97**, § 3.4] and [**Schm–77a**, § 5]).

Observe that the natural projection $p : \widetilde{X} \to X$ is Γ-equivariant and that $p_*(\widetilde{\mu}) = \mu$. So, $(X, \mu) \curvearrowleft \Gamma$ is a factor of the action $(\widetilde{X}, \widetilde{\mu}) \curvearrowleft \Gamma$. In particular, if $(\widetilde{X}, \widetilde{\mu}) \curvearrowleft \Gamma$ is ergodic, then $(X, \mu) \curvearrowleft \Gamma$ is ergodic.

The type of ergodic action $(X, \mu) \curvearrowleft \Gamma$ can be read off properties of its Maharam extension (see Theorem 5.2 in [**Schm–77a**]). We will only need the following very special result.

PROPOSITION 13.B.6. *Assume that the Maharam extension $(\widetilde{X}, \widetilde{\mu}) \curvearrowleft \Gamma$ is ergodic.*

Then the action $(X, \mu) \curvearrowleft \Gamma$ is of type III_1.

PROOF. Since $r(X, \Gamma, \mu)$ is closed in $\overline{\mathbf{R}_+^\times}$, it suffices to show that $\mathbf{R}_+^\times \subset r(X, \Gamma, \mu)$.

Let $\lambda \in \mathbf{R}_+^\times$ and set $a = \log \lambda$. Let $\varepsilon > 0$ and let A be a measurable subset of X with $\mu(A) > 0$. Then

$$\widetilde{A} := A \times (-a - \varepsilon/2, -a + \varepsilon/2)$$

is a measurable subset of $\widetilde{X}$ with $\widetilde{\mu}(\widetilde{A}) > 0$. The set

$$\widetilde{A_0} := \bigcup_{\gamma \in \Gamma} (A \times (-\varepsilon/2, \varepsilon/2))\gamma$$

is a Γ-invariant measurable subset of $\widetilde{X}$ with $\widetilde{\mu}(\widetilde{A_0}) > 0$. It follows from the ergodicity of $(\widetilde{X}, \widetilde{\mu}) \curvearrowleft \Gamma$ that $\widetilde{\mu}(\widetilde{X} \smallsetminus \widetilde{A_0}) = 0$. Since Γ is countable and $\widetilde{\mu}(\widetilde{A_0}) > 0$, there exists an element $\gamma \in \Gamma$ such that

$$\widetilde{\mu}(\widetilde{A} \cap (A \times (-\varepsilon/2, \varepsilon/2))\gamma^{-1}) > 0.$$

Choose $t \in (-a - \varepsilon/2, -a + \varepsilon/2)$ such that, for the t-section

$$B := \left\{x \in A \mid (x,t) \in \widetilde{A} \cap (A \times (-\varepsilon/2, \varepsilon/2))\gamma^{-1}\right\}$$

of $\widetilde{A} \cap (A \times (-\varepsilon/2, \varepsilon/2))\gamma^{-1}$, we have $\mu(B) > 0$.

Let $x \in B$. There exists $(y,s) \in A \times (-\varepsilon/2, \varepsilon/2)$ such that

$$x = y\gamma^{-1} \quad \text{and} \quad t = s + \log c_\mu(y, \gamma^{-1}).$$

Since

$$\log c_\mu(y, \gamma^{-1}) = \log c_\mu(x\gamma, \gamma^{-1}) = \log c_\mu(x, \gamma)^{-1} = -\log c_\mu(x, \gamma),$$

we have

$$|\log c_\mu(x,\gamma) - a| \le |\log c_\mu(x,\gamma) + t| + |-t - a| = |s| + |t - (-a)| \le \varepsilon.$$

Observe that $x\gamma = y \in A$. So, we have $B\gamma \subset A$ and

$$|\log c_\mu(x,\gamma) - \log \lambda| \le \varepsilon \quad \text{for all } x \in B.$$

This shows that $\lambda \in r(X, \Gamma, \mu)$. $\square$

We return to the situation of Example 13.A.19. We identify the real projective line $X = \mathbf{P}^1(\mathbf{R})$ with $\overline{\mathbf{R}} = \mathbf{R} \cup \{\infty\}$ and denote by μ the Lebesgue measure on X. The group $G = \mathrm{SL}_2(\mathbf{R})$ acts on X from the right by Möbius transformations given by

$$xg = \frac{ax + c}{bx + d} \quad \text{for} \quad g = \begin{pmatrix} a & b \\ c & d \end{pmatrix} \in G, \; x \in X.$$

The associated Radon–Nikodym cocycle $c_\mu : X \times G \to \mathbf{R}_+^\times$ is given, for $g \in G$ as above, by

$$c_\mu(x, g) = \frac{1}{(bx+d)^2} \quad \text{for} \quad x \in X.$$

We identify in the following proposition the Maharam extension of the restriction of this action to any countable subgroup of G.

PROPOSITION 13.B.7. *Let Γ be a countable subgroup of $G = \mathrm{SL}_2(\mathbf{R})$. The Maharam extension $(\widetilde{X}, \widetilde{\mu})$ of the action of Γ on $X = \mathbf{P}^1(\mathbf{R})$ as above can be identified with $(\mathbf{R}^2 \smallsetminus \{(0,0)\}, \lambda)$, where λ is the Lebesgue measure and the action of Γ on $\mathbf{R}^2 \smallsetminus \{(0,0)\}$ is the usual linear action from the right. More precisely, there exists a Borel isomorphism*

$$\Phi : \mathbf{R}^2 \smallsetminus \{(0,0)\} \to \widetilde{X}$$

with the following properties:

(i) *$\Phi_*(\lambda) = \widetilde{\mu}$, for an appropriate normalization of λ;*
(ii) *for λ-almost every $(u,v) \in \mathbf{R}^2 \smallsetminus \{(0,0)\}$ and every $\gamma \in \Gamma$, we have $\Phi((u,v)\gamma) = \Phi(u,v)\gamma$.*

PROOF. Define $\Phi : \mathbf{R}^2 \smallsetminus \{(0,0)\} \to \overline{\mathbf{R}} \times \mathbf{R}$ by

$$\Phi(u,v) = \begin{cases} (u/v, -\log v^2) & \text{if } v \neq 0, \\ (\infty, -\log u^2) & \text{if } v = 0. \end{cases}$$

It is immediately checked that Φ is a Borel isomorphism.

The image under Φ of the Lebesgue measure $d\lambda(u,v) = dudv$ on $\mathbf{R}^2$ is $\frac{1}{2}d\widetilde{\mu}(x,t) = e^{-t}dxdt$; indeed, since the Jacobian of Φ at (u,v) is $-2/v^2$, we have, for every measurable function $f : \widetilde{X} \to \mathbf{R}_+$,

$$\begin{aligned} \int_{\mathbf{R}^2} f(\Phi(u,v))dudv &= \int_{\mathbf{R}^2} f(u/v, -\log v^2)dudv \\ &= \frac{1}{2}\int_{\mathbf{R}^2} f(x,t)e^{-t}dxdt. \end{aligned}$$

Let $\gamma = \begin{pmatrix} a & b \\ c & d \end{pmatrix} \in \Gamma$. On the subset of $\mathbf{R}^2 \smallsetminus \{(0,0)\}$ of full Lebesgue measure consisting of vectors $(u,v) \in \mathbf{R}^2 \smallsetminus \{(0,0)\}$ with $v \neq 0$ and $bu + dv \neq 0$, we have

$$\begin{aligned} \Phi((u,v)\gamma) &= \Phi(au+cv, bu+dv) \\ &= \left(\frac{au+cv}{bu+dv}, -\log(bu+dv)^2\right) \\ &= \left(\frac{a(u/v)+c}{b(u/v)+d}, -\log v^2 - \log(b(u/v)+d)^2\right) \\ &= \left(\frac{au+cv}{bu+dv}, -\log v^2 + \log(c_\mu(u/v,\gamma))\right) \\ &= \Phi(u,v)\gamma. \end{aligned}$$

□

We are now ready to conclude the proof that the crossed product von Neumann algebras as in Examples 13.A.19 are factors of type III.

COROLLARY 13.B.8. *Let Γ be a lattice in $G = \mathrm{SL}_2(\mathbf{R})$ and $\overline{\Gamma}$ its image in $\mathrm{PSL}_2(\mathbf{R})$. Let $X = \mathbf{P}^1(\mathbf{R})$ be equipped with the Lebesgue measure μ.*

The action of Γ on (X,μ) by Möbius transformations is of type III_1. Therefore $L^\infty(X,\mu) \rtimes \overline{\Gamma}$ is a factor of type III_1.

PROOF. In view of Proposition 13.B.6 and Proposition 13.B.7, it suffices to show that the linear action of Γ on $(\mathbf{R}^2 \smallsetminus \{(0,0)\}, \lambda)$ is ergodic.

Observe that G acts transitively on $\mathbf{R}^2 \smallsetminus \{(0,0)\}$ and the stabilizer of the point $(1,0)$ is the subgroup

$$N = \left\{\begin{pmatrix} 1 & 0 \\ c & 1 \end{pmatrix} \;\middle|\; c \in \mathbf{R}\right\}.$$

Therefore $\mathbf{R}^2 \smallsetminus \{(0,0\}$ can be identified, as a G-space, with $N\backslash G$. Under this identification the Lebesgue measure λ on $\mathbf{R}^2 \smallsetminus \{(0,0\}$ corresponds to the unique (up to positive scalar multiples) G-invariant measure on the Borel subsets of $N\backslash G$.

Since Γ is a lattice in G and since N is not relatively compact in G, the claim follows from Moore's duality theorem and Moore's ergodicity theorem (compare with Example 13.A.19). □

The following corollary, of independent interest, is an immediate consequence of Corollary 13.B.8 and Proposition 13.B.4.

COROLLARY 13.B.9. *Let* Γ *be a lattice in* $\mathrm{SL}_2(\mathbf{R})$. *There exists no* Γ*-invariant* σ*-finite measure on* $\mathbf{P}^1(\mathbf{R})$ *which is equivalent to the Lebesgue measure.*

REMARK 13.B.10. The proof of Proposition 13.B.6 can easily be extended to show the following more general result. Let G be a simple noncompact connected Lie group, P a minimal parabolic subgroup in G, and Γ a lattice in G. Then the action of Γ on $(G/P, \mu)$ is of type III_1, where μ is a G-quasi-invariant Radon measure on G/P (see Example 4.3.15 in [**Zimm–84**]). For further results establishing that such "boundary actions" are of type III_1, see [**Kaim–00**] and [**Spat–87**].

CHAPTER 14

Construction of factor representations for some semi-direct products

Following [**Guic–63**], *in this chapter we first describe factor representations of a semi-direct product* $G = H \ltimes N$, *where*

- H *is countable discrete subgroup;*
- N *is a second-countable locally compact abelian normal subgroup of* G.

The factor representations we construct are induced representations $\widetilde{\pi}_\mu = \mathrm{Ind}_N^\Gamma \pi_\mu$, *where* μ *is a convenient* σ*-finite* H*-quasi-invariant measure on* $\widehat{N}$ *and* π_μ *is the canonical representation of* N *on* $L^2(\widehat{N}, \mu)$. *The proof that* $\widetilde{\pi}_\mu$ *is indeed factorial as well as the identification of its type rely on results on the measure space construction of Murray and von Neumann from Section* 13.A.

The particular case where μ *is the Haar measure on* $\widehat{N}$ *is of special interest as the associated representation* $\widetilde{\pi}_\mu$ *is equivalent to the regular representation* λ_G. *Criteria are given ensuring that* λ_G *is a factor representation. We give examples of groups* G *for which* λ_G *is a factor representation of type* II_∞ *or* III.

The representation $\widetilde{\pi}_\mu$ *of a group* $G = H \ltimes N$ *as above is a normal factor representation of type* I_∞ *or of type* II_∞, *provided* μ *is a Radon measure on an open and* H*-invariant subset* U *of* $\widehat{N}$ *with the following properties:*

- μ *is* H*-invariant;*
- μ *is infinite and nonatomic;*
- *the action of* H *on* (U, μ) *is essentially free and ergodic.*

In this case, the character associated with $\widetilde{\pi}_\mu$ *admits an expression in terms of* μ. *In Section* 14.E, *we will give examples of groups for which we can show that measures* μ *as above exist; in this way we obtain normal factor representations of type* I_∞ *for the Baumslag–Solitar group* $\mathrm{BS}(1, p)$, *Proposition* 14.E.2, *and normal factor representations of type* I_∞ *and of type* II_∞ *for the lamplighter group* $\mathbf{Z} \wr (\mathbf{Z}/2\mathbf{Z})$, *Corollaries* 14.E.6 *and* 14.E.12.

As a consequence of the construction of normal factor representations of type I_∞ *for* $\Gamma = \mathrm{BS}(1, p)$ *or* $\Gamma = \mathbf{Z} \wr (\mathbf{Z}/2\mathbf{Z})$, *we will see that the natural map from* $E(\Gamma)$ *to* $\mathrm{Prim}(\Gamma)$ *is neither injective nor surjective, in contrast to the case of nilpotent groups (Proposition* 11.D.2 *and Corollary* 12.A.3*), the affine group over a field (Corollary* 12.B.4*), or the general linear group over an infinite algebraic extension of a finite field (Corollary* 12.E.2*).*

14.A. Crossed product von Neumann algebras for semi-direct products

Let $G = H \ltimes N$ be a semi-direct product of a countable discrete subgroup H with a second-countable locally compact abelian normal subgroup N.

Recall that H acts on the right on $\widehat{N}$ by $(\chi, h) \mapsto \chi^h$, where $\chi^h(n) = \chi(hnh^{-1})$ for $\chi \in \widehat{N}$, $h \in H$, and $n \in N$.

CONSTRUCTION 14.A.1. Consider a group $G = H \ltimes N$ as above, and a nonzero σ-finite Borel measure μ on the second-countable LC space $\widehat{N}$. We assume that μ is quasi-invariant by the action of H (we do *not* assume that μ is finite).

Let π_μ be the canonical representation of N associated with μ, as in Construction 2.A.1; recall that π_μ is defined on the Hilbert space $L^2(\widehat{N}, \mu)$ by

$$(\pi_\mu(n)f)(\chi) = \chi(n)f(\chi) = \widehat{n}(\chi)f(\chi) \quad \text{for all } n \in N, \ f \in L^2(\widehat{N}, \mu), \ \chi \in \widehat{N}.$$

Denote by $\widetilde{\pi}_\mu$ the induced representation $\mathrm{Ind}_N^G \pi_\mu$. Since H is a right transversal for N in G, i.e., since $G = \bigsqcup_{h \in H} Nh$, we can realize $\widetilde{\pi}_\mu$ on the Hilbert space $\mathcal{H}_\mu := \ell^2(H, L^2(\widehat{N}, \mu))$ by

$$\begin{aligned} (\widetilde{\pi}_\mu(h, n)F)(h')(\chi) &= (\pi_\mu(h' \cdot n)F(h'h))(\chi) \\ &= \chi^{h'}(n)F(h'h)(\chi), \end{aligned}$$

for $h, h' \in H$, $n \in N$, $F \in \mathcal{H}_\mu$, and $\chi \in \widehat{N}$ (see Construction 1.F.4(2)).

We denote by

$$\mathcal{M} := \widetilde{\pi}_\mu(G)'' \subset \mathcal{L}(\mathcal{H}_\mu)$$

the von Neumann algebra generated by $\widetilde{\pi}_\mu(G)$.

We first identify $\mathcal{M}$ with $L^\infty(\widehat{N}, \mu) \rtimes H$, the crossed product von Neumann algebra associated as in Section 13.A with the action $(\widehat{N}, \mu) \curvearrowleft H$.

Recall that, for $h \in H$ and $\chi \in \widehat{N}$,

$$c(\chi, h) = \frac{d(h^{-1})_*(\mu)}{d\mu}(\chi)$$

denotes the Radon–Nikodym cocycle of the action $(\widehat{N}, \mu) \curvearrowleft H$, as in Appendix A.3.

PROPOSITION 14.A.2. *Let $G = H \ltimes N$, μ, $\mathcal{H}_\mu$, $\widetilde{\pi}_\mu$, and $\mathcal{M}$ be as in the previous construction.*

The von Neumann algebras $\mathcal{M}$ and $L^\infty(\widehat{N}, \mu) \rtimes H$ inside $\mathcal{L}(\mathcal{H}_\mu)$ are unitarily equivalent. More precisely, for the unitary operator $U : \mathcal{H}_\mu \to \mathcal{H}_\mu$ defined by

$$(UF)(h')(\chi) = c(\chi, h')^{1/2}F(h')(\chi^{h'}) \quad \textit{for all } h' \in H, \ F \in \mathcal{H}_\mu, \ \chi \in \widehat{N},$$

we have

$$\mathcal{M} = U^{-1}\left(L^\infty(\widehat{N}, \mu) \rtimes H)\right)U.$$

Moreover, if ρ_μ denotes the conjugate representation $(h, n) \mapsto U^{-1}\widetilde{\pi}_\mu(h, n)U$ of G, we have

$$(\rho_\mu(h, n)F)(h')(\chi) = c(\chi, h)^{1/2}\chi(n)F(h'h)(\chi^h),$$

for $h, h' \in H$, $n \in N$, $F \in \mathcal{H}_\mu$, and $\chi \in \widehat{N}$.

PROOF. It is routine to check, using the cocycle relation for c, that the inverse of U is given by

$$(U^{-1}F)(h')(\chi) = c(\chi, h'^{-1})^{1/2}F(h')(\chi^{h'^{-1}}),$$

and, using the change of variable formula $(***)$ of Appendix A.3, that U is unitary. To check that ρ_μ is given by the formula stated in the proposition, recall that

$$(\widetilde{\pi}_\mu(h,n)F)(h')(\chi) = \chi^{h'}(n)F(h'h)(\chi),$$

and compute

$$\begin{aligned}(U^{-1}\widetilde{\pi}_\mu(h,n)UF)(h')(\chi) &= c(\chi,h'^{-1})^{1/2}(\widetilde{\pi}_\mu(h,n)UF)(h')(\chi^{h'^{-1}}) \\ &= c(\chi,h'^{-1})^{1/2}\chi(n)(UF(h'h))(\chi^{h'^{-1}}) \\ &= c(\chi,h'^{-1})^{1/2}\chi(n)c(\chi^{h'^{-1}},h'h)^{1/2}F(h'h)(\chi^h) \\ &= c(\chi,h)^{1/2}\chi(n)F(h'h)(\chi^h) \\ &= (\rho_\mu(h,n)F)(h')(\chi),\end{aligned}$$

for $h,h'\in H,\ n\in N,\ F\in\mathcal{H}_\mu$, and $\chi\in\widehat{N}$.

Next, we claim that $U^{-1}\mathcal{M}U$ and $L^\infty(\widehat{N},\mu)\rtimes H$ coincide, that is,

$$\rho_\mu(G)'' = L^\infty(\widehat{N},\mu)\rtimes H.$$

The Koopman representation $h\mapsto u_h$ of H on $L^2(\widehat{N},\mu)$ is given by

$$u_h(f)(\chi) = c(\chi,h)^{1/2}f(\chi^h) \quad \text{for all } f\in L^2(\widehat{N},\mu),\ \chi\in\widehat{N}$$

(see Definition 13.A.5). Let $h\in H$. We have

$$(\rho_\mu(h,e)F)(h')(\chi) = c(\chi,h)^{1/2}F(h'h)(\chi^h) \quad \text{for all } h'\in H,\ F\in\mathcal{H}_\mu,\ \chi\in\widehat{N}$$

and hence

$$(\rho_\mu(h,e)F)(h') = u_h(F(h'h)) \quad \text{for all } h'\in H,\ F\in\mathcal{H}_\mu.$$

It follows that $\rho_\mu(h,e) = U_h$, where $h\mapsto U_h$ is the representation of H on $\mathcal{H}_\mu = \ell^2(H,L^2(\widehat{N},\mu))$ involved in the definition of the crossed product von Neumann algebra $L^\infty(\widehat{N},\mu)\rtimes H$ (see Definition 13.A.7).

Let $n\in N$. We have

$$(\rho_\mu(e,n)F)(h')(\chi) = \chi(n)F(h')(\chi) = \widehat{n}(\chi)F(h')(\chi) \quad \text{for all } F\in\mathcal{H}_\mu,\ h'\in H,\ \chi\in\widehat{N},$$

where $\widehat{n}\in C^b(\widehat{N})$ is defined by $\widehat{n}(\chi)=\chi(n)$. It follows that $\rho_\mu(e,n) = M(\widehat{n})$, where $\varphi\mapsto M(\varphi)$ is the representation of $L^\infty(\widehat{N},\mu)$ involved in the definition of $L^\infty(\widehat{N},\mu)\rtimes H$. As a consequence, we see that $\mathcal{M}$ is contained in $L^\infty(\widehat{N},\mu)\rtimes H$.

It remains to show that $L^\infty(\widehat{N},\mu)\rtimes H$ is contained in $\mathcal{M}$. Since $L^\infty(\widehat{N},\mu)\rtimes H$ is generated by

$$\{U_h \mid h\in H\}\cup\{M(\varphi)\mid \varphi\in L^\infty(\widehat{N},\mu)\}$$

and since we have shown that $\{U_h\mid h\in H\}\subset\mathcal{M}$, it suffices to show that

$$\{M(\varphi)\mid\varphi\in L^\infty(\widehat{N},\mu)\}$$

is contained in $\mathcal{M}$.

By what we have just shown, the algebra $\mathrm{Trig}(\widehat{N})$ of trigonometric polynomials on $\widehat{N}$ (see Appendix A.7) in contained in $\mathcal{M}$.

Now, $\mathrm{Trig}(\widehat{N})$ is dense in $L^\infty(\widehat{N},\mu)$ for the weak*-topology (see Proposition A.7.7); moreover, the map

$$L^\infty(\widehat{N},\mu)\to\mathcal{L}(\mathcal{H}_\mu),\quad \varphi\mapsto M(\varphi)$$

is continuous, when $L^\infty(\widehat{N},\mu)$ is equipped with the weak*-topology and $\mathcal{L}(\mathcal{H}_\mu)$ with the weak operator topology, by Proposition 1.H.3.

It follows that $\{M(\varphi) \mid \varphi \in \mathrm{Trig}(\widehat{N})\}$ is dense in $\{M(\varphi) \mid \varphi \in L^\infty(\widehat{N},\mu)\}$ for the weak operator topology. Therefore, $\{M(\varphi) \mid \varphi \in L^\infty(\widehat{N},\mu)\}$ is contained in $\mathcal{M}$. □

14.B. Some factor representations of semi-direct products

As in Section 14.A, let $G = H \ltimes N$ be a semi-direct product of a countable discrete subgroup H with a second-countable locally compact abelian normal subgroup N. We will use Proposition 14.A.2, in combination with results from Section 13.A on the crossed product von Neumann algebras, in order to construct various factor representations of G.

Recall from Section 14.A that we have associated a representation $\widetilde{\pi}_\mu$ of G in $\ell^2(H, L^2(\widehat{N},\mu))$ to every σ-finite Borel measure μ on $\widehat{N}$ which is quasi-invariant by the action of H.

THEOREM 14.B.1. *Let $G = H \ltimes N$ be as above. Let μ be a nonzero σ-finite Borel measure on $\widehat{N}$ which is quasi-invariant by the action of H and $\widetilde{\pi}_\mu$ the associated representation of G as in Section* 14.A. *Assume that the action $(\widehat{N},\mu) \curvearrowleft H$ is essentially free and ergodic.*

(1) *The representation $\widetilde{\pi}_\mu$ is a factor representation of G.*
(2) *If μ is an atomic measure, then $\widetilde{\pi}_\mu$ is a factor representation of type I.*
(3) *If μ is an H-invariant nonatomic finite measure, then $\widetilde{\pi}_\mu$ is a factor representation of type II_1.*
(4) *If μ is an H-invariant nonatomic infinite measure, then $\widetilde{\pi}_\mu$ is a factor representation of type II_∞.*
(5) *Assume that there does not exist any H-invariant σ-finite measure on $\widehat{N}$ which is equivalent to μ; then $\widetilde{\pi}_\mu$ is a factor representation of type III.*

PROOF. By Proposition 14.A.2, $\mathcal{M} := \widetilde{\pi}_\mu(G)''$ is unitarily equivalent to the crossed product von Neumann algebra $L^\infty(\widehat{N},\mu) \rtimes H$. Since the action $(\widehat{N},\mu) \curvearrowleft H$ is essentially free and ergodic, Items (1) to (5) follow from Items (1) to (5) in Theorem 13.A.13, respectively. □

EXAMPLE 14.B.2. We can use Theorem 14.B.1 in order to produce factor representations of type II_∞ and of type III for any one of our favorite groups Γ (with the exception of the general group $\mathrm{GL}_n(\mathbf{K})$) as well as recover some of the irreducible representations and representations of type II_1 of Γ considered in previous sections. Here we will consider only the case where Γ is the Heisenberg group over $\mathbf{Z}$. Similar results hold for the Heisenberg group and the affine group over a countable infinite field, for the Baumslag–Solitar groups $\mathrm{BS}(1,p)$, and for the lamplighter group.

Recall from Section 3.B that elements in $\Gamma = H(\mathbf{Z})$ can be written as triples (a,b,c) of integers, and that Γ is the semi-direct product $H \ltimes N$, where

$$H = \{(a,0,0) \in \Gamma \mid a \in \mathbf{Z}\} \quad \text{and} \quad N = \{(0,b,c) \in \Gamma \mid b,c \in \mathbf{Z}\}.$$

Fix $\theta \in [0,1[$ irrational. Let ψ_θ be the unitary character of the centre Z of Γ defined by $\psi_\theta(0,0,c) = e^{2\pi i\theta c}$ for all $c \in \mathbf{Z}$. Let $X \subset \widehat{N}$ be the H-invariant set of unitary characters χ of N such that $\chi|_Z = \psi_\theta$. Then X is homeomorphic to $\mathbf{T}$ and the action of H on X corresponds to the action of the irrational rotation

R_θ on the circle group $\mathbf{T}$ (see Corollary 3.B.14). In particular, the action of H on X is free. So, the action of H on (X, μ) is automatically essentially free for any H-quasi-invariant measure μ on X.

- Let μ be the counting measure on an R_θ-orbit $\mathcal{O}$ in X. Then the representation π_μ is equivalent to the direct sum $\bigoplus_{\chi \in \mathcal{O}} \chi$, and hence $\widetilde{\pi}_\mu = \mathrm{Ind}_N^\Gamma \pi_\mu$ is equivalent to
$$\bigoplus_{\chi \in \mathcal{O}} \mathrm{Ind}_N^\Gamma \chi.$$
As seen in Corollary 3.B.14, the representations $\mathrm{Ind}_N^\Gamma \chi$ are irreducible and mutually equivalent. As a result, $\widetilde{\pi}_\mu$ is a multiple of the irreducible representation $\mathrm{Ind}_N^\Gamma \chi$ for any $\chi \in \mathcal{O}$. Thus, we recover the fact proved in Theorem 14.B.1 that $\widetilde{\pi}_\mu$ is a factor representation of type I; in this case, $\widetilde{\pi}_\mu$ is of type I_∞.
- Let μ be the Lebesgue measure on X. Then the representation π_μ is equivalent to the induced representation $\mathrm{Ind}_Z^N \psi_\theta$, and hence $\widetilde{\pi}_\mu$ is equivalent to $\mathrm{Ind}_Z^\Gamma \psi_\theta$. This is the representation of type II_1 associated with the character $\psi_\theta \in E(\Gamma)$ from Corollary 12.A.8.
- By Theorem 5.F.2, there exist uncountably many nonequivalent infinite and σ-finite nonatomic measures on X which are T_θ-invariant and ergodic. For every such measure μ, the representation $\widetilde{\pi}_\mu$ of Γ is a factor representation of type II_∞.

  We do not know whether nonequivalent measures as above always yield non-quasi-equivalent factor representations. See however Proposition 14.C.4 below.
- Theorem 5.F.2 has been extended in [**Krie–76**] to cover the existence of quasi-invariant but noninvariant measures, at least in the case of $\mathbf{Z}$-actions. This result implies that there exist uncountably many nonequivalent nonatomic probability measures on X which are quasi-invariant by R_θ and for which there is no equivalent σ-finite R_θ-invariant measure. For every such measure μ, the representation $\widetilde{\pi}_\mu$ of Γ is a factor representation of type III.

14.C. Some normal factor representations

As in the previous sections, $G = H \ltimes N$ is a semi-direct product of a countable discrete subgroup H with a second-countable locally compact abelian normal subgroup N.

Let μ be a σ-finite Borel measure on $\widehat{N}$ which is quasi-invariant, essentially free, and ergodic by the action of H. As seen in Theorem 14.B.1, the associated representation $\widetilde{\pi}_\mu$ of G is then a factor representation. We will show below that, under a further assumption, $\widetilde{\pi}_\mu$ is a *normal* factor representation.

Let ν_N denote a Haar measure on N. For every $h \in H$, the Borel measure $h_*(\nu_N)$ is invariant under N; by uniqueness of the Haar measure, there exists $\Delta(h) > 0$ such that $h_*(\nu_N) = \Delta(h)\nu_N$, and $\Delta : H \to \mathbf{R}_+^\times$ is a homomorphism. A Haar measure ν_G on G is given by $\nu_G = (\Delta \nu_H) \otimes \nu_N$, where ν_H is the counting measure on H. We have
$$\int_G f(g)\, d\nu_G(g) = \sum_{h \in H} \Delta(h) \int_N f(h,n)\, d\nu_N(n)$$

for every measurable function $f : G \to [0, \infty]$. It is straightforward to check that ν_G is left and right invariant, hence that G is unimodular.

For a function $f : G \to \mathbf{C}$, we denote by f_e the function on N defined by

$$f_e(n) = f(e, n) \qquad \text{for all } n \in N.$$

Observe that $f_e \in L^1(N, \nu_N)$ if $f \in L^1(G, \nu_G)$. We set

$$L^1(G, \nu_G)_+ = \{f^* * f \mid f \in L^1(G, \nu_G)\}.$$

(As G is unimodular, the involution on $L^1(G, \nu_G)$ is given here by $f^*(g) = \overline{f(g^{-1})}$.) Observe that every $f \in L^1(G, \nu_G)_+$ is positive as element of the C*-algebra $C^*_{\max}(G)$.

The following technical lemma will be used in the proofs of Theorem 14.C.2 and Proposition 14.C.4 below. As in Appendix A.1, we denote by $C^{c,K}(\widehat{N})$ the space of complex-valued continuous functions on $\widehat{N}$ with support contained in some compact subset K of $\widehat{N}$.

LEMMA 14.C.1. *Let U be an open subset of $\widehat{N}$ and K a compact subset of U. Then there exists a compact subset K' of U with $K \subset K'$ such that $C^{c,K}(\widehat{N})$ is contained in the closure of the linear span of*

$$\{\overline{\mathcal{F}}(f_e) \mid f \in L^1(G, \nu_G)_+\} \cap C^{c,K'}(\widehat{N})$$

for the topology of uniform convergence on $\widehat{N}$.

PROOF. Since U is open and K compact, we can find a compact neighbourhood V of $\{e\}$ such that KV is contained in U. Let $K' = KV$ and set

$$A_{K'} := \{\overline{\mathcal{F}}(f_e) \mid f \in L^1(G, \nu_G)_+\} \cap C^{c,K'}(\widehat{N}).$$

Let $\varphi \in C^{c,K}(\widehat{N})$. We claim that φ is in the closure of the linear span of $A_{K'}$.

Upon considering the positive and negative parts of the real and imaginary parts of φ, we can assume that $\varphi \geq 0$.

We assume that the Haar measures ν_N on N and $\nu_{\widehat{N}}$ on $\widehat{N}$ are so normalized that the Fourier transform

$$\mathcal{F} : L^2(N, \nu_N) \to L^2(\widehat{N}, \nu_{\widehat{N}})$$

is an isomorphism of Hilbert spaces (Plancherel's Theorem A.7.6). Then this is also true for the inverse Fourier transform

$$\overline{\mathcal{F}} : L^2(N, \nu_N) \to L^2(\widehat{N}, \nu_{\widehat{N}}).$$

Choose an approximate identity supported in V, that is, a sequence $(\psi_i)_i$ in $C^c(\widehat{N})$ such that $\mathrm{supp}(\psi_i) \subset V$ and $\lim_i \psi_i * \psi = \psi$ uniformly on $\widehat{N}$ for every $\psi \in C^c(\widehat{N})$. Set

$$\psi := \varphi^{1/2}, \qquad r := \overline{\mathcal{F}}^{-1}(\psi), \qquad \text{and} \qquad r_i := \overline{\mathcal{F}}^{-1}(\psi_i).$$

Since $r_i \in L^2(N, \nu_N)$ and $r \in L^2(N, \nu_N)$, we have $r_i r \in L^1(N, \nu_N)$. Moreover,

$$\overline{\mathcal{F}}(r_i r) = \overline{\mathcal{F}}(r_i) * \overline{\mathcal{F}}(r) = \psi_i * \psi$$

and hence $\lim_i \overline{\mathcal{F}}(r_i r) = \psi$ uniformly on $\widehat{N}$.

Set

$$s_i := (r_i r)^* * r_i r \in L^1(N, \nu_N).$$

We claim that $\lim_i \overline{\mathcal{F}}(s_i) = \varphi$ uniformly on $\widehat{N}$ and that $\overline{\mathcal{F}}(s_i) \in A_{K'}$, for every i.

Indeed, we have

$$\overline{\mathcal{F}}(s_i) = |\overline{\mathcal{F}}(r_i r)|^2$$

and hence

$$\lim_i \overline{\mathcal{F}}(s_i) = |\psi|^2 = \varphi$$

uniformly on $\widehat{N}$.

Moreover, since $\mathrm{supp}(\psi_i) \subset V$ and $\mathrm{supp}(\psi) \subset K$, we have $\mathrm{supp}(\psi_i * \psi) \subset KV = K'$. As $\overline{\mathcal{F}}(s_i) = |\psi_i * \psi|^2$, it follows that

$$\mathrm{supp}(\overline{\mathcal{F}}(s_i)) = \mathrm{supp}(\psi_i * \psi) \subset K'.$$

Therefore $\overline{\mathcal{F}}(s_i) \in C^{c,K'}(\widehat{N})$.

It remains to show that every s_i is of the form $(f^{(i)})_e$ for some $f^{(i)} \in L^1(G, \nu_G)_+$. Let $g_i \in L^1(G, \nu_G)$ be the trivial extension of $r_i r$ to G, defined by

$$g_i(h,n) = \begin{cases} (r_i r)(n) & \text{if } h = e, \\ 0 & \text{if } h \neq e. \end{cases}$$

For $n \in N$, we have

$$\begin{aligned} (g_i^* * g_i)_e(n) &= (g_i^* * g_i)(e,n) \\ &= \sum_{h' \in H} \Delta(h') \int_N g_i^*((h',n')^{-1})\, g_i((h',n')(e,n))\, d\nu_N(n') \\ &= \sum_{h' \in H} \Delta(h') \int_N \overline{g_i(h',n')}\, g_i(h', n'(h' \cdot n))\, d\nu_N(n') \\ &= \int_N \overline{g_i(e,n')}\, g_i(e,n'n)\, d\nu_N(n') \\ &= \int_N \overline{(r_i r)(n')}\, (r_i r)(n'n)\, d\nu_N(n') \\ &= \int_N (r_i r)^*(n'^{-1})\, (r_i r)(n'n)\, d\nu_N(n') \\ &= ((r_i r)^* * (r_i r))(n) \\ &= s_i(n). \end{aligned}$$

So, for $f^{(i)} := g_i^* * g_i \in L^1(G, \nu_G)_+$, we have $s_i = f_e^{(i)}$. Therefore for every i, we have $\overline{\mathcal{F}}(s_i) \in A_{K'}$. □

THEOREM 14.C.2. *Let $G = H \ltimes N$ be as above. Let μ be a nonzero H-invariant Radon measure on an open H-invariant subset U of $\widehat{N}$. Assume moreover that the action $(\widehat{N}, \mu) \curvearrowleft H$ is essentially free and ergodic.*

(1) *The representation $\widetilde{\pi}_\mu$ associated with μ is a normal factor representation; moreover, the corresponding character t is given by*

$$t(f) = \int_{\widehat{N}} \overline{\mathcal{F}}(f_e)(\chi)\, d\mu(\chi) \quad \textit{for all } f \in L^1(G, \nu_G)_+,$$

where $\overline{\mathcal{F}}$ denotes the inverse Fourier–Stieltjes transform, and $f_e : N \to \mathbf{C}$ the function defined by $f_e(n) = f(e,n)$ for all $n \in N$.

(2) *The factor representation $\widetilde{\pi}_\mu$ is of type I_∞ if μ is an infinite atomic measure.*

(3) *The factor representation $\widetilde{\pi}_\mu$ is of type II_∞ if μ is an infinite nonatomic measure.*

PROOF. (1) By Theorem 14.B.1, the von Neumann algebra $\mathcal{M} := \widetilde{\pi}_\mu(G)''$ is a semi-finite factor. We have to show that $\widetilde{\pi}_\mu$ is a trace representation of G.

Upon conjugating $\mathcal{M}$ by a unitary operator, we can assume that $\mathcal{M}$ coincides with the crossed product von Neumann algebra $L^\infty(\widehat{N}, \mu) \rtimes H$ and that $\widetilde{\pi}_\mu$ is the representation ρ_μ of Proposition 14.A.2; thus, we can assume that

$$(\widetilde{\pi}_\mu(h,n)F)(h')(\chi) = \chi(n)F(h'h)(\chi^h),$$

for $h, h' \in H$, $n \in N$, $F \in \mathcal{H}_\mu$, and $\chi \in \widehat{N}$.

Let τ be the unique (up to positive scalar multiples) semi-finite faithful normal trace on the factor $\mathcal{M}$. By Proposition 10.D.2, it suffices to show there exists an element $f \in L^1(G)$ such that

$$0 < \tau(\widetilde{\pi}_\mu(f^* * f)) < +\infty.$$

Set $\mathcal{H}_\mu := \ell^2(H, L^2(\widehat{N}, \mu))$ and let $E : \mathcal{M} \to L^\infty(\widehat{N}, \mu)$ be the conditional expectation onto $L^\infty(\widehat{N}, \mu)$. Recall (see the observation before Lemma 13.A.10) that, for $T \in \mathcal{M}$, the function $E(T) \in L^\infty(\widehat{N}, \mu)$ is determined by

$$\int_{\widehat{N}} E(T)(\chi) f_1(\chi)\overline{f_2(\chi)}\, d\mu(\chi) = \langle T(\delta_e \otimes f_1) \mid \delta_e \otimes f_2\rangle \quad \text{for all } f_1, f_2 \in L^2(\widehat{N}, \mu),$$

where $\delta_e \otimes f_i \in \mathcal{H}_\mu$ is defined by $(\delta_e \otimes f_i)(h,n) = f_i(n)$ if $h = e$ and $(\delta_e \otimes f_i)(h,n) = 0$ otherwise. By Proposition 13.A.11(2) (see also Remark 13.A.12), we have

$$\tau(T) = \mu \circ E(T) = \int_{\widehat{N}} E(T)(\chi)\, d\mu(\chi) \quad \text{for all } T \in \mathcal{M}.$$

Fix $f \in L^1(G, \nu_G)_+$ and set

$$T := \widetilde{\pi}_\mu(f) \in \mathcal{L}(\mathcal{H}_\mu).$$

For $F, F' \in \mathcal{H}_\mu$, we have

$$\begin{aligned}
\langle T(F) \mid F'\rangle &= \int_G f(g)\langle \pi_\mu(g)F \mid F'\rangle\, d\nu_G(g) \\
&= \sum_{h\in H} \Delta(h) \int_N f(h,n)\langle \widetilde{\pi}_\mu(h,n)F \mid F'\rangle\, d\nu_N(n) \\
&= \sum_{h,h'\in H} \Delta(h) \int_N \int_{\widehat{N}} f(h,n)\chi(n)F(h'h)(\chi^h)\overline{F'(h')(\chi)}\, d\nu_N(n) d\mu(\chi);
\end{aligned}$$

in particular, we have

$$\begin{aligned}
\langle T(\delta_e \otimes f_1) \mid \delta_e \otimes f_1\rangle &= \int_N \int_{\widehat{N}} f(e,n)\chi(n) f_1(\chi)\overline{f_2(\chi)}\, d\nu_N(n)\, d\mu(\chi) \\
&= \int_{\widehat{N}} \left(\int_N f(e,n)\chi(n) d\nu_N(n) \right) f_1(\chi)\overline{f_2(\chi)}\, d\mu(\chi)
\end{aligned}$$

for all $f_1, f_2 \in L^2(\widehat{N}, \mu)$. This shows that

$$E(T)(\chi) = \int_N f(e,n)\chi(n)\, d\nu_N(n) = \overline{\mathcal{F}}(f_e)(\chi).$$

We therefore have

$$\tau(\widetilde{\pi}_\mu(f)) = \int_{\widehat{N}} \overline{\mathcal{F}}(f_e)(\chi) d\mu(\chi) \qquad \text{for all } f \in L^1(G)_+.$$

Since μ is a Radon measure on U, it takes finite values on $C^{c,K}(U)$ for every compact subset K of U. Moreover, since μ is nonzero, there exists a compact subset K of U and $\varphi \in C^{c,K}(U)$ such that

$$\int_{\widehat{N}} \varphi(\chi)\, d\mu(\chi) > 0.$$

It then follows from Lemma 14.C.1 that there exist a compact subset K' of U with $K \subset K'$ and a function $f \in L^1(G, \nu_G)_+$ such that $\overline{\mathcal{F}}(f_e) \in C^{c,K'}(U)$ and such that

$$\int_{\widehat{N}} \overline{\mathcal{F}}(f_e)(\chi)\, d\mu(\chi) > 0.$$

Since $\overline{\mathcal{F}}(f_e) \in C^{c,K'}(U)$, we also have

$$\int_{\widehat{N}} \overline{\mathcal{F}}(f_e)(\chi)\, d\mu(\chi) < +\infty.$$

So, $\widetilde{\pi}_\mu$ is a normal factor representation and we have identified its character $t = \tau \circ \widetilde{\pi}_\mu$ on $L^1(G, \nu_G)_+$.

(2) Assume that μ is an infinite atomic measure. Then, by Theorem 14.B.1, $\widetilde{\pi}_\mu$ is of type I. We have to check that $\mathcal{M}$ is an infinite factor.

By invariance and ergodicity, μ takes the same value $\alpha > 0$ on every χ in the support S of μ. For every subset $A \subset S$, we have, viewing $\mathbf{1}_A$ as a projection in $L^\infty(\widehat{N}, \mu) \subset \mathcal{M}$,

$$\tau(\mathbf{1}_A) = \int_{\widehat{N}} \mathbf{1}_A(\chi)\, d\mu(\chi) = \alpha|A|.$$

Since S is infinite, this shows that τ takes infinitely many values on projections. Therefore $\mathcal{M}$ is a factor of infinite type.

Item (3) follows from Theorem 14.B.1. □

REMARK 14.C.3. Let μ be a measure as in Theorem 14.C.2; assume that μ is atomic. Then the support S of μ is an H-orbit, by ergodicity of μ. Since μ is a Radon measure, S has to be a discrete subset of $\widehat{N}$.

For instance, let S be an orbit of an irrational rotation on $\mathbf{T}$. Then S is dense in $\mathbf{T}$ and so the corresponding counting measure μ does not fulfill the assumption of Theorem 14.C.2. The associated representation $\widetilde{\pi}_\mu$, which is a factor representation of type I_∞ of the Heisenberg group $H(\mathbf{Z})$ as in Example 14.B.2, is *not* normal. Indeed, $H(\mathbf{Z})$ has no normal factor representation of infinite type at all (Proposition 11.D.2).

Next, we show that nonequivalent measures μ as in Theorem 14.C.2 yield non-quasi-equivalent factor representations $\widetilde{\pi}_\mu$.

PROPOSITION 14.C.4. *Let G be as in Theorem 14.C.2. Let μ and μ' be nonzero H-invariant Radon measures on open H-invariant subsets U and U' of $\widehat{N}$. Assume moreover that the actions $(\widehat{N}, \mu) \curvearrowleft H$ and $(\widehat{N}, \mu') \curvearrowleft H$ are essentially free and ergodic. The following properties are equivalent:*

(i) *the factor representations $\widetilde{\pi}_\mu$ and $\widetilde{\pi}_{\mu'}$ are quasi-equivalent;*

(ii) *the restrictions of* μ *and* μ' *to* $U \cap U'$ *are proportional to each other and nonzero.*

PROOF. Set $U'' := U \cap U'$. Since μ is ergodic and U'' is H-invariant, we have either $\mu(U'') = 0$ or $\mu(\widehat{N} \smallsetminus U'') = 0$. Similarly, we have either $\mu'(U'') = 0$ or $\mu'(\widehat{N} \smallsetminus U'') = 0$.

Assume that $\mu|_{U''}$ and $\mu'|_{U''}$ are proportional to each other and nonzero. Then, we necessarily have

$$\mu(\widehat{N} \smallsetminus U'') = \mu'(\widehat{N} \smallsetminus U'') = 0.$$

It follows from Theorem 14.C.2 that the restrictions to $L^1(G, \nu_G)_+$ of the characters t and t' associated with $\widetilde{\pi}_\mu$ and $\widetilde{\pi}_{\mu'}$ coincide. Since $L^1(G, \nu_G)$ is dense in $C^*_{\max}(G)$, this implies (see Chap. I, §2, Proposition 5 in [**Guic–63**]) that t and t' are proportional to each other. Therefore $\widetilde{\pi}_\mu$ and $\widetilde{\pi}_{\mu'}$ are quasi-equivalent, by Proposition 10.D.10.

Conversely, assume that $\widetilde{\pi}_\mu$ and $\widetilde{\pi}_{\mu'}$ are quasi-equivalent. Then the characters t and t' described in Theorem 14.C.2 are proportional to each other (Proposition 10.D.10), that is, there exists $\alpha > 0$ such that

$$(*) \qquad \int_{\widehat{N}} \overline{\mathcal{F}}(f_e)(\chi)\, d\mu(\chi) = \alpha \int_{\widehat{N}} \overline{\mathcal{F}}(f_e)(\chi)\, d\mu'(\chi) \qquad \text{for all } f \in L^1(G, \nu_G)_+,$$

Assume, by contradiction, that $\mu(U'') = 0$.

Then $\mu(U') = 0$, since $\mu(\widehat{N} \smallsetminus U) = 0$ and $U'' = U \cap U'$. Lemma 14.C.1, applied to the Radon measure μ' on U', shows that we can find $f \in L^1(G, \nu_G)_+$ such that $\overline{\mathcal{F}}(f_e) = 0$ on $\widehat{N} \smallsetminus U'$ and such that

$$\int_{\widehat{N}} \overline{\mathcal{F}}(f_e)(\chi)\, d\mu'(\chi) > 0.$$

On the one hand, we then have, by $(*)$,

$$\begin{aligned} \int_{U'} \overline{\mathcal{F}}(f_e)(\chi)\, d\mu(\chi) &= \int_{\widehat{N}} \overline{\mathcal{F}}(f_e)(\chi)\, d\mu(\chi) \\ &= \alpha \int_{\widehat{N}} \overline{\mathcal{F}}(f_e)(\chi)\, d\mu'(\chi) > 0. \end{aligned}$$

On the other hand, we have

$$\int_{U'} \overline{\mathcal{F}}(f_e)(\chi)\, d\mu(\chi) = 0,$$

since $\mu(U') = 0$. This contradiction shows that $\mu(\widehat{N} \smallsetminus U'') = 0$.

Similarly, we have $\mu'(\widehat{N} \smallsetminus U'') = 0$. As a result, we may assume that both μ and μ' are Radon measures on the open set U''.

In view of Lemma 14.C.1, it follows from $(*)$ that μ agrees with $\alpha\mu'$ on $C^{c,K}(U'')$, for every compact subset K of U''. Therefore $\mu = \alpha\mu'$ on U''. □

REMARK 14.C.5. One cannot expect that every normal factor representation of $G = H \ltimes N$ is quasi-equivalent to a representation $\widetilde{\pi}_\mu$ for a measure μ as in Theorem 14.C.2.

Indeed, as is easily checked, the equivalence class of measures attached to $\widetilde{\pi}_\mu|_N$ as in Proposition 2.F.3 is the class $[\mu]$. If π is a factor representation which is quasi-equivalent to $\widetilde{\pi}_\mu$, then the class attached to $\pi|_N$ is $[\mu]$. However, there are examples of factor representations for which the action of H on $(\widehat{N}, \mu')$ is not essentially free, for some (equivalently, for any) measure μ' in the class attached to $\pi|_N$. Indeed, if π is a normal factor representation of H lifted to G, then the Dirac measure at the group unit of N is attached to $\pi|_N$; however, $(\widehat{N}, \mu')$ is not essentially free (when H and N are nontrivial).

The answer to the following question, which is more sensible to ask, seems to be unknown; for some partial results, see Chap. II, §2, Proposition 1 in [**Guic–63**].

QUESTION. Let π be a normal factor representation of G and μ an ergodic and quasi-invariant probability measure on $\widehat{N}$ attached to $\pi|_N$. Assume that the action of H on $(\widehat{N}, \mu)$ is essentially free. Is μ necessary as in Theorem 14.C.2 (that is, is μ equivalent to a nonzero H-invariant Radon measure on an open H-invariant subset U of $\widehat{N}$) and π quasi-equivalent to $\widetilde{\pi}_\mu$?

14.D. The von Neumann algebra of the regular representation as a crossed product

As in the previous sections, $G = H \ltimes N$ is a semi-direct product of a countable discrete subgroup H with a second-countable locally compact abelian normal subgroup N.

We apply the results of Sections 14.A, 14.B, and 14.C to the special case where μ is a Haar measure on $\widehat{N}$. As already mentioned in Section 14.C (for N instead of $\widehat{N}$), μ is relatively invariant (and hence quasi-invariant) by the action of H: for every $h \in H$, there exists a constant $\Delta(h) > 0$ such that $h_*(\mu) = \Delta(h)\mu$.

PROPOSITION 14.D.1. *Let $G = H \ltimes N$ be as above. Let μ be a Haar measure on $\widehat{N}$ and let $\widetilde{\pi}_\mu$ be the associated representation as in Section* 14.A.

(1) *The representation $\widetilde{\pi}_\mu$ is equivalent to the regular representation λ_G of G.*
(2) *The crossed product von Neumann algebra $L^\infty(\widehat{N}, \mu) \rtimes H$ is unitarily equivalent to the von Neumann algebra $\mathcal{L}(G) = \lambda_G(G)''$ generated by the regular representation of G.*

PROOF. (1) With π_μ as in Section 14.A, we have $\widetilde{\pi}_\mu = \mathrm{Ind}_N^G \pi_\mu$. Since λ_G is equivalent to $\mathrm{Ind}_N^G \lambda_N$, it suffices to show that λ_N is equivalent to π_μ.

We can assume that the Haar measures ν on N and μ on $\widehat{N}$ are so normalized that the Fourier transform $\overline{\mathcal{F}} : L^2(N, \nu) \to L^2(\widehat{N}, \mu)$ is a Hilbert space isomorphism. As already mentioned Example 1.G.10, $\overline{\mathcal{F}}$ intertwines λ_N and π_μ, hence λ_N is equivalent to π_μ.

(2) By (1), the von Neumann algebra $\widetilde{\pi}_\mu(G)''$ is unitarily equivalent to $\lambda_G(G)''$. The claim now follows from Proposition 14.A.2. □

We can now give sufficient conditions for the regular representation of G to be a factor representation, and determine its type.

COROLLARY 14.D.2. *Let $G = H \ltimes N$ be as in Proposition* 14.D.1 *and let μ be a Haar measure on $\widehat{N}$.*

(1) *If N is compact and $N \neq \{e\}$, then λ_G is not a factor representation.*

Assume for the following items that the action $(\widehat{N}, \mu) \curvearrowleft H$ is essentially free and ergodic.

(2) *If N is discrete and infinite, then λ_G is a factor representation of type II_1.*

(3) *If N is neither compact nor discrete and μ is H-invariant, then λ_G is a factor representation of type II_∞.*

(4) *If μ is not H-invariant, then λ_G is a factor representation of type III.*

PROOF. (1) Assume that N is compact. Then $\widehat{N}$ is discrete and so μ is the counting measure on $\widehat{N}$. In particular, we have $\mu(\{e\}) > 0$. Since the action of H on $\widehat{N}$ fixes e, it cannot be ergodic (as $N \neq \{e\}$), and so $\lambda_G(G)''$ is not a factor, by Proposition 13.A.14.

(2) Assume that N is discrete and infinite. Then $\widehat{N}$ is compact and infinite. Therefore μ is a finite nonatomic measure, which can be normalized to be a probability measure. It follows that the action of H on $\widehat{N}$ preserves μ, by uniqueness of μ. The claim follows from Theorem 14.B.1.

(3) Assume that N is neither compact nor discrete and μ is H-invariant. Then $\widehat{N}$ is neither discrete nor compact. Therefore μ is an infinite nonatomic measure. The claim follows from Theorem 14.B.1.

(4) Assume μ is not H-invariant. So, the modular function $\Delta : H \to \mathbf{R}^\times_*$ defined by $h_*(\mu) = \Delta(h)\mu$ is not trivial. We claim that there does not exist a σ-finite H-invariant measure on $\widehat{N}$ which is equivalent to μ. Once this is proved, Item (4) will follow from Theorem 14.B.1.

To prove the claim, assume, by contradiction, that there exists a σ-finite H-invariant measure ν on $\widehat{N}$ which is equivalent to μ. Then $\nu_H \otimes \nu$ is a left (and right) invariant σ-finite measure on the LC group $H \ltimes \widehat{N}$, where ν_H is the counting measure on H. Moreover, $\nu_H \otimes \nu$ is equivalent to $\nu_H \otimes \mu$, which is itself equivalent to the Haar measure $(\Delta\nu_H) \otimes \mu$ on $H \ltimes \widehat{N}$. Therefore $(\Delta\nu_H) \otimes \mu = f(\nu_H \otimes \nu)$ for a measurable function $f : \widehat{N} \to \mathbf{R}^\times_+$. Since $(\Delta\nu_H) \otimes \mu$ and $\nu_H \otimes \nu$ are both H-invariant, f is an H-invariant function. It follows from the ergodicity assumption that f is constant $\nu_H \otimes \nu$-almost everywhere. This shows that Δ is constant and hence trivial on H, and this is a contradiction. □

REMARK 14.D.3. The condition that $(\widehat{N}, \mu) \curvearrowleft H$ is ergodic is necessary for λ_G to be a factor representation, by Proposition 13.A.14. However, it can happen that λ_G is a factor representation even when $(\widehat{N}, \mu) \curvearrowleft H$ is not essentially free (see the example in Remark 13.A.15).

Assuming that λ_G is factorial, we now ask: when is λ_G a normal factor representation?

COROLLARY 14.D.4. *Let $G = H \ltimes N$ be as in Proposition* 14.D.1 *and let μ be a Haar measure on $\widehat{N}$. Assume that μ is nonatomic and that $(\widehat{N}, \mu) \curvearrowleft H$ is essentially free and ergodic.*

(1) *The regular representation λ_G is a normal factor representation if and only if μ is H-invariant.*

(2) *Assume that μ is H-invariant. The character t_G of λ_G is given by*

$$t_G(f^* * f) = (f^* * f)(e) \quad \textit{for all } f \in C^c(G),$$

for a convenient normalization of μ.

PROOF. (1) Assume that μ is H-invariant. Since μ is a Radon measure on $\widehat{N}$, it follows from Proposition 14.D.1 and Theorem 14.C.2 that λ_G is a normal factor representation.

Conversely, assume that λ_G is a normal factor representation. Then $\lambda_G(G)''$ is semi-finite and Corollary 14.D.2 shows that μ is H-invariant.

(2) By Theorem 14.C.2, the character t_G of the normal factor representation λ_G is given by

$$t_G(f) = \int_{\widehat{N}} \overline{\mathcal{F}}(f_e)(\chi)\, d\mu(\chi) \qquad \text{for all } f \in C^c(G).$$

We can choose normalizations of μ and of the Haar measure ν on N so that $\overline{\mathcal{F}}$ is a Hilbert space isomorphism between $L^2(N,\nu)$ and $L^2(\widehat{N},\mu)$. Then the inversion formula holds (see Proposition 4 in [**BTS1–2**, II, § 4]):

$$\varphi(e) = \int_{\widehat{N}} \overline{\mathcal{F}}(\varphi)(\chi)\, d\mu(\chi)$$

for every $\varphi \in A(N)$, where $A(N)$ is the linear span of

$$\{\varphi_1 * \varphi_2 \mid \varphi_1, \varphi_2 \in C^c(N)\}.$$

One checks that $(f^* * f)_e \in A(N)$ for $f \in C^c(G)$ and the claim follows. $\square$

REMARK 14.D.5. Assume that G is discrete and that λ_G is factorial. As has to be the case, the trace t_G from Corollary 14.D.4 coincides with $\tau_G \circ \lambda_G$, where τ_G is the canonical trace on $\mathcal{L}(G)$ as in Example 7.B.6.

In order to apply the results from the previous sections, we have to ensure that the action $(\widehat{N},\mu) \curvearrowleft H$ is essentially free and ergodic.

For $h \in H$, set

$$N_h := \{(h\cdot n)n^{-1} \mid n \in N\}.$$

Observe that, since N is abelian, N_h is a subgroup of N. The annihilator of N_h is the subgroup $N_h^\perp := \{\chi \in \widehat{N} \mid \chi(x) = 1 \text{ for all } x \in N_h\}$ of the dual of N.

PROPOSITION 14.D.6. *Let μ be a Haar measure on $\widehat{N}$.*

1. *For $h \in H$, the set $\mathrm{Fix}(h)$ of fixed points of h in $\widehat{N}$ coincides with the annihilator $N_h^\perp$ of N_h in $\widehat{N}$.*
2. *The action $(\widehat{N},\mu) \curvearrowleft H$ is essentially free if and only if $\mu(N_h^\perp) = 0$, for every $h \in H \smallsetminus \{e\}$.*
3. *Assume that N is discrete. The action $(\widehat{N},\mu) \curvearrowleft H$ is essentially free if and only if N_h is infinite, for every $h \in H \smallsetminus \{e\}$.*
4. *Assume that N is discrete. The action $(\widehat{N},\mu) \curvearrowleft H$ is ergodic if and only if the H-orbit $\{h\cdot n \mid h \in H\}$ of every $n \in N \smallsetminus \{e\}$ is infinite.*

PROOF. (1) For $\chi \in \widehat{N}$, we have

$$\begin{aligned} \chi \in \mathrm{Fix}(h) &\iff \chi^h = \chi \\ &\iff \chi((h\cdot n)n^{-1}) = 1 \text{ for all } n \in N \\ &\iff \chi \in N_h^\perp. \end{aligned}$$

(2) follows immediately from Item (1).

(3) Assume that N is discrete and let $h \in H$. By Pontryagin duality, the dual group of N_h can be identified with $\widehat{N}/N_h^\perp$. Therefore, in view of (2), we have to prove that $\mu(N_h^\perp) > 0$ if and only if $N_h^\perp$ has finite index in $\widehat{N}$. This is indeed the

case, since μ is a translation-invariant probability measure on $\widehat{N}$ and since $\widehat{N}$ is a disjoint union of cosets of $N_h^{\perp}$.

(4) Since N is discrete, $\widehat{N}$ is compact, the Haar measure μ on $\widehat{N}$ can be normalized to be a probability measure, and μ is preserved under the action of H.

Since, by Pontryagin duality, the dual group of $\widehat{N}$ is N, the claim then follows from a classical result in ergodic theory (see [**BeMa–00**, Proposition 1.5]). $\square$

EXAMPLE 14.D.7. Using Corollary 14.D.2 and Proposition 14.D.6, we give some examples of groups G for which the regular representation λ_G is a factor representation, and its type.

(1) Let H be an arbitrary countable infinite group and let $N = \bigoplus_{h\in H} \mathbf{Z}/2\mathbf{Z}$. Then H acts on N by shifting the coordinates:

$$h \cdot (\bigoplus_{h'\in H} x_{h'}) = \bigoplus_{h'\in H} x_{h^{-1}h'} \quad \text{for all } h \in H, \bigoplus_{h'\in H} x_{h'} \in N.$$

We identify the dual group $\widehat{N}$ with the direct product $\prod_{h\in H} \mathbf{Z}/2\mathbf{Z}$. The Haar measure on $\widehat{N}$ is $\mu = \bigotimes_{h\in H} \nu$, where ν is the equidistribution on $\mathbf{Z}/2\mathbf{Z}$. The action of H on $\widehat{N}$ is given by shifting the coordinates again.

We claim that $(\widehat{N}, \mu) \curvearrowleft H$ is essentially free. Indeed, let $h \in H \smallsetminus \{e\}$. In view of Proposition 14.D.6, we have to prove that the subgroup N_h appearing there is infinite.

We can choose an infinite subset L of H such that

$$h^{-1}L \cap L = \emptyset.$$

For every $l \in L$, let $a^{(l)} := \bigoplus_{h'\in H} x_{h'} \in N$ be defined by $x_l = x_{h^{-1}l} = 1$ and $x_{h'} = 0$ if $h' \notin \{h^{-1}l, l\}$. Then clearly,

$$h \cdot a^{(l)} + a^{(l)} \neq h \cdot a^{(l')} + a^{(l')} \quad \text{for } l \neq l'.$$

Therefore N_h is infinite.

We claim that $(\widehat{N}, \mu) \curvearrowleft H$ is ergodic. Indeed, let $\bigoplus_{h'\in H} x_{h'} \in N$, with $x_{h'}$ not all 0. Then, obviously, the H-orbit of $\bigoplus_{h'\in H} x_{h'}$ is infinite. The claim follows now from Proposition 14.D.6.

Therefore λ_G is a factor representation of type II_1. Equivalently, the crossed product algebra $L^\infty(\prod_{h\in H} \mathbf{Z}/2\mathbf{Z}, \mu) \rtimes H$ is a factor of type II_1.

(2) The group $H = \mathrm{SL}_2(\mathbf{Z})$ acts naturally on $N = \mathbf{Z}^2$. The dual group $\widehat{N}$ is the torus $\mathbf{T}^2$, with Lebesgue measure μ as the Haar measure.

The action $(\widehat{N}, \mu) \curvearrowleft H$ is essentially free. Indeed, for every $h \in \mathrm{SL}_2(\mathbf{Z}) \smallsetminus \{I\}$, the corresponding subgroup $N_h = \{(h - I)(x) \mid x \in \mathbf{Z}^2\}$ as in Proposition 14.D.6 is clearly infinite.

The action $(\widehat{N}, \mu) \curvearrowleft H$ is ergodic. Indeed the $\mathrm{SL}_2(\mathbf{Z})$-orbit of every vector $x \in \mathbf{Z}^2 \smallsetminus \{0\}$ is infinite.

Therefore λ_G is a factor representation of type II_1. Equivalently, the crossed product algebra $L^\infty(\mathbf{T}^2, \mu) \rtimes \mathrm{SL}_2(\mathbf{Z})$ is a factor of type II_1.

(3) Consider the group $H = \mathrm{SL}_2(\mathbf{Q})$ acting on $N = \mathbf{R}^2$ in the usual way. Let μ denote the Haar measure on $\widehat{N} = \mathbf{R}^2$, i.e., the Lebesgue measure on $\mathbf{R}^2$. As in the previous examples, one checks that the action $(\widehat{N}, \mu) \curvearrowleft H$ is essentially free

and ergodic. Since μ is an infinite measure, λ_G is a factor representation of type II_∞.

(4) Take Example (3) with $H = \mathrm{GL}_2(\mathbf{Q})$ instead of $\mathrm{SL}_2(\mathbf{Q})$. Similarly, one shows that $(\widehat{N}, \mu) \curvearrowleft H$ is essentially free and ergodic. Since the Lebesgue measure μ on $\mathbf{R}^2$ is not $\mathrm{GL}_2(\mathbf{Q})$-invariant, it follows that λ_G is a factor representation of type III.

REMARK 14.D.8. (1) It is easy to check that the groups of (1) & (2) in Example 14.D.7 are icc discrete groups. That their regular representations are factorial of type II_1 also follows from Propositions 7.A.1 and 7.C.5.

(2) Example 14.D.7(4) is attributed to Godement (for example in [**Suth–78**, page 247]).

(3) For examples of subtypes of type III (see Remark 13.B.5), consider moreover the triangular group $\mathrm{T}_2(\mathbf{Q}) = \begin{pmatrix} \mathbf{Q}^\times & \mathbf{Q} \\ 0 & \mathbf{Q}^\times \end{pmatrix}$, its subgroup

$$\mathrm{ST}_2(\mathbf{Q}) = \left\{ \begin{pmatrix} a & b \\ 0 & d \end{pmatrix} \in \mathrm{T}_2(\mathbf{Q}) \;\middle|\; ad = 1 \right\},$$

and, for every $\lambda \in]0,1[$, its subgroup $\mathrm{S}_\lambda\mathrm{T}_2(\mathbf{Q})$ generated by $\mathrm{ST}_2(\mathbf{Q}) \cup \begin{pmatrix} \sqrt{\lambda} & 0 \\ 0 & \sqrt{\lambda} \end{pmatrix}$; each of these groups acts naturally on $\mathbf{R}^2$. Then, as shown in [**Suth–78**]:

- the regular representation of $\mathrm{ST}_2(\mathbf{Q}) \ltimes \mathbf{R}^2$ is a factor representation of type II_∞;
- the regular representation of $\mathrm{T}_2(\mathbf{Q}) \ltimes \mathbf{R}^2$ is a factor representation of type III_1;
- the regular representation of $\mathrm{S}_\lambda\mathrm{T}_2(\mathbf{Q}) \ltimes \mathbf{R}^2$ is a factor representation of type III_λ.

(4) In [**Blac–77**], there are other constructions of groups G for which $\lambda_G(G)''$ are factors of type III_λ, that are moreover infinite tensor products of factors of type I.

(5) For a countable group H and an action of H by homeomorphism on a locally compact space X, there is a related notion of a reduced crossed product C*-algebra $C^0(X) \rtimes_{\mathrm{red}} H$. When H acts on N, and therefore on $\widehat{N}$, as above, the restriction of the isomorphism between $L^\infty(\widehat{N}, \mu) \rtimes H$ and $\mathcal{L}(G)$ that appears in Proposition 14.D.1 provides an isomorphism from the crossed product $C^0(X) \rtimes_{\mathrm{red}} H$ to the reduced C*-algebra of $G = H \ltimes N$. When moreover G is amenable, i.e., when H is amenable, so that the reduced C*-algebra of G coincides with its maximal C*-algebra, this provides examples of descriptions of maximal C*-algebras that can be added to those of Section 8.B.

14.E. Examples of normal factor representations

In Chapter 12, we determined the characters of finite type as well as the associated factor representations for our favorite groups (the two-step nilpotent groups, the affine group, the Baumslag–Solitar groups $\mathrm{BS}(1,p)$, the lamplighter group, and the general linear group over an infinite field). So far, we had no example of normal factor representations of type I_∞ and of type II_∞ for any one of our examples.

Using Theorem 14.C.2, we will give an example of a normal factor representation of type I_∞ for the Baumslag–Solitar group $\mathrm{BS}(1,p)$ and of a normal factor representation type of II_∞ for the lamplighter group $\mathbf{Z} \wr (\mathbf{Z}/2\mathbf{Z})$.

14.E.a. Normal factor representations of type I_∞ of solvable Baumslag–Solitar groups. Let p be a prime. Let $\Gamma = A \ltimes N \approx \mathbf{Z} \ltimes \mathbf{Z}[1/p]$ be the Baumslag–Solitar group $\mathrm{BS}(1,p)$ as in Sections 3.D, 4.B, 9.D, and 12.C. Recall that we can identify $\widehat{N}$ with the p-adic solenoid

$$\mathbf{Sol}_p = (\mathbf{Q}_p \times \mathbf{R})/\Delta \quad \text{for } \Delta = \{(a,-a) \mid a \in \mathbf{Z}[1/p]\}.$$

The action of A on $\widehat{N}$ is given by the transformation $T : \mathbf{Sol}_p \to \mathbf{Sol}_p$ induced by the multiplication by p.

The crucial tool in our construction is the following lemma which shows a peculiarity of the structure of T-orbits in $\mathbf{Sol}_p$.

LEMMA 14.E.1. *There exists a point $s \in \mathbf{Sol}_p$ for which the T-orbit $\{T^n s \mid n \in \mathbf{Z}\}$ is an infinite and* discrete *subset of $\mathbf{Sol}_p$.*

PROOF. Let $s = [(1,0)] \in \mathbf{Sol}_p = (\mathbf{Q}_p \times \mathbf{R})/\Delta$. We claim that we have

$$\lim_{n\to+\infty} T^n s = [(0,0)] \quad \text{and} \quad \lim_{n\to+\infty} T^{-n} s = [(0,0)].$$

Indeed, we have for every $n \in \mathbf{N}$,

$$T^n s = [(T^n 1, 0)] = [(p^n, 0)]$$

and

$$T^{-n} s = [(T^{-n}1,0)] = [(p^{-n},0)] = [(p^{-n},0) + (-p^{-n},p^{-n})] = [(0,p^{-n})].$$

Since $\lim_{n\to+\infty} p^n = 0$ in $\mathbf{Q}_p$ and $\lim_{n\to+\infty} p^{-n} = 0$ in $\mathbf{R}$, the claim is proved. It follows that the T-orbit of s is an infinite and discrete subset of $\mathbf{Sol}_p$. □

COROLLARY 14.E.2. *Let s be a point in $\mathbf{Sol}_p$, as in Lemma* 14.E.1. *Let μ be the measure $\sum_{n\in\mathbf{Z}} \delta_{T^n s}$ on $\widehat{N} = \mathbf{Sol}_p$.*

Then the representation $\widetilde{\pi}_\mu$ of $\Gamma = \mathrm{BS}(1,p)$ associated to μ as in Section 14.A *is a normal factor representation of type I_∞.*

PROOF. Indeed, since the T-orbit of s is a discrete subset of $\mathbf{Sol}_p$, the atomic T-invariant measure $\mu = \sum_{n\in\mathbf{Z}} \delta_{T^n s}$ is a Radon measure on $\widehat{N} = \mathbf{Sol}_p$. Moreover, μ is infinite, since s is not a T-periodic point. Observe that the action of A on $(\widehat{N},\mu)$ is essentially free (since s is not periodic). Therefore by Theorem 14.C.2, $\widetilde{\pi}_\mu$ is a normal factor representation of type I_∞. □

REMARK 14.E.3. Let $s \in \mathbf{Sol}_p$ with an infinite T-orbit and let χ_s be the associated unitary character of N. Recall from Proposition 3.D.6 that the induced representation $\mathrm{Ind}_N^\Gamma \chi_s$ corresponding to s is irreducible. Corollary 14.E.2 implies that, if the T-orbit of s is a discrete subset of $\mathbf{Sol}_p$, the representation $\mathrm{Ind}_N^\Gamma \chi_s$ is normal. We suspect that the converse statement is true: every normal infinite-dimensional irreducible representation of Γ is equivalent to a representation of the form $\mathrm{Ind}_N^\Gamma \chi_s$, where the T-orbit of s is an infinite discrete subset of $\mathbf{Sol}_p$.

The following corollary shows that the Thoma dual $E(\Gamma)$ does not provide a parametrization of $\mathrm{Prim}(\Gamma)$ for $\Gamma = \mathrm{BS}(1,p)$, in contrast to the case of nilpotent groups (Proposition 11.D.2 and Corollary 12.A.3), the affine group over a field (Corollary 12.B.4), or the general linear group over an infinite algebraic extension of a finite field (Corollary 12.E.2). Here we denote by κ the restriction to $\mathrm{QD}(\Gamma)_{\mathrm{fin}}$ of the map $\kappa_{\mathrm{prim}}^{\mathrm{qd}} : \mathrm{QD}(G) \to \mathrm{Prim}(G)$ of Corollary 8.D.4, hence also the restriction of the map $\kappa_{\mathrm{prim}}^{\mathrm{norm}} : \mathrm{QD}(G)_{\mathrm{norm}} \to \mathrm{Prim}(G)$ of Section 11.D.

COROLLARY 14.E.4. *Let $\Gamma = \mathrm{BS}(1,p)$ be the Baumslag–Solitar group and consider the map*

$$\kappa : \mathrm{QD}(\Gamma)_{\mathrm{fin}} \to \mathrm{Prim}(\Gamma).$$

(1) *κ has at least one uncountable fiber; in particular κ is not injective;*
(2) *κ is not surjective.*

PROOF. (1) Recall that $\mathrm{EP}(\mathbf{Sol}_p)$ denotes the set of ergodic T-invariant nonatomic probability measures on $\mathbf{Sol}_p$. By Corollary 12.D.5, there exist uncountably many measures in $\mathrm{EP}(\mathbf{Sol}_p)$ with support $\mathbf{Sol}_p$. Let μ_1, μ_2 be two such measures. For $i = 1,2$, consider $\Phi(\mu_i) \in E(\Gamma)$, with Φ as in Theorem 12.C.1, and $\pi_i = \pi_{\Phi(\mu_i)} \in \mathrm{QD}(\Gamma)_{\mathrm{fin}}$, as in Theorem 11.C.5.

We claim that $\kappa(\pi_1) = \kappa(\pi_2)$. Indeed, the support S_i of the projection-valued measure associated with the restriction $\pi_i|_N$ coincides with the support of μ_i. Therefore $S_1 = S_2 = \mathbf{Sol}_p$. Theorem 9.D.1 (see first step of the proof) then shows that π_1 and π_2 are weakly equivalent, that is, $\kappa(\pi_1) = \kappa(\pi_2)$.

(2) Let $s \in \mathbf{Sol}_p$ be a point for which the T-orbit $\{Ts \mid n \in \mathbf{Z}\}$ is an infinite and discrete subset of $\mathbf{Sol}_p$ (such a point exists by Lemma 14.E.1). The induced representation $\pi = \mathrm{Ind}_N^\Gamma \chi_s$ corresponding to s is irreducible (see Proposition 3.D.5).

We claim that $\mathrm{C}^*\mathrm{ker}(\pi)$ is not in the range of κ.

Assume, by contradiction, that $\mathrm{C}^*\mathrm{ker}(\pi)$ lies in the image of κ. Since the T-orbit of s is infinite, π is infinite-dimensional and therefore there exists $\mu \in \mathrm{EP}(\mathbf{Sol}_p)$ such that the factor representation π_μ associated with $\Phi(\mu)$ is weakly equivalent to π.

On the one hand, the support S of the projection-valued measure associated with the restriction $\pi|_N$ coincides with the closure of the T-orbit of s. On the other hand, the support of the projection-valued measure associated with the restriction $\pi_\mu|_N$ coincides with the support of μ. Since π_μ is weakly equivalent to π, it follows that the support of μ coincides with S, that is, with the closure of the T-orbit of s.

However, since the T-orbit of s is a discrete subset of $\mathbf{Sol}_p$, there exists a neighbourhood U of s such that $U \cap S = \{s\}$. Since s belongs to the support S of μ, we have $\mu(U) \neq 0$. It follows that $\mu(\{s\}) \neq 0$. This contradicts the fact that μ is nonatomic. □

14.E.b. Normal factor representations of type I_∞ for the lamplighter group. Recall from Sections 3.E, 4.B, 9.E, and 12.D that the lamplighter group is the semi-direct product $\Gamma = A \ltimes N$ of $A = \mathbf{Z}$ with $N = \bigoplus_{k \in \mathbf{Z}} \mathbf{Z}/2\mathbf{Z}$, where the action of $\mathbf{Z}$ on $\bigoplus_{k \in \mathbf{Z}} \mathbf{Z}/2\mathbf{Z}$ is given by shifting the coordinates. Recall also that $\widehat{N}$ can be identified with $X = \prod_{k \in \mathbf{Z}} \{0,1\}$, the dual action $\mathbf{Z}$ on $\widehat{N}$ being given by the shift transformation T on X. Moreover, the normalized Haar measure on $\widehat{N} = X$ is the measure $\mu = \bigotimes^{\mathbf{Z}} \nu$, where ν is the uniform probability measure on $\{0,1\}$.

The following lemma is similar to Lemma 14.E.1 and will allow us to construct a countable infinite number of normal factor representations of type I_∞ for the lamplighter group Γ.

We denote by $\mathbf{0}$ the element in X with all coordinates equal to 0.

LEMMA 14.E.5. *Let $(x_n)_{n\in\mathbf{Z}} \in X$ be such that $x_n = 1$ for finitely many $n \in \mathbf{Z}$. Then the T-orbit of $(x_n)_{n\in\mathbf{Z}} \in X$ is an infinite and* discrete *subset of X.*

PROOF. It is clear that we have

$$\lim_{n\to+\infty} T^n x = \mathbf{0} \quad \text{and} \quad \lim_{n\to+\infty} T^{-n} x = \mathbf{0}.$$

The claim follows. □

As in the case of the Baumslag–Solitar group (Corollary 14.E.2), Lemma 14.E.5 has the following consequence.

COROLLARY 14.E.6. *Let x be a point in X as in Lemma* 14.E.5. *Let μ be the measure $\sum_{n\in\mathbf{Z}} \delta_{T^n x}$ on $\widehat{N} = X$.*

Then the representation $\widetilde{\pi}_\mu$ of the lamplighter group associated to μ as in Section 14.A *is a normal factor representation of type I_∞.*

The following corollary shows that the Thoma dual $E(\Gamma)$ does not provide a parametrization of $\mathrm{Prim}(\Gamma)$. Its proof is similar to the proof of the corresponding result (Corollary 14.E.4) for the Baumslag–Solitar group $\mathrm{BS}(1,p)$.

COROLLARY 14.E.7. *Let Γ be the lamplighter group and consider the map*

$$\kappa : \ \mathrm{QD}(\Gamma)_{\mathrm{fin}} \to \mathrm{Prim}(\Gamma).$$

(1) *κ has at least one uncountable fiber; in particular, κ is not injective;*
(2) *κ is not surjective.*

Next, we aim to construct a normal factor representation of type II_∞ for the lamplighter group Γ. In view of Theorem 14.C.2, it suffices to exhibit an ergodic T-invariant Radon measure μ on an open and T-invariant subset U of X which is infinite and nonatomic.

14.E.c. Another construction of infinite invariant measures. Let X be a locally compact space and $T : X \to X$ a homeomorphism. The method from Section 5.F to construct infinite T-invariant measures on X always seems to produce *nonregular* measures. We will introduce another well-known construction (see [**Aaro–97**, § 1.5]) of an infinite invariant measure, associated with a given probability measure on a subset A of X, which is invariant for the return map to A.

Let $(X, \mathcal{B})$ be a Borel space, $T : X \to X$ a Borel automorphism, and A a measurable subset of X, fixed throughout this subsection. We endow A with the induced σ-algebra

$$\mathcal{B}_A = \{A \cap B \mid B \in \mathcal{B}\}.$$

DEFINITION 14.E.8. The **first return time** to A is the map

$$r_A : \ A \to \mathbf{N}^* \cup \{+\infty\}$$

defined by $r_A(x) := \min\{n \geq 1 \mid T^n x \in A\}$ if there exists $n \geq 1$ such that $T^n x \in A$ and $r_A(x) = +\infty$ otherwise.

The **induced transformation** by T to A (also known as the first return transformation or the Poincaré map) is the map $T_A : A \to A$ defined by

$$T_A(x) = \begin{cases} T^{r_A(x)}(x) & \text{if } r_A(x) < +\infty, \\ x & \text{if } r_A(x) = +\infty. \end{cases}$$

The induced transformation T_A is measurable. Indeed, the sets

$$r_A^{-1}(\{N\}) = A \cap \Big(\bigcap_{1 \le n < N} (X \smallsetminus T^{-n}(A)) \Big) \cap T^{-N}(A)$$

for $N \in \mathbf{N}^*$ and

$$r_A^{-1}(\{+\infty\}) = A \cap \Big(\bigcap_{n \in \mathbf{N}^*} (X \smallsetminus T^{-n}(A)) \Big)$$

are measurable. Therefore

$$T_A^{-1}(A \cap B) = \Big(\bigcup_{N \in \mathbf{N}^*} r_A^{-1}(\{N\}) \cap T^{-N}(A \cap B) \Big) \cup \big(r_A^{-1}(\infty) \cap B \big)$$

is measurable for every measurable subset B of X.

Set

$$A_N := r_A^{-1}(\{N\}) \text{ for } N \in \mathbf{N}^* \quad \text{and} \quad A_\infty := r_A^{-1}(\{+\infty\}).$$

Observe that we have a measurable partition

$$A = A_\infty \sqcup \bigsqcup_{N \in \mathbf{N}^*} A_N,$$

such that $T_A(x) = x$ when $x \in A_\infty$ and $T_A(x) = T^N(x)$ when $x \in A_N$.

THEOREM 14.E.9. *Let A be a measurable subset of X and μ_A a T_A-invariant probability measure on $(A, \mathcal{B}_A)$. Assume that $\mu_A(A_\infty) = 0$. Let μ be the measure on X defined by*

$$\mu(B) = \sum_{N=1}^{\infty} \sum_{k=0}^{N-1} \mu_A\big(T^{-k}(B) \cap A_N\big) = \sum_{N=1}^{\infty} \sum_{k=0}^{N-1} \mu_A\big(T^{-k}(B \cap T^k(A_N))\big)$$

for every $B \in \mathcal{B}$.

(1) *The measure μ on X is σ-finite and T-invariant.*

(2) *The measure μ is infinite if and only if*

$$\sum_{N=1}^{\infty} N \mu_A(A_N) = +\infty.$$

(3) *If μ_A is T_A-ergodic, then μ is T-ergodic.*

PROOF. (1) Observe that, by definition of the return map, the sets

$$T^k(A_N) \quad \text{for } N \in \mathbf{N}^* \text{ and } k \in \{0, \dots, N-1\}$$

are pairwise disjoint. Since μ_A is a measure on A, it follows from the formula for μ that μ is σ-additive. Moreover, we have

$$\mu(T^k(A_N)) = \mu_A(A_N) < +\infty \quad \text{for all } N \in \mathbf{N}^* \text{ and } k \in \{0, \dots, N-1\}.$$

Since

$$\mu(X) = \sum_{N=1}^{\infty} \sum_{k=0}^{N-1} \mu(T^k(A_N)),$$

we see that μ is σ-finite.

Let us show that μ is T-invariant. Let $B \in \mathcal{B}$. We have

$$\begin{aligned}\mu(T^{-1}(B)) &= \sum_{N=1}^{\infty} \sum_{k=0}^{N-1} \mu_A\left(T^{-(k+1)}(B) \cap A_N\right) \\ &= \sum_{N=1}^{\infty} \sum_{k=1}^{N} \mu_A\left(T^{-k}(B) \cap A_N\right) \\ &= \sum_{N=1}^{\infty} \sum_{k=1}^{N-1} \mu_A\left(T^{-k}(B) \cap A_N\right) + \sum_{N=1}^{\infty} \mu_A\left(T^{-N}(B) \cap A_N\right).\end{aligned}$$

For every $N \in \mathbf{N}^*$, we have by definition of the return map, $T_A = T^N$ on A_N and hence

$$T_A^{-1}(B \cap A) = (B \cap A_\infty) \sqcup \bigsqcup_{N^* \in \mathbf{N}} (T^{-N}(B) \cap A_N),$$

since $A = A_\infty \sqcup \bigsqcup_{N^* \in \mathbf{N}} A_N$. As $\mu_A(A_\infty) = 0$, it follows that

$$\sum_{N=1}^{\infty} \mu_A\left(T^{-N}(B) \cap A_N\right) = \mu_A\left(T_A^{-1}(B \cap A)\right).$$

Using the T_A-invariance of μ_A, we therefore have

$$\sum_{N=1}^{\infty} \mu_A\left(T^{-N}(B) \cap A_N\right) = \mu_A(B \cap A).$$

Since

$$\mu_A(B \cap A) = \sum_{N=1}^{\infty} \mu_A(B \cap A_N),$$

we obtain

$$\begin{aligned}\mu(T^{-1}(B)) &= \sum_{N=1}^{\infty} \sum_{k=1}^{N-1} \mu_A\left(T^{-k}(B) \cap A_N\right) + \mu_A(B \cap A) \\ &= \sum_{N=1}^{\infty} \sum_{k=0}^{N-1} \mu_A\left(T^{-k}(B) \cap A_N\right) = \mu(B).\end{aligned}$$

This proves the T-invariance of μ.

(2) The claim follows from the fact that

$$\mu(X) = \sum_{N=1}^{\infty} \sum_{k=0}^{N-1} \mu_A(A_N) = \sum_{N=1}^{\infty} N \mu_A(A_N).$$

(3) Let B be a T-invariant measurable subset of X such that $\mu(B) > 0$. We have to show that $\mu(X \setminus B) = 0$.

It follows from the definition of μ that there exist $N \in \mathbf{N}^*$ and $k \in \{0, \ldots, N-1\}$ such that $\mu_A(T^{-k}(B) \cap A_N) > 0$. By T-invariance of B, we therefore have

$$\mu_A(B \cap A) \geq \mu_A(B \cap A_N) = \mu_A\left(T^k(B) \cap A_N\right) > 0.$$

Since $T_A = T^N$ on A_N, we have, again by T-invariance of B,

$$\begin{aligned} T_A^{-1}(B \cap A) &= (B \cap A_\infty) \cup \bigcup_{N^* \in \mathbf{N}} (T^{-N}(B) \cap A_N) \\ &= (B \cap A_\infty) \cup \bigcup_{N^* \in \mathbf{N}} (B \cap A_N) \;=\; B \cap A. \end{aligned}$$

So, $B \cap A$ is T_A-invariant. Since $\mu_A(B \cap A) > 0$, it follows from the T_A-ergodicity of μ_A that $\mu_A((X \setminus B) \cap A) = 0$. Since $X \setminus B$ is T-invariant, we therefore have

$$\mu(X \setminus B) \;=\; \sum_{N=1}^{\infty} \sum_{k=0}^{N-1} \mu_A\left((X \setminus B) \cap A_N\right) \;=\; 0,$$

and this concludes the proof. □

14.E.d. Normal factor representations of type II_∞ for the lamplighter group. Let $X = \prod_{k \in \mathbf{Z}} \{0,1\}$ and T the shift transformation on X, as in Subsection 14.E.b. In order to construct normal factor representations of type II_∞ for the lamplighter group, we have to find appropriate T-invariant infinite measures on X (see Theorem 14.C.2). For this, we will use Theorem 14.E.9. As an intermediate step, we will first consider shift spaces over an infinite alphabet.

Consider a countable set S (a "state" space or "alphabet"), the product space

$$\Sigma_S \;:=\; S^{\mathbf{Z}} \;=\; \{(s_n)_{n \in \mathbf{Z}} \mid s_n \in S\},$$

and the shift transformation $\sigma : \Sigma_S \to \Sigma_S$ given by

$$\sigma((s_n)_{n \in \mathbf{Z}}) \;=\; (s_{n+1})_{n \in \mathbf{Z}} \qquad \text{for all } (s_n)_{n \in \mathbf{Z}} \in \Sigma_S.$$

Let $\mathcal{B}$ be the σ-algebra on Σ_S generated by the cylinders

$$C^{n_1,\ldots,n_k}_{a_1,\ldots,a_k} \;=\; \{(s_n)_{n \in \mathbf{Z}} \in \Sigma_S \mid s_{n_i} = a_i \text{ for } i = 1, \ldots, k\},$$

for integers $n_1 < n_2 < \cdots < n_k$ in $\mathbf{Z}$ and (not necessarily distinct) elements $a_1, a_2, \ldots, a_k$ in S.

Let $\mathbf{p}$ be a probability measure on S. The product measure

$$\nu_{\mathbf{p}} \;:=\; \mathbf{p}^{\otimes \mathbf{Z}}$$

is the unique probability measure on $(\Sigma_S, \mathcal{B})$ such that, for every cylinder $C^{n_1,\ldots,n_k}_{s_1,\ldots,s_k}$, we have

$$\nu_{\mathbf{p}}(C^{n_1,\ldots,n_k}_{a_1,\ldots,a_k}) \;=\; p_{a_1} p_{a_2} \cdots p_{a_k},$$

with $p_a = \mathbf{p}(\{a\})$. One checks that the shift transformation

$$\sigma : \Sigma_S \to \Sigma_S$$

is measurable with respect to $\mathcal{B}$. Since

$$\sigma^{-1}(C^{n_1,\ldots,n_k}_{a_1,\ldots,a_k}) \;=\; C^{n_1+1,\ldots,n_k+1}_{a_1 \ldots, a_k},$$

we have

$$\nu_{\mathbf{p}}(\sigma^{-1}(C^{n_1,\ldots,n_k}_{a_1,\ldots,a_k})) \;=\; p_{a_1} p_{a_2} \cdots p_{a_k} \;=\; \nu_{\mathbf{p}}(C^{n_1,\ldots,n_k}_{a_1,\ldots,a_k}),$$

and it follows that $\nu_{\mathbf{p}}$ is σ-invariant.

For a proof of the following classical fact, see e.g. Theorem 1.12 in [**Walt–82**]; the proof given there for a finite set S carries over without change to a countable infinite set S.

PROPOSITION 14.E.10. *Let S be a countable set and $\mathbf{p}$ a probability measure on S.*

The product measure $\nu_{\mathbf{p}}$ on Σ_S is ergodic under the shift transformation σ. □

Coming back to the space $X = \prod_{k\in\mathbf{Z}}\{0,1\}$, we set

$$U := X \setminus \{\mathbf{0}\}.$$

THEOREM 14.E.11. *There exists an ergodic T-invariant Radon measure μ on the open and T-invariant subset U of X which is infinite and nonatomic.*

PROOF. Consider the open and closed subset

$$A := \{(x_n)_{n\in\mathbf{Z}} \mid x_0 = 1\}$$

of U. For $S := \mathbf{N}$, we define a map $\Phi : \Sigma_S \to A$ as follows: for $\mathbf{n} = (n_k)_{k\in\mathbf{Z}} \in \Sigma_S$, we set $\Phi(\mathbf{n}) = (x_n)_{n\in\mathbf{Z}} \in A$, where

$$x_n = \begin{cases} 1 \text{ if } n = n_0 + \cdots + n_k + k + 1 \text{ for some } k \in \mathbf{N}, \\ 1 \text{ if } n = -(n_{-1} + \cdots + n_{-k} + k) \text{ for some } k \in \mathbf{N}^*, \\ 0 \text{ otherwise;} \end{cases}$$

thus,

$$\Phi(\mathbf{n}) = (\ldots, 1, [n_{-k}], 1, \ldots, 1, [n_{-1}], \underset{*}{1}, [n_0], 1, [n_1], 1, \ldots, 1, [n_k], 1, \ldots),$$

where $[n] = \underbrace{(0, \ldots, 0)}_{n-\text{times}}$ for $n \in \mathbf{N}$ and $*$ denotes the 0-th position in the sequence.

• *First step.* We claim that Φ is measurable and that $\Phi \circ \sigma = T_A \circ \Phi$.

Indeed, to show that Φ is measurable, it suffices to show that $\Phi^{-1}(C_\varepsilon^N)$ is a measurable subset of Σ_S for every $N \in \mathbf{Z}$, where

$$C_\varepsilon^N = \{(x_n)_{n\in\mathbf{Z}} \mid x_N = \varepsilon\}$$

and $\varepsilon \in \{0,1\}$. This is the case for $\Phi^{-1}(C_1^N)$ for $N \geq 0$, since

$$\Phi^{-1}(C_1^N) = \bigcup_{0\leq k\leq N} \ \bigcup_{(n_0,n_1,\ldots,n_k)\in S_{N,k}} C_{n_0,n_1,\ldots,n_k}^{0,1,\ldots,k},$$

where $S_{N,k}$ is the finite subset of $\{0,\ldots,N\}^k$ consisting of all k-tuples $(n_0, n_1, \ldots, n_k)$ such that $n_0 + \cdots + n_k + k + 1 = N$. The case $N < 0$ is treated similarly. It follows that we also have $\Phi^{-1}(C_0^N) \in \mathcal{B}$, since $C_0^N = A \setminus C_1^N$.

Let $\mathbf{n} = (n_k)_{k\in\mathbf{Z}} \in \Sigma_S$. Then

$$\begin{aligned} T_A(\Phi(\mathbf{n})) &= (\ldots, 1, [n_{-1}], 1, [n_0], \underset{*}{1}, [n_1], 1, [n_2], 1 \ldots) \\ &= \Phi((n_{k+1})_{k\in\mathbf{Z}}) = \Phi(\sigma(\mathbf{n})). \end{aligned}$$

This proves the claim.

Let $\mathbf{p}$ be a probability measure on $\mathbf{N}$ such that $p_n := \mathbf{p}(\{n\}) < 1$ for every $n \in \mathbf{N}$. Below we will make a more precise choice of $\mathbf{p}$.

Let

$$\mu_A := \Phi_*(\mathbf{p})$$

be the image of $\mathbf{p}$ under Φ.

• *Second step.* We claim that μ_A is a T_A-invariant nonatomic probability measure on A which is ergodic.

Indeed, $\mathbf{p}$ is nonatomic: since $\sum_{n\in\mathbf{N}} p_n = 1$ and $p_n < 1$, there exists $0 \leq q < 1$ such that $p_n \leq q$ for every $n \in \mathbf{N}$ and hence

$$\mathbf{p}(\{(n_k)_{k\in\mathbf{Z}}\}) = \lim_{k\to+\infty} \prod_{i=-k}^{k} p_{n_i} \leq \lim_{k\to+\infty} q^{2k+1} = 0.$$

Moreover, $\mathbf{p}$ is σ-invariant and ergodic by Proposition 14.E.10. Since $\mu_A = \Phi_*(\mathbf{p})$ and since $\Phi \circ \sigma = T_A \circ \Phi$ by the first step, the claim follows.

• *Third step.* We claim that $\mu_A(r_A^{-1}(\{+\infty\})) = 0$ and that $\mu_A(r_A^{-1}(\{N\})) = p_N$ for every $N \in \mathbf{N}^*$.

Indeed, we have

$$r_A^{-1}(\{+\infty\}) = \{(x_n)_{n\in\mathbf{Z}} \in A \mid x_n = 0 \text{ for all } n \geq 1\};$$

hence $\Phi^{-1}(r_A^{-1}(\{+\infty\})) = \emptyset$, and so

$$\mu_A(r_A^{-1}(\{+\infty\})) = \mathbf{p}(\Phi^{-1}(r_A^{-1}(\{+\infty\}))) = 0.$$

For $N \in \mathbf{N}^*$, we have, by definition of Φ,

$$\Phi^{-1}(r_A^{-1}(\{N\})) = \{(n_k)_{k\in\mathbf{Z}} \in \Sigma_S \mid n_0 = N-1\}$$

and hence

$$\mu_A(r_A^{-1}(\{N\})) = \mathbf{p}(\Phi^{-1}(r_A^{-1}(\{N\}))) = p_{N-1}.$$

Let μ be the measure on U associated with the T_A-invariant probability measure μ_A as defined in Theorem 14.E.9.

• *Fourth step.* We claim that μ is an ergodic T-invariant Borel measure on U which is nonatomic and σ-finite.

Indeed, the fact that μ is T-invariant, ergodic, and σ-finite follows from Theorem 14.E.9 and the second step (observe that $\mu_A(r_A^{-1}(\{+\infty\})) = 0$, by the third step). Since μ_A is nonatomic, the formula defining μ shows that μ is nonatomic.

Choose the probability measure $\mathbf{p}$ on $\mathbf{N}$ such that

$$\sum_{n\in\mathbf{N}} (n+1)p_n = +\infty.$$

• *Fifth step.* We claim that the associated measure μ on U is infinite.

Indeed, this follows from Theorem 14.E.9(2) and the fact that $\mu_A(r_A^{-1}(\{N\})) = p_{N-1}$ for every $N \in \mathbf{N}^*$ (see the third step).

Assume now that the probability measure $\mathbf{p}$ on $\mathbf{N}$ is chosen so that, moreover,

$$\sum_{n\in\mathbf{N}} \sqrt{n+1}\,p_n < +\infty.$$

(To fulfill all requirements, we could take $p_n = \dfrac{6}{\pi^2(n+1)^2}$ for every $n \in \mathbf{N}$.)

• *Sixth step.* We claim that μ is a Radon measure on U.

Indeed, since U is a second-countable locally compact space, it suffices to show that μ is finite on every compact subset of U (see Theorem A.4.1).

Let B be a compact subset of $U = X \smallsetminus \{\mathbf{0}\}$. Then there exists an open neighbourhood V of $\mathbf{0}$ in X such that $B \cap V = \emptyset$. Choose $N_0 \in \mathbf{N}$ so that the cylinder

$$C := \{(x_n)_{n\in\mathbf{Z}} \in X \mid x_n = 0 \quad \text{for all } n \text{ such that } -N_0 \leq n \leq N_0\}$$

is contained in V.

Let $N > N_0^2$. We claim that

$$T^{-k}(B) \cap A_N = \emptyset \quad \text{for all } \sqrt{N} \leq k \leq N - \sqrt{N}.$$

Indeed, for k such that $\sqrt{N} \leq k \leq N - \sqrt{N}$, let $y = (y_n)_{n \in \mathbf{Z}} \in T^{-k}(B)$. There exists $x = (x_n)_{n \in \mathbf{Z}} \in B$ with $T^k(x) = y$. Since $B \cap C = \emptyset$, we can find $i \in \{-N_0, \ldots, N_0\}$ such that $x_i = 1$. Then

$$i + k \leq N_0 + N - \sqrt{N} < N \quad \text{and} \quad i + k \geq -N_0 + \sqrt{N} > 0,$$

that is, $0 < i + k < N$. Since $y_{i+k} = x_i$, we have $y_{i+k} = 1$. This shows that $T^{i+k}y \in A$. Since $0 < i + k < N$, this implies that y does not belong to A_N and the claim is proved.

It follows that

$$\begin{aligned}
\mu(B) &= \sum_{N=1}^{\infty} \sum_{k=0}^{N-1} \mu_A(T^{-k}(B) \cap A_N) \\
&= \sum_{N=1}^{N_0^2} \sum_{k=0}^{N-1} \mu_A(T^{-k}(B) \cap A_N) \\
&\quad + \sum_{N > N_0^2} \left(\sum_{k=0}^{\sqrt{N}-1} \mu_A(T^{-k}(B) \cap A_N) + \sum_{k=N-\sqrt{N}}^{N-1} \mu_A(T^{-k}(B) \cap A_N) \right) \\
&\leq \sum_{N=1}^{N_0^2} \sum_{k=0}^{N-1} \mu_A(A_N) + \sum_{N > N_0^2} \left(\sum_{k=0}^{\sqrt{N}-1} \mu_A(A_N) + \sum_{k=N-\sqrt{N}}^{N-1} \mu_A(A_N) \right) \\
&= \sum_{N=1}^{N_0^2} N \mu_A(A_N) + \sum_{N > N_0^2} 2\sqrt{N} \mu_A(A_N).
\end{aligned}$$

Since, by assumption, the series $\sum_{n \in \mathbf{N}} \sqrt{n+1} p_n$ converges, we see that $\mu(B) < +\infty$. □

The following corollary is a direct consequence of Theorem 14.E.11 and Theorem 14.C.2.

Corollary 14.E.12. *Let $\Gamma = \mathbf{Z} \ltimes N$ be the lamplighter group, with $N = \bigoplus_{k \in \mathbf{Z}} \mathbf{Z}/2\mathbf{Z}$. Identify the dual $\widehat{N}$ of N with $X = \prod_{k \in \mathbf{Z}} \mathbf{Z}/2\mathbf{Z}$ and the action of the generator 1 of $\mathbf{Z}$ with the two-sided shift T of X.*

Let μ be a measure on X as in Theorem 14.E.11. Let $\widetilde{\pi}_\mu$ be the corresponding representation of Γ, as in Construction 14.A.1.

Then $\widetilde{\pi}_\mu$ is a normal factor representation of Γ of type II_∞.

CHAPTER 15

Separating families of finite type representations

Finite type representations of a topological group G have been introduced in Section 11.B and several examples of such representations were given in Chapter 12, mainly in the case where G is discrete.

In this chapter, we will deal with the problem of deciding when G has "enough" finite type representations. More precisely, we study the following two questions:

- *when does the family of finite type representations of G separate the points of G?*
- *when does the family of finite type representations of a discrete group G separate the points of the maximal C^*-algebra $C^*_{\max}(G)$ of G?*

It is shown in Section 15.A that a faithful finite type representation of G gives rise to a bi-invariant metric on G. Among the class of compactly generated locally compact groups, the groups with a separating family of finite type representations are characterized as being the SIN groups (Section 15.B). The class of SIN groups also coincides with the class of locally compact groups for which the von Neumann algebra generated by the regular representation is finite (Section 15.C). It follows that a connected locally compact group G has a separating family of finite type representations if and only if G is the product of a compact group with $\mathbf{R}^n$. Moreover, the class of compactly generated totally disconnected locally compact groups with a separating family of finite type representations agrees with the class of projective limits of discrete groups (Section 15.D).

Concerning the second question, which is deeper, only partial results are available. We show in Section 15.E that the answer to this question is negative for $G = \mathrm{SL}_n(\mathbf{Q})$ for $n \geq 2$. In contrast to this, the answer is positive for amenable groups as well as for free groups. In fact, we will prove that free groups satisfy even two stronger properties, which we call the C^*-MAP property and the C^*-residual finiteness property; this last property means that the representations with finite image separate the points of the maximal C^*-algebra of a free group.

In Section 15.F, we show that every so-called random invariant subgroup of a discrete group G gives rive to a trace of G. As a consequence, if G is character rigid in the sense that $E(G)$ consists only of 1_G and δ_e, then every action of G by measure-preserving transformations of a probability space is essentially free.

15.A. Finite type representations and bi-invariant metrics

Let G be a topological group. Recall from Definition 11.B.1 that a representation π of G is of finite type if the von Neumann algebra $\pi(G)''$ has a faithful finite normal trace. We denote by $\mathrm{Rep}_{\mathrm{FT}}(G)$ the set of finite type representations of G, up to quasi-equivalence.

We show that a finite type representation of G defines a bi-invariant metric on an appropriate quotient of G. Recall that a pseudo-metric or a metric d on a group G is **bi-invariant** if

$$d(gx, gy) = d(xg, yg) = d(x, y) \quad \text{for all } x, y, g \in G.$$

PROPOSITION 15.A.1. *Let G be a topological group and π a representation of G. Let τ be a normalized finite normal trace on $\pi(G)''$, let $\varphi = \tau \circ \pi$ be the normalized trace on G associated with π, and define*

$$d_{(\pi,\tau)} : G \times G \to \mathbf{R}, \qquad (x, y) \mapsto \sqrt{\tau\big((\pi(x) - \pi(y))^*(\pi(x) - \pi(y))\big)}.$$

(1) *We have*

$$d_{(\pi,\tau)}(x, y) = \sqrt{2 - 2\mathrm{Re}\, \varphi(x^{-1}y)} \quad \text{for all } x, y \in G.$$

(2) *$d_{(\pi,\tau)}$ is a bi-invariant continuous pseudo-metric on G, and its diameter is at most 2, i.e., $\sup_{x,y\in G} d_{(\pi,\tau)}(x, y) \leq 2$.*

(3) *Assume that τ is faithful. Then $d_{(\pi,\tau)}$ factorizes into a bi-invariant metric on $G/\ker \pi$.*

PROOF. Claim (1) follows from a straightforward computation.

(2) Let $(\pi', \mathcal{H}', \xi')$ be a GNS triple associated with the normalized function of positive type $\varphi = \tau \circ \pi$, so that

$$\tau(\pi(g)) = \varphi(g) = \langle \pi'(g)\xi' \mid \xi' \rangle \quad \text{for all } g \in G.$$

We have

$$d_{(\pi,\tau)}(x, y) = \sqrt{2 - 2\mathrm{Re}\, \varphi(x^{-1}y)} = \|\pi'(x)\xi' - \pi'(y)\xi'\|$$

for all $x, y \in G$. The fact that $d_{(\pi,\tau)}$ is a left-invariant pseudo-metric follows immediately from this equality. Since φ is conjugation invariant and continuous, $d_{(\pi,\tau)}$ is right invariant and continuous.

(3) If τ is faithful, then $d_{(\pi,\tau)}(x, y) = 0$ if and only if $\pi(x) = \pi(y)$, that is, $x^{-1}y \in \ker \pi$. This shows that $d_{(\pi,\tau)}$ factorizes by $G/\ker \pi$ and defines a metric on this quotient group. □

Recall from Definition 1.B.18 that a family $\mathcal{F}$ of representations of G *separates the points of G*, or is *separating*, if, for every $g \in G$ such that $g \neq e$, there exists $\pi \in \mathcal{F}$ such that $\pi(g) \neq \mathrm{Id}$.

The following corollary is an immediate consequence of Proposition 15.A.1.

COROLLARY 15.A.2. *Let G be a topological group. Assume that $\mathrm{Rep}_{\mathrm{FT}}(G)$ contains a separating sequence $(\pi_n)_{n\geq 1}$. For every $n \geq 1$, let τ_n be a normalized faithful finite normal trace on $\pi_n(G)''$.*

Then

$$d := \sum_{n\geq 1} \frac{1}{2^n} d_{(\pi_n,\tau_n)}$$

is a continuous bi-invariant metric on G.

REMARK 15.A.3. Groups as in Corollary 15.A.2 are related to a class of topological groups introduced by Popa in [**Popa–07**] and defined as follows.

A Polish group G is said to be of **finite type** if G is isomorphic, as a topological group, to a closed subgroup of the unitary group $\mathcal{U}(\mathcal{M})$ of a finite von Neumann algebra $\mathcal{M}$ acting on a separable Hilbert space, where $\mathcal{U}(\mathcal{M})$ is equipped with the strong operator topology.

It is shown in [**AnMa–12**, Theorem 2.7] that a Polish group G is of finite type if and only if $\mathrm{Rep}_{\mathrm{FT}}(G)$ contains a separating sequence and the topology of G is generated by the metric d as in Corollary 15.A.2.

15.B. Groups for which finite type representations are separating

We characterize below compactly generated locally compact groups G for which $\mathrm{Rep}_{\mathrm{FT}}(G)$ is separating.

DEFINITION 15.B.1. A topological group G is a **SIN group** if G has a basis of neighbourhoods of e which are invariant under conjugation, that is, neighbourhoods V of e such that $gVg^{-1} = V$ for all $g \in G$. The acronym SIN stands for "Small Invariant Neighbourhoods".

The class of SIN topological groups includes discrete groups, abelian LC groups, and compact groups. It is closed under direct products.

LEMMA 15.B.2. *Let G be a locally compact SIN group.*
Then G is unimodular.

PROOF. Let μ_G be a left Haar measure on G and let V be a conjugation-invariant compact neighbourhood of e. For every $g \in G$, we have, on the one hand, $\mu_G(g^{-1}Vg) = \mu(V) > 0$ and, on the other hand, $\mu_G(g^{-1}Vg) = \Delta_G(g)\mu(V)$, where Δ_G is the modular function of G (see Appendix A.6). It follows that $\Delta_G(g) = 1$. □

The following result is Théorème 10 in [**Gode–54**].

THEOREM 15.B.3. *Let G be a locally compact SIN group.*
Then $\mathrm{Rep}_{\mathrm{FT}}(G)$ separates the points of G.

PROOF. Let μ_G be a left Haar measure on G. By Lemma 15.B.2, μ_G is also a right Haar measure.

Let $g \in G$ be such that $g \neq e$. We can find a compact neighbourhood V of e which is invariant under conjugation, such that $V = V^{-1}$, and $g \notin V^2$. Set

$$f_V := \frac{\mathbf{1}_V}{\sqrt{\mu_G(V)}} \in L^2(G, \mu_G)$$

and let φ_V be the normalized continuous function of positive type on G defined by

$$\varphi_V(x) = \langle \lambda_G(x) f_V \mid f_V \rangle = \mu_G(V)^{-1}\mu_G(xV \cap V) \quad \text{for every } x \in G.$$

Then, for every $x, y \in G$, we have

$$\begin{aligned}\varphi_V(yxy^{-1}) &= \mu_G(V)^{-1}\mu_G(yxy^{-1}V \cap V) = \mu_G(V)^{-1}\mu_G(xy^{-1}Vy \cap y^{-1}Vy) \\ &= \mu_G(V)^{-1}\mu_G(xV \cap V) = \varphi_V(x),\end{aligned}$$

since μ_G and V are invariant under conjugation by y. Therefore $\varphi_V \in \mathrm{Tr}_1(G)$. Moreover, since $g \notin V^2 = V^{-1}V$, we have $gV \cap V = \emptyset$, and so $\varphi_V(g) = 0$.

Let π_V be the cyclic subrepresentation of λ_G defined on the closure of the linear span of $\{\lambda_G(x) f_V \mid x \in G\}$. Observe (see Proposition 11.B.7) that $\pi_V \in \mathrm{Rep}_{\mathrm{FT}}(G)$, since $\varphi_V \in \mathrm{Tr}_1(G)$; moreover, $\pi_V(g) \neq \mathrm{Id}$ since $\varphi_V(g) = 0$. □

We now show that the converse of Theorem 15.B.3 is true, under the additional assumption that G is compactly generated. This result appears in [**Dixm–C***, 17.3.7]; see also [**Gode–51b**, Théorème 6].

THEOREM 15.B.4. *Let G be a compactly generated locally compact group. Assume that* $\mathrm{Rep}_{\mathrm{FT}}(G)$ *separates the points of G.*

Then G is a SIN group.

PROOF. • *First step.* Let C be a compact subset of G such that $e \notin C$. We claim that there exists a neighbourhood U of e which is disjoint from C and invariant under conjugation.

Indeed, let $g \in C$. Since $g \neq e$, there exists $\pi \in \mathrm{Rep}_{\mathrm{FT}}(G)$ such that $g \notin \ker \pi$. Let τ be a normalized finite faithful normal trace on $\pi(G)''$ and $d = d_{(\pi,\tau)}$ the associated pseudo-metric on G as in Proposition 15.A.1. Since d is continuous, since $d(g, e) > 0$, and since d is bi-invariant, the set

$$V_g = \{x \in G \mid d(x, e) \geq d(g, e)/2\}$$

is a closed neighbourhood of g such that $e \notin V_g$ and V_g is invariant under conjugation.

By compactness of C, there exist finitely many elements $g_1, \dots, g_n$ of C such that $C \subset \bigcup_{i=1}^n V_{g_i}$. The set

$$U := G \smallsetminus \bigcup_{i=1}^{n} V_{g_i}$$

is a neighbourhood of e with the required properties.

• *Second step.* We claim that G is a SIN group.

Indeed, let K be a compact generating set for G and let V be an open neighbourhood of e with compact closure. Upon replacing K by $(K \cup \overline{V}) \cup (K \cup \overline{V})^{-1}$, we can assume that $V \subset K$ and that $K = K^{-1}$.

Set $C := K^3 \cap (G \smallsetminus V)$. Then C is a compact subset of G such that $e \notin C$. Therefore, by the first step, there exists a conjugation-invariant neighbourhood U of e such that $U \cap C = \emptyset$.

Let $g \in K$ and $x \in V \cap U$. Then $gxg^{-1} \in K^3$; moreover, $gxg^{-1} \in U$ and therefore $gxg^{-1} \notin C$. Hence, $gxg^{-1} \in V \cap U$. This shows that $V \cap U$ is invariant under conjugation by every element of K. Since K generates G, it follows that $V \cap U$ is invariant under conjugation by every element of G. As $V \cap U$ is a neighbourhood of e contained in V, the claim is proved. □

When G is a MAP group (see Subsection 4.C.b), it is obvious that $\mathrm{Rep}_{\mathrm{FT}}(G)$ separates the points of G. The following corollary is therefore an immediate consequence of Theorem 15.B.4.

COROLLARY 15.B.5. *Let G be a compactly generated locally compact MAP group.*

Then G is a SIN group.

The assumption about the compact generation of G in Corollary 15.B.5 and *a fortiori* in Theorem 15.B.4 is necessary, as shown by the following example which was communicated to us by P.-E. Caprace.

EXAMPLE 15.B.6. For every $n \in \mathbf{N}$, let $A_n = \mathrm{Alt}(5)$ be the alternating group on 5 symbols and let C_n be any proper subgroup of A_n such that $C_n \neq \{e\}$. The product

$$A := \prod_{n \in \mathbf{N}} A_n,$$

equipped with the product topology, is a compact group containing $C := \prod_{n \in \mathbf{N}} C_n$ as a closed subgroup. Let G be the *restricted product* of the A_n's with respect to the C_n's; recall that G is the subgroup of A defined as follows:

- $G = \bigcup_F \left(\prod_{n \in F} A_n \times \prod_{n \in \mathbf{N} \smallsetminus F} C_n\right)$, where F runs over the finite subsets of $\mathbf{N}$;
- G is equipped with the unique group topology for which C is an open compact subgroup of G.

The group G is a MAP group; this follows from Proposition 4.C.6 because the canonical injection of G into the compact group A is obviously continuous. (More explicitly, for every $g \neq e$ in G, there exists $n \in \mathbf{N}$ such that $p_n(g) \neq e$, where $p_n : G \twoheadrightarrow A_n$ is the canonical projection, and p_n can be checked to be continuous; if π_n is a finite-dimensional faithful representation of $A_n = \mathrm{Alt}(5)$, then $\pi_n \circ p_n(g) \neq \mathrm{Id}$.)

We are going to show that G is not a SIN group.

We first claim that every open normal subgroup of G has finite index in G. Indeed, let N be an open normal subgroup of G. Then $N \cap C$ is an open subgroup of the compact group C. By definition of the product topology on C, the neighbourhood $N \cap C$ of $\{e\}$ in C contains a subgroup H of the form

$$H = \prod_{n \in F} \{e\} \times \prod_{n \in \mathbf{N} \smallsetminus F} C_n,$$

for some finite subset F of $\mathbf{N}$. Since N is normal, N contains all G-conjugates of H. As A_n is simple and $C_n \neq \{e\}$ for all n, it follows that N contains the subgroup

$$G \cap \left(\prod_{n \in F} \{e\} \times \prod_{n \in \mathbf{N} \smallsetminus F} A_n \right),$$

which has finite index in G.

Assume now, by contradiction, that G is a SIN group. Then C, which is a neighbourhood of $\{e\}$ in G, contains a G-invariant neighbourhood U of $\{e\}$ such that $U \neq C$. It follows that

$$N := \bigcap_{g \in G} gCg^{-1}$$

is a normal subgroup of G containing U. Hence, N is open in G. By what we have seen above, this implies that N has finite index in G. This is a contradiction, since N is contained in C and since C has infinite index in G.

15.C. Locally compact groups with a finite von Neumann algebra

In this section, we will show that the class of LC groups with a finite group von Neumann algebra coincides with the class of SIN groups.

We will need the following elementary result.

LEMMA 15.C.1. *Let G be an LC group and $\tau : \mathcal{L}(G) \to \mathbf{C}$ a normal positive linear functional on the von Neumann algebra $\mathcal{L}(G) = \lambda_G(G)''$ generated by the regular representation of G.*

Then the function $\tau \circ \lambda_G : G \to \mathbf{C}$ belongs to $C^0(G)$.

PROOF. Recall that λ_G has vanishing matrix coefficients (see Section 12.E). Since τ is normal, τ is continuous with respect to the ultraweak topology on

$\mathcal{L}(L^2(G))$; it follows (see Théorème 1 in [**Dixm–vN**, Chap. I, §4, no. 2]) that there exists a sequence $(f_i)_{i\geq 1}$ in $L^2(G)$ with $\sum_{i=1}^{\infty} \|f_i\|^2 < +\infty$ such that

$$\tau(T) = \sum_{i=1}^{\infty} \langle Tf_i \mid f_i\rangle \quad \text{for every } T \in \mathcal{L}(G).$$

For $N \geq 1$, the function φ_N, defined by $\varphi_N(g) = \sum_{i=1}^{N} \langle \lambda(g) f_i \mid f_i\rangle$, belongs to $C^0(G)$ and we have

$$\begin{aligned} |\varphi_N(g) - \tau(\lambda(g))| &\leq \sum_{i=N+1}^{\infty} |\langle \lambda(g) f_i \mid f_i\rangle| \\ &\leq \sum_{i=N+1}^{\infty} \|f_i\|^2, \end{aligned}$$

for every $\in G$. This implies that $(\varphi_N)_{N\geq 1}$ converges to $\tau \circ \lambda_G$ uniformly on G and the claim follows. □

REMARK 15.C.2. A more precise version of Lemma 15.C.1 is in fact true: every normal positive linear functional on $\mathcal{L}(G)$ is a matrix coefficient of λ_G (see [**Eyma–64**, Théorème p. 218]).

The following result is proved in [**Dixm–C***, 13.10.5] under the additional assumption that G is unimodular.

THEOREM 15.C.3. *Let G be a locally compact group. The following properties are equivalent:*

(i) *G is a SIN group;*
(ii) *$\mathcal{L}(G)$ is a finite von Neumann algebra.*

PROOF. Assume that G is a SIN group. Let $(V_i)_{i\in I}$ be a basis of compact neighbourhoods of e which are invariant under conjugation. For every $i \in I$, set

$$f_i := \frac{\mathbf{1}_{V_i}}{\sqrt{\mu_G(V_i)}}.$$

The function of positive type φ_i on G associated with (λ_G, f_i) is conjugation invariant (see the proof of Theorem 15.B.3).

Let τ_i be the linear extension of φ_i to $\mathcal{L}(G)$; thus, τ_i is the normal positive linear functional on $\mathcal{L}(G)$ defined by

$$\tau_i(T) = \langle Tf_i \mid f_i\rangle \quad \text{for every } T \in \mathcal{L}(G).$$

Since φ_i is invariant under conjugation, τ_i is a trace on $\mathcal{L}(G)$.

Let $T \in \mathcal{L}(G)$. Assume that $\tau_i(T^*T) = 0$ for every $i \in I$. Since

$$\tau_i(T^*T) = \langle Tf_i \mid Tf_i\rangle = \|Tf_i\|^2,$$

it follows that $Tf_i = 0$ for every $i \in I$. Observe that, since $(V_i)_{i\in I}$ is a basis of neighbourhoods of e, the family $(f_i)_{i\in I}$ is an approximate identity for the $L^1(G)$-module $L^2(G)$, that is,

$$\lim_i \|f_i * f - f\|_2 = 0 \quad \text{for every } f \in L^2(G).$$

As T commutes with right convolutions by elements of $C_c(G)$, we therefore have, for every $f \in C^c(G)$,

$$Tf = \lim_i T(f_i * f) = \lim_i T(f_i) * f = 0.$$

Hence $T = 0$. This shows that $\mathcal{L}(G)$ is a finite von Neumann algebra.

Assume conversely that $\mathcal{L}(G)$ is a finite von Neumann algebra. Let $\mathcal{T}$ be the set of finite normal traces τ on $\mathcal{L}(G)$ normalized by $\tau(I) = 1$ and set

$$\Phi := \{\tau \circ \lambda_G \mid \tau \in \mathcal{T}\}.$$

Observe that every $\varphi \in \Phi$ is a conjugation-invariant function of positive type on G.

We claim that Φ separates the points of G. Indeed, let $x, y \in G$ with $x \neq y$. For

$$T := \lambda_G(x) - \lambda_G(y) \in \mathcal{L}(G),$$

we have $T \neq 0$ and hence there exists $\tau \in \mathcal{T}$ such that $\tau(T^*T) \neq 0$. It follows from Proposition 15.A.1 that $\tau \circ \lambda_G(x) \neq \tau \circ \lambda_G(y)$.

Fix $\varphi_0 \in \Phi$ and set

$$K_0 := \{g \in G \mid |\varphi_0(g) - 1| \leq 1/2\}.$$

Then K_0 is a closed neighbourhood of e and is conjugation invariant. Moreover, K_0 is compact, since $\varphi_0 \in C^0(G)$, by Lemma 15.C.1.

Consider the product space $\prod_{\varphi \in \Phi} \mathbf{C}$ equipped with the product topology. The map

$$\psi : K_0 \to \prod_{\varphi \in \Phi} \mathbf{C}, \quad g \mapsto \varphi(g)$$

is continuous. Moreover, ψ is injective, since Φ separates the points of G. It follows from the compactness of K_0 that the inverse map

$$\psi^{-1} : \psi(K_0) \to K_0$$

is continuous. This implies that the family of the sets $U_{F,\varepsilon}$, defined by

$$U_{F,\varepsilon} = \{g \in K_0 \mid |\varphi(g) - 1| \leq \varepsilon \text{ for all } \varphi \in F\}$$

for a finite subset F of Φ and $\varepsilon > 0$, is a fundamental family of neighbourhoods of e. Since every set $U_{F,\varepsilon}$ is conjugation invariant, G is a SIN group. □

15.D. Connected and totally disconnected LC groups

We apply Theorem 15.B.4 to two classes of groups, the connected LC groups and the totally disconnected LC groups. For this, we will need to use some structure theorems for SIN groups contained in either one of these classes. The first one is due to Iwasawa ([**Iwas–51**, Theorem 3]). For other conditions equivalent to (i) and (ii) below, see Example 4.C.7(2).

THEOREM 15.D.1 (**Iwasawa's Theorem**). *Let G be a connected locally compact group. The following properties are equivalent:*

(i) *G is a SIN group;*
(ii) *G is isomorphic, as a topological group, to a direct product $K \times \mathbf{R}^n$ for some compact group K and some $n \geq 0$.*

Observe that a connected LC group is generated by any compact neighbourhood of its unit element. The following result from [**KaSi–52**] is an immediate consequence of Theorem 15.B.3, Theorem 15.B.4, and Iwasawa's Theorem; the result was previously proved for semi-simple real Lie groups in [**SevN–50**].

THEOREM 15.D.2. *Let G be a connected LC group. The following properties are equivalent:*

(i) $\mathrm{Rep}_{\mathrm{FT}}(G)$ *separates the points of G;*
(ii) *G is isomorphic, as a topological group, to a direct product $K \times \mathbf{R}^n$ for some compact group K and some $n \geq 0$.*

Theorem 15.D.2 has the following consequence for the finite characters of a connected LC group.

COROLLARY 15.D.3. *Let G be a connected LC group.*
Every factor representation of finite type of G is a multiple of a finite-dimensional irreducible representation; equivalently, $E(G)$ consists of the characters associated with the finite-dimensional irreducible representations of G.

PROOF. Let π be a factor representation of finite type of G. Then $\pi \circ p$ is a faithful factor representation of finite type of $G/\ker \pi$, where $p : G \to G/\ker \pi$ is the canonical epimorphism. Since $G/\ker \pi$ is a connected LC group, it follows from Theorem 15.D.2 that $G/\ker \pi$ is isomorphic to a direct product $K \times \mathbf{R}^n$ for some compact group K and some $n \geq 0$. Every factor representation of $K \times \mathbf{R}^n$ is a multiple of a finite-dimensional irreducible representation. □

We now turn to totally disconnected groups. Here, we will use van Dantzig's Theorem; for a proof, see [**HeRo–63**, Theorem 7.7] or [**CoHa–16**, Theorem 2.E.6].

THEOREM 15.D.4 (**van Dantzig's Theorem**). *Let G be a totally disconnected LC group.*
The compact open subgroups of G form a basis of neighbourhoods of the group unit e.

The following result was suggested to us by P.-E. Caprace.

THEOREM 15.D.5. *Let G be a* compactly generated *totally disconnected LC group. The following properties are equivalent:*

(i) $\mathrm{Rep}_{\mathrm{FT}}(G)$ *separates the points of G;*
(ii) *G is a SIN group;*
(iii) *the compact open* normal *subgroups of G form a basis of neighbourhoods of the group unit e;*
(iv) *G is the projective limit of discrete groups.*

PROOF. The fact that (i) and (ii) are equivalent follows from Theorem 15.B.4 and Theorem 15.B.3.

Assume that G is a SIN group and let U be a conjugation invariant neighbourhood of e. By van Dantzig's Theorem, U contains a compact open subgroup H of G. In turn, H contains a conjugation-invariant neighbourhood V of e, since G is a SIN group. Let

$$N := \bigcap_{g \in G} gHg^{-1}.$$

Then N is a compact normal subgroup of G and we have $V \subset N \subset U$. So, N is a neighbourhood of e contained in U. This proves that (ii) implies (iii). It is obvious that (iii) implies (ii).

The fact that (iii) and (iv) are equivalent follows from the definition of a projective limit of LC groups (see [**BTG1–4**, Chap. III, § 7]). □

REMARK 15.D.6. In Theorem 15.D.5, it is necessary to assume that G is compactly generated; indeed, the group G from Example 15.B.6 is a totally disconnected group which is a MAP group but not a SIN group.

The following corollary of Theorem 15.D.5 reduces the determination of the Thoma dual of a compactly generated totally disconnected LC group to that of its discrete quotient groups.

COROLLARY 15.D.7. *Let G be a compactly generated totally disconnected locally compact group.*

Then

$$E(G) = \bigcup_{N \in \mathcal{N}} E(G/N),$$

where $\mathcal{N}$ is the family of open normal subgroups of G.

In particular, if G has no proper open normal subgroup, then $E(G) = \{1_G\}$, that is, every representation of finite type of G is a multiple of the trivial representation 1_G.

PROOF. Let π be a factor representation of finite type of G. Then $\overline{G} := G/\ker \pi$ is a compactly generated totally disconnected LC group and $\pi = \overline{\pi} \circ p$ for a faithful factor representation of finite type $\overline{\pi}$ of $\overline{G}$, where $p : G \twoheadrightarrow \overline{G}$ is the canonical epimorphism.

It follows from Theorem 15.D.5 that $\overline{G}$ is a projective limit of discrete groups. Lemma 8.D.2 implies that $\overline{\pi}$ is trivial on an open normal subgroup $\overline{N}$ of $\overline{G}$. Then π is trivial on the open normal subgroup $p^{-1}(\overline{N})$ of G and the claim is proved. □

Corollary 15.D.7 shows that $E(G) = \{1_G\}$ for every compactly generated totally disconnected LC group G which is topologically simple; an example is the group $G(\mathbf{K})^+/Z$, where $G(\mathbf{K})^+$ is the subgroup of $G(\mathbf{K})$ generated by the unipotent radicals of $\mathbf{K}$-parabolic subgroups of a simple algebraic group G over a non-Archimedean local field $\mathbf{K}$ and Z the centre of $G(\mathbf{K})^+$. For other examples which include appropriate groups of tree automorphisms and some Kac–Moody groups, we refer to Appendix A in [**CaRW–17**].

There are compactly generated totally disconnected LC groups G which are not topologically simple and for which $E(G) = \{1_G\}$. Examples of such groups are $G = \mathrm{SL}_2(\mathbf{K})$ or $G = \mathrm{SL}_2(\mathbf{K}) \ltimes \mathbf{K}^2$ for a non-Archimedean local field $\mathbf{K}$; indeed, in both cases, G has no proper open normal subgroup.

15.E. Faithful traces on group C*-algebras

Let Γ be a discrete group. Observe that, since Γ is a SIN group, $\mathrm{Rep}_{\mathrm{FT}}(\Gamma)$ separates the points of Γ (Theorem 15.B.3). In fact, the regular representation λ_Γ is a representation of finite type (Proposition 7.A.3) and is obviously separating.

Recall from Section 8.B that every representation of Γ has a unique extension to a representation of the maximal C*-algebra $C^*_{\max}(\Gamma)$ of Γ and that every representation of $C^*_{\max}(\Gamma)$ is an extension of a representation of Γ. Separating families of representations for groups have been introduced in Definition 1.B.18. A similar notion can be defined for involutive algebras, and provides for group algebras a stronger condition:

DEFINITION 15.E.1. A family $\mathcal{F}$ of representations of an involutive algebra A is **separating** if, for every $x \in A$, $x \neq 0$, there exists $\pi \in \mathcal{F}$ such that $\pi(x) \neq 0$.

REMARK 15.E.2. Let A be a C*-algebra. For a family $\mathcal{F}$ of representations of A, the following conditions are equivalent:

(i) $\mathcal{F}$ is separating, i.e., for every $x \neq 0$ in A, there exists $\pi \in \mathcal{F}$ such that $\pi(x) \neq 0$;
(ii) the intersection $\bigcap_{\pi \in \mathcal{F}} \ker(\pi)$ is $\{0\}$;
(iii) the representation $\bigoplus_{\pi \in \mathcal{F}} \pi$ is faithful;
(iv) for every $x \in A$, we have $\|x\| = \sup_{\pi \in \mathcal{F}} \|\pi(x)\|$.

The equivalence of (i), (ii), and (iii), is obvious; indeed, this holds for any involutive algebra, not only for a C*-algebra. The equivalence of (iii) and (iv) follows from the fact that any faithful representation ρ of a C*-algebra A is norm preserving, i.e., is such that $\|\rho(x)\| = \|x\|$ for all $x \in A$ [**Dixm–C***, 1.3.7, 1.8.1].

REMARK 15.E.3. Let G be a locally compact group and $C^*_{\max}(G)$ its maximal C*-algebra.

(1) Let $\mathcal{F}$ be a family of representations of G, equally viewed as representations of $C^*_{\max}(G)$. If $\mathcal{F}$ is separating for $C^*_{\max}(G)$, then $\mathcal{F}$ is separating for G.

To show this, we suppose that $\mathcal{F}$ does not separate G, and we show below that $\mathcal{F}$ does not separate $C^*_{\max}(G)$.

Let $g \in G$ be such that $g \neq e$ and $\pi(g) = \pi(e)$ for all $\pi \in \mathcal{F}$. Let μ_G be a left Haar measure on G. For $\varphi \in C^c(G)$, denote by $\lambda_g \varphi$ the translated function defined by $\lambda_g \varphi(x) = \varphi(g^{-1}x)$ for all $x \in G$. For $\pi \in \mathcal{F}$, we have

$$\pi(\lambda_g \varphi) = \int_G \varphi(g^{-1}x)\pi(x)\, d\mu_G(x) = \int_G \varphi(y)\pi(gy)\, d\mu_G(y) = \pi(\varphi)$$

and therefore $\pi(\lambda_g \varphi - \varphi) = 0$. Let U be a relatively compact neighbourhood of e in G such that $U \cap gU = \emptyset$. There exists $\varphi : G \to \mathbf{R}_+$ in $C^c(G)$ supported in U and not zero. Since the support gU of $\lambda_g \varphi$ is disjoint from U, the function $\psi = \lambda_g \varphi - \varphi$ is not zero, and $\pi(\psi) = 0$ for all $\pi \in \mathcal{F}$. Hence $\mathcal{F}$ does not separate $C^c(G)$, and all the more separates neither $L^1(G, \mu_G)$ nor its completion $C^*_{\max}(G)$.

(2) A set $\mathcal{F}$ of representations of G separating G need not separate $C^*_{\max}(G)$.

Consider for example the cyclic group $G := \{1, \varepsilon\}$ of order 2 and the one-element family $\mathcal{F} = \{\chi\}$ where $\chi : G \to \{\pm 1\}$ is the nontrivial unitary character. Then $\mathcal{F}$ separates G. However, for any nonzero constant function $\varphi : G \to \mathbf{C}$ (which is an element of $C^*_{\max}(G)$), we have $\chi(\varphi) = \varphi(1) - \varphi(\varepsilon) = 0$.

In this section we will deal with the question of whether the family of finite type representations of Γ separates the points of $C^*_{\max}(\Gamma)$. This question may be rephrased as follows (Corollary 15.E.7):

QUESTION 15.E.4. Does $C^*_{\max}(\Gamma)$ admit a faithful finite trace?

The question above is of interest. For instance, it was shown in [**Kirc–93**, Proposition 8.1] that the existence of such a trace for $\Gamma = \mathrm{SL}_2(\mathbf{Z}) \times \mathrm{SL}_2(\mathbf{Z})$ is equivalent to a positive solution of the Connes embedding problem. This problem is an outstanding open question in operator algebra; its original formulation [**Conn–76**, Page 105] asks whether every factor of type II_1 acting on a separable Hilbert space can be embedded in an ultrapower R^ω of the hyperfinite factor R of type II_1. Solving the Connes embedding problem is equivalent to solving a variety of other important problems in operator algebras, most notably by the work of Kirchberg; see also [**Ozaw–13**].

We first observe that the existence of a faithful trace on the maximal C*-algebra of a group is inherited by each of its subgroups.

PROPOSITION 15.E.5. *Let Γ be a discrete group and H a subgroup. If $C^*_{\max}(\Gamma)$ admits a faithful finite trace, then so does $C^*_{\max}(H)$.*

PROOF. As Γ is discrete, the canonical map $\mathbf{C}[H] \hookrightarrow \mathbf{C}[\Gamma]$ extends to an injective homomorphism $C^*_{\max}(H) \hookrightarrow C^*_{\max}(\Gamma)$, by Theorem 8.C.12. The claim is now clear. □

Recall that $\mathrm{Tr}(\Gamma)$ is the space of traces on Γ, that is, the space of functions on Γ which are central of positive type. It was shown in Section 11.A that every trace on Γ extends to a unique finite trace of the maximal C*-algebra $C^*_{\max}(\Gamma)$ of Γ and that, conversely, every finite trace of $C^*_{\max}(\Gamma)$ defines by restriction a trace on Γ.

We relate the faithfulness of a trace on $C^*_{\max}(\Gamma)$ to the faithfulness of the associated GNS representation of $C^*_{\max}(\Gamma)$ as in Construction 8.A.8.

LEMMA 15.E.6. *Let $\varphi \in \mathrm{Tr}_1(\Gamma)$ and denote by t_φ its unique extension to a trace on $C^*_{\max}(\Gamma)$. Let π_φ be the GNS representation of G associated to φ.*
Then

$$\mathrm{C^*ker}(\pi_\varphi) \; = \; \{x \in C^*_{\max}(\Gamma) \mid t_\varphi(x^*x) = 0\}.$$

As a consequence, the following properties are equivalent:

(i) $\mathrm{C^*ker}(\pi_\varphi) = \{0\}$;
(ii) t_φ *is faithful.*

PROOF. Recall how the GNS representation $(\pi, \mathcal{H}, \xi)$ of $C^*_{\max}(\Gamma)$ associated to t_φ is defined (see Construction 8.A.8).

The set $J = \{x \in C^*_{\max}(\Gamma) \mid t_\varphi(x^*x) = 0\}$ is a closed left ideal of $C^*_{\max}(\Gamma)$, because t_φ is a positive functional, and is self-adjoint, because t_φ is a trace, hence it is a two-sided ideal. Let $\mathcal{H}$ be the Hilbert space completion of $C^*_{\max}(\Gamma)/J$ for the scalar product on $C^*_{\max}(\Gamma)/J$ defined by

$$\langle x + J \mid y + J \rangle \; = t_\varphi(y^*x) \qquad \text{for all } \; x, y \in C^*_{\max}(\Gamma).$$

The GNS representation π of $C^*_{\max}(\Gamma)$ on $\mathcal{H}$ is defined on $C^*_{\max}(\Gamma)/J$ by

$$\pi(x)(y + J) \; = \; xy + J \qquad \text{for all } \; x, y \in C^*_{\max}(\Gamma)$$

and has $\xi = 1 + J$ as a cyclic vector. One checks that the restriction of π to Γ is equivalent to the GNS representation π_φ of Γ associated to φ as in Construction 1.B.6.

Let $x \in C^*_{\max}(\Gamma)$. Then $x \in \ker \pi = \mathrm{C^*ker}(\pi_\varphi)$ if and only if $\pi(x)(y + J) = xy + J = J$ for every $y \in C^*_{\max}(\Gamma)$, that is, if and only if $x \in J$. The claim follows, since (by definition) the trace t_φ is faithful if and only if $J = 0$. □

We relate the existence of a family of finite type representations of Γ which separates the points of $C^*_{\max}(\Gamma)$ to the existence of a faithful trace of $C^*_{\max}(\Gamma)$, for a countable group Γ.

COROLLARY 15.E.7. *Let Γ be a countable group. The following properties are equivalent:*

(i) *The family* $\mathrm{Rep}_{\mathrm{FT}}(\Gamma)$ *of representations of finite type separates the points of $C^*_{\max}(\Gamma)$;*
(ii) *there exists a faithful finite trace on $C^*_{\max}(\Gamma)$.*

PROOF. Let $\mathcal{T}_1$ be the set of finite traces t on $C^*_{\max}(\Gamma)$ normalized by $\|t\| = t(1) = 1$.

Set $A := C^*_{\max}(\Gamma)$. Since A is separable, the unit ball $A^*_{\leq 1}$ of the topological dual A^* of A is a compact metrizable space for the weak*-topology. Since $\mathcal{T}_1 \subset A^*_{\leq 1}$, it follows that $\mathcal{T}$ is separable for the weak*-topology.

Let $(t^{(n)})_{n\geq 1}$ be a dense sequence in $\mathcal{T}_1$. Then

$$t := \sum_{n\geq 1} \frac{1}{2^n} t^{(n)}$$

is a trace on $A = C^*_{\max}(\Gamma)$ with $t(1) = 1$, that is, $t \in \mathcal{T}_1$.

We claim that $\mathrm{Rep}_{\mathrm{FT}}(\Gamma)$ separates the points of $C^*_{\max}(\Gamma)$ if and only if t is faithful.

Indeed, assume that t is faithful. Then the GNS representation π of $C^*_{\max}(\Gamma)$ associated to t, which belongs to $\mathrm{Rep}_{\mathrm{FT}}(\Gamma)$, is faithful by Lemma 15.E.6.

Conversely, assume that $\mathrm{Rep}_{\mathrm{FT}}(\Gamma)$ separates the points of $C^*_{\max}(\Gamma)$. Let $x \in C^*_{\max}(\Gamma)$, $x \neq 0$. It follows from Lemma 15.E.6 that there exists $t_x \in \mathcal{T}_1$ such that $t_x(x^*x) \neq 0$. Since $(t^{(n)})_{n\geq 1}$ is weak*-dense in $\mathcal{T}_1$, we have $t^{(n)}(x^*x) \neq 0$ for some $n \geq 1$. It follows that $t(x^*x) \neq 0$ and this shows that t is faithful. □

15.E.a. The case of amenable groups. We next show that Question 15.E.4 has a positive answer when Γ is an amenable group. Recall first that, for every discrete group Γ, the Dirac function δ_e is a trace on Γ.

PROPOSITION 15.E.8. *Let Γ be a discrete group. The following properties are equivalent:*

(i) *the unique extension t_Γ of δ_e to a trace on $C^*_{\max}(\Gamma)$ is faithful;*
(ii) *Γ is amenable.*

PROOF. The GNS representation corresponding to δ_e is the regular representation λ_Γ of Γ. Therefore, by Lemma 15.E.6, t_Γ is faithful if and only if $\mathrm{C^*ker}(\lambda_\Gamma) = \{0\}$. This happens if and only if Γ is amenable (see Theorem 1.C.3). □

15.E.b. C*-maximally almost periodic groups. The situation with regard to Question 15.E.4 is less understood in the case of nonamenable groups. We introduce a class of groups for which the answer is positive.

DEFINITION 15.E.9. A discrete group Γ is **C*-maximally almost periodic** group, or a **C*-MAP** group, if the finite-dimensional representations of Γ separate the points of $C^*_{\max}(\Gamma)$.

REMARK 15.E.10. Let A be a C*-algebra. It is common (see [**BrOz–08**, Definition 7.1.6]) to say that A is **residually finite-dimensional**, if the family of the finite-dimensional representations of A separates the points of A. With this terminology, a group Γ is a C*-MAP group if $C^*_{\max}(\Gamma)$ is a residually finite-dimensional C*-algebra.

We first give a reformulation of the C*-MAP property in terms of Fell's topology.

PROPOSITION 15.E.11. *Let Γ be a discrete group. Let $\widehat{\Gamma}_{\rm fd}$ be as in Section 4.C the set of equivalence classes of irreducible finite-dimensional representations of Γ. The following properties are equivalent:*

(i) *Γ is a C*-MAP group;*
(ii) *$\widehat{\Gamma}_{\rm fd}$ is dense in $\widehat{\Gamma}$ for the Fell topology;*
(iii) *for every $x \in C^*_{\max}(\Gamma)$, we have*

$$\|x\| = \sup_{\pi \in \widehat{\Gamma}_{\rm fd}} \|\pi(x)\|.$$

PROOF. Let π be a finite-dimensional representation of Γ; then we can decompose π as a direct sum $\pi = \bigoplus_{i=1}^n \pi_i$ of irreducible representations, which are of course finite-dimensional. We have

$$\|\pi(x)\| = \sup_{i=1,\dots,n} \|\pi_i(x)\| \qquad \text{for all } x \in C^*_{\max}(\Gamma).$$

This shows that (i) and (iii) are equivalent (see Condition (iv) in Remark 15.E.2).

Observe that (ii) holds if and only if

$$\bigcap_{\pi \in \widehat{\Gamma}_{\rm fd}} \mathrm{C^*ker}(\pi) = \bigcap_{\pi \in \widehat{\Gamma}} \mathrm{C^*ker}(\pi),$$

that is, if and only if

$$\bigcap_{\pi \in \widehat{\Gamma}_{\rm fd}} \mathrm{C^*ker}(\pi) = \{0\}.$$

This shows that (i) and (ii) are equivalent. □

Next, we observe that a C*-MAP group is a MAP group (definition in Section 4.C) and that its maximal C*-algebra has a faithful trace.

PROPOSITION 15.E.12. *Let Γ be a C*-MAP group. Then*

(1) *Γ is a MAP group;*
(2) *when Γ is countable, $C^*_{\max}(\Gamma)$ has a faithful finite trace.*

PROOF. (1) For $\gamma \in \Gamma \smallsetminus \{e\}$, we have $\delta_\gamma - \delta_e \in C^*_{\max}(\Gamma) \smallsetminus \{0\}$, hence, there exists a finite-dimensional representation $(\pi, \mathcal{H})$ of Γ such that $\delta_\gamma - \delta_e \notin \mathrm{C^*ker}(\pi)$, that is, $\pi(\gamma) \neq \mathrm{Id}_{\mathcal{H}}$.

(2) is an immediate consequence of Corollary 15.E.7. □

As we now show, the class of amenable MAP groups coincides with the class of amenable C*-MAP groups; the proof is based on the following result of independent interest.

PROPOSITION 15.E.13. *Let Γ be a MAP group. The regular representation λ_Γ is weakly contained in $\bigoplus_{\pi \in \widehat{\Gamma}_{\rm fd}} \pi$.*

PROOF. It suffices to show that the Dirac function δ_e can be pointwise approximated by functions of positive type associated to finite-dimensional representations of Γ.

Let $\gamma_1, \ldots, \gamma_n \in \Gamma \smallsetminus \{e\}$. Fix $i \in \{1, \ldots, n\}$. Since Γ is a MAP group, there exists a finite-dimensional representation $(\pi_i, \mathcal{H}_i)$ with $\pi_i(\gamma_i) \neq \mathrm{Id}_{\mathcal{H}_i}$. Let $\xi_i \in \mathcal{H}_i$ be a unit vector such that $\pi_i(\gamma_i)\xi_i \neq \xi_i$ and let φ_i be the function of positive type associated to (π_i, ξ_i).

Define a normalized function of positive type $\varphi : \Gamma \to \mathbf{C}$ by

$$\varphi(\gamma) = \frac{1}{n+1}\left(1 + \sum_{i=1}^{n} \varphi_i(\gamma)\right) \qquad \text{for all } \gamma \in \Gamma.$$

Observe that $|\varphi_i(\gamma_j)| \leq 1$ and $\varphi_i(\gamma_i) \neq 1$ for all $i, j \in \{1, \ldots, n\}$. It follows that $|\varphi(\gamma_i)| < 1$ and hence

$$\lim_k \varphi^k(\gamma_i) = 0 \qquad \text{for all } i \in \{1, \ldots, n\}.$$

Since φ^k is a *normalized* function of positive type associated to a finite-dimensional representation of Γ, the claim is proved. □

The following corollary is a direct consequence of Proposition 15.E.13, Theorem 1.C.3, and Proposition 15.E.11.

COROLLARY 15.E.14. *Every amenable MAP group is C*-MAP.*

REMARK 15.E.15. There are groups which are not C*-MAP but which have a maximal C*-algebra admitting a faithful finite trace. Indeed, if Γ is an amenable group which is not MAP, $C^*_{\max}(\Gamma)$ has a faithful finite trace (Proposition 15.E.8) and Γ is not C*-MAP. An example of such a group is $\mathrm{SL}_n(\mathbf{K})$, where $n \geq 2$ and $\mathbf{K}$ is an infinite algebraic extension of a finite field.

The next result will provide us with the first example of a nonamenable group Γ for which $C^*_{\max}(\Gamma)$ admits a faithful finite trace. For the proof, we need the following elementary fact from operator theory.

LEMMA 15.E.16. *Let $\mathcal{H}$ be a Hilbert space and $T \in \mathcal{L}(\mathcal{H})$ such that $\|T\| \leq 1$. The operator U on $\mathcal{H} \oplus \mathcal{H}$ defined by the matrix*

$$\begin{pmatrix} T & (I - TT^*)^{1/2} \\ (I - T^*T)^{1/2} & -T^* \end{pmatrix}$$

is unitary.

PROOF. One computes that $UU^* = U^*U = \begin{pmatrix} I & 0 \\ 0 & I \end{pmatrix}$. □

The next theorem is from [**Choi–80**].

THEOREM 15.E.17 (**Choi**). *Let $\Gamma = F_2$ be the free group on 2 generators. Then Γ is a C*-MAP group; in particular, $C^*_{\max}(\Gamma)$ admits a faithful finite trace.*

PROOF. Let $\pi : C^*_{\max}(\Gamma) \to \mathcal{L}(\mathcal{H})$ be a faithful representation of $C^*_{\max}(\Gamma)$ on a separable Hilbert space $\mathcal{H}$ (one may take, for instance, $\pi = \bigoplus_n \pi_n$ for a dense sequence $(\pi_n)_n$ in the second countable topological space $\widehat{\Gamma}$). Set $A = \pi(a)$ and $B = \pi(b)$, where a and b are free generators of Γ.

Let $(P_n)_{n\geq 1}$ be a sequence of orthogonal projections in $\mathcal{L}(\mathcal{H})$ such that $\dim P_n(\mathcal{H}) = n$ for every $n \geq 1$ and $\lim_n P_n = I$ in the strong operator topology. Set

$$S_n := P_n A P_n \quad \text{and} \quad T_n := P_n B P_n.$$

Then S_n, T_n may be viewed as operators on $\mathcal{H}_n := P_n(\mathcal{H})$ and we have $\|S_n\|, \|T_n\| \leq 1$. Define $A_n, B_n \in \mathcal{L}(\mathcal{H}_n \oplus \mathcal{H}_n)$ by

$$\begin{aligned} A_n &= \begin{pmatrix} S_n & (I - S_n S_n^*)^{1/2} \\ (I - S_n^* S_n)^{1/2} & -S_n^* \end{pmatrix}, \\ B_n &= \begin{pmatrix} T_n & (I - T_n T_n^*)^{1/2} \\ (I - T_n^* T_n)^{1/2} & -T_n^* \end{pmatrix}. \end{aligned}$$

By Lemma 15.E.16, A_n and B_n are unitary operators on $\mathcal{H}_n \oplus \mathcal{H}_n$. So, a finite-dimensional representation $\pi_n : \Gamma \to \mathcal{L}(\mathcal{H}_n \oplus \mathcal{H}_n)$ of Γ is defined by setting

$$\pi_n(a) := A_n \quad \text{and} \quad \pi_n(b) := B_n.$$

We claim that the sequence $(\pi_n)_{n\geq 1}$ separates the points of $C^*_{\max}(\Gamma)$.

Indeed, observe that, viewing A_n and B_n as operators on $\mathcal{H} \oplus \mathcal{H}$ by defining them to be 0 on $(\mathcal{H}_n \oplus \mathcal{H}_n)^\perp$, we have

$$\lim_{n\to+\infty} A_n = \begin{pmatrix} A & 0 \\ 0 & -A^* \end{pmatrix} \quad \text{and} \quad \lim_{n\to+\infty} B_n = \begin{pmatrix} B & 0 \\ 0 & -B^* \end{pmatrix}$$

and likewise

$$\lim_{n\to+\infty} A_n^* = \begin{pmatrix} A^* & 0 \\ 0 & -A \end{pmatrix} \quad \text{and} \quad \lim_{n\to+\infty} B_n^* = \begin{pmatrix} B^* 0 & \\ 0 & -B \end{pmatrix}$$

in the strong operator topology. Let $x \in \mathbf{C}[\Gamma]$; then x can be written as $x = P(a, a^{-1}, b, b^{-1})$ for a noncommutative polynomial P in 4 variables. Therefore, we have

$$\begin{aligned} \lim_{n\to+\infty} \pi_n(x) &= \lim_{n\to+\infty} P(A_n, A_n^*, B_n, B_n^*) \\ &= \begin{pmatrix} P(A, A^*, B, B^*) & 0 \\ 0 & P(-A^*, -A, -B^*, -B) \end{pmatrix} \end{aligned}$$

in the strong operator topology. It follows that

$$\left\| \begin{pmatrix} P(A, A^*, B, B^*) & 0 \\ 0 & P(-A^*, -A, -B^*, -B) \end{pmatrix} \right\| \leq \liminf_{n\to+\infty} \|\pi_n(x)\|.$$

Since $\|\pi(x)\| = \|P(A, A^*, B, B^*)\|$, we have

$$\|\pi(x)\| \leq \left\| \begin{pmatrix} P(A, A^*, B, B^*) & 0 \\ 0 & P(-A^*, -A, -B^*, -B) \end{pmatrix} \right\|.$$

We deduce that

$$\|\pi(x)\| \leq \liminf_{n\to+\infty} \|\pi_n(x)\|$$

and hence $\|\pi(x)\| = \sup_{n\geq 1} \|\pi_n(x)\|$ for every $x \in \mathbf{C}[\Gamma]$.

Therefore, the representation

$$C^*_{\max}(\Gamma) \to \mathcal{L}\Big(\bigoplus_n (\mathcal{H}_n \oplus \mathcal{H}_n)\Big), \qquad x \mapsto \bigoplus_{n\geq 1} \pi_n(x)$$

is isometric on the dense subspace $\mathbf{C}[\Gamma]$ and hence isometric on $C^*_{\max}(\Gamma)$. $\square$

More generally, for any extended integer $r \in \{1, 2, \ldots, \infty\}$, the free group F_r is C*-MAP. Indeed, since F_r is a subgroup of F_2, the maximal C*-algebra $C^*_{\max}(F_r)$ is a sub-C*-algebra of $C^*_{\max}(F_2)$, see Corollary 8.C.10, and sub-C*-algebras of residually finite-dimensional C*-algebras are obviously residually finite-dimensional. Note also that the modular group $\mathrm{SL}_2(\mathbf{Z})$ is C*-MAP. Indeed, it contains a free subgroup F_2 of finite index, and it is easy to show that a group is C*-MAP if it has a C*-MAP subgroup of finite index (details in [**BeLo–00**, Lemma 4.6]). It is known that the fundamental group of a closed surface is C*-MAP; see Remark 15.E.21 below.

15.E.c. C*-residually finite groups. We introduce a property stronger than the C*-MAP property from the previous subsection and show that it is shared by free groups.

DEFINITION 15.E.18. We say that a discrete group Γ is **C*-residually finite** if the representations of Γ with finite images separate the points of $C^*_{\max}(\Gamma)$.

We want to show that free groups are C*-residually finite; this strengthening of Theorem 15.E.17 is Theorem 2.2 in [**LuSh–04**], where C*-residually finite groups are called groups with property FD.

We will give a variation of the proof of [**LuSh–04**] using the fact that every bijective measure-preserving transformation of the interval $[0, 1[$ can be approximated by dyadic permutations [**Halm–60**, p.65]; the crucial tool we will use is Proposition 15.E.19, which is based on [**AlEd–79**, Theorem 2.1].

Let $n \geq 1$ be an integer. A *dyadic interval* of rank n is an interval of the form $[i/2^n, (i+1)/2^n[$ for some $i \in \{0, \ldots, 2^n - 1\}$. The *dyadic decomposition* of rank n of $[0, 1[$ consists of all dyadic intervals of length 2^{-n}. A *dyadic permutation* of rank n is a bijection from $[0, 1[$ to $[0, 1[$ which maps every dyadic interval of rank n into a dyadic interval of rank n by an ordinary translation.

Observe that every dyadic permutation preserves the Lebesgue measure on $[0, 1[$ and that the set $\mathcal{P}_n$ of dyadic permutations of rank n is a finite group isomorphic to the symmetric group on 2^n letters. More precisely, every $P \in \mathcal{P}_n$ is uniquely determined by a permutation σ of $\{1, \ldots, 2^n\}$ for which $P(I_i) = I_{\sigma(i)}$, where $I_i = [i/2^n, (i+1)/2^n[$.

PROPOSITION 15.E.19. *Let S be a finite set of measurable Borel automorphisms of $[0, 1[$ preserving the Lebesgue measure μ.*

For given $\delta > 0$ and $\varepsilon > 0$, there exist an integer $n \geq 1$ and a set $\{P^s \mid s \in S\}$ of dyadic permutations of rank n such that

$$\mu\left(\{x \in [0, 1[\mid |s(x) - P^s(x)| \geq \delta\}\right) < \varepsilon \quad \textit{for every } s \in S.$$

PROOF. Let $I_1, \ldots, I_{2^m}$ be a dyadic decomposition of $[0, 1[$ such that $\mu(I_i) = \dfrac{1}{2^m} < \delta$. For every $s \in S$ and every $i \in \{1, \ldots, 2^m\}$, we can choose a compact subset C_i^s of $s^{-1}(I_i)$ such that

$$\mu(C_i^s) \geq (1 - \varepsilon)\mu(s^{-1}(I_i)) = \frac{1 - \varepsilon}{2^m},$$

by regularity of μ. Let

$$c := \inf_{s \in S} \inf_{1 \leq i \neq j \leq m} d(C_i^s, C_j^s),$$

where $d(A, B) = \inf\{|x - y| \mid x \in A, y \in B\}$ for subsets A, B of $[0, 1[$. Since the C_i^s's are compact and since $C_i^s \cap C_j^s = \emptyset$ for $i \neq j$, we have $c > 0$.

Choose another dyadic decomposition $J_1, \dots, J_{2^n}$ of $[0,1[$ such that $\mu(J_j) < c$. For every $j \in \{1, \dots, 2^n\}$ and $s \in S$, we then have $J_j \cap C_i^s \neq \emptyset$ for at most one $i \in \{1, \dots, 2^m\}$.

For $i \in \{1, \dots, 2^m\}$, denote by τ_i^s the set of $j \in \{1, \dots, 2^n\}$ such that $J_j \cap C_i^s \neq \emptyset$. Since $C_i^s \subset s^{-1}(I_i)$, it is clear that

$$\mu\Big(\bigcup_{j\in\tau_i^s} J_j\Big) \leq \mu(s^{-1}(I_i)) = 1/2^m.$$

Fix $s \in S$. We are going to define a dyadic permutation P^s of rank n such that $P^s(J_j) \subset I_i$ for every $i \in \{1, \dots, 2^m\}$ and every $j \in \tau_i^s$.

For $i \in \{1, \dots, 2^m\}$, denote by τ_i the set of $j \in \{1, \dots, 2^n\}$ such that $(J_j)_{j\in\tau_i}$ are the dyadic intervals of length $1/2^n$ contained in I_i. Since $\mu(\bigcup_{j\in\tau_i^s} J_j) \leq 1/2^m = \mu(I_i)$, we can find an injective map $\sigma^s : \tau_i^s \to \tau_i$. As $\tau_i^s \cap \tau_{i'}^s = \emptyset$ for $i \neq i'$, we can extend σ^s to a permutation of $\{1, \dots, 2^n\}$. Let P^s be the associated dyadic permutation.

Set

$$A^s := \bigcup_{i=1}^{2^m} C_i^s = \bigcup_{i=1}^{2^m} \bigcup_{j\in\tau_i^s} (J_j \cap C_i^s).$$

Since the C_i^s are pairwise disjoint, we have

$$\mu(A^s) = \sum_{i=1}^{2^m} \mu(C_i^s) \geq \sum_{i=1}^{2^m} \frac{1-\varepsilon}{2^m} = 1 - \varepsilon.$$

Let $x \in A^s$. Then $x \in J_j \cap C_i^s$ for some $i \in \{1, \dots, 2^m\}$ and some $j \in \tau_i^s$. On the one hand, since $x \in C_i^s \subset s^{-1}(I_i)$, we have $s(x) \in I_i$. On the other hand, since $x \in J_j$ and $j \in \tau_i^s$, we also have $P^s(x) \in I_i$. It follows that

$$|s(x) - P^s(x)| \leq \frac{1}{2^m} < \delta$$

for all $x \in A^s$. □

We are now ready to prove that free groups are C*-residually finite.

THEOREM 15.E.20 (**Lubotzky–Shalom**). *Let $r \geq 1$ be an integer. The free group F_r on r generators is C*-residually finite.*

PROOF. Let π be a representation of $\Gamma = F_r$ in a separable Hilbert space. We have to show that π is weakly contained in a direct sum of representations of Γ with finite images (see Condition (ii) in Remark 15.E.2).

• *First step.* We may assume that π is the Koopman representation π_X on $L^2(X, \mu)$ associated with a measure-preserving action of Γ on a standard Borel space X, equipped with a probability measure μ.

Indeed, it is well known (see [**BeHV–08**, A.7.15] and [**Zimm–84**, 5.2.13]) that there exists a measure-preserving action of Γ on a standard probability space (X, μ) such that π is equivalent to a subrepresentation of the Koopman representation of Γ on $L^2(X, \mu)$. The claim follows immediately.

• *Second step.* Let $\Gamma \curvearrowright (X, \mu)$ be a measure-preserving action of Γ on a standard Borel space X equipped with a probability measure μ. We claim that the Koopman representation π_X is weakly contained in a direct sum of representations of Γ with finite images.

Assume first that the set $A \subset X$ of atoms of μ is not empty. Since μ is a probability measure, the subset A_α of atoms of μ-measure α is finite for every $\alpha > 0$ and A is at most countable. It follows that there exists a sequence $(\alpha_i)_{i\geq 1}$ of positive numbers such that $A = \bigcup_{i\geq 1} A_{\alpha_i}$. Every subset A_{α_i} is Γ-invariant. We therefore have a decomposition

$$\pi_X = \Big(\bigoplus_{i\geq 1} \pi_{A_{\alpha_i}}\Big) \oplus \pi_{X\smallsetminus A}.$$

Since Γ acts on A_{α_i} by permutations, the representation $\pi_{A_{\alpha_i}}$ factorizes by a finite quotient of Γ, for every $i \geq 1$. It follows that $\pi_A = \bigoplus_{i\geq 1} \pi_{A_{\alpha_i}}$ is a direct sum of representations of Γ with finite images.

It remains to show that $\pi_{X\smallsetminus A}$ is weakly contained in a direct sum of representations of Γ with finite images.

Since $X \smallsetminus A$ is a standard Borel space and μ has no atoms in $X \smallsetminus A$, we may assume that $X \smallsetminus A = [0,1]$ and μ is the Lebesgue measure (see Theorem A.3.8).

Let S be a finite set such that Γ is the free group over S. Let $f_1, \ldots, f_k$ be continuous functions on $[0,1]$ and let $\varepsilon > 0$. Let $\delta > 0$ be such that $|f_i(x) - f_i(y)| < \varepsilon$ for all $x, y \in [0,1]$ such that $|x-y| < \delta$ and all $i \in \{1,\ldots,k\}$.

By Proposition 15.E.19, there exists a set $\{P_s \mid s \in S\}$ of dyadic permutations of the *same* rank such that

$$\mu\left(\{x \in [0,1] \mid |s(x) - P^s(x)| \geq \delta\}\right) < \varepsilon \qquad \text{for every } \ s \in S.$$

The set $(P_s)_{s\in S}$ defines a measure-preserving action of the free group Γ on $([0,1],\mu)$, which factorizes by a finite quotient of Γ. Denote by ρ the Koopman representation on $L^2([0,1],\mu)$ associated with this action.

Now writing π for $\pi_{X\smallsetminus A}$, we have, for every $s \in S$ and every $i \in \{1,\ldots,k\}$,

$$\begin{aligned}
\|\pi(s^{-1})f_i - \rho(s^{-1})f_i\|_2^2 &= \int_0^1 |f_i(s(x)) - f_i(P^s(x))|^2 \, d\mu(x) \\
&= \int_{\{x \,:\, |s(x)-P^s(x)|\geq\delta\}} |f_i(s(x)) - f_i(P^s(x))|^2 \, d\mu(x) \\
&\quad + \int_{\{x \,:\, |s(x)-P^s(x)|<\delta\}} |f_i(s(x)) - f_i(P^s(x))|^2 \, d\mu(x) \\
&\leq 4 \max_{1\leq j\leq k} \|f_j\|_\infty^2 \varepsilon + \varepsilon^2.
\end{aligned}$$

It follows that there exists a sequence of representations $(\rho_n)_{n\geq 1}$ of Γ, every one factorizing by a finite quotient, such that

$$\lim_{n\to+\infty} \langle \rho_n(s) f_i \mid f_i\rangle = \langle \pi(s) f_i \mid f_i \rangle \qquad \text{for all } \ i \in \{1,\ldots,k\}.$$

Since $C([0,1])$ is dense in $L^2([0,1])$, this implies that π is weakly contained in $\bigoplus_{n\geq 1} \rho_n$. □

REMARK 15.E.21. It was shown in [**LuSh–04**] that some other lattices in rank one Lie groups, as for instance surface groups, are C*-residually finite.

15.E.d. Group C*-algebras with no faithful finite trace. Using Theorem 12.E.3, we now give examples of groups for which the maximal C*-algebra has no faithful finite trace.

Observe first that $\Gamma = \mathrm{GL}_n(\mathbf{K})$ or $\Gamma = \mathrm{SL}_n(\mathbf{K})$ with $n \geq 2$ is amenable if $\mathbf{K}$ is an algebraic extension of a finite field; so, in this case, $C^*_{\max}(\Gamma)$ has a faithful finite trace by Proposition 15.E.8.

THEOREM 15.E.22. *Let* $\mathbf{K}$ *be a field which is not an algebraic extension of a finite field. Let* $\Gamma = \mathrm{GL}_n(\mathbf{K})$ *or* $\Gamma = \mathrm{SL}_n(\mathbf{K})$ *for* $n \geq 2$.

The family $\mathrm{Rep}_{\mathrm{FT}}(\Gamma)$ *does not separate the points of* $C^*_{\max}(\Gamma)$*; equivalently (see Corollary* 15.E.7*),* $C^*_{\max}(\Gamma)$ *has no faithful finite trace.*

PROOF. If the characteristic of $\mathbf{K}$ is 0, then $\mathbf{K}$ contains $\mathbf{Q}$; if the characteristic of $\mathbf{K}$ is p, then $\mathbf{K}$ contains the purely transcendental extension $\mathbf{F}_p(T)$. In view of Proposition 15.E.5, it suffices to prove the result for $\Gamma = \mathrm{SL}_2(\mathbf{Q})$ and for $\Gamma = \mathrm{SL}_2(\mathbf{F}_p(T))$.

Let t be a finite normalized trace on $C^*_{\max}(\Gamma)$; we claim that t is not faithful. Let $\varphi \in \mathrm{Tr}_1(\Gamma)$ be the trace on Γ corresponding to t. By Lemma 15.E.6, it suffices to show that the GNS representation π_φ corresponding to φ is not faithful on $C^*_{\max}(\Gamma)$.

It follows from Theorem 12.E.3 that there exists $\alpha \in [0,1]$ and $\psi \in \mathrm{Tr}_1(\Gamma)$ such that $\psi = 0$ on $\Gamma \smallsetminus Z$ and

$$\varphi = \alpha \mathbf{1}_\Gamma + (1-\alpha)\psi,$$

where Z is the centre of Γ.

Since Z is amenable, the GNS representation π_ψ corresponding to ψ is weakly contained in λ_Γ. As a consequence, π_φ is weakly contained in the direct sum $1_\Gamma \oplus \lambda_\Gamma$. Therefore to show that π_φ is not faithful on $C^*_{\max}(\Gamma)$, it suffices to show that the representation $1_\Gamma \oplus \lambda_\Gamma$ is not faithful on $C^*_{\max}(\Gamma)$.

Observe that $\mathrm{SL}_2(\mathbf{Q})$ contains the subgroup $\mathrm{SL}_2(\mathbf{Z})$ which is a lattice in the LC group $\mathrm{SL}_2(\mathbf{R})$ and that $\mathrm{SL}_2(\mathbf{F}_p[T])$ can be embedded as a lattice in the LC group $\mathrm{SL}_2(\mathbf{F}_p((T^{-1})))$, where $\mathbf{F}_p((T^{-1}))$ is the local field of Laurent series over $\mathbf{F}_p$.

Set $\Lambda = \mathrm{SL}_2(\mathbf{Z})$ when $\Gamma = \mathrm{SL}_2(\mathbf{Q})$ and $\Lambda = \mathrm{SL}_2(\mathbf{F}_p[T])$ when $\Gamma = \mathrm{SL}_2(\mathbf{F}_p(T))$. We see that in either case Γ contains a lattice Λ in the LC group $G = \mathrm{SL}_2(\mathbf{k})$ for a local field $\mathbf{k}$.

Let π be an irreducible representation of G such that $\pi \neq 1_G$ and which is not weakly contained in λ_G (that is, π is a representation from the complementary series of G). We claim that $\pi|_\Gamma$ is not weakly contained in $1_\Gamma \oplus \lambda_\Gamma$. Once proved, this will imply that $1_\Gamma \oplus \lambda_\Gamma$ is not faithful on $C^*_{\max}(\Gamma)$.

Assume by contradiction that $\pi|_\Gamma$ is weakly contained in $1_\Gamma \oplus \lambda_\Gamma$. Then $\pi|_\Lambda$ is weakly contained in $1_\Lambda \oplus \lambda_\Lambda$, since λ_Λ is weakly equivalent to $\lambda_\Gamma|_\Lambda$.

Let X denote the set of all $\sigma \in \widehat{\Lambda}$ which are weakly contained in $\pi|_\Lambda$. By assumption, every $\sigma \in X \smallsetminus \{1_\Lambda\}$ is weakly contained in λ_Λ. Since Λ is not amenable, 1_Λ is not weakly contained in λ_Λ, and hence 1_Λ does not belong to the closure of $X \smallsetminus \{1_\Lambda\}$ in the Fell topology. It follows that either 1_Λ is *contained* in $\pi|_\Lambda$ or $X \subset \widehat{\Lambda} \smallsetminus \{1_\Lambda\}$.

By the Howe–Moore Theorem (see [**HoMo–79**, Theorem 5.2]), π has vanishing matrix coefficients; since Λ is a discrete infinite subgroup of G, this implies that 1_Λ is not contained in $\pi|_\Lambda$. So, $X \subset \widehat{\Lambda} \smallsetminus \{1_\Lambda\}$ and this means that $\pi|_\Lambda$ is weakly contained in λ_Λ.

It follows, by continuity of induction, that $\mathrm{Ind}_\Lambda^G \pi|_\Lambda$ is weakly contained in $\mathrm{Ind}_\Lambda^G \lambda_\Lambda \simeq \lambda_G$. However,

$$\mathrm{Ind}_\Lambda^G \pi|_\Lambda \simeq \pi \otimes \mathrm{Ind}_\Lambda^G 1_\Lambda$$

(see [**BeHV–08**, Proposition E.2.5]) and, since Λ is a lattice in G, the trivial representation 1_G is contained in $\mathrm{Ind}_\Lambda^G 1_\Lambda$. This implies that π is weakly contained in λ_G and contradicts the fact that π belongs to the complementary series of G. $\square$

REMARK 15.E.23. The groups of Theorem 15.E.22 are not residually finite (they are not even MAP). There are examples of residually finite groups which are not C*-MAP: it is shown in [**Bekk–99**] that $\mathrm{SL}_2(\mathbf{Z}[\sqrt{2}])$ is not C*-MAP and in [**Bekk–07**] that the maximal C*-algebra of $\mathrm{SL}_n(\mathbf{Z})$ has no faithful finite trace for $n \geq 3$.

15.F. Traces and Invariant Random Subgroups

In this section, we establish a link between traces of a discrete group Γ and invariant random subgroups of Γ.

DEFINITION 15.F.1. Let G be a locally compact group. The **space of closed subgroups of** G is the space

$$\mathrm{Sub}(G) = \{H \subset G \mid H \text{ is a closed subgroup of } G\},$$

equipped with the Chabauty topology.

Recall that a basis for this topology is given by sets of the form

$$\mathcal{U}(K; V_1, \dots, V_n) = \{H \in \mathrm{Sub}(G) \mid H \cap K = \emptyset \text{ and } H \cap V_i \neq \emptyset \text{ for } i = 1, \dots, n\},$$

where K is a compact subset and the V_i's are open subsets of G. The space $\mathrm{Sub}(G)$ is compact (see [**Fell–62b**, Lemma 1]) and conjugation gives rise to an action of G by homeomorphisms on $\mathrm{Sub}(G)$.

DEFINITION 15.F.2. An **invariant random subgroup** of G, abbreviated IRS, is a G-invariant probability measure on the Borel subsets of $\mathrm{Sub}(G)$. We denote by $\mathrm{IRS}(G)$ the set of IRS's of G.

The set $\mathrm{IRS}(G)$ is a convex compact subset of the set of probability measures on $\mathrm{Sub}(G)$, equipped with the weak*-topology. The extreme points of $\mathrm{IRS}(G)$ are the **ergodic** IRS's, that is, the G-invariant ergodic probability measures on $\mathrm{Sub}(G)$. The convex hull of the set of ergodic IRS's is dense in $\mathrm{IRS}(G)$, by the Krein–Milman Theorem.

IRS's generalize both closed normal subgroups and lattices of G, as shown by Example 15.F.3. The notion of an IRS was introduced in [**AbGV–14**], and appeared implicitly in [**StZi–94**]. See [**Gela–18**] for a survey on IRS's.

EXAMPLE 15.F.3. (1) Let N be a closed normal subgroup of G; the Dirac measure δ_N is an IRS of G. IRS's of this form are exactly the point measures in $\mathrm{IRS}(G)$.

(2) Let Γ be a lattice in the LC group G and let $\mu_{G/\Gamma}$ be the unique G-invariant regular probability measure on the Borel subsets of G/Γ. Let $\Phi_*(\mu_{G/\Gamma})$ be the image of $\mu_{G/\Gamma}$ under the measurable map

$$\Phi : G/\Gamma \to \mathrm{Sub}(G), \quad g\Gamma \mapsto g\Gamma g^{-1}.$$

Then $\Phi_*(\mu) \in \mathrm{IRS}(G)$.

(3) The construction of IRS's in (2) can be generalized as follows. Let G be a second countable LC group and $G \curvearrowright (X, \mathcal{B})$ a measurable action of G on a countably separated Borel space $(X, \mathcal{B})$. Then, for every $x \in X$, the stabilizer G_x of x in G is a closed subgroup of G (see Corollary 2.1.21 in [**Zimm–84**]).

Let μ be a probability measure on $(X, \mathcal{B})$ which is invariant by G. The image $\Phi_*(\mu)$ of μ under the measurable map

$$\Phi : X \to \mathrm{Sub}(G), \quad x \mapsto G_x$$

is an IRS of G.

(4) Let G be a second countable LC group. Every IRS of G arises as in (3) for some measure-preserving action of G on a countably separated probability space $(X, \mathcal{B}, \mu)$; see [**AbGV–14**, Proposition 14] in the discrete case and [**ABB+–17**, Theorem 2.6] in general.

(5) Let G be a connected simple Lie group with trivial centre and with real rank greater than two. It is a remarkable result of [**StZi–94**] (see also Theorem 1.14 in [**ABB+–17**]) that every IRS of G is a convex combination of δ_G, $\delta_{\{e\}}$, and IRS's associated with lattices in G as in (2).

(6) IRS's of the infinite symmetric group $\mathrm{Sym}_{\mathrm{fin}}(\mathbf{N})$ as in Example 11.C.7(1) have been classified in [**Vers–12**].

The following result was first established in [**Vers–11**], with a different proof; for the original proof, see Theorem 15.F.6 and Remark 15.F.7(1).

PROPOSITION 15.F.4. *Let Γ be a discrete group and $\nu \in \mathrm{IRS}(\Gamma)$.*
The function $\varphi_\nu : \Gamma \to \mathbf{C}$ defined by

$$\varphi_\nu(\gamma) \; = \; \nu\left(\{H \in \mathrm{Sub}(G) \mid \gamma \in H\}\right)$$

is a normalized trace of Γ, that is, $\varphi_\nu \in \mathrm{Tr}_1(\Gamma)$.

PROOF. The function φ_ν is central, since for $\gamma, \delta \in \Gamma$, we have, by Γ-invariance of ν,

$$\begin{aligned} \varphi_\nu(\delta^{-1}\gamma\delta) \; &= \; \nu\left(\{H \in \mathrm{Sub}(G) \mid \delta^{-1}\gamma\delta \in H\}\right) \\ &= \; \nu\left(\{\delta H \delta^{-1} \in \mathrm{Sub}(G) \mid \delta^{-1}\gamma\delta \in H\}\right) \\ &= \; \nu\left(\{H \in \mathrm{Sub}(G) \mid \gamma \in H\}\right) \; = \; \varphi_\nu(\gamma). \end{aligned}$$

Moreover, φ is normalized: $\varphi_\nu(e) = \nu(\mathrm{Sub}(G)) = 1$.

It remains to show that φ_ν is positive definite. By the definition of φ_ν, we have, for every $\gamma \in \Gamma$,

$$\varphi_\nu(\gamma) \; = \; \int_{\mathrm{Sub}(G)} \mathbf{1}_H(\gamma) \, d\nu(H).$$

Let $c_1, \ldots, c_n \in \mathbf{C}$ and $\gamma_1, \ldots, \gamma_n \in \Gamma$. Since $\mathbf{1}_H$ is a function of positive type (see Example 1.B.9 and Proposition 1.F.10) for every $H \in \mathrm{Sub}(G)$ and since ν is a positive measure on $\mathrm{Sub}(G)$, we have

$$\begin{aligned} \sum_{i,j=1}^n c_i \overline{c_j} \varphi_\nu(\gamma_j^{-1}\gamma_i) \; &= \; \sum_{i,j=1}^n c_i \overline{c_j} \int_{\mathrm{Sub}(G)} \mathbf{1}_H(\gamma_j^{-1}\gamma_i) \, d\nu(H) \\ &= \; \int_{\mathrm{Sub}(G)} \Big(\sum_{i,j=1}^n c_i \overline{c_j} \mathbf{1}_H(\gamma_j^{-1}\gamma_i) \Big) d\nu(H) \; \geq \; 0, \end{aligned}$$

as was to be shown. □

EXAMPLE 15.F.5. (1) Let N be a normal subgroup of the discrete group Γ. The trace φ_ν of Γ corresponding to the IRS $\nu = \delta_N$ is $\mathbf{1}_N$.

(2) Let Γ be a countable group and let $\Gamma \curvearrowright (X, \mathcal{B}, \mu)$ be a measure-preserving action of Γ on a countably separated Borel space $(X, \mathcal{B})$, equipped with a probability measure μ on $\mathcal{B}$. As shown in Example 15.F.3(3), this action defines an IRS ν of Γ; the associated trace φ_ν as in Proposition 15.F.4, which we now prefer to denote by φ_μ, is given by

$$\varphi_\mu(\gamma) = \mu(X^\gamma) \qquad \text{for all } \gamma \in \Gamma,$$

where $X^\gamma = \{x \in X \mid \gamma x = x\}$ is the set of fixed points of γ in X.

We are going to describe a concrete model for the GNS triple associated with the trace φ_μ. For this, we need to discuss a variant of the group measure space construction of Murray and von Neumann of Section 13.A.

Let $\mathcal{R}$ be the equivalence relation on X of which the equivalence classes are the Γ-orbits. We view $\mathcal{R}$ as a subset of $X \times X$:

$$(x, y) \in \mathcal{R} \quad \text{if and only if} \quad y \in \Gamma x.$$

For $A \subset \mathcal{R}$ and $x \in X$, we set $A_x = A \cap (\{x\} \times X)$, and let $|A_x|$ denote its cardinality. We define a σ-finite measure m on the Borel subsets A of $\mathcal{R}$ by

$$m(A) = \int_X |A_x| d\mu(x).$$

We define two representations π and ρ of Γ on the Hilbert space $L^2(\mathcal{R}, m)$ by

$$(\pi(\gamma)f)(x, y) = f(\gamma^{-1}x, y) \quad \text{and} \quad (\rho(\gamma)f)(x, y) = f(x, \gamma^{-1}y),$$

for $\gamma \in \Gamma$, $f \in L^2(\mathcal{R}, m)$, and $(x, y) \in \mathcal{R}$. These two representations commute: $\pi(\gamma_1)\rho(\gamma_2) = \rho(\gamma_2)\pi(\gamma_1)$ for all $\gamma_1, \gamma_2 \in \Gamma$.

Let $\mathbf{1}_\Delta$ be the characteristic function of the diagonal $\Delta = \{(x, x) \mid x \in X\}$ of $\mathcal{R}$.

The following result is Theorem 9 in [**Vers–11**].

THEOREM 15.F.6 (**Vershik**). *Let Γ be a countable group acting by measure-preserving transformations on a standard Borel space $(X, \mathcal{B})$ equipped with a probability measure μ. Let*

$$\varphi_\mu : \Gamma \to \mathbf{C}, \quad \gamma \mapsto \mu(\{x \in X \mid \gamma x = x\})$$

be the associated trace of Γ. Let R, m, π, ρ, Δ be as above.

Then $\mathbf{1}_\Delta$ belongs to $L^2(\mathcal{R}, m)$ and we have $\pi(\Gamma)\mathbf{1}_\Delta = \rho(\Gamma)\mathbf{1}_\Delta \subset L^2(\mathcal{R}, m)$.

Let $\mathcal{K}$ be the closed linear span of $\pi(\Gamma)\mathbf{1}_\Delta$ (equivalently, of $\rho(\Gamma)\mathbf{1}_\Delta$). Let $(\pi, \mathcal{K})$ and $(\rho, \mathcal{K})$ denote the subrepresentations of π and ρ in $\mathcal{K}$.

Then both $(\pi, \mathcal{K}, \mathbf{1}_\Delta)$ and $(\rho, \mathcal{K}, \mathbf{1}_\Delta)$ are GNS-triples for φ_μ.

PROOF. The characteristic function $\mathbf{1}_\Delta$ belongs to $L^2(\mathcal{R}, m)$ and $\|\mathbf{1}_\Delta\| = 1$ because

$$\int_{\mathcal{R}} |\mathbf{1}_\Delta(x, y)|^2 \, dm(x, y) = \int_X |\Delta_x| \, d\mu(x) = \mu(X) = 1.$$

A straightforward computation shows that

$$\pi(\gamma)\mathbf{1}_\Delta = \mathbf{1}_{\{(x, \gamma^{-1}x) | x \in X\}} \quad \text{and} \quad \rho(\gamma)\mathbf{1}_\Delta = \mathbf{1}_{\{(x, \gamma x) | x \in X\}}$$

for all $\gamma \in \Gamma$, so that

$$\pi(\Gamma)\mathbf{1}_\Delta = \rho(\Gamma)\mathbf{1}_\Delta.$$

It follows that $\mathcal{K}$ is invariant by both π and ρ, and that $\mathbf{1}_\Delta$ is a cyclic vector for both $(\pi, \mathcal{K})$ and $(\rho, \mathcal{K})$.

Moreover, for every $\gamma \in \Gamma$, we have

$$\begin{aligned}\langle \pi(\gamma)\mathbf{1}_\Delta, \mathbf{1}_\Delta \rangle &= \int_{\mathcal{R}} \mathbf{1}_\Delta(\gamma^{-1}x, y)\mathbf{1}_\Delta(x, y)\, dm(x, y) \\ &= \int_{\mathcal{R}} \mathbf{1}_{\{(x,x) \in X \times X \mid \gamma^{-1}x = x\}}\, dm(x, y) \\ &= \int_X \big|\{x \in X \mid \gamma x = x\}\big|\, d\mu(x) \\ &= \mu\left(\{x \in X \mid \gamma x = x\}\right) = \varphi_\mu(\gamma),\end{aligned}$$

and, by a similar computation,

$$\langle \rho(\gamma)\mathbf{1}_\Delta, \mathbf{1}_\Delta \rangle = \varphi_\mu(\gamma).$$

This shows that $(\rho, \mathcal{K}, \mathbf{1}_\Delta)$ and $(\pi, \mathcal{K}, \mathbf{1}_\Delta)$ are both GNS-triples for φ_μ. □

REMARK 15.F.7. (1) The equality $\varphi_\mu = \langle \pi(\cdot)\mathbf{1}_\Delta, \mathbf{1}_\Delta \rangle$, established during the proof of Theorem 15.F.6, provides another proof of Proposition 15.F.4.

(2) One may ask which properties of the action $\Gamma \curvearrowright (X, \mu)$ imply that the associated function $\varphi \in \mathrm{Tr}_1(\Gamma)$ is indecomposable, that is, $\varphi \in E(\Gamma)$. For a discussion of this question, see [**Vers–11**] and [**DuGr–18**].

Next, we introduce a notion which strenghtens the notion of simplicity of a discrete group.

DEFINITION 15.F.8. A discrete group Γ is **character rigid** if $E(\Gamma) = \{\delta_e, \mathbf{1}_\Gamma\}$.

REMARK 15.F.9. Let Γ be a character rigid group. Then Γ is simple; indeed, let N be a normal subgroup of Γ. Since $\mathbf{1}_N \in \mathrm{Tr}_1(\Gamma)$, it follows that $\mathbf{1}_N = t\delta_e + (1-t)\mathbf{1}_\Gamma$ for some $t \in [0, 1]$. This implies that $N = \{e\}$ or $N = \Gamma$.

The following result from [**DuMe–14**] and [**PeTh–16**] shows that the ergodic actions of character rigid groups are essentially free. Recall that essentially free group actions have been introduced in Definition 13.A.1.

THEOREM 15.F.10. *Let Γ be a countable character rigid group.*

Then every measure-preserving ergodic action $\Gamma \curvearrowright (X, \mu)$ of Γ on a standard Borel space X, equipped with a nonatomic probability measure μ, is essentially free.

PROOF. Since $\Gamma \curvearrowright (X, \mu)$ is ergodic and μ nonatomic, the orbit Γx of μ-almost every $x \in X$ is infinite, by Proposition A.5.10.

Assume, by contradiction, that $\Gamma \curvearrowright (X, \mu)$ is not essentially free. Then, since Γ is countable, there exists $\gamma \in \Gamma \smallsetminus \{e\}$ such that

$$\mu\left(\{x \in X \mid \gamma x = x\}\right) > 0.$$

This means that, for the associated trace $\varphi_\mu \in \mathrm{Tr}(\Gamma)$ as in Theorem 15.F.6, we have $\varphi_\mu \neq \delta_e$. Since $E(\Gamma) = \{\delta_e, \mathbf{1}_\Gamma\}$, it follows that

$$\varphi_\mu = t\mathbf{1}_\Gamma + (1-t)\delta_e$$

for some $t > 0$.

By Theorem 15.F.6, $(\rho, \mathcal{K}, \mathbf{1}_\Delta)$ is a GNS triple for φ for some closed subspace $\mathcal{K}$ of $L^2(\mathcal{R}, m)$. Therefore, there exists a nonzero $\rho(\Gamma)$-invariant function in $L^2(\mathcal{R}, m)$.

It follows that there exists a $\rho(\Gamma)$-invariant Borel subset A of $\mathcal{R}$ such that $0 < m(A) < \infty$.

Recalling that $A_x = \{y \in X \mid (x, y) \in \mathcal{R}\}$, set

$$Y := \{x \in X \mid A_x \neq \emptyset\}.$$

Since $m(A) > 0$, we have $\mu(Y) > 0$. Observe that the $\rho(\Gamma)$-invariance of A means that $A_x = \Gamma x$ for every $x \in Y$. Therefore $|A_x| = \infty$ for almost every $x \in Y$, as almost every Γ-orbit is infinite. It follows that

$$m(A) = \int_X |A_x| \, d\mu(x) = \infty \cdot \mu(Y) = \infty,$$

and this is a contradiction. □

REMARK 15.F.11. As proved in [**DuMe–14**] and [**PeTh–16**], the conclusion of Theorem 15.F.10 holds more generally if Γ is a countable icc group and if $E(\Gamma)$ consists only of δ_e and of countably many characters all associated with finite-dimensional irreducible representations.

We draw the following consequence of Theorem 15.F.10.

COROLLARY 15.F.12. *Let Γ be a countable character rigid group. Then the ergodic IRS's of Γ are δ_Γ and $\delta_{\{e\}}$.*

PROOF. We may clearly assume that Γ is not the group with one element.

Since Γ is countable, the compact space $\mathrm{Sub}(\Gamma)$ is metrizable and is therefore a standard Borel space. Let μ be an ergodic Γ-invariant probability measure on $\mathrm{Sub}(\Gamma)$.

We claim that μ is atomic. Indeed, otherwise, $\Gamma \curvearrowright (\mathrm{Sub}(\Gamma), \mu)$ would be essentially free (Theorem 15.F.10) and this is possible only if Γ is the group with one element.

Since μ is atomic, μ is a Dirac measure and hence $\mu = \delta_N$ for some normal subgroup N of Γ. It follows from the simplicity of Γ (see Remark 15.F.9) that $\mu = \delta_\Gamma$ or $\mu = \delta_{\{e\}}$. □

EXAMPLE 15.F.13. (1) Let $\Gamma = \mathrm{PSL}_n(\mathbf{K})$, where $\mathbf{K}$ is an infinite countable field and $n \geq 2$. Then Γ is a countable character rigid group (see Theorem 12.E.3). By Corollary 15.F.12, the ergodic IRS's of Γ are δ_G and $\delta_{\{e\}}$.

(2) Example (1) has been extended in [**Bekk–20**] to more general algebraic groups as follows. Let G be an algebraic group defined and simple over an infinite countable field $\mathbf{K}$, and let $G(\mathbf{K})^+$ be the subgroup of $G(\mathbf{K})$ generated by the unipotent radicals of parabolic subgroups of G defined over $\mathbf{K}$; then $\Gamma = G(\mathbf{K})^+/Z$ is character rigid, where Z is the centre of $G(\mathbf{K})^+$.

(3) Let Γ be the commutator subgroup of a group from the Higman–Thompson families $F_{n,r}$ or $G_{n,r}$. Then Γ is character rigid; see [**DuMe–14**].

Appendix

This book requires various notions that the potential reader may not know or may have forgotten. The purpose of the present Appendix is to recall a few definitions and state some "well-known facts" which will hopefully make it easier to read various parts of the book.

A.1. Topology

Properties of topological spaces. A topological space X is T_0 if, for any two points $x, y \in X$, $x \neq y$, there exists an open subset U in X such that $x \in U$ and $y \notin U$, or $x \notin U$ and $y \in U$. Equivalently: if $\overline{\{x\}} \neq \overline{\{y\}}$ for any points $x, y \in X$ such that $x \neq y$.

The space X is T_1 if, for any two points $x, y \in X$, $x \neq y$, there exist open subsets U, V in X such that $x \in U$, $y \notin U$, $x \notin V$, $y \in V$. Equivalently: if $\{x\}$ is closed in X for every $x \in X$.

A T_2 space, also known as a **Hausdorff space**, is a space X satisfying Hausdorff's separation axiom: for any two points $x, y \in X$, $x \neq y$, there exist disjoint open subsets U, V in X such that $x \in U$ and $y \in V$.

It follows from the definitions that a T_1 space is T_0, and a T_2 space is T_1.

Bourbaki calls a T_0 topological space a **Kolmogorov space** and a T_1 topological space an **accessible space**. See Exercice 2 of Chapter I, §1 and Exercice 1 of Chapter I, §8 in [**BTG1–4**]. We recall the following proposition, which appears in Exercice 27 of Chapter I, §8 of [**BTG1–4**], and in [**Will–07**, Section 6.2]. It shows that every topological space admits a T_0 topological quotient space satisfying an appropriate universal property.

PROPOSITION A.1.1. *Let X be topological space. There exists a pair (Y, p) consisting of a T_0 topological space Y and a surjective continuous map $p : X \twoheadrightarrow Y$ with the following property:*

for every pair (Z, f) consisting of a T_0 topological space Z and a continuous map $f : X \to Z$, there exists a continuous map $g : Y \to Z$ such that $f = g \circ p$.

PROOF. We consider the equivalence relation $\sim$ on X defined by

$$x \sim x' \quad \text{if} \quad \overline{\{x\}} = \overline{\{x'\}}.$$

Let Y be the quotient space $X/\sim$ equipped with the quotient topology and

$$p : X \twoheadrightarrow X/\sim$$

the canonical projection.

We claim that $X/\sim$ is a T_0 space. Indeed, let $x, x' \in X$ be such that $p(x) \neq p(x')$. We assume that $\overline{\{x\}} \not\subset \overline{\{x'\}}$ (the other case, $\overline{\{x'\}} \not\subset \overline{\{x\}}$, can be dealt with similarly). It is enough to check that there exists an open subset of $X/\sim$ which contains $p(x)$ and not $p(x')$.

Set $U := X \smallsetminus \overline{\{x'\}}$. Observe that every $v \in X$ such that $\overline{\{v\}} \not\subset \overline{\{x'\}}$ is necessarily a point in U.

We claim that U is saturated for the equivalence relation. Indeed, for $u \in U$ and $u' \in X$ such that $u \sim u'$, we have $\overline{\{u'\}} = \overline{\{u\}} \not\subset \overline{\{x'\}}$, so that $u' \in U$. It follows that $p(U)$ is an open subset of $X/\sim$ such that $p(x) \in p(U)$ and $p(x') \notin p(U)$. This shows that $X/\sim$ is a T_0 space.

Next, let Z be T_0 topological space Z and $f : X \to Z$ a continuous map. Let $x, x' \in X$ such that $x \sim x'$. Then

$$f(x') \in f(\overline{\{x'\}}) = f(\overline{\{x\}}) \subset \overline{\{f(x)\}},$$

by continuity of f, and hence $\overline{\{f(x')\}} \subset \overline{\{f(x)\}}$. Similarly, we have $\overline{\{f(x)\}} \subset \overline{\{f(x')\}}$. Therefore, $\overline{\{f(x)\}} = \overline{\{f(x')\}}$. Since Z is a T_0 topological space, it follows that $f(x) = f(x')$. Therefore the map $g : X/\sim \to Z, \;\; p(x) \mapsto f(x)$ is well defined, and $g \circ p = f$.

Moreover, g is continuous. Indeed, for every open subset U of Z, the set $p^{-1}(g^{-1}(U)) = f^{-1}(U)$ is open in X and hence $g^{-1}(U)$ is open in Y. $\square$

For a given topological space X, it follows from the universal property that the pair (Y, p) as in Proposition A.1.1 is unique in the following sense: if (Y', p') is a pair with the same properties as (Y, p), then there exists a homeomorphism $f : Y \to Y'$ such that $f \circ p = p'$. We will call (Y, p) or Y the **universal Kolmogorov quotient** of X.

Such spaces appear in Section 1.E, where $\mathrm{Prim}(G)$ is identified as the universal Kolmogorov quotient of $\widehat{G}$, and in Section 9.D, as quasi-orbit spaces.

Given a continuous action of a group G on a topological X (see Definition A.5.7), the **quasi-orbit space** of this action is the quotient space $X/\sim$ for the equivalence relation $\sim$ defined on X by

$$x \sim x' \quad \text{if} \quad \overline{Gx} = \overline{Gx'}.$$

An equivalence class for $\sim$ is called a **quasi-orbit** for G. The space $X/\sim$ equipped with the quotient topology is a T_0 topological space. In fact, let X/G be the space of G-orbits in X equipped with the quotient topology. Then

$X/\sim$ is the universal Kolmogorov quotient of X/G, as defined above.

A topological space is **quasi-compact** if every open covering contains a finite covering, and is **compact** if it is both quasi-compact and Hausdorff. It is **locally compact** if any of its points has a compact neighbourhood and if it is Hausdorff.

A topological space is a **Baire space** if countable intersections of open dense subspaces are dense. Locally compact spaces and completely metrizable spaces are Baire spaces (Baire Category Theorem [**BTG5–10**, Chap. 9, § 5]). The quotient of a Baire space by an open equivalence relation is a Baire space [**BTG5–10**, Chap. 9, § 5, Exercise 25].

A topological space is **Polish** if it is second-countable and homeomorphic to a complete metric space. A **Polish group** is a topological group of which the topology is Polish. For the following proposition, we refer to [**BTG5–10**, Chap. 9, § 6, no 1]; it is used in the proof of our Proposition 8.A.13. Recall that a G_δ set is the intersection of a countable family of open subspaces.

PROPOSITION A.1.2. *Let X be a Polish space.*

(1) *Every closed subspace of X is Polish,*
(2) *every open subspace of X is Polish,*
(3) *a subspace of X is Polish if and only if it is a G_δ set.*

Most topological spaces and groups of interest in this book are locally compact. At several places, they have to satisfy one or another of the following restrictive conditions.

A topological space is **σ-compact** if it is a countable union of compact subspaces. In a locally compact space X which is σ-compact, there exists a nested sequence of compact subspaces $K_0 \subset K_1 \subset \ldots \subset K_n \subset K_{n+1} \subset \ldots$ such that $X = \bigcup_{n \geq 0} K_n$, and K_n is contained in the interior of K_{n+1} for all $n \geq 0$ [**BTG1–4**, Page I.68].

A topological space is **second-countable** if its topology has a countable basis of open sets, and **separable** if it contains a countable dense subset. It is straightforward that a second-countable space is separable. The converse holds for metrizable spaces (Theorem A.1.3), but it does not hold in general.

For example, let F be a finite group with at least two elements and G the direct product of copies of F indexed by $[0, 1]$; then G is a compact group which is separable [**Enge–89**, Corollary 2.3.16] and which is not second-countable.

We record the following equivalences, which contain a particular case of the Urysohn Metrization Theorem (see for example [**BTG5–10**, Page IX.21]).

THEOREM A.1.3 (**Second-countable locally compact spaces**). *For a locally compact space X, the following conditions are equivalent:*

(1) *X is second-countable;*
(2) *X is metrizable and σ-compact;*
(3) *X is metrizable and contains a countable dense subset;*
(4) *X is Polish.*

Spaces which satisfy the conclusion of Proposition A.1.4 appear below in the proof of (1) of Theorem A.4.1. Observe that, in a σ-compact locally compact space, an open subset need not be σ-compact; indeed, any non-σ-compact locally compact space can be seen as an open subset of its one-point compactification. However:

PROPOSITION A.1.4. *Let X be a second-countable locally compact space. Then every open subset U of X is σ-compact.*

PROOF. By Theorem A.1.3, there exists a metric d on X which induces the topology of X. For any positive integer n, set

$$F_n = \{x \in X \mid d(x, X \smallsetminus U) \geq 1/n\}.$$

Then F_n is closed in X, hence F_n is σ-compact, and U is equal to the countable union $\bigcup_n F_n$. Therefore U is σ-compact. $\square$

Spaces of complex-valued continuous functions. For X a topological space, we denote in this book by

$C(X)$ the space of continuous complex-valued functions on X,
$C^b(X)$ the subspace of bounded functions in $C(X)$,
$C^c(X)$ the subspace of compactly supported functions in $C(X)$,
$C^{c,K}(X)$ the subspace of $C(X)$ of functions with support contained in a compact subspace K of X.

When X is locally compact, we denote by

$C^0(X)$ the subspace of $C(X)$ of functions vanishing at infinity.

All these spaces are commutative complex involutive algebras for the pointwise product and the involution defined by $f^*(x) = \overline{f(x)}$ for all $x \in X$. In special situations, they have other relevant structures; for example, when G is a locally compact group, with a given left-invariant Haar measure, $C^c(G)$ is also an algebra for the convolution product.

When X is locally compact, $C^0(X)$ is a commutative C*-algebra. Conversely, the Gel'fand representation states that every commutative C*-algebra A is isomorphic to $C^0(X)$, where X is the space of characters of A equipped the weak*-topology, which is a locally compact space [**Dixm–C***, 1.4.1]. (A character of A is a nonzero morphism from A to $\mathbf{C}$.)

A.2. Borel spaces

Let X be a set. A **σ-algebra** of subsets of X is a collection $\mathcal{B}$ of subsets of X such that $X \in \mathcal{B}$, $X \smallsetminus B \in \mathcal{B}$ for all $B \in \mathcal{B}$, and $\bigcup_{n\geq 1} B_n \in \mathcal{B}$ for all sequences $(B_n)_{n\geq 1}$ of subsets B_n in $\mathcal{B}$. A **measurable space** is a set X endowed with a σ-algebra $\mathcal{B}$ of subsets; a measurable space is also called a **Borel space**. Let $(X, \mathcal{B})$ and $(Y, \mathcal{C})$ be two measurable spaces; a map $f : X \to Y$ is a **measurable map** if $f^{-1}(C) \in \mathcal{B}$ for all $C \in \mathcal{C}$; a measurable map is also called a **Borel map**.

Let $(X, \mathcal{B})$ be a Borel space. A subset $\mathcal{S}$ of $\mathcal{B}$ is **separating** if, for every pair (x, x') of distinct points in X, there exists $S \in \mathcal{S}$ such that $x \in S$ and $x' \notin S$. The space $(X, \mathcal{B})$ is **countably separated** if there exists in $\mathcal{B}$ a countable subset which is separating, and **countably generated** if there exists in $\mathcal{B}$ a countable subset which generates $\mathcal{B}$ and which is separating. Note that a countably generated Borel space is countably separated, by definition; moreover, countable separation and countable generation are inherited by subspaces, as it is straightforward to check. (We follow [**Mack–57**] and much of the literature – note however that some authors define "countably generating" without requiring "separating"; see for example [**Kech–95**].)

PROPOSITION A.2.1. *Let $(X, \mathcal{B})$ be a Borel space.*

(1) *$(X, \mathcal{B})$ is countably separated if and only if there exists an injective Borel map $X \to [0, 1]$.*
(2) *$(X, \mathcal{B})$ is countably generated if and only if it is isomorphic to a subset of $[0, 1]$.*

REMARKS, AND REFERENCES FOR THE PROOF. First, the reader should mark the difference between "subset" here in (2) and "Borel subset" in (b) below. Then note that, because of Theorem A.2.2, the proposition could be stated with any standard Borel space instead of $[0, 1]$.

For the proof of (1), recall that $\{0, 1\}^{\mathbf{N}}$ is Borel isomorphic to the Cantor set, a subspace of $[0, 1]$; details of the argument can be found for example in [**Zimm–84**, Appendice A]. For the proof of (2), see [**Mack–57**, Theorem 2.1]. □

Let Y be a topological space. The **Borel σ-algebra** $\mathcal{B}_Y$, also denoted by $\mathcal{B}(Y)$, is the σ-algebra generated by the open subsets of Y.

For a Borel space $(X, \mathcal{B})$, the following conditions are equivalent:

(a) $(X, \mathcal{B})$ is Borel isomorphic to $(Y, \mathcal{B}_Y)$ for some Polish space Y;
(b) $(X, \mathcal{B})$ is Borel isomorphic to $(C, \mathcal{B}_C)$ for some Borel subset C of a Polish space Y.

The pair $(X, \mathcal{B})$ is a **standard Borel space** if these conditions are satisfied. A standard Borel space is countably generated, but the converse does not hold. For standard Borel spaces, see [**Mack–57**, Section 3].

It is obvious from the equivalence above that every Borel subset of a standard Borel space is a standard Borel space.

Standard Borel spaces can be classified as follows; see [**Kura–66**, Page 41] or [**Kech–95**, Theorem 15.6].

THEOREM A.2.2 (**Kuratowski**). *A standard Borel space is Borel isomorphic to one of the following spaces, depending on its cardinal:*

(1) *the interval* $[0, 1]$,
(2) *the countable infinite set* $\mathbf{N}$,
(3) *a finite space.*

A.3. Measures on Borel spaces and σ-finiteness

A **measure**, more precisely a **positive measure**, on a measurable space $(X, \mathcal{B})$ is a map $\mathcal{B} \to \overline{\mathbf{R}}_+$ such that $\mu(\bigcup_{n\geq 1} B_n) = \sum_{n\geq 1} \mu(B_n)$ for all sequences $(B_n)_{n\geq 1}$ of pairwise disjoint subsets B_n in $\mathcal{B}$. A **measure space** is a triple $(X, \mathcal{B}, \mu)$ where $(X, \mathcal{B})$ is a measurable space and μ a measure on $(X, \mathcal{B})$. Instead of "a measure on $(X, \mathcal{B})$", we also write "a measure on X", or "a measure on $\mathcal{B}$", when this is clear enough from the context.

A **finite measure** is a measure μ on $(X, \mathcal{B})$ such that $\mu(X) < \infty$, and a **probability measure** is one such that $\mu(X) = 1$.

A measure space $(X, \mathcal{B}, \mu)$ is **σ-finite** if there is a countable sequence $(B_n)_{n\geq 1}$ of subsets in $\mathcal{B}$ such that $X = \bigcup_{n\geq 1} B_n$ and $\mu(B_n) < \infty$ for all $n \geq 1$. This condition on a measure space is necessary for several important results, for example for the Fubini Theorem and the Radon–Nikodym Theorem.

An **atom** in a measure space $(X, \mathcal{B}, \mu)$ is a point $x \in X$ such that $\mu(\{x\}) > 0$. The measure μ is **nonatomic** if there does not exist any atom in X.

Given a measure μ on X, two measurable functions f, g defined on X are **equal μ-almost everywhere** if $\mu(\{x \in X \mid f(x) \neq g(x)\}) = 0$.

Measurable functions with values either in $\overline{\mathbf{R}}_+$ or in a Banach space can be integrated, as discussed in standard expositions, for example in [**Rudi–66**, Chapter 1] and [**Lang–93**, Chapter VI]. Integration of functions with values in Banach spaces are important in this book, for example for the definition of the representation of the group algebra $L^1(G)$ associated with a representation of a locally compact group G; see Section 8.B.

More precisely, let $(X, \mathcal{B}, \mu)$ be a measure space and let E be a Banach space, real or complex; the two cases $E = \mathbf{R}$ and $E = \mathbf{C}$ are particularly important. First, any measurable function $f : X \to \overline{\mathbf{R}}_+$ has an **integral**, denoted by $\int_X f d\mu$, or $\int_X f(x)d\mu(x)$, or $\mu(f)$, which is a number in $\overline{\mathbf{R}}_+$. Then, any measurable function $g : X \to E$ such that $\int_X \|g(x)\| d\mu(x) < \infty$ can be integrated, and the integral $\int_X g d\mu$ is a vector in E. The integrals of two measurable functions which are equal

μ-almost everywhere are equal. It is abusive, but often convenient, to identify a measurable function to its equivalence class modulo μ-almost everywhere equality.

Let $(X, \mathcal{B}, \mu)$ be a measure space and p a real number, $p \geq 1$. For a measurable function $f : X \to \mathbf{C}$, set $\|f\|_p := \big(\int_X |f|^p d\mu \big)^{1/p}$. Denote by $L^p(X, \mu)$, or simply by $L^p(X)$, the space of measurable functions $f : X \to \mathbf{C}$ (modulo μ-almost everywhere equality) such that $\|f\|_p < \infty$; then $L^p(X)$ is naturally a Banach space. When $p = 2$, the space $L^2(X)$ is naturally a Hilbert space. When $p = 1$, note that $f \mapsto \int_X f d\mu$ defines a continuous linear form on $L^1(X)$.

To define a Banach space $L^\infty(X)$, one may proceed as follows. Let $(D_i)_i$ be a countable basis of the topology of $\mathbf{C}$, consisting of open discs. Let $f : X \to \mathbf{C}$ be a measurable function. Let S_f be the union of the discs D_i for which $f^{-1}(D_i)$ has zero measure. The **essential range** of f is the complement R_f of S_f in $\mathbf{C}$; it is a closed subset of $\mathbf{C}$. The **essential supremum** $\|f\|_\infty$ of f is the supremum of the moduli $|z|$ of the points $z \in R_f$. The function f is **essentially bounded** if $\|f\|_\infty < \infty$. Denote by $L^\infty(X, \mu)$, or simply by $L^\infty(X)$, the space of (equivalence classes of) measurable functions $f : X \to \mathbf{C}$ such that $\|f\|_\infty < \infty$; then $L^\infty(X)$ is naturally a Banach space. See [**Rudi–66**, Chapter 3].

REMARK A.3.1. Let $p, q \in [1, \infty]$ be such that $\frac{1}{p} + \frac{1}{q} = 1$. For all $f \in L^p(X, \mu)$ and $g \in L^q(X, \mu)$, the function fg is in $L^1(X, \mu)$, hence there is a duality

$$L^p(X, \mu) \times L^q(X, \mu) \to \mathbf{C}, \qquad (f, g) \mapsto \int_X fg d\mu.$$

Consider $L^p(X, \mu)$ as a Banach space when $p \in [1, \infty[$, and with the topology defined by this duality when $p = \infty$ (i.e., a posteriori, with the weak*-topology). Denote by $L^p(X, \mu)'$ the dual space of $L^p(X, \mu)$. Assume that the measure μ is σ-finite. Then there is an isomorphism

$$L^q(X, \mu) \to L^p(X, \mu)', \qquad g \mapsto \Big(f \mapsto \int_X fg d\mu \Big).$$

When $p < \infty$, a proof can be read in [**Rudi–66**, Theorem 6.16]. When $p = \infty$, it follows from general principles (the dual of a locally convex topological vector space E with a topology $\sigma(E, F)$ can be identified with F; see [**BEVT1–5**, Chap. II, § 6, no 2, Proposition 3]).

Similarly, for E a Banach space, there is for each $p \in [1, \infty[$ a Banach space $L^p(X, \mu, E)$ of measurable functions $g : X \to E$ (up to equality μ-almost everywhere) such that $\|g\|_p := \big(\int_X \|g(x)\|_E d\mu(x) \big)^{1/p} < \infty$, and a Banach space $L^\infty(X, \mu, E)$ of measurable functions g such that the essential supremum $\|g\|_\infty$ of the function $X \ni x \mapsto \|g(x)\|_E \in \mathbf{R}_+$ is finite. In case E is a Hilbert space, $L^2(X, \mu, E)$ is also a Hilbert space; it appears for example in Construction 2.B.1.

Radon–Nikodym derivatives and equivalence of measures. Let μ_1 and μ_2 be two positive measures on a Borel space $(X, \mathcal{B})$. The measures μ_1 and μ_2 are **singular with each other**, or **disjoint**, if there exist disjoint sets $A_1, A_2 \in \mathcal{B}$ such that $\mu_1(X \smallsetminus A_1) = 0$ and $\mu_2(X \smallsetminus A_2) = 0$. The measure μ_2 is **absolutely continuous** with respect to μ_1 if $\mu_2(B) = 0$ for every $B \in \mathcal{B}$ such that $\mu_1(B) = 0$. The measures μ_1 and μ_2 are **equivalent** if μ_2 is absolutely continuous with respect to μ_1 and μ_1 absolutely continuous with respect to μ_2.

THEOREM A.3.2 (**Radon–Nikodym**). *Let μ_1, μ_2 be two σ-finite positive measures on $(X, \mathcal{B})$. Suppose that μ_2 is absolutely continuous with respect to μ_1.*

Then there exists a measurable function $f : X \to \mathbf{R}_+$ such that

$$\mu_2(B) = \int_B f(x)\, d\mu_1(x) \qquad \text{for all } B \in \mathcal{B}.$$

The function f is uniquely determined by μ_1 and μ_2 up to equality μ_1-almost everywhere.

Moreover, f is locally μ_1-integrable, i.e., $\int_B f(x)d\mu_1(x) < \infty$ for all $B \in \mathcal{B}$ such that $\mu_1(B) < \infty$.

We refer to [**Rudi–66**, Chapter 6]. In the situation of the theorem, the function f is called the **Radon–Nikodym derivative** of μ_2 with respect to μ_1, and it is denoted by $d\mu_2/d\mu_1$. We write $\mu_2 = f\mu_1$, or $d\mu_2(x) = f(x)d\mu_1(x)$.

Observe that $f = d\mu_2/d\mu_1$ is determined by the relation

$$(*) \qquad \int_X \varphi(x)\, d\mu_2(x) = \int_X \varphi(x) \frac{d\mu_2}{d\mu_1}(x)\, d\mu_1(x)$$

for every measurable function $\varphi : X \to \mathbf{R}_+$.

There is a chain rule: let μ_1, μ_2, μ_3 be three σ-finite positive measures on $(X, \mathcal{B})$ such that μ_3 is absolutely continuous with respect to μ_2 and μ_2 absolutely continuous with respect to μ_1; then $d\mu_3/d\mu_1 = (d\mu_3/d\mu_2)\,(d\mu_2/d\mu_1)$, the equality being μ_1-almost everywhere. It follows that, if μ_1 and μ_2 are equivalent, the Radon–Nikodym derivative $d\mu_2/d\mu_1$ can be viewed as a measurable function $X \to \mathbf{R}_+^\times$.

THEOREM A.3.3 (**Lebesgue decomposition theorem**). *Let μ_1, μ_2 be two σ-finite positive measures on $(X, \mathcal{B})$.*

There is a unique pair $(\mu_{2,a}, \mu_{2,s})$ of σ-finite positive measures on $(X, \mathcal{B})$ such that $\mu_2 = \mu_{2,a} + \mu_{2,s}$, with $\mu_{2,a}$ absolutely continuous with respect to μ_1 and $\mu_{2,s}$, μ_1 disjoint.

It follows from Theorem A.3.3 that two σ-finite positive measures μ_1 and μ_2 on $(X, \mathcal{B})$ are not singular with each other if and only if there exists a nonzero measure μ which is absolutely continuous with respect to both μ_1 and μ_2.

Radon–Nikodym cocycle. Let $(X, \mathcal{B})$ be a Borel space, equipped with a σ-finite measure μ. The **direct image** of μ by a Borel map $\theta : X \to X$ is the measure $\theta_*(\mu)$ on $(X, \mathcal{B})$ defined by

$$\theta_*(\mu)(B) = \mu(\theta^{-1}(B)) \qquad \text{for all } B \in \mathcal{B}.$$

Equivalently, $\theta_*(\mu)$ is defined by

$$(**) \qquad \int_X \varphi(x)\, d\theta_*(\mu)(x) = \int_X \varphi(\theta(x))\, d\mu(x)$$

for every measurable function $\varphi : X \to \mathbf{R}_+$.

Let $Aut(X, \mu)$ denote the set of Borel automorphisms $\theta : X \to X$ such that $\theta_*(\mu)$ is equivalent to μ. It is obvious that $\theta_1 \circ \theta_2^{-1} \in Aut(X, \mu)$ for $\theta_1, \theta_2 \in Aut(X, \mu)$, so that $Aut(X, \mu)$ is a group.

We define $c : Aut(X, \mu) \times X \to \mathbf{R}_+^\times$ by

$$c(\theta, x) = \frac{d(\theta^{-1})_*(\mu)}{d\mu}(x) \qquad \text{for all } x \in X,\ \theta \in Aut(X, \mu).$$

We have

$$(***) \qquad \int_X \varphi(\theta(x))c(\theta, x)\, d\mu(x) \;=\; \int_X \varphi(x)\, d\mu(x)$$

for every measurable function $\varphi : X \to \mathbf{R}_+$; indeed, $\int_X \varphi(\theta(x))d(\theta^{-1})_*(\mu)(x)$ is equal to the left-hand term by $(*)$ and to the right-hand term by $(**)$. The function $x \mapsto c(\theta, x)$ is determined by the identity $(***)$, as an identity for all measurable function $\varphi : X \to \mathbf{R}_+$.

PROPOSITION A.3.4. *Let $(X, \mathcal{B})$ be a Borel space and μ a σ-finite measure on $(X, \mathcal{B})$.*

The function $c\colon \mathrm{Aut}(X, \mu) \times X \to \mathbf{R}_+^\times$ introduced above satisfies the following cocycle relation:

$$(\mathrm{RN}) \qquad c(\theta_1 \circ \theta_2, x) \;=\; c(\theta_1, \theta_2(x))c(\theta_2, x)$$

for all $\theta_1, \theta_2 \in \mathrm{Aut}(X, \mu)$ and μ-almost all $x \in X$.

PROOF. Using relation $(***)$ above for θ_1 and for θ_2, we have

$$\begin{aligned}\int_X \varphi(x)\, d\mu(x) \;&=\; \int_X \varphi(\theta_1(x))c(\theta_1, x)\, d\mu(x)\\ &=\; \int_X \varphi(\theta_1(\theta_2(x))c(\theta_1, \theta_2(x))c(\theta_2, x)\, d\mu(x)\end{aligned}$$

for every measurable function $\varphi : X \to \mathbf{R}_+$ and this proves the claim. □

Using the terminology introduced in Definition 5.B.1, Proposition A.3.4 shows that $c\colon \mathrm{Aut}(X, \mu) \times X \to \mathbf{R}_+^\times$ is a cocycle over (X, μ) with values in $\mathbf{R}_+^\times$ for the tautological (left) action of the group $\mathrm{Aut}(X, \mu)$ on X.

For reference, we record the following elementary fact.

PROPOSITION A.3.5. *Let μ be a nonzero σ-finite measure on a Borel space $(X, \mathcal{B})$.*

Then there exists a probability measure ν on $(X, \mathcal{B})$ which is equivalent to μ.

PROOF. Let $(B_n)_n$ be a sequence in $\mathcal{B}$ such that X is the disjoint union $\bigcup_n B_n$, and $0 < \mu(B_n) < \infty$ for every n. For each n, let μ_n be the probability measure defined by

$$\mu_n(B) \;=\; \frac{1}{\mu(B_n)}\mu(B \cap B_n) \qquad \text{for all } B \in \mathcal{B}.$$

Let $(c_n)_n$ be a sequence of positive real numbers such that $\sum_n c_n = 1$. The measure $\nu = \sum_n c_n\mu_n$ is a probability measure on $(X, \mathcal{B})$. For $B \in \mathcal{B}$, we have $\nu(B) = 0$ if and only if $\mu(B \cap B_n) = 0$ for every $n \geq 1$, that is, if and only if $\mu(B) = 0$. Therefore ν is equivalent to μ. □

We state two classical results of von Neumann. The first one is a version of his selection theorem [**vNeu–49**, Lemma 5]; we refer to the proof of Theorem Z.2 in [**Mack–76**, Chap. 2, § 2.2].

A σ-finite measure μ on a Borel space $(X, \mathcal{B})$ is said to be a **standard measure** if there exists a Borel subset N of X with $\mu(N) = 0$ such that $X \smallsetminus N$ is a standard Borel space.

THEOREM A.3.6. *Let X be a Borel space equipped with a standard measure μ and let Y be a standard Borel space. Let Ω be a Borel subset of $X \times Y$ such that the restriction to Ω of the projection $X \times Y \twoheadrightarrow X$ is surjective.*

Then there exists a Borel subset X_1 of X with $\mu(X \smallsetminus X_1) = 0$ and a Borel map $\varphi : X_1 \to Y$ such that $(x, \varphi(x)) \in \Omega$ for every $x \in X_1$.

The second result is [**vNeu–32**, Satz 1] and shows that a standard probability space is characterized by its algebra of essentially bounded functions; we refer to the proof in [**Dixm–vN**, Appendix IV].

THEOREM A.3.7. *Let X and Y be standard Borel spaces, equipped with σ-finite measures μ and ν. Assume that there exists an isomorphism of involutive algebras*

$$F : L^\infty(X, \mu) \to L^\infty(Y, \nu).$$

Then there exist Borel subsets N, M of X, Y respectively such that $\mu(N) = 0$, $\nu(M) = 0$ and a Borel isomorphism $f : X \smallsetminus N \to Y \smallsetminus M$ such that $f_*(\mu)$ is equivalent to ν and such that $F(\varphi)(f(x)) = \varphi(x)$ for all $\varphi \in L^\infty(X, \mu)$ and $x \in X \smallsetminus N$.

For reference, we record the following fact, which is sometimes called the isomorphism theorem for measures on standard Borel spaces; we refer to the elementary proof in [**Kech–95**, Theorem]. Recall (Theorem A.2.2) that an uncountable standard Borel space is Borel isomorphic to the interval $[0, 1]$.

THEOREM A.3.8. *Let $(X, \mathcal{B})$ be a standard Borel space equipped with a non-atomic probability measure μ.*

Then there exists a Borel isomorphism $f : X \to [0, 1]$ such that $f_*(\mu)$ is the Lebesgue measure on $[0, 1]$.

Complex measures. Unless explicitly written otherwise, measures in this book are positive measures. A **complex measure** on a Borel space $(X, \mathcal{B})$ is a map $\mu : \mathcal{B} \to \mathbf{C}$ such that $\mu(B) = \sum_i \mu(B_i)$ for all $B \in \mathcal{B}$ and all partitions $(B_i)_i$ of B such that $B_i \in \mathcal{B}$ for all i and $B_i \cap B_j = \emptyset$ for all i, j such that $i \neq j$. The **total variation** of such a complex measure μ is the finite positive measure $|\mu|$ defined by

$$|\mu|(B) = \sup \sum_i |\mu(B_i)|$$

for all $B \in \mathcal{B}$, where the supremum is taken over all Borel partitions $(B_i)_i$ of B as above. The space of complex measures on $(X, \mathcal{B})$ is a complex vector space, with a norm defined by $\|\mu\| = |\mu|(X)$.

There is a version of the Radon–Nikodym Theorem which implies that there exists a measurable function $h : X \to \mathbf{C}$ such that $|h(x)| = 1$ for all $x \in X$, and $\mu = h|\mu|$, i.e., $\mu(B) = \int_B h d|\mu|$ for all $B \in \mathcal{B}$. For a measurable function $f \in L^1(X, \mathcal{B}, |\mu|)$, the integral $\int_X f d\mu$ is defined to be $\int_X f h d|\mu|$. For complex measures, see [**Rudi–66**, Chapter 6].

A.4. Radon measures on locally compact spaces

Let X be a topological space; recall that the Borel σ-algebra $\mathcal{B}_X$ has been defined in Section A.2. Let μ be a measure on $(X, \mathcal{B}_X)$ such that

$$(R_1) \qquad \mu(K) < \infty \quad \text{for every compact subset } K \subset X.$$

The measure μ is **outer regular** if

$$(R_2) \qquad \mu(B) = \inf\{\mu(U) \mid B \subset U \text{ and } U \text{ open}\}$$

for all $B \in \mathcal{B}_X$, and **inner regular** if

$$(R_3) \qquad \mu(B) = \sup\{\mu(K) \mid B \supset K \text{ and } K \text{ compact}\}$$

for all $B \in \mathcal{B}_X$ such that $\mu(B) < \infty$ and for all open subsets B in X. A **Radon measure** on a locally compact space X is a measure on $(X, \mathcal{B}_X)$ which satisfies Conditions (R_1), (R_2), and (R_3).

Note that, if X is a σ-compact locally compact space, a Radon measure on $(X, \mathcal{B}_X)$ is σ-finite, by definition.

THEOREM A.4.1. *Let X be a second-countable locally compact space. Let μ a Borel measure on $(X, \mathcal{B}_X)$ which is finite on every compact subset of X.*

(1) *The measure μ is a Radon measure.*
(2) *The Banach space $L^p(X, \mu)$ is separable for $1 \leq p < \infty$.*

PROOF. For Claim (1), we refer to [**Rudi–66**, Theorem 2.18]. Rudin assumes that X is a locally compact space in which every open subset is σ-compact; the latter condition holds when X is second-countable, by Proposition A.1.4.

Here is a proof for Claim (2). Since X is locally compact and second-countable, there exists a countable family $\{U_n\}_n$ of relatively compact open subsets of X which is a basis of the topology. We may assume that $\{U_n\}_n$ is closed under finite unions. We denote by $\mathbf{1}_B$ the characteristic function of a Borel subset B of X.

We are going to show that the linear span of $\{\mathbf{1}_{U_n}\}_n$ is dense in $L^p(X, \mu)$. For this, consider $f \in L^q(X, \mu)$ such that $\langle f \mid \mathbf{1}_{U_n} \rangle = 0$ for every n, where q is the conjugate exponent of p. It suffices to show that $f = 0$.

Let U be any open subset of X of finite measure. Since U is a union of some of the U_n's, it follows from Lebesgue dominated convergence that $\int_U f(x) d\mu(x) = \langle f \mid \mathbf{1}_U \rangle = 0$.

Let B be a Borel subset of X of finite measure. By outer regularity of μ, there exists a sequence of open sets $(V_i)_i$ such that $B \subset V_i$ and $\mu(V_i) \leq \mu(B) + \frac{1}{i}$ for all i. Then $\|\mathbf{1}_B - \mathbf{1}_{V_i}\|_p^p = \mu(V_i \smallsetminus B) \leq \frac{1}{i}$, and so $\lim_i \mathbf{1}_{V_i} = \mathbf{1}_B$ in $L^p(X, \mu)$. It follows that $\int_B f(x) d\mu(x) = \langle f \mid \mathbf{1}_B \rangle = 0$. Since the linear span of $\{\mathbf{1}_B \mid B \in \mathcal{B}_X, \ \mu(B) < \infty\}$ is dense in $L^p(X, \mu)$, the claim follows. □

REMARK A.4.2. Locally compact spaces which are second-countable are of special importance. This hypothesis of second-countability implies that probability measures are Radon measures and that Radon measures are σ-finite, so that various fundamental theorems can be applied (Fubini, Radon–Nikodym). Note also that the results of our Section 1.H require separable Hilbert spaces, so that applications to $L^2(\widehat{G}, \mu)$ for an LCA group G require $\widehat{G}$ (and therefore G) to be second-countable.

There are known examples of compact spaces, necessarily not second-countable ones, having probability measures which are not Radon measures [**Boga–07**, Volume II, Example 7.1.3].

Let X be a locally compact space, μ a Radon measure on $(X, \mathcal{B}_X)$, and $C^c(X)$ the algebra of complex-valued continuous functions on X with compact supports. Then the map

$$C^c(X) \to \mathbf{C}, \qquad f \mapsto \int_X f d\mu$$

is well-defined, linear, and positive in the sense that $\int_X f d\mu \geq 0$ whenever $f(x) \geq 0$

for all $x \in X$. Conversely:

THEOREM A.4.3 (**Riesz representation theorem**). *Let X be a locally compact space.*

Every positive linear form on $C^c(X)$ is of the form $f \mapsto \int_X f d\mu$ for a Radon measure μ on $(X, \mathcal{B}_X)$, for all $f \in C^c(X)$.

For all this, see [**Rudi–66**, Chapter 2].

Complex Radon measures. Since "measure" means "positive measure" in this book, we also agree that "Radon measure" means "positive Radon measure", unless explicitly written otherwise. A **complex Radon measure** on a locally compact space X is a complex measure on $(X, \mathcal{B})$ of which the total variation is a finite Radon measure. We denote by $M^b(X, \mathcal{B}_X)$, or for short by $M^b(X)$, the space of complex Radon measures on X. It is a complex vector space. For the norm defined by $\|\mu\| = |\mu|(X)$ as in Section A.2, it is a Banach space [**Rudi–66**, Exercice 6.3].

The following theorem is a version of the Riesz representation theorem for complex Radon measures. Since the total variation of a complex measure is a finite measure, it is convenient to use the space $C^0(X)$ of continuous functions on X vanishing at infinity, rather than the space $C^c(X)$ as in Theorem A.4.3; recall that $C^c(X)$ is dense in the Banach space $C^0(X)$. For a proof of this theorem, we refer to [**Rudi–66**, Theorem 6.19].

THEOREM A.4.4 (**Riesz representation theorem for complex measures**). *Let X be a locally compact space.*

To each bounded linear form $\Phi : C^0(X) \to \mathbf{C}$ there corresponds a unique complex Radon measure μ on $(X, \mathcal{B}_X)$ such that $\Phi(f) = \int_X f d\mu$ for all $f \in C^0(X)$. Moreover $\|\Phi\| = |\mu|(X)$.

The assignment $\Phi \mapsto \mu$ provides an isomorphism from the dual of the Banach space $C^0(X)$ onto the Banach space $M^b(X)$ of complex Radon measures on $(X, \mathcal{B})$.

A.5. Groups and actions

Let $\mathcal{P}$ and $\mathcal{Q}$ be group properties. A group Γ is $\mathcal{P}$-by-$\mathcal{Q}$ if it has a normal subgroup N with Property $\mathcal{P}$ such that the quotient Γ/N has Property $\mathcal{Q}$. When $\mathcal{P}$ is a property inherited by subgroups and $\mathcal{Q}$ is the property of being finite, a group with property $\mathcal{P}$-by-$\mathcal{Q}$ is also called a virtually $\mathcal{P}$ group. In particular, a **virtually abelian** group is a group which has an abelian subgroup of finite index.

For a group Γ, we denote by $\mathrm{FC}(\Gamma)$ its **FC-centre**, that is, the normal subgroup of elements in Γ with a finite conjugacy class. A group Γ is an **FC-group** if $\mathrm{FC}(\Gamma) = \Gamma$. The following result is classical.

PROPOSITION A.5.1. *Let Γ be group.*

1. *If Γ is an FC-group, then the derived group $D(\Gamma)$ is finite.*
2. *If Γ is finitely generated and the FC-centre* $\mathrm{FC}(\Gamma)$ *is of finite index in Γ, then Γ is virtually abelian.*

PROOF. The result of Claim (1) is due to Schur. We refer to [**Robi–96**, Theorem 10.1.4].

For Claim (2), note first that $\mathrm{FC}(\Gamma)$ is finitely generated, as a finite index subgroup of a finitely generated group. Choose a finite generating subset S. For every $s \in S$, the centralizer $Z_{\mathrm{FC}(\Gamma)}(s)$ is of finite index in $\mathrm{FC}(\Gamma)$. Then the subgroup $A := \bigcap_{s \in S} Z_{\mathrm{FC}(\Gamma)}(s)$ is of finite index in $\mathrm{FC}(\Gamma)$. Moreover $A = \bigcap_{\gamma \in \mathrm{FC}(\Gamma)} Z_{\mathrm{FC}(\Gamma)}(\gamma)$ is the centre of $\mathrm{FC}(\Gamma)$. It follows that A is a normal abelian subgroup of finite index in Γ. □

REMARK A.5.2. (1) The hypothesis of finite generation is necessary in Proposition A.5.1(2): see Example 7.D.7.

(2) A virtually abelian finitely generated group need not have a centre of finite index. For example, the infinite dihedral group is virtually abelian and has a trivial centre.

DEFINITION A.5.3. A group Γ is **icc**, or has infinite conjugacy classes, if it is not $\{1\}$ and if all its conjugacy classes distinct from $\{1\}$ are infinite; equivalently: if $\{1\} = \mathrm{FC}(\Gamma) \neq \Gamma$.

EXAMPLE A.5.4. (1) Let $\mathrm{Aff}(\mathbf{K})$ be the affine group over an infinite field $\mathbf{K}$, as in Section 3.C. This group is icc, as is easy to check

(2) Similarly, for any integer $n \geq 2$, the Baumslag–Solitar group $\mathrm{BS}(1, n)$ of Section 3.D is icc.

(3) For any infinite group H and any group $A \neq \{e\}$, the restricted wreath product $H \wr A$ is icc. This applies to the lamplighter group $\mathbf{Z} \wr (\mathbf{Z}/2\mathbf{Z})$ of Section 3.E.

(4) For an infinite field $\mathbf{K}$ and an integer $n \geq 2$, the general linear group $\Gamma = \mathrm{GL}_n(\mathbf{K})$ of Section 3.F is not type I. The quotient group $\mathrm{PGL}_n(\mathbf{K})$ can be shown to be icc, for example because it is Zariski dense in the appropriate algebraic group with trivial centre (compare [**BeHa–94**, Proposition 2]).

(5) Examples of icc groups are very abundant. Other than those listed above, let us mention nonabelian free groups and appropriate groups acting on trees [**Corn–09**], lattices in semisimple connected real Lie groups without centre and without compact factors (as a consequence of the Borel density theorem), and weakly branch groups [**Grig–11**, Theorem 9.17].

DEFINITION A.5.5. Let Γ be a group. Two subgroups H, H' of Γ are **commensurate** if $H \cap H'$ has finite index in both H and H'. The **commensurator** of H in Γ is the subgroup

$$\mathrm{Comm}_\Gamma(H) = \{\gamma \in \Gamma \mid H \text{ and } \gamma^{-1} H \gamma \text{ are commensurate}\}$$

of Γ.

PROPOSITION A.5.6. *Let Γ be a group and H a subgroup.*

(1) *The commensurator $\mathrm{Comm}_\Gamma(H)$ is a subgroup of Γ containing H.*

For $\gamma \in \Gamma$, the following properties are equivalent:

(2) *$\gamma \in \mathrm{Comm}_\Gamma(H)$, i.e., $H \cap \gamma^{-1} H \gamma$, is of finite index in both H and $\gamma^{-1} H \gamma$;*
(3) *$H \cap \gamma^{-1} H \gamma$ and $H \cap \gamma H \gamma^{-1}$ are both of finite index in H;*
(4) *the H-orbits of the classes $H\gamma$ and $H\gamma^{-1}$ in $H \backslash \Gamma$ are both finite;*

(5) *there exists a finite index subgroup K of H such that $H\gamma$ and $H\gamma^{-1}$ are fixed by K.*

(Note that (4) *and* (5) *refer to the canonical right action $H\backslash\Gamma \curvearrowleft H$.)*

Consider a right transversal T for H in Γ and the restriction to H of the right action $T \curvearrowleft \Gamma$ defined in Construction 1.F.4*(2). Recall from there that, for $t \in T$, the point $t^* \in T$ is defined by $Ht^{-1} = Ht^*$. Then, for $t \in T$:*

(6) *$t \in \mathrm{Comm}_\Gamma(H)$ if and only if the H-orbits of t and t^* in T are both finite.*

Moreover $\mathrm{Comm}_\Gamma(H)$ *depends on the commensurability class of H only:*

(7) *If H and H' are two commensurate subgroups of Γ,*
then $\mathrm{Comm}_\Gamma(H) = \mathrm{Comm}_\Gamma(H')$.

There are similar equivalent properties in terms of left actions $\Gamma \curvearrowright \Gamma/H$ and $\Gamma \curvearrowright S$, with S a left transversal for H in Γ.

PROOF. We show that (3) $\Leftrightarrow$ (4) $\Leftrightarrow$ (5), and leave the verification of the other claims to the reader.

Let $\gamma \in \Gamma$. For the action $H\backslash\Gamma \curvearrowleft H$, denote by $\mathrm{Stab}_H(H\gamma)$ the stabilizer in H of $H\gamma$, i.e., $\mathrm{Stab}_H(H\gamma) = H \cap \gamma^{-1}H\gamma$. The orbit $(H\gamma)H$ is in canonical bijection with the coset space $\mathrm{Stab}_H(H\gamma)\backslash H = (H \cap \gamma^{-1}H\gamma)\backslash H$, and the orbit $(H\gamma^{-1})H$ is in canonical bijection with $(H \cap \gamma H\gamma^{-1})\backslash H$. The equivalence between (3) and (4) follows.

A point $H\gamma$ in the space $H\backslash\Gamma$ has a finite H-orbit if and only if there exists a subgroup of finite index K in H which fixes $H\gamma$. The equivalence between (4) and (5) follows. □

Group actions.

DEFINITION A.5.7. An **action from the left** of a group G on a set X is a map $\alpha : G \times X \to X$, $(g,x) \mapsto gx$ such that $(g_1g_2)x = g_1(g_2x)$ and $ex = x$ for all $g_1, g_2 \in G$ and $x \in X$. Such an action is equivalently described by the group homomorphism $G \to \mathrm{Sym}(X)$ mapping $g \in G$ to the permutation $x \mapsto gx$ of X. As agreed in Notation 1.A.16, we refer to such an action by the notation

$$G \curvearrowright X.$$

An **action from the right** of G on X is a map $X \times G \to X$, $(x,g) \mapsto xg$ such that $x(g_1g_2) = (xg_1)g_2 = x(g_1g_2)$ and $xe = x$ for all $g_1, g_2 \in G$ and $x \in X$. Our corresponding notation is

$$X \curvearrowleft G.$$

For brevity, we often write "action" for "action from the left".

Let $(X, \mathcal{B})$ be a Borel space and G a topological group. A **measurable action** of G on $(X, \mathcal{B})$ is an action of G by measurable automorphisms of $(X, \mathcal{B})$ such that the map $G \times X \to X$, $(g,x) \mapsto gx$ is Borel, where G is equipped with the Borel structure induced by its topology and $G \times X$ with the product Borel structure.

DEFINITION A.5.8. Consider a measurable action of a topological group G on a measurable space $(X, \mathcal{B})$, and a measure μ on $(X, \mathcal{B})$.

- The measure μ is **invariant** by G if $\mu(gB) = \mu(B)$ for all $g \in G$ and $B \in \mathcal{B}$; equivalently: if the image $g_*(\mu)$ is equal to μ for all $g \in G$.

- The measure μ is **quasi-invariant** by G if the set of μ-null elements in $\mathcal{B}$ is G-invariant, that is, if we have $\mu(gB) = 0$ for all $g \in G$ and $B \in \mathcal{B}$ such that $\mu(B) = 0$; equivalently: if the image $g_*(\mu)$ is equivalent to μ for all $g \in G$.
- Suppose moreover that G is locally compact and second-countable, and that μ is σ-finite and quasi-invariant by G. The action of G on $(X, \mathcal{B}, \mu)$ is **ergodic**, and the measure μ is **ergodic** for the action of G, if, for every G-invariant subset $B \in \mathcal{B}$, we have either $\mu(B) = 0$ or $\mu(X \smallsetminus B) = 0$. Equivalently (see Proposition A.5.9), this action is ergodic if, for every $B \in \mathcal{B}$ such that $\mu(gB \Delta B) = 0$ for all $g \in G$, we have either $\mu(B) = 0$ or $\mu(X \smallsetminus B) = 0$.
- Suppose now that Γ is countable and that μ is nonatomic. The action is **aperiodic** if the orbit Γx is infinite for μ-almost all $x \in X$.

Given an action of a group G on a measure space $(X, \mathcal{B}, \mu)$, a measurable function $f : X \to \mathbf{C}$ is **essentially G-invariant** if, for every $g \in G$, the function $x \mapsto f(gx)$ is constant μ-almost everywhere.

PROPOSITION A.5.9. *Let G be a second-countable locally compact group acting measurably on a Borel space $(X, \mathcal{B})$ and let μ be a quasi-invariant σ-finite measure on $(X, \mathcal{B})$. The following properties are equivalent:*

(i) *the action of G on $(X, \mathcal{B}, \mu)$ is ergodic;*
(ii) *every essentially G-invariant measurable function $f : X \to \mathbf{C}$ is constant μ-almost everywhere.*

For the proof, see Theorem 1.3 in [**BeMa–00**], or Proposition 13.A.3 for the particular case of a countable group G.

The following fact is given here with a proof, for the reader's convenience.

PROPOSITION A.5.10. *Let Γ be a countable group acting measurably on a standard Borel space $(X, \mathcal{B})$ and let μ be a nonatomic probability measure on $(X, \mathcal{B})$ which is invariant and ergodic.*

Then the action $\Gamma \curvearrowright (X, \mathcal{B}, \mu)$ is aperiodic.

PROOF. For an integer $n \geq 1$, let X_n be set of elements $x \in X$ such that $|\Gamma x| \leq n$. It is clear that X_n is Γ-invariant.

We claim that $X_n \in \mathcal{B}$. Indeed, let $\mathcal{P}_n$ be the set of finite subsets Δ of Γ such that $|\Delta| \leq n$. We have

$$X_n = \bigcup_{\Delta \in \mathcal{P}_n} \bigcap_{\gamma \in \Gamma} \bigcup_{\delta \in \Delta} \{x \in X \mid \gamma \cdot x = \delta \cdot x\}.$$

Since Γ and $\mathcal{P}_n$ are countable and since $\{x \in X \mid \gamma \cdot x = \delta \cdot x\} \in \mathcal{B}$ for every $\gamma, \delta \in \Gamma$, the claim follows.

We have to prove that

$$\mu\Big(\bigcup_{n \geq 1} X_n \Big) = 0.$$

For this, it suffices to show that $\mu(X_n) = 0$ for every $n \geq 1$.

Assume, by contradiction, that $\mu(X_n) \neq 0$ for some $n \geq 1$. It follows from the ergodicity of the action $\Gamma \curvearrowright (X, \mathcal{B}, \mu)$ that $\mu(X_n) = 1$.

Since $(X, \mathcal{B})$ is a standard Borel space, we may assume that $X = [0,1]$ and that $\mathcal{B}$ is the σ-algebra of all Borel subsets of $[0,1]$. Define

$$B := \{x \in X_n \mid x \leq \gamma \cdot x \ \text{ for all } \ \gamma \in \Gamma\} = \bigcap_{\gamma \in \Gamma} \{x \in X_n \mid x \leq \gamma \cdot x\}.$$

Then $B \in \mathcal{B}$ and $X_n = \bigcup_{\gamma \in \Gamma} \gamma B$. As $\mu(X_n) > 0$, we have $\mu(B) > 0$ by invariance of μ. Since μ is nonatomic, therefore there exists a set $A \in \mathcal{B}$ such that $A \subset B$ and $0 < \mu(A) < \mu(B)$. Then

$$A' := \bigcup_{\gamma \in \Gamma} \gamma A$$

is a Γ-invariant set in $\mathcal{B}$ such that

$$0 < \mu(A') < \mu(X_n) = 1.$$

This is a contradiction to the ergodicity of the Γ-action on X. □

A.6. Locally compact groups

Let G be a locally compact group. There exists on G a nonzero Radon measure μ_G which is left-invariant, that is, such that $\mu_G(gB) = \mu_G(B)$ for all $g \in G$ and a Borel subset B of G, and such a measure is unique up to multiplication by a positive real number. This is called a **left Haar measure** on G, or often, abusively, the Haar measure on G.

For each $g \in G$, there is a positive number $\Delta(g)$ such that, for each Borel subset B of G, we have

$$\mu_G(Bg) = \Delta_G(g)\mu_G(B).$$

The map

$$\Delta_G : G \to \mathbf{R}_+^\times, \qquad g \mapsto \Delta_G(g)$$

is a continuous homomorphism called the **modulus**, or **modular function**, of G. For $\varphi \in C_c(G)$, we have

$$\int_G \varphi(xg^{-1}) \, d\mu_G(g) = \Delta_G(g) \int_G \varphi(x) \, d\mu_G(x) \qquad \text{for all } \ g \in G,$$

$$\int_G \varphi(x^{-1}) \, d\mu_G(g) = \int_G \varphi(x)\Delta_G(x^{-1}) \, d\mu_G(x).$$

The group G is **unimodular** if $\Delta(g) = 1$ for all $g \in G$, i.e., if the measure μ_G on G is right-invariant (and so left- and right-invariant). Examples of unimodular groups include compact groups, locally compact abelian groups, locally compact nilpotent groups, and semisimple Lie groups. More on modular functions and unimodular groups in [**BeHV–08**, A.3].

On the Banach space $M^b(G)$ of complex Radon measures on G, there is a **convolution product** $(\nu, \sigma) \mapsto \nu * \sigma$, where $\nu * \sigma$ is the direct image of the measure $\nu \times \sigma$ by the multiplication $G \times G \to G$; alternatively, using the Riesz representation Theorem A.4.4, we have

$$\int_G \varphi(x) \, d(\nu * \sigma)(x) = \int_G \int_G \varphi(yz) \, d\nu(y) d\sigma(z) \qquad \text{for all } \ \varphi \in C^0(G).$$

It is easy to check that $\|\nu * \sigma\| \leq \|\nu\| \, \|\sigma\|$. On $M^b(G)$, there is also an involution $\nu \mapsto \nu^*$ defined by

$$\nu^*(B) = \overline{\nu(B^{-1})} \qquad \text{for all Borel set } \ B \in \mathcal{B}_G.$$

This makes $M^b(G)$ a Banach involutive algebra called the **measure algebra** of the locally compact group G. It is an algebra with unit, which is the Dirac measure supported at the origin of G.

Let φ, ψ be two complex-valued measurable functions on G. Their **convolution** is the function defined on G by

$$(\varphi * \psi)(g) = \int_G \varphi(gx)\psi(x^{-1})\, d\mu_G(x) = \int_G \varphi(x)\psi(x^{-1}g)\, d\mu_G(x) \quad \text{for all } g \in G$$

whenever these integrals make sense. There is an **involution** defined on these functions by

$$\varphi^*(g) = \Delta(g^{-1})\overline{\varphi(g^{-1})}.$$

For $\varphi, \psi \in L^1(G, \mu_G)$, the value $(\varphi * \psi)(g)$ is defined for μ_G-almost all $g \in G$; the resulting (equivalence class of the) function $\varphi * \psi$ is again in $L^1(G, \mu_G)$. Similarly, φ^* is in $L^1(G, \mu_G)$. It is important to note that the convolution product $*$ depends on the choice of the Haar measure μ_G.

For $\varphi, \psi \in L^1(G, \mu_G)$, the measures $\varphi\mu_G$ and $\psi\mu_G$ are in $M^b(G)$. The convolution $\varphi * \psi$ can equivalently be defined by

$$(\varphi * \psi)\mu_G = (\varphi\mu_G) * (\varphi\mu_G),$$

and the involution by

$$\varphi^*\mu_G = (\varphi\mu_G)^*.$$

The Banach space $L^1(G, \mu_G)$, together with this convolution product and this involution, is called the **group algebra** of the locally compact group G. The map

$$L^1(G, \mu_G) \to M^b(G), \qquad \varphi \mapsto \varphi\mu_G$$

is an isometric morphism of Banach involutive algebras; it identifies $L^1(G, \mu_G)$ as a closed two-sided *-ideal of $M^b(G)$. More precisely, let $\nu \in M^b(G)$ and $\varphi \in L^1(G, \mu_G)$. Then $\nu * (\varphi\mu_G) = (\nu * \varphi)\mu_G$ and $(\varphi\mu_G) * \nu = (\varphi * \nu)\mu_G$ for the functions $\nu * \varphi$ and $\varphi * \nu$ in $L^1(G, \mu_G)$ given by

$$(\nu * \varphi)(g) = \int_G \varphi(x^{-1}g)\, d\nu(x)$$

and

$$(\varphi * \nu)(g) = \int_G \Delta(x^{-1})\varphi(gx^{-1}))\, d\nu(x)$$

for all $g \in G$.

For a discrete group Γ and the counting measure μ_Γ, we have $L^1(\Gamma, \mu_\Gamma) = M^b(\Gamma)$; the notation is then $\ell^1(\Gamma)$ more often than $L^1(\Gamma, \mu_\Gamma)$. For a locally compact group G which is not discrete, the group algebra $L^1(G, \mu_G)$ does not have a unit; in particular $L^1(G, \mu_G)$ does not coincide with $M^b(G)$. For an analysis of the complement $M^b(G) \smallsetminus L^1(G, \mu_G)$, see Theorems 19.20 and 20.25 in [**HeRo–63**].

For $\varphi, \psi \in C^c(G)$, note that $\varphi * \psi$ and φ^* have compact supports, so that $C^c(G)$ is also a convolution algebra, and a $*$-subalgebra of $L^1(G, \mu_G)$ and of $M^b(G)$. Also, for two functions φ, ψ with finite support, the convolution has finite support, so that $\mathbf{C}[G]$ is a convolution algebra. When G is discrete, $\mathbf{C}[G]$ is a $*$-subalgebra of the group algebra $\ell^1(G)$. But $\mathbf{C}[G]$ can be of use even when G is not discrete; see for example our Proposition 1.B.6.

The following result can be found in many references. For discussion and details, see, e.g., [**BeHV–08**, Appendix B]. Quasi-invariance has been defined in Definition A.5.8.

PROPOSITION A.6.1. *Let G be a locally compact group, H a closed subgroup, and G/H the corresponding G-space.*

(1) *There exists a nonzero Radon measure μ on G/H which is quasi-invariant by G. The support of such a measure is the whole of G/H. Two quasi-invariant Radon measures on G/H are equivalent to each other.*
(2) *There exists a nonzero Radon measure on G/H which is invariant by G if and only if the modular function of H is the restriction to H of the modular function of G.*

A.7. Locally compact abelian groups and duality

Let G be a locally compact abelian group, for short an **LCA group**. The dual group of G, or **Pontryagin dual** of G, is the group $\widehat{G}$ of all continuous homomorphisms from G to $\mathbf{T}$, with the compact-open topology; it is itself a locally compact abelian group. Let μ_G be a Haar measure on G.

REMARK A.7.1. The topology on $\widehat{G}$ admits another description. Elements of $\widehat{G}$ are bounded continuous functions from G to $\mathbf{C}$, so that $\widehat{G}$ embeds naturally in $L^\infty(G, \mu_G)$. The restriction to $\widehat{G}$ of the weak*-topology of $L^\infty(G, \mu_G)$ coincides with the previously defined topology of $\widehat{G}$ (see Proposition 1 in [**BTS1–2**, Chap. II, § 1, No 1]).

For $g \in G$, denote by $\widehat{g} \in \widehat{\widehat{G}}$ the unitary character of $\widehat{G}$ defined by

$$\widehat{g}(\chi) = \chi(g) \quad \text{for } \chi \in \widehat{G}.$$

THEOREM A.7.2 (**Pontryagin duality**). *Let G be a locally compact abelian group, and $\widehat{G}$ its dual.*

The map

$$G \mapsto \widehat{\widehat{G}}, \qquad g \mapsto \widehat{g}$$

is an isomorphism of topological groups.

REFERENCES FOR PROOFS. The theorem is essentially due to Pontryagin. A proof and historical notes can be found in Chapter VI of [**Weil–40**]. We also refer to Théorème 2 in [**BTS1–2**, Chap. II, § 1, No 5]. □

The theorem suggests that properties of G are reflected by properties of $\widehat{G}$. For example, and as shown in Proposition 1.D.6, a locally compact abelian group G is compact if and only if $\widehat{G}$ is discrete. Also:

PROPOSITION A.7.3. *Let G be a locally compact abelian group.*

Then G is second-countable if and only if $\widehat{G}$ is second-countable.

PROOF. By Pontryagin duality, it suffices to show that, if G is second-countable, then $\widehat{G}$ is second-countable.

We consider $\widehat{G}$ as a subspace of $L^\infty(G, \mu_G)$, and the latter with the weak*-topology; see Remark A.7.1. Fix a countable dense subset Q of $\mathbf{C}$. A basis for the topology of $\widehat{G}$ is given by finite intersections of sets of the form

$$U_{f,q,n} = \left\{ \chi \in \widehat{G} \mid |\chi(f) - q| < 1/n \right\},$$

with $f \in L^1(G, \mu_G)$, $q \in Q$, and an integer $n \geq 1$, where we have written $\chi(f)$ for $\int_G \chi(g) f(g) d\mu_G(g)$.

Since G is second-countable, $L^1(G, \mu_G)$ is separable; so, there exists a dense sequence $(f_k)_{k\geq 1}$ in $L^1(G, \mu_G)$. The family of finite intersections of the sets

$$U_{f_k,q,n} = \left\{\chi \in \widehat{G} \mid |\chi(f_k) - q| < 1/n\right\}$$

for k, $n \geq 1$, and $q \in Q$, is countable and is clearly a basis for the topology of $\widehat{G}$. □

Let μ be a complex Radon measure on G. The **Fourier–Stieltjes transform** of μ is the function on $\widehat{G}$, denoted by $\mathcal{F}(\mu)$ or $\widehat{\mu}$, defined by

$$\mathcal{F}(\mu)(\chi) = \int_G \overline{\chi(g)}\, d\mu(g) \qquad \text{for all } \chi \in \widehat{G}.$$

The **inverse Fourier–Stieltjes transform** of μ is the function on $\widehat{G}$, denoted by $\overline{\mathcal{F}}(\nu)$, defined by

$$\overline{\mathcal{F}}(\mu)(\chi) = \mathcal{F}(\mu)(\chi^{-1}) = \int_G \widehat{g}(\chi)\, d\mu(\chi) = \int_G \chi(g)\, d\mu(\chi).$$

(Our notation agrees with that of [**BTS1–2**].) Both $\mathcal{F}(\mu)$ and $\overline{\mathcal{F}}(\mu)$ are bounded continuous complex-valued functions on $\widehat{G}$. If $\check{\mu}$ denotes the direct image of a measure μ on G by the map $G \to G$, $g \mapsto g^{-1}$, observe that $\overline{\mathcal{F}}(\mu) = \mathcal{F}(\check{\mu})$.

Let $f \in L^1(G, \mu_G)$; observe that $f\mu_G$ is a complex measure on G. The **Fourier transform** of the function f, denoted by $\mathcal{F}(f)$ or $\widehat{f}$, is the Fourier–Stieltjes transform of the measure $f\mu_G$, that is,

$$\mathcal{F}(f)(\chi) = \int_G \overline{\chi(g)} f(g)\, d\mu_G(g) \qquad \text{for all } \chi \in \widehat{G}.$$

Similarly, the **inverse Fourier transform** $\overline{\mathcal{F}}(f)$ of f is the inverse Fourier transform of $f\mu_G$, that is,

$$\overline{\mathcal{F}}(f)(\chi) = \int_G \chi(g) f(g)\, d\mu_G(\chi) \qquad \text{for all } \chi \in \widehat{G}.$$

For the next proposition, we refer to [**BTS1–2**, Chap. II, § 1], or to [**Foll–16**, Chapter 4]. Recall that $M^b(G)$ is a commutative Banach involutive algebra for convolution, and $C^b(G)$ a commutative C*-algebra for the pointwise product.

PROPOSITION A.7.4. *Let G be a locally compact abelian group.*

(1) *For $\mu \in M^b(G)$ a complex Radon measure on G, the Fourier–Stieltjes transform of μ and the inverse Fourier–Stieltjes transform of μ are in $C^b(\widehat{G})$.*
(2) *The mappings $\mathcal{F} : M^b(G) \to C^b(\widehat{G})$ and $\overline{\mathcal{F}} : M^b(G) \to C^b(\widehat{G})$ are morphisms of Banach involutive algebras.*
(3) *For $f \in L^1(G, \mu_G)$, the Fourier–Stieltjes transform of f and the inverse Fourier–Stieltjes transform of f are in $C^0(\widehat{G})$.*
(4) *The mappings $\mathcal{F} : M^b(G) \to C^b(\widehat{G})$ and $\overline{\mathcal{F}} : M^b(G) \to C^b(\widehat{G})$ are injective.*

The next theorem characterizes the images of positive measures. We refer to [**Foll–16**, 4.23] or [**BeHV–08**, D.2.2].

THEOREM A.7.5 (**Bochner's Theorem**). *Let G be a locally compact abelian group.*

The map $\mu \mapsto \mathcal{F}(\mu)$ establishes a bijection from the space $M^b(\widehat{G})_+$ of finite positive Radon measures on the dual of G onto the space $P(G)$ of functions of positive type on G (as defined in Section 1.B*).*

Let $\mu_{\widehat{G}}$ be a Haar measure on $\widehat{G}$. For $f \in L^1(G, \mu_G) \cap L^2(G, \mu_G)$, the Fourier transform $\widehat{f}$ is in $L^1(\widehat{G}, \mu_{\widehat{G}}) \cap L^2(\widehat{G}, \mu_{\widehat{G}})$. Recall that $L^1(G, \mu_G) \cap L^2(G, \mu_G)$ is dense in $L^2(G, \mu_G)$. The map $f \mapsto \widehat{f}$ extends to a continuous linear map from $L^2(G, \mu_G)$ onto $L^2(\widehat{G}, \mu_{\widehat{G}})$, denoted again by $f \mapsto \widehat{f}$. Moreover, there is a choice of the Haar measure $\mu_{\widehat{G}}$ such that $\|\widehat{f}\|_{L^2(\widehat{G}, \mu_{\widehat{G}})} = \|f\|_{L^2(G, \mu_G)}$ for all $f \in L^2(G, \mu_G)$; the Haar measure $\mu_{\widehat{G}}$ on $\widehat{G}$ is said to be **associated** to the Haar measure μ_G on G. For this and for the next Theorem, see [**BTS1–2**, Chap. II, § 1, no 3].

THEOREM A.7.6 (**Plancherel's Theorem**). *Let G be a locally compact abelian group, and $\widehat{G}$ its dual. Let μ_G be a Haar measure on G and $\mu_{\widehat{G}}$ the Haar measure on $\widehat{G}$ which is associated with μ_G.*

Then the Fourier transform

$$L^2(G, \mu_G) \to L^2(\widehat{G}, \mu_{\widehat{G}}), \qquad f \mapsto \widehat{f},$$

is an isometry of the Hilbert space $L^2(G, \mu_G)$ onto the Hilbert space $L^2(\widehat{G}, \mu_{\widehat{G}})$.

Let G be a locally compact abelian group and $C^b(\widehat{G})$ the algebra of bounded continuous complex-valued function on its dual. The **algebra of trigonometric polynomials** on $\widehat{G}$ is the linear span of $\{\widehat{g} \mid g \in G\}$; it is a subalgebra of $C^b(\widehat{G})$.

Note that $\mathrm{Trig}(\widehat{G})$ is a dense subalgebra of the algebra $AP(\widehat{G})$ of almost periodic functions on $\widehat{G}$ for the topology of uniform convergence. Indeed, $\mathrm{Trig}(\widehat{G})$ is a separating $*$-subalgebra of the algebra $C(\mathrm{Bohr}(\widehat{G}))$ of continuous functions on the Bohr compactification of $\widehat{G}$, and the claim follows from the Stone–Weierstraß Theorem. (For $AP(\widehat{G})$ and $\mathrm{Bohr}(\widehat{G})$, see Section 4.C.)

Given a finite Radon measure μ on $\widehat{G}$, there is a canonical map from the algebra $C^b(\widehat{G})$ to the algebra $L^\infty(\widehat{G}, \mu)$. The image by this map of $\mathrm{Trig}(\widehat{G})$ is naturally a subalgebra of $L^\infty(\widehat{G}, \mu)$, and also a subspace of $L^p(\widehat{G}, \mu)$ for every $p \geq 1$.

The following proposition is used on several occasions, (2.A.6), Proposition 2.B.3, Proposition 14.A.2.

PROPOSITION A.7.7. *Let G be a second-countable locally compact abelian group and μ a probability measure on its dual $\widehat{G}$.*

(1) *The image of the algebra* $\mathrm{Trig}(\widehat{G})$ *is norm dense in $L^p(\widehat{G}, \mu)$ for every $p \in [1, +\infty[$.*
(2) *The image of the algebra* $\mathrm{Trig}(\widehat{G})$ *is weak*-dense in $L^\infty(\widehat{G}, \mu)$.*

PROOF. Given $p \in [1, +\infty]$, denote by $q \in [1, +\infty]$ the conjugate exponent, defined by $\frac{1}{p} + \frac{1}{q} = 1$. Let $L^p(\widehat{G}, \mu)$ be equipped with the norm topology when $p < +\infty$ and with the weak*-topology when $p = +\infty$. The topological dual space of $L^p(\widehat{G}, \mu)$ can be identified in the usual way with $L^q(\widehat{G}, \mu)$; see Remark A.3.1. To show both (1) and (2), using the Hahn–Banach Theorem, it suffices to show that a continuous linear form on $L^p(\widehat{G}, \mu)$ which vanishes on $\mathrm{Trig}(\widehat{G})$ is the zero form, i.e., to show the following.

CLAIM. If $f \in L^q(\widehat{G}, \mu)$ is such that

$$(\dagger) \qquad \int_{\widehat{G}} \widehat{g}(\chi) f(\chi)\, d\mu(\chi) = 0 \quad \text{for all } g \in G,$$

then $f = 0$.

Since μ is a probability measure, $f \in L^q(\widehat{G}, \mu) \subset L^1(\widehat{G}, \mu)$, and ($\dagger$) can be written $\overline{\mathcal{F}}(f\mu) = 0$. By Proposition A.7.4, we therefore have $f\mu = 0$, i.e., $f(\chi) = 0$ for μ-almost all $\chi \in \widehat{G}$, hence $f = 0$ in $L^q(\widehat{G}, \mu)$. □

A.8. Hilbert spaces and operators

In a Hilbert space, a Schauder basis, or simply a basis, is a maximal family of orthonormal vectors. Every Hilbert space has a basis (as a consequence of Zorn's lemma), and two bases have the same cardinality. By definition this cardinality is the **dimension** of the Hilbert space.

Let $\mathcal{H}$ be a Hilbert space. We denote by $\mathcal{L}(\mathcal{H})$ the involutive algebra of bounded linear operators on $\mathcal{H}$. The algebra $\mathcal{L}(\mathcal{H})$ has several useful topologies. The **strong topology** is the locally convex topology defined by the set of seminorms $x \mapsto \|x\xi\|$, for $\xi \in \mathcal{H}$. The **weak topology** is the locally convex topology defined by the set of seminorms $x \mapsto |\langle x\xi \mid \eta\rangle|$, for $\xi, \eta \in \mathcal{H}$. The **ultra-weak topology** is the locally convex topology defined by the set of seminorms $x \mapsto |\sum_{i=1}^{\infty} \langle x\xi_i \mid \eta_i\rangle|$, for sequences $(\xi_i)_{i\geq 1}$ and $(\eta_i)_{i\geq 1}$ of vectors in $\mathcal{H}$ such that $\sum_{i=1}^{\infty} \|\xi_i\|^2 < \infty$ and $\sum_{i=1}^{\infty} \|\eta_i\|^2 < \infty$.

In $\mathcal{L}(\mathcal{H})$, the unit ball is weakly compact; more generally, bounded subsets of $\mathcal{L}(\mathcal{H})$ which are weakly closed are weakly compact. On bounded subsets of $\mathcal{L}(\mathcal{H})$, the weak topology and the ultra-weak topology coincide; this implies that bounded subsets of $\mathcal{L}(\mathcal{H})$ which are ultra-weakly closed are ultra-weakly compact.

On $\mathcal{L}(\mathcal{H})$, there is also the **norm topology**, defined by the norm $\|x\| = \sup_{\xi\in\mathcal{H}, \|\xi\|\leq 1} \|x\xi\|$. This makes it a Banach algebra, and indeed a C*-algebra.

For all this, see [**Dixm–vN**, Chap. I, § 3].

Denote by

$$\mathcal{U}(\mathcal{H}) = \{x \in \mathcal{L}(\mathcal{H}) \mid x^*x = xx^* = \mathrm{Id}_{\mathcal{H}}\}$$

the group of unitary operators on $\mathcal{H}$. The restrictions to $\mathcal{U}(\mathcal{H})$ of the strong, weak, and ultra-weak topologies coincide, and they make $\mathcal{U}(\mathcal{H})$ a topological group; this is the topology to consider for the theory of unitary representations. The restriction to $\mathcal{U}(\mathcal{H})$ of the norm topology is of little interest for the subject of this book because, for many topological groups G, the only homomorphism $G \to (\mathcal{U}(\mathcal{H}), \text{norm topology})$ is the trivial homomorphism [**Sing–52**].

For a bounded linear operator on a Hilbert space, the following conditions are equivalent:

(P1) $x^* = x$ and the spectrum of x is contained in the set $\mathbf{R}_+$ of nonnegative real numbers;
(P2) there exists $y \in \mathcal{L}(\mathcal{H})$ such that $x = y^*y$;
(P3) there exists $z \in \mathcal{L}(\mathcal{H})$ such that $z^* = z$ and $x = z^2$.
(P4) $\langle x\xi|\xi\rangle \geq 0$ for all $\xi \in \mathcal{H}$.

The operator x is called **positive** if it satisfies these conditions. The set $\mathcal{L}(\mathcal{H})_+$ of positive operators is a closed convex cone such that $\mathcal{L}(\mathcal{H})_+ \cap (-\mathcal{L}(\mathcal{H})_+) = \{0\}$. There is an order on positive operators, defined by $x \leq y$ if $y - x$ is positive. For all this, see, e.g., [**Dixm–C***, 1.6].

For a positive operator x on $\mathcal{H}$, functional calculus provides a positive operator, denoted by $\sqrt{x}$, or $x^{\frac{1}{2}}$, of which the square is x; it is the **square root** of x. Note that, in $\mathcal{L}(\mathcal{H})$ with the strong topology, $\sqrt{x}$ is the limit of polynomials in x.

For $x \in \mathcal{L}(\mathcal{H})$, the square root of x^*x is the **absolute value** of x, denoted by $|x|$.

Let $\mathcal{H}_1, \mathcal{H}_2$ be two Hilbert spaces. A bounded linear operator $u : \mathcal{H}_1 \to \mathcal{H}_2$ is a **partial isometry** if there exists a closed subspace $\mathcal{K}$ of $\mathcal{H}_1$ such that the restriction of u to $\mathcal{K}$ is an isometry and such that $u = 0$ on $\mathcal{K}^\perp$. The subspace $\mathcal{K} = (\ker u)^\perp$ is the **initial space** and its image $u(\mathcal{K})$ is the **final space** of the partial isometry u. Note that u^*u is the orthogonal projection of $\mathcal{H}_1$ onto $\mathcal{K}$, and uu^* is the orthogonal projection of $\mathcal{H}_2$ onto $u(\mathcal{K})$.

Every operator $x : \mathcal{H}_1 \to \mathcal{H}_2$ has a **polar decomposition** $x = u|x|$, where $|x| = (x^*x)^{1/2}$ and where u is a partial isometry with initial space $(\ker x)^\perp$ and final space the closure of the image of x. The definition of u is as follows. First, $u\xi = 0$ for $\xi \in \ker(x) = \ker(|x|)$. Next, one has to observe that the linear map $|x|\xi \mapsto x\xi$ from the range of $|x|$ to the range of x is an isometry, and to define u on the closure $\overline{\text{range}(|x|)} = \ker(|x|)^\perp$ of the range of $|x|$ as the unique continuous linear extension of $|x|\xi \mapsto x\xi$. Then u is such that $u^*x = |x|$.

A.9. Projection-valued measures

Let $(X, \mathcal{B})$ be a Borel space and $\mathcal{H}$ be a Hilbert space. Denote by $\text{Proj}(\mathcal{H})$ the set of orthogonal projections in $\mathcal{L}(\mathcal{H})$. A **projection-valued measure** on $(X, \mathcal{B})$ and $\mathcal{H}$ is a map $E : \mathcal{B} \to \text{Proj}(\mathcal{H})$ such that

(1) $E(\emptyset) = 0$ and $E(X) = \text{Id}_{\mathcal{H}}$;
(2) $E(B \cap B') = E(B)E(B')$ for all $B, B' \in \mathcal{B}$;
(3) if $(B_i)_{i \in I}$ is a sequence of pairwise disjoint sets in $\mathcal{B}$, then

$$E(\bigcup_{i \in I} B_i) = \sum_{i \in I} E(B_i),$$

where the sum converges in the strong topology.

We also say that E is on X and $\mathcal{H}$ if $\mathcal{B}$ is clear from the context.

Projection-valued measures are also called spectral measures [**Halm–51**] and resolutions of the identity [**Rudi–73**]. It can easily be shown that Condition (2) is redundant, as it follows from (1) and (3); see, e.g., [**Halm–51**, § 36].

Let $E : \mathcal{B} \to \text{Proj}(\mathcal{H})$ be a projection-valued measure as above. For every $\xi, \eta \in \mathcal{H}$, a (signed) bounded Borel measure $\mu_{\xi,\eta}$ is defined on X by

$$\mu_{\xi,\eta}(B) = \langle E(B)\xi \mid \eta \rangle \quad \text{for all } B \in \mathcal{B}.$$

In the case of a locally compact space X and its σ-algebra $\mathcal{B}_X$ of Borel sets, it is customary to define a **regular** projection-valued measure by requiring that all these measures $\mu_{\xi,\eta}$ are Radon measures; this is automatic when X is a second-countable locally compact space, by Theorem A.4.1.

When $\xi = \eta$, the measure $\mu_{\xi,\eta}$ is denoted by μ_ξ. If $\|\xi\| = 1$, then μ_ξ is a probability measure.

Construction A.9.1. To a projection-valued measure $E : \mathcal{B} \to \text{Proj}(\mathcal{H})$, there is an associated functional calculus, i.e., a morphism of an appropriate algebra of functions on X to the algebra of operators $\mathcal{L}(\mathcal{H})$, defined as follows.

Denote by $\mathrm{Bor}^b(X)$ the commutative involutive algebra of complex-valued bounded measurable functions on X, with the norm defined by $\|f\| = \sup_{x \in X} |f(x)|$; it is a Banach involutive algebra. For every $f \in \mathrm{Bor}^b(X)$, there exists a unique operator $\Psi(f) \in \mathcal{L}(\mathcal{H})$ such that

$$\langle \Psi(f)\xi \mid \eta \rangle = \int_X f(x)\, d\mu_{\xi,\eta}(x) \qquad \text{for all } \xi, \eta \in \mathcal{H}.$$

Moreover, the map

$$\mathrm{Bor}^b(X) \to \mathcal{L}(\mathcal{H}), \qquad f \mapsto \Psi(f)$$

is a homomorphism of Banach involutive algebras. The operator $\Psi(f)$ is usually denoted by

$$\int_X f(x)\, dE(x).$$

For all this, see Chapter 12 in [**Rudi–73**].

EXAMPLE A.9.2. Let S be a normal operator on a Hilbert space $\mathcal{H}$. Let X denote its spectrum, which is a nonempty compact subset of the complex plane. One formulation of the spectral theorem is that there exists a unique projection-valued measure $E : \mathcal{B}_X \to \mathrm{Proj}(\mathcal{H})$ such that $S = \int_X z dE(z)$.

Let X be a topological space and let E be a projection-valued measure on $(X, \mathcal{B}_X)$ and some Hilbert space. The **support** of E is the set of $x \in X$ such that, for every neighbourhood U of x, we have $E(U) \neq 0$; equivalently, the support of E is the complement of the largest open set U of X such that $E(U) = 0$. For $x \in X$, we say that x is an **atom** of E if $E(\{x\}) \neq 0$. Obviously, every atom of E lies in the support of E.

A.10. C*-algebras

Though we expect from the reader some familiarity with C*-algebras, here we recall the following definitions. Several of them have already appeared in the present appendix.

An **involutive algebra** is a complex algebra A given with a conjugate linear involution, i.e., a conjugate linear map $A \to A$, $x \mapsto x^*$ such that $(x^*)^* = x$ and $(xy)^* = y^*x^*$ for all $x, y \in A$. A **Banach algebra** is a complex algebra A given with a norm $\| \cdot \| : A \to \mathbf{R}_+$ for which A is a Banach space, and such that $\|xy\| \leq \|x\| \, \|y\|$ for all $x, y \in A$. A **Banach involutive algebra** is an algebra A which is both an involutive algebra and a Banach algebra, and which is such that $\|x^*\| = \|x\|$ for all $x \in A$. A **C*-algebra** is a Banach algebra A with a conjugate linear involution $x \mapsto x^*$ such that $\|x^*x\| = \|x\|^2$ for all $x \in A$; it is easy to check that this implies $\|x^*\| = \|x\|$ for all $x \in A$.

Let A be a C*-algebra. A **projection** in A is an element $p \in A$ such that $p = p^* = p^2$.

For $x \in A$, the analogues of Conditions (P1), (P2), (P3) of Section A.8 are equivalent. An element $x \in A$ is **positive** if it satisfies these conditions. The set A_+ of positive elements in A is a closed convex cone such that $A_+ \cap (-A_+) = \{0\}$, and every positive element in A has a unique positive square root ion A (again see [**Dixm–C***, 1.6]). There is an order on positive elements of A, defined for $x, y \in A_+$ by $x \leq y$ if $y - x \in A_+$.

Let A be a complex normed algebra. An **approximate identity** in A is a family $(u_\iota)_{\iota \in I}$ of elements of A indexed by a directed set I such that

(1) $\|u_\iota\| \le 1$ for all $\iota \in I$;
(2) $\lim_{\iota \in I} \|u_\iota x - x\| = \lim_{\iota \in I} \|x u_\iota - x\| = 0$ for all $x \in A$.

For example, let $L^1(G, \mu_G)$ be the group algebra of an LC group G, with Haar measure μ_G. Let $\mathcal{V}$ be a basis of a neighbourhood of e in G. For each $V \in \mathcal{V}$, there exists a continuous function $u_V : G \to \mathbf{R}_+$ such that $u_V(g^{-1}) = u_V(g)$ for all $g \in G$, with support compact and contained in V, and such that $\int_G u_V(g) d\mu_G(g) = 1$. Then $(u_V)_V$ is an approximate identity in $L^1(G, \mu_G)$ [**Foll–16**, Proposition 2.44].

Let A moreover be a C*-algebra. An **increasing approximate identity** in A is an approximate identity $(u_\iota)_{\iota \in I}$ such that

(3) $u_\iota \ge 0$ for all $\iota \in I$,
(4) $u_\iota \le u_{\iota'}$ for all $\iota, \iota' \in I$ with $\iota \le \iota'$.

Every C*-algebra has an increasing approximate identity. More precisely, given a dense two-sided ideal $\mathfrak{m}$ in A, there exists an increasing approximate identity $(u_\iota)_{\iota \in I}$ in A such that $u_\iota \in \mathfrak{m}$ for all $\iota \in I$. If A is separable, there exists such an approximate identity indexed by $\mathbf{N}$ [**Dixm–C***, 1.7.1].

A.11. Von Neumann algebras

Let $\mathcal{H}$ be a Hilbert space, and $\mathcal{L}(\mathcal{H})$ the involutive algebra of bounded linear operators on $\mathcal{H}$. The **commutant** of a subset S of $\mathcal{L}(\mathcal{H})$ is $S' := \{x \in \mathcal{L}(\mathcal{H}) \mid xs = sx \text{ for all } s \in S\}$, and the **bicommutant** is $S'' := (S')'$. Note that S' and S'' are subalgebras of $\mathcal{L}(\mathcal{H})$, and that $S \subset S''$. For S, T two subsets of $\mathcal{L}(\mathcal{H})$, the inclusion $S \subset T$ implies $T' \subset S'$. In particular, $S \subset S''$ implies $(S'')' \subset S' \subset (S')''$, so that $S''' = S'$.

A **von Neumann algebra** acting on $\mathcal{H}$ is an involutive subalgebra $\mathcal{M} \subset \mathcal{L}(\mathcal{H})$ that is equal to its bicommutant $\mathcal{M}''$. For example, S' and S'' are von Neumann algebras for any selfadjoint subset S of $\mathcal{L}(\mathcal{H})$. In particular, $\mathcal{L}(\mathcal{H})$ and $\mathbf{C}\mathrm{Id}_{\mathcal{H}}$ are von Neumann algebras acting on $\mathcal{H}$.

A **homomorphism** of a von Neumann algebra $\mathcal{M}$ acting on $\mathcal{H}$ to a von Neumann algebra $\mathcal{N}$ acting on $\mathcal{K}$ is a linear map $\Phi : \mathcal{M} \to \mathcal{N}$ such that $\Phi(ST) = \Phi(S)\Phi(T)$ and $\Phi(S^*) = \Phi(S)^*$ for all $S, T \in \mathcal{M}$. For such a homomorphism Φ, it follows that $\Phi(\mathcal{M}_+) \subset \Phi(\mathcal{N}_+)$, that the image by Φ of a projection is a projection, that $\|\Phi(S)\| \le \|S\|$ for all $S \in \mathcal{M}$, and when Φ is injective that $\|\Phi(S)\| = \|S\|$ for all $S \in \mathcal{M}$ (Chap. I, §1, no. 5 in [**Dixm–vN**]).

Let $\mathcal{M}, \mathcal{N}$ be two von Neumann algebras. A linear map $\Phi : \mathcal{M} \to \mathcal{N}$ is positive if $\Phi(\mathcal{M}_+) \subset \mathcal{N}_+$, and **normal** if, moreover, $\Phi(x)$ is the least upper bound of $(\Phi(x_\iota))_{\iota \in I}$ for every increasing generalized sequence $(x_\iota)_{\iota \in I}$ in $\mathcal{M}_+$ with least upper bound x. A normal positive linear map is ultraweakly continuous [**Dixm–vN**, Chap I, §4, no. 3].

For the next theorem, see Chap I, §3, no. 4 in [**Dixm–vN**].

THEOREM A.11.1 (**von Neumann bicommutant theorem**). *Let $\mathcal{H}$ be a Hilbert space and $\mathcal{L}(\mathcal{H})$ the involutive algebra of bounded operators on $\mathcal{H}$. Let $\mathcal{M}$ be a selfadjoint subalgebra of $\mathcal{L}(\mathcal{H})$ such that the set*

$$\{T\xi \mid T \in \mathcal{M},\ \xi \in \mathcal{H}\}$$

is total in $\mathcal{H}$ (this is, for instance, the case if $\mathcal{M}$ contains the identity operator of $\mathcal{H}$).

The bicommutant $\mathcal{M}''$ of $\mathcal{M}$ is the closure of $\mathcal{M}$ for the weak operator topology as well as for the strong operator topology.

In particular, the following conditions are equivalent:

(i) *$\mathcal{M}'' = \mathcal{M}$, i.e., $\mathcal{M}$ is a von Neumann algebra in $\mathcal{L}(\mathcal{H})$;*
(ii) *$\mathcal{M}$ is closed in $\mathcal{L}(\mathcal{H})$ for the weak operator topology;*
(iii) *$\mathcal{M}$ is closed in $\mathcal{L}(\mathcal{H})$ for the strong operator topology.*

The von Neumann algebra generated by a selfadjoint subset S of $\mathcal{L}(\mathcal{H})$ is equal to the bicommutant S''.

DEFINITION A.11.2. A von Neumann subalgebra $\mathcal{M}$ of $\mathcal{L}(\mathcal{H})$ is a **factor** if the centre of $\mathcal{M}$ consists only of the scalar multiples of $\mathrm{Id}_{\mathcal{H}}$, that is, if

$$\mathcal{M} \cap \mathcal{M}' = \{\lambda \mathrm{Id}_{\mathcal{H}} \mid \lambda \in \mathbf{C}\}.$$

Observe that $\mathcal{M}$ is a factor if and only if $\mathcal{M}'$ is a factor.

EXAMPLE A.11.3. For each positive integer n, the algebra $\mathcal{L}(\mathbf{C}^n)$ is a factor, by definition a factor of type I_n; its dimension is finite, equal to n^2. For an infinite-dimensional Hilbert space $\mathcal{H}$, the algebra $\mathcal{L}(\mathcal{H})$ is a factor, by definition a factor of type I_∞.

It is an insight of Murray and von Neumann that there exist other types of factors, in particular factors of type II_1, which are infinite-dimensional factors, say $\mathcal{M}$, such that there exists a linear form $t : \mathcal{M} \to \mathbf{C}$ such that $t(xy) = t(yx)$ for all $x, y \in \mathcal{M}$ and $t(1) = 1$. There is more on the types of von Neumann algebras and of factors in Section 7.B.

PROPOSITION A.11.4. *Let $\mathcal{H}$ be a Hilbert space and $\mathcal{M}$ a subalgebra of $\mathcal{L}(\mathcal{H})$. Denote by $\mathcal{P}$ the set of the orthogonal projections $P \in \mathcal{L}(\mathcal{H})$ which belong to $\mathcal{M}$. Then:*

(1) *$\mathcal{M}$ is the von Neumann algebra generated by $\mathcal{P}$, i.e., $\mathcal{P}'' = \mathcal{M}$;*
(2) *$\mathcal{M}$ is abelian if and only if $PQ = QP$ for all $P, Q \in \mathcal{P}$;*
(3) *$\mathcal{M}$ is a factor if and only if 0 and $\mathrm{Id}_{\mathcal{H}}$ are the only projections in the centre of $\mathcal{M}$.*

PROOF. Since $\mathcal{P} \subset \mathcal{M}$, we have $\mathcal{P}'' \subset \mathcal{M}''$, hence $\mathcal{P}'' \subset \mathcal{M}$ by the von Neumann bicommutant theorem, Theorem A.11.1. To show (1), it remains to show that $\mathcal{M} \subset \mathcal{P}''$. We choose $S \in \mathcal{P}'$ and $T \in \mathcal{M}$, and we have to show that $ST = TS$. Upon considering the real part and the imaginary part of T (which both belong to $\mathcal{M}$), we may assume that T is a selfadjoint operator.

By the spectral theory of selfadjoint operators, we have

$$\{T\}'' = \{\mathbf{1}_B(T) \mid B \in \mathcal{B}(\mathbf{R})\}'',$$

where the projection $\mathbf{1}_B(T)$ is defined by functional calculus (see [**Rudi–73**, Theorem 12.23]; also, compare with (2) in the SNAG Theorem 2.C.3). We have $\mathbf{1}_B(T) \in \mathcal{P}$ for all $B \in \mathcal{B}(\mathbf{R})$; see [**Dixm–vN**, Chap. I, § 1, no. 1]. It follows that $S \in \{T\}'$.

Claims (2) and (3) follow from Claim (1) and the definitions. □

DEFINITION A.11.5. Let $\mathcal{H}$ be a Hilbert space. For an operator $T \in \mathcal{L}(\mathcal{H})$, the **support** of T is the projection E of $\mathcal{H}$ onto the orthogonal of the kernel of T.

Note that E is the smallest of the projections $E_1 \in \mathcal{L}(\mathcal{H})$ such that $TE_1 = T$ (projections are ordered by $E_1 \leq E_2$ if $E_2 - E_1$ is a positive operator).

Let F denote the projection of $\mathcal{H}$ onto the closure of the image $T(\mathcal{H})$. Then F is the smallest of the projections $F_1 \in \mathcal{L}(\mathcal{H})$ such that $F_1 T = T$. The last equality is equivalent to $T^* F_1 = F_1$; it follows that F is the support of T^*.

DEFINITION A.11.6. Let $\mathcal{M}$ be a von Neumann algebra on $\mathcal{H}$, and $\mathcal{Z}$ its centre. Let T be in the commutant $\mathcal{M}'$ of $\mathcal{M}$. For a projection $E_1 \in \mathcal{Z} = \mathcal{M}' \cap \mathcal{M}$ such that $TE_1 = T$, we have $T^* E_1 = E_1 T^* = (TE_1)^* = T^*$. It follows that the smallest projection E in $\mathcal{Z}$ which is larger than the support of T is also the smallest projection in $\mathcal{Z}$ which is larger than the support of T^*. Thus E is the **central support** of T.

Let P be a projection in $\mathcal{M}'$; set $\mathcal{K} = P(\mathcal{H})$. The subspace $\mathcal{K}$ of $\mathcal{H}$ is invariant by $\mathcal{M}$. The central support E of P belongs to $\mathcal{Z} = \mathcal{M}' \cap \mathcal{M}$, and is the smallest of the projections E_1 in $\mathcal{Z}$ such that $P \leq E_1$. The set

$$\{PS|_{\mathcal{K}} \mid S \in \mathcal{M}\} \cong P\mathcal{M}P$$

is a von Neumann subalgebra of $\mathcal{L}(\mathcal{K})$, denoted by $\mathcal{M}_P$ and called the **induced von Neumann algebra**, or the von Neumann algebra induced by $\mathcal{M}$ on $\mathcal{K}$. The **induction**

$$\mathcal{M} \to \mathcal{M}_P, \quad S \mapsto PS|_{\mathcal{K}}$$

is a homomorphism of von Neumann algebras. For another projection $Q \in \mathcal{M}'$ such that $P \leq Q$, we have inductions $\mathcal{M} \to \mathcal{M}_Q$ and $\mathcal{M}_Q \to \mathcal{M}_P$, and their composition is the induction $\mathcal{M} \to \mathcal{M}_P$.

For the next propositions, see in [**Dixm–vN**, Chap. I], Proposition 2 of §2, no. 1, and Theorem 3 of §4, no. 4.

PROPOSITION A.11.7. *Let $\mathcal{M}$ be a von Neumann algebra acting on $\mathcal{H}$. Let P be a projection in $\mathcal{M}'$.*

If the central support of P is $\mathrm{Id}_{\mathcal{H}}$, then the induction $\mathcal{M} \to \mathcal{M}_P$ is an isomorphism.

PROPOSITION A.11.8. *Let $\mathcal{M}, \mathcal{N}$ be two von Neumann algebras acting on Hilbert spaces $\mathcal{H}, \mathcal{K}$ respectively, and $\Phi : \mathcal{M} \to \mathcal{N}$ an isomorphism.*

Then there exist a Hilbert space $\mathcal{H}_2$, a projection $P \in (\mathcal{M} \otimes \mathbf{C}\mathrm{Id}_{\mathcal{H}_2})' \subset \mathcal{L}(\mathcal{H} \otimes \mathcal{H}_2)$, with range some closed subspace $\mathcal{K}_2$ of $\mathcal{H} \otimes \mathcal{H}_2$, and an isomorphism of Hilbert spaces $U : \mathcal{K}_2 \to \mathcal{K}$, such that

$$\Phi(T) = UP\big(T \otimes \mathrm{Id}_{\mathcal{H}_2}\big)PU^{-1} \quad \textit{for every } T \in \mathcal{M}.$$

[Observe that $\mathcal{K}_2$ is invariant under $\mathcal{M} \otimes \mathrm{Id}_{\mathcal{H}_2}$, because $P \in (\mathcal{M} \otimes \mathrm{Id}_{\mathcal{H}_2})'$.]

Bibliography

[Aaro–97] J. Aaronson, *An introduction to infinite ergodic theory*, Mathematical Surveys and Monographs, vol. 50, American Mathematical Society, Providence, RI, 1997. MR1450400. 368, 390

[AbEL–11] F. Abdelmoula, M. Elloumi, and J. Ludwig, *The C^*-algebra of the motion group* $\mathrm{SO}(n) \ltimes \mathbb{R}^n$, Bull. Sci. Math. **135** (2011), no. 2, 166–177. MR2773395. 246

[AbGV–14] M. Abért, Y. Glasner, and B. Virág, *Kesten's theorem for invariant random subgroups*, Duke Math. J. **163** (2014), no. 3, 465–488. MR3165420. 416, 417

[ABB+–17] M. Abert, N. Bergeron, I. Biringer, T. Gelander, N. Nikolov, J. Raimbault, and I. Samet, *On the growth of L^2-invariants for sequences of lattices in Lie groups*, Ann. of Math. (2) **185** (2017), no. 3, 711–790. MR3664810. 417

[AlBe–95] J. L. Alperin and R. B. Bell, *Groups and representations*, Graduate Texts in Mathematics, vol. 162, Springer-Verlag, New York, 1995. MR1369573. 130, 277

[AlEd–79] S. Alpern and R. D. Edwards, *Lusin's theorem for measure-preserving homeomorphisms*, Mathematika **26** (1979), no. 1, 33–43. MR557124. 412

[Ambr–44] W. Ambrose, *Spectral resolution of groups of unitary operators*, Duke Math. J. **11** (1944), 589–595. MR11172. 82

[AnMa–12] H. Ando and Y. Matsuzawa, *On Polish groups of finite type*, Publ. Res. Inst. Math. Sci. **48** (2012), no. 2, 389–408. MR2928146. 399

[ArKa–97] R. J. Archbold and E. Kaniuth, *On the primitive ideal space of a finitely generated nilpotent group*, Forum Math. **9** (1997), no. 2, 193–210. MR1431120. 331

[Arti–57] E. Artin, *Geometric algebra*, Interscience Publishers, Inc., New York-London, 1957. MR0082463. 144

[Arve–76] W. Arveson, *An invitation to C^*-algebras*, Graduate Texts in Mathematics, No. 39, Springer-Verlag, New York-Heidelberg, 1976. MR0512360. 71

[AuMo–66] L. Auslander and C. C. Moore, *Unitary representations of solvable Lie groups*, Mem. Amer. Math. Soc. **62** (1966), 199. MR207910. 204, 228

[AuKo–71] L. Auslander and B. Kostant, *Polarization and unitary representations of solvable Lie groups*, Invent. Math. **14** (1971), 255–354. MR293012. 209

[Bagg–68] L. Baggett, *A description of the topology on the dual spaces of certain locally compact groups*, Trans. Amer. Math. Soc. **132** (1968), 175–215. MR409720. 38

[Bagg–72] L. Baggett, *A separable group having a discrete dual space is compact*, J. Functional Analysis **10** (1972), 131–148. MR0346090. 35, 36

[BaMe–86] L. Baggett and K. Merrill, *Representations of the Mautner group and cocycles of an irrational rotation*, Michigan Math. J. **33** (1986), no. 2, 221–229. MR837580. 165

[BaSu–81] L. Baggett and T. Sund, *The Hausdorff dual problem for connected groups*, J. Functional Analysis **43** (1981), no. 1, 60–68. MR639797. 38

[BaPa–94] L. Baggett and J. Packer, *The primitive ideal space of two-step nilpotent group C^*-algebras*, J. Funct. Anal. **124** (1994), no. 2, 389–426. MR1289356. 271

[Bana–83] W. Banaszczyk, *On the existence of exotic Banach-Lie groups*, Math. Ann. **264** (1983), no. 4, 485–493. MR716262. 27

[Bana–91] W. Banaszczyk, *Additive subgroups of topological vector spaces*, Lecture Notes in Mathematics, vol. 1466, Springer-Verlag, Berlin, 1991. MR1119302. 27

[Barg–47] V. Bargmann, *Irreducible unitary representations of the Lorentz group*, Ann. of Math. (2) **48** (1947), 568–640. MR21942. 8, 121

[BaLS–64] H. Bass, M. Lazard, and J.-P. Serre, *Sous-groupes d'indice fini dans* $\mathbf{SL}(n, \mathbf{Z})$ (French), Bull. Amer. Math. Soc. **70** (1964), 385–392. MR161913. 143

[Bedo–94] E. Bédos, *On the C^*-algebra generated by the left regular representation of a locally compact group*, Proc. Amer. Math. Soc. **120** (1994), no. 2, 603–608. MR1181157. 260

[Bekk–99] B. Bekka, *On the full C^*-algebras of arithmetic groups and the congruence subgroup problem*, Forum Math. **11** (1999), no. 6, 705–715. MR1725593. 416

[Bekk–07] B. Bekka, *Operator-algebraic superridigity for* $\mathrm{SL}_n(\mathbb{Z})$, $n \geq 3$, Invent. Math. **169** (2007), no. 2, 401–425. MR2318561. 326, 331, 416

[Bekk–19] B. Bekka, *Infinite characters on* $GL_n(\mathbf{Q})$, *on* $\mathrm{SL}_n(\mathbf{Z})$, *and on groups acting on trees*, J. Funct. Anal. **277** (2019), no. 7, 2160–2178. MR3989142. 313

[Bekk–20] B. Bekka, *Character rigidity of simple algebraic groups*, Math. Ann. **378** (2020),no. 3–4, 1223–1243. 326, 420

[BeLS–92] B. Bekka, A. T. Lau, and G. Schlichting, *On invariant subalgebras of the Fourier-Stieltjes algebra of a locally compact group*, Math. Ann. **294** (1992), no. 3, 513–522. MR1188135. 251

[BeHa–94] B. Bekka and P. de la Harpe, *Représentations d'un groupe faiblement équivalentes à la représentation régulière* (French, with English and French summaries), Bull. Soc. Math. France **122** (1994), no. 3, 333–342. MR1294459. 225, 261, 290, 432

[BeVa–95] B. Bekka and A. Valette, *Lattices in semi-simple Lie groups, and multipliers of group C^*-algebras*, Recent advances in operator algebras (Orléans, 1992), Astérisque **232** (1995), 67–79. MR1372525. 251

[BeLo–00] B. Bekka and N. Louvet, *Some properties of C^*-algebras associated to discrete linear groups*, C^*-algebras (Münster, 1999), Springer, Berlin, 2000, pp. 1–22. MR1796907. 412

[BeMa–00] B. Bekka and M. Mayer, *Ergodic theory and topological dynamics of group actions on homogeneous spaces*, London Mathematical Society Lecture Note Series, vol. 269, Cambridge University Press, Cambridge, 2000. MR1781937. 108, 112, 117, 118, 168, 275, 386, 434

[BeCu–03] B. Bekka and R. Curtis, *On Mackey's irreducibility criterion for induced representations*, Int. Math. Res. Not. **38** (2003), 2095–2101. MR1994777. 51

[BeHV–08] B. Bekka, P. de la Harpe, and A. Valette, *Kazhdan's Property (T)*, Cambridge Univ. Press, 2008. [An erratum is available on the web, and contains in particular a corrected proof of Proposition F.2.7, quoted on several occasions in the present book. `https://perso.univ-rennes1.fr/bachir.bekka/Erratum-Kazhdan.pdf`] 17, 19, 21, 25, 26, 27, 28, 29, 31, 33, 38, 40, 43, 44, 52, 86, 148, 151, 172, 190, 214, 249, 254, 265, 267, 289, 356, 413, 416, 435, 436, 438

[BeEc–20] B. Bekka and S. Echterhoff, *On unitary representations of algebraic groups over local fields*, `arXiv:2003.02297v1`, 4 Mar 2020. 208

[BeFr–20] B. Bekka and C. Francini, *Characters of algebraic groups over number fields*, `arXiv:2002.07497v1`, 18 Feb 2020. 326

[BYTs–16] I. Ben Yaacov and T. Tsankov, *Weakly almost periodic functions, model-theoretic stability, and minimality of topological groups*, Trans. Amer. Math. Soc. **368** (2016), no. 11, 8267–8294. MR3546800. 27

[BCRZ–16] V. Bergelson, J. C. Christopherson, D. Robertson, and P. Zorin-Kranich, *Finite products sets and minimally almost periodic groups*, J. Funct. Anal. **270** (2016), no. 6, 2126–2167. MR3460236. 145

[Bind–93] M. W. Binder, *Induced factor representations of discrete groups and their types*, J. Funct. Anal. **115** (1993), no. 2, 294–312. MR1234393. 225

[Bern–74] I. N. Bernšteĭn, *All reductive $\mathfrak{p}$-adic groups are of type I* (Russian), Funkcional. Anal. i Priložen. **8** (1974), no. 2, 3–6. MR0348045. 204

[Blac–77] B. E. Blackadar, *The regular representation of restricted direct product groups*, J. Functional Analysis **25** (1977), no. 3, 267–274. MR0439979. 387

[Blat–63] R. J. Blattner, *Positive definite measures*, Proc. Amer. Math. Soc. **14** (1963), 423–428. MR147580. 46

[BLSV–78] J. Boidol, H. Leptin, J. Schürman, and D. Vahle, *Räume primitiver Ideale von Gruppenalgebren* (German), Math. Ann. **236** (1978), no. 1, 1–13. MR498971. 261

[Boga–07] V. I. Bogachev, *Measure theory. Vol. I, II*, Springer-Verlag, Berlin, 2007. MR2267655. 430

[Bore–86] A. Borel, *Hermann Weyl and Lie groups*, Hermann Weyl, 1885–1985, Eidgenössische Tech. Hochschule, Zürich, 1986, pp. 53–82. MR875268. 7

[Boul–16] M. Bouljihad, *Existence of rigid actions for finitely-generated non-amenable linear groups*, `arXiv:1606.05911v1`, 19 Jun 2016. 364

[BA1–3] N. Bourbaki, *Algèbre, chapitres 1 à 3*, Bourbaki, 1970. 124

[BTG1–4] N. Bourbaki, *Topologie générale, chapitres 1 à 4*, Hermann, 1971. 32, 39, 138, 254, 405, 421, 423

[BTG5–10] N. Bourbaki, *Topologie générale, chapitres 5 à 10*, Hermann, 1971. 31, 422, 423

[BEVT1–5] N. Bourbaki, *Espaces vectoriels topologiques, chapitres 1 à 5*, N. Bourbaki et Springer, 2007. 362, 426

[BTS1–2] N. Bourbaki, *Théories spectrales, chapitres 1 et 2*, Hermann, 1967. 25, 100, 115, 116, 385, 437, 438, 439

[BoHo–19] R. Boutonnet and C. Houdayer, *Stationary characters on lattices of semisimple Lie groups*, `arXiv:1908.07812v1`, 21 Aug 2019. 326

[Boye–83] R. P. Boyer, *Infinite traces of AF-algebras and characters of* $\mathrm{U}(\infty)$, J. Operator Theory **9** (1983), no. 2, 205–236. MR703808. 326

[BoMa–77] R. Boyer and R. Martin, *The group C^*-algebra of the de Sitter group*, Proc. Amer. Math. Soc. **65** (1977), no. 1, 177–184. MR473718. 246

[Broo–88] R. Brooks, *Constructing isospectral manifolds*, Amer. Math. Monthly **95** (1988), no. 9, 823–839. MR967343. 51

[Brow–73a] I. D. Brown, *Representation of finitely generated nilpotent groups*, Pacific J. Math. **45** (1973), 13–26. MR352324. 165

[Brow–73b] I. D. Brown, *Dual topology of a nilpotent Lie group*, Ann. Sci. École Norm. Sup. (4) **6** (1973), 407–411. MR352326. 110

[BrOz–08] N. P. Brown and N. Ozawa, *C^*-algebras and finite-dimensional approximations*, Graduate Studies in Mathematics, vol. 88, American Mathematical Society, Providence, RI, 2008. MR2391387. 408

[BrTi–72] F. Bruhat and J. Tits, *Groupes réductifs sur un corps local* (French), Inst. Hautes Études Sci. Publ. Math. **41** (1972), 5–251. MR327923. 190

[Bump–97] D. Bump, *Automorphic forms and representations*, Cambridge Studies in Advanced Mathematics, vol. 55, Cambridge University Press, Cambridge, 1997. MR1431508. 52

[BuHa–97] M. Burger and P. de la Harpe, *Constructing irreducible representations of discrete groups*, Proc. Indian Acad. Sci. Math. Sci. **107** (1997), no. 3, 223–235. MR1467427. 49

[Burn–11] W. Burnside, *The theory of groups of finite order*, 2nd Edition. Cambridge Univ. Press, 1911. [Reprint, Dover, 1955.] 49

[CaRW–17] P.-E. Caprace, C. D. Reid, and G. A. Willis, *Locally normal subgroups of totally disconnected groups. Part II: Compactly generated simple groups*, Forum Math. Sigma **5** (2017), e12, 89. MR3659769. 405

[CaLe–16] A. Carderi and F. Le Maître, *More Polish full groups*, Topology Appl. **202** (2016), 80–105. MR3464151. 327

[CaMo–84] A. L. Carey and W. Moran, *Characters of nilpotent groups*, Math. Proc. Cambridge Philos. Soc. **96** (1984), no. 1, 123–137. MR743708. 320, 326, 329

[Choi–80] M. D. Choi, *The full C^*-algebra of the free group on two generators*, Pacific J. Math. **87** (1980), no. 1, 41–48. MR590864. 410

[Conn–76] A. Connes, *Classification of injective factors. Cases II_1, II_∞, III_λ, $\lambda \neq 1$*, Ann. of Math. (2) **104** (1976), no. 1, 73–115. MR454659. 224, 407

[Conr] K. Conrad, *The character group of* $\mathbf{Q}$, Unpublished note, available on `http://www.math.uconn.edu/~kconrad/blurbs/` 105

[Corn–09] Y. Cornulier, *Infinite conjugacy classes in groups acting on trees*, Groups Geom. Dyn. **3** (2009), no. 2, 267–277. MR2486799. 432

[CoHa–16] Y. Cornulier and P. de la Harpe, *Metric geometry of locally compact groups*, EMS Tracts in Mathematics, vol. 25, European Mathematical Society (EMS), Zürich, 2016. Winner of the 2016 EMS Monograph Award. MR3561300. 114, 254, 404

[Corw–75] L. Corwin, *Induced representations of discrete groups*, Proc. Amer. Math. Soc. **47** (1975), 279–287. MR387487. 51

[CoSt–91] M. Cowling and T. Steger, *The irreducibility of restrictions of unitary representations to lattices*, J. reine angew. Math. **420** (1991), 85–98. MR1124567. 261, 290

[Cuntz–77] J. Cuntz, *Simple C^*-algebras generated by isometries*, Comm. Math. Phys. **57** (1977), no. 2, 173–185. MR467330. 314

[Dixm–vN] J. Dixmier, *Les algèbres d'opérateurs dans l'espace hilbertien (algèbres de von Neumann)*, deuxième édition, revue et augmentée. Gauthier-Villars, 1969. [First Edition 1957, English translation North-Holland 1981.] xi, 12, 53, 54, 56, 59, 60, 64, 80, 184, 195, 211, 218, 219, 220, 222, 230, 259, 298, 304, 306, 307, 308, 309, 311, 312, 315, 321, 353, 361, 362, 402, 429, 440, 443, 444, 445

[Dixm–C*] J. Dixmier, *Les C^*-algèbres et leurs représentations*, deuxième édition. Gauthier–Villars, 1969. [First Edition 1964, English translation North-Holland 1977.] xi, 9, 12, 19, 21, 30, 31, 37, 52, 53, 55, 56, 71, 132, 133, 135, 136, 184, 192, 193, 195, 196, 199, 200, 201, 203, 204, 208, 220, 230, 232, 233, 234, 235, 236, 237, 238, 243, 244, 245, 258, 260, 298, 301, 307, 309, 312, 328, 399, 402, 406, 424, 440, 442, 443

[Dixm–57] J. Dixmier, *Sur les représentations unitaires des groupes de Lie algébriques* (French), Ann. Inst. Fourier (Grenoble) **7** (1957), 315–328. MR99380. 208

[Dixm–59] J. Dixmier, *Sur les représentations unitaires des groupes de Lie nilpotents. V* (French), Bull. Soc. Math. France **87** (1959), 65–79. MR115097. 204, 208

[Dixm–60a] J. Dixmier, *Sur les C^*-algèbres* (French), Bull. Soc. Math. France **88** (1960), 95–112. MR121674. 228, 240, 254, 259

[Dixm–60b] J. Dixmier, *Opérateurs de rang fini dans les représentations unitaires* (French), Inst. Hautes Études Sci. Publ. Math. **6** (1960), 13–25. MR136684. 261

[Dixm–60c] J. Dixmier, *Sur les structures boréliennes du spectre d'une C^*-algèbre* (French), Inst. Hautes Études Sci. Publ. Math. **6** (1960), 5–11. MR124755. 259

[Dixm–61] J. Dixmier, *Sur le revêtement universel d'un groupe de Lie de type I* (French), C. R. Acad. Sci. Paris **252** (1961), 2805–2806. MR133407. 210

[Dixm–63] J. Dixmier, *Traces sur les C^*-algèbres* (French), Ann. Inst. Fourier (Grenoble) **13** (1963), no. fasc., fasc. 1, 219–262. MR149317. 217

[Dixm–64a] J. Dixmier, *Traces sur les C^*-algèbres. II* (French), Bull. Sci. Math. (2) **88** (1964), 39–57. MR181906. 314

[Dixm–64b] J. Dixmier, *Utilisation des facteurs hyperfinis dans la théorie des C^*-algèbres* (French), C. R. Acad. Sci. Paris **258** (1964), 4184–4187. MR161174. 58

[Dixm–66] J. Dixmier, *Existence de traces non normales* (French), C. R. Acad. Sci. Paris Sér. A-B **262** (1966), A1107–A1108. MR196508. 215, 218

[Dixm–69] J. Dixmier, *Sur la représentation régulière d'un groupe localement compact connexe* (French), Ann. Sci. École Norm. Sup. (4) **2** (1969), 423–436. MR260933. 227

[Dudk–11] A. Dudko, *Characters on the full group of an ergodic hyperfinite equivalence relation*, J. Funct. Anal. **261** (2011), no. 6, 1401–1414. MR2813475. 327

[DuMe–14] A. Dudko and K. Medynets, *Finite factor representations of Higman-Thompson groups*, Groups Geom. Dyn. **8** (2014), no. 2, 375–389. MR3231220. 320, 326, 419, 420

[DuGr–17a] A. Dudko and R. Grigorchuk, *On irreducibility and disjointness of Koopman and quasi-regular representations of weakly branch groups*, Modern theory of dynamical systems, Contemp. Math., vol. 692, Amer. Math. Soc., Providence, RI, 2017, pp. 51–66. MR3666066. 290

[DuGr–17b] A. Dudko and R. Grigorchuk, *On spectra of Koopman, groupoid and quasi-regular representations*, J. Mod. Dyn. **11** (2017), 99–123. MR3627119. 290

[DuGr–18] A. Dudko and R. Grigorchuk, *On diagonal actions of branch groups and the corresponding characters*, J. Funct. Anal. **274** (2018), no. 11, 3033–3055. MR3782986. 419

[Dye–59] H. A. Dye, *On groups of measure preserving transformations. I*, Amer. J. Math. **81** (1959), 119–159. MR131516. 165

[EcKl–12] S. Echterhoff and H. Klüver, *A general Kirillov theory for locally compact nilpotent groups*, J. Lie Theory **22** (2012), no. 3, 601–645. MR3012147. 268

[Effr–63] E. G. Effros, *A decomposition theory for representations of C^*-algebras*, Trans. Amer. Math. Soc. **107** (1963), 83–106. MR146682. 9, 242

[Effr–65] E. G. Effros, *Transformation groups and C^*-algebras*, Ann. of Math. (2) **81** (1965), 38–55. MR174987. 259

[Elli–77] G. A. Elliott, *The Mackey Borel structure on the spectrum of an approximately finite-dimensional separable C^*-algebra*, Trans. Amer. Math. Soc. **233** (1977), 59–68. MR454648. 294, 295

[Enge–89] R. Engelking, *General topology*, 2nd ed., Sigma Series in Pure Mathematics, vol. 6, Heldermann Verlag, Berlin, 1989. Translated from the Polish by the author. MR1039321. 423

[EnSc–92] M. Enock and J.-M. Schwartz, *Kac algebras and duality of locally compact groups*, Springer-Verlag, Berlin, 1992. With a preface by Alain Connes; With a postface by Adrian Ocneanu. MR1215933. 10

[Erne–64] J. Ernest, *A new group algebra for locally compact groups*, Amer. J. Math. **86** (1964), 467–492. MR167566. 260

[Erne–65] J. Ernest, *A new group algebra for locally compact groups. II*, Canadian J. Math. **17** (1965), 604–615. MR182676. 260

[Eyma–64] P. Eymard, *L'algèbre de Fourier d'un groupe localement compact* (French), Bull. Soc. Math. France **92** (1964), 181–236. MR228628. 402

[Eyma–72] P. Eymard, *Moyennes invariantes et représentations unitaires* (French), Lecture Notes in Mathematics, Vol. 300, Springer-Verlag, Berlin-New York, 1972. MR0447969. 51

[FeMo–77] J. Feldman and C. C. Moore, *Ergodic equivalence relations, cohomology, and von Neumann algebras. I*, Trans. Amer. Math. Soc. **234** (1977), no. 2, 289–324. MR578656. 292, 365

[Fell–60a] J. M. G. Fell, *The dual spaces of C^*-algebras*, Trans. Amer. Math. Soc. **94** (1960), 365–403. MR146681. 8, 30, 121

[Fell–60b] J. M. G. Fell, *C^*-algebras with smooth dual*, Illinois J. Math. **4** (1960), 221–230. MR124754. 259

[Fell–61] J. M. G. Fell, *The structure of algebras of operator fields*, Acta Math. **106** (1961), 233–280. MR164248. 9, 121, 246

[Fell–62] J. M. G. Fell, *Weak containment and induced representations of groups*, Canadian J. Math. **14** (1962), 237–268. MR150241. 110, 113

[Fell–62b] J. M. G. Fell, *A Hausdorff topology for the closed subsets of a locally compact non-Hausdorff space*, Proc. Amer. Math. Soc. **13** (1962), 472–476. MR139135. 416

[Fell–64] J. M. G. Fell, *Weak containment and induced representations of groups. II*, Trans. Amer. Math. Soc. **110** (1964), 424–447. MR159898. 251

[Foll–16] G. B. Folland, *A course in abstract harmonic analysis*, 2nd ed., Textbooks in Mathematics, CRC Press, Boca Raton, FL, 2016. MR3444405. 21, 35, 40, 42, 46, 54, 56, 92, 106, 110, 113, 121, 128, 152, 165, 203, 243, 438, 443

[FoSW] B. E. Forrest, N. Spronk, and M. Wiersma, *Existence of tracial states on reduced group C^*-algebras*, `arXiv:1706.05354v2`, 19 Jul 2017. 215

[Freu–36] H. Freudenthal, *Topologische Gruppen mit genügend vielen fastperiodischen Funktionen* (German), Ann. of Math. (2) **37** (1936), no. 1, 57–77. MR1503269. 136

[Frob–Col] F. G. Frobenius, *Gesammelte Abhandlungen. Bände I, II, III* (German), Herausgegeben von J.-P. Serre, Springer-Verlag, Berlin-New York, 1968. MR0235974. 7, 9

[Fuc73] L. Fuchs, *Infinite abelian groups. Vol. II*, Pure and Applied Mathematics. Vol. 36-II, Academic Press, New York-London, 1973. MR0349869. 106

[Furs–61] H. Furstenberg, *Strict ergodicity and transformation of the torus*, Amer. J. Math. **83** (1961), 573–601. MR133429. 154, 165

[Gela–18] T. Gelander, *A lecture on invariant random subgroups*, New directions in locally compact groups, London Math. Soc. Lecture Note Ser., vol. 447, Cambridge Univ. Press, Cambridge, 2018, pp. 186–204. MR3793288. 416

[Gelf–41] I. M. Gel'fand, *Normierte Ringe*, Mat. Sbornik **51** (1941), 3–24 [**Gelf–Col**, Volume I, 181–201]. 8

[Gelf–50] I. M. Gel'fand, *Spherical functions on symmetric Riemannian spaces*, Doklady Akad. Nauk SSSR (N.S.) **70** (1950), 5–8 [**Gelf–Col**, Volume II, 31–35]. 187

[Gelf–Col] I. M. Gel'fand, *Collected papers. Volumes I, II, III*, Springer, 1987, 1988, 1989. 451, 452

[GeRa–43] I. M. Gel'fand and D. Raikov, *Irreducible unitary representations of locally bicompact groups*, Rec. Math. [Mat. Sbornik] N.S. **13(55)** (1943), 301–316 [**Gelf–Col**, Volume II, 3–17]. 8, 27

[GeNa–43] I. M. Gel'fand and M. A. Naimark, *On the imbedding of normed rings into the ring of operators in Hilbert space*, Mat. Sbornik **54** (1943), 197–213 [**Gelf–Col**, Volume I, 241–257]. 8

[GeNa–47a] I. M. Gel'fand and M. A. Naimark, *Unitary representations of the group of linear transformations of the straight line*, Dokl. Akad. Nauk SSSR (N.S.) **55** (1947), 567–570 [**Gelf–Col**, Volume II, 18–21]. 8

[GeNa–47b] I. M. Gel'fand and M. A. Naimark, *Unitary representations of the Lorentz group*, Izvestiya Akad. Nauk SSSR. Ser. Mat., **11** (1947), 411–504. [**Gelf–Col**, Volume II, 41–123]. 8, 9, 121

[GeNa–48] I. M. Gel′fand and M. A. Naĭmark, *Normed rings with involutions and their representations* (Russian), Izvestiya Akad. Nauk SSSR. Ser. Mat. **12** (1948), 445–480. MR0026763. 21

[GeNa–57] I. M. Gelfand and M. A. Neumark, *Unitäre Darstellungen der klassischen Gruppen* (German), Akademie-Verlag, Berlin, 1957. MR0085262. 9

[GGPS–69] I. M. Gel'fand, M. I. Graev, and I. I. Pyatetskii-Shapiro, *Representation theory and automorphic functions*, translated from the Russian by K. A. Hirsch. W. B. Saunders Company, 1969 [Russian original 1966]. 52, 113, 208

[Glas–98] E. Glasner, *On minimal actions of Polish groups*, 8th Prague Topological Symposium on General Topology and Its Relations to Modern Analysis and Algebra (1996), Topology Appl. **85** (1998), no. 1-3, 119–125. MR1617456. 27

[Glim–61a] J. Glimm, *Type I C^*-algebras*, Ann. of Math. (2) **73** (1961), 572–612. MR124756. 8, 258, 259, 291

[Glim–61b] J. Glimm, *Locally compact transformation groups*, Trans. Amer. Math. Soc. **101** (1961), 124–138. MR136681. 291

[Gode–44] R. Godement, *Sur une généralisation d'un théorème de Stone* (French), C. R. Acad. Sci. Paris **218** (1944), 901–903. MR14589. 82

[Gode–48] R. Godement, *Les fonctions de type positif et la théorie des groupes* (French), Trans. Amer. Math. Soc. **63** (1948), 1–84. MR23243. 30

[Gode–51a] R. Godement, *Sur la théorie des représentations unitaires* (French), Ann. of Math. (2) **53** (1951), 68–124. MR38571. 53, 59, 211

[Gode–51b] R. Godement, *Mémoire sur la théorie des caractères dans les groupes localement compacts unimodulaires* (French), J. Math. Pures Appl. (9) **30** (1951), 1–110. MR41857. 9, 227, 399

[Gode–52] R. Godement, *A theory of spherical functions. I*, Trans. Amer. Math. Soc. **73** (1952), 496–556. MR52444. 204, 208

[Gode–54] R. Godement, *Théorie des caractères. II. Définition et propriétés générales des caractères* (French), Ann. of Math. (2) **59** (1954), 63–85. MR58880. 9, 310, 312, 399

[GoSV–16] M. Göll, K. Schmidt, and E. Verbitskiy, *A Wiener lemma for the discrete Heisenberg group*, Monatsh. Math. **180** (2016), no. 3, 485–525. MR3513217. 165

[GoRo–79] E. C. Gootman and J. Rosenberg, *The structure of crossed product C^*-algebras: a proof of the generalized Effros-Hahn conjecture*, Invent. Math. **52** (1979), no. 3, 283–298. MR537063. 263

[Gorb–80] N. V. Gorbachëv, *C^*-algebra of the Heisenberg group* (Russian), Uspekhi Mat. Nauk **35** (1980), no. 6(216), 157–158. MR601762. 246

[GoKV–14] V. Gorin, S. Kerov, and A. Vershik, *Finite traces and representations of the group of infinite matrices over a finite field*, Adv. Math. **254** (2014), 331–395. MR3161102. 326

[Grig–11] R. I. Grigorchuk, *Some topics of the dynamics of group actions on rooted trees* (Russian, with Russian summary), Tr. Mat. Inst. Steklova **273** (2011), no. Sovremennye Problemy Matematiki, 72–191; English transl., Proc. Steklov Inst. Math. **273** (2011), no. 1, 64–175. MR2893544. 290, 432

[GrKO–15] R. Grigorchuk, R. Kravchenko, and A. Ol'shanskii, *Constructions of torsion-free countable, amenable, weakly mixing groups*, Enseign. Math. **61** (2015), no. 3-4, 321–342. MR3539841. 145

[GrMo–71a] S. Grosser and M. Moskowitz, *Compactness conditions in topological groups*, J. reine angew. Math. **246** (1971), 1–40. MR284541. 136

[GrMo–71b] S. Grosser and M. Moskowitz, *Harmonic analysis on central topological groups*, Trans. Amer. Math. Soc. **156** (1971), 419–454. MR276418. 38

[Guic–63] A. Guichardet, *Caractères des algèbres de Banach involutives* (French), Ann. Inst. Fourier (Grenoble) **13** (1963), 1–81. MR147925. 9, 263, 312, 314, 331, 373, 382, 383

[Guic–65] A. Guichardet, *Utilisation des sous-groupes distingués ouverts dans l'étude des représentations unitaires des groupes localement compacts* (French), Compositio Math. **17** (1965), 1–35 (1965). MR188348. 263, 273

[Haa–33] A. Haar, *Der Massbegriff in der Theorie der kontinuierlichen Gruppen* (German), Ann. of Math. (2) **34** (1933), no. 1, 147–169. MR1503103. 7

[HaKL–90] L. A. Harrington, A. S. Kechris, and A. Louveau, *A Glimm-Effros dichotomy for Borel equivalence relations*, J. Amer. Math. Soc. **3** (1990), no. 4, 903–928. MR1057041. 291

[Halm–51] P. R. Halmos, *Introduction to Hilbert Space and the theory of Spectral Multiplicity*, Chelsea Publishing Company, New York, N. Y., 1951. MR0045309. 71, 80, 441

[Halm–60] P. R. Halmos, *Lectures on ergodic theory*, Chelsea Publishing Co., New York, 1960. MR0111817. 412

[Halp–75] H. Halpern, *Quasi-equivalence classes of normal representations for a separable C^*-algebra*, Trans. Amer. Math. Soc. **203** (1975), 129–140. MR367669. 9, 314

[Hari–47] Harish–Chandra, *Infinite irreducible representations of the Lorentz group*, Proc. Roy. Soc. London, Ser. A **189** (1947), 372–401 [**Hari–Col**, Volume I, 162–191]. 121

[Hari–53] Harish–Chandra, *Representations of a semisimple Lie group on a Banach space. I*, Trans. Amer. Math. Soc. **75** (1953), 185–243 [**Hari–Col**, Volume I, 391–449]. 204

[Hari–54] Harish–Chandra, *Representations of a semisimple Lie group. III*, Trans. Amer. Math. Soc. **76** (1954), 234–253 [**Hari–Col**, Volume I, 490–509]. 6, 317

[Hari–56] Harish–Chandra, *The characters of semisimple Lie groups*, Trans. Amer. Math. Soc. **83** (1956), 98–163 [**Hari–Col**, Volume II, 156–221]. 9

[Hari–Col] Harish–Chandra, *Collected papers. Volumes I, II, III, IV*, Edited by V.S. Varadarajan. Springer, 1984. 8, 453

[Harp–07] P. de la Harpe, *On simplicity of reduced C^*-algebras of groups*, Bull. Lond. Math. Soc. **39** (2007), no. 1, 1–26. MR2303514. 289

[Haue–86] W. Hauenschild, *Zwei Dualräume der unendlichen, symmetrischen Gruppe* (German), Arch. Math. (Basel) **46** (1986), no. 3, 211–215. MR834838. 328

[Haus–14] F. Hausdorff, *Grundzüge der Mengenlehre*, Veit, 1914. 143

[Hawk–71] T. Hawkins, *The origins of the theory of group characters*, Arch. History Exact Sci. **7** (1971), no. 2, 142–170. MR1554141. 8

[Hawk–72] T. Hawkins, *Hypercomplex numbers, lie groups, and the creation of group representation theory*, Arch. History Exact Sci. **8** (1972), no. 4, 243–287. MR1554156. 8

[Hawk–74] T. Hawkins, *New light on Frobenius' creation of the theory of group characters*, Arch. History Exact Sci. **12** (1974), 217–243. MR0446837. 8

[HeRo–63] E. Hewitt and K. A. Ross, *Abstract harmonic analysis. Vol. I: Structure of topological groups. Integration theory, group representations*, Die Grundlehren der mathematischen Wissenschaften, Bd. 115, Academic Press, Inc., Publishers, New York; Springer-Verlag, Berlin-Göttingen-Heidelberg, 1963. MR0156915. 31, 404, 436

[HeRo–70] E. Hewitt and K. A. Ross, *Abstract harmonic analysis. Vol. II: Structure and analysis for compact groups. Analysis on locally compact Abelian groups*, Die Grundlehren der mathematischen Wissenschaften, Band 152, Springer-Verlag, New York-Berlin, 1970. MR0262773. 10, 86

[HiHi–05] T. Hirai and E. Hirai, *Positive definite class functions on a topological group and characters of factor representations*, J. Math. Kyoto Univ. **45** (2005), no. 2, 355–379. MR2161697. 322

[HoRa–19] C. Houdayer and S. Raum, *Locally compact groups acting on trees, the type I conjecture and non-amenable von Neumann algebras*, Comment. Math. Helv. **94** (2019), no. 1, 185–219. MR3941470. 209

[Howe–77a] R. E. Howe, *On representations of discrete, finitely generated, torsion-free, nilpotent groups*, Pacific J. Math. **73** (1977), no. 2, 281–305. MR499004. 271, 326, 329

[Howe–77b] R. E. Howe, *The Fourier transform for nilpotent locally compact groups. I*, Pacific J. Math. **73** (1977), no. 2, 307–327. MR492059. 99, 268

[HoMo–79] R. E. Howe and C. C. Moore, *Asymptotic properties of unitary representations*, J. Functional Analysis **32** (1979), no. 1, 72–96. MR533220. 415

[HoRo–89] R. E. Howe and J. Rosenberg, *The unitary representation theory of* GL(n) *of an infinite discrete field*, Israel J. Math. **67** (1989), no. 1, 67–81. MR1021362. 276

[Hump–80] J. E. Humphreys, *Arithmetic groups*, Lecture Notes in Mathematics, vol. 789, Springer, Berlin, 1980. MR584623. 142

[Isma–69] R. S. Ismagilov, *The irreducible cycles connected with a dynamic system* (Russian), Funkcional. Anal. i Priložen. **3** (1969), no. 3, 92–93. MR0249571. 154, 165

[Iwas–51] K. Iwasawa, *Topological groups with invariant compact neighborhoods of the identity*, Ann. of Math. (2) **54** (1951), 345–348. MR43106. 136, 403

[Jaco–45] N. Jacobson, *A topology for the set of primitive ideals in an arbitrary ring*, Proc. Nat. Acad. Sci. U.S.A. **31** (1945), 333–338. MR13138. 8

[KaSi–52] R. V. Kadison and I. M. Singer, *Some remarks on representations of connected groups*, Proc. Nat. Acad. Sci. U.S.A. **38** (1952), 419–423. MR48456. 227, 404

[Kaim–00] V. A. Kaimanovich, *Ergodic properties of the horocycle flow and classification of Fuchsian groups*, J. Dynam. Control Systems **6** (2000), no. 1, 21–56. MR1738739. 371

[Kaji–83] T. Kajiwara, *Group extension and Plancherel formulas*, J. Math. Soc. Japan **35** (1983), no. 1, 93–115. MR679078. 228

[KaKo–44] S. Kakutani and K. Kodaira, *Über das Haarsche Mass in der lokal bikompakten Gruppe* (German), Proc. Imp. Acad. Tokyo **20** (1944), 444–450. MR14401. 254

[Kall–70] R. R. Kallman, *Certain topological groups are type I*, Bull. Amer. Math. Soc. **76** (1970), 404–406. MR255725. 210

[Kall–73] R. R. Kallman, *Certain topological groups are type I. II*, Advances in Math. **10** (1973), 221–255. MR338256. 209

[Kani–69] E. Kaniuth, *Der Typ der regulären Darstellung diskreter Gruppen* (German), Math. Ann. **182** (1969), 334–339. MR260896. 224

[Kani–79] E. Kaniuth, *Primitive ideal spaces of groups with relatively compact conjugacy classes*, Arch. Math. (Basel) **32** (1979), no. 1, 16–24. MR532844. 40

[Kani–80] E. Kaniuth, *Ideals in group algebras of finitely generated FC-nilpotent discrete groups*, Math. Ann. **248** (1980), no. 2, 97–108. MR573341. 329

[Kani–82] E. Kaniuth, *On primary ideals in group algebras*, Monatsh. Math. **93** (1982), no. 4, 293–302. MR666831. 268

[KaTa–13] E. Kaniuth and K. F. Taylor, *Induced representations of locally compact groups*, Cambridge Tracts in Mathematics, vol. 197, Cambridge University Press, Cambridge, 2013. MR3012851. 38, 40, 92, 121, 152, 271

[KaWe–72] Y. Katznelson and B. Weiss, *The construction of quasi-invariant measures*, Israel J. Math. **12** (1972), 1–4. MR316679. 164

[Kapl–51a] I. Kaplansky, *Group algebras in the large*, Tohoku Math. J. (2) **3** (1951), 249–256. MR48712. 225

[Kapl–51b] I. Kaplansky, *The structure of certain operator algebras*, Trans. Amer. Math. Soc. **70** (1951), 219–255. MR42066. 9, 259

[Kazh–67] D. A. Kazhdan, *Connection of the dual space of a group with the structure of its closed subgroups* (Russian), Funkcional. Anal. i Priložen. **1** (1967), 71–74. MR0209390. 38

[Kean–71] M. Keane, *Sur les mesures quasi-ergodiques des translations irrationnelles* (French), C. R. Acad. Sci. Paris Sér. A-B **272** (1971), A54–A55. MR272982. 164

[Kech–95] A. S. Kechris, *Classical descriptive set theory*, Graduate Texts in Mathematics, vol. 156, Springer-Verlag, New York, 1995. MR1321597. 424, 425, 429

[KeRa–17] M. Kennedy and S. Raum, *Traces on reduced group* C**-algebras*, Bull. Lond. Math. Soc. **49** (2017), no. 6, 988–990. MR3743482. 215

[Khal–74] I. Khalil, *Sur l'analyse harmonique du groupe affine de la droite* (French), Studia Math. **51** (1974), 139–167. MR350330. 203

[Kirc–93] E. Kirchberg, *On nonsemisplit extensions, tensor products and exactness of group* C^**-algebras*, Invent. Math. **112** (1993), no. 3, 449–489. MR1218321. 406

[Kiri–62] A. A. Kirillov, *Unitary representations of nilpotent Lie groups* (Russian), Uspehi Mat. Nauk **17** (1962), no. 4 (106), 57–110. MR0142001. 8, 99, 110, 204, 317

[Kiri–65] A. A. Kirillov, *Positive-definite functions on a group of matrices with elements from a discrete field* (Russian), Dokl. Akad. Nauk SSSR **162** (1965), 503–505. MR0193183. 331, 346

[Kiri–67] A. A. Kirillov, *Dynamical systems, factors and group representations* (Russian), Uspehi Mat. Nauk **22** (1967), no. 5 (137), 67–80. MR0217256. 154

[Kiri–73] A. A. Kirillov, *Representation of the infinite-dimensional unitary group*, Dokl. Akad. Nauk SSSR **212** (1973), 288–290 [English translation: Soviet Math. Dokl. **14** (1973), no. 5, 1355–1358]. 122, 209

[Kiri–76] A. A. Kirillov, *Elements of the theory of representations*, Grundlehren der math. Wiss. **220**, Springer, 1976 [Russian original 1972]. 259

[Kiri–99] A. A. Kirillov, *Merits and demerits of the orbit method*, Bull. Amer. Math. Soc. (N.S.) **36** (1999), no. 4, 433–488. MR1701415. 110

[Klep–61] A. Kleppner, *On the intertwining number theorem*, Proc. Amer. Math. Soc. **12** (1961), 731–733. MR131783. 51

[KlLi–72] A. Kleppner and R. L. Lipsman, *The Plancherel formula for group extensions. I, II*, Ann. Sci. École Norm. Sup. (4) **5** (1972), 459–516; ibid. (4) 6 (1973), 103–132. MR342641. 113

[KlLi–73] A. Kleppner and R. L. Lipsman, *The Plancherel formula for group extensions. I, II*, Ann. Sci. École Norm. Sup. (4) **5** (1972), 459–516; ibid. (4) 6 (1973), 103–132. MR342641. 228

[Knap–02] A. W. Knapp, *Lie groups beyond an introduction*, 2nd ed., Progress in Mathematics, vol. 140, Birkhäuser Boston, Inc., Boston, MA, 2002. MR1920389. 189, 205

[KoKa–43] K. Kodaira and S. Kakutani, *Normed ring of a locally compact Abelian group*, Proc. Imp. Acad. Tokyo **19** (1943), 360–365. MR14094. 260

[Koop–31] B. O. Koopman, *Hamiltonian systems and transformations in Hilbert space*, Proc. Nat. Acad. Sci. U.S.A. **17** (1931), 315–318. 356

[Krie–71] W. Krieger, *On quasi-invariant measures in uniquely ergodic systems*, Invent. Math. **14** (1971), 184–196. MR293060. 164

[Krie–76] W. Krieger, *On Borel automorphisms and their quasi-invariant measures*, Math. Z. **151** (1976), no. 1, 19–24. MR417389. 377

[Kura–66] K. Kuratowski, *Topology. Vol. I*, New edition, revised and augmented, Academic Press, 1966. [First edition in French, 1933.] 425

[Lang–85] S. Lang, $SL_2(\mathbf{R})$, Graduate Texts in Mathematics, vol. 105, Springer-Verlag, New York, 1985. Reprint of the 1975 edition. MR803508. 205

[Lang–93] S. Lang, *Real and functional analysis*, 3rd ed., Graduate Texts in Mathematics, vol. 142, Springer-Verlag, New York, 1993. MR1216137. 425

[LeLu–94] H. Leptin and J. Ludwig, *Unitary representation theory of exponential Lie groups*, De Gruyter Expositions in Mathematics, vol. 18, Walter de Gruyter & Co., Berlin, 1994. MR1307383. 110

[LeRo–68] H. Leptin and L. Robertson, *Every local compact MAP group is unimodular*, Proc. Amer. Math. Soc. **19** (1968), 1079–1082. MR230839. 136

[Lieb–72] A. Lieberman, *The structure of certain unitary representations of infinite symmetric groups*, Trans. Amer. Math. Soc. **164** (1972), 189–198. MR286940. 209

[Lips–70] R. L. Lipsman, *The dual topology for the principal and discrete series on semisimple groups*, Trans. Amer. Math. Soc. **152** (1970), 399–417. MR269778. 121

[Lips–72] R. L. Lipsman, *Representation theory of almost connected groups*, Pacific J. Math. **42** (1972), 453–467. MR327975. 208

[LiMo–74] J. Liukkonen and R. Mosak, *The primitive dual space of $[FC]^-$ groups*, J. Functional Analysis **15** (1974), 279–296. MR0346088. 40

[LuTu–11] J. Ludwig and L. Turowska, *The C^*-algebras of the Heisenberg group and of thread-like Lie groups*, Math. Z. **268** (2011), no. 3-4, 897–930. MR2818736. 246

[LuSh–04] A. Lubotzky and Y. Shalom, *Finite representations in the unitary dual and Ramanujan groups*, Discrete geometric analysis, Contemp. Math., vol. 347, Amer. Math. Soc., Providence, RI, 2004, pp. 173–189. MR2077037. 412, 414

[Mack–51] G. W. Mackey, *On induced representations of groups*, Amer. J. Math. **73** (1951), 576–592. MR42420. 48, 49, 51, 58

[Mack–52] G. W. Mackey, *Induced representations of locally compact groups. I*, Ann. of Math. (2) **55** (1952), 101–139. MR44536. 8, 40, 55, 113

[Mack–53] G. W. Mackey, *Induced representations of locally compact groups. II. The Frobenius reciprocity theorem*, Ann. of Math. (2) **58** (1953), 193–221. MR56611. 211

[Mack–57] G. W. Mackey, *Borel structure in groups and their duals*, Trans. Amer. Math. Soc. **85** (1957), 134–165. MR89999. 2, 258, 424, 425

[Mack–58] G. W. Mackey, *Unitary representations of group extensions. I*, Acta Math. **99** (1958), 265–311. MR98328. 8, 40, 42, 91, 92, 152

[Mack–61] G. W. Mackey, *Induced representations and normal subgroups*, Proc. Internat. Sympos. Linear Spaces (Jerusalem, 1960), Jerusalem Academic Press, Jerusalem; Pergamon, Oxford, 1961, pp. 319–326. MR0139687. 228

[Mack–63] G. W. Mackey, *Infinite-dimensional group representations*, Bull. Amer. Math. Soc. **69** (1963), 628–686. MR153784. 203

[Mack–76] G. W. Mackey, *The theory of unitary group representations*, Chicago Lectures in Mathematics, University of Chicago Press, Chicago, Ill.-London, 1976. Based on notes by James M. G. Fell and David B. Lowdenslager of lectures given at the University of Chicago, Chicago, Ill., 1955. MR0396826. 8, 27, 132, 169, 172, 183, 209, 220, 428

[Mack–78] G. W. Mackey, *Unitary group representations in physics, probability, and number theory*, Mathematics Lecture Note Series, vol. 55, Benjamin/Cummings Publishing Co., Inc., Reading, Mass., 1978. MR515581. 169

[Mal'c–40] A. I. Mal'cev, *On the faithful representation of infinite groups by matrices*, Amer. Math. Soc. Transl. (2) **45** (1965) 1–18 [Russian original: Mat. SS.(N.S.) 8(50) (1940), pp. 405–422]. 136

[Mare–75] O. Maréchal, *Une remarque sur un théorème de Glimm*, Bull. Sci. Math. (2) **99** (1975), no. 1, 41–44. MR430797. 259

[Maut–50a] F. I. Mautner, *Unitary representations of locally compact groups. I*, Ann. of Math. (2) **51** (1950), 1–25. MR32650. 53, 211

[Maut–50b] F. I. Mautner, *Unitary representations of locally compact groups. II*, Ann. of Math. (2) **52** (1950), 528–556. MR36763. 211, 227, 228

[Maut–50c] F. I. Mautner, *The structure of the regular representation of certain discrete groups*, Duke Math. J. **17** (1950), 437–441. MR39727. 221, 223, 224

[Megr–01] M. G. Megrelishvili, *Every semitopological semigroup compactification of the group $H_+[0,1]$ is trivial*, Semigroup Forum **63** (2001), no. 3, 357–370. MR1851816. 27

[Menn–65] J. L. Mennicke, *Finite factor groups of the unimodular group*, Ann. of Math. (2) **81** (1965), 31–37. MR171856. 143

[Mili–71] D. Miličić, *Topological representation of the group C^*-algebra of* SL(2, R) (English, with Serbo-Croatian summary), Glasnik Mat. Ser. III **6(26)** (1971), 231–246. MR308795. 121, 246

[Miln–71] P. Milnes, *Identities of group algebras*, Proc. Amer. Math. Soc. **29** (1971), 421–422. MR274661. 245

[Moor–65] C. C. Moore, *Decomposition of unitary representations defined by discrete subgroups of nilpotent groups*, Ann. of Math. (2) **82** (1965), 146–182. MR181701. 208

[Moor–72] C. C. Moore, *Groups with finite dimensional irreducible representations*, Trans. Amer. Math. Soc. **166** (1972), 401–410. MR302817. 132, 210, 254

[MoRo–76] C. C. Moore and J. Rosenberg, *Groups with T_1 primitive ideal spaces*, J. Functional Analysis **22** (1976), no. 3, 204–224. MR0419675. 39, 40

[Mott–12] L. Motto Ros, *On the complexity of the relations of isomorphism and bi-embeddability*, Proc. Amer. Math. Soc. **140** (2012), no. 1, 309–323. MR2833542. 291

[MuvN–36] F. J. Murray and J. von Neumann, *On rings of operators*, Ann. of Math. (2) **37** (1936), 116–229 [**vNeu–Col**, Volume III, 6–119]. 353, 362

[MuvN–43] F. J. Murray and J. von Neumann, *On rings of operators, IV*, Ann. of Math. (2) **44** (1943), 716–808 [**vNeu–Col**, Volume III, 229–321]. 211

[Naim–43] M. Naimark [also written Neumark], *Positive definite operator functions on a commutative group*, Bull. Acad. Sci. URSS Sér. Math. [Izvestia Akad. Nauk SSSR] **7** (1943), 237–244. 82

[Neeb–14] K.-H. Neeb, *Unitary representations of unitary groups*, Developments and retrospectives in Lie theory, Dev. Math., vol. 37, Springer, Cham, 2014, pp. 197–243. MR3329940. 209

[vNeu–31] J. von Neumann, *Die Eindeutigkeit der Schrödingerschen Operatoren*, Math. Ann. **104** (1931), no. 1, 570–578 [**vNeu–Col**, Volume II, 220–229]. 8, 109

[vNeu–32] J. von Neumann, *Einige Sätze über messbare Abbildungen*, Ann. of Math. (2) **33** (1032), 574–586 [**vNeu–Col**, Volume II, 294-306]. 429

[vNeu–34] J. von Neumann, *Almost periodic functions in a group. I*, Trans. Amer. Math. Soc. **36** (1934), 445–492 [**vNeu–Col**, Volume II, 454–501]. 133, 144

[vNeu–40] J. von Neumann, *On rings of operators. III*, Ann. of Math. (2) **41** (1940), 94–161 [**vNeu–Col**, Volume III, 161–228]. 362

[vNeu–49] J. von Neumann, *On rings of operators. Reduction theory*, Ann. of Math. (2) **50** (1949), 401–485, the article was written in 1937–38 [**vNeu–Col**, Volume III, 400–484]. 53, 59, 428

[vNWi–40] J. von Neumann and E. P. Wigner, *Minimally almost periodic groups*, Ann. of Math. (2) **41** (1940), 746–750 [**vNeu–Col**, Volume IV, 220–224]. 144

[vNeu–Col] J. von Neumann, *Collected works. Volumes I to VI*, General editor: A. H. Taub. Pergamon Press, 1961–1969. 456, 457, 459

[vNeu–Col] J. von Neumann, *Collected works. Vol. III: Rings of operators*, General editor: A. H. Taub, Pergamon Press, New York-Oxford-London-Paris, 1961. MR0157873. 8

[Obat–89] N. Obata, *Some remarks on induced representations of infinite discrete groups*, Math. Ann. **284** (1989), no. 1, 91–102. MR995384. 49

[O'Do–81] D. P. O'Donovan, *Irreducible representations of the irrational rotation covariance C^*-algebra*, Proc. Roy. Irish Acad. Sect. A **81** (1981), no. 2, 231–236. MR654822. 271

[Ol's–77] G. I. Ol'shanskii, *Classification of the irreducible representations of the automorphism groups of Bruhat–Tits trees*, Funct. Anal. Appl. **11** (1977), no. 1, 26–34. 209

[Ol's–78] G. I. Ol'shanskii, *Unitary representations of the infinite-dimensional classical groups $U(p,\infty), \mathrm{SO}(p,\infty), Sp(p,\infty)$ and of the corresponding motion groups*, Funct. Anal. Appl. **12** (1978), no. 1, 185–195. 122, 209

[Ol's–80] G. I. Ol'shanskii, *New "large" groups of type I*, Akad. Nauk. SSSR **16** (1980), 31–52. [English translation: J. Soviet Math. **18** (1992), 22–39.] 209

[OrWe–80] D. S. Ornstein and B. Weiss, *Ergodic theory of amenable group actions. I. The Rohlin lemma*, Bull. Amer. Math. Soc. (N.S.) **2** (1980), no. 1, 161–164. MR551753. 165

[OsAd–76] B. Osofsky and S. Adams, *Problems and Solutions: Advanced Solutions: 6102*, Amer. Math. Monthly **85** (1978), no. 6, 504–505. MR1538750. 143

[Ozaw–13] N. Ozawa, *About the Connes embedding conjecture: algebraic approaches*, Jpn. J. Math. **8** (2013), no. 1, 147–183. MR3067294. 407

[Palm–78] T. W. Palmer, *Classes of nonabelian, noncompact, locally compact groups*, Rocky J. Math. **8** (1978), no. 4, 683–741. Updated and expanded in Sections 12.5 and 12.6 of [**Palm–01**]. 38, 40, 136

[Palm–01] T. W. Palmer, *Banach algebras and the general theory of $*$-algebras. Vol. 2: $*$-algebras*, Encyclopedia of Mathematics and its Applications, vol. 79, Cambridge University Press, Cambridge, 2001. MR1819503. 457

[Pede–79] G. K. Pedersen, *C^*-algebras and their automorphism groups*, London Mathematical Society Monographs, vol. 14, Academic Press, Inc. [Harcourt Brace Jovanovich, Publishers], London-New York, 1979. MR548006. 217

[Pest–07] V. Pestov, *The isometry group of the Urysohn space as a Levÿ group*, Topology Appl. **154** (2007), no. 10, 2173–2184. MR2324929. 27

[PeWe–27] F. Peter and H. Weyl, *Die Vollständigkeit der primitiven Darstellungen einer geschlossenen kontinuierlichen Gruppe*, Math. Ann. **97** (1927), no. 1, 737–755 [**Weyl–Col**, Volume III, 58–75]. 7

[Pete–15] J. Peterson, *Character rigidity for lattices in higher-rank groups*, Preprint 2015. 326

[PeTh–16] J. Peterson and A. Thom, *Character rigidity for special linear groups*, J. reine angew. Math. **716** (2016), 207–228. MR3518376. 320, 351, 419, 420

[Pfef–95] C. Pfeffer Johnston, *Primitive ideal spaces of discrete rational nilpotent groups*, Amer. J. Math. **117** (1995), no. 2, 323–335. MR1323678. 329

[Pogu–82] D. Poguntke, *Discrete nilpotent groups have a T_1 primitive ideal space*, Studia Math. **71** (1981/82), no. 3, 271–275. MR667315. 39, 329

[Pogu–83] D. Poguntke, *Simple quotients of group C^*-algebras for two-step nilpotent groups and connected Lie groups*, Ann. Sci. École Norm. Sup. (4) **16** (1983), no. 1, 151–172. MR719767. 254, 328

[Pont–39] L. Pontrjagin, *Topological Groups*, Princeton Mathematical Series, v. 2, Princeton University Press, Princeton, 1939. Translated from the Russian by Emma Lehmer. MR0000265. 8

[Popa–07] S. Popa, *Cocycle and orbit equivalence superrigidity for malleable actions of w-rigid groups*, Invent. Math. **170** (2007), no. 2, 243–295. MR2342637. 398

[Powe–75] R. T. Powers, *Simplicity of the C^*-algebra associated with the free group on two generators*, Duke Math. J. **42** (1975), 151–156. MR374334. 313

[Puka–71] L. Pukanszky, *Unitary representations of solvable Lie groups*, Ann. Sci. École Norm. Sup. (4) **4** (1971), 457–608. MR439985. 227

[Puka–74] L. Pukanszky, *Characters of connected Lie groups*, Acta Math. **133** (1974), 81–137. MR409728. 9, 328, 329

[Puka–99] L. Pukánszky, *Characters of connected Lie groups*, Mathematical Surveys and Monographs, vol. 71, American Mathematical Society, Providence, RI, 1999. With a preface by J. Dixmier and M. Duflo. MR1707323. 328, 329

[Raeb–82] I. Raeburn, *On group C^*-algebras of bounded representation dimension*, Trans. Amer. Math. Soc. **272** (1982), no. 2, 629–644. MR662057. 38

[Rams–76] A. Ramsay, *Nontransitive quasi-orbits in Mackey's analysis of group extensions*, Acta Math. **137** (1976), no. 1, 17–48. MR460531. 147, 162

[RiZa–10] L. Ribes and P. Zalesskii, *Profinite groups*, 2nd ed., Ergebnisse der Mathematik und ihrer Grenzgebiete. 3. Folge. A Series of Modern Surveys in Mathematics [Results in Mathematics and Related Areas. 3rd Series. A Series of Modern Surveys in Mathematics], vol. 40, Springer-Verlag, Berlin, 2010. MR2599132. 139

[Rief–74] M. A. Rieffel, *Induced representations of C^*-algebras*, Advances in Math. **13** (1974), 176–257. MR353003. 251

[Rief–81] M. A. Rieffel, *C^*-algebras associated with irrational rotations*, Pacific J. Math. **93** (1981), no. 2, 415–429. MR623572. 271

[Robe–78] A. Robert, *Exemples de groupes de Fell* (French, with English summary), C. R. Acad. Sci. Paris Sér. A-B **287** (1978), no. 8, A603–A606. MR514536. 214

[Robe–83] A. Robert, *Introduction to the representation theory of compact and locally compact groups*, London Mathematical Society Lecture Note Series, vol. 80, Cambridge University Press, Cambridge-New York, 1983. MR690955. 175

[Robe–00] A. M. Robert, *A course in p-adic analysis*, Graduate Texts in Mathematics, vol. 198, Springer-Verlag, New York, 2000. MR1760253. 116

[Robe–69] L. C. Robertson, *A note on the structure of Moore groups*, Bull. Amer. Math. Soc. **75** (1969), 594–599. MR245721. 132

[Robi–96] D. J. S. Robinson, *A course in the theory of groups*, 2nd ed., Graduate Texts in Mathematics, vol. 80, Springer-Verlag, New York, 1996. MR1357169. 137, 141, 431

[Rose–76] J. Rosenberg, *The C^*-algebras of some real and p-adic solvable groups*, Pacific J. Math. **65** (1976), no. 1, 175–192. MR447467. 246

[Rose–04] J. Rosenberg, *A selective history of the Stone-von Neumann theorem*, Operator algebras, quantization, and noncommutative geometry, Contemp. Math., vol. 365, Amer. Math. Soc., Providence, RI, 2004, pp. 331–353. MR2106827. 8

[Rudi–62] W. Rudin, *Fourier analysis on groups*, Interscience Tracts in Pure and Applied Mathematics, No. 12, Interscience Publishers (a division of John Wiley and Sons), New York-London, 1962. MR0152834. 142

[Rudi–66] W. Rudin, *Real and complex analysis*, 3rd ed., McGraw-Hill Book Co., New York, 1987. MR924157. 367, 425, 426, 427, 429, 430, 431

[Rudi–73] W. Rudin, *Functional analysis*, McGraw-Hill Series in Higher Mathematics, McGraw-Hill Book Co., New York-Düsseldorf-Johannesburg, 1973. MR0365062. 32, 441, 442, 444

[Saka–66] S. Sakai, *On a characterization of type I C^*-algebras*, Bull. Amer. Math. Soc. **72** (1966), 508–512. MR192363. 259

[Saka–67] S. Sakai, *On type I C^*-algebras*, Proc. Amer. Math. Soc. **18** (1967), 861–863. MR213887. 259

[Schm–77a] K. Schmidt, *Cocycles on ergodic transformation groups*, Macmillan Lectures in Mathematics, Vol. 1, Macmillan Company of India, Ltd., Delhi, 1977. MR0578731. 168, 365, 366, 367, 368

[Schm–77b] K. Schmidt, *Infinite invariant measures on the circle*, Symposia Mathematica, Vol. XXI (Convegno sulle Misure su Gruppi e su Spazi Vettoriali, Convegno sui Gruppi e Anelli Ordinati, INDAM, Rome, 1975), Academic Press, London, 1977, pp. 37–43. MR0476996. 165, 166

[Schm–84] K. Schmidt, *Asymptotic properties of unitary representations and mixing*, Proc. London Math. Soc. (3) **48** (1984), no. 3, 445–460. MR735224. 144

[Schu–24] I. Schur, *Neue Anwendungen der Integralrechnung auf Probleme der Invariantentheorie, I-III*, Sitzungberichte der Akademie der Wiss. zu Berlin (1924), 189–208; Ges. Abh. **II**, pp. 440–459. 7

[Sega–47] I. E. Segal, *Irreducible representations of operator algebras*, Bull. Amer. Math. Soc. **53** (1947), 73–88. MR20217. 21

[Sega–50a] I. E. Segal, *An extension of Plancherel's formula to separable unimodular groups*, Ann. of Math. (2) **52** (1950), 272–292. MR36765. 227

[Sega–50b] I. E. Segal, *The two-sided regular representation of a unimodular locally compact group*, Ann. of Math. (2) **51** (1950), 293–298. MR36764. 227

[Sega–51] I. E. Segal, *Decompositions of operator algebras. I*, Mem. Amer. Math. Soc. **9** (1951), 67. MR44749. 211

[Sega–57] I. E. Segal, *The structure of a class of representations of the unitary group on a Hilbert space*, Proc. Amer. Math. Soc. **8** (1957), 197–203. MR84122. 122

[SevN–50] I. E. Segal and J. von Neumann, *A theorem on unitary representations of semisimple Lie groups*, Ann. of Math. (2) **52** (1950), 509–517 [**vNeu–Col**, Volume IV, 332–340]. 144, 404

[Serr–77] J.-P. Serre, *Arbres, amalgames,* SL_2 (French), Société Mathématique de France, Paris, 1977. Avec un sommaire anglais. Rédigé avec la collaboration de Hyman Bass. Astérisque, No. 46. MR0476875. 189

[Shod–33] K. Shoda, *Über die monomialen Darstellungen einer endlichen Gruppe*, Proc. Phys. Math. Soc. Japan (3) **15** (1933), 249–257. 51

[Shte–71] A. I. Shtern (= Štern), *The connection between the topologies of a locally bicompact group and its dual space*, Funkcional Anal. i Priložen. **5** (1971), no. 4, 56–63. 35, 36

[Sing–52] I. M. Singer, *Uniformly continuous representations of Lie groups*, Ann. of Math. (2) **56** (1952), 242–247. MR49201. 440

[Skud–76] H.-L. Skudlarek, *Die unzerlegbaren Charaktere einiger diskreter Gruppen* (German), Math. Ann. **223** (1976), no. 3, 213–231. MR463356. 326

[Smir–66] S. V. Smirnov, *Positive definite functions on algebraic nilpotent groups over a discrete field* (Russian), Dokl. Akad. Nauk SSSR **170** (1966), 524–525. MR0201573. 326

[Spat–87] R. J. Spatzier, *An example of an amenable action from geometry*, Ergodic Theory Dynam. Systems **7** (1987), no. 2, 289–293. MR896799. 371

[Stew–76] J. Stewart, *Positive definite functions and generalizations, an historical survey*, Rocky Mountain J. Math. **6** (1976), no. 3, 409–434. MR430674. 19

[Ston–30] M. H. Stone, *Linear transformations in Hilbert space. III. Operational methods and group theory*, Proc. Nat. Acad. Sci. U.S.A. **16** (1930), 172–175. 8, 82, 109

[StZi–94] G. Stuck and R. J. Zimmer, *Stabilizers for ergodic actions of higher rank semisimple groups*, Ann. of Math. (2) **139** (1994), no. 3, 723–747. MR1283875. 416, 417

[StVo–75] S. Stratila and D. Voiculescu, *Representations of AF-algebras and of the group $U(\infty)$*, Lecture Notes in Math. **486**, Springer, 1975. 245, 326

[Suna–86] T. Sunada, *Riemannian coverings and isospectral manifolds*, Ann. of Math. (2) **121** (1985), no. 1, 169–186. MR782558. 51

[Sund–87] V. S. Sunder, *An invitation to von Neumann algebras*, Universitext, Springer-Verlag, New York, 1987. MR866671. 367

[Sund–97] V. S. Sunder, *Functional analysis: Spectral theory*, Birkhäuser Advanced Texts: Basler Lehrbücher. [Birkhäuser Advanced Texts: Basel Textbooks], Birkhäuser Verlag, Basel, 1997. MR1646508. 71, 77, 80

[Suth–78] C. E. Sutherland, *Type analysis of the regular representation of a nonunimodular group*, Pacific J. Math. **79** (1978), no. 1, 225–250. MR526681. 387

[Suth–83] C. E. Sutherland, *On a construction of unitary cocycles and the representation theory of amenable groups*, Ergodic Theory Dynam. Systems **3** (1983), no. 1, 129–133. MR743031. 162, 293

[Take–57] O. Takenouchi, *Sur la facteur-représentation d'un groupe de Lie résoluble de type (E)* (French), Math. J. Okayama Univ. **7** (1957), 151–161. MR97464. 208

[Take–79] M. Takesaki, *Theory of operator algebras. I*, Springer-Verlag, New York-Heidelberg, 1979. MR548728. 353

[Tayl–76] K. F. Taylor, *The type structure of the regular representation of a locally compact group*, Math. Ann. **222** (1976), no. 3, 211–224. MR425007. 227

[Thom–64a] E. Thoma, *Über unitäre Darstellungen abzählbarer, diskreter Gruppen*, Math. Ann. **153** (1964), 111–138. 7, 9, 21, 221, 324
Translation by A. Törnquist: *On unitary representations of countable discrete groups*, October 2006, `www.math.toronto.edu/asger/thoma.pdf`

[Thom–64b] E. Thoma, *Über positiv-definite Klassenfunctionen abzählbarer Gruppen* (German), Math. Z. **84** (1964), 389–402. MR170217. 7, 9, 324, 331

[Thom–64c] E. Thoma, *Die unzerlegbaren, positiv-definiten Klassenfunktionen der abzählbar unendlichen, symmetrischen Gruppe* (German), Math. Z. **85** (1964), 40–61. MR173169. 9, 326

[Thom–68] E. Thoma, *Eine Charakterisierung diskreter Gruppen vom Typ I* (German), Invent. Math. **6** (1968), 190–196. MR248288. 221

[Thos–15] S. Thomas, *A descriptive view of unitary group representations*, J. Eur. Math. Soc. (JEMS) **17** (2015), no. 7, 1761–1787. MR3361728. 291, 295, 296

[Tsan–12] T. Tsankov, *Unitary representations of oligomorphic groups*, Geom. Funct. Anal. **22** (2012), no. 2, 528–555. MR2929072. 209

[Vale–84] A. Valette, *Notes on the structure and the K-theory of the C^*-algebras associated to* $SL_2(\mathbf{R})$, Bull. Soc. Math. Belg. Sér. B **36** (1984), no. 1, 29–56. MR885552. 121, 246

[Vale–86] A. Valette, *Representations of minimally almost periodic groups*, J. Austral. Math. Soc. Ser. A **41** (1986), no. 3, 366–375. MR855403. 144

[Veec–69] W. A. Veech, *Strict ergodicity in zero dimensional dynamical systems and the Kronecker-Weyl theorem* mod 2, Trans. Amer. Math. Soc. **140** (1969), 1–33. MR240056. 165

[VeKe–81] A. M. Vershik and S. V. Kerov, *Asymptotic theory of the characters of a symmetric group* (Russian), Funktsional. Anal. i Prilozhen. **15** (1981), no. 4, 15–27, 96. MR639197. 9, 326

[Vers–11] A. M. Vershik, *Nonfree actions of countable groups and their characters* (Russian, with English and Russian summaries), Zap. Nauchn. Sem. S.-Peterburg. Otdel. Mat. Inst. Steklov. (POMI) **378** (2010), no. Teoriya Predstavleniĭ, Dinamicheskie Sistemy, Kombinatornye Metody. XVIII, 5–16, 228; English transl., J. Math. Sci. (N.Y.) **174** (2011), no. 1, 1–6. MR2749291. 9, 417, 418, 419

[Vers–12] A. M. Vershik, *Totally nonfree actions and the infinite symmetric group* (English, with English and Russian summaries), Mosc. Math. J. **12** (2012), no. 1, 193–212, 216. MR2952431. 417

[Vile–68] N. Ja. Vilenkin, *Special functions and the theory of group representations*, Translated from the Russian by V. N. Singh. Translations of Mathematical Monographs, Vol. 22, American Mathematical Society, Providence, R. I., 1968. MR0229863. 52

[Voic–76] D. Voiculescu, *Représentations factorielles de type II1 de* $U(\infty)$ (French), J. Math. Pures Appl. (9) **55** (1976), no. 1, 1–20. MR442153. 326

[Wall–92] N. R. Wallach, *Real reductive groups. II*, Pure and Applied Mathematics, vol. 132, Academic Press, Inc., Boston, MA, 1992. MR1170566. 12

[Walt–82] P. Walters, *An introduction to ergodic theory*, Graduate Texts in Mathematics, vol. 79, Springer-Verlag, New York-Berlin, 1982. MR648108. 342, 393

[Wang–75] P. S. Wang, *On isolated points in the dual spaces of locally compact groups*, Math. Ann. **218** (1975), no. 1, 19–34. MR384993. 35

[Warn–72] G. Warner, *Harmonic analysis on semi-simple Lie groups. I*, Springer-Verlag, New York-Heidelberg, 1972. Die Grundlehren der mathematischen Wissenschaften, Band 188. MR0498999. 203

[Weav–03] N. Weaver, *A prime C^*-algebra that is not primitive*, J. Funct. Anal. **203** (2003), no. 2, 356–361. MR2003352. 240

[Weil–40] A. Weil, *L'intégration dans les groupes topologiques et ses applications* (French), Actual. Sci. Ind., no. 869, Hermann et Cie., Paris, 1940. [This book has been republished by the author at Princeton, N. J., 1941.] MR0005741. 8, 133, 136, 249, 437

[Weyl–24] H. Weyl, *Zur Theorie der Darstellung der einfachen kontinuierlichen Gruppen*, (Aus einem Schreiben an Herrn I. Schur.) Sitzungsber. Akad. Wiss. Berlin, (1924), pp. 338–345, letter dated 28. November 1924 [**Weyl–Col**, Volume II, 453–460]. 7

[Weyl–Col] H. Weyl, *Gesammelte Abhandlungen. Bände I, II, III, IV* (German), Herausgegeben von K. Chandrasekharan, Springer-Verlag, Berlin-New York, 1968. MR0230597. 7, 9, 457, 461

[Wign–39] E. Wigner, *On unitary representations of the inhomogeneous Lorentz group*, Ann. of Math. (2) **40** (1939), no. 1, 149–204. MR1503456. 8

[Will–07] D. P. Williams, *Crossed products of C^*-algebras*, Mathematical Surveys and Monographs, vol. 134, American Mathematical Society, Providence, RI, 2007. MR2288954. 245, 421

[Wils–98] J. S. Wilson, *Profinite groups*, London Mathematical Society Monographs. New Series, vol. 19, The Clarendon Press, Oxford University Press, New York, 1998. MR1691054. 137

[Wolf–07] J. A. Wolf, *Harmonic analysis on commutative spaces*, Mathematical Surveys and Monographs, vol. 142, American Mathematical Society, Providence, RI, 2007. MR2328043. 191

[Yosh–51] H. Yoshizawa, *Some remarks on unitary representations of the free group*, Osaka Math. J. **3** (1951), 55–63. MR41854. 58

[Zimm–84] R. J. Zimmer, *Ergodic theory and semisimple groups*, Monographs in Mathematics, vol. 81, Birkhäuser Verlag, Basel, 1984. MR776417. 27, 108, 150, 365, 371, 413, 417, 424

Notation Index

The following table contains a list of symbols that are frequently used throughout the book. The numbers indicate the pages of an appearance or a definition of each symbol. In a few cases, the same symbol may have different meanings, depending on the context.

Notation	Meaning	Page
$\simeq$	equivalence of representations	15
$\simeq_\Gamma$	equivalence of representations of Γ (occasionally)	291
$\preceq$	weak containment of representations	27
$\sim$	weak equivalence of representations	28
$\approx$	quasi-equivalence of representations	170
$\approx$	isomorphism of sets, i.e., bijection	2
$\approx$	isomorphism of other structures (Hilbert spaces, C"-algebras, ...)	
$\mathbf{N} = \{0, 1, 2, 3, \dots\}$	nonnegative integers	
$\mathbf{N}^* = \{1, 2, 3, \dots\}$	positive integers	
$\overline{\mathbf{N}^*} = \{1, 2, \dots, \infty\}$	extended positive integers	73
$\mathbf{Z}$	rational integers	10
$\mathbf{Q}$	rational numbers	31
$\mathbf{Z}_p$	p-adic integers	35
$\mathbf{Q}_p$	p-adic numbers	35
$\mathbf{R}$	real numbers	17
$\mathbf{R}_+ = [0, \infty[$	nonnegative real numbers	112
$\mathbf{R}_+^\times =]0, \infty[$	positive real numbers	112
$\overline{\mathbf{R}_+^\times} =]0, \infty] = \mathbf{R}_+^\times \cup \{\infty\}$	extended positive real numbers	366
$\overline{\mathbf{R}} = \mathbf{R} \cup \{\infty\} = \mathbf{P}^1(\mathbf{R})$	real projective line	369
$\mathbf{Sol}_p = (\mathbf{Q}_p \ltimes \mathbf{R})/\Delta$	p-adic solenoid	116
$\mathbf{C}$	complex numbers	14
$\mathbf{T} = \mathbf{R}/\mathbf{Z} = \{z \in \mathbf{C} \mid \lvert z \rvert = 1\}$	circle group	17
$\mathbf{K}$	field	36
$\mathbf{F}_p$	finite field of order p	105

$\mathcal{H}$, or $\mathcal{K}$	Hilbert space	13
$\mathcal{U}(\mathcal{H})$	unitary group of $\mathcal{H}$	14
$\mathcal{L}(\mathcal{H}_1, \mathcal{H}_2)$	bounded linear operators from $\mathcal{H}_1$ to $\mathcal{H}_2$	14
$\mathcal{L}(\mathcal{H})$	bounded linear operators on $\mathcal{H}$	14
$I = \mathrm{Id}_{\mathcal{H}} \in \mathcal{L}(\mathcal{H})$	identity operator on $\mathcal{H}$	24
$\mathrm{Proj}(\mathcal{H})$	set of projections in $\mathcal{L}(\mathcal{H})$	441
$\ell^2(T, \mathcal{K})$	Hilbert space	43
$L^2(X, \mu, \mathcal{K})$	Hilbert space	54
$L^2(\widehat{G}, \mu, \mathcal{K})$	Hilbert space	73
$L^p(X, \mu)$ or $L^p(X)$	Banach space	426
$\mathbf{1}_U$	characteristic function of U	430
G	topological group	14
e	unit element in a group	14
μ_G	left Haar measure on an LC group G	2
Δ_G	modular function of an LC group G	435
$G_{\mathrm{ab}} = G/\overline{[G, G]}$	abelianization of a group G	31
G_{disc}	topological group G made discrete	26
$G = H \ltimes N$	semi-direct product	124
$\check{\varphi}$	$\check{\varphi}(g) = \check{\varphi}(g^{-1})$ for $\varphi : G \to \mathbf{C}$	19
φ^*	$\varphi^*(g) = \Delta(g^{-1})\overline{\varphi(g^{-1})}$ for $\varphi : G \to \mathbf{C}$	436
1_G	unit representation of G on $\mathbf{C}$	14
π, ρ, σ	(unitary) representations of a topological group	14
λ_G	left regular representation of an LC group G	26
ρ_G	right regular representation of an LC group G	26
$\mathrm{Hom}_G(\pi_1, \pi_2)$	space of intertwiner operators	14
$\mathbf{C}[G]$	group algebra of a group G	20
$L^1(G, \mu_G)$ or $L^1(G)$	group algebra of an LC group G	242
$C^*_\lambda(G)$	reduced C*-algebra of an LC group G	243
$C^*_{\mathrm{max}}(G)$	maximal C*-agebra of an LC group G	239
$\mathcal{L}(\Gamma)$	von Neumann algebra of λ_Γ	218
$\widehat{G}$	(unitary) dual of a topological group G	18
$\widehat{G}_{\mathrm{fd}}$	finite-dimensional part of $\widehat{G}$	31
$\mathrm{Hom}(G, \mathbf{T})$	group of continuous homomorphisms $G \to \mathbf{T}$	18
$P(G), P_{\leq 1}(G), P_1(G)$	functions of positive type on G	19
$\mathrm{Extr}(P_1(G))$	indecomposable elements in $P_1(G)$	23
$\mathrm{Prim}(G)$	primitive dual of a topological group G	38
$\mathrm{QD}(G)$	quasi-dual of a topological group G	192
$\mathrm{QD}(G)_{\mathrm{norm}}$	normal quasi-dual of an LC group G	314
$\mathrm{QD}(G)_{\mathrm{fin}}$	finite part of $\mathrm{QD}(G)$	319

$\mathrm{Char}(G)$	equivalence classes of characters of an LC group G	314
$\mathrm{Char}(G)_{\mathrm{fin}}$	finite part of $\mathrm{Char}(G)$	319
$\mathrm{Tr}(G), \mathrm{Tr}_{\leq 1}(G), \mathrm{Tr}_1(G)$	traces on G	320
$E(G) = \mathrm{Extr}(\mathrm{Tr}_1(G))$	Thoma's dual of a topological group G	324
$\widehat{G}$	Pontryagin dual of an LCA group G	437
$\mathcal{F}$	Fourier transform	438
$\overline{\mathcal{F}}$	inverse Fourier transform	438
A	C*-algebra	214
A_+	positive cone in A	214
$\widehat{A}$	spectrum of A	233
$\mathrm{Prim}(A)$	primitive ideal space of A	233
$A^*_+, A^*_{+,\leq 1}, A^*_{+,1}$	positive linear forms on a C*-algebra A	234
$\mathrm{P}(A) = \mathrm{Extr}(A^*_{+,1})$	set of pure states of A	235
$\mathcal{J}(A)$	set of closed two-sided ideals of A	238
$\mathcal{N}(A)$	C*-semi-norms on a C*-algebra A	238
$\mathcal{E}(A)$	extremal C*-semi-norms on a C*-algebra A	239
$t : A_+ \to [0, \infty]$	trace on A	214
$\mathfrak{m}_t$	ideal of definition of t	298
$\mathfrak{n}_t$	$\{x \in A \mid t(x^*x) < \infty\}$	297
$\mathcal{M}$	von Neumann algebra	218
$\mathcal{M}_+$	positive cone in $\mathcal{M}$	218
$\mathcal{A}$	Hilbert algebra	298
σ-compact	sigma-compact topological space	423
$C(X)$	space of continuous functions on X	423
$C^b(X)$	space of bounded continuous functions on X	423
$C^c(X)$	subspace of $C(X)$ of functions with compact support	423
$C^{c,K}(X)$	subspace of $C^c(X)$ of functions with support in K	423
$C^0(X)$	subspace of $C(X)$ of functions vanishing at infinity	423
$M^b(X)$	space of bounded complex measures on $(X, \mathcal{B})$	431

Index

Absolute value of an operator on a Hilbert space, 441
Accessible topological space, T_1, 421
Action
  2-transitive, 49
  aperiodic, 434
  ergodic, 353, 434
  essentially free, 353
  essentially transitive, 353
  nonsingular, 353
  type II, III_λ, 366
Action of G on X, respectively $G \curvearrowright X$ and $X \curvearrowleft G$, 18, 433
Affine group
  $\mathrm{Aff}(\mathbf{R})$, $\mathrm{Aff}(\mathbf{R})_0$, 41, 113, 202, 228
  $\mathrm{Aff}(\mathbf{Q})$, nonuniqueness of direct integral decomposition in irreducible, 59
  $\mathrm{Aff}(\mathbf{Q})$, quasi-regular representation of type III, 225
  $\mathrm{Aff}(\mathbf{K})$ of a field $\mathbf{K}$, 110
Affine group $\mathrm{Aff}(\mathbf{K})$ of an infinite field
  examples of irreducible representations, 110
  finite-dimensional representations, 129
  primitive dual, 271
  examples of irreducible representations, 165
  icc group, 432
  Thoma dual, 336
Algebras
  continuous functions, $C(X)$, $C^b(X)$, $\mathbf{C}^c(X)$, $C^0(X)$, 19, 215, 424
  trigonometric polynomials on $\widehat{G}$, 375, 439
  convolution algebras, $C^c(G)$, $C^0(G)$, $L^1(G)$, 242, 259, 305
  Hilbert algebra, 298
  C*-algebra, 442
  von Neumann algebra, 443
Almost periodic function, 135
Amenable
  group, 28
  homogeneous space, 51
Aperiodic action, 434
Approximate identity, 443
Atom
  of a measure, 425
  of a projection-valued measure, 442

Baire space, 238, 422
Banach involutive algebra, 442
Baumslag–Solitar group $\mathrm{BS}(1, p)$
  examples of irreducible representations, 114, 165
  finite-dimensional representations, 130
  icc groop, 432
  normal factor representations, 388
  primitive dual, 272
  Thoma dual, 338
Bi-invariant pseudo-metric on a group, 398
Bihomomorphism, 99
Bohr compactification, 134
Borel equivalence relation, 290
  Borel equivalent, 291
  Borel reducible, 291
  smooth, 290
Borel map, 424
Borel space, 424
  countably generated, 424

countably separated, 424
standard, 424
Bounded element of a Hilbert algebra, 298
Bruhat decomposition
of $\mathrm{GL}_n(\mathbf{K})$, 277
of $\mathrm{SL}_n(\mathbf{K})$, 285

C*-algebra, 442
commutative C*-algebra, 424
cartesian product, 245
restricted product, 245
C*-kernel, 244
C*-maximally almost periodic group, or C*-MAP group, 408
C*-residually finite group, 412
C*-semi-norm, 238
Canonical representation π_μ of an LCA group G on $L^2(\widehat{G}, \mu)$, 68
Cartesian product of C*-algebras, 245
CCR
representation, 202, 265
group, 202
C*-algebra, 257
Central character, 17, 257
Central direct integral decomposition, 196
Central function = class function, 1
Central support of an element in a von Neumann algebra, 445
Character
unit 1_G, 14
unitary character, 17
central, 17, 257
of an LC group, 314
finite, 319, 325
of a normal factor representation of an LC group, 315
Class function = central function, 1
Cocycle, 42, 149
Radon–Nikodym, 148, 365
cohomologous, 149, 152
irreducible, 152
Commensurate subgroups, 51, 432
Commensurator, 48, 432
Commutant, 14, 443
Commutation Theorem for Hilbert algebras, 299
Commutator $[a, b] := a^{-1}b^{-1}ab$, 31
Commutator subgroup = derived group, 31
Conditional expectation, 359
Conjugate representation, 18, 44
Connes embedding problem, 406
Continuity of induction, 44
Convolution
of two functions on an LC group, 436
of two measures on an LC group, 435
Crossed product, 357, 387
Cyclic representation, 20

Decomposable operator over a measure space, 54, 59
Decomposable von Neumann algebra, 64
Decomposition of representations
into irreducible representations for type I groups, 199
into irreducible representations, 57
into factor representations, 195
Derived group = commutator subgroup, 31
Diagonal matrix coefficient, 20
Diagonalisable operator, 55
Dihedral group D_∞
dual, 95
Thoma dual, 326
Dimension
of a Hilbert space, 440
Direct integral
of Hilbert spaces, 53
of representations, 55
of von Neumann algebras, 64
Direct sum of representations, 15
Disjoint
measures, 426
representations, 15, 170
Dual
short for unitary dual, 18
surjection on primitive dual, 38
finite-dimensional part, 31
injection in quasi-dual, 192

Enveloping C*-algebra of measure algebra, 248
Equivalent
  representations, 15
  cyclic representations, 22
  measures, 426
  projection-valued measures, 82
Ergodic
  action, 353, 434
  IRS, 416
  measure, 434
Essential range
  of a cocycle, 366
  of a function, 426
Essential supremum, 426
Essentially free action, 353
Essentially transitive action, 353
Extremal semi-norm, 239

Factor, 89, 444
Factor representation, 89, 173, 179
Faithful trace, 214
FC-centre FC(Γ) of a group Γ, 431
FC-group, 431
Fell topology, 29, 245
Field of Hilbert spaces over a measure space, 53
Finite part of the quasi-dual, 319
Finite trace, 214
Finite type, for a Polish group, 398
Finite-dimensional part of the dual, 31
First return time, 390
Fourier transform, 438
Fourier–Stieltjes transform, 438
Free group
  C*-MAP group, 410
  C*-residually finite group, 413
  icc, 432
  inside $\mathrm{SL}_2(\mathbf{Q})$ and $\mathrm{SL}_2(\mathbf{F}_p(T))$, 282
  nonnormal irreducible infinite-dimensional representation, 313
  nonuniqueness of direct integral decomposision in irreducible, 58
Full Hilbert algebra, 300
Function of positive type, 19
  associated with a representation, 20
Fundamental family of measurable vector fields over a Borel space, 53

GCR
  group, 200
  C*-algebra, 257
Gel'fand pair, 187
Gel'fand–Naimark–Segal representation, 20, 234, 321
  with $\varphi \in P(G)$, 20
  with $\varphi \in A_+^*$, 234
  with $\varphi \in \mathrm{Tr}(G)$, 321
Gel'fand–Raikov Theorem, 27
General linear group $\mathrm{GL}_n(\mathbf{K})$
  Bruhat decomposition, 277
  examples of irreducible representations, 119
  finite-dimensioinal irreducible representations, 145
  not MAP, 137
  primitive dual when $\mathrm{GL}_n(\mathbf{K})$ is amenable, 276
  principal series, 277
  quotient $\mathrm{PGL}_n(\mathbf{K})$ icc, 432
  Thoma dual, 346
Glimm's Theorem, 258
GNS, short for Gel'fand–Naimark–Segal, 20
Group measure space construction, see crossed product, 357

Haar measure, 435
Hausdorff topological space, T_2, 421
Heisenberg group
  $H(R)$ over a commutative ring R, 102
  $H(\mathbf{K})$ over a field $\mathbf{K}$, 105
  $H(\mathbf{Z})$, 106
  $H(\mathbf{R})$, 109, 246
  $H_\infty(\mathbf{F}_p)$, 227
Heisenberg group $H(\mathbf{K})$
  examples of irreducible representations, 105
  finite-dimensional representations, 128
  primitive dual, 268

Thoma dual, 334
Heisenberg group $H(\mathbf{Z})$
factor representations of type I_∞, II_1, II_∞, and III, 376
examples of irreducible representations, 106
finite-dimensional representations, 129
no normal representation of infinite type, 329, 381
primitive dual, 270
representations involving cocycles, 164
Thoma dual, 335
Hilbert algebra, 298
bounded element, 298
full, 300
natural trace, 301
Homogeneous representation, 80

Icc group, 211, 432
Ideal of definition of a trace, 297
Indecomposable
for $\varphi \in P_{\leq 1}(G)$, 23
for $\varphi \in A^*_{+,\leq 1}$, 235
for $\varphi \in \mathrm{Tr}_{\leq 1}(G)$, 324
Induced representation, 40
Induced transformation, 391
Induced von Neumann algebra, 445
Integral, 425
Intertwiner
between representations of a group, 14
between cocycles on a group, 152
Invariant measure, 433
Invariant random subgroup, or IRS, 416
Involutive algebra, 442
Involutive algebra $\mathcal{L}(\mathcal{H})$, 14
Irreducible representation, 16

Jacobson topology, 233

Kazhdan Property (T), 38
Kolmogorov topological space, T_0, 421
Koopman representation, 356

Lamplighter group
examples of irreducible representations, 118, 165
finite-dimensional representations, 131
icc group, 432
normal factor representations, 389, 393
primitive dual, 275
Thoma dual, 341
Large compact subgroup, 203
LC for "locally compact", 36
LCA for "locally compact abelian", 67
Left regular representation, 26, 212, 224, 243
Liminal
locally compact group, 202
representation, 202
C*-algebra, 257
Locally quasi-compact topological space, 32
Lower semi-continuous trace, 214

Mackey machine, 92
Mackey–Borel structure, 193, 255
Maharam extension, 368
Matrix coefficient, 20
Mautner group $M := \mathbf{C}^2 \rtimes \mathbf{R}$, 228
Maximal C*-algebra of an LC group, 243
Maximally almost periodic group, or MAP group, 133, 400
Measurable
space, 424
map, 424
action $G \curvearrowright (X, \mathcal{B})$, 433
Measurable field
measurable vector field over a Borel space, 53
measurable field of Hilbert spaces over a Borel space, 53
measurable field of operators over a measure space, 54
measurable field of representations, 55
measurable field of von Neumann algebras, 63
Measure
σ-finite, 425

absolutely continuous, 426
algebra, 436
complex, 429
disjoint, 426
equivalent, 426
ergodic, 434
invariant, 433
nonatomic, 425
quasi-invariant, 434
Radon, 430
singular, 426
total variation, 429
Measure algebra, 436
Minimally almost periodic group, or m.a.p. group, 143
Modular function of a locally compact group, 435
Monomial representation, 46
Moore group, 132
Motion group, 187
Multiplicity of a representation, 184
Multiplicity-free representation, 75, 175, 181, 185

Natural trace of a Hilbert algebra, 301
Nondegenerate representation of an involutive algebra, 232
Nonsingular action, 353
Normal factor representation, 312
Normal positive linear map from a von Neumann algebra to another, 443
Normal quasi-dual, 314
Normal trace, 218

Orbit equivalence of group actions, 165

Partial isometry, 441
initial space, 441
final space, 441
Polar decomposition of an operator, 441
Polish space, Polish topology, 422
Pontryagin dual, 437
Positive definite function, see functions of positive type, 19
Positive element in a C*-algebra
square root, 442
Positive element in a C*-algebra, 440, 442
square root, 441
Positive type
for a function, 19
for a Radon measure, 309
Postliminal
C*-algebra, 257
Prime ideal, 239
Primitive dual
of a topological group, 38, 244
Primitive ideal, 233
Primitive ideal space
of a C*-algebra, 233
Principal series
of $\mathrm{GL}_n(\mathbf{K})$, 120, 277
of $\mathrm{SL}_n(\mathbf{K})$, 285
Profinite completion, 138
Profinite group, 137
Projection in a C*-algebra, 442
Projection-valued measure, 441
atom, 442
equivalent, 82
regular, 441
support, 442
Projective kernel, 256
Projective limit of finite groups, 137
Properly infinite
von Neumann algebra, 218
representation, 220
Property
$\mathcal{P}$-by-$\mathcal{Q}$, 431
Pure state on a C*-algebra, 235
Purely infinite
von Neumann algebra, 218
representation, 220
C*-algebra, 314

Quasi-compact topological space, 422
Quasi-dual, 192
type I part, 192
normal, 314
finite part, 319
Quasi-equivalent
representations, 170, 177
trace representations of an LC group, 308

finite type representations of a topological group, 322
Quasi-invariant measure, 434
Quasi-orbit, quasi-orbit space, 273, 422
Quasi-regular representation, 43, 49, 190, 191, 225

Radon measure, 430
complex, 431
Radon–Nikodym
cocycle, 148, 365
derivative, 427
Reduced C*-algebra of an LC group, 243
Reduced Enveloping C*-algebra of measure algebra, 248
Regular projection-valued measure, 441
Representation
C*-kernel, 244
canonical (for an LCA group), 68
CCR, 202, 265
conjugate, 18, 44
cyclic, 20
direct sum, 15
disjoint, 15, 170
equivalent $\simeq$, 15
factor, 89, 173, 179
finite type, 220, 319
homogeneous, 80
induced, 40
irreducible, 16, 232
Koopman, 356
left regular, 26
monomial, 46
multiplicity of a representation, 184
multiplicity-free, 175, 181, 185
nondegenerate, 232
normal, 312
of $C^*_{\max}(G)$ associated to a representation of G, 244
of $L^1(G)$ associated to a representation of G, 243
of an involutive algebra, of a C*-algebra, 232
quasi-equivalent $\approx$, 170, 177, 308, 322
quasi-regular, 43, 49, 190, 191, 225
right regular, 26
standard, 308
subordinate, 170
trace, 308
traceable, 309
trivial (for a C*-algebra), 232
type I, 175, 181
type II, II_1, II_∞, III, 220
unit, 14
weakly contained $\preceq$, 27, 244
weakly equivalent $\sim$, 28, 244
with vanishing coefficients, 346
Representation = unitary group representation, 14
Representation of finite type, 320
Residually finite group, 136
Residually finite-dimensional C*-algebra, 408
Restricted product of C*-algebras, 245
Right regular representation, 26

Schröder–Bernstein property for representations, 171
Schur's lemma, 17
Second-countable topological space, 423
Semi-direct product, 124
Semi-finite trace, 214
Separable topological space, 423
Separating family
of Borel subsets of a space, 424
of continuous functions on a space, 439
of representations of a group, 26
of representations of an involutive algebra, 406
Separation axioms T_0, T_1, T_2, 421
Sigma-compact topological space, 423
SIN group, 136, 399
Solenoid $\mathbf{Sol}_p$, 116
Special linear group $\mathrm{SL}_n(\mathbf{K})$
Bruhat decomposition, 285
minimally almost periodic, 144
principal series, 285
$(\mathrm{SL}_n(\mathbf{R}), \mathrm{SO}(n))$ is a Gelfand pair, 189

$(\mathrm{SL}_n(\mathbf{Q}_p), \mathrm{SL}_n(\mathbf{Z}_p)$ is a Gelfand pair, 189
$\mathrm{SL}_2(\mathbf{R})$ is a CCR group, 204
Thoma duals of $\mathrm{SL}_n(\mathbf{K})$ and $\mathrm{PSL}_n(\mathbf{K})$, 351, 420
Spectrum of a C*-algebra, 233
Square root of a positive operator, 441
Square-integrable vector field over a measure space, 53
Standard
representation of $(\mathcal{M}, \tau)$, 308
Borel space, 424
measure on a Borel space, 428
State on a C*-algebra, 234
Strong topology on $\mathcal{L}(\mathcal{H})$ and $\mathcal{U}(\mathcal{H})$, 14, 440
Subordinate representation, 170
Subrepresentation, 14
Support of a projection-valued measure, 442
Support of an operator, 444

Tame group, 259
Thoma's dual, 324
Topological space
σ-compact, 423
accessible, 421
Baire, 238, 422
compact, 422
Kolmogorov, 421
locally compact, 422
quasi-compact, 422
second-countable, 423
separable, 423
separation axioms T_0, T_1, T_2, 421
Trace
on a C*-algebra, 214, 297
domination, 214
on a von Neumann algebra, 218
ideal of definition, 297
natural trace on the von Neumann algebra associated with a Hilbert algebra, 301
on a topological group, 320
Trace representation, 308
quasi-equivalent, 308, 322
Traceable representation, 309
Transversal T, and $G = \bigsqcup_{t \in T} tH$, 40
Trigonometric polynomials on $\widehat{G}$, 375, 439
Trivial extension of a function defined on a subgroup, 21
Trivial representation, 232
Two-step nilpotent groups
every normal factor representation of $H(\mathbf{K})$ is of finite type, 329
examples of irreducible representations, 99
primitive dual, 265
Thoma dual, 331
Type I
representation, 175, 181
part of the quasi-dual, 192
group, 196, 258
von Neumann algebra, 218
Type II, III
von Neumann algebra, 218
representation, 220
action, 366
III_λ, 366, 387

Ultra-weak topology on $\mathcal{L}(\mathcal{H})$ and $\mathcal{U}(\mathcal{H})$, 440
Unimodular group, 435
Unit representation or unit character 1_G, 14
Unitary character, 17
Unitary dual, 18
Unitary representation = representation, 14
Universal Kolmogorov quotient, 422

Vanishing coefficients (representation), 346
von Neumann algebra, 443
factor, 89, 444
type I, 218
continuous, 218
finite, 218
semi-finite, 218
type II_1, 218
type II_∞, 218
type III, 218
properly infinite, 218
purely infinite, 218
crossed product, 357
enveloping, 260

group measure space construction, 357
von Neumann algebra $\mathcal{L}(\mathbf{G})$ of an LC group G, 212
von Neumann kernel of a topological group, 135
Weak topology on $\mathcal{L}(\mathcal{H})$ and $\mathcal{U}(\mathcal{H})$, 14, 440
Weakly
contained (representation), 27, 244
equivalent (representation), 28, 244

SELECTED PUBLISHED TITLES IN THIS SERIES

250 **Bachir Bekka and Pierre de la Harpe,** Unitary Representations of Groups, Duals, and Characters, 2020

249 **Nikolai M. Adrianov, Fedor Pakovich, and Alexander K. Zvonkin,** Davenport–Zannier Polynomials and Dessins d'Enfants, 2020

248 **Paul B. Larson and Jindrich Zapletal,** Geometric Set Theory, 2020

247 **István Heckenberger and Hans-Jürgen Schneider,** Hopf Algebras and Root Systems, 2020

246 **Matheus C. Bortolan, Alexandre N. Carvalho, and José A. Langa,** Attractors Under Autonomous and Non-autonomous Perturbations, 2020

245 **Aiping Wang and Anton Zettl,** Ordinary Differential Operators, 2019

244 **Nabile Boussaïd and Andrew Comech,** Nonlinear Dirac Equation, 2019

243 **José M. Isidro,** Jordan Triple Systems in Complex and Functional Analysis, 2019

242 **Bhargav Bhatt, Ana Caraiani, Kiran S. Kedlaya, Peter Scholze, and Jared Weinstein,** Perfectoid Spaces, 2019

241 **Dana P. Williams,** A Tool Kit for Groupoid C^*-Algebras, 2019

240 **Antonio Fernández López,** Jordan Structures in Lie Algebras, 2019

239 **Nicola Arcozzi, Richard Rochberg, Eric T. Sawyer, and Brett D. Wick,** The Dirichlet Space and Related Function Spaces, 2019

238 **Michael Tsfasman, Serge Vlăduţ, and Dmitry Nogin,** Algebraic Geometry Codes: Advanced Chapters, 2019

237 **Dusa McDuff, Mohammad Tehrani, Kenji Fukaya, and Dominic Joyce,** Virtual Fundamental Cycles in Symplectic Topology, 2019

236 **Bernard Host and Bryna Kra,** Nilpotent Structures in Ergodic Theory, 2018

235 **Habib Ammari, Brian Fitzpatrick, Hyeonbae Kang, Matias Ruiz, Sanghyeon Yu, and Hai Zhang,** Mathematical and Computational Methods in Photonics and Phononics, 2018

234 **Vladimir I. Bogachev,** Weak Convergence of Measures, 2018

233 **N. V. Krylov,** Sobolev and Viscosity Solutions for Fully Nonlinear Elliptic and Parabolic Equations, 2018

232 **Dmitry Khavinson and Erik Lundberg,** Linear Holomorphic Partial Differential Equations and Classical Potential Theory, 2018

231 **Eberhard Kaniuth and Anthony To-Ming Lau,** Fourier and Fourier-Stieltjes Algebras on Locally Compact Groups, 2018

230 **Stephen D. Smith,** Applying the Classification of Finite Simple Groups, 2018

229 **Alexander Molev,** Sugawara Operators for Classical Lie Algebras, 2018

228 **Zhenbo Qin,** Hilbert Schemes of Points and Infinite Dimensional Lie Algebras, 2018

227 **Roberto Frigerio,** Bounded Cohomology of Discrete Groups, 2017

226 **Marcelo Aguiar and Swapneel Mahajan,** Topics in Hyperplane Arrangements, 2017

225 **Mario Bonk and Daniel Meyer,** Expanding Thurston Maps, 2017

224 **Ruy Exel,** Partial Dynamical Systems, Fell Bundles and Applications, 2017

223 **Guillaume Aubrun and Stanisław J. Szarek,** Alice and Bob Meet Banach, 2017

222 **Alexandru Buium,** Foundations of Arithmetic Differential Geometry, 2017

221 **Dennis Gaitsgory and Nick Rozenblyum,** A Study in Derived Algebraic Geometry, 2017

220 **A. Shen, V. A. Uspensky, and N. Vereshchagin,** Kolmogorov Complexity and Algorithmic Randomness, 2017

219 **Richard Evan Schwartz,** The Projective Heat Map, 2017